CHRISTOPHER JON BJERKNES

THE MANUFACTURE
AND SALE OF
SAINT EINSTEIN

I

Christopher Jon Bjerknes

The Manufacture and Sale of Saint Einstein

Volume I

Published by
Omnia Veritas Ltd

www.omnia-veritas.com

© Omnia Veritas Ltd - Christopher Jon Bjerknes - 2019.

All Rights Reserved. No part of this publication may be reproduced, distributed, or transmitted in any form or by any means, including photocopying, recording, or other electronic or mechanical methods, without the prior written permission of the publisher, except in the case of brief quotations embodied in critical reviews and certain other noncommercial uses permitted by copyright law.

TABLE OF CONTENTS:

1 EINSTEIN DISCOVERS HIS RACIST CALLING --------- 7

1.1 INTRODUCTION --------- 7
1.2 THE MANUFACTURE AND SALE OF ST. EINSTEIN --------- 8
   1.2.1 Promoting the "Cult" of Einstein --------- 10
   1.2.2 The "Jewish Press" Sanctifies a Fellow Jew --------- 11
1.3 IN A RACIST ERA --------- 31

2 THE DESTRUCTIVE IMPACT OF RACIST JEWISH TRIBALISM --------- 129

2.1 INTRODUCTION --------- 129
2.2 DO NOT BLASPHEME THE "JEWISH SAINT" --------- 135
2.3 HARVARD UNIVERSITY ASKS A FORBIDDEN QUESTION --------- 152
2.4 AMERICANS REACT TO THE INVASION OF EASTERN EUROPEAN JEWS --------- 157
   2.4.1 Jewish Disloyalty --------- 164
   2.4.2 In Answer to the "Jewish Question" --------- 175

3 ROTHSCHILD, *REX IVDÆORVM* --------- 181

3.1 INTRODUCTION --------- 181
3.2 JEWISH MESSIANIC SUPREMACISM --------- 191
3.3 THE "EASTERN QUESTION" AND THE WORLD WARS --------- 195
   3.3.1 Dönmeh Crypto-Jews, The Turkish Empire and Palestine --------- 198
   3.3.2 The World Wars—A Jewish Antidote to Jewish Assimilation --------- 205
3.4 ROTHSCHILD WARMONGERING --------- 217
   3.4.1 Inter-Jewish Racism --------- 219
      3.4.1.1 Rothschild Power and Influence Leads to Unbearable Jewish Arrogance --------- 230
      3.4.1.2 Jewish Intolerance and Mass Murder of Gentiles --------- 242
   3.4.2 The Messiah Myth --------- 255
3.5 JEWISH DOGMATISM AND CONTROL OF THE PRESS STIFLES DEBATE --------- 280
   3.5.1 Advertising Einstein in the English Speaking World --------- 320
   3.5.2 Reaction to the Unprecedented Einstein Promotion --------- 332
   3.5.3 The Berlin Philharmonic—The Response in Germany --------- 340
   3.5.4 Jewish Hypocrisy and Double Standards --------- 394
3.6 THE MESSIAH ROTHSCHILDS' WAR ON THE GENTILES—AND THE JEWS --------- 414

4 EINSTEIN THE RACIST COWARD --------- 479

4.1 INTRODUCTION --------- 479
4.2 THE POWER OF JEWISH TRIBALISM INHIBITS THE PROGRESS OF SCIENCE AND DELIBERATELY PROMOTES "RACIAL" DISCORD --------- 480
4.3 A JEW IS NOT ALLOWED TO SPEAK OUT AGAINST A JEW --------- 481
4.4 THE BAD NAUHEIM DEBATE --------- 491
   4.4.1 Einstein Desires a "Race" War Which Will Exterminate the European

Esau ---------- 504
4.4.2 Genocidal Judaism—Pruning the Branches of the Human Family Tree ---------- 516
4.4.3 Crypto-Jews ---------- 544
4.4.4 The Gentiles Must be Exterminated Lest God Cut Off the Jews ---------- 561
4.4.5 Jewish Dualism and Human Sacrifice—Evil is Good ---------- 576
4.4.6 Gentiles are Destined to Slave for the Jews, Then the Slaves Will be Exterminated ---------- 583

1 EINSTEIN DISCOVERS HIS RACIST CALLING

In 1919, Albert Einstein rose to international fame for predicting that the gravitational field of the sun would deflect rays of light. Eclipse observations confirmed this prediction. Newspapers around the world covered the story and declared that Albert Einstein had surpassed the genius of Copernicus, Kepler, Galileo and Newton. It seemed that all was right with the world—but then everything went tragically wrong.

"Probably Professor Einstein does not realize how sensationally and cunningly he has been advertised. From the point of view of awakening popular curiosity, his press-notices could hardly have been improved. The newspapers first announced his discovery as revolutionizing science. This sounds well, but its meaning, after all, is rather vague. Then they printed a series of entertaining oddities, supposedly deducible from his hypothesis, although most of them could have been equally well deduced from the conclusions of Lorentz or Poincaré: for example, moving objects are shortened in the direction of their motion."—GERTRUDE BESSE KING

"If anyone should ask how Einstein managed to get such vast publicity in the matter of relativity, we may observe that he has the habit of a promoter."—THOMAS JEFFERSON JACKSON SEE[1]

"While he lived in Germany, however, Einstein seems to have accepted the then-prevalent racist mode of thought, often invoking such concepts as 'race' and 'instinct,' and the idea that the Jews form a race."—JOHN STACHEL[2]

1.1 Introduction

Racist physicist Albert Einstein became internationally famous in 1919 when newspapers around the world reported that he had correctly predicted that the gravitational field of the sun would deflect rays of light. The press promoted the virulently racist and segregationist Zionist, Albert Einstein, as if he were the world's greatest mind, a mind that had surpassed the genius of Copernicus, Galileo and Newton.

In April of 1921, Albert Einstein took advantage of his newly found fame and traveled to America. He promoted racist Zionism to the Jews of America, while raising money for the Eastern European Zionists who had made him famous. Einstein championed the racist doctrine of Theodor Herzl, that Jews were a distinct race of human beings, who could not assimilate into any Gentile society and therefore ought to segregate themselves and form a nation in Palestine. Einstein also believed that there ought to be a world government.

---

[1]. T. J. J. See, "EINSTEIN A TRICKSTER?", *The San Francisco Journal*, (27 May 1923).

[2]. J. Stachel, "Einstein's Jewish Identity", *Einstein from 'B' to 'Z'*, Birkhäuser, Boston, Basel, Berlin, (2002), pp. 57-83, at 68.

However, Einstein thought that Israel ought to be a distinct nation. Though he described himself as non-religious, Einstein's racist views, and his concurrent call for a world government and a segregated "Jewish State" mirrored Jewish Messianic prophecies.

Einstein raised money in America for the Hebrew University in Jerusalem. He also tried to popularize the racist Zionist cause. The news media enthusiastically covered his trip to the United States. Mainstream news media claimed that all of Einstein's critics were anti-Semites, but did not criticize Einstein for his rabid racism or his segregationist politics.

Prof. Arvid Reuterdahl of St. Thomas College, in St. Paul, Minnesota, responded to Einstein's aggressive self-promotion. With reference to the notorious circus promoter P. T. Barnum, Prof. Reuterdahl dubbed Albert Einstein the "Barnum of the Scientific World". He publicly challenged Einstein to a debate over the merits of the theory of relativity and publicly accused Einstein of plagiarism.

Einstein refused to debate Reuterdahl. Einstein stated that his sole purpose for coming to America was to raise money for the Hebrew University in Jerusalem and that he could not be bothered with issues related to "his" theories. Even before coming to America, Einstein had earned an international reputation for hiding from his critics. His favorite tactic to avoid debate was to accuse his critics of being "anti-Semites", while refusing to address their legitimate accusations of his, Einstein's, irrationality and plagiarism. Like most bullies by bluff, Einstein was a coward, who hid behind the power of the racist Jews who attempted to shield him from criticism through well-orchestrated smear campaigns in the international press.

In spite of this, or perhaps because of this, Einstein generally had a hard time in America. Due to his incompetence, and the tribalistic racism he and his Jewish friends exhibited, Einstein faced scandal after scandal. Though Einstein had arrived to a triumphant welcome in New York City, he left the United States an utter disgrace. Though Einstein had accepted many honors from American universities, he publicly ridiculed American scholars and Americans in general in a widely published interview he gave after he had returned to Europe. The grapes had turned to sour gripes.

1.2 The Manufacture and Sale of St. Einstein

Isaac Newton believed that light is composed of matter converted into tiny "corpuscles". Newton predicted that the gravitational attraction of other matter would attract light corpuscles, just as it attracted everything else made up of matter. Einstein repeated Newton's prediction that gravitational fields would deflect light.

Like countless others before him, Einstein had proposed a non-Newtonian law of gravity. In Einstein's gravitational theory the deflection of light rays was twice as great as in Newton's gravitational theory.

In 1918-1920, the British astronomers Frank Watson Dyson, Charles Davidson and Arthur Stanley Eddington collaborated with Albert Einstein, and his friends Alexander Moszkowski, Max Born, Erwin Freundlich and Hendrik Antoon Lorentz to promote and sensationalize contrived reports that eclipse observations had confirmed Einstein's prediction. The astronomers had attempted to photograph stars which could be seen near the edge of the Sun during a full eclipse. The images of these stars might indicate that the path of the rays of light coming from stars behind the Sun had curved when passing near the Sun, thereby displacing the images of the stars from the position they would otherwise have had on the pictures, had not the gravitational field of the sun altered the path of light coming from the stars behind the Sun. Johann Georg von Soldner (in 1801) and Albert Einstein (in late 1915) predicted that the deflection would be twice the amount the Newtonian theory of gravitation predicted. This factor of two distinguished their theories from Newton's. Though it was Newton who first predicted the effect, and it was Soldner who first correctly predicted the amount of the deflection for light rays, it was Einstein who took credit for both predictions.

Dyson, Davidson, Eddington and Einstein misrepresented the photographic evidence, which was of poor quality and, therefore, inconclusive. They falsely claimed that the photographs taken during eclipse of the Sun proved not only that the deflection of light had occurred, but that it was twice the Newtonian value, in accord with Einsteinian (Soldnerian) theory. However, this is not what the photographs had shown, and it is doubtful that the photographs could in any case have been conclusive. The effect was exceedingly small and the equipment the astronomers employed was primitive and did not have the precision needed to accurately record the predicted effect.

The press promoted these falsified reports and told the general public that Newtonian theory had been overthrown and that Einstein was a great genius, who was at least the equal of Copernicus, Kepler, Galileo and Newton. Newspapers asserted that Einstein had introduced a new world view, one that was true no matter how strange it appeared to be, with its "warped space-time", "hundred foot poles in fifty foot barns", and other "paradoxes". The press reported that Einstein's unique insight was so sophisticated and enlightened that only twelve men in the world could understand it. Reporters told the people of the world that a dramatic revolution in science had taken place—though this magnificent and unprecedented revolution, so deserving of international attention and praise, had changed nothing in their lives and they had no need, nor reason, to try to understand it.

The sensational reports created a mass hysteria for Einstein in America, one which culminated in Einstein's visit to the United States in the spring of 1921. Einstein's trip came shortly after Einstein had endured a series of public humiliations in the scientific community in Germany in 1920. He was hiding from the German scientists who had informed the public that he was a fraud. Whenever Einstein faced overwhelming problems in Germany, he wisely traveled to other nations, in part for publicity purposes to promote Zionism— which gave him undeserved publicity and paid for his trips—and which gave him the means to hide from his many critics. Einstein went to Spain and to Japan, continually promoting himself by being seen in the company of royalty, heads of state and international celebrities.

In spite of all the humiliating defeats Einstein met in the scientific world, a pro-Einstein press stuck by him and unfairly smeared those who legitimately criticized him. Some of his critics were highly respected Nobel Prize winning physicists, but this did not inhibit the pro-Einstein press from attacking their reputations merely because they had dared to disagree with the racist Zionist Albert Einstein, on purely scientific matters.

1.2.1 Promoting the "Cult" of Einstein

In an epiphany of Saint Einstein, Jewish journalist Alexander Moszkowski wrote to Albert Einstein on 1 February 1917,

> "Regardless of what happens, I would like to continue the 'cult'; for you it is secondary, for me it is of paramount importance in life. Additionally, I have the encouraging feeling that, with my modest writing abilities, I may also serve the cause once in a while."[3]

Moszkowski used his writing talents to make Einstein a superstar. In October of 1919, Moszkowski fulfilled his promise to Einstein to promote the "cult" of Einstein, and began the international "Einstein mania", which peaked in November and December of 1919.

Einstein knew that the newspaper hype was disingenuous and distasteful, but he blamed the public for the hype his racist Jewish friends had manufactured. In mid-December, 1919, Einstein wrote to his friend and confidant Heinrich Zangger,

> "The newspaper drivel about me is pathetic; this kind of exaggeration meets a certain need among the public. Really, a harmless ideology."[4]

On 24 December 1919, Einstein wrote to Zangger and justified the lies as "harmless tomfoolery",

> "[T]his business reminds one of the tale of 'The Emperor's New Clothes,' but it is harmless tomfoolery. [***] The disparity between what you are and what others believe, or at least, say about you, is far too great."[5]

---

[3]. A. Moszkowski to A. Einstein, translated by A. M. Hentschel, *The Collected Papers of Albert Einstein*, Volume 8, Document 292, Princeton University Press, (1998), p. 281.
[4]. Letter from A. Einstein to H. Zangger of 15 or 22 December 1919, English translation by A. Hentschel, *The Collected Papers of Albert Einstein*, Volume 9, Document 217, Princeton University Press, (2004), pp. 185-186, at 186.
[5]. Letter from A. Einstein to H. Zangger of 24 December 1919, English translation by A. Hentschel, *The Collected Papers of Albert Einstein*, Volume 9, Document 233, Princeton University Press, (2004), pp. 197-198.

When Albert Einstein's critic physicist Ernst Gehrcke made similar statements, Einstein called him "anti-Semitic". Zangger received yet another letter from Albert Einstein dated 3 January 1920, in which Einstein stated, among other things,

"As for me, since the light deflection result became public, such a cult has been made out of me that I feel like a pagan idol."[6]

When Einstein's critic Ernst Gehrcke made similar statements, Einstein called him "anti-Semitic".

The press claimed that Einstein was the greatest and most original thinker that the world had ever seen. No one knew better than Einstein himself that the press was deliberately lying to the public. Albert Einstein wrote to Hendrik Antoon Lorentz on 19 January 1920,

"Nevertheless, unlike you, nature has not bestowed me with the ability to deliver lectures and dispense original ideas virtually effortlessly as meets your refined and versatile mind. [***] This awareness of my limitations pervades me all the more keenly in recent times since I see that my faculties are being quite particularly overrated after a few consequences of the general theory stood the test."[7]

1.2.2 The "Jewish Press" Sanctifies a Fellow Jew

Adapting his title from a poem by Adelbert von Chamisso,[8] Kurt Joël promoted

---

[6]. Letter from A. Einstein to H. Zangger of 3 January 1920, English translation by A. Hentschel, *The Collected Papers of ALbert Einstein*, Volume 9, Document 242, Princeton University Press, (2004), pp. 204-205, at 204.

[7]. Letter from A. Einstein to H. A. Lorentz of 19 January 1920, English translation by A. Hentschel, *The Collected Papers of Albert Einstein*, Volume 9, Document 265, Princeton University Press, (2004), p. 220.

[8]. Adelbert von Chamisso:

Die Sonne bringt es an den Tag

Gemächlich in der Werkstatt saß
Zum Frühtrunk Meister Nikolas,
Die junge Hausfrau schenkt' ihm ein,
Es war im heitern Sonnenschein. —
Die Sonne bringt es an den Tag.
Die Sonne blinkt von der Schale Rand,
Malt zitternde Kringeln an die Wand,
Und wie den Schein er ins Auge faßt,
So spricht er für sich, indem er erblaßt :
"Du bringst es doch nicht an den Tag" —

"Wer nicht? was nicht?'. die Frau fragt gleich,
"Was stierst du so an? was wirst du so bleich?"
Und er darauf: "Sei still, nur still !
Ich's doch nicht sagen kann noch will.
Die Sonne bringt's nicht an den Tag."

Die Frau nur dringender forscht und fragt,
Mit Schmeicheln ihn und Hadern plagt,
Mit süßem und mit bitterm Wort;
Sie fragt und plagt ihn Ort und Ort :
"Was bringt die Sonne nicht an den Tag?"

"Nein nimmermehr!" — "Du sagst es mir noch."
"Ich sag es nicht." — "Du sagst es mir doch."
Da ward zuletzt er müd und schwach
Und gab der Ungestümen nach. —
Die Sonne bringt es an den Tag.

"Auf der Wanderschaft, 's sind zwanzig Jahr,
Da traf es mich einst gar sonderbar.
Ich hatt nicht Geld, nicht Ranzen, noch Schuh,
War hungrig und durstig und zornig dazu. —
Die Sonne bringt's nicht an den Tag.

Da kam mir just ein Jud in die Quer,
Ringsher war's still und menschenleer,
'Du hilfst mir, Hund, aus meiner Not!
Den Beutel her, sonst schlag ich dich tot!'
Die Sonne bringt's nicht an den Tag.

Und er: 'Vergieße nicht mein Blut,
Acht Pfennige sind mein ganzes Gut!'
Ich glaubt ihm nicht und fiel ihn an ;
Er war ein alter, schwacher Mann —
Die Sonne bringt's nicht an den Tag.

So rücklings lag er blutend da;
Sein brechendes Aug in die Sonne sah;
Noch hob er zuckend die Hand empor,
Noch schrie er röchelnd mir ins Ohr.
'Die Sonne bringt es an den Tag!'

Ich macht ihn schnell noch vollends stumm
Und kehrt ihm die Taschen um und um:
Acht Pfenn'ge, das war das ganze Geld.
Ich scharrt ihn ein auf selbigem Feld —
Die Sonne bringt's nicht an den Tag.

Dann zog ich weit und weiter hinaus,

Albert Einstein in the *Vossische Zeitung* morning edition on 29 May 1919.

> "Die Sonne bringt es an den Tag?
> Eine Himmelsentscheidung in der
> Relativitätstheorie.
>
> Von
> Kurt Joël.

Sonnenfinsternisse sind sicherlich nichts Seltenes. Wiederholt sind in den letzten hundert Jahren wissenschaftliche Expeditionen ausgerüstet worden, um sie zu beobachten und die Ergebnisse dieser Beobachtung zu verarbeiten. Und doch sieht man der Verfinsterung unseres Zentralgestirns, die heute, am 29. Mai, eintritt und 3 Stunden 17 Minuten währt, mit besonderer Spannung entgegen. Nicht etwa wegen der langen Dauer dieser Finsternis, die mit der schmalen Zone ihrer Totalität das nördliche Brasilien und Mittelafrika durchschreitet und zu deren Erforschung von England aus zwei Unternehmungen — die eine mit dem Standort in Sobral (Brasilien), die andere nach der Insel Isla do Principe, etwa 180 Kilometer von der afrikanischen Küste — ausgerüstet worden sind. Nicht bloß die Astronomen, auch Physiker, Mathematiker, selbst Philosophischen harren auf die endgültigen Ergebnisse dieser Himmelsbeobachtung, da sie mittelbar helfen sollen, eine der wichtigsten neueren physikalischen, ja erkenntnistheoretischen Fragen, die Einsteinsche Gravitationstheorie, zu

---

Kam hier ins Land, bin jetzt zu Haus. —
Du weißt nun meine Heimlichkeit,
So halte den Mund und sei gescheit!
Die Sonne bringt's nicht an den Tag.

Wann aber sie so flimmernd scheint,
Ich merk es wohl, was sie da meint,
Wie sie sich müht und sich erbost, —
Du, schau nicht hin und sei getrost :
Sie bringt es doch nicht an den Tag."

So hatte die Sonn eine Zunge nun,
Der Frauen Zungen ja nimmer ruhn. —
"Gevatterin, um Jesus Christ!
Laßt Euch nicht merken, was Ihr nun wißt!" —
Nun bringt's die Sonne an den Tag.

Die Raben ziehen krächzend zumal
Nach dem Hochgericht, zu halten ihr Mahl.
Wen flechten sie aufs Rad zur Stund?
Was hat er getan? wie ward es kund?
Die Sonne bracht es an den Tag.

beantworten.

Nach der Einsteinschen Relativitätstheorie muß ein Strahl, der von einem Stern aus tangential zur Sonne verläuft, um $1,74''$ abgelenkt werden und die Ablenkung für andere Sterne umgekehrt proportional diesem Abstand vom Mittelpunkt der Sonne sein. Beeinflußt nun wirklich die Sonne den Lichtstahl und damit die scheinbaren Oerter der Sterne? Diese Frage sollte bereits im August 1914, wo ebenfalls eine Sonnenfinsternis stattfand, entschieden werden, jedoch hat der Krieg die Arbeit der meisten Expeditionen gestört. Welche Entscheidung wird nun der Himmel für Einsteins Theorien bringen?

Schon einmal hat dieser Forscher den Himmel zum Zeugen für die Richtigkeit seiner Theorie angerufen. Es handelte sich um die Perihelbewegung des Merkur, die bis dahin den Erklärungsversuchen der Physiker und Astronomen getrotzt hatte. Das Perihel (der Punkt der Sonnennähe) erfährt im Sinne der Bewegung des Planeten eine sehr geringe, aber ganz sicher nachgewiesene Bewegung, die in hundert Jahren auf den freilich nicht übermäßigen Betrag von 43 Bogensekunden wächst, sich aber aus den Grundlagen der von Newton begründeten klassischen Mechanik nicht hat ableiten lassen. Der Astronom Leverrier hat durch Rechnung gezeigt, daß diese Abweichung der Beobachtung von der Rechnung bei Zugrundelegung der Newtonschen Mechanik nur durch die Annahme unbekannter Massen erklärt werden könne. Aber nach solchen Massen hat man bisher vergeblich gesucht. Da verband Albert Einstein die Gravitation mit seiner Relativitätstheorie; die gewonnenen Bewegungsgleichungen lieferten in ganz überraschender Weise für den Umlauf eines Planeten um die Sonne eine Bewegung des Perihels, die für den Merkur vollständig mit der beobachteten übereinstimmt, während sie bei den entfernteren Planeten einen so geringen Betrag ausmacht, daß sie auch da mit den nicht mit völliger Sicherheit ermittelten kleinen Bewegungen übereinstimmen würde.

Bevor wir uns der hohen wissenschaftlichen Bedeutung der heutigen Sonnenfinsternis zuwenden, wollen wir in wenigen Sätzen das Wesen des Relativitätsprinzips erläutern. Unstreitig sind alle Beobachtungen und Wahrnehmungen relativ, d. h. abhängig von den Bewegungs- und Geschwindigkeitsunterschieden, die zwischen dem beobachteten Vorgang und dem Beobachter bestehen. Betrachten wir z. B. den freien Fall eines Körpers auf der Erde und nehmen wir an, daß diesen Vorgang einmal jemand beobachtet, der ruhig auf der Erde steht, und das andere Mal jemand, der sich etwa mit 100 000 Kilometer in der Sekunde von der Erde fortbewegt. Dann ist es ohne weiteres klar, daß beide Beobachter verschiedene Fallzeiten und Räume feststellen würden. Einstein hat nun gezeigt, daß eine Zeitangabe niemals etwas Absolutes und für alle Orte in gleicher Weise Zutreffendes ist, sondern nur in Verbindung mit dem Bewegungszustande eines Körpers einen bestimmten Sinn haben kann.

Nachdem er so klargelegt hatte, daß man den Begriff der Zeit und der Länge relativieren, d. h. abhängig von dem Bezugsystem annehmen muß, ist er weiter dazu übergegangen, auf den Zusammenhang zwischen

Gravitation und Trägheit im Lichte dieser Relativitätstheorie hinzuweisen. Er veranschaulicht das durch folgende Betrachtungen. Wenn ein irgendwo in der Welt in einem geschlossenen Kasten befindlicher Physiker beobachtete, daß alle sich selbst überlassenen Gegenstände in eine bestimmte Beschleunigung geraten, etwa stets mit konstanter Beschleunigung auf den Boden des Kastens fallen, so könnte er diese Erscheinung auf zwei Arten erklären: Erstens könnte er annehmen, daß sein Kasten auf einem Himmelskörper ruhe, und den Fall der Gegenstände auf dessen Gravitationswirkung zurückführen. Zweitens aber könnte er auch annehmen, daß der Kasten sich mit konstanter Beschleunigung nach „oben" bewegt; dann wäre das Verhalten der „fallenden" Gegenstände durch ihre Trägheit erklärt. Beide Erklärungen sind genau gleich möglich, jener Physiker hat kein Mittel, zwischen ihnen zu entscheiden. Nimmt man an, daß alle Beschleunigungen relativ sind, daß also ein Unterscheidungsmittel prinzipiell fehlt, so läßt sich dies verallgemeinern: an jedem Punkt des Universums kann man die beobachtete Beschleunigung eines sich selbst überlassenen Körpers entweder als Trägheitswirkung auffassen oder als Gravitationswirkung, d. h. man kann entweder sagen: „das Bezugsystem, von dem aus ich den Vorgang beobachte, ist beschleunigt" oder: „der Vorgang findet in einem Gravitationsfelde statt". Die Identität der trägen und der gravitierenden Masse ist, wie M. Schlick in seinem Schriften „Raum und Zeit in der gegenwärtigen Physik" ausführt, der eigentliche Erfahrungsgrund, der uns erst das Recht gibt zu der Annahme oder der Behauptung, daß die Trägheitswirkungen, die wir an einem Körper beobachten, auf den Einfluß zurückzuführen sind, den er von anderen Körpern erleidet. Einstein ist es nun wirklich gelungen, ein Grundgesetz aufzustellen, das Trägheits- und Gravitationserscheinungen in gleicher Weise umfaßt.

Denken wir wieder an den beschleunigten Kasten und nehmen an, daß er an seiner Seitenwand ein Loch habe. Welchen Weg legt nun ein Lichtstrahl, der senkrecht zur Bewegungsrichtung in den Kasten fällt, gegenüber dem Kasten zurück? In einem gleichförmig bewegten System läuft er geradlinig, in einem beschleunigten System wird ein quer zur Bewegungsrichtung lausender Lichtstrahl demnach zurückbleiben. Sind nun die Gesetze der Schwerefelder wie die bewegter Systeme, so muß auch im Schwerefelde der Lichtstrahl in der Richtung der Schwerkraft aus der geraden Bahn abgelenkt werden. Das folgt aus Einsteins Theorien, und diese Folgerung hat auch der Forscher gezogen. Auf der Erde selbst ist eine solche Messung nicht durchzuführen, da ihr Gravitationsfeld nicht stark genug ist. Wohl aber könnte das Gravitationsfeld der Sonne dazu ausreichen. Das Licht eines Sternes, das sehr nahe an der Sonne vorbeikommt, müßte durch ihr Gravitationsfeld um 1,74" aus seiner Bahn abgelenkt werden. Die Beobachtungen der Astronomen bei der heutigen Sonnenfinsternis — die Sonne ist infolgedessen genügend abgeblendet, um eine Beobachtung des reichen Feldes von Sternen in ihrer Nähe zuzulassen — sollen nun den Beweis erbringen, ob Einsteins Voraussage richtig ist. Damit wäre zugleich eine neue experimentelle Stütze für die Relativitätstheorie geschaffen, die berufen ist, unsere bisherigen Raum- und Zeitbegriffe wesentlich zu beeinflussen."

Carrying on the tradition of the literary tributes paid to Newton in Edmund Halley's *Ode to Newton*,[9] and Voltaire's *Letters Concerning the English Nation*, Alexander Moszkowski promoted the cult of Einstein with a tribute to Albert Einstein in the *Berliner Tageblatt* (which Jewish racist Zionist Theodor Herzl called a "Jewish paper"[10]), Volume 48, Number 476, on 8 October 1919,

"Die Sonne bracht' es an den Tag!
Von
Alexander Moszkowski.

Sie wurde befragt, sie hat Antwort gegeben, und das Echo ihres Orakels wird durch die Jahrhunderte klingen. Wir Menschen von heute stehen dem Ereignis selbst noch zu nahe, als daß wir dessen weitreichende Bedeutung vollkommen ermessen könnten. Aber wir erinnern uns der Ansage des Goetheschen Ariel:

>Phöbus' Räder rollen prasselnd,
>Welch Getöse bringt das Licht!
>Es trometet, es posaunet,
>Auge blinzt und Ohr erstaunet!

Es wird des Erstaunens kein Ende sein über diese Sonnenbotschaft, die sich an das Zentrum menschlichen Denkens wandte. Wir wollten wissen: Ist die Verfassung der Welt begreiflich? Und Phöbus sprach: Sie ist es, ist dem menschlichen Verstand zugänglich, wenn die neue allgemeine Relativitätslehre Einsteins aller Betrachtung zugrunde gelegt wird.

---

[9]. E. Halley, in Newton's *Principia* in the translation by A. Motte, revised and annotated by F. Cajori, "Ode to Newton", *Principia*, Volume 1, University of California Press, Berkeley, Los Angeles, London, (1962), pp. XIII-XV.

[10]. Racist political Zionist Theodor Herzl wrote on 12 June 1895,

> "Jewish papers! I will induce the publishers of the biggest Jewish papers (*Neue Freie Presse, Berliner Tageblatt, Frankfurter Zeitung,* etc.) to publish editions over there, as the *New York Herald* does in Paris."—T. Herzl, English translation by H. Zohn, R. Patai, Editor, *The Complete Diaries of Theodor Herzl*, Volume 1, Herzl Press, New York, (1960), p. 84.

THE DEARBORN INDEPENDENT, praised the *New York Herald*. "When Editors Were Independent of the Jews", THE DEARBORN INDEPENDENT, (5 February 1921). *See also:* T. Herzl, English translation by H. Zohn, R. Patai, Editor, *The Complete Diaries of Theodor Herzl*, Volumes 1 and 2, Herzl Press, New York, (1960), pp. 37, 97, 170, 455, 457, 480. *See also:* A. Elon, *Herzl*, Holt, Rinehart and Winston, New York, (1975), pp. 167-168. *See also: The Collected Papers of Albert Einstein*, Volume 7, Document 35, Princeton University Press, (2002), pp. 296-297, note 8.

Am 29. Mai dieses Jahres wurde die Sonne zur Zeit einer totalen Bedeckung befragt. Ihre Antwort bestand zunächst nur in einigen Lichtpunkten auf photographischen Platten. Aber in diesen Punkten lag die Erklärung des Geheimnisses beschlossen. Es bedurfte noch allerfeinster Messungen, um diese Punktierschrift in eine gültige physikalische Erklärung zu übersetzen. Zwei englische Expeditionen, nach Brasilien und nach Innerafrika, hatten es übernommen, dies zu entwickeln, zu messen und auszudeuten. Vor wenigen Tagen traf die Bestätigung ein: Die Lichtbotschaft steht in vollstem Einklang mit der Annahme jenes Weltsystems, wie es von Einsteins Lehre gefordert wird. Und diese selbst, aus Gedankenexperimenten entsprossen, ist nunmehr auch durch das sinnlich erfaßbare, astronomische Experiment unerschütterlich bewiesen.

Nur mit wenigen Worten sei das Wesen dieses Experimentes andeutungsweise erläutert. Nach Einstein begeben sich die kosmischen Ereignisse in einer vierdimensionalen Raumzeitwelt, innerhalb deren die Newtonsche Bewegungslehre der Himmelskörper nur eine Annäherung darstellt. Zur Erfassung der allgemeinen Vorgänge bedarf es der Einführung einer Ueber-Euklidischen Geometrie, deren Ermittelung von „Weltlinien" im Raumzeitlichen und der Aufgabe jeder Fernwirkung, deren Annahme eigentlich dem menschlichen Denken widerspricht. Die zuerst so verwirrende, mathematisch verwickelte und deshalb überaus schwierige Lehre verwandelt sich, je mehr man in sie eindringt, in die denkbar lichtvollste Vereinfachung des gesamten Weltbildes, in eine wirklich restlose Erfassung der letzten kosmischen Fragen.

Schon einmal hatte diese Lehre in einem früheren Stadium ihrer Entwicklung eine sichtbare Kreuzprobe bestanden, damals, als es ihr gelang, gewisse, sonst ganz unerklärliche Anomalien in der Bahn des Planeten Merkur als durchaus normal und mit der Berechnung übereinstimmend zu erweisen. Aber hinter dieser Kreuzprobe stand eine zweite, die den Lichtstrahl selbst auf seiner Wanderung durch die Welt verfolgen sollte. Eine Ungeheuerlichkeit tat sich auf: Bestand diese Lehre zu Recht, dann mußte sich in sehr starken Gravitationsfeldern — also etwa beim Durchgang in Sonnennähe — eine merkliche Krümmung der Lichtstrahlen herausstellen. Und eben hierauf waren die Anstrengungen der beiden englischen Expeditionen gerichtet. Es galt die Abbiegung der Lichtstrahlen zu erweisen, die, von Fixsternen ausgesendet, an der verdunkelten Sonne vorbeistreichen, um unser Auge oder — experimentell sicherer — die photographische Platte zu erreichen. Fand diese Abbiegung wirklich statt, so mußte sich dies dadurch offenbaren, daß auf der Platte die Sterne weiter auseinanderstanden, als man nach ihrer wirklichen Position erwarten konnte.

Um wieviel wohl? Die Berechnung verlangte unglaubliche Feinheiten des Ausmaßes. Man stelle sich den ganzen Himmelsbogen vor, in Grade eingeteilt: dann ergibt eine Mondbreite etwa einen halben Grad. Hiervon der dreißigste Teil, eine Bogenminute, ist noch gut vorstellbar. Aber hiervon wiederum der sechzigste Teil, die Bogensekunde, entzieht sich nahezu aller sinnlichen Erfaßbarkeit. Und auf dieses Kleinmaß kam es an: denn die in reiner Gedankenarbeit entwickelte Theorie sagte eine Ablenkung von ein und sieben Zehntel Bogensekunde an. So stand diese Größenordnung auf dem Papier, vorläufig ohne Bewahrheitung durch astronomische Praxis, aber festverankert in einem System unheimlicher Gleichungen, die in ihrer Gesamtheit die wahre Ordnung des bewegten Universums verkünden.

Wirklich, es war etwas viel verlangt von den fernen Welten, denen nunmehr ein blinkendes Zeugnis abverlangt wurde. Sie hatten sich zur Zeit einer totalen Sonnenfinsternis so rundum zu gruppieren, daß sie eben noch leuchtende Lichtpünktchen entwarfen, deren Stellung mit Ja und Nein für die vorausberechnete Größenordnung einstehen sollte. Und zwar mit einem Zeugnis, das im Bejahungsfall eine durch Jahrtausende überlieferte Grundanschauung des Menschenhirns überwältigte.

Wie denn? Ein Sternstrahl soll krumm werden können? Widerstreitet daß nicht dem Elementarbegriff der geraden, der kürzesten Linie, für die wir ja keine anschaulichere Vorstellung besitzen, als eben im Strahl? Hatte doch Leonardo da Vinci die Gerade direkt so definiert, so benannt als die „*linea radiosa*"!

Aber für diese vermeintliche Selbstverständlichkeit ist in der vom Forschergeist Einsteins durchstrahlten Welt kein Platz mehr. Die am 29. Mai befragte Konstellation hat die Entscheidung geliefert. Mehr als ein Vierteljahr hat es gedauert, ehe die Punktrunen genügend entziffert waren. Jetzt ist die Bestätigung eingetroffen: die Sternstrahlen werden tatsächlich im Schwerefelde der Sonne abgelenkt, sie zeigen eine Krümmung mit der Hohlseite zur Sonne gewendet, so daß sich der scheinbare Abstand der geprüften Sterne vergrößert: und dies innerhalb gewisser Beobachtungsgrenzen, die Einsteins vorausgesagter Größenordnung entsprechen. Was nur dann möglich ist, wenn das Fundamentalgerüst Einsteins, die allgemeine Relativitätstheorie, als die wahre Verfassung des Universums angesprochen wird.

Galt dies dem mathematischen Denker, dem strengen Physiker schon vorher als Gewißheit, so wird fortan auch für den Erkenntnistheoretiker der letzte Zweifel die letzte Zuflucht zu räumen haben. Ja, man darf voraussagen, daß der größte Gewinn aus der jetzt völlig sichergestellten Einsicht dereinst dem Philosophen zufallen wird, der darauf ausgeht erkenntnistheoretisch das allereinfachste, mit allen Beobachtungstatsachen restlos harmonierende Weltbild zu entwerfen. Er wird auf Kant fußend, aber über Kant hinauswachsend die Idealformen der Anschauung in Raum und Zeit erhöhen und emporläutern zum vierdimensionalen Ordnungsschema, in welchem der letzte Restsinnlicher Schlacke abzufallen hat vor der reinen Erkenntnis des wahren raum-zeitlichen Weltgefüges. Wenn dereinst ein bestimmter Augenblick bezeichnet werden soll als historisches Zeichen für die große Wandlung in menschlicher Anschauung gegenüber dem Universum, so wird manch einer den zuvor genannten Tag als das deutlichste Merkdatum wählen. Und wenn er ihn nennt, so wird er hinzufügen, daß eine letzte Wahrheit entschleierbar war über Galilei und

Newton, über Kant hinaus, bestätigt durch einen Orakelspruch aus der Tiefe des Himmels, in lesbarer Strahlschrift. Das Uebereinstimmen einer Menschenforschung mit der Wirklichkeit des Weltgeschehens — „Die Sonne bracht' es an den Tag!'"''

Shortly after this article appeared, Heinrich Zangger wrote to Albert Einstein on 22 October 1919,

"I already filled the official's heads with the bent light, years ago.— Proclaimed Galileo-Newton-Einstein—so if you want the appointment, or keep it, resp., it would be a joy to all."[11]

Friedrich Karl Wiebe[12] alleged in 1939, that the press in post-World War One Germany, and with it public opinion, was largely controlled by traitorous Jews who cheapened the medium with sensationalism—by Jews who allegedly only cared about Jewish interests and who would pursue those perceived self-interests at the expense of other Germans. Jews have long been noted for making judgments based on selfish interests to the exclusion of broader societal interests, or pure principles, or a sense of fairness, as is typified by the common racist Jewish expression, "Is it good for the Jews?"

Though Wiebe only incidentally mentions the publisher Julius Springer, a man who was very influential in promoting Einstein and who sought to discredit Einstein's critics, Wiebe does name the publishing house of the Jewish brothers Ullstein, and the publishing house of the "Eastern Jew" Rudolf Mosse. Wiebe states that the *Berliner Morgenpost*, which he alleged had the largest circulation of any German newspaper, was controlled by Jews, as was the politically influential *Vossische Zeitung*, under editor-in-chief Geog Bernhard. The *Berliner Tageblatt*, which served as spokesman for Germany abroad and was often quoted in America and England, was led by editor-in-chief Theodor Wolff, and the *Acht-Uhr-Abendblatt* also had a Jewish chief editor. One might, together with Theodor Herzl,[13] add the *Frankfurter Zeitung* to the list of "Jewish

---

11. Letter from H. Zangger to A. Einstein of 22 October 1919, English translation by A. Hentschel, *The Collected Papers of Albert Einstein*, Volume 9, Document 148, Princeton University Press, (2004), pp. 126-128, at 127.

12. F. K. Wiebe, *Deutschland und die Judenfrage*, M. Müller & Sohn, Hrsg. im Auftrage des Instituts zum Studium der Judenfrage, Berlin, (1939); **English** translation, *Germany and the Jewish Problem*, Published on behalf of the Institute for the Study of the Jewish Problem, Berlin, (1939); **French** translation, *L'Allemagne et la Question Juive*, Berlin, Edité sous les auspices de l'Institut pour l'étude de la question juive, (1939); **Spanish** translation, *Alemania y la Cuestión Judía*, Publicado por encargo del Instituto para el Estudio de la Cuestión Judía, Berlín, (1939).

13. Racist political Zionist Theodor Herzl wrote on 12 June 1895,

"Jewish papers! I will induce the publishers of the biggest Jewish papers (*Neue Freie Presse, Berliner Tageblatt, Frankfurter Zeitung*, etc.) to publish editions over there, as the *New York Herald* does in Paris."—T. Herzl, English translation by H. Zohn, R. Patai, Editor, *The Complete Diaries of Theodor*

newspapers". Many of these papers promoted Einstein and personally attacked his critics. Wiebe alleged that Jews ran the *Reichverband der Deutschen Presse* and the *Verein Berliner Presse*. Wiebe names Georg Bernhard, Theodor Wolff and Maximilian Harden as Jews who had "stabbed Germany in the back" following World War One. He noted that historian Friedrich Thimme dubbed Harden, "the Judas of the German people". Harden was a politically active Zionist Jew.[14]

Germany had been very good to the Jews. German Jews were the wealthiest people in the world. In the years following the First World War, the Germans resented the fact that the Jews, Einstein being their chief spokesman, had stabbed the Germans in the back during the war, and then twisted the knife at the peace negotiations in France, where a large contingent of Jews decided Germany's fate, and reneged on Woodrow Wilson's Fourteen Points, one of which assured Germany that it would lose no territory. The Germans had thought that Wilson's pledge would be honored after the Germans had surrendered in good faith. Had not the Germans received this promise of the Fourteen Points, they would not have surrendered and were in a position to continue the war. The promise was broken by Jews and their agents.

In addition, the Allies insisted that Germany pay draconian war reparations that would forever ruin the nation. Leading Jews in Germany sided with the Allies against their native land. It was obvious that leading Jews were profiteering from the war in every way possible, at the expense of the German nation and its People. Jewish leaders instigated crippling strikes in the arms industry, which left German troops without adequate armaments. Jewish revolutionaries took advantage of Germany's weakened state, which Jews had deliberately caused for the purpose, and created a Soviet Republic in Bavaria and overthrew the monarchy. Civil war and revolutions were always a Jewish strategy to turn Gentile brother against brother (*Judges* 7:22. *Haggai* 2:22). German-Jewish bankers cut off Germany's access to funds. German-Jewish Zionists moved to London and brought America into the war on the side of the British at the very moment Germany was about to win the war. Those arms which were produced were often substandard and were peddled by Jews to Jews in the German Government, which also left the German troops without adequate arms, while making Jews immensely wealthy. German-Jewish bankers

---

Herzl, Volume 1, Herzl Press, New York, (1960), p. 84.

THE DEARBORN INDEPENDENT, praised the *New York Herald*. "When Editors Were Independent of the Jews", THE DEARBORN INDEPENDENT, (5 February 1921). *See also:* T. Herzl, English translation by H. Zohn, R. Patai, Editor, *The Complete Diaries of Theodor Herzl*, Volumes 1 and 2, Herzl Press, New York, (1960), pp. 37, 97, 170, 455, 457, 480. *See also:* A. Elon, *Herzl*, Holt, Rinehart and Winston, New York, (1975), pp. 167-168. *See also:* The Collected Papers of Albert Einstein, Volume 7, Document 35, Princeton University Press, (2002), pp. 296-297, note 8.

14. S. E. Weltmann, "Germany, Turkey, and the Zionist Movement, 1914-1918", *The Review of Politics*, Volume 23, Number 2, (April, 1961), pp. 246-269, at 266.

conspired with German arms manufacturers to produce weapons for both sides. The German-Jewish press, which had initially beat the war drums louder than anyone else, teamed up with leading Jews in the German Government at the end of the war and demanded that Germany submit to the demands of the Allies, give up vast territories and make the reparations payments. The German-Jewish press and Jews in the German Government, many of whom were the same persons who had most boisterously called upon the German People to go to war, insisted that the Germans accept responsibility for causing the war, though they had not caused it. Etc. Etc. Etc.

England was not immune to the same processes of Socialism which brought about the ruin of Germany and Russia at the hands of the Jewish bankers. Socialists had long attacked British industrialization and sought to undermine British society so that they could overthrow the British Government. On 17 March 1919, *The London Times* reported on page 18,

"AN ALBERT HALL SPEECH.
SOCIALIST'S DEFENCE.

At Bow-street Police Court on Saturday, before Sir John Dickinson, WILLIAM FORSTER WATSON, 37, turner's engineer, of Enderwick-road, Hornsey, and Featherstone-buildings, Holborn, was charged, on remand, under the Defence of the Realm Regulations, with making seditious utterances at a meeting, convened by the British Socialist Party, held at the Albert Hall on February 8. In a speech the defendant, it was alleged, urged the audience to seize upon every little bit of industrial unrest, and to make demands upon the employers with which they could not comply.

Sir Archibald Bodkin conducted the case on behalf of the Director of Public Prosecutions.

Chief Inspector Parker, of the Special Branch at Scotland Yard, produced some documents found in the possession of the defendant, and in cross-examination said the fact that the defendant had recently started a paper had nothing whatever to do with this prosecution.

At the close of the case for the prosecution the defendant pointed to the few persons in the public part of the Court and asked that some of the large crowd waiting outside might be admitted. He gave an assurance that, so far as he had any influence, the untoward demonstration in Court last week would not be repeated.

The Magistrate said that such a demonstration would never be allowed again in any Court. He refused to permit the admission of any of the public other than a few persons whom the defendant had specially mentioned.

For the defence, Mr. Edward Charles Fairchild, Chairman of the Albert Hall meeting, said that the impression left upon his mind by the defendant's speech was that if there should be continual encroachments upon liberty, the workers would be ultimately entitled to resist, but there was no specific call to workers to arm themselves for purposes of immediate violence.

The Rev. Cavendish Moxon, a curate of the Church of England, said that he was not in favour of aggressive violence in any movement and was not an extreme pacifist. The defendant's speech, taken as a whole, did not impress him as being an incitement to violence. One of his phrases, 'Arm yourselves if necessary,' meant, in the witness's view, that if the worst came to the worst, the workers would have to arm themselves in self-defense.

The Magistrate quoted from the transcript of the defendant's speech, and asked the witness if he considered it right to make demands upon the employing class for such conditions as would make it impossible for them to carry on.

The witness replied that that was the Socialist view, and he agreed with it in the sense of substituting the control of the workers for the control of the masters.

Ex-Inspector John Syme, who said he was now engaged in 'Exposing the Home Office,' expressed the opinion that the defendant's speech was not meant to be taken literally. The defendant certainly did not create the impression that he was advocating the immediate purchase of revolvers, guns, and such things.

The defendant.—There are plenty doing that to-day without my advocating it.

Other evidence for the defence having been given, the defendant was again remanded on bail in two sureties of £100 each.

On leaving the Court the defendant was loudly cheered by a large crowd of sympathizers."

Michael B. Laughlin wrote of the French Socialists, who sought to cripple their nation in time of war with general strikes,

"In 1901 an obscure French history professor named Gustave Hervé attained instant notoriety by writing an article in the regional socialist press which seemed to include the image of the French tricolour planted in a pile of manure. Hervé soon founded an important and diverse anti-militarist movement called Hervéism which attempted to end war and promote socialism by revolutionary means, including a military insurrection and a workers' general strike in the event of war.[1] When France's various socialist parties united in 1905 to form the SFIO (Section Française de l'Internationale Ouvrière), the Insurrectional Socialist component of Hervéism became its most extreme faction. In 1906 Hervé and his heterogeneous band founded a weekly newspaper, *La Guerre sociale*, which attempted to unite socialists, syndicalists and anarchists around his anti-militarist programme. For the next six years Hervéists employed an array of tactics, including sensational press campaigns and the formation of provocative conspiratorial organizations to generate interest and to unite France's divided revolutionary factions.²"[15]

---

15. M. B. Loughlin, "Gustave Hervé's Transition from Socialism to National Socialism:

Infamous British Communist John Spargo admitted in 1929 that Socialists were always out to destroy society so as to leave it ripe for revolution, and one might add that they blamed the ills that they deliberately caused on those who were trying to prevent them—they covertly caused the People to suffer in the name of a new "Utopia" to come,

> "[T]he sooner the process of degradation is effected the better, for the sooner will the agony be over and the glorious consummation of Socialism be realized. [***] Haters of All Social Reforms. That logic controlled the policy of British Socialism in the days of my youth. That is why we busied ourselves distributing leaflets bearing the significant title, 'To Hell With Trade Unionism!' and appropriately printed in red. That also is why we inveighed against life insurance in our propaganda with all the bitterness of which we were capable. Life insurance was a protective device against poverty, an ameliorative measure designed to avert the poverty and degradation without which our Utopia could not be reached. In the same spirit and under the compulsion of the same Marxian dogma we opposed every form of thrift, all philanthropy and social reforms calculated to lessen social misery and improve the conditions of life and labor. We regarded all these things with the hate and horror which religious fanatics might feel towards deliberate human thwarting of the clearly manifested design of God."[16]

While millions of Germans were starving to death, top Jews in Germany had never known better times. Whenever anyone revealed the truth of what was happening, the Jewish press immediately smeared them by calling them "anti-Semites". The situation was similar to, though even worse than, the situation in America today.

In 1933, the Jews Abraham Myerson and Isaac Goldberg alleged many of the same facts Wiebe would later allege, though they offered an entirely different perspective on the same issues. Myerson and Goldberg wrote, in 1933, in their book, *The German Jew: His Share in Modern Culture*,

> "The circles of criticism and of journalism in Germany were, up to the incursions of Hitler, predominantly Jewish. Julius Bab, Alfred Kerr, Fritz Engel, Felix Holländer, Felix Salten (author of *Bambi*), Siegmund Freund, Emil Faktor... the roster is long; nor have we mentioned critics from the professorial fold, such as Richard M. Meyer.

Another Example of French Fascism?", *Journal of Contemporary History*, Volume 36, Number 1, (January, 2001), pp. 5-39, at 5.

16. J. Spargo, "Why I Am No Longer a Socialist", *Nation's Business*, Volume 17, (February, 1929), pp. 15-17, 96, 98, 100; (March, 1929), pp. 29-31, 168, 170; at pages 96 and 98 of the February issue. Reprinted: *Why I Am No Longer a Socialist*, Chamber of Commerce of the United States, Washington, D.C., (1929).

Publishing in Germany has largely been built up by a Jewish passion for commercial pursuits that parallels the passion of intellect so freely evidenced in the Jew. Through such powerful interests as those of the Lachmann-Mosse family and the estate of Leopold Ullstein, the largest publishing firm in Germany, the press and the magazine world have been controlled by German Jews. Before it was 'coordinated' into the Nationalist régime, the house of Ullstein employed almost eight thousand persons, and issued almost a hundred newspapers and periodicals. Ullstein (1826-99) passed the fast-growing business on to five industrious sons.

Rudolph Mosse (1843-1920) founded the *Berliner Tageblatt* in 1872. It was, until the descent of Hitler upon the Jews, one of the great newspapers of the world, known to all journalists as a palladium of liberalism... . Naturally, although these newspapers and their allied interests employed a host of Gentile workers, there were countless Jews in their offices. Among editors and journalistic powers were to be found such gifted paladins as Maximilian Harden and Theodor Wolff. The statistical fact is that the Jewish mind, for reasons that have impelled it to the other artistic and literary pursuits, engages naturally in journalism and criticism. Even so anti-Semitic a writer as Friedrich von Oppeln-Bronowski has been quoted as blaming, not the Jews, but the inertia of his fellow-Germans. 'The outcry of the conservative press against the literary incursions of the Jew reminds me of the clamour raised by the inferior business man against his more clever, ' unfair' competitor. Instead of making complaint, it had better improve itself. If it is true that the Jews have assumed so disproportionate a role in journalism, we can undoubtedly connect the fact with their exclusion under the old régime from the higher governmental positions.' [*Footnote:* See I. E. Poritzky: 'The Jew in the Intellectual Life of Germany,' *Menorah Journal*, Vol. XII, No. 6 (1926). I refer to this article those who are in search of many Jewish names.]

In book-publishing the Jew has become a power in Germany since 1910. It is interesting to observe that at about this same time the Jew in the United States was entering upon a distinguished career in the publication of belles-lettres. In Germany the house of S. Fischer, founded in 1886, may stand for a quasi-hegemony that includes such important firms as Drei Masken, Bruno Cassirer, Kurt Wolff, Paul Zsolnay, Felix Bloch Erben, and Oesterheld & Company.

Incidentally, the famous Universal Edition, Vienna, publisher of modernist scores, though by no means confining itself to the musical advance guard, is presided over by Dr. Alfred Kalmus.

One can, therefore, understand the exaggerated outcry of Herr Bartels—though hardly sympathize with his bigoted implications—when, after descanting upon the prominence of Jews in the art and the business of letters, he is suddenly led to exclaim: 'There is no doubt that on the eve of the war our entire German life was no longer German in temper.' The situation, to him, appeared so critical that, instead of commending the universality of outlook displayed by all these Jewish publishers—can it be only a commercial accident that the Jewish firms in other countries display a like interest in publishing works of international spirit and origin?— Bartels hinted at some sort of apostasy on the part of those Gentile writers who allowed themselves to be published by Jews. These leading publishers were not only providers of books; at times they were the supporters of movements.

It is only half metaphorical to declare that, whether in the higher reaches of literature or in the forum of journalism, the German Jew has mingled his blood with printer's ink in the service of German culture. The cruelty of a régime may hold the Jew at once excommunicated and incommunicado; not by fiat, not by a conflagration of books, can it exterminate the past. Books burn; men burn; passions and ideas are immortal."[17]

With Einstein's blessing, the Jewish litterateur Alexander Moszkowski published a sensationalistic and hagiographic book, which advertised Einstein to the public in an unprecedented and shameless way: *Einstein Einblicke in seine Gedankenwelt Gemeinverständliche Betrachtungen über die Relativitätstheorie und ein neues Weltsystem Entwickelt aus Gesprächen mit Einstein*, Hoffman und Campe, Hamburg, (1921); in English translation, *Einstein: The Searcher*, E. P. Dutton, New York, (1921). This self-aggrandizing book recorded Moszkowski's conversations with Einstein, and presented Einstein to the public as if he were a god condescending to speak to mere mortals.

The public was vulnerable to such hype. Heike Kamerlingh Onnes wrote to Albert Einstein on 8 February 1920, as if Einstein were the law giver Moses,

"In my imagination I can already see you at our university's venerable rostrum that was born of the struggle for freedom of conscience,[2] smiling down at us and telling us about your communion with the gods and about the fine interplay of harmony by which hints of Nature's laws are revealed, your kind eyes sparkling with delight!"[18]

Though Jewish litterateurs were infamous for overrating Spinoza's philosophy, Mendelssohn's music, Marx's and Lasalle's political philosophies, Theodor Lessing's *Nathan der Weise*, Bergson's philosophy, etc.; that shameless self-glorification did not begin to approach the magnitude and the absurdity of the promotion of the Jewish racist Albert Einstein. Many leading scientists found such unprecedented advertising for Einstein distasteful. In 1924, Ernst Gehrcke preserved conclusive evidence that Moszkowski's book was promoted in the daily newspapers as part of an overall plan to promote Albert Einstein to the

---

17. A. Myerson and I. Goldberg, *The German Jew: His Share in Modern Culture*, A. A. Knopf, New York, (1933), pp. 140-142.
18. Letter from H. K. Onnes to A. Einstein of 8 February 1920, A. Hentschel, translator, *The Collected Papers of Albert Einstein*, Volume 9, Document 304, Princeton University Press, (2004), pp. 254-255, at 255.

gullible public through intensive advertising.[19]

As revealed in their letters to Albert Einstein,[20] the Jewish physicist Max Born and his Jewish wife Hedwig knew that this unprecedented and tasteless self-promotion would occur and that it would vindicate Einstein's critics. The Borns, who were apostate Jews, went to the extremes of threatening Einstein in order to prevent the publication of Moszkowski's book. Max Born even requested permission from Einstein to sue Moszkowski in order to block the publication of his book. The Borns had experience with Moszkowski in the past, and they knew that he would shamelessly hype Einstein for personal profit—profits the Borns wanted all to themselves. The Borns knew that Moszkowski's book would serve as proof for the outspoken Einstein critics Paul Weyland, Ernst Gehrcke and Philipp Lenard that Einstein was advertising himself to the public. The Borns, who were peddling a book of their own, *Einstein's Theory of Relativity*,[21] and who were themselves seeking to profiteer off of the Einstein brand, failed in their efforts to prevent the release of Moszkowski's work.

The press and elements of the Physics community did indeed create an "Einstein 'brand'" which has lasted. Peter Rogers, editor of *Physics World*, stated in his editorial in the August, 2004, issue of *Physics World*,

> "His legacy as the greatest physicist of all time is guaranteed, despite the regular claims that 'Einstein was wrong' or that he stole his ideas from someone else. The real opportunity presented by 2005 is the chance to sell Einstein and physics to the young. Physicists have to realize that physics needs the 'outside world' more than it needs physics. [***] Physics as a subject is lucky in having Einstein as a 'brand'[.]"[22]

Rodgers wrote, in September of 2003,

> "[...]Einstein developed the special theory of relativity in 1905. This potted history is true, of course, but it overlooks the contributions of Poincare and Lorentz. However, if every article had to give full credit for every advance in the history of physics, there would be little room for what is going on today."[23]

Rodgers also stated, in November of 2003,

---

19. E. Gehrcke, *Die Massensuggestion der Relativitätstheorie: Kuturhistorisch-psychologische Dokumente*, Berlin, Hermann Meusser, (1924), pp. 19-22, 25, 56.
20. English translation by I. Born, *The Born-Einstein Letters*, Walker and Company, New York, (1971), pp. 34-52.
21. M. Born, *Die Relativitätstheorie Einsteins und ihre physikalischen Grundlagen: gemeinverständlich dargestellt*, J. Springer, Berlin, (1920).
22. P. Rogers, "Another *Annus Mirabilis?*", *Physics World*, (August, 2004); posted on *Physics Web*, <http://physicsweb.org/articles/world/17/8/1>
23. P. Rogers, "History Revisited", *Physics World*, (September, 2003); posted on *Physics Web*, <http://physicsweb.org/articles/world/16/9/1>

"Fabrication, plagiarism and a range of other offences—duplicate submissions, conflicts of interest and referee misconduct—were among the topics discussed at a recent workshop on scientific misconduct [***] Failure to cite the work of others adequately is also an offence [***] [J]ust one more major case of fabrication or plagiarism would be very bad news for our subject."[24]

The Einstein brand was already established and used to market products in January of 1920, shortly after the press hyped Einstein and the theory of relativity in November and December of 1919. Alexander Eliasberg, a Jew who wore his Jewishness on his sleeve, wrote to Albert Einstein on 27 January 1920,

"This new type of monthly, which will serve a very large readership, is characterized by its emphasis on the sciences—of which your illustrious name serves as a symbol[.]"[25]

In letter to Albert Einstein, Paul Epstein described Alexander Eliasberg, who was Epstein's cousin, in the following terms, in the hopes that it would impress the Jewish racist and segregationist Albert Einstein,

"Eliasberg is a Jew of nationalistic bent, who stresses his Jewishness at every opportunity that presents itself. His name is emblazoned on the cover of the Jewish monthly *Jüdische Monatshefte*; furthermore, he has published a library's worth of translations from Yiddish."[26]

The Borns had a vested interest in maintaining the "Einstein myth". Einstein, himself, wrote,

"There you [Max Born] are, giving relativity lectures to stave off bankruptcy of the institute[.]"[27]

Hedwig Born's father delighted in the attention paid to Einstein in the press, because it made him proud as a Jew and as a German to see the world's scientists

---

[24]. P. Rogers, "Do's and don'ts [sic] for authors", *Physics World*, (November, 2003); posted on *Physics Web*, <http://physicsweb.org/articles/world/16/11/1>
[25]. Letter from A. Eliasberg to A. Einstein of 27 January 1920, A. Hentschel, translator, *The Collected Papers of Albert Einstein*, Volume 9, Document 286, Princeton University Press, (2004), pp. 238-239, at 239.
[26]. Letter from P. Epstein to A. Einstein of 31 January 1920, A. Hentschel, translator, *The Collected Papers of Albert Einstein*, Volume 9, Document 290, Princeton University Press, (2004), pp. 240-241.
[27]. Letter from A. Einstein to H. and M. Born of 27 January 1920, A. Hentschel, translator, *The Collected Papers of Albert Einstein*, Volume 9, Document 284, Princeton University Press, (2004), pp. 235-238, at 236.

bow down to Einstein. Viktor G. Ehrenberg, Hedwig's father, wrote to Einstein on 23 November 1919,

> "So it uplifts the heart and strengthens one's faith in the future of mankind when one sees the researchers of all nations prostrating themselves before a man of Jewish blood, who thinks and writes in the German language, in full recognition of his greatness."[28]

Paul Oppenheim also took pride in the fact that a Jew and a German was receiving a great deal of positive public attention. He wrote Albert Einstein,

> "The purpose of these lines is to congratulate you from the bottom of my heart and to express quite artlessly the pure joy that we have such a man among 'us'—in the double sense."[29]

Alexander Moszkowski was a Jewish litterateur and journalist. It had often been alleged that Jews were guilty of self-advertisement, sought to control professorships in Germany and dominate entire fields of research through corrupt means, and that there was alliance between literary and journalistic Jews—like Moszkowski—and professors—like Einstein—to market themselves to the public. For example, the primary exponent of the modern racial anti-Jewish sentiments that evolved among Hegelian revolutionaries, Zionists, Socialists and Communists in the Eighteenth and Nineteenth Centuries;[30] Eugen Karl Dühring wrote in the 1880's, decades before Moszkowski published his hagiographic book sanctifying Einstein:

> "The harmony of professors and Jews is characteristic for both parts. Incidentally, the Jews also press industriously towards university professorships; for they know that there is in this sphere something corrupt to capitalise on. Ruin allures them here too, as everywhere. In turn, the professors make use of the Jews to let the rotten structure be displayed through bold advertisement as a most highly upright and strong one. They even flirt with the literary Jews and flatter them already so that the latter may, through their press and their journals, give to the little professorial authority the varnish which these people appointed to the lectern need very much indeed. The Jews for their part, however, make a business once again

---

28. Letter from V. G. Ehrenberg to A. Einstein of 23 November 1919, English translation by A. Hentschel, *The Collected Papers of Albert Einstein*, Volume 9, Document 173, Princeton University Press, (2004), p. 145.

29. Letter from P. Oppenheim to A. Einstein of 27 November 1919, English translation by A. Hentschel, *The Collected Papers of Albert Einstein*, Volume 9, Document 179, Princeton University Press, (2004), pp. 153-154, at 153. *See also:* Editor's note 3 in the German ed.

30. P. L. Rose, *Revolutionary Antisemitism in Germany from Kant to Wagner*, Princeton University Press, (1990).

through this habitation in society. In this way they exploit for themselves not only the parties but also one of the most important branches of administration in which they become most harmful, namely that of higher education. [***] But the Germans would, however, indeed not like to forget, in the long run, their ancient forests in which they settled affairs with the Romans, to dutifully let Sinai and the Jewish blood rule. They have too much organic politics of action, and the politics of the Jews consists always only of one thing, namely of the advertisement for their people. This has revealed itself even in Messieurs Gambetta and Disraeli. [***] If the Jews in the newspapers cannot push any longer for the bad products of their people and of their comrades into the advertisement-organs and, at the same time, silence the good and suppress it through distortion, the Jewish or judaised literature will no longer appear anywhere with its wretchedness. It must, as an artificial product of the Jewish advertisement, fall into nothing, if the support of this insolent Jewish advertisement is removed which, where it suits it, raises the most inadequate daily publication to the heavens. Such Jewish advertisement manages to proclaim a subordinate Jewish litterateur or parliamentarian as a great publicist or politician, who exercises a most decisive influence on the development of at least an entire field if not indeed of the entire culture. In general, all other advertisements are strongly affected if the newspaper Jews do not have them any longer in their hands. What sort of advertisement has not been made by the latter in the newspapers, for example, for the most recent German legislation procedure of Jewish stamp, and how these press-Jews have glorified everything to the public before its introduction and, afterwards, when everybody could grasp tangibly its uselessnesses, extenuated it according to their ability! If the newspaper power remains a Jewish power, then in literature and politics, indeed even in the actual science, the most shameless advertisement is made for everything which emerges either from the Jews themselves or from those who side with the Jews, thus from actual Jewish comrades. On the contrary, the really preferable and in general everything good and honorable—to which the Jews already have an aversion from inherited instinct even when it does not have the least to do with pro or con in relation to the Jews—is basically and in an artificial way thrown aside. That however which produced from the character of the modern peoples and so is an especial honour for the nations is in every case devalued where it cannot be silenced. If the nations therefore wish that among them a public word may still be possible for the appropriate evaluation of their best people, they must free themselves from the Jewish press."[31]

Dühring gave his accounts credence by citing Jewish British Prime Minister

---

31. E. K. Dühring, *Die Judenfrage als Racen-, Sitten- und Culturfrage: mit einer weltgeschichtlichen Antwort*, H. Reuther, Karlsruhe, (1881); English translation by A. Jacob, *Eugen Dühring on the Jews*, Nineteen Eighty Four Press, Brighton, England, (1997), pp. 133-134, 138-139, 178-179.

Benjamin Disraeli, who knew in 1844 that the European revolutions of 1848 were about to occur under Jewish leadership. Disraeli wrote,

> "'You never observe a great intellectual movement in Europe in which the Jews do not greatly participate. The first Jesuits were Jews; that mysterious Russian Diplomacy which so alarms Western Europe is organized and principally carried on by Jews; that mighty revolution which is at this moment preparing in Germany, and which will be, in fact, a second and greater Reformation, and of which so little is as yet known in England, is entirely developing under the auspices of Jews, who almost monopolize the professorial chairs of Germany. Neander the founder of Spiritual Christianity, and who is Regius Professor of Divinity in the University of Berlin, is a Jew. Benary, equally famous, and in the same University, is a Jew. Wehl, the Arabic Professor of Heidelberg, is a Jew. Years ago, when I was in Palestine, I met a German student who was accumulating materials for the History of Christianity, and studying the genius of the place; a modest and learned man. It was Wehl; then unknown, since become the first Arabic scholar of the day, and the author of the life of Mahomet. But for the German professors of this race, their name is Legion. I think there are more than ten at Berlin alone.[']"[32]

Einstein's correspondence is filled with discussions about professorships and other positions of influence—as one would expect from a very well-connected professor, regardless of his or her ethnic origin. However, Einstein, who was a racist Zionist, stated that he preferred Jews for his friends and he also stated that he considered all Jews to be his brothers.[33]

Stephen G. Bloom wrote in his book *Postville: A Clash of Cultures in Heartland America*,

> "Yet despite the lack of Jewish worship and observance, and my family's total assimilation into everything American and secular, we were thoroughly Jewish. Our perspective was Jewish, as was our very essence. The world was split into two distinct halves: Jews and gentiles. Jews were always sought in business or social dealings over gentiles. A common expression used by Jews to describe a slow, dense person was—and still is—'He's got a *goyisher kop*,' which literally means 'He's got a gentile head' but figuratively means 'slow-witted.' First question when I came home and boasted of making a new friend always was 'Is he Jewish?' 'God forbid!' (my father's expression) if I should ever go out with a gentile girl, and *'Oy vey!'* (which literally means 'Oh pain!') if I ever got serious with her. All my parents' friends were Jews. They all shared the same role

---

32. B. Disraeli, *Coningsby; or, The New Generation*, H. Colburn, London, (1844), here quoted from The Century Co. edition of 1904, New York, pp. 231-232.
33. M. Born, *The Born-Einstein Letters*, Walker and Company, New York, (1971), p. 16. A. Einstein, *The World As I See It*, Citadel Press, New York, (1993), p. 89.

models: Sandy Koufax, Bernard Baruch, Bess Meyerson, Sam Levinson, Hank Greenberg, Arthur Goldberg, Golda Meir, Albert Einstein—these were people to be admired. And that poet with the beard, Allen Ginsberg, so smart, but the *faygeleh* (homosexual) business, such a waste!"[34]

In 1930, some German Jews recognized the danger of Zionist racism and demanded that Albert Einstein stop using his scientific fame to promote racism, disloyalty and "interracial" strife. *The New York Times* reported on 7 December 1930 on page 11,

> "The National German-Jewish Union, a small group of extreme nationalist and anti-Zionist Jews, protested against Professor Einstein using his world-fame as a scientist for 'propagating Zionism.'"

After the Second World War, Jews again criticized Einstein for his nationalistic Zionism. Einstein responded,

> "In my opinion condemning the Zionist movement as 'nationalistic' is unjustified. [***] Thus already our precarious situation forces us to stand together irrespective of our citizenship."[35]

Einstein believed that "affirmative action" was needed and justified to balance the discrimination Jews faced in Europe. He was especially concerned that a "Jewish university" be founded in Palestine to provide an opportunity for higher education to the Jews of Eastern Europe. Einstein and his friends attempted to fill universities, and the editorial staff of publications, with Jewish professors and lecturers who would be agreeable to his personal scientific and political views. Einstein agreed with Dühring that "Jews" exercised an undue influence in the press and Einstein stated that relativity theory was advertised, or rejected, in the press based on political bias. Leading Jews in the press and at the universities had organized to silence Dühring and to destroy his career. They did the same to composer Richard Wagner. The campaign to muzzle Dühring only legitimized Dühring's beliefs and fueled him on to publish several very influential works against Jews.

1.3 In a Racist Era

There was a panic in the western world following the violent Bolshevik Revolution in Russia in 1917. *The New York Times* in the late teens and early twenties published numerous articles warning of the dangers of Bolshevism.

---

[34]. S. G. Bloom, *Postville: A Clash of Cultures in Heartland America*, Harcourt, Inc., New York, (2000), pp. 63-64.
[35]. H. Dukas and B. Hoffmann, *Albert Einstein: The Human Side*, Princeton University Press, (1979), p. 55.

Many conservative German newspapers also tried to rouse public apprehensions over the dangers of the Communist revolution and Einstein was widely seen as an anarchist and a Communist.[36] Max Born wrote, "Einstein was well known to be politically left-wing, if not 'red'."[37] Einstein put his name to Communist and Socialist causes and both groups actively sought his support, with varying degrees of success.[38] When Einstein wanted to visit the United States in the early 1930's many protested against his admission into the country on the grounds that he was a Communist, an anarchist and a Socialist. *The New York Times*, on 4 December 1932, on the front page, stated,

> "The board of the National Patriotic Council in a statement today termed Dr. Einstein 'a German Bolshevik' and said his original theory 'was of no scientific value or purpose, not understandable because there was nothing there to understand.'"

The Patriot of 22 December 1932 published an article "The Visa of Professor Einstein" detailing the objections raised to the granting of a visa to Albert Einstein,

> "Professor Einstein has informed the world, through the Press, of his difficulty in getting an American visa in Berlin, owing to the U. S. Consul having been warned that he is an undesirable alien by the American Women's Patriotic Association. In the end the professor got his visa, and chuckled over the fact that the sentries of America had not given heed to 'the wise, patriotic ladies,' but had forgotten the occasion when 'the Capitol of mighty Rome was once saved by the cackling of its faithful geese.' The fact is that the patriotic American women had as substantial a reason for giving warning as had the Roman geese. *The Patriot* has given many instances in which Americans had as much right to object to the meddling of Professor Einstein in revolutionary movements on his visits to the U. S. as we have to protest against the Bolshevik finger in the preparation of revolution by British Communists."[39]

The *Patriot* article continued with extracts from the law and from the charges, which proved that Einstein was a member of several Communist front organizations and encouraged illegal activities, and that he could not be lawfully

---

36. M. Janssen, *et al.*, Editors, *The Collected Papers of Albert Einstein*, Volume 7, Note 7, Princeton University Press, (2002), pp. 124-125.
37. M. Born, *My Life: Recollections of a Nobel Laureate*, Charles Scribner's Sons, New York, (1975), p. 185.
38. *See, for example: The Collected Papers of Albert Einstein*, Volume 9, Documents 44 and 64, Princeton University Press, (2004).
39. D. Fahey, *The Mystical Body of Christ in the Modern World*, Browne and Nolan Limited, London, (1935), pp. 273-275, *see also:* 275-280, *especially* points 2 and 16, at pp. 277-279.

admitted into the United States of America. Einstein had influential friends and his record was ignored. The protests that he should not be allowed a visa to come to the United States were ultimately unsuccessful.[40] Einstein expressed himself in Marxist terms and his friends as well as his foes recognized the Socialistic tones in his statements in the early 1920's.[41] In 1949, Einstein published an article in the *Monthly Review* in which he advocated Socialism.[42] Since both world wars weakened the nations of the world, both wars created an atmosphere where Communism could flourish.

There were vocal advocates of anarchism, Communism, and Socialism in many Jewish communities. Many such individuals were romantic, very good-natured humanitarian people who sought social justice for the poor, and we today enjoy many benefits from their sacrifices. Others were mere opportunists who used Communism as a front to promote themselves into positions of dictatorial power. Perhaps most outside of Bolshevik dominated countries were not the murderous material that the genocidal tyrants Lenin and Stalin were. However, in many circles all Communists were seen as dangerous propagandists for imposed atheism, murderous revolution and a conspiracy to rule the world in a unified reign of tyranny led by the Jews.

There certainly were Communist elements in the world striving for the horrific goals of imposed atheism, murderous revolution and a conspiracy to rule the world by a "proletariat" which was in reality an obedient army of the subjugated. Mass murderers like Vladimir Ilyich Lenin, Béla Kun, Joseph Stalin and Mao Tse-Tung, did the biding of Jewish financiers who placed them in power to ruin Gentile nations, destroy Gentile religions and capture Gentile wealth. These assertions will be proven further on in this text. These were murderers whom Einstein admired for their political savvy, while disagreeing with some of their ideals.[43] Though the lower level Communists can be forgiven as deceived Utopian idealists, the upper levels Jews who financed and directed them were out to fulfill horrific Jewish prophecies, and the childish ideals of Communism were but bait in a vile trap. The worst of the Communists were those directly under the control of Jewish bankers, the openly genocidal Bolsheviks who had already slain tens of millions of Slavic Christians by the early 1920's. Einstein wrote to Hedwig and Max Born on 27 January 1920 that he found the Bolshevists not unappealing.[44]

Bolshevik atrocities shocked the free world. The Bolsheviks mass murdered tens of millions of innocent people and criminalized Christianity. The Bolsheviks were conspicuously and predominantly led and financed by Jews.

---

40. "Consul Investigated Charge", *The New York Times*, (6 December 1933), p. 6.
41. J. Stachel, *Einstein from 'B' to 'Z'*, Birkhäuser, Boston, (2002), p. 71.
42. A. Einstein, "Why Socialism?", *Monthly Review*, (May, 1949); reprinted in *Ideas and Opinions*, Crown, New York, (1954), pp. 151-158.
43. M. Janssen, *et al.*, Editors, *The Collected Papers of Albert Einstein*, Volume 7, Princeton University Press, Volume 7, Note 7 (2002), pp. 124-145.
44. Letter from A. Einstein to the Borns of 27 January 1920, *The Collected Papers of Albert Einstein*, Volume 9, Document 284, Princeton University Press, (2004).

Many have tied the dogmatism and cruelty of Communism to the dogmatism and cruelty of Judaism. The primitive and dogmatic dictator cults of personality, which are common to Communist régimes, mirror obeisance to a vengeful and jealous Jewish God and the ascendence of the Jewish King as the Messiah.

Jews have been praying for thousands of years for a Jewish Messiah to arrive and wipe out the Gentile nations, religions, cultures, and, eventually, peoples. The fact that leading Jews were accomplishing these Jewish Messianic ends through Communism concerned many people around the world. Just as the Jewish religion asserts that there can only be one God to rule the universe, the Jews have chosen themselves to rule over mankind and to destroy it. The relevant religious passages which evince these facts will be quoted later on in this text. When responsible persons voiced their legitimate concerns about Jewish Bolshevik destruction, they were often smeared in the Jewish press around the world as if "anti-Semites".

However, Jewish Bolshevik Zionist apologists were free to publicly identify the identities of Bolshevism, Christianity and their common source, genocidal Judaism, with its prophetic myths—as did "Mentor" in 1919. Like many other Zionists, Mentor forecast the Second World War shortly after the First had ended in *The Jewish Chronicle* on 28 March 1919 on pages 9 and 10,

"PEACE, WAR— AND BOLSHEVISM.

By MENTOR.

SOON after the armistice was signed, a contribution appeared in this column to which the caption, 'The Oath of the Peoples,' was rendered. It depicted something of the horrors of modern warfare. Yet ghastly, terrible, as were the facts which it presented, it was manifest that only a tiny corner of the veil was lifted by it which hid from the average man the Jazz Dance of Hell that careered across so much of the world for upwards of four years. It was necessary, in a subsequent article, to declare that although the war was suspended, it was not yet ended, and that, therefore, the prevailing condition of this and the other belligerent lands was still one of War—War suspended but not ended. It was a necessary reservation that was then made, because it was a reminder that circumstances still obtained which could be met by human beings by no other method than warfare; because, to human beings, there has until now been practically revealed no other. The reminder to which I have referred, if it was necessary—and it was—at the time was made, is even more important at this moment. For four-and-a-half months, representatives of the chief belligerent Powers and delegates of several neutral nations have been foregathering in conference at Paris. The war was constantly heralded as a war to end War. And now, as has been well said, the Peace Conference threatens to produce a peace that will end Peace.

THIS explains to some extent why the war went on as long as it did. Hateful as War must have been to those responsible for it in all the countries engaged in the struggle, they doubtless feared even more than War, once they engaged in it, the laying-down of arms because of the menace which Peace would bring to the future peace of the world. In the four-and-a-half months that have elapsed since the Peace Conference foregathered, the aspirations and ideals, the finely-spun purposes and the nobly-conceived objects which were to be compassed by the Conference, seem gradually to have crumbled like the Dead Sea fruit of the cities of Sin. The great French historian, Lavisse, in an address the other day, described to his pupils at the *Ecole Normale* what has happened. He declared:—

> You are following the discussions of the Conference of all the world. The most different voices speak there. Ancient quarrels revive, and visions, egotisms, hatreds, legacies from the past obstruct the future. Yet we hope that the Conference will be able, despite all these difficulties, to secure some articles of the creed of a humanity which is still without doubt at a great distance.

'Some articles!' 'the creed of a humanity, still at a great distance!' The war which was to end War, is being followed—it is feared—by a peace that will end Peace.

EVERY one of us, even those in whom normally and naturally to them the vein of pessimism runs, hopes that from the *impasse* into which the Conference has been drawn by circumstances which they could not control, conditions which they did not foresee, and events which they could not overcome, may emerge somehow with a better message to mankind than M. Lavisse prognosticates. For the condition of affairs throughout the world to-day is unmatched by any of which, though we search through all history, we can find any parallel. There have been long and exhausting wars ere this, and the belligerents at the end of them have lain prone, under the burden which War entails. This is not the first time that at the end of a long and wearying struggle, in which hundreds and thousands of the world's youth have been sacrificed to the demons which implant blood-lust in the hearts of men—this is not the first time when great nations have been crippled by war and at the end found the result of it all so much less than the objects which they sought in beginning the enterprise. But in previous conflicts, there has been just this difference: it was the Dynasts, the Kings, the Emperors, the Tsars, who carried on the war. It was their armies which were employed as instruments of their sovereign will. To-day, all that is altered. When countries go to war now, it is the peoples of those countries that are involved. And there is all the difference in the world between a conflict of Dynasts and a conflict of peoples. War is not ended now at the will of Dynasts and diplomatists. Therein in truth lies the danger of the spirit which has been manifesting itself among the delegates at Paris, and of which M. Lavisse has spoken. Because that spirit is dictated by great popular feelings and passions which Conferences may interpret, but cannot control. There was much force, in the quotation from the great statesman Burke, which was printed in this week's *Jewish World* upon the same point. 'Nothing is more common,' said Burke, 'than for men to wish and call loudly too, for reformation, who, when it arrives, do by no means like the severity of its aspect. Reformation is one of those pieces which must be put at some distance in order to please. Its greatest favourers love it better in the abstract than in the substance.' This was said of individuals. It is proving true also of peoples, and the proceedings at the Conference in Paris are an exemplification of Burke's works,

WHATEVER the faults may be, whether they be in fundamental construction, in spirit, in temper, or merely in method and procedure, which have brought the Paris Conference to its present dilemma, it is perfectly clear that the wild rejoicings of Armistice Day were premature and misplaced, if those engaged in them imagined that the Armistice had brought Peace to the world and that the war had ended War. For we are even now face to face with a war the extent and seriousness of which no man can foresee, and the ultimate effect of which no man can foretell. Bolshevism is the aftermath of the war that has not yet ended though it is suspended; as that in its turn was the catastrophic harvest which the world reaped for generations of political, social, and economic iniquity. The ideas and the ideals of the Western world collided with those of which Tsarist Russia and the Prussia which Bismarck made, were the most conspicuous and the most awful examples. The world of liberalism revolted against the world of retrogression, the world of freedom against the world of oppression, the world of liberty against the world of militarism. That was the conflict for which the two main elements in the war took up arms, and for the prevalence, one way or another, of which, they determined to measure their respective strength; and the fact that Tsarist Russia was opposed to militarist Prussia was only a political accident which does not in the least modify the real meaning of the world-struggle. The instant that Russia joined the *Entente*, Tsarism was to all intents and purposes dead. If the *Entente* did not mean that Tsarism should die, as surely as it meant that Prussian militarism should, then the Russian alliance was an absurdity. But when Russian Tsarism died, the *Entente* looked upon the fact as a defeat rather than a victory. Shortsightedly, it counted bayonets instead of hearts and machine guns instead of souls. It counted armies instead of principles, and measured battalions instead of the spirit that animates men. With this attitude of the *Entente* towards the Russian Revolution, another struggle for re-birth became inevitable. Bolshevism means the revolution of the people against itself—the revolution of the people against a system for which the people itself became responsible, when War ceased to be the concern merely of Dynasts and Kings and their armies, and became that of the whole of the belligerent peoples who engaged in it. That fact, it is to be feared, was not duly taken into account when the *personnel* of the Conference, which was to end War and initiate the reign of Peace, was chosen; and to that fact, it is probable, must be attributed much of the position in which the Conference now finds itself.

THERE is no need to descant upon the dangers of Bolshevism from many points of view or upon the ruinous upset which its prevalence must mean to society. There is no need to point to Bolshevism as a creed that is detestable, because it is the negation of democracy, meaning as it does the ruling by a single class instead of the government of the people by the people, for the people. But we do not get any nearer to understanding the phenomenon of Bolshevism by merely abusing it, not by calling down imprecations upon the outrageous conduct of those who are leading this strange, wild movement of the masses. It is, to be sure, a *bouleversement* of the ideas that have ruled hitherto, when Bolshevism declares that the man or woman who earns his or her bread by the sweat of his or her brow, is to have first consideration—that he who labours must have preferential treatment by the State. But is no more ridiculous than the system which gives first consideration to those who are idle because they are rich, to those who, however themselves incapable of work, live upon the sweat of the brows of others. It is, as I say, easy to denounce the cruelties, the wicked demoniacal cruelties, if half or quarter of what has been reported of Bolshevists in Russia be true. But if what has been reported be the fact, is it all really any worse than—is it, to be frank, as bad as—the outrages in Russia for which Tsarism was responsible, the infamous wickedness of the Ochrana, or such abominations as the wholesale evacuation of a quarter of a million of our people under the guise of military necessity, to which, early in the war, it was my painful duty to call attention? The Conference at Paris seems disposed to try to stamp out Bolshevism by military force. But Bolshevism is precisely a protest against military force and all social and economic forces upon which militarism relies. It would seem therefore that the application of further military force is more likely to increase the hold of Bolshevism upon the minds of people rather than to eliminate it.

AND here I must break off—as they say in the House of Commons, I must adjourn and ask leave to sit again. For Bolshevism has now, and will have increasingly in the future, a particular interest for us Jews, which it were ridiculous and short-sighted for us to ignore. Because Bolshevism is rightly unpopular and because all men and women of right thinking loathe and abominate the outrage and the murder, the injustice and the terrorism associated with Bolshevism, it were absurd to suppose that we have said the last word about it as Jews by making some such declaration, as I observe Major Lionel de Rothschild ventured the other day, when he said he very much doubted whether any good Jews, any believing Jews, were Bolshevists. This, of course, is, in fact, mere moonshine. The gallant Major was evidently unaware that, to give only one instance, one of the men who stands and has stood as a great Jewish religious force, in America, a 'believing' Jew if ever there was one, an earnest high-minded man, although it may be somewhat *bizarre*, had declared publicly his sympathy with Bolshevism. It must be taken for granted that a man like Dr. J. L. Magnes [Magnes was a lecherous agent of Jacob Schiff—the Jewish banker, and Rothschild agent, behind the Russian Revolution.] before so proclaiming himself, was satisfied that Bolshevism and Judaism are not as entirely incompatible as Major de Rothschild evidently thinks. In any case we Jews cannot airily dissociate all Jews from Bolshevism by declaring that to be a Bolshevist is necessarily to be a bad Jew. The ranging himself of Dr. J. L. Magnes as a Bolshevist—to say nothing of the many excellent Jews who are Bolshevists in Eastern Europe to-day—proves the futility of the Major's observation. No folly could be greater than for us Jews to show the white feather of cowardice in pretending what is untrue, and to declare that the political creed of Bolshevism and the religious creed of Judaism are incompatible merely because the association of Jews with an unpopular movement may be awkward for us. The truth in the long run is our surest buckler. It will never in the end fail us. It were well, then, to examine what the exact meaning of the portent we call

Bolshevism is, and why Jews have become associated with it. That I propose to attempt, as the novels say, in the next chapter."

Note that Mentor sophistically blames the *Entente*, the Allies, for the conditions which precipitated the Second World War, which war Jewish leadership had planned before it began the First. Mentor blames the Czar for Bolshevist atrocities, atrocities which the Czar sought to prevent. Mentor—already in 1919—blamed the Allies for creating the Second World War by rejecting Bolshevism.

However, if the Allies had truly fought against Bolshevism over the objections of vocal and influential Jews like Mentor and Israel Zangwill who asked the Allies to leave Bolshevism to its work,[45] there would have been no Second World War, and there would have been no Bolshevik Nazis and the lives of tens of millions of Slavs the Bolsheviks—Nazi and Soviet—mass murdered would have been spared. Note that Mentor focuses on abuses the Czar allegedly committed specifically against Jews, and Mentor makes it clear that Bolshevism was an act of retaliation by Jews against the Russian People—and ultimately against all non-Jews—"the people against itself"—the controlled self-destruction of the Gentile Peoples as an act of Jewish revenge. In the name of "peace", Mentor petitioned the Allies to passively allow Bolshevism to wage war against the world and mass murder innocent civilians—Gentile civilians.

Mentor wrote in *The Jewish Chronicle* on 4 April 1919 on page 7,

"PEACE, WAR— AND BOLSHEVISM.

By MENTOR.

WHAT is written here is pendent to what appeared in this column last week. As I intimated, I propose to revert to the subject then referred to.
BOLSHEVISM is at once the most serious menace to, and the best hope of, Civilisation. Paradoxical as this may sound, but a little thought will show it to be abundantly true. The menace of Bolshevism is manifest. It pulls down what, until now, it has shown itself unable efficiently to replace. In the name of freedom, it imposes galling slavery. In the name of humanity, it inflicts the direst evil upon the men, women, and children who come under its sway. It protests against class domination and itself imposes the domination of class wherever it can obtain power. It knows no bounds either in justice or in liberty. It murders, imprisons and tortures with the ruthlessness of an autocracy drunk with new-found authority. It is ruthless, relentless, all-engulfing. It falls upon the country it infects like a dire pestilence which casts people prone. It is a political disease, an economic infliction, a social disaster.
YET, none the less, in Bolshevism there lies, to-day, the hope of Humanity. For in essence, it is the revolt of peoples against the social state, against the evil; the

---

45. "Socialists at the Albert Hall", *The London Times*, (10 February 1919), p. 10.

iniquities—and the inequalities—that were crowned by the cataclysm of the War under which the world groaned for upwards of four years. It is a revolution against a social state which suffered Tsarism to exist in Russia and militarism in Prussia and which still allows, alas, so many a crying wrong in countries that plume themselves on their freedom and boast of their liberty. Bolshevism is the signal to mankind to halt in its social, political, and economic ways of old; to stay and examine them in the light of the sacrifice of the millions of youth who have gone down to darkness eternal, of the millions of treasure which war has wasted, and to ponder them in the light of the incalculable, ineffable burden which the years of struggle have placed upon Society, and, heaviest of all, upon the poor—in light of the war which was proof in all surety that the old order was doomed if civilisation was to survive. That Bolshevism broke out first in the country most oppressed is nothing for wonder; it is merely natural. For centuries Russia had been the forcing ground of every infamy imposed by power and every wickedness done in the name of Government. That the creed has spread to a country whose national aspirations were for generations crushed, and where autocracy ruled, is nothing for wonder. Nor is the protest of Bolshevism merely a matter for Russia and Hungary, or a menace only to bayonet-ridden Germany. It is a challenge to the world—not least to the nations of freedom and liberty. It is a challenge to all the nations including the peoples who nourish liberty and freedom as precious principles, but who have passively allowed a state of affairs to grow and putrefy into the infamies of Russian Tsarism, the iniquity of Hungary, and the wickedness of German militarism; to the world that has suffered Society to fester into these and to break out into the prurient, gaping, sloughing, agonising tumour of such a war as that which is not ended, though it is suspended. And the fact that this protest has been made is the world's best hope. It is a demand for another order of things, for a social state which will render humanity immune from the wickedness and such evil as resulted in the greatest war mankind has ever known. It asks for some guarantee against a system which dragged peoples innocent of any intention of killing, slaying, and slaughtering into the vortex of War—peacefully intentioned peoples who loathed and hated War (such as was England before that fateful day in August, 1914)—from which even the most innocent of belligerents, and even those who stood aside from the contest are suffering to-day; though none were wholly guiltless of it, because for generations all passively concurred in the system. If the world, as a result of the War, had received no such warning as Bolshevism, the evil would, in all probability have gone on, deepening in its wrong, becoming ever blacker. Bolshevism is a social fever which indicates a high blood temperature. It gives the warning of mischief that may be fatal. A wise doctor takes note of the fever and seeks to remove the cause. He does not call the fever ugly names or denounce it, nor is he so stupid as to confuse the patient's consequent delirium with his normal condition, as so many are confusing the delirium of Bolshevism with the normal state of the countries in which it is finding vogue.

ALL such indications on the part of the body politic that there is a disease that must be removed, else the patient must go under, are as unpleasant, as inimical, as is the delirium of the fever-stricken patient distressing. The French Revolution drowned Paris in blood. Its excesses were far greater than anything that even the most malicious has attributed to Bolshevism. It instituted a Reign of Terror. It massacred Royalty. It condemned men and women day by day to the tumbril; so commonly indeed, that the men and women walking in the streets of Paris hardly looked round when some victim of the Jacobins was being taken to the Guillotine. Nothing and nobody was safe from the raging, tearing fever of the Revolution. For years it inflicted upon France a series of infamies, of torture, of horror, of bloodshed almost unparalleled in history. Yet, at the end of it all, and notwithstanding its reaction in Napoleonism, a great English writer declared that there had been nothing greater and more glorious in all history than the French Revolution. By common consent what liberty, equality, and fraternity—liberty, equality, and fraternity which the French Revolution never gained, and which in seeking after it demeaned and disgraced—the rest of the world possesses to-day, it draws in large measure from the days in which France was bathed in the anarchy of revolt. That is because the motive-spring which set the French Revolution into being was an ideal for the betterment of mankind, a protest against the social, political and economic infamies which will for ever be associated with the *régime* of the Bourbons, a striving for a social state that would not allow unbridled luxury, lascivious prodigality, selfish extravagence, inhuman carelessness, to thrive in the Court and to go on side by side with poverty, hunger, a life of groaning and moaning in the alleys hard by. And, even now, while the terror of Bolshevism is in full swing, a writer in an English Daily paper is brought to declare, as one did the other day, that at root Bolshevism in ideal has nothing comparable to it since the teachings which Jesus of Nazareth gave to the world. The writer had, there is little doubt, recollected the parable of the rich man, torn with suffering in Hell, pleading to Lazarus, the beggar whose sores the dogs licked, resting in the bosom of Abraham in Heaven. It is the parable of the ideals of Bolshevism.

IT is not difficult to see why a people which has managed to subsist through Tsardom, because of the religious ideals and ideas which it nourished throughout all its classes, and not least among its peasantry, has been attacked by the ideals of Bolshevism, and why, released from Tsardom, it has, pendulum-like, swung into the arms of Lenin, looking to the ideals of his creed, and not to its wickedness or its excesses. The same reason obtains for the number of Jews who are to be found in the Bolshevist ranks. The Jew is an idealist. He will give much for an ideal. He thirst for idealism as a goal of life. This may seem strange to those who associate the Jew with materialism. But the capacity of the Jew for idealism is such that he notoriously idealises even the material. The fact that there are so many of our people who have associated themselves with the ideals of Bolshevism, even although as Jews its excesses must be repugnant to them, has to be placed in conjunction with another fact. These men will be found for the most part unassociated with or dissociated from the Synagogue. In the ordinary way of speaking they are not observing Jews. Is it not patent that the Synagogue, having failed to attract them by its idealism, and no other ideal, not even a material ideal, having been provided for them—for they are not men of wealth and substance, such as are usually to be found among the *bourgeoisie*—they have ranged themselves on the side of Bolshevism, because here was no Jewish ideal to which these Jews could devote their sentiments and their energies? I cannot understand how people who for generations have, unprotesting, allowed the Jew, particularly in Eastern Europe, in Russia, to suffer pogroms, to be massacred and ill-treated, and tortured and murdered, and for two thousand years have kept our people outside the ambit of the most potent source of idealism that can appeal to men—that associated with National being—now have the hypocrisy, the soulless impertinence, to complain that so many of our people are Bolshevists! That Jews have been chosen to the extent they have to take a leading part in the movement in Russia and in Hungary, is merely because they are heavily endowed with intellectualism and capacity, as compared with the rest of the population. But the world must not surprised that the Jew, who is an idealist or nothing, has turned to the idealism of Bolshevism, which a British writer has declared to be comparable to the idealism preached by the founder of Christianity. It were surprising, really, were it otherwise. You cannot keep a people out of their rightful place amid the nations of the world, and then complain because they take the leading part which their abilities entitle them to in the nations among whom you have scattered them. The fact that a timorous millionaire afraid, and doubtless with good cause, of Bolshevism, which he probably has never taken the trouble, or perhaps has not the capacity to appreciate in full measure, places a ban of religious excommunication upon those Jews who are Bolshevists, is a thing for the gods to laugh at!

THERE is much in the fact of Bolshevism itself, in the fact that so many Jews are Bolshevists, in the fact that the ideals of Bolshevism at many points are consonant with the finest ideals of Judaism, some of which went to form the basis of the best teachings of the founder of Christianity—these are things which the thoughtful Jew will examine carefully. It is the thoughtless one who looks upon Bolshevism only in the ugly repulsive aspects which all social revolutions assume and which make it so hateful to the freedom-loving Jew—when allowed to be free. It is the thoughtless one that thus partially examines the greatest problem the modern world has been set, and as his contribution to the solution dismisses it with some exclamation made in obedient deference to his own social position, and to what for the moment happens to be conventionally popular."

Mentor falsely blamed the Czar for the hardships of the Russian People, which Jewish leaders had deliberately caused so as to make the Russian People clamor for a revolution—a revolution which would put Jews into power—if not on the throne, then behind it. Jewish leaders deliberately ruined the Russian economy by obstructing Russia's access to investment capital, by provoking a war with Japan and funding the Japanese while cutting off Russia's access to funds,[46] by conducting massive strikes, by assassinations and attempted

---

[46]. R. Smethurst, "Takahashi Korekiyo, the Rothschilds and the Russo-Japanese War, 1904-1907", *The Rothschild Archive: Review of the Year April 2005 to March 2006*, London, (2007), pp. 20-25. <http://www.rothschildarchive.org/ib/articles/AR2006.pdf>

revolutions, by attempting to discredit the Russian Government in the press around the world, by instigating the First World War, etc. For example, John Hays Hammond[47] wrote in *The New York Times* on 18 November 1911, on page 2,

> "I, however, convinced them that there was no lack of friendliness toward Russians on the part of Americans, who remembered Russia's friendship to us at the time of our civil war. [***] Mr. Jacob H. Schiff has done more to accentuate the troubles of his co-religionists in Russia than any other one man, because of his boastful statement that the money of Jewish bankers had made it possible for Japan to wage a successful war against Russia."

The reason why Russia was the first and the primary target of Jewish Bolshevism was that Russia had the world's largest Jewish population and the Zionists wanted to export these Jews against their will to Palestine. The Czar, far from directing racism at the Jews, asked the Jews not to segregate and prohibited racist Zionist Nationalism in order to sponsor Jewish integration with the other Peoples of the Empire, in order that all Peoples in the Empire would live together in harmony and peace. For this act of kindness, Jewish leadership heaped ruin upon Russia and murdered the Czar and his family. Hungary also had a very large racist Jewish population and it, too, fell victim to Jewish Bolshevism and its murderous savagery, as did Poland, with its very large Jewish population. Will the United States be next?

The Ladies' Literary Cabinet, Being a Repository of Miscellaneous Literary Productions in Prose and in Verse, Volume 1, Number 4, (5 June 1819), p. 29, wrote,

"THE JEWS.

> In the year 1290, in the reign of Edward I., the property of all the Jews in England was confiscated to the use of the crown; 280 of them were hanged in one day, charged with adulterating the coin. Above fifteen thousand of these unfortunate people, in that reign, were plundered of all their wealth, and banished the kingdom. In the year 1811, in the reign of George III. Mr. Rothschild, a celebrated Jew, was at the head of most of the loans to the European kings and emperors. How remarkably do these facts speak in favour of the progress of liberal and enlightened opinions in that country."

Under the heading "Foreign Articles", the following statement appeared in *Niles' Weekly Register*, Volume 17, Number 427, (13 November 1819), p. 169,

---

47. *See also:* J. H. Hammond, "Russia", *The Autobiography of John Hays Hammond; Illustrated with Photographs*, In Two Volumes, Chapter 23, Farrar & Rinehart, incorporated, New York, (c1935); *reprinted:* Arno Press, New York, (1974), pp. 454-478.

"Mr. Rothschild, the great London banker, indignant at the persecution of his Jewish brethren in Germany, has refused to take bills upon any of the cities in which they are persecuted; and great embarrassments to trade have been experienced in consequence of his determination. ☞It is intimated that the persecution of the Jews is in part owing to the fact, that Mr. Rothschild and his brethren were among the chief of those who furnished the 'legitimates,' with money to forge chains for the people of Europe."

In an article entitled "The Jews", *The Knickerbocker; or New York Monthly Magazine*, Volume 53, Number 1, (January, 1859), pp. 41-51, at 44-45, wrote,

"Yet the Jews of the Ottoman Empire, notwithstanding their degradation, exhibit a certain intellectual tendency. They live in an ideal world, frivolous and superstitious though it be. The Jew who fills the lowest offices, who deals out *raki* all day long to drunken Greeks, who trades in old nails, and to whose sordid soul the very piastres he bandies have imparted their copper haze, finds his chief delight in mental pursuits. Seated by a taper in his dingy cabin, he spends the long hours of the night in poring over the Zohar, the Chaldaic book of the magic Cabala, or, with enthusiastic delight, plunges into the mystical commentaries on the Talmud, seeking to unravel their quaint traditions and sophistries, and attempting, like the astrologers and alchymists, to divine the secrets and command the powers of Nature. 'The humble dealer, who hawks some article of clothing or some old piece of furniture about the streets; the obsequious mass of animated filth and rags which approaches to obtrude offers of service on the passing traveller, is perhaps deeply versed in Talmudic lore, or aspiring, in nightly vigils, to read into futurity, to command the elements, and acquire invisibility.' Thus wisdom is preferred to wealth, and a Rothschild would reject a family alliance with a Christian prince to form one with the humblest of his tribe who is learned in Hebrew lore.

The Jew of the old world, has his revenge:

'THE pound of flesh which I demand of him
Is dearly bought, is mine, and I will have it.'

Furnishing the hated Gentiles with the means of waging exterminating wars, he beholds, exultingly, in the fields of slaughtered victims a bloody satisfaction of his 'lodged hate' and 'certain loathing,' more gratifying even than the golden Four-per-cents on his princely loans. Of like significance is the fact that in many parts of the world the despised Jews claim as their own the possessions of the Gentiles, among whom they dwell. Thus the squalid *Yeslir*, living in the Jews' quarter of Balata or Haskeni, and even more despised than the unbelieving dogs of Christians, traffics secretly in the estates, the palaces and the villages of the great Beys and Pachas, who would regard his touch as pollution. What, apparently, can be more absurd? Yet these assumed possessions, far more valuable, in fact, than the best 'estates in Spain,' are bought and sold for money, and inherited from generation to generation."

A philo-Semitic article entitled "The Jews in the United States", *The World's Work*, Volume 11, Number 3, (January, 1906), pp. 7030-7031; stated,

"In European capitals there are Hebrew bankers who dictate certain international relations because they hold the purse-strings of governments; and every European country owes much to the men of great genius that the race has contributed to the arts and to statecraft."

Jewish bankers and their agents deliberately ruined the economies of target nations like Russia. They then used their disproportionate influence in the press to blame the current government for the hardships they themselves had deliberately caused, thereby creating resentment between the People and their government and preventing the People from realizing the true cause of their misery. Jewish leadership instigated: the English Revolution, which made their agent Oliver Cromwell a dictator; the French Revolution, which made their agents Robespierre and then Napoleon dictators; the "Young Turk" Revolution, which made their crypto-Jewish agents Talaat Pasha[48] and Atatürk[49] dictators; the Bolshevik Revolution, which made their agent Vladimir Ilyich Lenin a dictator; the Nazi Revolution, which made their agent Adolf Hitler a dictator; the Spanish Revolution, which made their agent Francisco Franco a dictator; etc. etc. etc.

In America today, Jewish propagandists are blaming George Bush for the problems Jewish leadership have caused America. They are also attempting to discredit the American system of government in general by pointing out that the Founding Fathers were Freemasons and were influenced by the ideas of the Illuminati, but without mentioning that these institutions were each subservient to the Jewish bankers and were a means used by them to obtain compromised Gentile leaders who had divulged all their dark secrets in order to gain admission into these secret societies. However, the root problem is not the American system of government, but rather the deliberate corruption of that system by Jewish leaders, Jewish racism and Jewish tribalism. Changing the form of government will only worsen that problem, because the same Jews who are decrying the system—either directly or through their agents—are those who have corrupted it. If their calls for revolution and a gold based currency are heeded, they will take it over completely and deliberately ruin the nation. It is the Jewish bankers who own the gold and who want to sell it to the American Government—and to a large extent this is gold they first stole from the American People, which they

---

48. R. De Nogales, *Four Years Beneath the Crescent*, Charles Scribner's Sons, New York, (1926), pp. 26-27.
49. "The Sort of Man Mustafa Kemal Is", *The Literary Digest*, Volume 75, Number 2, Whole Number 1695, (14 October 1922), pp. 50, 52-53, at 50.

desire to sell back to America at an immense profit, so that they can again steal it at a discount and leave America without its own independent money supply. In the name of "reform", Jewish leaders will lead America into a Soviet-style nightmare and perpetual world war. In the name of defending American sovereignty, they will deliver America into a world government and war with America's neighbors.

Jewish leaders are teaching Americans to distrust American leadership, without exposing the fact that Jewish leaders are deliberately causing America's problems. Americans are being primed for a revolution which will put an anti-Semitic dictator into power who will then do the bidding of Jewish leadership, as happened in Germany when Hitler rose to power lifted up on golden strings held in the Jewish bankers' hands, and the German economy grew as if by magic on the monies which poured in from Jewish bankers who were fattened on the American economy at the expense of the American People. Germany then collapsed when those monies mysteriously dried up and unnecessary war led into more unnecessary war—as Hitler and Stalin deliberately destroyed Germany and Eastern Europe, and Japan deliberately destroyed China as it had helped the Jewish bankers to destroy Russia.

These Jewish instigated revolutions and wars followed a common model. After actively provoking revolutions with the false premise that revolution was necessary to free the People from their government, the Cabalistic Jews deliberately collapsed the economy of the overthrown State, or otherwise deliberately brought chaos and general panic to the public. They then used their disproportionate influence in the press to promote the false message that only a dictator would be capable of restoring order to the land. The Cabalistic Jews thereby caused the People to enslave themselves with the trap of a revolution promising "liberty, equality and fraternity" that resulted instead in chaos and panic, only to offer up the promise of order and prosperity under a dictator of their choosing, who will supposedly restore order, then resign from office. Of course, it was the Jewish bankers who had deliberately made conditions unbearable in the first place, so as to create the necessary climate and needed conditions for revolution and war. They planned for dictatorship from the very beginning and their revolutions were based from the outset on deliberate lies and ill-intentions.

In the Eighteenth and Nineteenth Centuries, the Rothschild clan made a high art out of deliberately provoking wars and revolutions, which resulted in dictatorships of their manufacture and under their ultimate control. This furthered the Jewish Messianic goal of destroying the Gentile nations and supplanting them with universal Jewish rule. It also enabled the Rothschilds to further the Jewish Messianic goal of concentrating the wealth of the world in Jewish hands. By the 1870's, the Rothschilds had accumulated at least $3,400,000,000.00USD non-adjusted,[50] through wars and revolutions which they had fomented and financed, and from which they profited in perpetuity.

---

[50]. "The Rothschilds", The Chicago Tribune, (27 December 1875), p. 8.

The Rothschilds openly sought to become King of the Jews in the Nineteenth Century. The King of the Jews is, by definition, the Messiah, or anointed, of the Jews. The Old Testament teaches the Jews that their Messiah will rule the world—that in the "end times", after a terribly destructive world war, the Jewish King will lead a world government from Jerusalem (*Exodus* 34:11-17. *Psalm* 2; 72. *Isaiah* 2:1-4; 9:6-7; 11:4, 9-10; 42:1; 61:6. *Jeremiah* 3:17. *Joel* 3:16-17. *Micah* 4:2-3. *Zechariah* 8:20-23; 14:9).

The Jewish bankers used the tactic of perpetual war as a trap to ensnare the Gentile nations into surrendering their national sovereignty and accepting Jewish world government. After making the world weary of wars the Rothschilds had intentionally caused and lengthened, Cabalistic Jews used their disproportionate influence in the press to promote the myth that a world government would herald the end of war, because there would be no nations left to fight wars against each other. The false assertion that a world government was necessary to prevent war was a common theme in Jewish Bolshevik propaganda. Jewish leaders deliberately caused the People of the world to suffer, and then offered themselves up as the resolution to the problems the Jewish leaders had deliberately caused, but which the Jewish leaders falsely blamed on Gentile government and religions.

If successful, the Jewish bankers' plan to fulfill Jewish Messianic prophecy through political means will ultimately result in universal tyranny, and then the extermination of non-Jews and assimilated Jews. The process of creating war to make the world weary of war, while promoting the myth that the loss of national sovereignty will mean the end of war, is a trap used by Cabalistic Jews to ensnare non-Jewish Peoples into fulfilling the Jewish Messianic prophecy that Jews will rule a world government in the Messianic Age. Jewish Messianic prophecy predicts that only "righteous Jews" will be left alive in the "end times"—that the Jewish Messiah will judge and then exterminate the "wicked", all non-Jews and assimilated Jews (*Isaiah* 11. *Jeremiah* 3:17; 10:10-11; 23:5-8. *Sanhedrin* 105*a*. *Zohar*).

Psalm 110 says of the murderous Jewish King, whom the Jews intend to anoint as "Messiah",

> "The LORD said unto my Lord, Sit thou at my right hand, until I make thine enemies thy footstool. 2 The LORD shall send the rod of thy strength out of Zion: rule thou in the midst of thine enemies. 3 Thy people *shall be* willing in the day of thy power, in the beauties of holiness from the womb of the morning: thou hast the dew of thy youth. 4 The LORD hath sworn, and will not repent, Thou *art* a priest for ever after the order of Melchizedek. 5 The Lord at thy right hand shall strike through kings in the day of his wrath. 6 He shall judge among the heathen, he shall fill *the places with* the dead bodies; he shall wound the heads over many countries. 7 He shall drink of the brook in the way: therefore shall he lift up the head."[51]

---

51. *Cf.* N. De Manhar, *Zohar: Bereshith—Genesis: An Expository Translation from*

The *Zohar* informs us of the beliefs of Cabalistic Jews and their racist genocidal hatred of non-Jews.

The *Zohar*, I, 28b-29a, states,

> "At that time the mixed multitude shall pass away from the world [***] The mixed multitude are the impurity which the serpent injected into Eve. From this impurity came forth Cain, who killed Abel. [***] for they are the seed of Amalek, of whom it is said, 'thou shalt blot out the memory of Amalek' [***] Various impurities are mingled in the composition of Israel, like animals among men. One kind is from the side of the serpent; another from the side of the Gentiles, who are compared to the beasts of the field; another from the *mazikin* (goblins), for the souls [29a] of the wicked are literally the *mazikin* (goblins) of the world; and there is an impurity from the side of the demons and evil spirits; and there is none so cursed among them as Amalek, who is the evil serpent, the 'strange god'. He is the cause of all unchastity and murder, and his twin-soul is the poison of idolatry, the two together being called Samael (lit. poison-god). There is more than one Samael, and they are not all equal, but this side of the serpent is accursed above all of them."[52]

The *Zohar* I, 47a, states,

> "SAID Rabbi Abba: 'Nephesh hahaya' (living soul) truly denote the souls of Israel. They are the children of the Holy One and holy in his sight, but the souls of the heathen and idolatrous nations whence come they?'
>
> Said Rabbi Eleazar: 'They emanate from the left side of the sephirotic tree of life, which is the side of impurity, and therefore they defile all that come into contact with them. It is written, 'Let the earth bring forth the living creature after his kind, and creeping thing and beast of the earth after his kind' (Gen. 1-24). Wherefore does the word 'lemina' (after his kind) occur twice? It is to confirm what has lust been stated, that the souls of Israel are pure and holy, but the souls of the heathen being impure and unholy are symbolized by the creeping thing and beast of the earth, and therefore, like the foresaken in circumcision, are cut off."[53]

The *Zohar*, II, 219b, states,

> "So they went nearer and they heard him saying: 'Crown, crown, two sons are kept outside, and there will be no peace or rest until the bird is thrown

---

Hebrew, Third Revised Edition, Wizards Bookshelf, San Diego, (1995), p. 177.

52. H. Sperling and M. Simon, *The Zohar*, Volume 1, The Soncino Press, New York, (1933), pp. 108-110.

53. N. De Manhar, *Zohar: Bereshith—Genesis: An Expository Translation from Hebrew*, Third Revised Edition, Wizards Bookshelf, San Diego, (1995), p. 203.

down in Cæsarea.' R. Jose wept and said: 'Verily the *Galuth* is drawn out, and therefore the birds of heaven will not depart until the dominion of the idolatrous nations is removed from the earth, which will not be till the day when God will bring the world to judgement.'"[54]

The *Zohar*, III, 19*b*, states,

"It is, however, as R. Abba has said: all the other days are given over to the angelic principalities of the nations, but there is *one* day which will be the day of the Holy One, blessed be He, in which He will judge the heathen nations, and when their principalities shall fall from their high estate."[55]

The *Zohar*, III, 43*a*, states,

"To these He appointed as ministers Samael and all his groups—these are like clouds to ride upon when He descends to earth: they are like horses. That the clouds are called 'chariots' is expressed in the words, 'Behold the Lord rideth upon a swift cloud, and shall come into Egypt' (Isa. XIX, I). Thus the Egyptians saw their Chieftain like a horse bearing the chariot of the Holy One, and straightaway 'the idols of Egypt were moved at His presence, and the heart of Egypt melted in the midst of it' (*Ibid.*), i. e. they were 'moved' from their faith in their own Chieftain. AND EVERY FIRSTLING OF AN ASS THOU SHALT REDEEM WITH A LAMB, AND IF THOU WILT NOT REDEEM IT... THOU SHALT BREAK HIS NECK."[56]

The *Zohar*, III, 282*a*, states,

"From the side of idolatry Shabbethaj (Saturn) is called Lilith [*Footnote:* Lilith is a female demon, comp. Is. XXXIV. 14 and Weber, *Altsynagogale palästinische Theologie*, p. 246.], mixed dung, on account of the filth mixed from all kinds of dirt and worms, into which they throw dead dogs and dead asses, the sons of 'Esau and Ishma'el, and there (read הבו) Jesus and Mohammed, who are dead dogs, are buried among them. She (Lilith) is the grave of idolatry, where they bury the uncircumcised, (who are) dead dogs, abomination and bad smell, soiled and fetid, a bad family. She (Lilith) is the ligament [*Footnote:* אכדם is a fibre attached to the lungs] which holds fast the 'mixed multitude' (Ex. xii. 38), which is mixed among Israel, and which holds fast bone and flesh, that is, the sons of 'Esau and Ishma'el, dead bone and unclean flesh torn of beasts in the field, of which it is said (Ex. xxii. 31):

---

54. H. Sperling and M. Simon, *The Zohar*, Volume 2, The Soncino Press, New York, (1933), p. 311.
55. H. Sperling and M. Simon, *The Zohar*, Volume 3, The Soncino Press, New York, (1933), p. 63.
56. H. Sperling and M. Simon, *The Zohar*, Volume 3, The Soncino Press, New York, (1933), p. 132.

'Ye shall cast it to the dogs.'"[57]

Wanting for God's intervention, the Jewish bankers played the rôle of the Jewish Messiah and used Old Testament prophecies, the Talmud and Cabalistic writings as a plan they set out to artificially fulfill by their own intentional actions without any help from God. They have been highly successful, much to the detriment of mankind. They have given us Bolshevism, Nazism, Zionism, etc., each as an artificial political means to place a Jewish King at the head of the world.

Cabalistic Jews set yet another trap for the Gentile nations. They deliberately caused specific economies to grow and accumulate the wealth of the world by increasing the money supply in a target nation, or empire. They then deliberately collapsed the economy of the target nation by restricting the money supply and by running the target nation or empire into debt through deliberately mismanaged economic policy and perpetual war. Cabalistic Jews then used their disproportionate influence in the press to make the People clamor for banking reforms—usually a move toward the gold standard and a centralized privately owned bank, which operated under a fractional reserve system and a debt based issuance of currency, all of which profited the Jewish bankers who invariably and inevitably ran the system and profited from the debts of the nation the same Jewish bankers deliberately caused.

To summarize, there were three primary traps which Cabalistic Jews set for their non-Jewish neighbors in order to cause them to unwittingly fulfill Jewish Messianic prophecy by artificial political means. Jewish financiers used their agents to promote revolutions on the false promise that revolution would bring about freedom and democracy. After carrying out a revolution and deliberately creating a climate of fear and chaos, the Jewish financiers then installed a dictator of their choosing to subvert the freedoms of Gentile nations and bring them into perpetual war and perpetual debt. Jewish financiers deliberately caused perpetual wars to make the People of the world clamor for peace, and then proposed the false notion that world government was the only means to achieve an end to war—world government Jews have intended to lead from ancient times. Jewish financiers deliberately caused banking scandals in order to make Peoples clamor for banking reforms, but then subverted the reform process by instituting the very policies they had always sought—disastrous policies for the People, which syphoned off the wealth of the nation and the world into the coffers of the Jewish bankers.

Congressman Charles A. Lindbergh Sr. was very aware of the fact that the bankers had deliberately caused the panic in 1907 in order to make the public

---

57. G. Dalman, *Jesus Christ in the Talmud, Midrash, Zohar, and the Liturgy of the Synagogue*, Deighton Bell, Cambridge, (1893), p. 40. Though work is given an ancient attribution by its "discoverer", the Muhammadans are also mentioned in *Zohar*, II, 32a. Some consider the author to have been divinely inspired, some say the work evolved over time, some say the work is a fabrication—in any event, it is an now a very old writing and was very influential in Jewish political movements like the Frankists.

clamor for banking reforms, banking reforms the bankers would draft which would give them complete control over the money supply and wipe out the lower level, but numerous, competing banks,

"When the Aldrich-Vreeland Emergency Currency Bill was sprung on the House in its finished draft and ready for action to be taken, the debate was limited to three hours and Banker Vreeland placed in charge. It took so long for copies of the bill to be gotten that many members were unable to secure a copy until within a few minutes of the time to vote. No member who wished to present the people's side of the case was given sufficient time to enable him to properly analyze the bill. I asked for time and was told that if I would vote for the bill it would be given me, but not otherwise. Others were treated in the same way.

Accordingly, on June 30, 1908, the Money Trust won the first fight and the Aldrich-Vreeland Emergency Law was placed on the statute books. Thus the first precedent was established for the people's guarantee of the rich man's watered securities, by making them a basis on which to issue currency. It was the entering wedge. We had already guaranteed the rich men's money, and now, by this act, the way was opened, and it was intended that we should guarantee their watered stocks and bonds. Of course, they were too keen to attempt to complete, in a single act, such an enormous steal as it would have been if they had included all they hoped ultimately to secure. They knew that they would be caught at it if they did, and so it was planned that the whole thing should be done by a succession of acts. The first three have taken place.

Act No. 1 was the manufacture, between 1896 and 1907, through stock gambling, speculation and other devious methods and devices, of tens of billions of watered stocks, bonds, and securities.

Act No. 2 was the panic of 1907, by which those not favorable to the Money Trust could be squeezed out of business and the people frightened into demanding changes in the banking and currency laws which the Money Trust would frame.

The Act No. 3 was the passage of the Aldrich-Vreeland Emergency Currency Bill, by which the Money Trust interests should have the privilege of securing from the Government currency on their watered bonds and securities. But while the act contained no authority to change the form of the bank notes, the U. S. Treasurer (in some way that I have been unable to find a reason for) implied authority and changed the form of bank notes which were issued for the banks on government bonds. These notes had hitherto had printed on them, 'This note is secured by bonds of the United States.' He changed it to read as follows: 'This note is secured by bonds of the United States or other securities.' 'Or other securities' is the addition that was secured by special interests. The infinite care the Money Trust exercises in regard to important detail work is easily seen in this piece of management. By that change it was enabled to have the form of the money issued in its favor on watered bonds and securities, the same as bank notes secured on government bonds, and, as a result, the people do not know whether they get one or the other. None of the $500,000,000 printed and lying in the U. S. Treasury ready to float on watered bonds and securities has yet (April, 1913) been used. But it is there, maintained at a public charge, as a guarantee to the Money Trust that it may use it in case it crowds speculation beyond the point of its control. The banks may take it to prevent their own failures, but there is not even so much as a suggestion that it may be used to help keep the industries of the people in a state of prosperity.

The main thing, however, that the Money Trust accomplished as a result of the passing of this act was the appointment of the National Monetary Commission, the membership of which was chiefly made up of bankers, their agents and attorneys, who have generally been educated in favor of, and to have a community interest with, the Money Trust. The National Monetary Commission was placed in charge of the same Senator Nelson W. Aldrich and Congressman Edward B. Vreeland, who respectively had charge in the Senate and House during the passage of the act creating it.

The act authorized this commission to spend money without stint or account. It spent over $300,000 in order to learn how to form a plan by which to create a greater money trust, and it afterwards recommended Congress to give this proposed trust a fifty-year charter by means of which it could rob and plunder all humanity. A bill for that purpose was introduced by members of the Monetary Commission, and its passage planned to be the fourth and final act of the campaign to completely enslave the people.

The fourth act, however, is in process of incubation only, and it is hoped that by this time we realize the danger that all of us are in, for it is the final proposed legislation which, if it succeeds, will place us in the complete control of the moneyed interests. History records nothing so dramatic in design, nor so skillfully manipulated, as this attempt to create the National Reserve Association,—otherwise called the Aldrich plan,—and no fact nor occurrence contemplated for the gaining of selfish ends is recorded in the world's records which equals the beguiling methods of this colossal undertaking. Men, women, and children have been equally unconscious of how stealthily this greatest of all giant octopuses,—a greater Money Trust,—is reaching out its tentacles in its efforts to bind all humanity in perpetual servitude to the greedy will of this monster.

I was in Congress when the Panic of 1907 occurred, but I had previously familiarized myself with many of the ways of high financiers. As a result of what I discovered in that study, I set about to expose the Money Trust, the world's greatest financial giant. I knew that I could not succeed unless I could bring public sentiment to my aid. I had to secure that or fail. The Money Trust had laid its plans long before and was already executing them. It was then, and still is, training the people themselves to demand the enactment of the Aldrich Bill or a bill similar in effect. Hundreds of thousands of dollars had already been spent and millions were reserved to be used in the attempt to bring about a condition of public mind that would cause demand of the passage of the bill. If no other methods succeeded, it was planned to bring on a violent panic and to rush the bill through during the distress which would result from the panic. It was figured that the people would demand new banking and currency laws; that it would be impossible for them to get a definitely practical plan before Congress when they were in an excited state and that, as a result, the Aldrich plan would slip safely through. It was designed to pass that bill in the fall of 1911 or 1912." [58]

Jewish bankers used their financial influence to ruin Gentile Peoples, then Jewish bankers used their political influence and controlled press to blame Gentile governments and religions for the ruin Jewish bankers had deliberately caused. Beware of the agents of Cabalistic Jews bearing the "gifts" of revolution, banking reform and world government. Remember that it is these same Jewish leaders who are deliberately causing the pains and poverty of the world and who intend to lead gullible non-Jews into such severe suffering that they will gladly hand over all their power to the Jews who are perpetual portraying themselves in the media as the worst victims of conflict and most moral people—people who can deliver us all from the problems of life—with a bullet to the back of the head. Beware of Utopian promises and easy schemes to unseat the powerful from power. Beware of revolutionaries, especially anti-Semitic revolutionaries. Beware of those who point out the corruption of Jewish leadership, but then offer up solutions which will ultimately serve the interests of Jewish leadership. Jewish leaders have always used outrage against their outrages as a trap to put their own agents into power.

Of course, to solve the problems Jewish bankers were causing and blaming on their victims, the Peoples needed to know who was at fault and how to remedy the situation. This, too, proved to be an opportune situation for the Jewish bankers, who were highly racist and who desired to keep the "Holy Jews" segregated form the all the "inferior races", while maintaining control over Gentile societies.

Jews have, like all human beings, tended to integrate into the societies where they have lived. Jewish leaders have always chastised and punished assimilatory Jewry with death. After ruining nations and cultures with large Jewish populations, Jewish leaders often put anti-Semitic leaders into power, who then falsely blamed all Jews for the actions of Jewish leadership, and who proposed highly destructive "solutions" to the problems Jewish leaders had caused. In this way, Jewish leaders maintained their control over both Jews and non-Jews, and forced assimilating Jews back into segregation, thereby preserving the "divine Jewish race" from the dissolution of good natured integration.

The Jewish bankers then forced the Jews to flee to another nation, taking with them the wealth of their previous homeland. The new target nation or empire then grew with the influx of investment capital, drawing unto itself the wealth of the world, which ultimately filtered into the hands of the Jewish

---

58. C. A. Lindbergh, *Banking and Currency and The Money Trust*, National Capital Press, Washington, D.C., (1913), pp. 92-98.

bankers, who loaned it out at interest to finance wars they had caused and to pay for the disastrous economic policies they covertly implemented. It was not only important to Jewish leaders to accumulate the world's wealth so that they would be wealthy, but also to oppress non-Jews and inhibit their progress so as to prevent any future challenges to Jewish power (*Proverbs* 1:13-14). The perpetual debt of the Gentile nations Jewish leaders caused became a perpetual source of revenue for Jewish bankers. As economies collapsed, Jewish leadership gained wealth and had the means to buy up politicians, royalty, churches, businesses, real estate, arms, valuables and manufacturing capital at reduced prices.

These Jews used all of the ancient corrupt tactics of organized crime. They burned down nations and offered the protection racket of "world government" as if a solution to the problem of war, war which they had covertly caused. They loaned out monies secured by nations' taxes, then ensured that the borrowers could not repay the debts, then they took over entire economies. In prior times when the majority of the world's citizens were farmers, they ensured that the farms would fail so that they could collectivize the farms and force the Peoples of the world into slavery on lands they had stolen from the farmers.

The best means to dissolve Jewish power is to welcome Jews into other communities. Anti-Semitism has always only increased Jewish power by increasing Jewish racism and tribalism and by providing Jewish leaders with a means to put their agents into power on a political platform centered on shallow and counterproductive Jew-baiting. These "anti-Semitic" Jewish agents then deliberately ruin the anti-Semitic nation they have created. Jew-baiting is trap that ensnares Gentile Peoples and increases the power of Jewish leadership. Racist political Zionist leader Theodor Herzl wrote in his book *The Jewish State*,

> "Oppression and persecution cannot exterminate us. No nation on earth has survived such struggles and sufferings as we have gone through. Jew-baiting has merely stripped off our weaklings; the strong among us were invariably true to their race when persecution broke out against them. This attitude was most clearly apparent in the period immediately following the emancipation of the Jews. Later on, those who rose to a higher degree of intelligence and to a better worldly position lost their communal feeling to a very great extent. Wherever our political well-being has lasted for any length of time, we have assimilated with our surroundings. I think this is not discreditable. Hence, the statesman who would wish to see a Jewish strain in his nation would have to provide for the duration of our political well-being; and even Bismarck could not do that. [***] The Governments of all countries scourged by Anti-Semitism will serve their own interests in assisting us to obtain the sovereignty we want. [***] Great exertions will not be necessary to spur on the movement. Anti-Semites provide the requisite impetus. They need only do what they did before, and then they will create a love of emigration where it did not previously exist, and strengthen it where it existed before. [***] I imagine that Governments will, either voluntarily or under pressure from the Anti-Semites, pay certain attention to this scheme; and they may perhaps actually receive it here and there with a sympathy

which they will also show to the Society of Jews."⁵⁹

Adolf Hitler and Joseph Stalin were both agents of the Jewish bankers and both performed the valuable services of segregating the Jews and increasing Jewish hatred of non-Jews. Hitler and Stalin, who were both Bolshevik Zionists, brought the German People and the Russian People into war with each other, and helped the Jewish bankers to discredit and ruin Gentile government and to move the world towards a universal world government led by Jews—towards the "New World Order" or "Jewish Utopia" prophesied in *Isaiah* 65:17 and 66:22.

Jewish leaders deliberately caused Gentile Peoples to hate all Jews, then they used their controlled press and their disproportionate wealth to finance supposedly anti-Semitic leaders, who then deliberately destroyed the Gentile nations and caused war and famine by proposing the easy "solutions" of dictatorship, gold-backed currencies, "defensive" "preemptive"—truly aggressive—wars, and the segregation and expulsion of the Jews. Jewish leaders followed the example of Joseph found in *Genesis* 47, in which story the Jews steal the wealth of Egypt and take it with them on their way out; and the story of Esau and Jacob (*Genesis* 25:23; 27:38-41. *The Zohar*, Volume 2, Tol'doth, folios 138*b*-139*b*, Soncino, London, (1933), pp. 45-46.), where Jacob provokes Esau to "anti-Semitism", which "anti-Semitism" causes Esau and his descendants to eternally slave and soldier for Jacob—anti-Semitism causes the Gentiles to become the slaves of the Jews. In the Hebrew Bible, Jews justify their theft and genocide of other Peoples based on the anti-Jewish feelings they have deliberately provoked. In the Old Testament, and throughout history, Jews justify their racism and segregationist tribalism by deliberately provoking other Peoples to hate them. They forever blame others for the problems they themselves have caused. The *Zohar*, II, 160*a*, states, and note that the "evil side" is the allegedly sub-human Gentile world,

> "R. Hizkiah said: 'Assuredly it is so. Happy is he whose portion is firmly established on the good side, and who does not incline himself to the other side, but is delivered from them.' Said R. Judah: 'Assuredly it is so, and happy is he who is able to escape that side, and happy are those righteous who are able to wage war against that side.' R. Hizkiah asked: 'How?' R. Judah, in reply, began to discourse on the verse: *For by wise guidance thou shalt make thy war, etc.* (Prov. xxiv, 6). 'This war', he said, 'alludes to the war against the evil side, which man must combat and overcome, so as to be delivered from it. It was in this way that Jacob dealt with Esau, who was on the other side, so as to outwit him by craft, as was necessary in order to keep the upper hand of him from the beginning to the end, as befitted.'"⁶⁰

---

59. T. Herzl, *A Jewish State: An Attempt at a Modern Solution of the Jewish Question*, The Maccabæan Publishing Co., New York, (1904), pp. 5-6, 25, 68, 93.
60. H. Sperling and M. Simon, *The Zohar*, Volume 2, The Soncino Press, New York, (1933), pp. 114-115.

The process, by which Jewish leadership lead Gentile nations into self-destruction through artificial and controlled anti-Semitism and false promises of a Utopian society to come, is one of deliberate false diagnosis and contrived improper treatment. Jewish leaders covertly claim through their agents that all Jews are a cancer on the nation and the cure is the segregation of the Jews. But it is the patient—the non-Jews—who receive the fatal treatment of revolution, war, economic ruin and cultural degradation—a lethal dose of unneeded radiation. Racist Jewish leaders regularly sacrifice a few of their own and walk away with the wealth of other nations, and the contrived status of a blameless victim who must remain segregated for the sake of self-defense.

The solution to the problem is for non-Jews to recognize that the core problem is not Jewish people in general, but rather genocidal Judaism and corrupt Jewish leadership who view Jewish genocidal prophecy as a plan they must carry out at all costs, including the sacrifice of large numbers of innocent Jews. The solution is to welcome Jews in general into the broader community and to expose the methods and intentions of corrupt and racist Jewish leadership. Jews must in their turn abandon genocidal Judaism and abandon their virulent racism and corrupt tribalism. Jews must cease to hypocritically insist upon their own segregation, while demanding that the rest of the world integrate into a world government led by Jews.

Jewish Messianic prophecy is a plan too dangerous to ignore. It threatens to destroy human life on Earth. In the 1500's, Martin Luther wrote, among other things,

> "Further, they presume to instruct God and prescribe the manner in which he is to redeem them. For the Jews, these very learned saints, look upon God as a poor cobbler equipped with only a left last for making shoes. This is to say that he is to kill and exterminate all of us Goyim through their Messiah, so that they can lay their hands on the land, the goods, and the government of the whole world. And now a storm breaks over us with curses, defamation, and derision that cannot be expressed with words. They wish that sword and war, distress and every misfortune may overtake us accursed Goyim. They vent their curses on us openly every Saturday in their synagogues and daily in their homes. They teach, urge, and train their children from infancy to remain the bitter, virulent, and wrathful enemies of the Christians."[61]

Since Luther's time, many Jews have stated that the Jewish People and politics are the Jews' Messiah. Jewish Bolshevism accomplished, and sought to accomplish, many of the Jews' Messianic goals.

---

[61]. M. Luther, *Von den Juden und ihren Lügen*, Hans Lufft, Wittenberg, (1543); Reprinted, Ludendorffs, München, (1932); English translation by Martin H. Bertram, "On the Jews and Their Lies", *Luther's Works*, Volume 47, Fortress Press, Philadelphia, (1971), pp. 123-306, at 264.

The Jewish book of *Deuteronomy* 7:2 states,

"And when the LORD thy God shall deliver them before thee; thou shalt smite them, *and* utterly destroy them; thou shalt make no covenant with them, nor shew mercy unto them:"

Jüdische Rundschau, Number 82/83, (14 October 1921), pp. 595-596 (front page and second page of the issue), covered speeches by Zionist leaders in Berlin on Sunday, 9 October 1921, in Blüthner Hall welcoming back Nachum Sokolow, President of the Executive,

"Begrüssung für Sokolow
Zionistische Massendemonstration in Berlin

Wie bereits kurz gemeldet, fand am Sonntag, den 9. d. M. im überfüllten Blüthnersaal in Berlin ein großes Massenmeeting zur Begrüßund des Präsidenten der Exekutive, Herrn Nahum Sokolow, statt. Die Versammlung war ein lebendiger Beweis der Wertschätzung und Verehrung für den zionistischen Führer, der nach langjähriger Abwesenheit wieder zu kurzem Aufenthalt nach Berlin zurückgekehrt ist. Herr Sokolow war Gegenstand lebhafter Ovationen, die ein Ausdruck des Dankes für die große Arbeit waren, die Sokolow im Dienste des jüdischen Volkes mit hingebungsvoller Energie geleistet hat. Was dazu zu sagen ist, haben die Redner der Feier gesagt. Wir können uns daher auf die Wiedergabe ihrer Reden beschränken.

Die Versammlung wurde eröffnet vom Vorsitzenden der B. Z. V.,
Dr. Egon Rosenberg,
der es als glückliches Schicksal pries, daß dem jüdischen Volk in der schweren Zeit des Krieges zwei Männer vom politischen Ingenium und von der Tatkraft Weizmanns und Sokolows geschenkt wurden. Er begrüßt außer Sokolow noch die Herren Jabotinsky, Dr. Halpern und Dr. Scharja Levin, die lebhaft akklamiert wurden.

Als erster Redner spricht der Vorsitzende der Z. V. f. D.,
Feliz Rosenblüth,
der etwa folgendes ausführt:

Als der Zionismus zum ersten Male der Welt sein Programm verkündete, da hat man überall in der Welt und vielleicht nirgends lauter und hohnvoller als in Deutschland die Frage aufgeworfen, wie es möglich sein sollte, die zerstreuten Teile der Diasporajudenheit wieder zu einer nationalen Einheit als Staatsvolk zusammenzuschmieden. Man berief sich auf den jüdischen Individualismus, der jeder Einordnung und Führung spottet. Man hat dem jüdischen Volk die inneren Fähigkeiten abgesprochen, wieder ein nationales Gemeinwesen mit staatlich-sozialer Gliederung aufzubauen. Man hat bei uns jene sozialen Tagenden verneint, die eben erst aus einer zusammenhanglosen Masse von Menschen ein organisch verbundenes Volk machen. Wenn an diesem Vorwurf etwas richtig gewesen sein sollte, so können wir sagen, daß auch hier das Wort Theodor Herzls zutrifft, daß schon das Wandern auf dem Wege zum Ziele uns zu neuen, zu besseren Menschen gemacht hat. Wir haben alle schon oft erlebt, daß der Zionismus mit jener wunderbaren Kraft der Antizipation das Wunder einer inneren Wandlung an uns vollzogen hat, daß wir gelernt haben, uns ideell im vorhinein als Bürger unseres werdenden Gemeinwesens zu empfinden, das heißt, als Menschen mit der Verantwortlichkeit und den Pflichten des einzelnen gegenüber der höheren Ordnung der Gemeinschaft. Es ist in diesem Jahrzehnt der Arbeit des politischen Zionismus in der Tat so etwas wie ein zionistisches Staatsvolk entstanden, ein Vortrupp des werdenden Palästastaatsvolkes, eine Gemeinschaft mit eigentümlichen Kriterien der Ordnung und Gliederung, die sich beispielsweise im Zionistenkongreß eine parlamentarische Körperschaft mit eigenartiger gesetzgeberischer Kraft geschaffen hat. Dieser Prozeß der Staatsvolkswerdung aber, meine Damen und Herren, ist unlösbar verknüpft mit einem Phänomen, das auch erst durch den Zionismus wieder neu im jüdischen Volk geschaffen wurde, mit dem Phänomen des Führertums. Wir wollen hier nicht untersuchen, ob diese Wandlung vielleicht überhaupt erst möglich geworden ist dadurch, daß im Zionismus Führerpersönlichkeiten mit natürlicher Uebergeordnetheit entstanden sind, oder ob diese Menschen zu Führern einporgewachsen sind aus dem Drange dieses Umwandlungsprozesses. Aber wir wissen, daß erst der Zionism dem jüdischen Volk wieder Führer geschenkt hat, und wir betrachten dieses Führertum als Symbol der Regenerationsbewegung, in der wir stehen. Erst in den Tagen des Zionismus ist es wieder möglich geworden, daß jüdische Männer überall in der Welt von dem gleichen Gruß aus jüdischen Herzen als Führer empfangen wurden, und wir erkennen diese Erscheinung als sichtbaren Beweis dafür, daß wir heute in einer Zeit leben, in der unser Volk neu erwacht ist und seine Kraft neu sammeln will. Der Zionismus hat uns wieder Führer und Repräsentanten gegeben, auf die das jüdische Volk alles überträgt, was an Hoffnungen und Zukunftswillen in ihm lebt. Diese Männer können stark sein, weil sie sich als Träger dieses Volkswillens fühlen. Der Zionismus hat uns wieder zentrale Persönlichkeiten gegeben, und das ist der hoffnungsvollste Beweis dafür, daß im jüdischen Volk zentripetale, aufbauende, sammelnde Kräfte leben. Deshalb wollen wir, wenn wir in diese Begrüßungsfeier eintreten, uns bewußt sein, daß diese Feier keine äußere Demonstration ist, sondern eine Manifestation des Lebenswillens der jüdischen Nation, der nach Konzentration und nach Vereinheitlichung strebt und für den zentrale Führerpersönlichkeiten ein Symbol oder vielleicht sogar ein Beweis sind. In diesem Sinne begrüßt die Zionistische Vereinigung für Deutschland am heutigen Tage Herrn Nahum Sokolow, den Präsidenten der Exekutive, als den Repräsentanten unserer Bewegung, als den Mann, der zusammen mit Weizmann das Recht des jüdischen Volkes auf Palästina verkündet und verteidigt hat und der im Kampf für unser Ideal unser anerkannter Führer wurde. Wir grüßen in unserer Mitte Herrn Sokolow, und in diesem Gruß erleben wir unsere Uebereinstimmung mit der Judenheit der ganzen Welt, die Einheit der jüdischen Nation. (Lebh. Beifall.)

Dr. Schmarja Levin

sagt in seiner Rede u. a.: „Bei einer zionistischen Veranstaltung hat ein großer englischer Staatsmann, Sir Robert Cecil, gesagt, daß die einzigen Errungenschaften des Krieges die Balfour-Deklaration und die League of Nations sind. An der Balfour-Deklaration sind wir alle interessiert, der Bund der Nationen könnte uns aber als etwas Fernliegens und Fremdes erscheinen. Ich glaube aber, Robert Cecil hat den Zusammenhang zwischen diesen beiden Dingen tiefer erfaßt. Es ist nicht Uebertreibung noch Ueberhebung, wenn ich die These aufstelle, daß die Verwirklichung des Zionismus vom Siege der zweiten Idee bedingt ist. Denn in ihr liegt die Garantie der Dauerhaftigkeit. Noch vor dem Waffenstillstand hat sich in Amerika ein Mann gefunden, der Vertreter von 120 Millionen Menschen, der diese idee aufnahm. Es ist keine neue idee, es ist die alte jüdische Idee der Propheten. Wenn Sie die jüdische Psyche an den klassischen Denkmälern studieren, so werden Sie finden, daß kein Wort für den Begriff ‚Menschheit' vorhanden ist, sondern diese Werke sprechen immer von dem Verband aller Nationen. In einem Worte spiegelt sich eine Weltanschauung, und es ist kein Zufall, daß die hebräische Sprache, die bereits im Altertum ein solch hohe Entwicklung erreicht hat, kein besonderes Wort für den Begriff Menschheit geprägt hat. Denn sie haben den Sinn des historischen Prozesses tief begriffen, und es ist ihnen klar, daß die Nation das Höchste ist, was die Geschichte hervorbringt. Nicht das Verschwinden der Nationen, noch deren Verschmelzung zu einer Einheit hat ihnen vorgeschwebt, sondern das harmonische Leben sämtlicher Nationen und Völkerschaften. Sie waren zu ernst, um sich Illusionen hinzubringen und Phantomen zu dienen, deshalb galt ihre Predigt immer dem Bund der Nationen und nicht dem verschwommenen Begriff einer abstrakten Menschheit. Wilson, der diese Idee predigte, hatte kein Glück. Aber vielleicht ist es der Gang der Geschichte, daß die ‚erste Auflage' einer Idee zerbrochen wird und daß die zweite Auflage erscheinen muß, um zur Geltung zu kommen. Wir haben dafür ein krasses Beispiel in den zehn Geboten. Die ersten Gesetzestafeln wurden zerbrochen, und erst in der zweiten Auflage feierte die Idee, die ihnen zugrunde lag, ihre Auferstehung. Es ist unsere Sache, die Idee des Völkerbundes aufzunehmen, sie zu verbreiten, bis sie Wirklichkeit wird. Man kann sich nie auf eine einzelne Nation verlassen, mag sie auch die beste und edelste sein. Denn auch die besten und edelsten werden manchmal in ihren Handlungen von egoistischen Motiven geleitet. Das Gleichgewicht der Welt kann nur durch eine Körperschaft reguliert werden, die alle Nationen repräsentiert und den Interessen aller Rechnung trägt. Der Zionismus ist mit dieser großen Idee verknüpft, und es ist deshalb unsere Aufgabe, uns ihrer mit aller Energie anzunehmen. Wir können ihr in manchen Bezeihungen zum Siege verhelfen, denn wir haben schon manche Idee in der Welt populär gemacht. Es ist kein Zufall, daß gerade aus Palästina weltbefruchtende und weltbeherrschende Ideen ausgingen.

Es kann sein, daß unsere Unzufriedenheit, die uns nach Palästina treibt, gerade darin liegt, daß wir nach einem Platz für die Verwirklichung von neuen Ideen trachten. Denn das letzte Wort ist noch nicht gesprochen, und lange wird noch der mensch herumirren, bis er aus dem Labyrinth seinen Ausweg findet. Der richtige Ort für die Verwirklichung der einstweilen nur geahnten Idee ist weder in Genf noch in Haag zu suchen. Ein jüdischer Denker, der aber nicht nur strenger Logiker, wie mancher es glaubt, sondern auch ein großer Ahner unserer Zukunft ist, Achad Haam, hat von einem Tempel auf dem Berge Zion geträumt wo die Verteterschaft aller Nationen dem ewigen Frieden einen Tempel weihen wird. Und ich benutze gerade diese Gelegenheit, von der Idee der Völkerverbrüderung zu sprechen, weil sie mit der Persönlichkeit Sokolows verbunden ist. Sokolow hat es verstanden, den Zionismus in seiner Totalität aufzufassen, und deshalb war er ebenso energisch als Präsident der jüdischen Delegation wie in seiner rein zionistischen Tätigkeit, wobei er die glänzendste Gelegenheit hatte, mit den Vertretern der verschiedenen Nationen in beständigen Kontakt zu kommen und gar manchen vielleicht unbewußten Einfluß auf die Gestaltung solcher Beziehungen, die die Idee des Völkerbundes um einen Schritt weiterbringen, auszuüben.''

Kurt Blumenfeld

begrüßt darauf in kurzen Worten Herrn Sokolow. Er weist darauf hin, daß Herr Sokolow die Fülle des Wissens und die Fähigkeit, den Maßstab der Jahrhunderte anzulegen, mit der Kraft verbindet, dem Augenblick gerecht zu werden. Die zionistische Bewegung, die im Gegensatz zu dem kurzatmigen Revolutionen anderer Völker eine ,,Revolution mit langem Atem'' sei, brauche eine solche Persönlichkeit an führender Stelle. Nicht durch Tageserfolge sei die zionistische Sache zu fördern, sondern durch unverdrossene, stetige Arbeit. Die Energie, die im Augenblick erfordert wird, dürfe nicht aus einer Desperadostimmung kommen, sie brauche vielmehr die freudige Tat von Menschen, die von der Unzerstörbarkeit der zionistischen Sache überzeugt sind. Herr Blumenfeld sprach in diesem Zusammenhang über die Notwendigkeit, die Erkenntnis des wahren Zustandes der zionistischen Bewegung zur Grundlage unserer Arbeit zu machen.

Auf alle diese Reden antwortet sodann

Nahum Sokolow:

Herr Vorsitzender, meine Damen und Herren, der Zionistenkongreß liegt hinter uns. Wir gehen jetzt mit den Kongreßresolutionen in die Welt hinaus, um sie in die Tat umzusetzen. Schon einer der Herren Vorredner, Dr. Levin, bemerkte, daß Personenkultus keine jüdische Sache ist. Er hat Recht. Wenn diese Versammlung dazu bestimmt wäre, der Ausdruck eines persönlichen Kultus zu sein, so würde ich mit Dank ablehnen. Doch ich habe den Eindruck, daß keiner unter Ihnen diese Versammlung als eine persönliche Ehrung für mich betractet. Ich bin für Sie in diesem Augenblick der Vertreter einer Idee, der Repräsentant einer Organisation. Sie ehren nicht mich, sondern Sie ehren die Idee, zu deren Wortführern ich zu gehören den Vorzug habe. Ich möchte hier ein gut jüdische Wort zitieren: Hilf ihm, wenn er unter der Last zusammenbricht. Ich breche schier zusammen unter der Last der Komplimente, der wohlgemeinten, der weit übertriebenen, die an meine Adresse gerichtet sind. Helfen Sie mir, mich unter dieser Las aufzurichten. Ich werde Ihnen Gleiches mit Gleichem vergelten. Es wäre weder mir noch meinen Kollegen möglich gewesen, irgend etwas zu erreichen, wäre nicht unserer Arbeit eine Arbeit vorausgegangen, die hier, von Euch gemacht worden ist, die von Euch noch immer gemacht wird, von Euch, Zionisten Berlins, von Euch, Zionisten Deutschlands, von Euch, der zionistischen Jugend Deutschlands, die wir in allen Ländern als Vorbild zitieren. Wäre diese Arbeit nicht gemacht und entwickelt worden, und wurde diese Arbeit nicht jetzt einer größeren Zukunft entgegengehen, so wäre unsere Arbeit nicht möglich. Ich beglückwünsche Sie zu Ihrer Arbeit, zu Ihrer Begeisterung und Opferfreudigkeit, von der wir, die Zionisten der Welt, viel Großes erwarten. Ich bin unter Euch, und es ist mir wie ein Traum. Noch vor drei, vier, fünf Jahren hätte ich es nicht geahnt. Mir beweist dies, daß der Zionismus stärker ist als der Moloch des Weltkrieges, und daß wir jetzt enger vereinigt sind, als uns die äußeren Umstände trennen konnten. Es ist für mich ein Feiertag, daß ich hier unter Euch bin und von Euch empfangen werden kann. Das ist der Sieg der zionistischen Einheit.

Und nun ein Wort zu den Erfolgen. Wenn man Erfolge erzielt—und ich will nicht zu bescheiden sein und in Abrede stellen, daß wir politische Erfolge erzielt haben—so muß man immer darauf achten, welchen Methoden diese Erfolge zu verdanken sind. Dies ist nicht nur eine historische Betrachtung und soll nicht nur dazu dienen, irgendein Rätsel der Vergangenheit zu lösen, sondern sie soll auch als Anweisung für die weitere Tätigkeit dienen. Ein Wort zu den politischen Erfolgen. Sie bestehen, wie allen Zionisten bekannt ist, in dem, was wir im Laufe der Jahre angestrebt und was wir erreicht haben: Die internationale Anerkennung und die internationale Bestätigung unseres Ideals, der nationalen Heimstätte in Palästina. Hiermit stehen die Namen, unseres Präsidenten Dr. Chaim Weizmann und meine Wenigkeit, in Verbindung. Auch möchte ich bei dieser Gelegenheit eines teuren unvergeßlichen Namens gedenken, Dr. Tschlenow, der uns in der ersten Periode unserer Arbeit geholfen hat. Es sind keine Berufsdiplomaten, die diese Erfolge erzielt haben. Lange vor der Friedenskonferenz tauchten Juden auf, die versuchten, sich mit der Welt in Verbindung zu setzen, und in London, Paris, Rom und anderen politischen Zentren Propoganda zu machen. Wir sprachen mit den Machthabern der Welt dir Sprache ehrlicher Leute. Es gibt in England Tausende von Juden, die einflußreicher und bekannter sind als Weizmann, der aus Pinsk gebürtig ist. In Paris, dem Zentrum aller jüdischen Kapazitäten, war es meine Wenigkeit. Unter all diesen Leuten erscheint ein fremder Jude, der höchstens auf eine literarische Karriere in hebräischer Sprache im Osten Europas zurückschauen kann. Das ist alles, was ich in meinem Tornister trug, den Marschallstab eines europäischen Diplomaten trug ich nie in meinem Koffer. Wir sind die Schüler des ersten jüdischen Politikers, Theodor Herzls. Ich könnte nicht sagen, daß ich zu seinen Füßen saß. Ich saß viel früher zu den Füßen so manchen Rabbiners in Polen. Aber ich bemühte mich in den wenigen Jahren, die uns vergönnt waren, neben Herzl zu arbeiten, in seinen Geist einzudringen. Wir sprachen zu den Diplomaten der Welt in Namen des jüdischen Volkes und der Zionistischen Organisation. Wir sprachen die Sprache des nationalen Zionismus, die Sprache der nationalen Idee. Die Welt war auch vor dem Kriege national eingerichtet, aber sie wollte sich während des Krieges noch viel nationaler einrichten. Sie wollte die politische Geographie mit den Grenzen der nationalen Ethnographie womöglich in Einklang bringen. Deshalb, als sich die Völker beim Aeropag der Mächte mit ihren Ansprüchen meldeten, sagten wir uns, daß auch für uns die Zeit gekommen sei. Wenn wir uns jetzt nicht melden, so werden unsere Ansprüche der Verjährung verfallen. Wir erhoben also unsere Ansprüche auf unsere alte Heimat. Da sagte man uns: Wir sind entschlossen, die nationale Selbständigkeit der Völker, die sie seit einem Jahrhundert eingebüßt haben, wieder herzustellen, aber Eure Sache ist viel zu alt. Euch ist vor 2000 Jahren Unrecht geschehen. So historisch kann man nicht sein. Darauf erwiderten wir: Wir haben ein stärkeres Recht als andere Nationen, die seit 100 Jahren unter dem Verlust ihrer Selbständigkeit leiden, denn wir leiden schon seit 2000 Jahren. Darauf sagte man uns, die Politik richtet sich nach Analogien und Tatsachen. Darauf wiesen wir hin auf die Analogie des griechischen Volkes. Man sagte uns: Die Juden sind ja gar nicht in Palästina. Wir erwiderten: Die Griechen waren ja in Griechenland auch nicht da. Oeffnen Sie das Buch der Geschichte, so werden Sie sehen, daß das Land, das jetzt von Griechen bewohnt ist, von allen möglichen Mischstämmen bevölkert war, die nach und nach begannen, sich zu den Griechen zu bekennen. So hängt die Frage des heimatlichen Palästina mit dem jüdischen Volk in der ganzen Welt zusammen. Einen großen Teil meiner Zeit mußte ich diesen Verhandlungen widmen. Wir verlangten Minderheitsrechte für die Juden in allen Ländern, wo sie in großen Massen leben. Diese Forderung ist vorläufig auf dem Papier erfüllt worden. Auch den anderen Minderheiten sind Minderheitsrechte zugebilligt worden. Aber es existiert sonst keine einzige Minderheit, die nicht irgendwo in der Welt eine Mehrheit ist. Die Garantie der Minderheitsrechte hat nur insofern Wert, als zu gleicher Zeit dieses Volkselement in irgendeinem Lande in der Welt konzentriert ist und eine Mehrheit darstellt. Deshalb besteht ein tiefer logischer Zusammenhang zwischen der Diaspora und Palästina. Das jüdische Volk will nach Zion zurückkehren, das jüdische Volkstum wird sein Zentrum in Palästina haben. Große Teile des Judentums werden als jüdische Peripherien in der Welt leben, es muß für sie gesorgt werden, ihre Würde und ihre nationalen Rechte müssen gesichert werden. Zwischen diesen beiden Postulaten besteht kein Widerspruch, ist kein Widerspruch in der politischen Welt gefunden worden, weil wir im Namen des jüdischen Volkes sprachen, weil wir die Romantik von Palästina für uns hatten, weil wir die Romantik eines alten Volkes für uns hatten, das wieder jung zu werden beginnt. Wir sagten offen und ehrlich, was wir für Palästina und was wir für die Diaspora beanspruchen, so daß es als einheitliches System der Vernunft der Staatsmänner erschien. Deshalb haben wir das erreicht, was zu erreichen war. Es ist Tatsache geworden: wir sind in das Stadium eines Volkslebens eingetreten, wir sind schon in der Welt das anerkannte jüdische Volk, für welches ein Heim in Palästina gebaut wird. Wir haben diesen Bau schon begonnen. Ich kann Ihnen nicht auskalkulieren, wie wir es errichten werden, wieviel es kosten wird. Wenn wir Monumente in der Welt sehen, uns an ihnen ergötzen und an ihnen lernen, sie weiter zu schaffen und wenn in diesem Augenblick ein Rechenmeister mich fragt, wieviel es gekostet hat und wo man das Geld hergenommen hat, so könnte ich diese Fragen nicht beantworten. Dafür werden wir Rechenmeister haben, denn ohne Rechenmeister geht es nicht. Wir dürfen uns aber nicht von vornherein nur auf diesen Rechenstandpunkt stellen. Man muß sehr oft die Zahlen vergessen und sich hineinstürzen in eine große Sache. Wir, das jüdische Volk, sind auf Leben und Tod in diese Sache eingetreten. Wir müssen für das jüdische Volk das Nationalheim bauen, und da gibt es kein Rechen mehr. Jeder Jude muß eintreten mit seiner ganzen Person, mit all seiner Kraft, das ist unser Reichtum. Das übrige wird sich von selbst ergeben. Wenn Sie von politischen Erfolgen gesprochen haben, dürfen Sie nicht vergessen, daß diese Erfolge nur der Anfang sind, der Anfang einer Arbeit, die jetzt mit noch größerer Energie geleistet werden muß. Das Mandat ist noch nicht ratifiziert. Ich gebe zu, daß es mangelhaft ist, aber wir müssen diese Lücken ausfüllen. Sie wissen selbst, welche Möglichkeiten einer Interpretation gegeben sind, und wir müssen dafür sorgen, daß es so interpretiert wird, wie es unserer Sache dienlich ist. Die freie, nicht Immigration, sondern Repatriierung, muß vor sich gehen. Das muß ruhig und maßvoll gemacht werden. Nicht in aufreizender, provokatorischer Form, sondern ruhig, Schritt für Schritt, so muß Palästina unser werden. Ich glaube daran, ich bin überzeugt davon, daß Palästina in wenigen Jahren unser wird, und ich will hoffen, daß wir alle, die wir hier anwesend sind, es noch erleben werden, daß in Palästina eine auferstandene Welt zu sehen ist. Die Pioniere, die wir jetzt dort sehen, das ist die Rückkehr des jüdischen Volkes nach Erez-Israel. So sind die Juden auch aus Babylon zurückgekehrt, in Gruppen, in Familien, deren Namen angegeben werden. Und so werden auch wir zurückkehren. Mit Arbeit werden wir Palästina gewinnen, nicht erobern, sondern gewinnen, nicht nur für uns, sondern für die ganze Menschheit, und wir werden das goldene Jerusalem wieder zur Leuchte der Welt machen.

Sokolow schloß mit folgenden Worten:

„Ich bitte Sie, tragen Sie, die Zionisten, die Botschaft hinaus in das jüdische Volk. Wir haben im Namen das jüdischen Volkes und für das jüdische Volk Palästina bewilligt bekommen, es liegt an uns, in Palästina die Heimstätte zu errichten. Was ich unter Euch sehen will, ist Begeisterung. Wir stehen vor Jom Kippur. Und da kommt mir in Erinnerung ein Wort, das ein Wunderrabbi geäußert haben soll, als er vor Kol Nidre die Schule betrat. Er kam und fand all Leute in großer Andacht. Die großen Wachskerzen brannten, und alles war regelrecht zu Kol Nidre eingerichtet. Aber er fühlte, daß etwas fehlte und da sagte er: ‚Das Feuer ist nicht da!' Und als er das sagte, verbreitete sich eine Wärme in der Schule und durchdrang die Herzen und die Gemüter aller Andächtigen. Werde ich ein solcher Wunderrabbi sein?..."

Nach der mit einem großen Beifallssturm aufgenommenen Rede Sokolows verlangte die Versammlung spontan unter stürmischen Kundgebungen, daß auch der anwesende.

Jabotinsky

spreche. Jabotinsky sprach hierauf einige anfeuernde Worte. Er sagte u. a.: „Die Begeisterung hat nur Wert, wenn sie imstande ist, sich in menschliche Energie umzusetzen, in eine Energie, die Tag für Tag einen Schritt vorwärts geht, und wenn dieser Schritt nicht gelingt, ihn am nächsten Morgen von neuem versucht, es muß eine Energie sein, die sich in schöpferische Tat verwandelt. Unsere Parole muß sein: Arbeit in Palästina, Gold im Galuth, Blut, wenn es gilt, letzte Opfer zu bringen. Das ist, glaube ich, der Sinn der heutigen Versammlung und die Anregung, mit der wir heute Berlin verlassen. Berlin war immer das Vorbild der guten Organisation, und die Organisation besteht darin, daß man Tatsachen schafft. Gehen Sie weiter auf diesem Wege, dann wird man das Recht haben zu sagen, daß diese Versammlung ein großer Schritt vorwärts war." (Stürmischer Beifall.)

Die Versammlung nahm zum Schluß die nachstehende Resolution an:

Resolution.

„Die in Berlin am 9. Oktober 1921 tagende zionistische Festversammlung spricht dem Präsidenten der zionistischen Exekutive, Herrn Nahum Sokolow, den tiefsten Dank aus für seine Arbeit, die zur Anerkennung des historischen Rechtes des jüdischen Volkes geführt hat. Sie erneuert mit dem Ausdruck des Dankes das Gelöbnis, alle Kräfte anzuspannen, um der zionistischen Leitung den Aufbau Erez-Israels auf der durch die politischen Erfolge geschaffenen Grundlage zu ermöglichen. In der Erkenntnis, daß der Aufbau Palästinas das zentrale Problem der jüdischen Gegenwart ist, fordert sie jeden Juden auf, sich opferbereit an dieser Aufgabe zu beteiligen."

*

Montag, den 10. d. M., sprach Sokolow in einem Kreise geladener jüdischer Persönlichkeiten. Zu dieser Veranstaltung war die Einladung seitens eines Komitees ergangen, dem u. a. die Herren Prof. Einstein, Rabb. Dr. Baeck, Generalkonsul Landau, Dr. Alfred Apfel, Prof. Sobernheim sowie mehrere Zionisten angehörten. Die Ausführungen Sokolows, der die Prinzipien zionistischer Politik und die Erfahrungen seiner Arbeit darlegte, fanden bei den zahlreichen Anwesenden aufmerksamstes Interesse."

These Jewish Zionist leaders, who represented great power, but few Jews, revealed that the First World War was an act of human sacrifice to "Moloch", a holocaust which had strengthened the Zionists and unified them, and which was intended to make the Peoples of the world clamor for small ethically segregated nations. The Zionist Jews planned long before the First World War that if they could provoke a world war, then they could petition at the inevitable peace conferences they would control to steal Palestine from its indigenous populations on the false and racist basis that they were a pure race in need of a segregated land to call their own. The Jewish nationalism of the Balfour Declaration and the internationalism of the Zionist League of Nations—the loss of sovereignty of Gentile nations and concurrent creation of a Jewish sovereignty—were praised by Zionist leaders as the fulfillment of Jewish prophecy, which prophecy calls for the disappearance of Gentile government and the emergence of the Jewish nation as the exclusive ruler of the entire world. Though the Jewish bankers' agent President Woodrow Wilson had failed to unite the nations in world government after the contrived holocaust of the First World War, Zionist Jews intended to try and try again until the Peoples of the world capitulated to the Judaic prophecies.

They planned more world wars and Bolshevik takeovers in order to soften the will of the Peoples to protect their own sovereignty, such that they would gladly surrender to Jewish power as a supposed means to end their suffering. As Jabotinsky said, "Arbeit in Palästina, Gold im Galuth, Blut, wenn es gilt, letzte Opfer zu bringen." One of the most influential of Zionist Jews, Achad Ha'am, saw Zionism as the fulfillment of Jewish Messianic prophecy and believed Jerusalem would become the capital of the world, as was foretold and planned by Jewish "prophets" in antiquity—note that when these Jews speak of "eternal world peace" they are referring to the Jewish prophecy that the Jewish Messiah will obliterate the Gentile Peoples and rule the world—a world which will know no more war, because the Jewish Messiah will have killed off the enemies of the Jews—all Peoples but the Jews will have perished at the hands of the Jews. These Jews were deceiving the Gentiles into destroying themselves in the euphemistic name of "peace", which to these Jews meant the extermination of non-Jews. Remember that "eternal peace" to Cabalistic Jews meant the death of the Gentiles and they deliberately tried to lead Gentiles into welcoming this fate, this Utopia of "eternal peace"—their own extinction.

World famous aviator Charles A. Lindbergh, Jr. warned that the Jews, the British, and the Roosevelt administration were planning a Pearl Harbor type event, in a speech Lindbergh delivered on 11 September 1941 in Des Moines, Iowa.[62] Lindbergh was viciously smeared in the press, so viciously, that few dared to defend him. After the Pearl Harbor attack, any who might otherwise have said, "I told you so!" would have been branded a traitor and a Nazi. It is further interesting to note that Adolf Hitler declared war against America immediately after the United States declared war on Japan—this in the full knowledge that America's entrance into the war had cost Germany victory in the First World War—then Hitler declared war on the Soviets, thereby ensuring the destruction of Germany. It has since been proven that FDR did have foreknowledge of the Pearl Harbor attack.[63]

62.
<http://www.pbs.org/wgbh/amex/lindbergh/filmmore/reference/primary/desmoinesspeech.html>

On 2 April 1917, while petitioning the American Congress for war against Germany, President Woodrow Wilson, who was an agent of Zionist Jewish bankers, stated that he would be good to the Germans and attack them without provocation so that the First World War would accomplish world peace by means of world war—which happened to be an ancient Jewish plan, war in the name of peace, genocide for the benefit of the righteous Jew, tyranny and slavery in the name of democracy,

"We are glad, now that we see the facts with no veil of false pretence about them, to fight thus for the ultimate peace of the world and for the liberation of its peoples, the German peoples included: for the rights of nations great and small and the privilege of men everywhere to choose their way of life and of obedience. The world must be made safe for democracy. Its peace must be planted upon the tested foundations of political liberty. We have no selfish ends to serve. We desire no conquest, no dominion. We seek no indemnities for ourselves, no material compensation for the sacrifices we shall freely make. We are but one of the champions of the rights of mankind. We shall be satisfied when those rights have been made as secure as the faith and the freedom of nations can make them."[64]

According to Congressman Thorkelson, Lord Beaverbrook wrote an article entitled "A Military Alliance With England", which appeared in the *American Mercury* long before the attack on Pearl Harbor, in August of 1939, and which Congressman Thorkelson entered into the Congressional Record on 11 October 1939. This article revealed that some hoped for another world war which would empower the League of Nations,

"An attack by the Japanese on the Pacific coast of the United States would

---

63. R. B. Stinnett, *Day of Deceit: The Truth about FDR and Pearl Harbor*, Free Press, New York, (2000). *See also:* C. B. Dall, FDR, *My Exploited Father-in-Law*, Christian Crusade Publications, Tulsa, Oklahoma, (1967); **and** *A Tribute to Lincoln, Our Money-Martyred President: An Address in Springfield, Illinois*, Omni Publications, Hawthorne, California, (1970); **and** *Amerikas Kriegspolitik: Roosevelt und seine Hintermänner*, Grabert-Verlag, Tübingen, (1972); **and** A. J. Hilder and C. B. Dall, *The War Lords of Washington (Secrets of Pearl Harbor); an Interview with Col. Curtis Dall*, Educator Publications, Fullerton, California, (1972); **and** C. B. Dall, *Who Controls Our Nation's Federal Policies — and Why?*, Noontide Press, Los Angeles, (1973); **and** C. B. Dall and B. Freedman, *Israel's Five Trillion Dollar Secret*, Liberty Bell Publications, Reedy, West Virginia, (1977); **and** C. B. Dall, *Col. Dall Reports to the Board*, Liberty Lobby, Washington, D.C., Serial Publication, (1900's); **and** C. B. Dall and R. M. Bartell, *Liberty Lobby Progress Report*, Serial Publication, Liberty Lobby, Washington, D.C., (1970's) ; **and** C. B. Dall, *Colonel Dall Reports*, Serial Publication Liberty Lobby Washington, D.C., (1900's); **and** C. B. Dall and C. M. Dunn, *Ephemeral Materials*, 1957-, Liberty Lobby, Washington, D.C.
64. W. Wilson, "War Message", Sixty-Fifth Congress, First Session, Senate Document Number 5, Serial Number 7264, Washington, D.C., (1917) pp. 3-8.

certainly have to deal with a serious obstacle in Hawaii, although an assault on Pearl Harbor would not compare in danger with an assault on Singapore. [***] We have not got so far as that on this occasion. But we have had an English archbishop telling us that it may be necessary to have another great and horrible war to establish the efficacy of the League of Nations. 'This generation or the next will probably have to be sacrificed,' said the distinguished ecclesiastic. But there is good reason to suppose that this is a passing mood of the people, not a fixed attitude. It has sprung up swiftly during days of excitement, and generous, although misguided, emotion. The cause of 'Little Abyssinia' appealed very much as the cause of the Cuban rebels did to the people of the United States 40 years ago. And these storms of passion rarely, if ever, have an influence in shaping permanent policy. The mood changes too swiftly. Certainly the change in viewpoint is very marked compared with the situation we had in 1922. At that time I was able to take part in a movement which brought down the Prime Minister, Mr. Lloyd George, and destroyed his government. And what was the charge against him? What was the crime he had committed in the eyes of the public? Simply that he had threatened to use military sanctions against the Turks for an offense against a peace treaty, and therefore against the League, every bit as glaring as the Italian invasion of Ethiopia."[65]

Jews have often duped Gentiles with contrived "Christian" Utopian beliefs like that of the "Rapture". They have some Christians eagerly awaiting, and even deliberately seeking to provoke another world war and a nuclear holocaust, because Cabalistic Jews have led them to believe that the genocide of non-believers will bring back Christ. They are taught by Cabalist Jews, and these Jews' agents, that they will be privileged by their faith in disaster, and will be whisked away to safety while the rest of us are mass murdered at their behest. These Jews have sophistically tied fabricated and false Zionist propaganda to Christianity.

Jews have also duped many Gentiles with the Utopian lie of Communism. "Mentor" acknowledged that Bolshevism was a Jewish movement and Mentor saw Bolshevism and Zionism, which in tandem fulfill Jewish Messianic prophecy by political means, as the salvation of mankind—meaning the salvation of the Jews—to Cabalistic Jews, Gentiles are sub-human. When Mentor wrote, the world knew well the horrific nature of the Red Terror, which Mentor defended as a means to an end. Mentor wanted to defend the Bolsheviks from the Allies who were threatening to defend the Russian People from the Jewish bankers. *The World's Work* published the following article in March of 1919, which cannot even begin to capture the horrors of the Jewish bankers' Bolshevism,

---

65. Lord Beaverbrook, "A Military Alliance With England", *American Mercury*; as quoted in: *Congressional Record: Proceedings and Debates of the 76th Congress: Second Session*, Volume 85, Part 1, United States Government Printing Office, Washington, D. C., (21 September 1939- 31 October 1939), pp. 302-304, at 303.

"THE RED TERROR IN RUSSIA

An Eye-Witness's Story of the Mass Murders in Petrograd Directed by Lenine and Trotzky

BY

ARNO DOSCH-FLEUROT

I WAS passing before the Chinese Gate of the old Tartar city in Moscow one afternoon last summer when I got a mental snapshot of the red terror that has made a lasting impression on me. The incident was commonplace enough, but the composition of the picture seized the overwrought, terror-held imagination which I in common with everyone, even including the Bolsheviki, was suffering from in Russia.

The ancient Chinese Gate, ever remindful of the soft yielding of the Russians to outside, strange, particularly Oriental influences, was in the background. Before it, conspicuous among the lazy movements of the half-eastern, half-western crowd, passed a tall Mongolian soldier in the common Russian uniform, a bare automatic stuck in his belt flat on his stomach. He walked with a masterly stride like the other Mongolians who passed in and out of that gate hundreds of years ago among the same motley crowd of Russian peasants. And well he might feel his power, for he was one of the executioners hired by the Bolsheviki to take their prisoners—officers, bourgeois, peasants who objected to their dictatorship, anybody they did not like—and, forcing them to kneel in dark corners, to put that same automatic behind their ears and blow their heads off.

Just as he passed a load of his victims came gliding by. A modern police van, smooth-running, its dark green paint barely scratched, the only neat-looking thing left in Moscow, slipped silently across the square into the picture—bound for the Kremlin. It held ordinarily perhaps thirty persons, but was so tightly crowded I could see several heads through the tiny grating at the rear. Among them I recognized a young officer, who was soldier and nothing more. He was arrested simply because he was an officer, taken as a 'hostage,' and, as he was on his way to the Extraordinary Commission Against Counter-revolution, Speculation, and Sabotage, I did not have the slightest expectation of ever seeing him again. I never even knew his fate, nor did his family. He took a ride in the Bolshevist 'tumbril,' and that was all any one ever knew. That is one of the most terrible things about the red terror.

The next most terrible thing about the terror is that it was undertaken by the Bolsheviki as a political move. They put it into execution coldly, tried it out as an experiment on what the great Socialist newspaper, the *Vorwaerts*, referred to 'as the living body of society.' Recently in Copenhagen, I met a Bolshevik from Moscow and I asked him about the terror. 'Most of us think now it was a mistake,' he replied, calmly. 'A fine time to discover your mistake,' I replied, 'after you have murdered between 25,000 and 50,000 people.' It was in Copenhagen I made this bitter comment. In Moscow, I should not have dared.

The spirit of the red terror was obvious in Russia from the moment of the original revolution. The soldiers who killed their officers, the sailors who drowned their commanders, were terrorists. On the third and fourth day of the original revolution I expected any moment to hear the mass-slaughter of the civilians had begun. But the situation flattened out, and, except for the usual isolated killings of property owners by peasants, the amount of murder actuated by hatred in Russia was extraordinarily small during the spring and summer of 1917. It looked as if Russia might have something like permanent political freedom, and even the Jewish pogroms ceased.

The body which has been responsible for much of the red terror since the Revolutionary Tribunal, was organized immediately after the Bolshevist revolution and was anything but terrorist to begin with. For one thing it was then in the hands of Russian workmen, and not dominated by international adventurers. I remember well its first trial. Countess Panin, a kindly little woman known to all Russia as a philanthropist, had had charge of the hospitals and orphans under Kerensky, and, following the Bolshevist *coup d'état*, refused to give her funds to the usurpers. I think the charge was high treason, but the charge was a mere matter of words. She had opposed the Bolsheviki; that was the real crime. The court, Petrograd workmen, a mixture of Slav ferocity and gentleness, listened sagely to the testimony, which, of course, was very biased, and decided to dismiss the little countess with public rebuke! The second trial was that of Pouriskkevitch, a violent monarchist and a fool. He was caught in some absurd monarchistic plot, and the evidence was good. The court sentenced him to four years' hard labor, and then, because he was sick, really because he was an ass, sent him on his way.

The Revolutionary Tribunal did not last very long in such hands. That was not the kind of court planned by Lenine and Trotzky. They Soon put it in the hands of their obedient lieutenant, the little Ukrainian, Krylenko, the sublieutenant who was commander-in-chief of the Russian Army in the days when it demobilized itself and ignored his orders. He is president of the Revolutionary Tribunal yet. It is easy enough to get hireling soldiers, whether Letts or Chinamen, to execute your political enemies.

The real terror did not begin until after the signing of the treaty of Brest-Litovsk, long after in fact. Up to that time the Boisheviki had things their own way. The demand for peace in Russia, any kind of peace, shameful if necessary, was so strong among the uneducated Russian masses, that counter revolution had no chance. There was a Chouan movement that never died, and never has died, among the Cossacks, but it was powerless. And, if there was any shame in the mass of the Russian army for deserting its Allies, Trotzky had plenty of sophistical words to prove that the only possible shame was to fight another day.

So it was only after Russia felt herself out of the war that opposition worth mentioning began menacing the doctrinaire leaders of the Bolsheviki, who had proved from the start their inability to organize anything constructive. Opposition to them everywhere throughout the country had never ceased, and to combat it they organized the Extraordinary Commission against Counter-revolution, Speculation, and Sabotage. With a government based on usurped power, influential only until it got the country out of war, and from that time on backed by a very small minority of the population, this Extraordinary Commission had an opportunity to do as it liked. It had no laws whatsoever to check it, and as soon as it had been in the exercise of its power a short time, it was no longer even bound by the government.

During April and May, 1918, when the Extraordinary Commission began exercising its arbitrary power, I was in Sweden, but I returned to a Russia in June and remained until September, the period during which the red terror developed into a concrete movement. Meanwhile Petrograd, not liking the moving of the central government to Moscow, thus depriving the Petrograd workmen of the power to which they had become used, had formed the Commune of the North which pretended to govern northern Russia, but only succeeded in governing Petrograd with the terror inspired by its own Extraordinary Commission. Moscow had the chief Extraordinary Commission which reached out its long arm into all parts of Russia not strong enough to combat it, but Petrograd maintained its independence of action.

When I left Petrograd two months previously the local government of Petrograd was in the hands of the Soviet, which governed badly but with a certain laziness only sporadically ferocious which made life possible for those who did not come directly under its displeasure. Its president, the Bolshevist Zinoviev, placed there by Lenine, was forever laying every ill at the door of the bourgeoisie and trying by every art of a mediocre demagogue to induce the people to rise against the bourgeoisie, but he could not succeed. It took the single-handed power of Ouritzky, the adventurer, who became president of Petrograd's Extraordinary Commission, to give the bourgeoisie and all other enemies of the Bolsheviki, among them by this time most of the peasants, a due fear of the dictature of the proletariat. Ouritzky was himself a mere adventurer, who openly led a riotous life in Petrograd, made a great fortune himself by bribes and speculation, got most of it into foreign banks, but was shot before he got away. His more recent accumulations, 4,500,000 rubles, were discovered after his death in Petrograd, and nationalized solemnly by the Petrograd Soviet, but the Petrograd Soviet was unable to give back the lives of the '512 bourgeois hostages' who were shot in vengeance for his death.

The red terror really began with Ouritzky's death, that is to say, began on a scale that attracted foreign attention. But from the moment the Extraordinary Commission came into being several months previously it began exercising an arbitrary rule and terrorized everyone who fell under its displeasure. It would be more correct to say the red terror began with the dictature of the proletariat, but that the mass murders began only when the Bolsheviki felt their power threatened after the Fifth All-Russian Soviet at Moscow, July 5th, when the fanatic little Maria Spiridonovo made Lenine quail before her stinging words by saying that the Bolsheviki had failed, that the peasants were all against them, only a small portion of the workmen were with them, and that they were backed by the hooligans and the worst elements in the population. For that little Spiridonovo has been in jail ever since, though the charge against her is that she was in the plot that resulted in the murder of the German Ambassador Mirbach.

As Spiridonovo was the leader of the Left Social Revolutionists who helped the Bolsheviki stabilize their power during the winter and joined with them in driving out the Constitutional Assembly, the disaffection of the mad little woman was a severe blow to them. It meant that eventually all the peasants would be against them, and some immediately. They could not count on remaining dictators of Russia more than a few weeks without extraordinary procedures. Then they adopted the terror programme. Trotzky, Zinoviev, Carl Radek, Svertloff, all with consciences as hard as nails, had, long been for it, and now they were able to talk down the rest whose consciences were no better but who were inclined to believe that those who live by the sword are likely to die by the sword. I have often heard a distinction made in favor of Lenine in this respect, but it is undeserved. He supported all the decrees of the terror.

Incidents of actual terrorism are to me all intertwined with parallel examples of Bolshevist mentality, also explicative of the state of mind which could declare a terror. Zinoviev, President of the Petrograd Soviet, for instance, in the same days of July, when the mass arrests of 'bourgeois hostages' were taking place, began intensifying his campaign to rouse the workmen to go out and slaughter the rest of the citizens where found. He had been at it for months, but the Petrograd workmen, played upon as they had been for years by these furious fanatics, would not go out and kill the bourgeoisie in cold blood. Then, in July, came the cholera, intensified by the long, slow starvation to which Bolshevist disorganization had subjected the whole of Petrograd. It came violently, a thousand cases in one day, nearly half dying. The city was stricken, every doctor was in the hospitals or working night and day with the sick. That particular night I knew the Soviet was going to meet to take action and I was interested to go because I knew the burning question of free commerce to relieve the food situation and end the absurd unsuccessful food nationalization was bound to come up. But I could not go because my friend, with whom I lived, was attacked by the cholera. I knew a dozen doctors but could not get one. Finally by telephone I got one at a hospital and he authorized a drug store by telephone to sell me tincture of opium for him and with that we were able, by working all night, to save his life.

In the morning, relieved that the crisis was past, I walked out to quiet my nerves and bought a copy of the official newspaper, the *Communa*. In it was the report of the night's meeting. The food monopolization question had been raised, I found, but Zinoviev, seeing the danger of losing the Bolshevist grip, turned the thoughts of these simple men from the point at issue, as he had done a hundred times before, by delivering a passionate demagogic address, laying the cholera epidemic at the doors of the bourgeoisie, saying it was their doing. That was to be expected of him, but then he went on to say something for which this earth has no fitting punishment. He said that 'we,' the workmen, would put a stop to the epidemic, and if the bourgeois doctors did not do their duty, they would be shot on the spot. Emphasizing his point, evidently feeling he had nearly passed his political crisis, he said: 'Any workman who finds a doctor is not doing his duty right must kill him.' As the deaths were inevitable, this was a call to the assassination of every doctor in Petrograd. To the credit of the Petrograd workmen I must add I heard of no doctor being killed, but that does not let off Zinoviev. As if he did not know doctors always do their duty, especially in Russia where in times of epidemic their heroism is classic. In the country if the epidemic does not kill them, the peasants do. Politics knows nothing more contemptible than this effort to make political capital out of a common calamity.

I cannot write about the terror coldly because I lived it, my friends were victims of it. Night after night I lay and waited for them to come and take me, too. For some reason, not quite clear though, they left us Americans alone. I have no idea what help or shelter they could have expected from the 'imperial American Government.'

Life under these conditions in Russia was not bearable, and individuals set about fighting terror with terror. One young man killed Ouritzky. A young woman tried to kill Lenine. 'The White Terror,' cried the Bolsheviki, 'we must fight it with the Red Terror.' The same old dishonest way of turning things. They had by this time a goodly number of hostages, not only in Moscow and Petrograd, but in the provincial cities and the small towns everywhere and killed hundreds in vengeance. Most of these murdered hostages had never seen or heard of the attempted assassination. The record of terrorism in the provinces of Russia never can be told.

THE BOURGEOIS HOSTAGES

As I am here in Berlin, with none of my documentary proofs, I cannot cite from the Bolshevist papers. But in the month of September, these official organs were full of the lists of hostages killed 'to fight the White Terror.' The Bolsheviki, blind with their own rage, set down in their own official organs, the *Pravda* and *Isvestia* of Moscow, and the *Communa*, and *Pravda* Petrograd, the records of their own killings. I can only give out of my memory the one definite figure, 512, shot to avenge the death of Ouritzky, the scoundrel, whose rascality they later discovered. But when they discovered it, there was no regret at the hostages slaughtered because they wanted to kill them as 'boorjooy' hostages anyhow. It was indifferent to them whether they killed them because Ouritzky, or Ouritzky's dog, was killed.

Then, in September, came the culminating act of the Bolshevist Government, the manifest of September, written by Carl Radek, the most terrible document of which the brain of man was ever guilty. 1 will not attempt to quote it as I have not the manifest before me, but the tense of it was that every workman or peasant was immediately to kill, without parley, any one whom he suspected of counter-revolutionary tendencies. This threw down every bar, laid the way wide open to personal vengeance, plunder, and anarchy. The death and suffering that has occurred in Russia on account of this sweeping manifest passes all possibility of reckoning. It ended the last bit of justice between man and man in Russia. It turned loose anarchy in a situation filled with hate. It turned every man against his neighbor, made every house a fortress, and assured the deaths of tens of thousands of the only people who could possibly reconstruct Russia.

The Extraordinary Commission did its best to reduce the capable portion of the Russian population. It set about it systematically, even arresting people by occupations. The Russian engineers, for instance, are essential to the carrying on of that vast, scattered country, so the Bolsheviki began in September arresting them on any flimsy excuse and executing them out of hand. There was little pretence of trial, the Tribunal under Krylenko, and the Extraordinary Commission, presided during the worst of the Terror by a little Lett fanatic named Peters, divided up the work of signing death warrants, and were only occasionally interrupted in the orderly procedure of their assassinations by persistent pleaders for mercy, but the automatic pistols worked in the cellars of the Lubianka and the other prisons of Russia without ceasing. There is no use trying to give figures. The actual deaths from the Red Terror must surpass all estimates. By one kind of terrorism or another, the deaths in Russia in the autumn of 1918 must have averaged a thousand a day. As the total deaths of the French Revolution from the fall of the Bastille to the beheading of Robespierre was only about ten thousand, the difference is noticeable. Except for the affair of the Conciergerie, there was also in France some pretence at trial. Nor was there anything to match the manifest of September, the product of Radek, the Austrian.

But violent death was not enough. Fifty to a hundred thousand victims even is only a fraction of ten millions. So the Bolsheviki had to think of a more general terror, and they decided to starve people to death. By trying to run a food supply which they were incapable of organizing they had already practically starved the city populations of all classes, but now they set about finally to starve everyone except actual workmen. They had long had a system of cards by which the city populations were divided into four groups. Category No. 1 contained only men who worked hard with their hands. Category No. 2 contained those who worked less hard. Category No. 3 contained the liberal professions. I, as writer, had cards of the third category. The fourth category contained all who had an income from property or invested money. The plan was, and is, to make the third and fourth categories die of starvation. They cannot go to work with their hands, and thus get cards of first or second category. There is nothing for them to do, according to the plan, except to die. They are educated wrong, so they must die.

Of course, they did not all die off in a few days of starvation. They evaded the law and peasants, who were also openly disobeying the law, risked being shot by the Red Guard and came into the cities with their produce. So they live on, somehow, many dying slowly and all with their vitality and chances of recuperation greatly reduced. They are forbidden to buy anything, and the Red Guards are in the markets to see that the purchasers have only cards of the first and second categories. But the simple Russian people are themselves not so cruel as the Bolsheviki who are trying to lead them, somehow it is arranged, though with trouble. Since July 26th the fourth category has had only two herrings daily, and the third category was put on the same diet a few weeks later. I was supposed to be so nourished, but, in point of fact, I never ate a herring in Russia. I got food, illegally. But, as the first category gets from 50 to 100 grams of bread a day and the second category but 25 to 50 grams, there has not been much to choose between being a member of the bourgeoisie or of the proletariat. All have had to buy illegally or starve.

The Terror is having a certain success. It is gradually killing off all the culture there was in Russia, and, if it could go on long enough, there would be simply an aggregation of villages, some at peace, others at war. The cities have steadily disintegrated, and, after a year in power, the Bolsheviki have not one constructive act to their credit. But they are still in power, late in November as I write, and while they remain in power the Red Terror will continue."[66]

On 30 October 1939, Congressman Thorkelson warned the American Congress that some Jews were out to destroy America with another world war and by seeding Mexico with Communist revolutionaries—an old Eighteenth Century Rothschild plan, which is still in the Communists' works and is a real and present danger to America's security,

> "If House Joint Resolution 306, the present Neutrality Act, is passed as it is, it is my firm belief that such action on our part will bring about civil war in the United States, which may well terminate in the ultimate destruction of those in the invisible Government who sponsored this legislation and who are the silent promoters of the present war in Europe.
>
> As the first step in consideration of this so-called Neutrality Act of 1939, please ask yourself, Who is it that wants war? It certainly is not the people that want war, and it is their wish that we must consider, as we are their Representatives in Congress.
>
> Have any of your constituents asked you to vote for war, so that their children may be sent forth to drown in the Atlantic or die in the trenches of Europe? Are there any Members of Congress who want war? I do not believe so. Have you ever stopped to think, or have you tried to identify

---

[66]. A. Dosch-Fleurot, "The Red Terror in Russia", *The World's Work*, Volume 37, Number 5, (March, 1919), pp. 566-569.

those whose greatest ambition is to aline this country in war on the side of England? I have not found anyone that wants war except those who harbor hatreds toward Hitler, and strange as it may seem, they are the same people who approved of Stalin.

Is it logical or reasonable that all Christian civilized nations, such as the United States, England, Canada, Australia, France, Germany, Austria, and other European nationalities, must engage in internecine conflict or war of extermination, so that this group of haters may get even with one man? Shall we sacrifice millions of our young men from 18 to 30 years of age to appease personal hatreds of a small group of international exploiters? I think not. I do not believe that there is any one person worth such sacrifice, whether he be king, prince, or dictator.

Let me now carry this argument a little further, for I want to call your attention to the fact that this same group that now hates Hitler was pro-German during the World War, and it is the same group that ruled and directed Germany's military machine before and during the World War. It is the same group that brought about inflation and exploited the German people, and it is the same group that furnished the money that brought about revolution in Russia and eliminated the Russian Army when its aid was needed to win the World War. This same group of internationalists paid and promoted the bloody invasion of Hungary, in which the invaders destroyed life and property with utter disregard for civilized warfare or even decency. It is this same group that has spread and nourished communism throughout the whole world and that sponsored the 'red' revolution in Spain. It is the same communistic group which is now concentrated south of us in Mexico, waiting to strike when the time is ripe.

Please ask yourselves if you are justified in giving the President the power set forth in this Neutrality Act, and are you justified in repealing the arms-embargo clause, when you know it is for no other reason except to aline the United States with Gr€at Britain in another war as senseless as the World War. In considering this remember that there are no hatreds among the common people of the nations of the world, and for that reason no desire to destroy either life or property. Is it not time that we, the common people, learn a lesson—yes; a lesson in self-preservation instead of fighting for the 'invisible government'? Let us marshal this personnel into an army of their own and ship them some place to fight it out among themselves. It will be a blessing to civilization.

This contemplated war will not save the world for democracy because we have that now in the fullest measure; it is fully entrenched within the Government itself and in many organizations. We need no further evidence of that than the recent exposé of the League for Peace and Democracy, with its many members employed in strategic positions within the Federal Government, to further the cause of democracy and communism. No; this war will not be fought for so-called democracy or communism, for it is here, and is an evil that we will eventually be called upon to destroy or else be destroyed by it.

If the present agitation in Europe should terminate in an active war, its purpose will be to place all Christian civilized nations under the domination of an international government that expects to rule the world by the power of money and the control of fools who sit in the chairs of governments. I do not believe this will happen here, for the people are too well informed about this evil blight that is keeping the world at odds, and which is spreading dissension and hatreds by confusion and international intrigue. Let us shake off this evil, put our shoulders to the wheel, and push the carriage of state back on the road to sound constitutional government. Do not forget, if attack comes, it will be delivered by the Communists within the United States and next by the Communists who are waiting beyond our borders. Let us, therefore, give undivided attention to the Communists within our midst, for they have no place within a republican government. We should not tolerate foreign or hyphenated groups that, for reasons best known to themselves, cannot or will not assimilate to become Americans. For our own preservation we must get rid of those who cannot subscribe to the fundamental principles of this Republic, as set forth in the Constitution of the United States."[67]

On 22 September 1922, when the Jewish bankers had succeeded in obtaining the Palestine Mandate, but the majority of Jews did not wish to go to Palestine, and in the bankers' minds, the Jews needed another world war and an anti-Semitic dictator to convince Jews in general of the wisdom of being racist and murderous Jews, a Jewish Bolshevist Zionist who published under the pseudonym "Mentor" offered the Trojan Horse of "peace" to the world as bait for the nations to surrender their sovereignty to Jewish bankers and perish from the Earth,

Mentor wrote in *The Jewish Chronicle* on 15 September 1922 on pages 9 and 10,

"'Live Together or Die Together.'

By MENTOR.

DAY by day, almost hour by hour, the claim that the Great War was a war to end war appears to leer at us with a grim grin of ever deepening ironical mockery. It seems clear that of all the vain and illusory estimates that were made of the horrible disaster which fell upon mankind in August, 1914, none was so vain and illusory as that it was a war that would end war. Day by day, and almost hour by hour, fresh evidences crop up showing that the spirit of combat is as deeply ingrained in the nations of the world as ever. There are signs which cannot be

---

[67]. *Congressional Record: Proceedings and Debates of the 76th Congress: Second Session*, Volume 85, Part 1, United States Government Printing Office, Washington, D. C., (1939), p. 1068.

mistaken, indeed, that as a direct result of the war there were set going the intrigues of diplomats, the underhand workings of politicians, the selfish devices of statesmen, all of them forming a net-work of live wires, which, at some mere touch, may send once again into a great conflagration all the vile elements that go to constitute war. And this, notwithstanding the chorus of protestation that peace and concord, and only peace is the goal towards which all the nations are striving. There is as much truth in the protestation now as there was in like assurances during the fatal months before August, 1914, when Russia and Germany and France and Great Britain, the foremost combatants in the epic tragedy, vied with each other in their declaration of peace and good-will among men. To-day, as then, all the talk of peace and the prevention of war is in reality nothing more than a manœuvring for position, precisely in the manner of prize fighters about to enter into contest. There are strivings for alliances and *ententes* and understandings and interests, which those who do not forget the history of the world before the breaking out of the Great War, feel as sure are premonitions of another great catastrophe of a like sort, as are the Italian peasants that an earthquake is imminent when they hear the low rumbles of the tremulous earth. The other day, Mr. J. A. Spender published a striking article in the *Westminster Gazette*, the burden of which was the essential interdependence and unity not alone of England and France but of all the nations of Europe, and not alone, of all the nations of Europe but of all the continents of the world. Ha speaks of 'the next war,' and does not hesitate to say that as the world is going, though it may be uncertain which nations will be opposed to which in such a war, that any nation will be out of it is scarcely to be thought of. And he concludes that the nations of the world have therefore now to make up their minds that they must all live together or die together.

Limiting Armaments.

Live together or die together! It is in very surety for humanity at large a case of life or death. If War and the spirit of War be not eliminated, and War be allowed to develop in the sense in which the *Westminster Gazette* article contemplates 'the next war,' then it is not merely a question of life or death, as all wars are to the combatants engaged in them. It is a question of life or death to the nations of the world, life or death to civilisation. As I write these words, the Assembly of the League of Nations is meeting at Geneva, and good men are making strenuous efforts to secure that nation shall not lift up sword against nation, and that they shall learn war no more [*Isaiah* 2:4]. Limitation of armaments seems to be the one practical means that hitherto has suggested itself for accomplishing the peace of the world. But I cannot help thinking that this method is open to the gravest illusion, and may in fact prove to be in itself fraught with much danger of War. Because people are likely to rely upon it and neglect every other method and means, while all the time it may after all be a mere curtain hiding an intensive cultivation, instead of a limitation of warlike material. A country, for instance, may limit its naval equipment, and by thus saving millions may be able to devote so much the larger sum to some far more deadly form of warfare. A few. months ago the United States summoned a Conference at

Washington for the purpose of limiting armaments, and certain resolutions were come to for the limiting of navies. The average man and woman, just because America has taken this foremost lead in disarmament, doubtless conclude that America is bent entirely upon ways of peace, and is devoting herself exclusively to a national life that is humane in its policy. Yet the reports of the American Chemical Warfare Service for the three years ending 1920 show how disarmament as a policy may be as deceptive and as fatal as the placing of the stumbling block before the eyes of the blind.
Poison Gas.

We learn from this document that before the last war had ended sixty-three kinds of poison gas were in use, and that the Warfare Service of the American Government was engaged in research problems comprising some eight or nine gases that were said to be far more deadly than any that had hitherto been employed. We read of one gas that is capable of making the soil upon which it is cast as barren as once was Pharaoh's Egypt, and for the like period as that during which the famine raged in his land. Another gas is so deadly that a few whiffs of it are sufficient to cause a tree to wither and become pulverised. Upwards of eight hundred tons of these gases are being turned out by the United States weekly, and the cost is stated to be 100,000,000 dollars *per annum*, requiring forty-eight thousand men in the service. So successful—save the mark!—is this abominable business of wholesale slaughter, that it is being extended, and we hear of a kind of radio-activity whereby, at the finger touch of one man, death can be spread over a vast area. At the same time malignant disease germs are mentioned, which could be dropped from aeroplanes, or spread over an enemy's country by specially cultivated rats and fleas. What devilish work the aeroplane can do, God—if it be not blasphemous to use His name in such a connection—alone knows! This American report, for example, speaks of aeroplanes, one of which could poison in the course of a flight every living soul within an area some seven or eight miles long and a hundred feet wide! It needs no gift of imagination to think what a 'covey' of these dastardly productions could do if let loose upon an enemy country. The report acknowledges that a hundred of them could, in a single night, convert a great city into an necropolis, a huge Gomorrah of corpses.

The Next War.

But pray let me not be mistaken. I happen to have lighted, through reading these facts in a paper the other day, upon these particulars of what America is doing behind the screen of limitation of armaments. I do not suppose, however, that she is doing any worse, even if she is doing much more, in the direction of mass slaughter, than are other peoples. We read of wondrous air engines being made in this country, which are designed to be capable of annihilating the largest men-o'-war afloat, together with the whole of their crews by one fell swoop. If they can do that, the destruction that they could wreak on land can be better conceived than described. It is manifest that the air raids of the Great War, to give just one instance of the multiplying of this murder enterprise, compared with the air raids of the next war, will seem as a popgun compared with a rifle. Since the last war the problem of distance has been so modified that an aeroplane, carrying I know not how many tons of death-dealing bombs, can travel easily a mile in some three-quarters of a minute. The carrying capacity of the aeroplane has also enormously increased; while, weather and atmospheric conditions, which were so often a shield against invasion in the last war, will be no bar in the next. Nor is it only in the region of air engines of war that huge strides towards greater and more ruthless destruction have been made. Submarine instruments that proved so deadly in the last war, and so nearly came to crippling this country and defeating her, have been rendered many more times efficacious. So have the older instruments of warfare. Thus we learn of 'Big Berthas' that, planted at the Channel Ports of France, say, at Calais or Boulogne, could easily storm London, and might send their death-dealing contents far further into the land. Mr. Spender is right. The next war between the great nations of the world will mean that those nations will die together. It will be the alternative to their having refused to live together.

We see, then, how delusive limitation of armaments may be as a means of eliminating war, when America gathers together a Conference for the very purpose, agrees to a limitation of its navy, secures a limitation of the navies of other countries, and yet proceeds with the demoniacal manufacture of poison to be utilised by aeroplanes against any who may become the enemies of the United States. We see the futility of relying upon disarmament when we know that every country that is crying out for disarmament, England included, is at the same time using (or rather misusing) its best brains for devising methods whereby men and women can be shuffled off to death, because two or three men in one State cannot agree with two or three men in another—for that in its origin is what war really means; the nation's part comes in afterwards. It seems to me that disarmament, to be of any value, should be consequential. I mean that the mere laying down of arms will not ensure peace, if the spirit of war be not first exorcised. Great Britain was to all intents and purposes 'disarmed'— she was, in fact, unarmed, speaking comparatively—when she entered into the Great War. But the spirit of War became strong within her, and it was not long before she had vast armaments under her control. It is quite conceivable that a country without armaments could yet take its part, and a very sanguinary part, in a war. For armaments are quickly improvised, and to-day are cheap, for aeroplanes or submarines are very cheap when compared with such armaments as wars needed some years ago. An aeroplanes and submarines would be potent weapons to go on with anyway, by any nation, engaged in modern warfare. Indeed, many experts declare that those engines of destruction alone will decide the next war.

No; limitation of armaments must come, as disarmament must come as a result of man's feeling of disgust, and horror, and detestation of war. Man's disgust and horror, and detestation of war, will not comes as a result of disarmament; and so long as the feeling of war, the sentiment of its glory, the mirage of its beauty and grandeur are implanted in men's minds and souls, the possibility of war must be ever present and can always at very short notice overcome lack of arms. Perhaps, however, an even more potent guard against War would be the discovery of some means of national security, so that nations could be sure that others nations did not mean to attack them. Men do not go about armed in civilised society, not because arms are unobtainable nor so much because of their detestation and horror of killing or injuring a neighbour, who has insulted or annoyed or attacked them. It is because men feel that they are moderately safe. But, first and foremost, men must understand the reality of war, the meaning of it. For that reason it is perhaps not altogether to be deplored that war no longer is a matter only of the trained armies taking their part in it. When war is declared between two nations now, every man, woman and child of each of those nations is liable to be maimed and slaughtered and not only the fighting men who volunteer, or are compelled to do service. Indeed, as things are tending, it is not unduly paradoxical to suggest that the day may come when in war the safest place for the peoples engaged in it will be the battlefield. Soldiers who go out to fight will be dug in in trenches, or provided with elaborate security which it would be impossible to render to the whole of the population whom they leave behind, and who will be at the tender mercy of such horrors as the American report I have quoted details. So that war is coming home much more narrowly to every individual than even did the last war. Thus, uniformed or not uniformed, the shirker as well as the man who goes to 'do his bit,' the man who sees war only as a means for profiteering, as well as the man who sees in war glory and a road to honor—all will be equally liable to suffer the hellish damnations which are now involved in war. And this certainly creates a possibility that nations will not be quite so ready to embark on war, and statesmen will not find it so easy to obtain the wherewithal, financial and human, for carrying on war in the future as in the past.

'The Paths of Glory.'

War, said a writer whom I was reading the other day, is a madness that seizes peoples, and they are unable to restrain themselves when the passion and craze overtake them. To set a prophylactic against that madness nothing, it seems to me, could be more effectual than an intensive campaign telling of the realities of war as it has been in the past, and picturing what a war in the future must be. As an aid to this, nothing that has been published, I think, could be more assistant than a collection of poems written during the Great War, mostly by soldiers and entitled 'The Paths of Glory.' It is edited by Mr. Bertram Lloyd and published by Messrs. George Allen and Unwin, Ltd. It is really difficult to select out of such a collection (which includes, it is interesting to note, some contributions from the pens of Jews), and one which will the more surely convey to the reader something of War, as it appears to the man who has gone through it, of War stripped of its unreality, of what one contributor to this volume calls its gilded cozenings, its trappings, and its hideous jewels. And let me say, parenthetically, that the same writer has in this book a line that grips. 'Blood,' he says, 'will not build the new Jerusalem.' There is a world of admonition and teaching, of reproach and warning, in that line: 'Blood will not build the new Jerusalem.' But there is one poem in the book that, it seems to me, will appear remarkable, not only in itself, but because it was written by a German soldier, the product of German militarism and of a culture to which we applied, in the hate that was so carefully induced in so many of us, for nearly five years, the omnibus term of 'Hun.' It is a little poem called 'The Brothers,' and its translation reads thus:

> Before our wire there lay for long a dead man full in view:
> The sun burned down upon him, he was cooled by wind and dew.
>
> Day after day upon his pallid face I used to stare,
> And ever grew more certain: 'twas my brother lying there.
>
> And often as I looked at him outstretched before my gaze,
> I seemed to hear his merry voice from far-off peaceful days.
>
> And in my dreams I heard him crying out and weeping sore,
> 'Ah, brother, dearest brother, do you love me then no more?'
>
> At last I risked the bullets and the shrapnel-rain, and ran
> And fetched him in, and buried... an unknown fellow-man.
>
> My eyes deceived me, but my heart proclaimed the truth to me:
> In every dead man's countenance a brother's face I see.

If we could comprehend that it is a brother's face with which nation by nation is confronted when international quarrels occur—well, fratricide is not unknown, but it is rare, and war would be all the rarer if men called it fratricide—the murder of brother by brother.

What Are Jews Doing?

The work before all right-thinking men to-day, the chief work, the work that is more urgent than any other—that is abundantly clear—is war against war. A campaign against wholesale murder, so that humanity may be spared 'the next war,' and civilisation may be saved from the utter ruin and damnation which a war of any extent must bring upon it. The nations of the world are now, I believe, manœuvring for position with all their talk of peace and disarmament, of *ententes* and alliances. It is the rumbling of the volcano, the premonition of yet another disaster, a crowning disaster for mankind. The King the other day declared that the only war worth waging to-day is a war against war. And this holy war, this really glorious campaign, this battle of honour, veritably of Right over Might, this war for Peace, for the ideal preached by the Jewish prophets of old, and nourished by every Jew throughout the ages, and prayed for in his most solemn moments of converse between him and his God—in *this* war what are Jews doing and what are they going to do? How are Jews going to play *their* part. For Jews, if they be true to everything that makes Judaism worthy of them and makes them worthy of Judaism, must play their part in a great endeavour, so that the nations of the world may live together and not die together."

In *The Jewish Chronicle* on 22 September 1922 on pages 13 and 14, Mentor continued the plea to a world made weary by war and Bolshevism instigated by Jewish bankers, that the nations must surrender their sovereignty in order to obtain "peace",

"'What are the Jews Doing?'

By MENTOR.

WHEN I wrote in this column last week, I had no idea that the premonitions to which I alluded, of another great catastrophe of like sort to the war that began in 1914, would so soon be justified. Within a few hours of my words appearing in print a document was issued by the British Government, threatening the beginning of a war of which, once started, no man could foretell the end. Hardly was the last issue of the *Jewish Chronicle* published than we seemed whirled back in a sudden instant to the time eight years ago that preluded the terrible world-struggle that lasted through nearly five years. There were rumours of war; there were ominous movements of politicians from the four corners of the kingdom, which newspapers interpreted as meaning all sorts of things. The evil birds of Militarism were foregathering. Like vultures they flew to gather their prey. Stories were bruited abroad, craftily designed to work upon the sentiments and the emotions of the people. Reasons and excuses, arguments and assurances, were cleverly designed, so that when the dogs of war were unleashed, proof of the inevitability and the justification for starting wholesale murder, for man going out to kill his fellow man, might be prudently provided beforehand. As I write, the situation—as it is termed—seems, if anything, a good deal less dangerous than it did at the beginning of the week. That is because those who were for war, those who were willing if not anxious to resort to arms in order to fight about a dispute instead of adjusting it by negotiation, have not received the encouraging response from the country which they had evidently hoped would come to them. Once bit twice shy! All the conventional paraphernalia of diplomats and politicians were again employed by the men of war as they were used eight years ago. Then their assurances were accepted, and men believed they could by war accomplish a great deal. Now, some of the public at least are wiser, and recollect the fraud, the chicanery, the double-dealing, the falsity, and the two-facedness which were so largely responsible for the determination of this country to enter into war eight years ago. They know that the same people are up to the same dodges, that the like people are bent on the like wiles, and the country this time has put a large discount upon all the mongering for War. The experience of the Great War has thus not been wholly lost, and there seems a healthy disposition, in more than one quarter, to regard the Minister who leads this country into war as *ipso facto* unfitted to hold the trust he has dishonoured by muddlement. There is proved to be now a looking upon war as the crowning disaster of any nation, not as its glory, as a visitation and not as a proud happening.

Jewish Doctrine and Christian.

If war is averted, if those responsible for the Government of the country finding war 'no go,' because the people will have none of it, have to seek other means for adjusting international differences, then the incident which looked so grave at the beginning of the week will have been of advantage. For it will have shown at least one Government that the way of war is not the easiest at hand for them for settling any disputes that may arise. So far, so good; and if that spirit of antagonism to and hatred and—if you will—fear of war be maintained, so that men, beginning by disliking it, will go on to loathe and detest it, then we shall have made a long stride to the abolition of war and the arbitrament of the sword, and towards that condition which is the Jewish ideal; when man shall no longer lift up sword against man, nor learn war any more. [*Isaiah* 2:4] I call that the Jewish ideal, but we Jews have not a monopoly of it. Peace is a Christian ideal, too. Indeed, Christianity goes much farther, and is a doctrine of non-resistence to evil. Judaism does not teach that; it is far more practical and far more human. But if Christianity were really practised and the Christian spirit were truly in the souls of those who profess Christianity, war would be impossible. But a Jew is here writing for Jews, and it is because peace is a Jewish ideal that I revert to this question here and now—now, because we are on the threshold of the most sacred days in the Jewish calender, when the Jew, if ever, is brought into close contact with the Almighty, when, if ever, he feels strong upon him the duty which is his as a Jew.

The Jewish Mission.

And I ask: What are the Jews doing in the war against war, the war which the King himself the other day said is the only war worth while; the war for Civilisation, for salving Humanity, for making the life of men and women in the world tolerable and bearable; the war against one of the most fertile roots of poverty with its fruits of hunger, and vice, and disease—what are the Jews doing in the war for which the King of Kings long ago conscripted certainly every Jew? I suppose the answer will reach me that Jews ought not, as such and of themselves, to be expected to take any definite part in such a campaign. I shall be told that war is really a political matter, and that Jews have no politics of their own, they share in the politics of the nations of which they are citizens. But this argument, carried to its logical conclusion, would place the Jew in such a position that the whole of the claim which he has made concerning his place in the world, and in respect to the Judaism he professes, would have to be seriously overhauled. How can a Jew be true to Jewish teachings, to the teachings of the Prophets, to Rabbinical teaching, to all that Judaism connotes for the Jew, unless Peace on earth and Goodwill among men be believed in by him and hoped for by him? How can he pray, as he constantly prays, from year end to year end, and day by day, for peace, and yet not mean it and not wish it? And if he means it and wishes it, then how can he place even his duty to the State (if it is conceivable that his duty to the State can involve war as a principle) before his duty to his God? The Christian does it. He worships a Divinity that he hails as the emblem of peace. He invokes the one whom he regards as Messiah, the harbinger of peace. He subscribes to the doctrine of Peace enunciated by the great Founder of his faith, and yet he contrives instruments of violence, engines of slaughter, and all the hellish devices for maintaining War on earth and illwill towards men. But that is a matter for Christians. That they do thus is no reason, and assuredly no justification for Jews doing likewise. Following the multitude to do evil is not Jewish work. And so I ask again, just as we are slipping into yet another New Year: What are the Jews doing so that war shall cease from the earth, so that peace may reign and goodwill prevail among the children of men?

Our Separateness.

What are the Jews doing? It is a pertinent and not an impertinent question; because it asks, though not in those words, how is the Jew justifying his existence? We elect to remain a separate people. In every country and in every land we segregate ourselves from our fellow-citizens, and throughout the ages we have obstinately (as our enemies term it), faithfully as we believe, kept ourselves apart as a separate people. For what? Some Jews will tell you that we have refused to assimilate in the sense of losing ourselves in the multitudes surrounding us, because we have all along been conscious of being a separate national entity. So we have maintained our separateness in the hope that some day our national being would be restored. This, put very broadly, is the attitude of Zionists and Jewish Nationalists. But all Jews are not one or the other. The majority are neither, or at least care not at all for either striving. Their idea of Jewish separateness is altogether another. They say that we Jews have kept apart in order to carry on, amid the nations of the world, a Jewish Mission. That mission, so it is claimed, comprises our weaning other peoples away from error of thought and sin of action to a true conception of God. It means that we have to urge the breaking up of all idols and securing allegiance alone to the Almighty Governor of the universe. Very well, let us accept, for the purpose of argument, the contention of these fellow Jews that their separateness is maintained alone for the Mission potentialities of our people. Then I would ask: What are they doing in the way of propagating that Mission? Some of them argue that although it is true they are not actively engaged in spreading the message of Israel, or in preaching its truths to those of other faiths, they are doing service to the mission passively in the living of their lives. Their example, they say, is even better than precept. Surely this is a paltering with the question; it is an excuse, a subterfuge, and it makes the whole idea of the Mission of Israel not alone the sham that it is with those who thus argue, but a ridiculous parody of every idea of the purpose and the object which any mission worthy of the name must have.

The Jew's Contribution.

This paltry excuse for neglect of the call of the Mission of Israel does not rob us of the right to ask: What is the Jew doing in pursuance of what he believes to be his mission to Mankind? The answer must be: precious little. We are standing at the dawn of a New Year. We are about to reach another milestone in our history. Is the Jew to go on year by year in the same meaningless, chaotic existence, just living, just existing without a worthy purpose as Jew; for mere material selfish objects, as a people without an ideal, without an aspiration? Broadly speaking, there are only two possible ideals for Jews, the National ideal and the Mission of Israel ideal. They are not antagonistic or even mutually exclusive. For the Jewish Nationalist also believes—believes very strongly—in the Mission of Israel, but believes, too, that it is impossible of accomplishment without national existence in a Jewish land. But taking the Jewish position as it is, either aspiration, if the Jew be true to it, will justify his separateness among the nations of the world. But if he nourish neither of those ideals, as is the way with thousands and thousands of Jews, then the *raison d'être* of his existence is *nil*, the part he plays in the world is a mirage. He is a mere parasite, and he justifies nothing so much as the indictment that is made by some enemies of our people. They denounce us because we remain separate as a people, and yet take no count of any service which we should do as Jews for the common benefit of Mankind. Well, if there be any reality in the Mission of Israel ideal, then I ask again: What are the Jews doing? What part are they taking in the war against war, in leading men from violence and slaughter and murder in the wholesale, back or rather forward to ways of peace, to ways of goodwill and happiness among men. We are doing precious little, even as individual Jews. As a Jewish people, we are doing nothing.

Here surely, as I have more than once suggested, is a great and glorious opportunity for the Jewish People. They do not want to be a separate nation. They wish to be separate among the nations of the world. Very well, then let them justify that aspiration. All the trouble Jews encounter is traceable to nothing so surely as to the fact that they are despised. And they are despised, not as individuals—as individuals even anti-Semites respect Jews—but because, however commendable individual Jews may be, whatever service individual Jews may have done for the world and for civilisation—and Dr. Joseph Jacobs left a posthumous work showing how great had been the service of individual Jews in that respect—as a people Jews contribute nothing to the service of mankind. We do not cultivate a Jewish culture; we are not known for any great or enduring office which we perform. But suppose we carried on our mission, our God-given mission as the bringers and the promoters of peace, as the bearer of that great ideal, is it not palpable that there would be something we should be doing by which we should win the respect of mankind? Because sooner or later, after misunderstanding had passed away and misrepresentation and vituperation had evaporated, the world would come to acknowledge itself our debtors for the good we should have effected. It seems to me that in the times in which we live—with the constant menace and danger of war, with the ineffable wickedness which allows great talent and scientific attainment to be misused and misapplied, as they are being misused and misapplied in devising means for carnage, for bloodshed, for violence, for all the indescribable horror comprised in war—and particularly at this hour when we are entering into the most solemn moments of conclave—the Jew with his God—it is not inapt to ask: What are the Jews doing in the war that alone matters, the war against war? I ask it here and now, because the hearts of my fellow-Jews, attuned at this season to higher thoughts and loftier aspirations, may bethink themselves that there is a great evil in the world, the greatest evil that mankind and civilisation have to contend against. And mayhap there will arise in their souls a determination, each one as he can and where he can, to do what he can—thus making it a Jewish mission—so as to roll away the menace of war from the path that humanity is treading."

If the "Jewish Mission" were truly to convince the Peoples of the world that monotheism is the most rational choice among extant religions, then Jews would be applying themselves to this task, but they are not. Instead, it appears that where Jews involve themselves in religious questions, they are most often ridiculing other religions. Far from inviting other Peoples to join Judaism, Jewish leaders instead attempt through their disproportionate control of media and education to destroy all religious beliefs in other Peoples, including the monotheism of Christianity and Islam—save the false beliefs they have instilled in Dispensationalist Christian Zionists, who serve as their slavish and gleefully suicidal "Esau", to their deceptive and deceitful "Jacob".

Wealthy Jews have long sponsored flamboyant Christian Zionists, who dupe Christians into slaving and soldiering for Jews who despise them. Wealthy Jews created the many mythologies of Dispensational Evangelical Christianity and popularized them through such characters as John Nelson Darby, including the myth of the "Rapture", mythologies which were originally crafted by crypto-Jewish Jesuits in the 1500's,[68] and the Zionist mythologies which Cyrus Ingerson Scofield incorporated into his notations found in the heavily promoted Scofield Bible.

The true nature of the "Jewish Mission" is made obvious by the actions of Jewish leaders and is spelled out in Jewish religious literature. It is to destroy other cultures, religions, nations and "races". It is not a mission of peace and tolerance, rather it is a mission of segregation, "race" hatred, Jewish supremacy, war and genocide. As the Jewish book of *Exodus* 34:11-17 states, the "Jewish Mission" is to:

"11 Observe thou that which I command thee *this* day: behold, I drive out before thee the Amorite, and the Canaanite, and the Hittite, and the Perizzite, and the Hivite, and the Jebusite. 12 Take heed to thyself, lest thou make a

---

68. J. W. O'Malley, *The First Jesuits*, Harvard University Press, Cambridge, Massachusetts, (1993). Rev. D. McDougall, *The Rapture of the Saints*, Artisan Publishers, (1998).

covenant with the inhabitants of the land whither thou goest, lest it be for a snare in the midst of thee: 13 But ye shall destroy their altars, break their images, and cut down their groves: 14 For thou shalt worship no other god: for the LORD, whose name *is* Jealous, *is* a jealous God: 15 Lest thou make a covenant with the inhabitants of the land, and they go a whoring after their gods, and do sacrifice unto their gods, and *one* call thee, and thou eat of his sacrifice; 16 And thou take of their daughters unto thy sons, and their daughters go a whoring after their gods, and make thy sons go a whoring after their gods. 17 Thou shalt make thee no molten gods. [King James Version]"

And as the Jewish book of *Obadiah* states, and note that Edom and Esau signify Gentiles, and that Judah and Jacob signify the Jews, the "Jewish Mission" is to destroy the nations and exterminate the subhuman Gentile "cattle":

"1 The vision of Obadiah. Thus saith the Lord GOD concerning Edom: We have heard a message from the LORD, and an ambassador is sent among the nations: 'Arise ye, and let us rise up against her in battle.' 2 Behold, I make thee small among the nations; thou art greatly despised. 3 The pride of thy heart hath beguiled thee, O thou that dwellest in the clefts of the rock, thy habitation on high; that sayest in thy heart: 'Who shall bring me down to the ground?' 4 Though thou make thy nest as high as the eagle, and though thou set it among the stars, I will bring thee down from thence, saith the LORD. 5 If thieves came to thee, if robbers by night—how art thou cut off!—would they not steal till they had enough? If grape-gatherers came to thee, would they not leave some gleaning grapes? 6 How is Esau searched out! How are his hidden places sought out! 7 All the men of thy confederacy have conducted thee to the border; the men that were at peace with thee have beguiled thee, and prevailed against thee; they that eat thy bread lay a snare under thee, in whom there is no discernment. 8 Shall I not in that day, saith the LORD, destroy the wise men out of Edom, and discernment out of the mount of Esau? 9 And thy mighty men, O Teman, shall be dismayed, to the end that every one may be cut off from the mount of Esau by slaughter. 10 For the violence done to thy brother Jacob shame shall cover thee, and thou shalt be cut off for ever. 11 In the day that thou didst stand aloof, in the day that strangers carried away his substance, and foreigners entered into his gates, and cast lots upon Jerusalem, even thou wast as one of them. 12 But thou shouldest not have gazed on the day of thy brother in the day of his disaster, neither shouldest thou have rejoiced over the children of Judah in the day of their destruction; neither shouldest thou have spoken proudly in the day of distress. 13 Thou shouldest not have entered into the gate of My people in the day of their calamity; yea, thou shouldest not have gazed on their affliction in the day of their calamity, nor have laid hands on their substance in the day of their calamity. 14 Neither shouldest thou have stood in the crossway, to cut off those of his that escape; neither shouldest thou have delivered up those of his that did remain in the day of distress. 15 For the day of the LORD is near upon all the nations; as thou hast done, it shall be done unto thee; thy dealing shall return upon thine own head. 16 For as ye have drunk upon My holy mountain, so shall all the nations drink continually, yea, they shall drink, and swallow down, and shall be as though they had not been. 17 But in mount Zion there shall be those that escape, and it shall be holy; and the house of Jacob shall possess their possessions. 18 And the house of Jacob shall be a fire, and the house of Joseph a flame, and the house of Esau for stubble, and they shall kindle in them, and devour them; and there shall not be any remaining of the house of Esau; for the LORD hath spoken. 19 And they of the South shall possess the mount of Esau, and they of the Lowland the Philistines; and they shall possess the field of Ephraim, and the field of Samaria; and Benjamin shall possess Gilead. 20 And the captivity of this host of the children of Israel, that are among the Canaanites, even unto Zarephath, and the captivity of Jerusalem, that is in Sepharad, shall possess the cities of the South. 21 And saviours shall come up on mount Zion to judge the mount of Esau; and the kingdom shall be the LORD'S. [version of the Jewish Publication Society]"

Mentor refers to the other "ideal" of Judaism—other than the destruction of Gentile nations and peoples—and that other ideal is the establishment of Jewish State in Palestine. To the Jews, the establishment of the Jewish State heralds the appearance of the Jewish Messiah and the Jews' prophesied complete dominance over all other Peoples followed by the other Peoples' judgement and then extermination (*Obadiah* 21). Mentor is right to assert that for Jews there is no conflict in supremacist Judaism between these two Jewish "ideals" of Jewish Nationalism and the concurrent destruction of Gentile Nationalism. The establishment of a Jewish Kingdom to rule and ruin the Earth is the expressed purpose of Judaism and the attainment of these goals is the only reason that racist Jews have kept their people segregated from the rest of humanity for some two thousand five hundred years.

They remain separate so that they can eventually rule and utterly destroy every other group of human beings. It is their "divine" wish and sole purpose. And they believe that when they have accomplished these horrific goals, God will bless them with a "new earth" and a new spirit and a new heart and will cover their dry bones with a new flesh, as promised in the Jewish apocalyptic nightmares of *Isaiah* and *Ezekiel*. This "new earth" will not suffer Gentile life. These Cabalistic Jews, and their Christian dupes who have been schooled to believe in the "Rapture", intend to destroy the world so as to provoke God to create the "new earth". They do not fear the genetic harm they are deliberately causing humanity, nor do they fear the environmental harm they are causing, because they believe that these will hasten the arrival of the Jewish Messiah and the appearance of a "new earth". The *Zohar*, I, 28*a-b*, states,

"At that time every Israelite will find his twin-soul, as it is written, 'I shall give to you a new heart, and a new spirit I shall place within you' (Ezek. XXXVI, 26), and again, 'And your sons and your daughters shall prophesy'

(Joel III, 1); these are [28b] the new souls with which the Israelites are to be endowed, according to the dictum, 'the son of David will not come until all the souls to be enclosed in bodies have been exhausted', and then the new ones shall come."[69]

Mentor came like Greeks bearing gifts, gifts that would destroy those who received them. Recall that Mentor acknowledged that Bolshevism was a Jewish deception that enslaved whole Peoples in the name of freedom, and yet Mentor claimed that Bolshevism was the salvation of Mankind. In another Jewish deception, Mentor sought to destroy all Gentile nations in the name of "peace", "goodwill and happiness among men" and to establish Israel as a lone nation to rule the world. Mentor knew that Jews had caused the First World War, though Mentor blamed Christians, Tsarism, and everyone but the Jews who were truly responsible. Mentor knew that there were no benefits to Gentiles under Jewish world rule. Mentor knew that Jewish world rule signaled an end to war, because it signaled the end of Gentile life. Mentor knew that it was deceitful to lure the Gentiles into surrendering their sovereignty to Jewish world rule for the sake of "peace", because Mentor knew that it would be a peace which meant the assured destruction of the Gentiles. The solution to war was to bring an end to the tribal Jewish corruption which created it, not to destroy the national sovereignty of all Gentile nations and concurrently and artificially create a "Jewish State" to rule over all and then mass murder all Gentiles.

As with all Jewish promises to the Gentiles of Utopia, Mentor's offer was a trap set to lure the Gentile nations into destroying themselves. This wretched deceit should come as no surprise to the reader, because it is the central purpose of Judaism and Jewish tribalism to lure Gentiles into self-destruction with false promises of an Utopian society, which they promise will follow the end of the world. The reason Mentor was pleading with the Jews to petition for the "end times" peace prophesied in *Isaiah* 2:4 following the devastating First World War, was that Mentor wanted the Jews to sponsor the power of the Zionist League of Nations, which had recently issued the Palestine Mandate, but which had not yet convinced masses of Jews to move from their homes in Europe and America to Palestine. Mentor's plea for peace was in fact a plea for Jewish world rule and the formation of a Jewish State in Palestine following the prophesied World War that the Jews had brought about, in forced and artificial fulfillment of Jewish Messianic prophecy.

The Jewish bankers had largely succeeded in their plan to discredit Gentile government and inaugurate Jewish world rule through the Zionist League of Nations, which they had created. They had also succeeded in stealing Palestine from the Turkish Empire and its indigenous population. But they failed to convince the vast majority of Jews to ruin the Earth in the name of peace, and to follow their effort to fulfill the promises of the Jewish prophets through their

---

69. H. Sperling and M. Simon, *The Zohar*, Volume 1, The Soncino Press, New York, (1933), p. 108.

own devilish intervention in world affairs.

Since the First World War failed to convince the Jews to go to Palestine, it could not have been the final most horrific war of prophecy, and the Jewish bankers would see to it that a yet worse world war would take place, and then again test the Jews to see if they would need a third and still worse war to convince them to flee to the "Promised Land" and stay. Since Israel is today falling apart, and since Jews in America and Russia are again turning toward assimilation, sanity and humanity; there are likely plans in the works for a still worse world war than the Second World War, which racist Zionists believe will finally fulfill Jewish genocidal "end times" prophecy by means of the nuclear incineration of Gentile nations beginning with the Iranians.

Zionist Jews have requisitioned the nuclear arsenal of the United States through the use of disloyal Jewish agents in America, and by deputizing millions of Dispensationalist Zionist Christian dupes who hope to sacrifice the world for the sake of the Jews and who will gladly and madly kill off humanity in the insane belief that God will whisk them away to safety. These highly dangerous Christian Zionist beliefs were created and promoted by Zionist Jews, and were yet another deceitful Jewish trap set for the Gentiles to lure them into destroying themselves, as will be shown further on in this text. Jacob has yet again tricked Esau.

Though in the period immediately after the First World War most Jews hated the crazed Zionists, there were, however, a minority of Jews who went along with the Jewish bankers and called for the governments of the world to step aside and surrender their national sovereignty to the open the way to the universal and open rule of the Jewish bankers, who could then claim the throne of the Messiah. Jewish bankers created a devastating war in part to make the Gentile Peoples war weary. The Jewish bankers then spread the lie that world government alone could prevent war, knowing that world government would be run by them, at first covertly, and later openly. They would then kill off Esau. It must be stressed that the Jewish bankers covertly and deliberately caused war in order to make the world weary of war so that they could step in and offer themselves as the solution to war.

One such plea for the rule of the world by Jewish bankers and their coterie followed Mentor's column on the same page of *The Jewish Chronicle* on 22 September 1922 on page 14,

"The Remedy for War.

From Mr. JOSEPH FINN.
TO THE EDITOR OF THE JEWISH CHRONICLE.

SIR,—'Mentor,' in his article 'Live together or die together,' has rendered a great service to the cause of peace by showing how the terrible war with its consequent peace (which is as bad as was the war), was the result of two or three men in one State disagreeing with two or three men in another. But, 'Mentor,' like the other writers on the same subject, stops short at the remedy.

To eliminate the spirit of war by preaching against, and pointing out the horrors of war, is impossible. The fighting instinct in man cannot be eradicated. Take away the causes which awaken that spirit, and the chances of war would become nil.

For the past twelve years I have tried to convince leading pacifists that mere preaching against the horrors of war will not stop them, so long as nations will have to compete against each other for material gain, like individuals within the nations. [Hebrew deleted.] If it were not for fear of the law, even individuals would war against each other, because of the pressure of the struggle for material gain. As there is no international law strong enough to keep nations in check, the result is war. My pacifist friends argue that whilst the material and economic elements have something to do with the case, the moral element is the chief factor. When the war broke out, not because the various nations wanted to fight, but because the intrigues of diplomats dragged the nations into it, then those very pacifists forgot in a moment all that they had preached against war, and became the most bloodthirsty patriots. The same thing will repeat itself when the diplomats and statesmen bring on another war.

If we are to live and not die together, we must first of all take the great problem of the world's peace, prosperity, and security out of the hands of politicians, statesmen, and diplomats. They are psychologically unfit to solve that problem. Anyone, however slightly acquainted with history, must admit that governments and their agents can only destroy. Throughout history, capital, labour, science, and art have *built*, whilst statesmen and politicians have *destroyed*. That is not mere rhetoric; it is hard fact. Since we must have governments, we have to put up with politicians and statesmen; but when in the history of nations a state of conditions as at present prevailing is reached, when the more the statesmen and diplomats ostensibly try to drag us out of the mire, the more they push us into it, then it is high time for the various nations involved to ask these gentlemen to step aside for a time, and to let us help ourselves.

What then is to be done? As a first step, I suggest the calling of a world conference of all the nations—the delegates to such a conference to be sent from the following bodies: chambers of commerce, bankers' institutions, manufacturers' associations, traders' associations, universities, art institutions, churches, trade unions, co-operative societies, friendly societies, and hospitals. Politicians, statesmen, diplomats, and journalists should not be eligible as candidates. The 'Reconstruction of the world' should be the problem which such a conference should undertake to solve. Whilst that conference proceeds, the various governments should confine their activities to the administration of the common law and the performance of police duties. All international politics and diplomacy of any sort must cease during the sitting of that World Conference.

For the moment, I will say nothing about the programme. Suffice it to say for the present, that such a Conference would do more to reconstruct the world in one month, than the statesmen and diplomats could do in a century.

The war has shown that we Jews must always suffer more than other people when the world is in a state of upheaval. It behooves us, therefore, more than others, to strive for universal peace, security, and prosperity. We cannot find security in some corner in Palestine, while the nations are trying to destroy each other. Our welfare and happiness is dependent on the welfare of all the other nations. If we really believe that we have a mission in the world, then that mission can only be to help on—nay, to push on—the general advancement of the nations, even at the risk of temporary unpleasantness. Our true [Hebrew deleted.] will not be found in having our own politicians, statesmen, diplomats, generals, and soldiers. We will reach our [Hebrew deleted.] when all wars—military and commercial—shall cease, and in consequence thereof the nations become truly civilised and refined, when they begin to feel sorrow because of the wrongs they have done to us throughout the centuries. Then will *our* day come, when the nations will be eager to compensate us for the wrongs we are suffering and have suffered. Blessed be those who live to see that day!

Yours faithfully,
JOSEPH FINN.
10, Windsor Road, Forest Gate, E.7."

Finn speaks of the revenge of the Jews upon the Gentiles for the "Controversy of Zion"—of the prophesied Messianic Age when the Jews will enslave and then exterminate the Gentiles, after the Jewish Messiah passes judgment on non-Jews and assimilated Jews (*Isaiah* 11. *Jeremiah* 3:17; 10:10-11; 23:5-8). The Jewish book of *Zechariah* 8:23 promises the Jews that ten Gentiles will gladly slave for every Jew,

"Thus saith the LORD of hosts; In those days *it shall come to pass*, that ten men shall take hold out of all languages of the nations, even shall take hold of the skirt of him that is a Jew, saying, We will go with you: for we have heard *that* God *is* with you."

The Jewish Talmud at *Shabbath* 32*b* increases the number of Gentile slaves per Jew to 2,800 (note that the Talmud was written by the "the scribes and Pharisees" whom Jesus condemned in the New Testament, *Matthew* 23. *Mark* 7; 12:38-40. *Luke* 11. *John* 8.),

"Resh Lakish said: He who is observant of fringes will be privileged to be served by two thousand eight hundred slaves, for it is said, *Thus saith the Lord of hosts: In those days it shall come to pass, that ten men shall take hold, out of all the languages of the nations, shall even take hold of the skirt of him that is a Jew, saying, We will go with you,* etc."[70]

---

[70]. I. Epstein, Shabbath 32*b*, *The Babylonian Talmud*, Volume 7, The Soncino Press,

The Jewish book of *Genesis* 25:23; 27:38-41 promises the Gentiles to the Jews as their slaves and slave soldiers, and gives the Jews an incentive to exterminate the Gentiles, simply because the Gentiles dare to be angry at the Jews for deceiving them and using them as slaves,

"25:23 And the LORD said unto her, Two nations *are* in thy womb, and two manner of people shall be separated from thy bowels; and *the one* people shall be stronger than *the other* people; and the elder shall serve the younger. [***] 27:38 And Esau said unto his father, Hast thou but one blessing, my father? bless me, *even* me also, O my father. And Esau lifted up his voice, and wept. 27:39 And Isaac his father answered and said unto him, Behold, thy dwelling shall be the fatness of the earth, and of the dew of heaven from above; 27:40 And by thy sword shalt thou live, and shalt serve thy brother; and it shall come to pass when thou shalt have the dominion, that thou shalt break his yoke from off thy neck. 27:41 And Esau hated Jacob because of the blessing wherewith his father blessed him: and Esau said in his heart, The days of mourning for my father are at hand; then will I slay my brother Jacob."

The Jewish Talmud, in the book of *Sanhedrin* folios 56*a*-60*b*, sets forth seven laws which it insists non-Jews must follow on pain of death (*see also: Sanhedrin* 99*a*). These are called the "Noahide Laws". Rabbinical scholars fabricated these laws in part as a response to Christianity. Jewish leaders have already introduced these laws into the laws of the United States of America.[71] Christians, Hindus, Moslems, Buddhists, etc. who refuse to abandon their religious beliefs and become slaves of the Jews will all be killed. Anyone who questions Judaism will be killed. Anyone who disobeys the commands of the Jews will be killed.

Jewish leaders are attempting to persuade Gentiles to adopt the "Noahide Laws". These laws forbid Gentiles to worship idols. Jews view Christianity, Islam, Hinduism, Buddhism, and all other religions as forbidden religions (*Zechariah* 14:9). Jews are planning a nuclear war, which will turn night into day, and will eat the flesh off your bone, as prophesied in *Zechariah* 14:7, 12-13. Jews are planning to steal all of the wealth of the world (*Deuteronomy* 6:10-11; 11:24-25. *Joshua* 1:2-5; 6:19; 24:13. *Proverbs* 1:13-14. *Isaiah* 2:1-4; 23:17-18; 40:15-17, 22-24; 54:1-4; 60:5, 8-12, 16-17; and 61:5-6. *Obadiah. Haggai* 2:7-8. *Zechariah* 14:14. *Baba Kamma* 38*a*).

Jews insist that Gentiles obey the seven Noahide Laws—the seven Mitzvoth of the *Bnei Noach*. Jews plan to cut off the heads of Gentiles, Christians, Hindus, Moslems, Buddhists, etc. who refuse to bow down and worship the Jewish King, and who refuse to renounce their religions and obey the Noahide Laws and the

---

London, (1938), p. 149.
71. House Joint Resolution 104-102-14 enacted 20 March 1991.

Rabbis (*Exodus* 34:11-17. *Deuteronomy* 7:1-6. *Isaiah* 49:7, 23. *Zechariah* 8:23; 14:9, 16-20. *See also: Shabbath* 32*b*. *Sanhedrin* 56*a*-60*b*, 88*b*. *Erubin* 21*b*). *Zechariah* chapter 14 states,

"Behold, the day of the LORD cometh, and thy spoil shall be divided in the midst of thee. 2 For I will gather all nations against Jerusalem to battle; and the city shall be taken, and the houses rifled, and the women ravished; and half of the city shall go forth into captivity, and the residue of the people shall not be cut off from the city. 3 Then shall the LORD go forth, and fight against those nations, as when he fought in the day of battle. 4 And his feet shall stand in that day upon the mount of Olives, which is before Jerusalem on the east, and the mount of Olives shall cleave in the midst thereof toward the east and toward the west, *and there shall be* a very great valley; and half of the mountain shall remove toward the north, and half of it toward the south. 5 And ye shall flee *to* the valley of the mountains; for the valley of the mountains shall reach unto Azal: yea, ye shall flee, like as ye fled from before the earthquake in the days of Uzziah king of Judah: and the LORD my God shall come, *and* all the saints with thee. 6 And it shall come to pass in that day, *that* the light shall not be clear, *nor* dark: 7 But it shall be one day which shall be known to the LORD, not day, nor night: but it shall come to pass, *that* at evening time it shall be light. 8 And it shall be in that day, *that* living waters shall go out from Jerusalem; half of them toward the former sea, and half of them toward the hinder sea: in summer and in winter shall it be. 9 And the LORD shall be king over all the earth: in that day shall there be one LORD, and his name one. 10 All the land shall be turned as a plain from Geba to Rimmon south of Jerusalem: and it shall be lifted up, and inhabited in her place, from Benjamin's gate unto the place of the first gate, unto the corner gate, and *from* the tower of Hananeel unto the king's winepresses. 11 And *men* shall dwell in it, and there shall be no more utter destruction; but Jerusalem shall be safely inhabited. 12 ¶And this shall be the plague where*with* the LORD will smite all the people that have fought against Jerusalem; Their flesh *shall* consume away while they stand upon their feet, and their eyes shall consume away in their holes, and their tongue shall consume away in their mouth. 13 And it shall come to pass in that day, *that* a great tumult from the LORD shall be among them; and they shall lay hold every one on the hand of his neighbour, and his hand shall rise up against the hand of his neighbour. 14 And Judah also shall fight at Jerusalem; and the wealth of all the heathen round about shall be gathered together, gold, and silver, and apparel, in great abundance. 15 And so shall be the plague of the horse, of the mule, of the camel, and of the ass, and of all the beasts that shall be in these tents, as this plague. 16 And it shall come to pass, *that* every one that is left of all the nations which came against Jerusalem shall even go up from year to year to worship the King, the LORD of hosts, and to keep the feast of tabernacles. 17 And it shall be, *that* whoso will not come up of *all* the families of the earth unto Jerusalem to worship the King, the LORD of hosts, even upon them shall be no rain. 18 And if the family of Egypt go not up, and come not, *that* have no *rain*; there shall be the plague, where*with* the LORD will smite the heathen that come not up to keep the feast of tabernacles. 19 This shall be the punishment of Egypt, and the punishment of all nations that come not up to keep the feast of tabernacles. 20 In that day shall there be upon the bells of the horses, HOLINESS UNTO THE LORD; and the pots in the LORD's house shall be like the bowl's before the altar. 21 Yea, every pot in Jerusalem and in Judah shall be holiness unto the LORD of hosts: and all they that sacrifice shall come and take of them, and seethe therein: and in that day there shall be no more the Canaanite in the house of the LORD of hosts."

The Jewish prayer of Kiddush for the New Year summarizes the genocidal ambitions of the Jewish faith. It states, inter alia,

> "Blessed art thou, O Lord our God, King of the universe, who hast chosen us from all peoples and exalted us above all tongues, and sanctified us by thy commandments. [***] For thou hast chosen us and hast sanctified us above all nations; and thy word is truth and endureth for ever. Blessed art thou, O Lord, King over all the earth, who sanctifiest [the Sabbath and] Israel and the Day of Memorial. [***] We therefore hope in thee, O Lord our God, that we may speedily behold the glory of thy might, when thou wilt remove the abominations from the earth, and the idols will be utterly cut off, when the world will be perfected under the kingdom of the Almighty, and all the children of flesh will call upon thy name, when thou wilt turn unto thyself all the wicked of the earth. Let all the inhabitants of the world perceive and know that unto thee every knee must bow, every tongue must swear. Before thee, O Lord our God, let them bow and fall; and unto thy glorious name let them give honour; let them all accept the yoke of thy kingdom, and do thou reign over them speedily, and forever and ever. For the kingdom is thine, and to all eternity thou wilt reign in glory; as it is written in thy Law, The Lord shall reign for ever and ever."[72]

The Jewish Talmud teaches Jews that Jesus is an "idol" and that Christians are "idol worshipers". The Jewish Talmud instructs Jews to behead idol worshipers. Racist Jews intend to impose the Noahide Laws in the End Times and exterminate all Christians, pursuant to the Jewish Talmud in the book of *Sanhedrin*, folios 56*a*-60*b*, and 99*a* (note that the Talmud was written by "the scribes and Pharisees" whom Jesus condemned in the New Testament: *Matthew* 22; 23. *Mark* 7; 12:38-40. *Luke* 11. *John* 8).

Jews believe that Jesus was the son of a prostitute and a Roman soldier, who learned witchcraft in Egypt, and who beguiled Jews to worship him as an idol. While this myth is more fully enunciated in the *Toledoth Jeshua (also: Yeshua)*,

---

72. English translation by S. Singer, "Kiddush for New Year", *The Authorized Daily Prayer Book of the United Hebrew Congregations of the British Empire*, Ninth American Edition, Hebrew Publishing Co., New York, (not dated), pp. 243-255, at 243, 247-248.

the Talmud also iterates similar beliefs in *Shabbath* 104*b*,[73] in *Sanhedrin* 67*a*,[74] in *Sotah* 47*a*,[75] in *Sanhedrin* 43*a*, in *Sanhedrin* 107*b*,[76] and in *Sanhedrin* 106*a*-106*b*,

> "*Balaam also the son of Beor, the soothsayer,* [*did the children of Israel slay with the sword*].³ A soothsayer? But he was a prophet! — R. Johanan said: At first he was a prophet, but subsequently a soothsayer.⁴ R. Papa observed: This is what men say, 'She who was the descendant of princes and governors, played the harlot with carpenters.'⁵ *Did the children of Israel slay with the sword among them that were slain by them.*⁶ Rab said: They subjected him to four deaths, stoning, burning, decapitation and strangulation.⁷"[77]

Jesus ridiculed the scribes and Pharisees (*Matthew* 22; 23. *Mark* 7; 12:38-40. *Luke* 11. *John* 8). That was just one of the reasons why the Jews killed Jesus in the New Testament story, and that is one of the reasons why the Jewish Talmud says that Jesus is boiling in excrement in hell. The Jewish Talmud, in the book of *Erubin* 21*b*, states (*see also: Sanhedrin* 88*b*),

> "[A]s to the laws of the Scribes, whoever transgresses any of the enactments of the Scribes incurs the penalty of death. [***] This⁸ teaches that he who scoffs at the words of the Sages will be condemned to boiling excrements. [***] [F]or [neglecting] the words of the Rabbis³ one deserves death[.]"[78]

The Jewish Talmud tells Jews in *Gittin*, folio 57*a*, that Jesus is boiling in hell in hot excrement and semen, which is his curse for questioning Jewish authority,

> "He then went and raised Balaam by incantations. He asked him: Who is in repute in the other world? He replied: Israel. What then, he said, about joining them? He replied: *Thou shalt not seek their peace nor their prosperity all thy days for ever.*[*Footnote:* Deut. XXIII, 7.] He then asked:

---

[73]. I. Epstein, Editor, "Shabbath 104*b*", *The Babylonian Talmud*, Volume 8, The Soncino Press, London, (1938), pp. 502-505, at 504, *see especially footnote 2.*

[74]. I. Epstein, Editor, "Sanhedrin 67*a*", *The Babylonian Talmud*, Volume 27, The Soncino Press, London, (1935), pp. 454-461, at 456-457, *see especially footnote 5.*

[75]. I. Epstein, Editor, "Sotah 47*a*", *The Babylonian Talmud*, Volume 20, The Soncino Press, London, (1936), pp. 245-249, at 246-247, *see especially* footnote 3 on p. 246, and footnotes 7, 10 and 11 on p. 247.

[76]. I. Epstein, Editor, "Sanhedrin 107*b*", *The Babylonian Talmud*, Volume 28, The Soncino Press, London, (1935), pp.733-738, at 735-736, *see especially* footnote 4 on page 735, and footnote 2 on page 736.

[77]. I. Epstein, "Sanhedrin 106*a*-106*b*, *The Babylonian Talmud*, Volume 28, The Soncino Press, London, (1935), pp. 721-729, at 725.

[78]. I. Epstein, "Erubin 21*b*", *The Babylonian Talmud*, Volume 9, The Soncino Press, London, (1938), pp. 148-151, at 149-150.

What is your punishment? He replied: With boiling hot semen.[*Footnote:* Because he enticed Israel to go astray after the daughters of Moab. V. Sanh. 106a.] He then went and raised by incantations the sinners of Israel.[*Footnote:* {MS. M. Jesus}.] He asked them: Who is in repute in the other world? They replied: Israel. What about joining them? They replied: Seek their welfare, seek not their harm. Whoever touches them touches the apple of his eye. He said: What is your punishment? They replied: With boiling hot excrement, since a Master has said: Whoever mocks at the words of the Sages is punished with boiling hot excrement. Observe the difference between the sinners of Israel and the prophets of the other nations who worship idols."[79]

The Jewish Talmud states in *Rosh Hashanah*, folio 17a, that all Christians are going to burn in hell forever,

"But as for the minim[11] and the informers and scoffers,[1] who rejected the Torah and denied the resurrection of the dead, and those who abandoned the ways of the community,[2] and those who 'spread their terror in the land of the living',[3] and who sinned and made the masses sin, like Jeroboam the son of Nebat and his fellows—these will go down to Gehinnom and be punished there for all generations, as it says, *And they shall go forth and look upon the carcasses of the men that have rebelled against me*[4] etc."[80]

The Jewish Talmud in the book of *Shabbath*, folio 116a, instructs Jews to burn all Christian books,

"Come and hear: The blank spaces[5] and the Books of the Minim[6] may not be saved from a fire, but they must be burnt in their place, they and the Divine Names occurring in them. Now surely it means the blank portions of a Scroll of the Law? No: the blank spaces in the Books of Minim. Seeing that we may not save the Books of Minim themselves, need their blank spaces be stated?—This is its meaning: And the Books of Minim are like blank spaces."[81]

The Jewish Talmud also teaches Jews to cover up their acts of murder with a covering blanket of lies. Jews are taught to make it appear that they are doing a good deed for the community, when they are in fact deliberately murdering innocent human beings. The Talmud in the Jewish book of *Abodah Zarah*, folio 26a-26b, teaches Jews that when they murder someone by casting that person

---

79. I. Epstein, "Gittin 57a", *The Babylonian Talmud*, Volume 21, The Soncino Press, London, (1936), pp. 261-265, at 261.
80. I. Epstein, "Rosh Hashanah 17a", *The Babylonian Talmud*, Volume 13, The Soncino Press, London, (1938), pp. 64-67, at 64-65.
81. I. Epstein, Editor, "Shabbath 116a", *The Babylonian Talmud*, Volume 8, The Soncino Press, London, (1938), pp. 567-571, at 569.

into a pit, they should scape off any steps in the pit, remove all ladders from the pit, and cover the pit with a stone, so as to ensure the death of their victim. Should someone catch the murderous Jew scraping away the steps in a pit, making off with a ladder, or covering the pit with a stone, the Jewish Talmud recommends that the Jew should deceive the witness and claim that the Jew needs the ladder to rescue his son from a rooftop, or is covering the pit with a stone so that cattle do not fall in, or that he is scraping off the steps so that cattle are not lured into the pit. The Jewish Talmud in the book of *Abodah Zarah* 26a-26b states that it is permissible for Jews to kill Christians,

> "R. Joseph further had in mind to say, in regard to what has been taught that in the case of idolaters and shepherds of small cattle one is not obliged to bring them up [from a pit] though one must not cast them in it[2]—that for payment one is obliged to bring them up on account of ill feeling. Abaye, however, said to him: He could offer such excuses as, 'I have to run to my boy who is standing on the roof', or, 'I have to keep an appointment at the court.'
>
> R. Abbahu recited to R. Johanan: 'Idolaters and [Jewish] shepherds of small cattle need not be brought up [26b] though they must not be cast in, but *minim*,[3] informers, and apostates may be cast in, and need not be brought up.' Whereupon R. Johanan remarked: I have been learning that the words, *And so shalt thou do with every lost thing of thy brother's* [*thou mayest not hide thyself*],[4] are also applicable to an apostate, and you say he may be thrown down; leave out apostates! Could he not have answered that the one might apply to the kind of apostate who eats carrion meat to satisfy his appetite,[5] and the other to an apostate who eats carrion meat to provoke?— In his opinion, an apostate eating carrion meat to provoke is the same as a *min*.[6]
>
> It has been stated: [In regard to the term] apostate there is a divergence of opinion between R. Aha and Rabina; one says that [he who eats forbidden food] to satisfy his appetite, is an apostate, but [he who does it] to provoke is a *'min'*; while the other says that even [one who does it] to provoke is merely an apostate.—And who is a *'min'*?—One who actually worships idols.[1]
>
> An objection was raised: If one eats a flea or a gnat he is an apostate. Now such a thing could only be done to provoke, and yet we are taught that he is merely an apostate!—Even in that case he may just be trying to see what a forbidden thing tastes like.
>
> The Master said: 'They may be cast in and need not be brought up'—if they may be cast in need it be said that they need not be brought up?—Said R. Joseph b. Hama in the name of R. Shesheth: What is meant to convey is that if there was a step in the pit-wall, one may scrape it away, giving as a reason for doing so, the prevention of cattle being lured by the step to get unto the pit. Raba and R. Joseph both of them said: It means to convey that if there is a stone lying by the pit opening, one may cover the pit with it, saying that he does it for [the safety] of passing animals. Rabina said: It is

meant to convey that if there is a ladder there, he may remove it, saying, I want it for getting my son down from a roof."[82]

Cabalistic Jews consider themselves to be the divine presence of God on Earth. They believe that their King, the Jewish Messiah, will be God ruling the Earth. Gentiles will have to serve him and each Jew or face death. The infamous Nazi Zionist film *Der ewige Jude* quotes an older rendition of the Benediction of the Sabbath about 55 minutes into the film, which recalls the sentiments of Rabbi Simon ben Yohai,[83]

> "Praise be the Lord, who has set apart the Holy and the common nations, Israel and the other Peoples. The heathens, who do not keep your laws, you have made enemies to be exterminated. God's wrath is upon them, and He says, 'I will kill even the best among the heathens.' Since all among the peoples of the world are blasphemers, none are good, but the sons of Israel are all righteous."

Some argue that Jews in general have an indoctrinated tendency to stifle progress and restrict disputes to dogmatism. This is an effect of Judaism, which demands obedience to an arbitrary and absolute law. One cannot speak out against, or argue with, the one true God, or with those chosen to represent him and chosen to kill off the unchosen. Some, including Eugen Karl Dühring, Friedrich Nietzsche and Houston Stewart Chamberlain, have argued that Judaism is a slavish religion which inhibits human creativity. The ancient religion has little respect for personal choice and places in its stead absolute obedience to God and to God's laws, and to God's chosen people. Since Judaism is more political than it is religious, the effects of this authoritarianism lingered even in the writings of many German Jews who were supposedly atheists, including Karl Marx, Moses Hess and Ferdinand Lasalle.

This same charge was also made by philo-Semites like the famous cultural Zionist Ha'am. Ha'am wrote of the Jews as a slavish "people of the book" who suffered under the "long-standing disease" of the "tyranny of the written word" which forbade individual thought for the sake of absolute obedience to arbitrary dogmatic laws.[84] Chaim Nachman Bialik's speech at the opening of the "Hebrew University" provides us with a good example of the religious zealotry and of the dogmatic and intolerant worship of the Torah and Talmud of some Jews—

---

82. I. Epstein, Editor, "Abodah Zarah 26a-26b", *The Babylonian Talmud*, Volume 29, The Soncino Press, London, (1935), pp. 129-133, at 131-132.
83. "Gentile", *The Jewish Encyclopedia*, Volume 5, Funk and Wagnalls Company, New York, (1903), pp. 615-626, at 617. *See also:* A. Cohen, "Soferim 41a", *The Minor Tractates of the Talmud Massektoth Ketannoth in Two Volumes*, Volume 1, The Socino Press, London, (1965), pp. 287-288, *especially* note 50.
84. A. Ha-Am, "The Law of the Heart", in A. Hertzberg, *The Zionist Idea*, Harper Torchbooks, New York, (1959), pp. 251-255.

probably a very small percentage of Jews today.[85] Jewish children learn Hebrew and Judaism through a process of mindless repetition, which inhibits their ability to reason and think independently. Jewish leaders are often arrogant, absolutist, intolerant and dogmatic. In 1944, David Ben-Gurion cried out "for absolute allegiance to the Jewish revolution", which he defined in the Messianic terms of *"the complete ingathering of the exiles into a socialist Jewish state."*[86] Ben-Gurion believed that Jews should lead the Gentiles of the world to adopt Jewish religious mythologies and conduct "world revolution". Violent revolution, and the dictatorships imposed under the illusion of Utopian dreams, have been longstanding Jewish traditions. Reality and science give way to religion and childish delusion.

Like many before him, Albert Einstein believed that Jews had lived in darkness while Gentile Europeans had born reborn.[87] Judaism had inhibited the progress of science among Jews, who attempted to stifle free thought among their own people. When the Jewish community marketed the new Jewish heroes Karl Marx, Albert Einstein and Sigmund Freud to the general public in the Twentieth Century, the old habits remained and a new international dogmatism, like that of the old lawgiver Moses, emerged. No one dare question the pseudo-Messiahs, who had allegedly found ultimate truths that were not open to debate. The old Jewish traditions of hero worship and dogmatism carried on in a new age of mass suggestion through intensive advertising and a controlled and propagandistic press. To question a Jewish hero was to question a Jewish God, and therefore to be anti-Semitic, *per se*.[88]

This largely ended free and open debate, and with it normal scientific progress in these fields. Several nations were forced into the slavery of Communism under the false promise and childish premise of a Jewish Utopia to come. Physics degenerated into mysticism. Psychology reaped tremendous profits for its practitioners, while doing little for its patients that time alone would have otherwise accomplished. Each of these mythologies and advertised heroes could only survive in a climate where dissent was suppressed, and suppression and dogmatism were ancient traditions in the Jewish community. Anatole Leroy-Beaulieu wrote in the 1893,

> "Far from emanating from the Synagogue, the new ideas had great difficulty in making their way into it. The Synagogue had, so to speak, stopped up all the chinks and crannies in its traditions; in Poland, Hungary, and even in Germany, in fact, almost everywhere it had proceeded after the

---

85. H. N. Bialik, "Bialik on the Hebrew University", in A. Hertzberg, *The Zionist Idea*, Harper Torchbooks, New York, (1959), pp. 281-288, at 282-283.
86. D. Ben-Gurion, *Ba-Maarachah*, Volume 3, Tel-Aviv, (1948), pp. 200-211, English translation in A. Hertzberg, *The Zionist Idea*, Harper Torchbooks, New York, (1959), pp. 606-619, at 618.
87. A. Einstein, *Ideas and Opinions*, Crown, New York, (1954), p. 181.
88. "Prof. Einstein Here, Explains Relativity", *The New York Times*, (3 April 1921), pp. 1, 13, at 1.

fashion prevalent in cold countries, where at the beginning of winter the windows are fastened down with cement to keep the outer air from entering. Its most illustrious children were anathematised by the Synagogue; the *Herem*, with its awful imprecations, was hurled at whoever attempted innovations. Baruch Spinoza was excommunicated in the eighteenth century by the most enlightened community on earth; Moses Mendelssohn, who served as a model for Lessing's *Nathan the Wise*, had in that same century to see his German Pentateuch and Psalms condemned by German and Polish rabbis. The synagogue of Berlin rejected books written in the vernacular; it expelled one of its members for having read a German book. The bulk of Jews of both classes, the *Askenazim* and the *Sephardim*, abhorred the philosophers and their precepts. They held profane sciences in suspicion. [*Footnote:* See, especially, the autobiography of the rabbi-philosopher, Solomon Maimon, published in 1792-93, by R. P. Moritz, under the name: *Salomon Maimon's Lebensgeschichte*. Cf. Arvède Barine's *Un Juif Polonaise* (*Revue des Deux Mondes*, of October 15, 1889).] While the salons of Paris were discussing the philosophy of Descartes, or the approaching regeneration of man, the Jewish communities of Eastern and Central Europe were dreaming of cabalistic utopias, yielding themselves up to the craze of Hassidism, and growing fanatical over the rival claims of false Messiahs, such as Franck and Sabbatai. [*Footnote:* The Seventeenth and eighteenth centuries were, in fact, the age of false Messiahs and also of the diffusion of Hassidism or neo-cabalism, still prevalent in a number of communities. See Graetz's *Geschichte der Juden*, vol. x., chap. vi.-xi.]

III.

Everywhere, in the East as well as the West, it was from the outside, and thanks only to the lamps of the *goim*, that the new ideas, 'the light,' penetrated into the alleys of the Ghetto and pierced the gloom of the *Judengasse*. Could it, indeed, have been otherwise, after centuries of sequestration and debasement! However great may be Israel's elasticity, her mainspring seemed to have been broken. She was weighed down by the double load of her heavy talmudic traditions, and the hatred of a hostile society."[89]

Communists, Zionists and Nazis likewise have been notorious opponents of personal choice and viciously punished dissent and free speech. Each of these movements were led and financed by Jews and by crypto-Jews. The hero worship of figures like Einstein, Freud and Marx, which has led in many instances to a dogmatic stagnation of science and to fanatical personal attacks on dissenters, has been called a "Jewish trait"—the continuance of a persistent habit of intolerance after the abandonment of one religious Jewish creed for another,

---

[89]. A. Leroy-Beaulieu, *Israel chez les nations: Les Juifs et l'antisémitisme*, C. Lévy, Paris, (1893); English translation by F. Hellman, *Israel among the Nations: A Study of the Jews and Antisemitism*, G. P. Putnam's Sons, New York, W. Heinemann, London, (1895), pp. 60-61.

and the shameless perpetuation and proselytizing of a childish religious creed through the obstruction of open debate, and the self-aggrandizing advertising of Jewish cult figures.

On the other hand, many leading Jews have been very cosmopolitan and cultured people, who were eager to assimilate. They, too, fall victim to a fairly large contingent of racist Jews who wish to quash disagreement with their views by slandering and libeling anyone who brings the facts to the fore by calling them "anti-Semitic" for daring to argue with Jewish racists.[90] This is a highly vocal and well-organized minority in the Jewish community, which is mostly composed of racist Zionist Jews. Albert Einstein, who was himself a vocal racist, is a hero to other racist Jews. Racist Jews often have no regard for individual rights or democratic principles. They insist that everyone obey them, or face death. This charge is not made lightly or whimsically, and a good deal of evidence will be presented in this work to justify this accusation. Other Jews are by no means immune to the attacks of racist Jews, in fact they are the most common target of racist Jewish intolerance, totalitarianism and violence.

Many of the early Communist and Socialist philosophers were proudly in the traditions of Plato, the early Christians, the American Revolution, and the French Revolution—a fact that troubled many critics of Judaism and Jews, who often saw the French Revolution as a Jewish Frankist-style plot to destroy the monarchies of the world in order to obtain Jewish emancipation and a Jewish nation, then to rule the world from Jerusalem as was prophesied in the Old Testament (*Exodus* 34:11-17. *Psalm* 2; 72. *Isaiah* 2:1-4; 9:6-7; 11:4, 9-10; 42:1; 61:6. *Jeremiah* 3:17. *Joel* 3:16-17. *Micah* 4:2-3. *Zechariah* 8:20-23; 14:9). The French Revolution resulted in the "Terror" and many predicted that a "world revolution" would be yet more terrible. Indeed, the "Red Terror" of the Bolshevik Revolution was far more terrible than the Terror of the French Revolution. The Old Testament calls on Jews to commit still worse acts against humanity than the atrocities of the Bolshevik—and Nazi—revolutions.

Though centuries of Jewish propaganda have blinded many to these facts, the world public was acutely aware of them after the First World War. Though Jewish propaganda has largely erased this history from the consciousness of the American People and has misrepresented the facts so as to make it appear that there were no legitimate grounds to be suspicious of Judaism and Jewish racism in the era when Einstein faced his harshest criticism, there were many legitimate reasons why courageous individuals fought back against the destruction of their nations and their cultures. Many of these individuals were of Jewish descent and

---

90. P. Findley, *They Dare to Speak Out: People and Institutions Confront Israel's Lobby*, Lawrence Hill, Westport, Connecticut, (1985); **and** *Deliberate Deceptions: Facing the Facts about the U.S.-Israeli Relationship*, Lawrence Hill Books, Chicago, (1993); **and** *Silent No More: Confronting America's False Images of Islam*, D : Amana Publications, Beltsville, Maryland, (2001). *See also:* R. I. Friedman, "Selling Israel in America: The Hasbara Project Targets the U.S. Media", *Mother Jones*, (February/March, 1987), pp. 1-9; reprinted "Selling Israel to America", *Journal of Palestine Studies*, Volume 16, Number 4, (Summer, 1987), pp. 169-179.

knew well the agenda of the Jewish financiers who fomented and funded the Jewish revolutionaries.

Some saw democracy as a very bad thing—a tyranny of the mediocre over the superior person, which inhibits progress and cheapens culture, science and the arts; allowing for collusive elements to commercialize and destroy culture by vulgarizing it for mass consumption. Some, including Aristotle, believed that democracy inevitably degenerates into plutocracy. Some of the critics of the Jews of Einstein's day, in chorus with many proud Jews, pointed out the commonality of Bolshevism and Judaism. This promoted general prejudice against Jews, most of whom opposed Bolshevism.

There were, however, large numbers of Jewish Bolsheviks. Jews led and financed the international revolutionary movement and it must therefore be viewed as a Jewish movement. Though many good natured people were duped through romanticism and idealism into joining the world revolutionary movement, it became very clear after the Russian and Hungarian revolutions that the Bolsheviks were out to destroy the Gentile nations in fulfillment of horrific Judaic prophecies.

The Bolsheviks used Utopianism as a political platform to lure in recruits. After they succeeded in their revolution, they subverted the very ideals they had promoted as a means to place themselves into power. When this became widely known, they then themselves created anti-Bolshevik organizations, including the Nazi Party, as a means to place crypto-Bolsheviks into power, who would conduct a Bolshevik revolution in the name of fighting Bolshevism, in countries which had learned of the dangers of Bolshevism.

Though many early Socialists helped to organize labor unions, which developed the middle class, and were pursuing their Utopian dreams before the genocidal purges of Lenin, Stalin, Kun, Mao and other "Communist revolutionaries" would forever stigmatize the political agenda and ideas of Socialism, Bolsheviks were rightfully seen as terribly dangerous in 1919 and Germany was one of their primary targets. Peter Michelmore wrote in his biography *Einstein: Profile of the Man*,

> "But there was another, more sinister, reason. November 7[, 1919] was the second anniversary of the Bolshevik Revolution in Russia and Communist Party agents all over the world had in their hands a secret manifesto saying that this was the day when workers should be incited to overthrow governments, assassinate public officials, bomb army barracks and establish dictatorships of the proletariat. Berlin was a prime target. The amateur republican government of former basketmakers and blacksmiths was in daily danger of collapse under pressure from both extreme left and extreme right."[91]

Einstein, himself, wrote to Emil Zürcher on 15 April 1919 that he knew for

---

91. P. Michelmore, *Einstein: Profile of the Man*, Dodd, Mead, New York, (1962), p. 3.

certain that Bolshevik leaders were stealing the wealth of the Russian Nation and were "systematically" mass murdering everyone who did "not belong to the lowest class."[92]

In the 1920's, there were many theories about Jewish people, even (one might say, *especially*) in the conservative academic community, who should have been more enlightened. Einstein happened to fulfill many stereotypes. One such stereotype was the belief that Jewish people were genetically incapable of profound intuitive thought, but could only think "logically", *i. e.*, repetitively, deductively and mathematically. Philipp Frank wrote,

> "The members of the Jewish community had often been compelled to hear and to read that while their race possessed a certain craftiness in business pursuits, in science it could only repeat and illuminate the work of others, and that truly creative talents were denied them."[93]

It might be possible, though it seems unlikely to your author, that Jews would selectively mate with persons who were obedient to authority and shunned original thought, and that Jewish society so strongly selects against the survival of strong and moral individuals that the Jews have created a clannish and ignoble breed. Anecdotally, your author has found that the opposite is the case, at least in the modern era. It seems more likely to your author that conditioned reactionary tribalism gives the false appearance of intellectual stupor and anti-social immorality. Jews would blindly support patently false notions and would deliberately lie in a nearly uniform chorus, not because they were truly blind, but rather because they were truly clannish. They were probably made so by social conditioning, not blood—though the possibility exists that they have bred themselves into an especially clannish, selfish and unethical type, as governed by the general standards of Western Civilization.

In the 1893, Anatole Leroy-Beaulieu wrote,

> "There are two opinions current with regard to the Jew. One ascribes to him a spirit, if not a genius, foreign and antagonistic to our race, and calls it the 'Semitic' spirit. The other—often held by the very same persons—asserts that the Jew is utterly lacking in individual genius, in originality. According to this opinion he has never invented anything, and is in art and science, as everywhere else, capable only of adjusting and adapting. 'Look at them,' said one of my friends to me, 'see how quickly and with what monkey- or squirrel-like agility they climb the first rungs of any ladder; sometimes they even succeed in scaling its top, but they never add to it a single round.' Granted; but how many of us really add a single round to that mysterious ladder which we have set up in vacant space, and which reaches

---

[92]. Letter from A. Einstein to E. Zürcher of 15 April 1919, English translation by A. Hentschel, *The Collected Papers of Albert Einstein*, Volume 9, Document 23, Princeton University Press, (2004), p. 19.

[93]. P. Frank, *Einstein: His Life and Times*, Alfred A. Knopf, New York, (1967), p. 145.

toward the Infinite?

Men who consider the remnants of Israel as an ethnic element distinct from all others, insist that they have never displayed any originality, either in art, poetry, or philosophy. The Jew, in their opinion, is utterly lacking in creative power. It is this that is said to distinguish the Semitic, from the Aryan, spirit. The Semite is sterile; neither his brain nor his hands can produce anything new. He is content to appropriate the labour of others, in order to put it to use; he makes the most of ideas and inventions, as of dollars; he combines them and puts them into circulation; in short, he always subsists on others; one might almost say that he is the parasite of arts and sciences.

This is, approximately, the theory of Wagner [*Footnote:* Wagner's *Das Judenthum in der Musik*.] with regard to music, the art most cultivated by the Jews; according to him, Jews like Mendelssohn, Meyerbeer, and Halévy, although indeed able to compose a German symphony or a French opera, have not been able to invent a new form in art. But, is it necessary to invent new forms in order to be an original artist? And does this lack necessarily imply that Jewish genius consists entirely in a faculty for combination? Absence of creative power, of spontaneity and of originality, is said to be the mark of the Jew everywhere. Israel, it is asserted, displays ill this respect something of a woman's nature. The Semites are said to be a feminine race, possessing to a high degree the gift of receptivity, always lacking in virility and procreative power. From which it would seem to follow that they are, after all, an inferior race.

If this be indeed so, it suggests a reflection: If the Jew is merely an imitator, a copyist, a borrower, how can his race possibly denationalise our strong Aryan races? But, are we justified in regarding this lack of originality as a racial feature, the stamp impressed on Israel and the Semite by the hand of ages? As for myself, I must confess that if any of the ancient races seemed to possess originality, it was this race. Even those who have denied it a creative imagination [*Footnote:* Renan's Histoire Générale des Langues Sémitiques: 'The eminently subjective character of Arabian and Hebrew poetry is due to another trait of the Semitic spirit, to its complete lack of creative imagination and to the consequent absence of fiction.] have agreed that it gave the world religion—an invention that holds its own with any other."[94]

Even some of Einstein's staunchest supporters believed in this theory that Jews were genetically inferior to the creative intellect of Gentiles. Einstein tried to portray himself in opposition to intuition and against inductive reasoning,

---

[94]. A. Leroy-Beaulieu, *Israel chez les nations: Les Juifs et l'antisémitisme*, C. Lévy, Paris, (1893); English translation by F. Hellman, *Israel among the Nations: A Study of the Jews and Antisemitism*, G. P. Putnam's Sons, New York, W. Heinemann, London, (1895), pp. 246-247.

which unscientific stance fit the stereotype of the Jewish mind.[95]

The following letter to the editor, which appeared in *The New York Times* in 1919, captures the spirit of the times, both the commonplaces of the time and the prevailing influence of racialist thought and nationalism in academic circles in the 1920's:

"Einstein and His Theory.
To the Editor of The New York Times:

On the first day of the Autumn meeting of the National Academy of Sciences (New Haven, Nov. 10) Einstein's relativity theory was discussed by two brilliant men from Massachusetts. Perhaps some of your readers may be interested in two remarks made by the speakers. The first speaker, a brilliant mathematician, came to the conclusion that Einstein's theory is mere philosophy, which he explained by the fact that Einstein is a Jew. The second speaker, whom, as he said humorously, physicists look upon as a

---

[95]. On the myth among Einstein supporters, *see:* D. E. Rowe, "'Jewish Mathematics' at Göttingen in the Era of Felix Klein", *Isis*, Volume 77, Number 3, (September, 1986), pp. 422-449; **and** "Science in Germany: The Intersection of Institutional and Intellectual Issues", *Osiris*, Series 2, Volume 5, (1989), pp. 186-213. *See also:* A. Fölsing, *Albert Einstein: A Biography*, Viking, New York, (1997), p. 203. On the myth among Einstein's adversaries, *see:* K. Hentschel, *Physics and National Socialism: An Anthology of Primary Sources*, Basel, Boston, Birkhäuser, (1996). On Einstein's anti-intuition / anti-induction stance, *see:* A. Moszkowski, *Einstein: The Searcher*, E. P. Dutton, New York, (1921), pp. 179-182. *See also:* A. Einstein, "Antrittsreden des Hrn. Einstein", *Sitzungsberichte der Königlich Preussischen Akademie der Wissenschaften zu Berlin*, (1914), pp. 739-742; English translation by A. Engel, "Inaugural Lecture of Mr. Einstein", *The Collected Papers of Albert Einstein*, Volume 6, Document 3, Princeton University Press, (1997), pp. 16-18; **and** "Motive des Forschers", *Zu Max Plancks sechzigstem Geburtstag. Ansprachen, gehalten am 26. April 1918 in der Deutschen Physikalischen Gesellschaft von E. Warburg, M. v. Laue, A. Sommerfeld und A. Einstein*, C. F. Müllersche Hofbuchhandlung, (1918), pp. 29-32; English translation by A. Engel, "Motives for Research", *The Collected Papers of Albert Einstein*, Volume 7, Document 7, Princeton University Press, (2002), pp. 41-45; **and** "Time, Space, and Gravitation / Theories of Principle", *London Times*, (28 November 1919), p. 13-14, English translation corrected: "Einstein on His Theory", *London Times*, (2 December 1919), p. 17; **and** "Induktion und Deduktion in der Physik", *Berliner Tageblatt*, Morning Edition, 4. Beiblatt, (25 December 1919), p. 1; English translation by A. Engel, "Induction and Deduction", *The Collected Papers of Albert Einstein*, Volume 7, Document 28, (2002), pp. 108-109; "Physics and Reality", *The Journal of the Franklin Institute*, Volume 221, Number 3, (March, 1936), reprinted: A. Einstein, *Ideas and Opinions*, Crown Publishers, Inc., New York, (1954), pp. 290-323, *see especially:* Section 4, "The Theory of Relativity", p. 307. Maurice Solovine quotes Einstein as supporting intuition, "Physics,' he said, 'is essentially an intuitive and concrete science. Mathematics is only a means for expressing the laws that govern phenomena.'" Quoted in, *Einstein: A Centenary Volume*, International Commission on Physics Education, U. S. A., (1979), p. 9. *The New York Times* reported on 3 April 1921 on the front page, "One of his traveling companions described him as an 'intuitive physicist' whose speculative imagination is so vast that it senses great natural laws long before the reasoning faculty grasps and defines them."

mathematician and mathematicians consider a physicist, had a good word to say for the theory of Einstein, namely, that he, the speaker, heard in Paris that Einstein, who was and still is a member of the Kaiser Wilhelm's Academy in Berlin, expressed a laudable wish that the Germans should be beaten. Accordingly, Einstein's theory may be unscientific because Einstein is a Jew; on the other hand, the theory ought to be correct because Einstein was an anti-Hun. Undoubtedly the mental rays of some of our scientists suffered a more or less perceptible deviation from the normal, brought about by the course of Mars in the last four years.

SAMUEL JAMES MELTZER.

New York, Nov. 11, 1919."

Judaism, Jewish tribalism, and Jewish racism gave the Jews a bad name, and many confused these ethnic, cultural and religious traits with "racial" traits. However, Jews were often able to intimidate most scholars out of publicly condemning these behaviors, and from publishing examples of them, and conducting research into their causes. The tribalism itself provided racist Jews with a means to quash most public condemnation of Jewish racism and Jewish tribalism. Edward Alsworth Ross, a Professor of Sociology at the University of Wisconsin, wrote in his book, *The Old World in the New: The Significance of Past and Present Immigration to the American People*, The Century Co., New York, (1914), pp. 143-167, Chapter 7, "The East European Hebrews",

"CHAPTER VII
THE EAST EUROPEAN HEBREWS

IN his defense of Flaccus, a Roman governor who had 'squeezed' his Jewish subjects, Cicero lowers his voice when he comes to speak of the Jews, for, as he explains to the judges, there are persons who might excite against him this numerous, clannish and powerful element. With much greater reason might an American lower his voice to-day in discussing two million Hebrew immigrants united by a strong race consciousness and already ably represented at every level of wealth, power, and influence in the United States.

At the time of the Revolution there were perhaps 700 Jewish families in the colonies. In 1826 the number of Jews in the United States was estimated at 6000; in 1840, at 15,000; in 1848, at 50,000. The immigration from Germany brought great numbers, and at the outbreak of the Civil War there were probably 150,000 Jews in this country. In 1888, after the first wave from Russia, they were estimated at 400,000. Since the beginning of 1899, one and one-third millions of Hebrews have settled in this country.

Easily one-fifth of the Hebrews in the world are with us, and the freshet shows no signs of subsidence. America is coming to be hailed as the 'promised land,' and Zionist dreams are yielding to the conviction that it will be much easier for the keen-witted Russian Jews to prosper here as a free component in a nation of a hundred millions than to grub a living out of the baked hillsides of Palestine. With Mr. Zangwill they exult that:

'America has ample room for all the six millions of the Pale; any one of her fifty states could absorb them. And next to being in a country of their own, there could be no better fate for them than to be together in a land of civil and religious liberty, of whose Constitution Christianity forms no part and where their collective votes would practically guarantee them against future persecution.'

Hence the endeavor of the Jews to control the immigration policy of the United States. Although theirs is but a seventh of our net immigration, they led the fight on the Immigration Commission's bill. The power of the million Jews in the metropolis lined up the Congressional delegation from New York in solid opposition to the literacy test. The systematic campaign in newspapers and magazines to break down all arguments for restriction and to calm nativist fears is waged by and for one race. Hebrew money is behind the National Liberal Immigration League and its numerous publications. From the paper before the commercial body or the scientific association to the heavy treatise produced with the aid of the Baron de Hirsch Fund, the literature that proves the blessings of immigration to all classes in America emanates from subtle Hebrew brains. In order to admit their brethren from the Pale the brightest of the Semites are keeping our doors open to the dullest of the Aryans!

Migrating as families the Hebrews from eastern Europe are pretty evenly divided between the sexes. Their literacy is 26 per cent., about the average. Artisans and professional men are rather numerous among them. They come from cities and settle in cities—half of them in New York. Centuries of enforced Ghetto life seem to have bred in them a herding instinct. No other physiques can so well withstand the toxins of urban congestion. Save the Italians, more Jews will crowd upon a given space than any other nationality. As they prosper they do not proportionately enlarge their quarters. Of Boston tenement-house Jews Dr. Bushee testifies: 'Their inborn love of money-making leads them to crowd into the smallest quarters. Families having very respectable bank accounts have been known to occupy cellar rooms where damp and cold streaked the walls.' 'There are actually streets in the West End where, while Jews are moving in, negro housewives are gathering up their skirts and seeking a more spotless environment.'

The first stream of Russo-Hebrew immigrants started flowing in 1882 in consequence of the reactionary policy of Alexander III. It contained many students and members of scholarly families, who stimulated intellectual activity among their fellows here and were leaders in radical thought. These idealists established newspapers in the Jewish-German Jargon and thus made Yiddish (*Jüdisch*) a literary language. The second stream reached us after 1890 and brought immigrants who were not steeped in modern ideas but held to Talmudic traditions and the learning of the rabbis. The more recent flow taps lower social strata and is prompted by economic motives. These later arrivals lack both the idealism of the first stream and the religious culture of the second.

Besides the Russian Jews we are receiving large numbers from Galicia, Hungary, and Roumania. The last are said to be of a high type, whereas the Galician Jews are the lowest. It is these whom Joseph Pennell, the illustrator, found to be 'people who, despite their poverty, never work with their hands; whose town... is but a hideous nightmare of dirt, disease and poverty' and its misery and ugliness 'the outcome of their own habits and way of life and not, as is usually supposed, forced upon them by Christian persecutors.'

OCCUPATIONS

The Hebrew immigrants rarely lay hand to basic production. In tilling the soil, in food growing, in extracting minerals, in building, construction and transportation they have little part. Sometimes they direct these operations, often they finance them, but even in direst poverty they contrive to avoid hard muscular labor. Under pressure the Jew takes to the pack as the Italian to the pick.

In the '80's numerous rural colonies of Hebrews were planted, but, despite much help from outside, all except the colonies near Vineland, New Jersey, utterly failed. In New York and New England there are more than a thousand Hebrew farmers, but most of them speculate in real estate, keep summer boarders, or depend on some side enterprise—peddling, cattle trading or junk buying—for a material part of their income. The Hebrew farmers, said to number in all 6000, maintain a federation and are provided with a farmers' journal. New colonies are launched at brief intervals, and Jewish city boys are being trained for country life. Still, not over one Hebrew family in a hundred is on the land and the rural trend is but a trickle compared with the huge flow.

Perhaps two-fifths of the Hebrew immigrants gain their living from garment-making. Naturally the greater part of the clothing and dry goods trade, the country over, is in their hands. They make eighty-five per cent. of the cigars and most of the domestic cigarettes. They purchase all but an insignificant part of the leaf tobacco from the farmers and sell it to the manufacturers. They are prominent in the retailing of spirits, and the Jewish distiller is almost as typical as the German brewer.

None can beat the Jew at a bargain, for through all the intricacies of commerce he can scent his profit. The peddler, junk dealer, or pawn broker is on the first rung of the ladder. The more capable rise in a few years to be theatrical managers, bankers or heads of department stores. Moreover great numbers are clerks and salesmen and thousands are municipal and building contractors. Many of the second generation enter the civil service and the professions. Already in several of the largest municipalities and in the Federal bureaus a large proportion of the positions are held by keen-witted Jews. Twenty years ago under the spoils system the Irish held most of the city jobs in New York. Now under the test system the Jews are driving them out. Among the school teachers of the city Jewesses outnumber the women of any other nationality. Owing to their aversion to 'blind-alley' occupations Jewish girls shun housework and crowd into the factories, while those who can get training become stenographers, bookkeepers, accountants and private secretaries. One-thirteenth of the students in our seventy-seven leading universities and colleges are of Hebrew parentage. The young Jews take eagerly to medicine and it is said that from seven hundred to nine hundred of the physicians in New York are of their race. More noticeable is the influx into dentistry and especially into pharmacy. Their trend into the legal profession has been pronounced, and of late there is a movement of Jewish students into engineering, agriculture and forestry.

MORALS

The Jewish immigrants cherish a pure, close-knit family life and the position of the woman in the home is one of dignity. More than any other immigrants they are ready to assume the support of distant needy relatives. They care for their own poor, and the spirit of coöperation among them is very noticeable. Their temper is sensitive and humane; very rarely is a Jew charged with any form of brutality. There is among them a fine *élite* which responds to the appeal of the ideal and is found in every kind of ameliorative work.

Nevertheless, fair-minded observers agree that certain bad qualities crop out all too often among these eastern Europeans. A school principal remarks that his Jewish pupils are more importunate to get a mark changed than his other pupils. A settlement warden who during the summer entertains hundreds of nursing slum mothers at a country 'home' says: 'The Jewish mothers are always asking for *something extra* over the regular kit we provide each guest for her stay.' 'The last thing the son of Jacob wants,' observes an eminent sociologist, 'is a square deal.' A veteran New York social worker cannot forgive the Ghetto its littering and defiling of the parks. 'Look at Tompkins Square,' he exclaimed hotly, 'and compare it with what it was twenty-five years ago amid a German population!' As for the caretakers of the parks their comment on this matter is unprintable. Genial settlement residents, who never tire of praising Italian or Greeks, testify that no other immigrants are so noisy, pushing and disdainful of the rights of others as the Hebrews. That the worst exploiters of these immigrants are sweaters, landlords, employers and 'white slavers' of their own race no one gainsays.

The authorities complain that the East European Hebrews feel no reverence for law as such and are willing to break any ordinance they find in their way. The fact that pleasure-loving Jewish business men spare Jewesses but pursue Gentile girls excites bitter comment. The insurance companies scan a Jewish fire risk more closely than any other. Credit men say the Jewish merchant is often 'slippery' and will 'fail' in order to get rid of his debts. For lying the immigrant has a very bad reputation. In the North End of Boston 'the readiness of the Jews to commit perjury has passed into a proverb.' Conscientious immigration officials become very sore over the incessant fire of false accusations to which they are subjected by the Jewish press and societies. United States senators complain that during the close of the struggle over the immigration bill they were overwhelmed with a torrent of crooked statistics and misrepresentations by the Hebrews fighting the literacy test.

Graver yet is the charge that these East European immigrants lower standards wherever they enter. In the boot and shoe trade some Hebrew jobbers who, after sending in an order to the manufacturer, find the market taking an unexpected downward turn, will reject a consignment on some pretext in order to evade a loss. Says Dr. Bushee: 'The shame of a variety of underhanded methods in trade not easily punishable by law must be laid at the door of a certain type of Jew.' It is charged that for personal gains the Jewish dealer wilfully disregards the customs of the trade and thereby throws trade ethics into confusion. Physicians and lawyers complain that their Jewish colleagues tend to break down the ethics of their professions. It is certain that Jews have commercialized the social evil, commercialized the theatre, and done much to commercialize the newspaper.

The Jewish leaders admit much truth in the impeachment. One accounts for the bad reputation of his race in the legal profession by pointing out that they entered the tricky branches of it, viz., commercial law and criminal law. Says a high minded lawyer: 'If the average American entered law as we have to, without money, connections or adequate professional education, he would be a shyster too.' Another observes that the sharp practice of the Russo-Jewish lawyer belongs to the earlier part of his career when he must succeed or starve. As he prospers his sense of responsibility grows. For example, some years ago the Bar Association of New York opposed the promotion of a certain Hebrew lawyer to the bench on the ground of his unprofessional practices. But this same lawyer made one of the best judges the city ever had, and when he retired he was banqueted by the Association.

The truth seems to be that the lower class of Jews of eastern Europe reach here moral cripples, their souls warped and dwarfed by iron circumstance. The experience of Russian repression has made them haters of government and corrupters of the police. Life amid a bigoted and hostile population has left them aloof and thick-skinned. A tribal spirit intensified by social isolation prompts them to rush to the rescue of the caught rascal of their own race. Pent within the Talmud and the Pale of Settlement, their interests have become few, and many of them have developed a monstrous and repulsive love of gain. When now, they use their Old World shove and wile and lie in a society like ours, as unprotected as a snail out of its shell, they rapidly push up into a position of prosperous parasitism, leaving scorn and curses in their wake.

Gradually, however, it dawns upon this twisted soul that here there is no need to be weazel or hedgehog. He finds himself in a new game, the rules of which are made by *all* the players. He himself is a part of the state that is weakened by his law-breaking, a member of the profession that is degraded by his sharp practices. So smirk and cringe and trick presently fall away from him, and he stands erect. This is why, in the same profession at the same time, those most active in breaking down standards are Jews and those most active in raising standards are Jews—of an earlier coming or a later generation. 'On the average,' says a Jewish leader, 'only the third generation feels perfectly at home in American society.' This explains the frequent statement that the Jews are 'the limit'—among the worst of the worst and among the best of the best.

CRIME

The Hebrew immigrants usually commit their crimes for gain; and among gainful crimes they lean to gambling, larceny, and the receiving of stolen goods rather than to the more daring crimes of robbery and burglary. The fewness of the Hebrews in prison has been used to spread the impression that they are uncommonly law-abiding. The fact is it is harder to catch and convict criminals of cunning than criminals of violence. The chief of police of any large city will bear emphatic testimony as to the trouble Hebrew lawbreakers cause him. Most alarming is the great increase of criminality among Jewish young men and the growth of prostitution among Jewish girls. Says a Jewish ex-assistant attorney-general of the United States in an address before the B'nai B'rith: 'Suddenly we find appearing in the life of the large cities the scarlet woman of Jewish birth.' 'In the women's night court of New York City and on gilded Broadway the majority of street walkers bear Jewish names.' 'This sudden break in Jewish morality was not natural. It was a product of cold, calculating, mercenary methods, devised and handled by men of Jewish birth.' Says the president of the Conference of American Rabbis: 'The Jewish world has been stirred from the center to circumference by the recent disclosures of the part Jews have played in the pursuance of the white slave traffic.' On May 14, 1911, a Yiddish paper in New York said, editorially:

'It is almost impossible to comprehend the indifference with which the large New York Jewish population hears and reads, day after day, about the thefts and murders that are perpetrated every day by Jewish gangs—real bands of robbers—and no one raises a voice of protest, and no demand is made for the protection of the reputation of the Jews of America and for the life and property of the Jewish citizens.'

'A few years ago when Commissioner Bingham came out with a statement about Jewish thieves, the Jews raised a cry of protest that reached the heavens. The main cry was that Bingham exaggerated and overestimated the number of Jewish criminals. But when we hear of the murders, hold-ups and burglaries committed in the Jewish section by Jewish criminals, we must, with heartache, justify Mr. Bingham.'

Two weeks later the same paper said: 'How much more will Jewish hearts bleed when the English press comes out with descriptions of gambling houses packed with Jewish gamblers, of the blind cigar stores where Jewish thieves and murderers are reared, of the gangs that work systematically and fasten like vampires upon the peaceable Jewish population, and of all the other nests of theft, robbery, murder, and lawlessness that have multiplied in our midst.'

This startling growth reflects the moral crisis through which many immigrants are passing. Enveloped in the husks of medievalism, the religion of many a Jew perishes in the American environment. The immigrant who loses his religion is worse than the religionless American because his early standards are dropped along with his faith. With his clear brain sharpened in the American school, the egoistic, conscienceless young Jew constitutes a menace. As a Jewish labor leader said to me, 'the non-morality of the young Jewish business men is fearful. Socialism inspires an ethics in the heart of the Jewish workingman, but there are many without either the old religion or the new. I am aghast at the consciencelessness of the *Luftproletariat* without feeling for place, community or nationality.'

RACE TRAITS

If the Hebrews are a race certainly one of their traits is *intellectuality*. In Boston the milk station nurse gets far more result from her explanations to Jewish mothers than from her talks to Irish or Italian mothers. The Jewish parent, however grasping, rarely exploits his children, for he appreciates how schooling will add to their earning capacity. The young Jews have the foresight to avoid 'blind alley' occupations. Between the years of fourteen and seventeen the Irish and Italian boys earn more than the Jewish lads; but after eighteen the Jewish boys will be earning more, for they have selected occupations in which you can work up. The Jew is the easiest man to sell life insurance to, for he catches the idea sooner than any other immigrant. As philanthropist he is the first to appreciate scientific charity. As voter he is the first to repudiate the political leader and rise to a broad outlook. As exploited worker he is the first to find his way to a theory of his hard lot, viz. capitalism. As employer he is quick to respond to the idea of 'welfare work.' The Jewish patrons of the libraries welcome guidance in their reading and they want always the best; in fiction, Dickens, Tolstoi, Zola; in philosophy, Darwin, Spencer, Haeckel. No other readers are so ready to tackle the heavy-weights in economics and sociology.

From many school principals comes the observation that their Jewish pupils are either very bright or distinctly dull. Among the Russo-Jewish children many fall behind but some distinguish themselves in their studies. The proportion of backward pupils is about the average for school children of non-English-speaking parentage; but the brilliant pupils indicate the presence in Hebrew immigration of a gifted element which scarcely shows itself in other streams of immigration. Teachers report that their Jewish pupils 'seem to have hungry minds.' They 'grasp information as they do everything else, recognizing it as the requisite for success.' Says a principal: 'Their progress in studies is simply another manifestation of the acquisitiveness of the race.' Another thinks their school successes are won more by intense application than by natural superiority, and judges Irish pupils would do still better if only they would work as many hours.

The Jewish gift for mathematics and chess is well known. They have great imagination, but it is the 'combinative' imagination rather than the free poetic fancy of the Celt. They analyze out the factors of a process and mentally put them together in new ways. Their talent for anticipating the course of the market, making fresh combinations in business, diagnosing diseases, and suggesting scientific hypotheses is not questioned. On the other hand, an eminent savant thinks the best Jewish minds are not strong in generalization and deems them clever, acute and industrious rather than able in the highest sense. On the whole, the Russo-Jewish immigration is richer in gray matter than any other recent stream, and it may be richer than any other large inflow since the colonial era.

Perhaps *abstractness* is another trait of the Jewish mind. To the Hebrew things present themselves not softened by an atmosphere of sentiment, but with the sharp outlines of that desert landscape in which his ancestors wandered. As farmer he is slovenly and does not root in the soil like the German. As poet he shows little feeling for nature. Unlike the German artisan who becomes fond of what he creates, the Jew does not love the concrete for its own sake. What he cares for is the *value* in it. Hence he is rarely a good artisan, and perhaps the reason why he makes his craft a mere stepping-stone to business is that he does not relish his work. The Jew shines in literature, music and acting—the arts of expression—but not often is he an artist in the manipulation of materials. In theology, law and diplomacy—which involve the abstract—the Jewish mind has distinguished itself more than in technology or the study of nature.

The Jew has *little feeling for the particular*. He cares little for pets. He loves man rather than men, and from Isaiah to Karl Marx he holds the record in projects of social amelioration. The Jew loves without romance and fights without hatred. He is loyal to his purposes rather than to persons. He finds general principles for whatever he wants to do. As circumstances change he will make up with his worst enemy or part company with his closest ally. Hence his wonderful adaptability. Flexible and rational the Jewish mind cannot be bound by conventions. The good will of a Southern gentleman takes set forms such as courtesy and attentions, while the kindly Jew is ready with any form of help that may be needed. So the South looked askance at the Jews as 'no gentlemen.' Nor have the Irish with their strong personal loyalty or hostility liked the Jews. On the other hand the Yankees have for the Jews a cousinly feeling. Puritanism was a kind of Hebraism and throve most in the parts of England where, centuries before, the Jews had been thickest. With his rationalism, his shrewdness, his inquisitiveness and acquisitiveness, the Yankee can meet the Jew on his own ground.

Like all races that survive the sepsis of civilization, the Hebrews show great *tenacity of purpose*. Their constancy has worn out their persecutors and won them the epithet of 'stiff-necked.' In their religious ideas our Jewish immigrants are so stubborn that the Protestant churches despair of making proselytes among them. The sky-rocket careers leading from the peddler's pack to the banker's desk or the professor's chair testify to rare singleness of purpose. Whatever his goal—money, scholarship, or recognition—the true Israelite never loses sight of it, cannot be distracted, presses steadily on, and in the end masters circumstance instead of being dominated by it. As strikers the Jewish wage earners will starve rather than yield. The Jewish reader in the libraries sticks indomitably to the course of reading he has entered on. No other policy holder is so reliable as the Jew in keeping up his premiums. The Jewish canvasser, bill collector, insurance solicitor, or commercial traveler takes no rebuff, returns brazenly again and again, and will risk being kicked down stairs rather than lose his man. During the Civil War General Grant wrote to the war department regarding the Jewish cotton traders who pressed into the South with the northern armies: 'I have instructed the commanding officer to refuse all permits to Jews to come South, and I have frequently had them expelled from the department, but they come in with their carpet sacks in spite of all that can be done to prevent it.' Charity agents say that although their Hebrew cases are few, they cost them more than other cases in the end because of the unblushing persistence of the applicant. Some chiefs of police will not tolerate the Hebrew prostitute in their city because they find it impossible to subject her to any regulations.

THE RACE LINE

In New York the line is drawn against the Jews in hotels, resorts, clubs, and private schools, and constantly this line hardens and extends. They cry 'Bigotry' but bigotry has little or nothing to do with it. What is disliked in the Jews is not their religion but certain ways and manners. Moreover, the Gentile resents being obliged to engage in a humiliating and undignified scramble in order to keep his trade of his clients against the Jewish invader. The line is not yet rigid, for the genial editor of *Vorwaerts*, Mr. Abram Cahan, tells me that he and his literary brethren from the Pale have never encountered Anti-Semitism in the Americans they meet. Not the socialist Jews but the vulgar upstart parvenus are made to feel the discrimination.

This cruel prejudice—for all lump condemnations are cruel—is no importation, no hang-over from the past. It appears to spring out of contemporary experience and is invading circle after circle of broad-minded. People who give their lives to befriending immigrants shake their heads over the Galician Hebrews. It is astonishing how much of the sympathy that twenty years ago went out to the fugitives from Russian massacres has turned sour. Through fear of retaliation little criticism gets into print; in the open the Philo-semites have it all their way. The situation is: Honey above, gall beneath. If the Czar, by keeping up the pressure which has already rid him of two million undesired subjects, should succeed in driving the bulk of his six million Jews to the United States, we shall see the rise of the Jewish question here, perhaps riots and anti-Jewish legislation. No doubt thirty or forty thousand Hebrews from eastern Europe might be absorbed by this country each year without any marked growth of race prejudice; but when they come in two or three or even four times as fast, the lump outgrows the leaven, and there will be trouble.

America is probably the strongest solvent Jewish separatism has ever encountered. It is not only that here the Jew finds himself a free man and a citizen. That has occurred before, without causing the Jew to merge into the general population. It is that here more than anywhere else in the world *the future is expected to be in all respects better than the past.* No civilized people ever so belittled the past in the face of the future as we do. This is why tradition withers and dies in our air; and the dogma that the Jews are a 'peculiar people' and must shun intermarriage with the Gentiles is only a tradition. The Jewish dietary laws are rapidly going. In New York only one-forth of the two hundred thousand Jewish workmen keep their Sabbath and only one-fifth of the Jews belong to the synagogue. The neglect of the synagogue is as marked as the falling away of non-Jews from the church. Mixed marriages, although by no means numerous in the centers, are on the increase, and in 1909 the Central Conference of Jewish Rabbis resolved that such marriages 'are contrary to the tradition of the Jewish religion and should therefore be discouraged by the American Rabbinate.' Certainly every mixed marriage is, as one rabbi puts it, 'a nail in the coffin of Judaism,' and free mixing would in time end the Jews as a distinct ethnic strain.

The 'hard shell' leaders are urging the Jews in America to cherish their distinctive traditions and to refrain from mingling their blood with Gentiles. But the liberal and radical leaders insist that in this new, ultra-modern environment nothing is gained by holding the Jews within the wall of Orthodox Judaism. As a prominent Hebrew labor leader said to me: 'By blending with the American the Jew will gain in physique, and this with its attendant participation in normal labor, sports, athletics, outdoor life, and the like, will lessen the hyper-sensibility and the sensuality of the Jew and make him less vain, unscrupulous and pleasure-loving.'

It is too soon yet to foretell whether or not this vast and growing body of Jews from eastern Europe is to melt and disappear in the American population just as numbers of Portuguese, Dutch, English, and French Jews in our early days became blent with the rest of the people. In any case the immigrant Jews are being assimilated outwardly. The long coat, side curls, beard and fringes, the 'Wandering Jew' figure, the furtive manner, the stoop, the hunted look, and the martyr air disappear as if by magic after a brief taste of American life. It would seem as if the experience of Russia and America in assimilating the Jews is happily illustrated by the old story of the rivalry of the wind and the sun in trying to strip the traveler of his cloak."

Tragically, Einstein's racism and tribalism provoked a "racial" debate over his personality and the theory of relativity. Counterattacks predictably followed Einstein's ethnic slurs and Einstein's reckless and racist defamations of his legitimate critics. For example,

<h3 style="text-align:center">"NOTES BRÈVES</h3>

Einstein, plagiaire.
  Le Juif d'Allemagne Einstein est un plagiaire. La presse américaine en avait déjà (v. n° 225) fourni la preuve. Le *Dearb. Independent*, 25.3, y revient avec de nouveaux documents.
  Il montre les « découvertes » du Juif suivant pas à pas, et ses publications suivant volume par volume, les découvertes et les publications d'Arvid Reuterdahl, Américain d'origine suédoise, doyen de l'Ecole

d'architecture et de mécanique au collège Saint Thomas (St. Paul, Minn.). Le *Raum-Zeit-Kontinuum*, les *Raum-Zeit-Funktionen* et *Raum-Zeit-Koordinaten* du Juif ne sont que des démarquages du *Space-Time Potential* de l'Américain, grossièrement camouflés.

Les Juifs ne sont jamais que des plagiaires. Mais la stupidité des *goyim* leur permet de s'introduire dans la peau des hommes de génie à la manière de Chéri-Bibi. Et la presse de tous les pays, moyennant une poignée de dollars ou de *crasseux*, assassine de silence les vrais savants pour revêtir de leur gloire le gorille du Ghetto."[96]

THE DEARBORN INDEPENDENT published an article on 30 July 1921, on page 14, (American Jews successfully organized many large fund raising drives, as represented in the pages of *The New York Times*, especially during the First World War) which ridicules Einstein's anti-American interview upon his return to Europe:

"*Relatively Unimportant, Extremely Typical*

ALBERT EINSTEIN, who maintained a pose of dignified silence in the face of his scientific accusers while in the United States, has broken into most undignified speech immediately upon his return to Europe.

Knowledge of what he is and the traditional ill-manneredness of which he is an heir, this exhibition of boorishness was not unexpected.

Disgust with Einstein is somewhat an old thrill, because his plagiarism is so manifest and his fame is so directly the result of the circus-advertising instinct of his race. But a new emotion divides it now: What about those nose-led Americans who, in obedience to the swarthy New York ruling race, bowed down and worshipped Einstein and chanted loudest in the chorus of his praise?

Their position is most humiliated. And rightly so. Every white man, who bows down to the swarthy ruling race of New York and elsewhere, gets his nose rubbed into it sooner or later. It is the traditional repayment which that race—and all inferior races—renders when a superior race makes a fool of itself.

Mr. Einstein was gloriously received in the United States. Even the cold photographs retain the glow of passionate occasions. Literally over 150,000 persons by comptometer count, swarmed round him on his arrival. He had not done anything for science, for the easement of human pain nor for the solution of life's pressing problems, yet he was received as a royalty of the realm of reason, while others who have found the way to healing or achievement for the common man have been allowed to enter and leave New York unheeded. Mr. Einstein, by the way, *left* New York unheeded—there were half a dozen persons on the piers —which should, perhaps, be borne

---

96. *La Vieille* (Paris), Number 272, (20 April 1922), p. 15.

in mind.

Mr. Einstein was given the freedom of New York, under protest, and was refused the freedom of Boston, but the universities received him gladly and decked him with their doctorates. The press, in response to swarthy local committees, shouted itself hoarse. Clothing lofts poured out their Red intellectuals by the thousand, and taking it all in all the publicity manager of Mr. Einstein's stunt did a good job—until—scientists began to ask Einstein questions.

The only recorded answer which Einstein made to any but adulatory remarks while in the United States, was, 'See my secretary.' American collegians and scientists, philosophers and literary men besought him; others with the 'goods on him' openly challenged him; but surrounded by a swarthy ring that made everybody believe that a slight to Einstein was equal to sacrilege against the Holy of Holies on Mt. Zion, he maintained his silence and, supposedly, his dignity. That last, however, is not known. He left the United States rather unexpectedly.

THE DEARBORN INDEPENDENT is glad to say that it was one—perhaps not the only one—of the papers that were not taken in by the Einstein publicity managers. It is glad to remember also that it gave much-needed space to a scientific critic of Einstein's theory, who had been refused space elsewhere. A roster of the publications which were afraid of the swarthy crowd around Einstein gives much food for publication.

Therefore, perhaps, THE DEARBORN INDEPENDENT is not so embarrassed as are the Einstein devotees by the attack upon America which the professor has made. Not so embarrassed as, say, the *Scientific American.*

Mr. Einstein's charges are as follows: (1) That America is too exaggerated in its enthusiasm. 'This exaggerated enthusiasm for me and my work struck me as being a genuinely and peculiarly American phenomenon'; (2) that Americans are bored; (3) that America suffers from poverty in intellectual things; (4) that most American men think of nothing but work; (5) that the rest of the men are mere lap dogs for indolent women; (6) 'that women dominate the entire life of America' ; (7) that our excitement over the theory of relativity was 'comic'; (8) that the only real American scientist lives in Chicago and is a Jew!

As complete a slap in the face as the swarthy tribe has ever handed a white people!

Mr. Arthur Brisbane, pen-sentinel of the tribe, who held Mr. Einstein up as an example too lofty for Americans to emulate, yet to be worshipfully gazed upon as a distant and unattainable star, was plainly up against it.

Many people think that Mr. Brisbane is himself distantly connected by racial ties with people of Mr. Einstein's type, but others are assured that he is not. It is unfortunate, if he is not, because his admiration of the tribe is so great that assertion of his belonging to it would not be construed by him as an insult, but rather as a high compliment. Some people have commented on the name 'Brisbane,' saying that its Hebrew form is Brith Ben, or 'son of the covenant.' The name Einstein is not as Hebrew as is Brisbane;

Einstein is German for 'one stone.'

It was rather hard, therefore, after standing sponsor for Einstein in all the Hearst papers and before the American public, to have Einstein hurl his insult across the sea. What did Mr. Brisbane do then, quoth the little bird? Did he turn to his ever-present Hebrew secretary for inspiration as he often has done before? History may never know.

But it is certain that something stirred within Mr. Brisbane's breast, something American, something angry and tipped with truth; and there hurtled through his mobile mind with the clarifying turbulence and light of an electric storm, this luminous thought: 'No wonder Einstein thinks thus of America; *all that Einstein saw of America was the Jews*!' (Wild shrieks of 'pogrom! pogrom!' ringing through the darker recesses of Brisbane's brain!) 'That's it—that's how to explain it; he didn't see America at all—he just saw Jews.'

Lest the reader should think that statement too great a strain on his credulity, we hasten to offer, what we always have on hand in these matters, the evidence. Behold it!

Today

Einstein's Views.
What of the 5,000,000?
Valuable 'Devil's Finger.'
Hopeful Mr. Herrick.
—By ARTHUR BRISBANE—
Copyright. 1921.

Prof. Einstein, of the relativity theory, returned home, says:

First, he is amused by the wild enthusiasm of the entire American nation in greeting him. What Prof. Einstein saw, without knowing it, was the extremely enthusiastic welcome of his co-religionists. Our citizens of Jewish blood delight at another demonstration, in Einstein's person, of the ability of their race. It was Jewish enthusiasm that the professor witnessed, and there is no greater enthusiasm than that. It is a good explanation of the whole Einstein criticism.

It is a good explanation of the whole Einstein criticism.

Moreover, it is true. Outside an occasional university president and Senate, the white mayors and governors en route, once the President of the United States, the professor did not meet many Americans. He did not greatly want to meet Americans. Americans are inclined to sit in judgment first, and that spelled danger.

He has simply made the same error which others have made, in not properly distinguishing between racial strains of blood.

Einstein's charge about the comic enthusiasm is absolutely true; scores of photographs confirm the facts. But who furnished the enthusiasm? A little more candor on Einstein's part would have made that clear. As a long, long benefit of the doubt, it may be agreed that perhaps Mr. Einstein may have mentioned his co-nationalists in this respect, and it may have been changed to 'Americans.' But probably not. If it had been changed to 'Americans' from an original other, it would have made it rather difficult for certain newspapers who bow the knee to the tribe; especially in view of the tact that 75 per cent of the advertising in United States newspapers is paid for by the tribe. Jack Lait once said, 'The department store is the bulwark of free speech!' And he ought to know.

The tribe did make fools of themselves over Einstein. They made a fool of him, too. Now he makes a fool of both by describing the tribal defects and ascribing them to 'Americans.' What a plot for screaming farce by Morry Gest!

Mr. Brisbane is right. He is wrong on nearly everything else he tries to say on the related subject, but he is right in his analysis of Mr. Einstein's sources: Mr. Einstein's opinion of America is the result of his having seen only Jews. Some foreign governments are suffering from the same mis-view of us.

The Brisbane explanation of the Einstein theory of Americans may be applied all down the line. 'The intellectual poverty' he noticed is also due to the fact that all he saw of American intellect was Jewish. The tribe does not originate ideas; it grabs them and exploits them. The tribe is not at home in the study, but on the stage. In art it simply steals ideas and elaborates them. In music it performs, but does not create. In law, it manipulates, but does not clarify great principles. In politics it is opportunist. Intellectual bankruptcy may coexist with a very pert knowledge of what the schools teach, and the tribe is quite expert at possessing itself of that—all white man's knowledge, by the way.

And so on through the charges. The Brisbane explanation is hereby unanimously adopted: Prof. Einstein thinks what he does and says what he did because what he saw was not America but Jews. He couldn't see America for the swarthy swarm that smothered him. And what is worse, hundreds of thousands—millions—of that swam have never seen America either, and never will, for the same reason.

The Jews are strangely silent on the criticism. Rabbi Stephen S. Wise— in the Yiddish papers they spell it correctly, Weisz—refuses to comment. The tribal elders of the New York Board of Aldermen who fought for the freedom of the city for Mr. Einstein just as boldly as they fight for legally imposed social equality where they are not wanted, don't like to discuss it either.

Prof. Rautenstrauch is rather gentle in his comment 'His visit to this country was of too brief duration and his contacts while here were too narrow.' Second half of answer is right. It doesn't take long to know Americans: 10 minutes is the average time for striking up a real human kind of acquaintance here, and Einstein was with us weeks and weeks— but—

'his contacts while here were too narrow.' For reference, Mr. Brisbane's comment again.

Einstein's tribalists cannot answer; it is an outbreak of bad manners, rank contemptuousness and untruth which is indefensible. Einstein never was a great scientist; now we know he is not even an ordinarily passable individual.

What puzzles the Washington *Post* is the reason for Einstein staying on in the country after he had found what a detestable place it was; and why he went on accumulating university degrees and other academic honors when he had formed so low an opinion of our institutions, and when the only scholar he could find in the United States was a Jew out in Chicago.

It's a somewhat honest wail the *Post* puts forth:

'Why did Prof. Einstein not discover after a few days' stay in America his impressions and then make a speedy return to his haven? Why did he accept the attentions and awards from municipalities and educational institutions if he questioned their sincerity?'

The answer is simple, but the *Post* doesn't give it.

The answer is given in 'blank' verse by a poet on this page.

Things One Cannot Print
(Obviously done in blank verse.)
in writing for the Editors
    Telling funny news,
Omit from all you chance to say
    Mention of the South Americans.

Whene'er you feel the writing urge
    Why write whate'er you choose,
Except you must not write at all
    About our friends the Italians.

If verses fill your soul with song
    Turn fondly to your Muse,
But do not let her lead you far;
    Sing not about the British.

If funny stories fill your head
    And you would but amuse,
Why keep them laughing by all means,
    But not about the Greeks.

Fill up the page with anecdotes,
    Tell anything that's new
But let no story that you tell
    Poke fun at any Syrian.

You'll only tire your massive brain.
 Your time you'll surely lose,
If you submit to Editors
 Stories on the French.

I'm greatly hampered in my work,
 My stuff they all refuse,
Because the stories that I tell
 Are often on the Swiss.

I should be paid for what I write,
 My lawyer says to sue,
And that is what so puzzles me
 For he, too, is a Belgian.
 —New York Herald, July 3, 1921.

    LATER BULLETIN—Word comes from Amsterdam that Prof. Einstein did not say it. He is still dazed by the good will of America, still has the glory of America in his eyes, and so on. The difference is that the first story came under the names of responsible correspondents and through the channels of responsible newspapers; while the second story comes orphaned— probably from the Jewish Telegraph Agency, which is the associated press of international Jewry. The agency has not been functioning very much of late, the principal reason being that it cannot send long and harrowing dispatches about 'pogroms' and be believed any more, because there are too many neutral observers in the 'pogrom zone.' There are no pogroms [*see:* "Pogroms in Poland", *The New York Times*, (23 May 1919), p. 10; where the report claims that Germans may have fabricated myths, and spread rumors of Jewish pogroms in order to vilify their enemies.—CJB], but there is this: There is the sale for money of goods bought by the charity of the American people, mostly the American church people. The agency, however, doesn't deal in facts of that kind.
    It is rather singular that none of the tribe's dailies doubted the first Einstein report. They knew how delightfully and characteristically racial it was, how perfectly natural. They took it for granted.
    However, the Einstein matter is a mere speck on the racing river of events yet it shows something of the tendency of the river. No one has a license to feel badly over it, except the scientific publications that didn't have the intestinal integrity to challenge the man in the name of science; the universities that did not dare keep him off their list of honors; the society people who fêted the rather mangy lion; and the plain and more honest members of the tribe who thought Einstein might generously reflect a little glory on them. He hasn't."

Einstein apparently did not respond directly to many of the genuinely race-based attacks made against him, such as those above, which were made in no uncertain terms. He preferred to mischaracterize some of the scientific objections to his theories, and the legitimate concerns raised about his plagiarism, as if they were "anti-Semitism" *per se*. When Einstein arrived at America's shores, *The New York Times* emphasized the fact that theory of relativity was widely criticized,

> "The man was Dr. Albert Einstein, propounder of the much-debated theory of relativity that has given the world a new conception of space, and time and the size of the universe."[97]

Before Einstein stepped off the ship, he lied and "played the race card" in order to smear anyone who would dare to criticize him in America,

> "Professor Einstein was reluctant to talk about relativity, but when he did speak he said most of the opposition to his theories was the result of strong anti-Semitic feeling."[98]

The article continued,

> "He was asked about those who oppose his theory, and said:
> 'No man of culture or knowledge has any animosity toward my theories. Even the physicists opposed to the theory are animated by political motives.'
> When asked what he meant, he said he referred to anti-Semitic feeling. He would not elaborate on this subject, but said the attacks in Berlin were entirely anti-Semitic."[99]

Among those highly knowledgeable and cultured physicists and philosophers who actively opposed relativity theory, as it was expressed by Einstein, many of whom were Jewish—who, according to Einstein's assertions, must have been uncultured, ignorant anti-Semites—we find Hendrik Antoon Lorentz, Max Abraham, Alfred North Whitehead, Ernst Mach, Albert Abraham Michelson, Friedrich Adler, Henri Bergson, Oskar Kraus, Melchior Palágyi, [etc. etc. etc.]. Clearly, Einstein lied about a very serious matter, and, what is worse, Einstein was himself a racist instigator and a political agitator; and, therefore, a hypocrite and a deliberate inciter of "racial" discord.

Einstein and his friends' (especially Max von Laue's) wanton and reckless charges of anti-Semitism only served to intensify and provoke it, as evinced above, which was their goal. Einstein expressed the bizarre belief commonly

---

[97]. "Prof. Einstein Here, Explains Relativity", *The New York Times*, (3 April 1921), pp. 1, 13, at 1.
[98]. "Prof. Einstein Here, Explains Relativity", *The New York Times*, (3 April 1921), pp. 1, 13, at 1.
[99]. "Prof. Einstein Here, Explains Relativity", *The New York Times*, (3 April 1921), pp. 1, 13.

held by racist Jews, that anti-Semitism was a positive thing because it kept Jews segregated from Gentiles. Einstein argued that Jews should not mix with Gentiles, due to "racial" differences. Responding to the truly race-based attacks would have tended to discredit anti-Semitism, and with it racist political Zionism. However, Einstein and Max von Laue's tactic of mischaracterizing legitimate arguments about science and priorities issues as if "anti-Semitism" only inspired anti-Jewish sentiment—much to their delight.

Einstein was obviously scarred by childhood traumas.[100] Being a coward by nature, he hid behind reckless defamations in order to avoid legitimate criticism.

Hubert Goenner observed,

"Nevertheless, Kleinert (1979, 501-6) and Elton (1986, 95) documented that [*Albert Einstein*] *was first* in referring to anti-Semitism in public, well before any of his adversaries in the campaign against him [*Footnote:* Einstein soon regretted his statement.] [... .]"[101]

Einstein's accusation that no one but an anti-Semite would disagree with him was a smear against dissent heard round the world—obviously meant to stifle the debate. It was an open threat to anyone who would challenge him on the facts—anyone who dared to tell the truth and expose him. These smears were accompanied by alarmist (and shifting) misrepresentations of the audience's actions, and the proceedings, at the Berlin Philharmonic when Paul Weyland and Ernst Gehrcke lectured against the theory of relativity. This had a chilling effect on the debate over the facts, with some fearing to challenge Einstein, knowing full well that they would be accused of "anti-Semitism" in the international press no matter what they actually did, said or thought. Einstein's tactics served to provoke and intensify extant anti-Jewish feelings and to numb the ears of the world when the truly rabid and murderous NSDAP rose to power.

As was his habit, Einstein used alarmist tactics and sought to alienate anyone, including Jews, who dared to disagree with him. Most Jews felt a deep love for, and loyalty to, their present nationality, and wanted nothing of what they thought of as Einstein's archaic Zionist bigotry.[102] Einstein was a simplistic person and he sought to narrowly define people of diverse backgrounds and

---

100. Letter from A. Einstein to P. Nathan of 3 April 1920, *The Collected Papers of Albert Einstein*, Volume 9, Document 366, Princeton University Press, (2004), p. 492. Also: *The Collected Papers of Albert Einstein*, Volume 1, Princeton University Press, (1987), p. *lx*, note 44. J. Stachel, "Einstein's Jewish Identity", *Einstein from 'B' to 'Z'*, Birkhäuser, Boston, Basel, Berlin, (2002), pp. 57-83, at 69. *See also:* P. A. Bucky, Einstein, and A. G. Weakland, *The Private Albert Einstein*, Andrews and McMeel, Kansas City, (1992), pp. 83, 86.

101. H. Goenner, "The Reaction to Relativity Theory. I: The Anti-Einstein Campaign in Germany in 1920", *Science in Context*, Volume 6, Number 1, (1993), pp. 107-133, at 112. "Kleinert (1979, 501-6) and Elton (1986, 95)" refers to: A. Kleinert, in H. Nelkowski, et. al. Editors, *Einstein Symposium Berlin 1979*, pp. 501-506; **and** L. Elton, "Einstein, General Relativity and the German Press", *Isis*, Volume 79, (1986), p. 95.

102. P. Michelmore, *Einstein: Profile of the Man*, Dodd, Mead, (1962), p. 87.

beliefs,[103] and he sought to intimidate everyone into following his course, by degrading Jews who sought to assimilate and intimating that they were somehow traitors to a religious cause—a religious cause which he, himself, truly found ridiculous. Einstein stated,

> "I am neither a German citizen, nor is there in me anything that can be described as 'Jewish faith.' But I am happy to belong to the Jewish people, even though I don't regard them as the Chosen People. Why don't we just let the Goy keep his anti-Semitism, while we preserve our love for the likes of us?"[104]

Einstein was reciting the Herzlian brand of racist Zionism he had embraced as a route to personal fame. Theodor Herzl revealed his core beliefs when recalling a conversation he had had with racist Zionist Max Nordau:

> "Never before had I been in such perfect tune with Nordau. [***] This has nothing to do with religion. He even said that there was no such thing as a Jewish dogma. But we are of one race. [***] 'The Jews,' he says, 'will be compelled by anti-Semitism to destroy among all peoples the idea of a fatherland.' Or, I secretly thought to myself, to create a fatherland of their own."[105]

Herzl and Nordau's plans were carried out. The Zionists created the Nazi Party and funded it, in order to discredit Gentile government, and in order to segregate Jews and force them into Palestine against their will. Though racist Jews were behind the Nazis and guided their destiny, these same racist Jews then criticized Gentiles for the atrocities these same racist Jews had caused the Nazis to commit against non-Zionist Jews. Racist Jewish poseurs to this day claim the moral high ground over European Gentiles based on the actions the Nazis took at the behest of racist Jewish Zionist financiers.

Jews have always scapegoated their victims. Messianic Jewish religious mythology teaches Jews to commit genocide against all other peoples. Jews are taught to ruin the religions, cultures, nations and lives of the rest of humanity. They often hide their evil acts by scapegoating their victims and blaming them

---

103. H. Dukas and B. Hoffmann, *Albert Einstein: The Human Side*, Princeton University Press, (1979), pp. 55-56.
104. A. Einstein quoted in A. Fölsing, English translation by E. Osers, *Albert Einstein, a Biography*, Viking, New York, (1997), p. 494; which cites speech to the *Central-Verein Deutscher Staatsbürger Jüdischen Glaubens*, in Berlin on 5 April 1920, in D. Reichenstein, *Albert Einstein. Sein Lebensbild und seine Weltanschauung*, Berlin, (1932). This letter from Einstein to the Central Association of German Citizens of the Jewish Faith of 5 April 1920 is reproduced in *The Collected Papers of Albert Einstein*, Volume 9, Document 368, Princeton University Press, (2004).
105. T. Herzl, English translation by H. Zohn, R. Patai, Editor, *The Complete Diaries of Theodor Herzl*, Volume 1, Herzl Press, New York, (1960), p. 196.

for the atrocities Jews have committed and are committing today.

The Jewish mythology of the scapegoat is found in the Hebrew Bible at *Leviticus* 16 and in the Jewish Talmud in the book of *Yoma*. On the day of atonement, Jews used to select two goats, one to be sacrificed to God in the Temple, the other to be sent into the wilderness to Azazel. All the sins of the Jews were placed on the goat which was sent into the wilderness and in this way Jews unburdened themselves of the guilt of their sins. About the time Jesus Christ was said to have been sacrificed to atone for the sins of mankind, the rituals of atonement in the Jewish Temple began to fail, as was predicted in *Daniel* 9:24, 27. Forty years later, the Romans destroyed the Jewish Temple. Some Christians believe that Jewish ritual sacrifices ended with the sacrifice of Jesus, and that the Jews ought not to rebuild a Temple and must not resume animal sacrifices, for such sacrifices would constitute a blasphemy against the sacrifice on the cross.

In recent times, organized Jewry have scapegoated the Czar of Russia for the crimes against the Russian People Jews have deliberately committed. In this way, criminal Jews were able to drive a wedge between the Russian People and the government which was desperately trying to protect the Russian People from organized Jewry which sought to destroy them. Racist Jews organized strikes and carried out pogroms against Jews, and then Jews defamed the Czar in the international press by falsely blaming him for the misery Jews were deliberately causing other Jews and Gentiles. Jews promoted war between the Japanese and the Russians, and financed the Japanese while blocking Russia's access to funds.[106] Jews heavily financed violent and destructive revolutionaries to create

---

106. R. Smethurst, "Takahashi Korekiyo, the Rothschilds and the Russo-Japanese War, 1904-1907", *The Rothschild Archive: Review of the Year April 2005 to March 2006*, London, (2007), pp. 20-25.

<http://www.rothschildarchive.org/ib/articles/AR2006.pdf>

See also: L. Wolf, "The Zionist Peril. (Letters to the Editor)", *The London Times*, (8 September 1903), p. 5; **and** "Is Russia Solvent?", *The London Times*, (11 March 1905), p. 10; **and** "Is Russia Solvent?", *The London Times*, (14 March 1905), p. 3; **and** "Russia's Gold Reserve. (Letters to the Editor)", *The London Times*, (24 March 1905), p. 7; **and** "The Russian Gold Reserve. (Letters to the Editor)", *The London Times*, (28 March 1905), p. 8; **and** "Russia's Gold Reserve. (Letters to the Editor)", *The London Times*, (5 April 1905), p. 10; **and** "The Russian Gold Reserve. (Letters to the Editor)", *The London Times*, (1 May 1905), p. 11; **and** "The Russian Gold Reserve. (Letters to the Editor)", *The London Times*, (27 May 1905), p. 7; **and** "The Bankers And The Peace Negotiations. (Letters to the Editor)", *The London Times*, (18 August 1905), p. 5; **and** "The Massacres Of Jews. (Letters to the Editor)", *The London Times*, (7 December 1905), p. 12; **and** "The Reign Of Terror In Odessa. (Letters to the Editor)", *The London Times*, (21 December 1905), p. 7; **and** "Is Russia Solvent? (Letters to the Editor)", *The London Times*, (6 October 1906), p. 12. ***See also:*** I. Zangwill , *The Problem of the Jewish Race*, Judean Publishing Company, New York, (1914), p. 14; which was first published as an article, "The Jewish Race", *The Independent*, Volume 71, Number 3271, (10 August 1911), pp. 288-295, at 292. ***See also:*** "Jacob H. Schiff Rejoices", *The New York Times*, (18 March

discontent and unrest among the Russian People. Jews made the Russian People suffer and blamed the Czar for the harm Jews had done. In this way, Jews caused the Russian People to destroy themselves and their government and hand over all their independence and power to organized Jewry. Jews were then able to carry out their ancient plans as they mass murdered tens of millions of Slavic Christians.

Zionist Jews then placed Adolf Hitler into power to further the spread of Communism and to drive reluctant Jews to Palestine against their will. Jews then blamed German Gentiles for the harm to Jews and other Europeans, which these same Jews had deliberately caused.

Zionist Jews placed George Bush into power in America. Jews had their agent George Bush bring America into perpetual war and perpetual debt. These same Jews now blame Bush and Gentile government in the United States for the harm Jews are deliberately causing Americans and the non-Jewish peoples of the Middle East. Jews even blame the United States for Israel's unprovoked aggression and genocide against Lebanon. In this way, Jews not only unburden themselves from their guilt, they discredit Gentile governments and bring Gentile governments into unnecessary war with each other, all of which furthers the ambitions of ancient Jewish Messianic goals.

Jews are presently also scapegoating Hezbullah and the Palestinian People for the barbaric Jewish genocide of the Palestinian and Lebanese Peoples. Israelis followed the same model to create a pretext for the Jewish mass murder of helpless Arabs. Israelis sent Jewish soldiers into foreign territory and then pretended that the capture of these soldiers constituted a casus belli for the genocidal wars Jews have been planning for 2,500 as found in the book of Ezekiel. Jews overrate the strength of Hezbullah so as to provide a pretext for the complete destruction of Lebanon Jews planned in *Ezekiel* Chapters 27 and 28. Jews offer false hope that their unprovoked aggression will soon end, then dash those hopes by intensifying their murder of helpless Lebanese children and babies and pour the blood of their victims back onto their victims by scapegoating them for Jewish atrocities.

Scapegoating is but one form of deceit which is deeply ingrained in the Judaic psyche. Another ancient Jewish deceit is the use of crypto-Jews to undermine Gentile societies and religions. Christian Zionists, often led by crypto-Jews, are desperate to commit genocide against at least two billion human beings in the false and utterly selfish hopes that by mass murdering these innocents Christian Zionist mass murderers will provoke Jesus to Rapture the Christian Zionists mass murderers into Heaven. Crypto-Jews created these false

---

1917), Section 2, p. 2. *See also:* "Pacifists Pester till Mayor Calls Them Traitors", *The New York Times*, (24 March 1917), pp. 1-2. *See also:* "Kahn Asks Army of 6,000,000 Men", *The New York Times*, (30 December 1917), p. 4. *See also:* E. Slater and R. Slater, "Jacob Schiff", *Great Jewish Men*, Jonathan David Publishers, New York, (2003), pp. 274-276, at 275-276. *See also:* "Schiff, Jacob Henry", *Encyclopaedia Judaica*, Volume 14 RED-SL, Encyclopaedia Judaica, Jerusalem, The Macmillan Company, New York, (1971), cols. 960-962, at 961.

beliefs (and others like them) and sponsor them today in an effort to trick Gentiles into killing one another off. Ancient Jews taught this behavior in the Hebrew Bible in the book of Esther. Jews celebrate the genocide of Gentiles, and the deceit of the crypto-Jew, once every year in the Jewish festival of Purim. It is the Jews' favorite holiday.

Crypto-Jews, Zionist Jews, and Israel firsters have infiltrated the mass media and governments of all the world. They are deliberately attempting to orchestrate a nuclear World War III in the hopes that it will kill off the Gentiles and leave only "righteous Jews" alive in the "End Times". The People of the World must take action to save themselves from the genocide racist Jews have been planning for 2,500 years and which they believe they must carry out now that they have created the Jewish Kingdom in Palestine. We cannot depend upon government or media to help. Both have been corrupted by genocidal Jewish influence.

This is part of a broader plan to fulfill Judaic prophecy by political action meant to discredit Gentile governments and religions and promote the myth that Judaism and Jews are innocent and highly moral. We see it today in the widespread attacks on Islam and Moslem nations, which are fomented by racist and highly unethical Jews. Just as Zionist Jews subverted German society with crypto-Jewish leaders who rose to power on a platform of anti-Semitism, Zionist Jews are subverting Moslem nations with crypto-Jewish leaders and Jewish agents who rise to power on an anti-Zionist platform. Jews covertly commit acts of terrorism against other Jews, which they blame on non-Jews, in order to create a climate of antagonism and distrust, where Jewish racists can spuriously claim the moral high ground and utter their hateful and false defamations against other peoples with impunity and apparent justification.

The rabid nationalism Herzl and Einstein embraced, and the anti-Semitism they believed benefitted the Jews by uniting and segregating them, began to become very dangerous in the 1920's—much to the delight of the Zionists. Einstein's hypocrisy, his anti-Nationalism versus his Zionism, remained yet to be resolved in the minds of the naïve. For those who grasped the import of Judaic Messianic myth, Einstein was consistently obedient to the racist and genocidal Jewish prophets. In 1938, Einstein stated in his essay "Our Debt to Zionism",

> "Rarely since the conquest of Jerusalem by Titus has the Jewish community experienced a period of greater oppression than prevails at the present time. [***] Yet we shall survive this period too, no matter how much sorrow, no matter how heavy a loss in life it may bring. A community like ours, which is a community purely by reason of tradition, can only be strengthened by pressure from without."[107]

Einstein continues in his essay in an effort to justify the illogical and immoral

---

[107]. A. Einstein, "Our Debt to Zionism", *Out of My Later Years*, Carol Publishing Group, New York, (1995), pp. 262-264, at 262.

conflicts in his political philosophy, but without success. Einstein also reveals that his early assertions of the racial purity of Jews were nonsense employed for political effect—the political effect of deliberately bringing the Nazis into power in order to herd up the Jews of Europe and chase them into Palestine—the political effect of discrediting Gentile nationalism, while justifying Jewish nationalism. Zionist are today using the same tactics to discredit Islamic nationalism and promote Jewish nationalism. They delight in the fact that they are killing off large numbers of innocent Gentiles in the process.

Einstein and the Zionist Fascists were carrying on a long tradition of European and Judaic ethnocentrism and racism spanning the middle ages and reaching far back into antiquity. The hatred was directed in both directions—much to the delight of racist Jews.

In Einstein's day, Jews and Gentiles were finally becoming integrated. Racist Einstein and his Zionist friends artificially created a rise in anti-Semitism and demanded segregation. Einstein thought that anti-Semitic attacks and segregation were the best means to preserve the "Jewish race" from the "fatal assimilation" brought on by better relations between Jews and non-Jews.

2 THE DESTRUCTIVE IMPACT OF RACIST JEWISH TRIBALISM

Jews have an ancient tradition of racism and of deliberately segregating themselves from all other peoples. Jews even segregate each other into separate subdivisions of Sephardim and Ashkenazim. Sephardim have traditionally considered themselves to be more "racially pure" than Ashkenazim, and, therefore, "racially" superior to Ashkenazim. Ashkenazim have traditionally viewed themselves as "racially" superior to Gentiles. Since they cannot claim "racial" superiority over the Sephardim, the Ashkenazim use tribalistic politics to kill them off.

"Jews have not troubled themselves to justify, on any rational ground, the tenacious fight of their race against the storms of nineteen centuries of persecution. The fight has been its own justification. Obviously, a race that has endured what theirs has withstood must have some glorious mission to perform; to define that mission would be an element of positive weakness, since their enemies would then have a chance to meet them on the ground of reason, where their peculiar virtues, tenacity, single-mindedness, and pliant heroism, would avail them nothing."—RALPH PHILIP BOAS

"The position of the Jews is unique. For them race, religion, and country are interrelated, as they are interrelated in the case of no other race, no other religion, and no other country on earth. By a strange and most unhappy fate it is this people of all others which, retaining to the full its racial self-consciousness, has been severed from its home, has wandered into all lands and has nowhere been able to create for itself an organized social commonwealth. Only Zionism—so at least Zionists believe—can provide some mitigation of this great tragedy."—ARTHUR JAMES BALFOUR

2.1 Introduction

In the United States in the early 1920's, scholars became increasing concerned by the invasion of racist and tribalistic "Russian or Polish Jews", who had been pouring into America since the 1880's. These immigrants allegedly sought to take over American universities and to Judaize American society. Harvard University opened the question of whether or not it was in the best interests of American society to allow Jews from Poland to obtain majority control over highly influential American colleges and universities.

In 1917, Ralph Philip Boas, who was himself Jewish, discussed the tribalistic, segregationist and racist attitudes common among Jews of the era—and throughout history,

"DESPITE the fact that we are ceasing to persecute people who disagree with us in religion or politics, we only dimly realize that one of the greatest evils of persecution is the fact that it saves its victims the trouble of justifying themselves. Persecution begets martyrdom, a glory as lacking in reason as its progenitor. Whether Sir Roger Casement was right or not is now only an academic question; his execution, by enshrining him forever in the Pantheon of Irish martyrs, makes the heart rather than the mind his judge. So it is with the Jews. Jews have not troubled themselves to justify, on any rational ground, the tenacious fight of their race against the storms of nineteen centuries of persecution. The fight has been its own justification. Obviously, a race that has endured what theirs has withstood must have some glorious mission to perform; to define that mission would be an element of positive weakness, since their enemies would then have a chance to meet them on the ground of reason, where their peculiar virtues, tenacity, single-mindedness, and pliant heroism, would avail them nothing.

It is, therefore, a happy chance for the American Jew that his age-long persecution has either ended or has degenerated into petty social discrimination. For he must now realize that the day has gone when he could justify himself by recalling his heroic miseries. In other days and other countries he faced only the problems of existence. New ideas and opportunities could not pass the walls of the ghetto; custom made adherence to old ceremonies and beliefs not only easy but imperative. The Sabbath was the one day on which the Jew could be a man instead of a thing; the recurrent holidays gave him his one outlet for the emotions rigidly suppressed in daily life; the study and analysis of the Law and the Talmud furnished the intellectual exercise that his eager mind was denied in the schools and the learned circles of the country which tolerated him. The very fact that he was confined within a pale, therefore, made it easy for him to keep his race a distinct entity.

But now, if he is unable to find a rational ground for his religious and racial unity, he will meet a foe more insidious than persecution—the gradual disintegration of race and religious consciousness within the faith. Ironically enough, what pales, pogroms, and ghettos could not accomplish, freedom promises to bring to pass. So the time has come when the Jew in America must decide what he is going to do with and for himself; his enemies can no longer save him the effort of decision.

[***]

What is true of Europe is true also of the United States: the Jew occupies a position the importance of which is out of all proportion to his numbers. Hence the problem of Judaism is of real interest in America, because the influence which the Jew can have upon social life and the current political and financial situation depends almost entirely upon his mode of life and manner of thought. [***] What the Jew is going to do with this self-consciousness may, to Christians, seem of little moment. It is not of that loyal kind which moves men to blow up munition factories, or to plant bombs in steamships. For others, doubtless, its implications are not of great importance. For himself, however, they are everything. His self-consciousness colors his whole point of view. It is not a simple thing. It is compounded of many factors. It is both racial and religious; it makes him both hopeful and despondent; it gives cause both for pride and for a feeling of inferiority; it makes him clannish, and it makes him long for a wider field of acquaintance. [***] Judaism is clannish. Jews undoubtedly hang together. The combination of persecution with its inevitable concomitant, self-justification, acts as a centripetal force in driving Jews upon themselves. Just as Jews have the almost grotesque notion that a man will make his philosophic and religious convictions 'jibe' with his birth, so they have the wholly grotesque notion that a man should choose his friends and his wife from the small group among whom he happens to be born, though later education and environment may move him a thousand miles away. The results of this clannishness are paradoxical. For instance, the average Jew is sure that the chief reason why Anti-Semitism is everywhere ready to show its ugly head, is jealousy of the splendid history and the extraordinary business ability of the race. At the same time he subconsciously assumes the inferiority which has long been attributed to him, covering his feelings, however, by uncalled-for justification and bitter opposition to all criticism. It is torture to him, for example, that *The Merchant of Venice* should be read in the public schools. Who can blame him? For Shylock, although undoubtedly an exaggerated character, nevertheless makes concrete those qualities the portrayal of which hurts because it bears the sting of truth.

The development of committees 'On Purity of the Press' in Jewish societies, and the extraordinary wire-pulling over the Russian treaty and the Immigration bill, show to what lengths this consciousness can go. It is impossible for the Jew to be entirely at ease in the world. He is introspective and suspicious, often unhappy, always sure that, for good or ill, he is a marked man among men.

There are three attitudes which Jews in this country take toward their problem—a few as a result of having thought it through, the majority as a result of the forces of inertia, environment, or chance, forces of which they themselves are perhaps not aware. Some Jews attempt to get rid of their self-consciousness by separating from the group. They deliberately set out to convince themselves that there is no difference between them and other men, and that they can act and live in all respects like other American citizens. A second group find their fellow Jews entirely satisfactory. They are conscious of a difference between themselves and others, but, living as they do in large cities where the Jewish community numbers hundreds of thousands, they feel no need of association with non-Jews other than that which they get in business. They are rich, or at least well-to-do; they have all the comforts that money can buy; they occupy fine streets and build expensive synagogues. They are willing, not only to accept their group-consciousness, but to develop it to the fullest extent by means of societies and fraternal orders. In the third place, there is a small group of Jews keenly conscious of their race, who would like to make Judaism vital as a great religion and a great tradition. They differ from the second group in that they not only accept their individuality but try to justify it. It is not sufficient for them that there should be enough Jewish organizations and undertakings to make a respectable year-book: they are interested in showing why such organizations should exist They not only *are* Jews, but they *want to be* Jews; they want to feel that Judaism really has a mission to fulfill and a message to carry to the questioning world.

The Jew who attempts to solve his problem by separating from his community must leave the great centres of Jewish life and go to some small town where he may make a fresh start. There he will find himself in an anomalous position. He will have neither the support that comes from rubbing elbows with one's own kind, nor the mental and moral stiffening that comes from active opposition. He will be simply an odd fish, and as such will be subject, not to antagonism, but to curiosity. What cordiality he meets with is the cordiality of curiosity. He is a strange creature, similar—on a far lower scale of interest—to a Chinese traveler or a Hindu student. He is engaged in conversation on the 'Jewish problem,' or Jewish customs and history, until he sickens with trading on the race-consciousness that he is striving to forget. With cruel kindliness his friends impress upon him that his Judaism 'makes no difference,' with the result that he finds himself anticipating every imminent friendship by a clear statement of his race, lest the friendship be built upon the sands of prejudice. His social relations must be above reproach. A hasty word, an ill-considered action, in other men to be put down to idiosyncracy, in him is attributed to his birth. Even when there exists the frankest and most open friendship, he is continually seeing difficulties. The fathers have eaten a sour grape and the children's teeth are set on edge. The self-consciousness that he learned in youth reappears in maturity. Whether he will or no, a Jew he remains.

If he finds his situation intolerable he may, of course, utterly and completely deny his Jewish affiliation. He may consort with Christians, join a Christian church, marry a Christian wife, and tread under foot the old associations that will occasionally cast a disagreeable shadow across his life Unfortunately for such a solution, a cloud still hangs about the idea of apostasy. Such a refuge seems to a man of honor despicable. It is a cowardly procedure, surely, to deny one's birth and sail under false colors, the more so since, though it does no harm to others, it gains advantage for one's self. Why ii should it be treason for a Jew to abandon his religion and forget his birth any more than for a Frenchman or a Swede to do so? Probably for the reason that no one cares whether a man was born in France or not, whereas in certain circles it makes a great deal of difference if a man was born in Jewry. Furthermore, Christians feel strongly that the Jew who forsakes the religion into which he was born, does so, not because his eyes have been opened upon the truth, but because he sees in apostasy definite material advantages. The Jew who would take this means of obtaining peace, therefore, would find himself cursed by an irrational idealism which can disturb while it cannot fortify and achieve.

If, however, he returns to some great centre of Jewish life and attempts to affiliate with his own people, he is in a perilous position. He is more than likely to meet with distrust where he seeks sympathy. Jews are so extremely sensitive to criticism and so keenly conscious of the social discrimination which they encounter from Christians, that they can hardly believe that a man who seems to have lived for several years on an equal footing with

Christians has not either denied his birth, in which case he has been a traitor, or has not certain qualities of mind which, since they have been palatable to Christians, must be severely critical of Jews.

And, indeed, they have, perhaps, a measure of justice in their position. It is impossible for a Jew to live apart from his race for several years without looking upon his people with a new light. For one thing, distance has enabled him to focus. He has learned to sympathize more than a little with those hotel-keepers whose ban upon Jews is a terrible thorn in the flesh of the man whose money ought to take him anywhere. He has come to see that the clannishness of Jews serves only to intensify what social discrimination may exist, and to make present in the imagination much that does not. He has realized that persecution is not necessarily justification, and that because a Jew was blackballed at a fashionable club does not prove that he was a man of first-rate calibre. And finally, he has perceived that there is an arrogance of endurance as well as an arrogance of persecution, and that for a man to be continually assuming that people are taking the trouble to despise him for his birth, is to postulate an importance that does not exist.

On the other hand, he has, because of his distance, idealized Judaism. In his retirement he studied the history of his people; he thrilled with their martyrdom; he marveled at their tenacity and their fortitude. He built up for himself on the cobweb foundation of boyhood memories, visions of the simple nobility of Jewish ritual and ceremonies, and vague ideals of an inspiring religious faith. He may, perhaps, have met, far more frequently than ill-will, a sentimental and unbalanced adulation of Jews. The cult of the new is with us, and the history, the folk-lore, the literature, and the customs of Judaism have, for many people who pride themselves on their social liberality, the fascination of novelty. It is the easiest thing in the world for a Jew to yield to this sentimental tolerance, and to view his people in a rosy light.

It is, therefore, something of a shock to him when he reënters a great Jewish community, for he finds that the great mass of American Jews have sunk into a comfortable materialism. What persecution could not accomplish, success in business has brought to pass. The innate qualities of the Jew could not save him from the fate of the Christian who has become rich in a hurry—grossness and self-conceit. That Jeshurun waxed fat and kicked is as true now as it ever was, and there is little reason to expect that the race which was hopelessly cankered by national prosperity in the days of Solomon can escape a similar fate in the twentieth century. [***] The sad result is that in prosperity the Jewish self-consciousness ceases to be religious and becomes merely racial.

[***]

The number of immigrants, or children of immigrants, from countries where for centuries they have been trained in an atmosphere of slavish cunning and worship of money, who become rich, is almost incredible. In Russia, Galicia, or Roumania, they cultivated a self-respect by rigid adherence to dignified and beautiful customs; in America the florid exuberance of newly acquired wealth cannot be dignified. Clannishness, exclusion from circles of good taste and good breeding, the infiltration of the parvenu East-European Jews, and imitation of the most obvious aspects of Americanism—its flamboyant and tasteless materialism—all combine to make the thoughtful Jew sadly question what hope lies in the bulk of the Jews who live in the great American cities.

[***]

[Zionism] is actuated by a spirit of helpfulness and by an ideal of racial unity. [***] Aided by persecution and poverty, [American Judaism] furnished admirable discipline to a race naturally stubborn and tenacious. Persecution, poverty, and discipline gone, what is left?—an indistinct monotheism joined to an ethical tradition never formulated into a system, and only vaguely defined. None of the great Jewish philosophers ever succeeded in establishing a Jewish creed; indeed, there was no need of one when common suffering wrought so effectual a bond. [***] At all events it must be remembered that, since the problem of Judaism comes from intense self-consciousness, persecution and sentimental tolerance are both bad for the Jew. The one saves him the trouble of seeking out his reason for existence; the other flatters him into a belief that there is no necessity for the search. If men will treat Jews like other people, instead of nourishing their age-long notions of peculiarity, they will make it easier for time to settle the Jewish problem as it settles all others."[108]

In 1845, an article appeared in *The North American Review*, which revealed that governments were concerned by Jewish Messianic aspirations and the resultant disloyalty of Jews,

"The Jews in Russian Poland have lately been subjected to military service; and to the soldier's oath the government has added, for Israelitish recruits, the following clause: 'I swear to be faithful to my standard, and never desert it, even should the Messiah come upon earth.'"[109]

Frankist Jews in Poland asserted in the 1700's and throughout their later history that the Messiah had arrived in the person of Jacob Frank. They formed revolutionary and destructive bands, which tore apart Polish society. Frank began a dynasty of Messiahs, whose soul alleged migrated from one Messiah to the next through the process of Metempsychosis. Jews have long believed in the transmigration of souls and the perpetual reign of the seed of King David through reincarnation, as opposed to purely through reproduction.[110] It was the duty of

---

108. R. P. Boas, "The Problem of American Judaism", *The Atlantic Monthly*, Volume 119, Number 2, (February, 1917), pp. 145-152.
109. "The Modern Jews", *The North American Review*, Volume 60, Number 127, (April, 1845), pp. 329-368, at 348.
110. "LURIA, ISAAC BEN SOLOMON, *Encyclopaedia Judaica*, Volume 11 LEK-MIL, Macmillan, Jerusalem, (1971), cols. 572-578, at 576. J. A. Eisenmenger, *The Traditions*

the Messiah to utterly destroy the Gentile world.

2.2 Do Not Blaspheme the "Jewish Saint"

When Einstein arrived in America in early April of 1921, shortly after Einstein, himself, declared that anyone who disagreed with him must *ipso facto* be anti-Semitic, the Board of Aldermen of the City of New York met to vote on a proposal to grant Chaim Weizmann and Albert Einstein the "freedom of the city". Alderman Bruce M. Falconer voted against the proposal and was immediately assaulted, threatened with severe retaliation and smeared as an "anti-Semite"—an accusation he emphatically denied. *The New York Times*, which was owned by a Jewish publisher named Adolph S. Ochs,[111] published Alderman Falconer's name, occupation, and home address, on the front page together with the charges of anti-Semitism, a description of the assault against him, and a report of the threats to destroy him, as well as his denials of any prejudice.

Several stories describing the spectacle appeared in *The New York Times*, beginning with 6 April 1921,

"HOLDS UP FREEDOM OF CITY TO EINSTEIN

Alderman Falconer Blocks Move
to Grant Official Honors
to Two Scientists.

NEVER HEARD OF HIS THEORY

Alderman Friedman Shakes Fist
in Face of Opponent and
Calls Action an Insult.

There is at least one man in New York who never heard of Professor Albert Einstein, whose theory of relativity has been discussed for many months in newspapers and magazines. He is Alderman Bruce M. Falconer, whose lack of acquaintance with Professor Einstein's fame caused a row in the Board of Aldermen yesterday and resulted in the freedom of the city being temporarily refused to both Professor Einstein and Professor Chaim Weizmann, chemist and inventor of the high explosive trinitrotoluol.

---

of the Jews, Contained in the Talmud and other Mystical Writings, Volume 1, J. Robinson, London, (1748), pp. 277-338.
111. B. J. Hendrick, "The Jews in America: II Do the Jews Dominate American Finance?", *The World's Work*, Volume 44, Number 3, (January, 1923), pp. 266-286, at 282.

At the request of Aldermanic President LaGuardia, Mayor Hylan has called a special meeting of the Board for next Friday at 1:30 P. M., to take action on the resolution.

'I am expressing the feeling of the entire Board when I ask you to call this meeting in order that the desires of the people of this city may be carried out in extending this call to these distinguished people,' he said to the Mayor.

Professor Weizmann is President of the International Zionist Organization, and, with Professor Einstein, M. M. Ussischkin and Dr. Benzion Mossinson, is here to confer with American Zionists. They were received at the City Hall yesterday by Mayor Hylan and a committee of citizens. More than 5,000 Zionists filled the plaza in front of the City Hall.

It was thought that the granting of the freedom of the city to the two visitors would be a mere formality. So it would have been but for Alderman Falconer, who is a lawyer and lives at 701 Madison Avenue. After the ceremony the Aldermen went to their Chamber and a resolution was introduced by Alderman Louis Zeltner, Moritz Graubard and Samuel R. Morris in honor of the visitors. Every one was ready to vote favorably when Alderman Falconer arose. He confessed that until yesterday he never had heard of either Professor Einstein or Professor Weizmann. He asked to be enlightened, but nobody offered to explain the theory of relativity. Mr. Falconer said that he thought the freedom of the city had been too often granted, and, although his objection had nothing to do with racial or religious prejudices, he believed that caution should be exercised.

A storm broke about Alderman Falconer's head. Laughter and protests came from every side, and several members tried to tell him the records of the two men, but their recital made little impression upon the Alderman.

Rules Committee Dodges.

A motion that the resolution be made a general order for next week when it could be passed over Alderman Falconer's protest precipitated a parliamentary row, and in a few minutes the board was tangled up in rulings. President LaGuardia came in and took the chair. He ruled that the point of order to make the resolution a general order was debatable, and about this time the Committee on Rules, led by Alderman Kenneally, slipped out of the room.

Alderman Falconer was obdurate, and at the end of the debate the Rules Committee came back and an attempt was made to get around his objection. It was moved to suspend the rules, when the resolution could be passed over his objection. But Alderman Falconer suspected the purpose of the motion, and objected. Alderman Friedman then asked that the resolution be withdrawn.

After the incident was officially closed there were angry arguments in the boardroom. Alderman Friedman shook his fist under Mr. Falconer's nose and said that his action was an insult and that he would carry the issue into Mr. Falconer's district. Judge Gustave Hartmann tried unsuccessfully to tell Mr. Falconer what Professor Einstein had done in science.

After the adjournment of the meeting Judge Hartman charged Alderman Falconer with having made his objection to the resolution because of purely anti-Semitic motives. This brought a denial from the Alderman and when Judge Hartman repeated his charge Mr. Falconer said: 'You're a liar, I am most certainly not opposed to the Jewish people as a race.'

'I will not let this matter drop,' said Judge Hartman. 'Not only will I bring the matter before the people of the city and the intelligent Jewry, but I will also press this matter in the council of the Republican Party. I am firmly convinced that your attitude in this matter was prompted by anti-Semitism, and I will not be satisfied until you are retired from public life.'

When Professors Weizmann and Einstein arrived at the City Hall, accompanied by their wives and other members of the delegation, they were escorted to the Mayor's office by James F. Sinnott, Secretary to Mayor Hylan, and the Committee of Welcome led by Magistrate Rosenblatt.

'As Mayor of this city, which is the home of more than one-third of all the Jews in America,' said Mayor Hylan, 'I gladly join in felicitating those who have already accomplished so much toward the restoration of Palestine. The success thus far achieved may be regarded as a happy augury that continued endeavor will result in the final and complete attainment of the hope and aspiration of the Zionist organization.

'May I say to Dr. Weizmann and Professor Einstein that in New York we point with pride to the courage and fidelity of our Jewish population, demonstrated so unmistakably in the World War.'

George W. Wickersham, former Attorney General, also spoke of the achievements of the two leaders of the delegation.

Professor Weizmann thanked the Mayor and Mr. Wickersham for their welcome, which he accepted as showing sympathy for the cause he represented.

Mrs. Einstein lost a gold lorgnette with a chain attached during the reception at the City Hall. It was an heirloom."

Intimidation, threats of retaliation and retaliatory actions are common practice among Einstein advocates. The judge threatening and smearing the attorney was and is not unique to the legal profession and political life. American Zionism was headed by United States Supreme Court Justice Louis Dembitz Brandeis and represented by Judge Julian William Mack of the United States Circuit Court of Appeals for the Seventh Circuit. There have been accusations of Jewish American judges allowing guilty Zionist criminals to go free and otherwise preventing justice.[112] The Talmud and other Judaic literature encourage Jews to favor one another at the expense of Gentiles and to forgive

---

112. R. I. Friedman, *The False Prophet: Rabbi Meir Kahane: from FBI Informant to Knesset Member*, Lawrence Hill Books, Brooklyn, New York, (1990), p. 38.

crimes Jews commit against Gentiles. For example, *Sanhedrin* 58*b* states that a Gentile who strikes a Jew must be killed, because striking a Jew is like striking God. Yet according to *Sanhedrin* 57*a*, a Jew who murders a Gentile without cause will not be put to death and is not civilly liable for the crime.[113] Furthermore, a Jew may steal from a Gentile and may keep the stolen goods with both criminal and civil immunity under some interpretations of Jewish law.

Numerous physicists of international renown have complained directly to your author that their works in opposition to relativity theory, and which expose Einstein's career of plagiarism, have been refused publication without grounds and are often met with angry personal attacks and threats of retaliation as well as reactionary and unjustified accusations of anti-Semitism. Some peer reviewed journals and scientific conferences regularly refuse to even consider works and lectures which question relativity theory, or Einstein's originality. Even Jewish opponents are attacked as if *ipso facto* anti-Semites for daring to utter a syllable of truth about Einstein's plagiarism and the fallibility of "his" theories. Helen Dukas (Einstein's secretary) and Bannesh Hoffmann wrote,

> "Einstein had become a figure of enormous symbolic importance to Jews. In 1923, when he visited Mount Scopus, the site on which the Hebrew University was to rise, he was invited to speak from 'the lectern that has waited for you two thousand years.'"[114]

Dennis Overbye tells the story of Ilse Einstein's letter to Georg Nicolai of 22 May 1918 in which she complains of Albert Einstein's perverse sexual advances towards her. Albert Einstein was conducting an incestuous and adulterous relationship with her mother, Else Einstein, at the time. Albert Einstein was related to his cousin Else through both his mother and his father. Einstein was perhaps dissuaded from his perverse wish to marry Ilse Einstein by his uncle Rudolf Einstein's (Rudolf Einstein was Elsa Einstein's father and Ilse Einstein's grandfather, as well as Albert Einstein's uncle and father-in-law) dowry of 100,000 Marks, which Albert Einstein accepted when he married his cousin Else—Albert thereby continued to have access to Ilse.[115] Albert Einstein was behaving like a Frankist Jew.

Overbye states that Wolf Zuelzer preserved the letter,

> "despite pressure from Margot Einstein, Helen Dukas, and lawyers representing the Einstein estate to surrender it or destroy it. The tale, an example of the difficulties scholars have faced in telling the Einstein story, is preserved in Zuelzer's correspondence in the American Heritage archive

---

113. *Cf.* "Gentile", *The Jewish Encyclopedia*, Funk and Wagnalls Company, New York, (1903), pp. 615-626, at 618.
114. H. Dukas and B. Hoffmann, *Albert Einstein: The Human Side*, Princeton University Press, (1979), p. 55.
115. Letter from A. Einstein to "Berlin-Schöneberg Office of Taxation" of 10 February 1920, *The Collected Papers of Albert Einstein*, Volume 9, Document 306, Princeton University Press, (2004), pp. 256-257, at 257.

at the University of Wyoming."[116]

It is rather embarrassing for an ethnic "Saint" and national hero to be exposed as a pervert and a plagiarist, and Einstein had become both an ethnic saint and a national hero for Jews. Bruno Thüring used these facts to characterize Einstein as a rabid nationalist, who used his pacifistic preaching as a front to promote his Zionist agenda. Thüring recounted that the *Jüdische Rundschau* quoted the Zionist David Yellin's welcoming address to Einstein in the name of Jerusalem on 15 March 1929 and Einstein's response:

> "„Du hast den Namen ‚Gaon' verdient, den das jüdische Volk seinen erwählten geistigen Führern gibt — dies aber nicht nur wegen deiner genialen Leistungen in der Wissenschaft, wiewohl wir sie recht zu schätzen wissen — noch mehr aber bist du uns ein Gaon, weil du die Fahne der nationalen Wiedergeburt hoch in der Hand hältst und die hebräische Universität in Jerusalem gefordert hast."
> Und Einstein antwortete darauf:
> „Der heutige Tag ist der größte meines Lebens. Heute ist das wichtigste Ereignis in meiner Lebensgeschichte geschehen. Im Laufe meines Lebens lernte ich die Verirrung der jüdischen Seele, die Sünde der Selbstverleugnung des Volks-Jüdischen kennen. Und so freue ich mich, daß Israel seine Bedeutung in der Welt wieder zu erkennen beginnt. Diese Tat, die Befreiung der jüdischen Seele, wurde von der zionistischen Bewegung vollbracht."[117]

Einstein wrote to Paul Ehrenfest on 12 April 1926,

"I do believe that in time this endeavor will grow into something splendid; and, Jewish Saint that I am, my heart rejoices."[118]

The German Consul General in New York reported on 21 March 1931, "Es ist ein Charakteristikum für die New Yorker Volkspsyche, daß die Persönlichkeit Einsteins, ohne daß deutlich erkennbare Gründe dafür anzuführen wären, Ausbrüche einer Art Massenhysterie auslöste, und zwar

---

116. D. Overbye, *Einstein in Love: A Scientific Romance*, Viking, New York, (2000), pp. 343, 404, note 22. *See:* A. Einstein to Ilse Einstein, *The Collected Papers of Albert Einstein*, Volume 8, Document 536, Princeton University Press, (1998); **and** Ilse Einstein to Georg Nikolai, *The Collected Papers of Albert Einstein*, Volume 8, Document 545, Princeton University Press, (1998).
117. B. Thüring, "Albert Einsteins Umsturzversuch der Physik und seine inneren Möglichkeiten und Ursachen", *Forschungen zur Judenfrage*, Volume 4, (1940), pp. 134-162, at 142. Republished as: *Albert Einsteins Umsturzversuch der Physik und seine inneren Möglichkeiten und Ursachen*, Dr. Georg Lüttke Verlag, Berlin, (1941).
118. H. Dukas and B. Hoffmann, *Albert Einstein: The Human Side*, Princeton University Press, (1979), p. 55.

nicht nur bei den hierfür besonders veranlagten Gruppen von „Friedensfreunden" und den schwärmerischen Phantasten neu entstandener mystischer Religionsgesellschaften, sondern auch in relativ so kühlen Kreisen, wie z. B. bei den amerikanischen Förderern des Palästinawerkes. Inwieweit hierbei der Umstand eine Rolle spielte, daß sich unter den sieben Millionen Einwohnern New Yorks annähernd zwei Millionen Juden befinden, und ob in der Wechselwirkung zwischen Presse und Publikum erstere ihre zahllosen Spezialartikel über Einstein brachte, weil die Leser sich begehrten, oder ob letztere sich hierfür interessierten, weil die Zeitungen dieses Interesse schon vor Einsteins Ankunft erweckten und alsdann wachhielten, wird schwer zu entscheiden sein. Nicht ganz belanglos erscheint in letzterer Beziehung aber vielleicht das Scherzwort eines Rundfunkredners zur Zeit des Höhepunktes der Einstein-Begeisterung, daß wohl nicht 50 Personen wüßten, warum der Gelehrte überhaupt hier sei ... Einsteins Ausführungen brachten die Anwesenden in einen Begeisterungstaumel, der sich auch darin äußerte, daß zahlreiche Personen Einsteins Hände und Kleidungsstücke küßten."[119]

Philipp Frank wrote,

"The Jewish population of America itself regarded Einstein's visit as the visit of a spiritual leader, which filled them with pride and joy. The Jews felt that their prestige among their fellow citizens was raised by the fact that a man of Einstein's generally recognized intellectual greatness publicly acknowledged his membership in the Jewish community and made their interests his own."[120]

Chaim Weizmann recalled his visit with Einstein to New York in 1921,

"We had reckoned—literally—without our host, which was, or seemed to be, the whole of New York Jewry. Long before the afternoon ended, delegations began to assemble on the quay and even the docks."[121]

The ethnic, racial and religious prejudice of Einstein and his followers, even if in the understandable and forgivable form of misguided pride, has no place in science. Many unscrupulous individuals have dishonored the victims of the Holocaust and Pogroms by disingenuously smearing any person who dares to

---

119. Quoted in B. Thüring, "Albert Einsteins Umsturzversuch der Physik und seine inneren Möglichkeiten und Ursachen", *Forschungen zur Judenfrage*, Volume 4, (1940), pp. 134-162, at 156-157. Republished as: *Albert Einsteins Umsturzversuch der Physik und seine inneren Möglichkeiten und Ursachen*, Dr. Georg Lüttke Verlag, Berlin, (1941).
120. P. Frank, *Einstein, His Life and Times*, Alfred A. Knopf, New York, (1947), pp. 182-183.
121. C. Weizmann, *Trial and Error: The Autobiography of Chaim Weizmann*, Harper & Brothers, New York, (1949), p. 266.

question Einstein or the theory of relativity as an "anti-Semite", in order to change the subject from the critic's legitimate arguments, to a disingenuous personal attack against the legitimate critic, which evokes powerful emotions. They not only dishonor those who were murdered, by invoking the memory of the dead to distract from Einstein's errors and misdeeds, they inhibit the progress of science and the accurate portrayal of history, in the names of those who were murdered at the behest of racist Zionist Jews.

The saga of Alderman Falconer's exercising of his rights to oppose the award of the "Freedom of the City" to Weizmann and Einstein continued across the pages of *The New York Times* and newspapers around the world. *The New York Times* reported 7 April 1921,

"RELATIVITY AT CITY HALL.

Alderman FALCONER wants everybody to understand that when he said he had never heard of Professor ALBERT EINSTEIN he didn't know it was the famous EINSTEIN, the destroyer of time and space. The Alderman's reasoning is intelligible even if its result was rather unhappy. Two gentlemen were coming up to be formally endowed with such freedom as can still be granted in this well regulated city. Who were they? Mr. EINSTEIN and Mr. WEIZMANN. And how was any one to know—unless he had read the papers—that this EINSTEIN was the celebrated EINSTEIN? He was coming to New York not as a scientist but as a Zionist, in which capacity he hasn't been working long enough to become celebrated. Any nobody would have suspected that a Mayor hostile to art artists would be asking the freedom of the city for a couple of mere science scientists.

So Alderman FALCONER was led into the blunder in which he is now trying to justify himself. He says EINSTEIN is a German. True, he is German-born, and recently he spent a year or two in Berlin. But genuine blown-in-the-glass Germans of the Reventlow type would fling their hands and howl if they heard EINSTEIN called a German. One of the reasons for his leaving Berlin, apparently, was the attacks made on him by some of the reactionary monarchist organs. They had three counts against EINSTEIN—he is a Swiss citizen, a Jew and a democrat. Nobody but the Staats-Zeitung can seriously believe that 'hatred of the Germans' is behind this opposition to EINSTEIN.

But the professor probably felt quite at home in the City Hall, with or without freedom. Relativity was being practiced in those quarters long before EINSTEIN discovered it as a theory. The rays of logic emanating from the Mayor's office are bent as badly as EINSTEIN'S rays of light. EINSTEIN proved that things are not where they seem to be, but that is no news to gentlemen elected on a program of economy who have raised the city budget to unheard-of figures. And a man who has annihilated space may be able to provide our municipal Government with some happy thoughts on the rapid-transit problem.

And perhaps Alderman FALCONER has done no real harm. Mrs. EINSTEIN, emerging from the crowd which had gathered for the reception at the City Hall, missed a valuable gold lorgnette; so no doubt she and her husband are vividly impressed, already, with the freedom of our city."

Einstein and his advocates would sometimes flip-flop on the issue of Einstein's citizenship over the course of many years, often to avoid fulfilling national or political duties, or purely to allege bigotry, arbitrarily changing Einstein's status to fit the accusation and to emphasize and aggravate social divides for political profit.[122] Einstein was also dishonest about his religious status and misrepresented it to suit the occasion and encouraged his friend Paul Ehrenfest to do the same. Ehrenfest had more character than Einstein and Ehrenfest stood by his convictions.[123]

Some Jews felt no obligation to honor their obligations, especially any obligation to a Gentile. During the evening service for Yom Kippur, Jews recite the following dispensation of vows three times. It is called the Kol Nidre,

"All vows, obligations, oaths or anathemas, pledges of all names, which we have vowed, sworn, devoted, or bound ourselves to, from this day of atonement, until the next day of atonement (whose arrival we hope for in happiness) we repent, aforehand, of them all, they shall all be deemed absolved, forgiven, annulled, void and made of no effect; they shall not be binding, nor have any power; the vows shall not be reckoned vows, the obligations shall not be obligatory, nor the oaths considered as oaths."

Many Jews believe that reciting the Kol Nidre on the Day of Atonement entitles them to lie, violate contracts, etc. Some Rabbis will swear that the Kol Nidre refers only to personal vows made to ones self, but if these selfsame Rabbis have recited the Kol Nidre, of what value is their oath?

The Kol Nidre is based on the Talmud in the book of *Nedarim* folio 23*b*,

"And he who desires that none of his vows made during the year shall be

---

122. *Compare, for example:* Letter from A. Einstein to the League of German Scholars and Artists of 13 January 1920, *The Collected Papers of Albert Einstein*, Volume 9, Document 258, Princeton University Press, (2004); *to:* A. Einstein, *The World As I See It*, Citadel Press, New York, (1993), p. 89. *See also:* G. J. Whitrow, Editor, *Einstein: The Man and his Achievement*, Dover, New York, (1967), pp. 17-18. H. Dukas and B. Hoffmann, *Albert Einstein: The Human Side*, Princeton University Press, (1979), pp. 6-11. A. Fölsing, *Albert Einstein: A Biography*, Viking, New York, (1997), pp. 30, 39-41, 52, 58, 80-82, 83, 273, 327, 334-335, 346, 394, 426, 502, 515, 539-541, 643, 661, 667, 687, 714. A. Pais, *Subtle is the Lord*, Oxford University Press, (1982), p.504. R. Schulmann, *et al.*, Editors, *The Collected Papers of Albert Einstein*, Volume 8, Part A, Note 3, Princeton University Press, (1998), pp. 166-167. Letter from A. Einstein to A. S. Eddington of 2 February 1920, *The Collected Papers of Albert Einstein*, Volume 9, Document 293, Princeton University Press, (2004), p. 245. Letter from A. Einstein to "Berlin-Schöneberg Office of Taxation" of 10 February 1920, *The Collected Papers of Albert Einstein*, Volume 9, Document 306, Princeton University Press, (2004), pp. 256-257, at 256.

123. J. Stachel, *Einstein from 'B' to 'Z'*, Birkhäuser, Boston, (2002), pp. 60-61.

valid, let him stand at the beginning of the year and declare, 'Every vow which I may make in the future shall be null.¹ [HIS VOWS ARE THEN INVALID,] PROVIDING THAT HE REMEMBERS THIS AT THE TIME OF THE VOW. But if he remembers, he has cancelled the declaration and confirmed the vow?²—Abaye answered: Read: providing that it is *not* remembered at the time of the vow. Raba said, After all, it is as we said originally.³ Here the circumstances are e.g., that one stipulated at the beginning of the year, but does not know in reference to what. Now he vows. Hence, if he remembers [the stipulation] and he declares: 'I vow in accordance with my original intention', his vow has no reality. But if he does not declare thus, he has cancelled his stipulation and confirmed his vow."[124]

The political Zionists had successfully vilified Germans, and America's participation in the war which resulted from this deliberate vilification intensified the ill-will. Political Zionists have, from the very beginnings of their movement, employed smear tactic as their preferred response to legitimate criticism. Nachman Syrkin stated in 1898,

"Only cowards and spiritual degenerates will term Zionism a utopian movement."[125]

At the Sixth Zionist Congress in 1903, Max Nordau stated,

"After barely [*sic*] than a year's activity it called this Congress into being, a body to which none, but a few crazy Jewish opponents, denies the quality of legitimately representing the Jewish people. All serious people recognise that we are the executive and deliberate representatives of the Jewish people."[126]

The New York Times reported on 16 January 1917 on page 3,

"'We protagonists of universalism,' said Dr. Philipson, 'are being laughed to scorn. Our claim that Israel is an international religious community is being held up to ridicule. We are told that Israel can only survive by stressing its separatistic nationalism; that only by drawing ourselves off from our fellow inhabitants in the lands in which we live as a separate nationalistic group can we perpetuate Jewish life.'"

---

124. I. Epstein, Nedarim 23*b*, *The Babylonian Talmud*, Volume 19, The Soncino Press, London, (1936), p. 68.
125. N. Syrkin, under the nom de plume "Ben Elieser", *Die Judenfrage und der socialistische Judenstaat*, Steiger, Bern, (1898); English translation in A. Hertzberg, *The Zionist Idea*, Harper Torchbooks, New York, (1959), pp. 333-350, at 347.
126. "Dr. Nordau's Review of the Zionist Movement", *Supplement to the Jewish Chronicle*, (28 August 1903), pp. xi-xii, at xi.

The New York Times published a statement by Professor Ralph Philip Boas on 16 December 1917, Section 4, page 4—not long after the Balfour Declaration. Boas stated, *inter alia*,

> "Moreover, Zionism is continually emphasizing the breach between Jew and Christian which most of us are trying to bridge. As the child of anti-Semitism, it thrives on persecution. Its central argument is that Jews can never be at home in a 'foreign' land. It makes capital of every instance of petty intolerance and nourishes itself upon the ill-will which Jews are prone to fancy even when it is not present. The chip which many Jews bear more or less ostentatiously now that the yellow badge has been removed, some Zionists magnify into a veritable Pilgrim's burden which can drop from the bent back only upon the soil of Palestine. Zionists are continually heaping abuse upon the non-separatist, upon the man who has no desire to be different from other human beings and is very grateful that he does not have to be a marked man among men."

The truth is that the vast majority of Jews rejected the political Zionists. Political Zionist smear tactic was routine for Einstein supporters. *The New York Times* reported,

"EINSTEIN TO HAVE FREEDOM OF STATE

Senate Passes Resolution Honoring Visiting Scientist—Measure Before Assembly Today.

Special to The New York Times.

ALBANY, April 6.—The Board of Alderman having failed yesterday to extend to Drs. Albert Einstein and Chaim Weizmann, the Zionist emissaries, the freedom of the City of New York, the Senate today, by unanimous vote, extended to the distinguished visitors the freedom of the entire State of New York.

The resolution on which action was taken was sponsored by Senator Nathan Straus Jr. of New York, who characterized the failure of the Alderman to act on the Zeltner resolution as 'a disgrace.'

The text of the Straus resolution follows:

'Whereas Albert Einstein of Switzerland and Chaim Wezmann of Great Britain are now visiting our State; and

'Whereas the purpose of their visit is to cement the bonds of unity between the United States and her neighbors abroad in the great struggle for human progress and happiness, and especially to unite the old world and the new in establishing a cultural centre for the Jews of the World in Palestine; and

'Whereas the achievements of Dr. Einstein in the spheres of physics and astronomy have commanded the attention and the admiration of the entire civilized world, and the record of Dr. Weizmann as a chemist during the World War has made the people of the allied and associated powers his debtors, and,

'Whereas it the desire of the Commonwealth of New York to make these distinguished visitors feel that every true American heart goes out to them in cordial welcome; therefore,

'Be it resolved that (if the Assembly concurs) the people of the State of New York extend to Dr. Einstein and Dr. Chaim Weizmann and their associates the handclasp of fellowship and a heartfelt welcome.'

Senator Bernard Downing, another Democrat member from New York City, warmly eulogized the two Zionists and extolled their services to science and to mankind.

The Assembly had adjourned for the day when the Straus resolution was adopted, but upon reconvening tomorrow will have the measure before it for concurrent action.

FALCONER IS DENOUNCED.

Owasco Club Condemns Alderman
for Blocking Welcome to Einstein.

Resolutions denouncing Alderman Bruce Falconer for his action in blocking the resolution in the Board of Aldermen offering the freedom of the city to Professor Albert Einstein and his colleagues were passed at a meeting of the Owasco Club, the Democratic organization of the Seventeenth Assembly District, yesterday.

'The conduct of Alderman Falconer manifests a spirit of bigotry, narrow-mindedness and intolerance, and displays him as a champion of anti-Semitism, which is only a stepchild of anti-Americanism,' said the resolution."

The Judge found political opportunists who sought to make good on his threats and repeat his smears. One can only conclude that such hysteria in New York, such vicious and highly publicized smears and vindictive opportunistic attacks, must have had a chilling effect on the debate over relativity theory and Einstein's alleged originality. Such was the ignoble birth of the modern myth of St. Einstein's infallibility and originality—opposition was too often shouted down by smear tactic and intimidation—even by formal decree.

Falconer tried to calm and reassure the hysterical mob, who defamed him and sought to destroy his life. *The New York Times* reported on 9 April 1921

"FREEDOM OF CITY GIVEN TO EINSTEIN

Alderman Honor Relativity Discoverer
and Prof. Weizmann
Despite Falconer's Protest.

HE DEFENDS ADVERSE VOTE

Cites Courtesies to Dr. Cook, De
Valera, Mannix and Mrs.
MacSwiney as Mistakes.

Professor Albert Einstein, the noted mathematician and discoverer of relativity, and Professor Chaim Weizmann, British chemist now have the freedom of New York City. It was voted to them yesterday at a special meeting of the Board of Aldermen, made necessary by the refusal of Alderman Bruce Falconer to consent to the passage of the resolution when it first came up on Tuesday, when the two scientists were welcomed by Mayor Hylan at City Hall.

Alderman Falconer cast the only negative vote yesterday, and in so doing said he was not actuated by race prejudice, but that he had in mind the dignity of the honor which has been given to some of the greatest Americans, and thought it should not be conferred on any one unless he were known to every person in the city. He said his first ancestor in this country came as secretary to Lord Cornbury, the first person to receive the freedom of the city, in 1702.

Alderman William T. Collins, leader of the Democrats, seized upon the mention of Alderman Falconer's ancestors with avidity and ridiculed it.

'We on this board are just as proud of our city and of the conferring of the freedom of the city on guests as is Alderman Falconer,' he said. 'It was only narrowness and bigotry that made the one member of this board object to granting the freedom of the city to Dr. Weizmann and Professor Einstein.'

Alderman Falconer said that Alderman Friedman did him a great injustice in saying that his objection was based on race prejudice, and said that his private physician is a Jew and that many of his friends are Jews.

'In 1909,' he said, 'the keys of the city were unfortunately given by the Board of Alderman to Dr. Cook, who pretended to have discovered the North Pole, but were afterward officially withdrawn from him. After that the freedom of the city was not again extended for ten years, until the second year of the Hylan Administration, when it was given to Eamon de Valera, at a meeting which occurred when I happened to be away from the city.

'Since that time it has been extended to Cardinal Mercier, King Albert of Belgium, the Prince of Wales, Archbishop Mannix and Mrs. MacSwiney. At the time the resolution was suddenly proposed in connection with Archbishop Mannix, I did not vote in favor of conferring the honor upon him.

'The next and last individual upon whom this honor was conferred was Mrs. MacSwiney. I did not vote for it, and if I had had a proper chance would have objected.

'I have been assured,' he said, 'that Professor Einstein was born in Germany and was taken to Switzerland, but returned to Germany prior to the war. He is consequently a citizen of Germany, of an enemy country, and might be regarded as an alien enemy.'

Alderman Friedman told Alderman Falconer that Professor Einstein was not a citizen of Germany, but of Switzerland, and Alderman Vladeck, leader of the Socialists, also said that Professor Einstein was far from being a German citizen.

Alderman Ferrand, the Republican leader, in moving the question, said:

'For what has occurred I make no apology to this board or to the citizens of the city. It can be charged to no party. It can only be charged to an individual who is arrogant and ignorant. We will have to take it from whence it comes.'

Professor Einstein visited the College of the City of New York yesterday, and attended a class in mathematics and physics, where he listened to an explanation of his theory by Prof. Edward Kasner of Columbia University. President Sidney Mezes, of the City College, and a number of advanced students were present. Prof. Einstein, who understands English, although he does not speak it well, complimented Prof. Kasner on his presentation of the subject, and later made a twenty minute talk.

It was announced at Princeton University yesterday that Professor Einstein would be the guest of the University from May 9 to 15 and would give five lectures in that time on relativity."

On 11 April 1921, *The New York Times* began to see that Falconer had made a good point,

"A Ceremony in Need of Revision.

Now that the implacable FALCONER has been beaten and Dr. EINSTEIN possesses formally and officially the 'freedom of the city' that actually is granted to anybody from almost anywhere, it might be well to abandon the use of a phrase that long since ceased to have any meaning even remotely related to the words composing it. Then the ground would be cleared for its replacement by a designation indicative of a special municipal welcome, accorded to visitors made worthy of it by great achievements or honorable services.

With the ancient ceremony thus revised and brought into accord with modern conditions, Dr. EINSTEIN certainly would be among those thus honored by an appreciation not less honorable to those who manifested it, and at least it is to be hoped that the honor less often would be cheapened, as 'the freedom of the city' has been cheapened several times in recent years, by giving it to persons who—well, to persons whose claims for admiration and respect, unlike his, were not firmly founded on the unanimous opinion of competent judges."

It is noteworthy that the same newspaper which had called Einstein's theory "much-debated" on the front page on 3 April 1921, claimed one week later that there was unanimous support for it.

When Einstein visited Boston, they refused to award him the freedom of the city. *The New York Times Index* does not name any stories covering this event under "Einstein". All they list were their articles of May 18th and 19th of 1921. From 18 May 1921:

"EINSTEIN SEES BOSTON; FAILS ON EDISON TEST"

Asked to Tell Speed of Sound He
Refers Questioner to Text
Books.

Special to The New York Times.

BOSTON, May 17.—There was a large crowd at the South Station this morning to greet Professor Einstein of relativity fame and his party. From the station the visitors made an unexpected automobile tour through the north and west ends, Boston's Jewish quarters, and then proceeded to the Copley Plaza Hotel, where they sat down to breakfast with Governor Cox, Mayor Peters and some 75 distinguished guests.

Mrs. Weisemann, wife of Dr. Chaim Weisemann, of the visiting party, surprised the party when it came time to pass around the cigars by calmly producing a cigarette and lighting it. Her action was welcomed by the men. They wanted to smoke but hesitated to do so in the presence of Mrs. Weisemann and Mrs. Einstein, the only women present. Mrs. Weisemann's action in 'lighting up' paved the way and the men lit their cigars.

Professor Einstein gave out through his secretary the following message for Bostonians:

'I am happy to be in Boston. I have heard of Boston as one of the most famous cities of the world and the centre of education. I am happy to be here and expect to enjoy my visit to this city and Harvard.'

Of course the famous visitor had run into the ever-present Edison questionnaire controversy. He did not tackle the whole proposition but so far as he went failed and thereby became one of us. He was asked through his secretary, 'What is the speed of sound?' He could not say off-hand, he replied. He did not carry such information in his mind but it was readily available in text books.

Professor Einstein took issue with the famous inventor's contention that a college education is of little value. Professor Einstein said he believed education was a good thing. If a man had ability, he thought, a college education helped him to develop it. He stated he had not had an opportunity to study the Edison list of questions. He had heard of the American inventor in connection with the invention of the phonograph and electrical appliances.

Mrs. Einstein said that while Edison was an inventor who dealt with practical and material things, her husband was a theorist who dealt with problems of space and of the universe."

Einstein's "secretary" was Simon Ginsburg (a. k. a. "Salomon Ginzberg" and "Schlomo Ginossar"), who was the son of "Usher Ginsburg" (a. k. a. "Asher Ginberg" and "Ahad Ha'am"), who published under the *nom de plume* "Achad Ha-am". Ginsburg the elder was the secretary for the Odessa Committee for Palestine.

On 19 May 1921, *The New York Times* reported,

"Einstein Honored at Boston.

BOSTON, May 18.—Professor Albert Einstein, the scientist, and his associate, Professor Chaim Weizmann, were guests of Governor Cox at luncheon today. Professor Einstein had spent the forenoon at Harvard University, where he was received informally by President Lowell and members of the faculty. At his request he was escorted through the various college laboratories and museums."

In marked contrast to the long front page story *The New York Times* published upon Einstein's arrival to America, the notices of his departure were far more humble. On 30 May 1921, *The New York Times* wrote on page 8,

"EINSTEIN SAILS TODAY.

Dr. Weizmann Will Remain In Interests of Zionism.

Professor and Mrs. Einstein will sail for Europe today on the Celtic, leaving behind them some puzzled academic minds. Since he came to this country several weeks ago in the interests of the proposed University of Jerusalem Professor Einstein has been the centre of attraction for scientists who have heard him lecture on his famous theory of relativity. He has spoken at several universities and had the order of Doctor of Science conferred on him by Princeton University.

Dr. Chaim Weizmann of the World Zionist Organization and other members of the commission will remain here for a short time. Mrs. Weizmann, who is President of the Women's International Zionist Organization, which is trying to raise $5,000,000 for welfare work among Jewish women and children in Palestine, appealed yesterday for Jewish women to contribute their jewels and treasure, 'gold and silver, new and old,' to the fund."

and buried back on page 14 of the *The New York Times* of 31 May 1921 was,

"Prof. EINSTEIN SAILS.

Says Relativity Theory Is Receiving 'Sympathetic Dealing' Here.

Professor Albert Einstein, who has been lecturing in the United States for several weeks on his theory of relativity, sailed for Liverpool yesterday on the Celtic. In lieu of an interview, he gave out a formal statement in which he said:

'I would like to add that the respect and admiration that I always felt for American scientists have been greatly increased as a result of my personal contact with them. I have seen a sympathetic dealing with the theory of relativity and a truly detached scientific interest in it.'

Professor Einstein announced that he had refused to accept an invitation to be the guest of Lord Haldane in London, but gave no reason for his action. Mrs. Clara Louise Weizmann, wife of Chaim Weizmann, President of the World Zionist Organization, also was a passenger. Others who sailed were P. S. Hill, President of the Universal Leaf Tobacco Company; Martin Vogel, formerly Assistant United States Treasurer; Toscha Seidel, violinist; Karonongse, Siamese Minister to the United States; M. Ussichkin, Secretary of the World Zionist Organization, and Dr. George E. Vincent, head of the Rockefeller Foundation."

The joke was on those who made such a show of defending Einstein's "honor" and who went to such extraordinary lengths to cater to Einstein during his visit to America. Instead of exhibiting due gratitude, Albert Einstein ridiculed them and slandered America upon his return to Europe. He specifically attacked the American scientists whom he had earlier praised in his apparently scripted press statement quoted immediately above.

This spectacle did not go unnoticed in the foreign press.

While it is true that some of Einstein's critics were closet (unknown to Einstein) or public anti-Semites, it is also true that many were proud Jews, or

Gentiles without any anti-Semitic feelings. While anti-Semitism, which was common in Europe and America in the 1920's—even Einstein was an anti-Semite, was likely to bias its adherents and foster resentment in them of Einstein's public success, it did not in and of itself render legitimate scientific and philosophical non-race related arguments wrong, nor should it render such legitimate arguments taboo. The very bias of "race" prejudice provided an incentive for some to expose Einstein and the exposure of Einstein's plagiarism and irrationality is a good thing, even though "race" prejudice is not.

Einstein should not be pardoned and science should not be stagnated merely because Einstein was criticized by some who may have had more than one motive for exposing him. If the racism of important historical figures, in word or deed, should make it impossible for present day scholars to rely upon their non-race related arguments, we must burn the Bible, the Constitution of the United States of America, the Declaration of Independence, as well as the other writings of many of the Founding Fathers of America, and the works of Aristotle, Herbert Spencer, Albert Einstein, and countless others. Any "race" prejudice some of Einstein's critics may have had did not grant Einstein the license to plagiarize and deceive the public. Nor did it grant him the privilege to hide from debate over the merits of the theory of relativity. Prejudice did not convert Einstein's plagiarism into non-plagiarism, nor did it turn Einstein's irrationality into rationality. In addition, nothing prevents a person who has expressed a racist bias on one occasion, from making a true statement on another occasion. Einstein, who was himself an anti-Jewish and anti-"Gojisch" racist and a complete hypocrite, took the coward's way out to cover up his misdeeds, but that does not mean that it was untrue when he claimed to have been descended from Jewish parents. It is certainly true that Einstein had no integrity as a scientist, as a man, or as a Jew.

While racist bias is a factor to be considered when weighing the value of an opinion expressed by an individual, it by no means excludes the possibility that a given expression of opinion or fact is legitimate, logical and factually correct. To pretend otherwise is to supplant logic and truth with reactionary and irrational emotion. To pretend otherwise is to be biased against reason and fact, and amounts to the irrational assertion that dislike of the messenger gives one a right to discount the truth when it is convenient to do so. A debtor might as easily and irrationally pretend that her dislike of a creditor gives her a right to refuse to pay off a legitimate debt. A true fact becomes no less of a true fact merely because it is iterated by someone with a bias or an ulterior motive for expressing it. A debt legitimately due is not paid back by a mere expression of dislike, even if the dislike is warranted.

Some well meaning individuals have been duped into believing that it is a good thing to suppress a legitimate criticism made by any person who has ever uttered an untoward word towards a "race", and to bar every other person from repeating the same legitimate criticism, or to ridicule the criticism itself as a matter of course, even if made before adopted by a person with a known bias. No doubt most of these dupes are rather selective in their sanctions, privileging and excusing some racists like Einstein, while exaggerating the degree and the impact of the statements of others. That aside, such dupes ought recognize the proven danger of excusing corrupt Jews from criticism by any method, including the method of pointing out that a given critic of corrupt Jews has iterated a generally anti-Jewish sentiment. This practice provides corrupt Jews with an incentive to create and sponsor anti-Semitism and to create a class of professional anti-Semites, whose pronouncements shield corrupt Jews from criticism. Ultimately, the practice of inhibiting the criticism of corrupt Jews, or any Jewish icon, or even any Jew, sponsors Jewish corruption and will inevitably lead to a severe and unjust backlash against all Jews.

It is not surprising that Jewish critics criticize obvious examples of corruption by Jews. That does not place Jews above criticism. Nor does it mean that a non-racist person becomes a racist by noticing and commenting upon the same corruption by a corrupt Jew, which a known anti-Semite has criticized. Nor does it mean that a non-racist criticism of a corrupt Jew becomes racist if noticed and encouraged by a racist. If such were the case, a corrupt Jew could hire another person to pose as an anti-Semite and criticize the corrupt Jew, and then be shielded for life from criticism. More broadly, corrupt Jewish leaders and corrupt Jewish organizations could hire stooges and *agents provocateur* to pose as anti-Semites and make ridiculous anti-Semitic statements, together with legitimate statements of fact, and thereby stigmatize legitimate expressions of criticism as if the expression of "race hatred", *per se*. Such things have happened. Corrupt Jewish financiers paid Hitler's way,[127] and many who have legitimately criticized corruption by Jewish financiers have been likened to Hitler, who was paid by those same corrupt Jewish financiers to criticize them. Are we forbidden to criticize the financing of Adolf Hitler?

2.3 Harvard University Asks a Forbidden Question

In 1921, Ralph Philip Boas discussed a proposed quota system meant to prevent Jews, a small minority in America, from obtaining majority control over leading American universities. Boas employed racist apartheid arguments favoring Jewish domination of the universities, by attributing Jewish success in the colleges and universities to the alleged superiority of the Jewish "race". Boas largely ignored the controlling effects of circumstance, religion and culture. Limiting Jewish enrollment to proportional numbers would have opened the door to more representation by Blacks and other minorities—whether or not those doors would have remained open is a separate issue. Boas wrote in his article, "Who Shall Go to College?", *The Atlantic Monthly*, Volume 130, Number 4, (October, 1922), pp. 441-448, at 443-448:

> "Such methods of admission have been in use in many of the larger colleges during the last few years, quietly and effectively; there is little reason to

---

[127]. "Text of Untermyer's Address", *The New York Times*, (7 August 1933), p. 4. *See also:* "Untermyer Back, Greeted in Harbor", *The New York Times*, (7 August 1933), p. 4.

believe that they would have roused public discussion, had not Harvard, with candor worthy of her motto, thrown her cards upon the table and invited the country to discuss openly the question, Who shall go to college?

[***]

III

With the later immigration, however, the case was different. The great Jewish immigration, which began in the eighteen-eighties and still continues to the limit of the law, settled chiefly in the Eastern cities, especially, as it chanced, in or near the very cities where were the largest colleges: Philadelphia, New York, New Haven, and Boston. They brought with them an inherited tradition of education, intellects trained for centuries in the sharpest analysis and dialectic, a natural bent toward the professions, and—what, perhaps, is most important—the repression for years of their attempts to give these desires and characteristics free play. In time they acquired the economic independence necessary to send their children to college; where financial independence was lacking, those children undertook the burden of self-support with the tenacity of the race. There were no Jewish colleges founded for Jewish boys and girls, as with the Catholics, because there was no organized religious body to undertake their founding, and also because Jews have no desire for separation in anything except race and religion.

Now, it happened that Jews began to flock to the colleges at precisely the time when the colleges began to grow unwieldy in numbers and ill-assorted in membership. With the turn of the century, the old college simplicity began to disappear. Old buildings were supplemented by costly modern edifices. The fraternity house and the private dormitory were established to ease the pressure upon the college building funds. Athletics began to develop their present overwhelming importance. Fraternities established hundreds of new chapters. It became necessary to harmonize the differences between rich and poor, between the yearning for scholarship and the cultivation of useful leisure. It was the time when the colleges were violently criticized for their organization, their curricula, and their student life.

Added, therefore, to a burden of cares, came the problem of racial equilibrium. The number of Jews in the eastern colleges gradually increased, until to-day Jews would, were they permitted, in many cases form as much as fifty per cent of the students. The problem of what to do with other groups—negroes, Armenians, Italians—is as nothing when compared with the problem of the Jews.

In the first place, other groups have not the Jewish desire for education. At one remove from the immigrant quarter, other groups do not go to college. Success does not come to them with great rapidity, nor have they the same racial background of learning and scholarship which is, in some degree, in every Jew's blood. Then, too, other groups have not the Jew's adaptability. The Ethiopian cannot change his skin; but Jewish boys and girls differ from their Gentile companions often only in a racial tie so faint that insistence upon it is but a galling reminder of a difference that seems almost academic. Moreover, Jews themselves are the most incoherent of racial groups, varying from the most cultivated, who have acquired the most conservative traditions of Americanism, to the most blatant, who know no traditions except those of oppression. And the urban environment of Eastern colleges has a full case of Jewish types, with the more noticeable, as always, setting the standard of judgment of the race as a whole. Finally, the Jew is the most successful of the newer groups in college. The success of Jews in scholarship is a byword. Rarely a list of honors appears which does not contain Jewish names. When a Jew puts his mind upon achievement, he usually secures what he aims for. He pursues success in scholarship with an intensity and a singleness of purpose which make him at least noticeable. What his hand finds to do, he does with all his might. Fatal gift! If only Jews would be content with mediocrity, the 'Jewish problem' might automatically disappear.

It is not the mere number of Jews, nor their undoubted prominence in scholarship, which complicates the problem. The American college is not, and never has been, an institution primarily for the acquisition of knowledge or the attainment of degrees. It is a social organization, with a very highly organized social structure. In most colleges this structure rests upon a basis of fraternities and clubs, with unwritten rules more rigid than those which govern the most exclusive society, administered with all the relentlessness of youth. It is hard to believe that young men have any inherent objection to their Jewish fellow students as individuals. But the organizations to which they belong have an inherent objection to Jews in the mass. In the admission of Jews they see the subtle undermining of a social prestige which they must preserve, or perish. So far as the classroom is concerned, Jewish students are one thing; but at the 'prom,' or the class-day tea, the presence of Jews and their relatives ruins the tone which must be maintained if social standing is not to collapse. The result of the presence of a large number of students who are themselves not any too welcome at college affairs, and whose relatives are positively impossible, is necessarily disunion and strife within the social life of the college. Jews are naturally clannish, and the social discrimination which they constantly feel makes them doubly so. Isolated as they are, at a time of life and in an environment where isolation is poison, they create a group always sore, always aloof, always a thorn in the side of deans and presidents, who want unity above everything. Where Jewish fraternities and clubs are permitted, the situation becomes worse. Discontent, the gnawing sense of being unjustly treated, the rancor of a brilliant mind forced into social inferiority—these things become articulate and even vociferous; a sense of injustice crystallizes. Then too, the Jewish fraternities necessarily exclude some Jews, and there is left a poor, struggling, often unpleasant remnant, suffering from an aggravated inferiority complex, which makes them mere hangers-on of the collegiate society; men who are using the college for the financial gain of a college degree, men who make neither useful citizens of the college community nor alumni of whom the college can be proud.

The thought which comes into the mind of every right-thinking person is the essential injustice of the situation. In most cases Jewish students are men of good character and fair scholarship. As far as can be learned, they give no trouble to the disciplinary officers. Being what they are, they are despised and rejected; and, being despised and rejected, they develop all their worst traits instead of their best. Were charity, friendliness, forbearance, and kindliness the outstanding characteristics of college men, students of unpleasant personality could be made better college men and better citizens. But these characteristics are no more true of college men than of any group of people. Rather less so, indeed, for young people are notoriously snobbish, hero-worshiping, and intolerant of eccentricity. College authorities, however good their will may be, have not the power to reform the social prejudices of college students. Hence arises a dilemma: either the social nature of a college body must be changed and a new point of view adopted—which seems impossible; or the groups of students who interfere with the harmonious functioning of this social nature must be limited—which rouses a storm of protest.

Those who know the colleges of the East will have little doubt of the outcome: it is easier to endure a storm of protest than to change a point of view. It must be remembered that the point of view has been the slow development of years, and is held alike by trustees, faculties, and alumni.

IV

If the American college were an institution which aimed to find the sharpest brains of the country and to cultivate them, the problem of the limitation of enrollment would be simple. Jews would have nothing to fear from such a system. The bright minds would be admitted; the dull minds would be rejected; and among the successful would unquestionably be the high percentage of Jews who always succeed in an open competition where brains count most.

But, for good or ill, the endowed colleges are not looking for the sharpest brains. In general they would probably like to think of themselves as worthy of Hilaire Behloc's praise:—

Here is a House that armours a man
    With the eyes of a boy and the heart of a
        ranger,
And a laughing way in the teeth of the world
    And a holy hunger and thirst for danger:
Balliol made me, Balliol fed me
    Whatever I had she gave me again:
And the best of Balliol loved and led me,—
    God be with you, Balliol men.

It is obvious that such a conception of college means a careful selection of students to form a type. It means scholarship, to be sure; but it means also, as the presidents of Brown and Bates have stated publicly, that scholarship shall be only one qualification for candidates. Character, personality, the chances of the student's being a leader in life, social adaptability, the power to make friends, eligibility to social circles, conformity to discipline and to accepted thoughts and usages—these formally become the important criteria of admission, as they have been informally, in many cases, for several years. It is needless to say that such a conception of educational eligibility would exclude a large proportion of Jewish students, all negroes, and most members of other immigrant groups; and, with an ever increasing number of candidates for admission, would put a premium upon training in the great private schools.

Once accepted, this idea marks an epoch in American education, the full significance of which most people can hardly recognize, especially when it is remembered that, as the college is, so are large numbers of schools. It means the abandonment of scholastic achievement as the criterion of collegiate success; it means the creation of 'gentlemen's' colleges, as we have had, for a long time, 'gentlemen's' schools; it means the establishment of state universities which will be consciously for the masses, as opposed to 'aristocratic' groups; and it means that the colleges which, though perhaps grudgingly and even unconsciously, have been a powerful agent in Americanization, will now give up that work.

The matter of justice does not enter into this discussion, provided state and municipal colleges are called into existence to give the education which is the right of every qualified youth in a democracy. It is education which counts as a right, not education in any specific college. If Harvard, Yale, Princeton, Columbia, and other endowed colleges feel that social homogeneity is the most important thing in the world for them, they have the right to secure that homogeneity, so long as they maintain no monopoly of college education. It may matter intensely to the alumnus of a great college that his son should go to that college in the same environment which he enjoyed; the young man of immigrant stock, to whom that environment means nothing, ought not make the gratification of that desire impossible, so long as he personally can get his education elsewhere, and so long as the great graduate schools are free to all comers who are properly qualified. It is the thing which matters, not the place in which the thing is obtained. If, for good or ill, colleges wish to stand apart from the incoherencies and the clashings of our changing social life, they have a right to do so, as long as they encourage the founding and maintenance of new institutions which will provide an education for all qualified candidates. It is well to remember, however, that in the past the endowed colleges have opposed the establishment of state universities, and that some of them have already undertaken a policy of exclusion of Jews without informing the public, and without giving a thought, apparently, to the question where the rejected students are to be educated. One of the bad features of the present discussion is the reticence of most college authorities, who permit rumors and sensational news reports to take the place of frank and open discussion, so that the public mind is befogged and confused by anybody who chooses to start a sensational story.

Though the question of justice may be put aside, the question of wisdom may properly enter into the discussion. The important thing is, after all, not what charters permit colleges to do, but what their self-respect, their desire to serve their students and their community, and their best interests in the future tell them they ought to do. Under a policy of exclusion of certain racial groups, of preferring the development of social qualities to active scholastic competition, the colleges are bound to lose more than they will gain. They may be pleasanter places to live in, but they will no longer really represent the eager, heterogeneous, varied amalgam which is America. Young men will be protected from the presence of new Americans at the very age when they ought to be making contacts which will give them real knowledge of actual civic life. There is something disquieting, too, in the thought that their enthusiasm for democracy is so slight that they demand shelter from its perplexities and from its dangers. American college life, surely, ought to be more than a pleasant interlude; it ought to be a stirring achievement.

Most disquieting of all, however, is the feeling that, in the perpetual fight against bigotry, superstition, racial intolerance, and inverted nationalism, the colleges seem to be abandoning the side of the angels. It may be hard to see one's college harboring strange men with alien ways, to see the happy spirit of youthful friendship weakening beneath the fierce and relentless pursuit of knowledge which, to these strangers, is the whole of college life; but it is harder to see one's college the fostering mother of hates and racial dissensions, the parent of bitterness which for years will be a canker in the minds of men. Colleges will doubtless say that, in selecting their students in their own way, they have no such purpose. However, what usually matters is not the purpose of an act, but its result."

2.4 Americans React to the Invasion of Eastern European Jews

The effects of the Eastern European Jews' influence on American society appeared not only in the universities, but in the motion picture industry, which Jews monopolized in the Teens of the Twentieth Century—a fact which is widely acknowledged and celebrated by Jews today.[128] They did not use their monopolization of that industry, which was largely built by Thomas Edison, then stolen from him, to promote strong moral values and collegiate aspirations in

---

128. "Motion Pictures", *Encyclopaedia Judaica*, Volume 12 MIN-O, Macmillan, Jerusalem, (1971), cols. 446-476. *See also:* "Television and Radio", *Encyclopaedia Judaica*, Volume 15 SM-UN, Macmillan, Jerusalem, (1971), cols. 927-931. *See also:* N. Gabler, *An Empire of Their Own: How the Jews Invented Hollywood*, Crown Publishers, New York, (1988). *See also:* M. Medved, "Jews Run Hollywood, So What?", *Moment*, (August, 1996). **See** *also: Hollywood: An Empire of Their Own*, Video Documentary by A&E, Directed by Simcha Jacobovici, Originally Aired as *Hollywoodism: Jews, Movies and the American Dream*, (1997).

their Gentile neighbors in America. They did not promote the dignity of Black Americans and encourage them to pursue higher education. On the contrary, the Eastern European Jews glorified crime, violence, perpetual war, and vice in the form of tobacco and alcohol consumption—industries dominated by Jews. Eastern European Jews created an intensely anti-intellectual spirit in American Gentile culture, which impacted most strongly and negatively upon American Blacks. Their apartheid anti-Black mythologies became self-fulfilling prophecies.

The Jewish movie moguls degraded Blacks,[129] while stealing their cultural achievements in dance and music. The Jewish movie moguls sexually exploited actors and actresses and prompted their use of drugs, and promoted cultural decadence in general. In addition, some Jews corruptly kept Blacks from reaping the profits of their own labors and talents in the music industry. Jews, long engaged in the slave trade,[130] were the first racists to fabricate religious racial myths which relegated Blacks specifically, and Gentiles in general, to a sub-human slave status. These movie moguls, who were mostly Eastern European Jews, taught American Gentiles to loathe wealth accumulation and promoted the Communist myth of the "working-class hero" as an ideal aspiration for American youth. They also promoted the Communist ideal of "race" mixing. Jews generally taught their own children to segregate and pursue higher education and the professions.

Frederick T. Gates used Rockefeller's money to finance institutions of higher learning which benefitted Jews, while promoting the idea that Gentile students should be readied for factory work and work as field hands and farmers.[131] Charlotte Thomson Iserbyt wrote in her book *The Deliberate Dumbing Down of America: A Chronological Paper Trail*, Conscience Press, Ravenna, Ohio, (1999), p. 9, which is available online: http://deliberatedumbingdown.com/MomsPDFs/DDoA.pdf,

"1913"

JOHN D. ROCKEFELLER, JR.'S DIRECTOR OF CHARITY FOR THE ROCKEFELLER FOUNDATION, Frederick T. Gates, set up the Southern Education Board (SEB), which was later incorporated into the General Education Board (GEB) in 1913, setting in motion 'the deliberate dumbing down of America.' *The Country School of Tomorrow: Occasional Papers No. 1* (General Education Board: New York, 1913) written by Frederick T. Gates contained a section entitled 'A Vision of the Remedy' in which he wrote the following:

---

129. J. J. Goldberg, *Jewish Power: Inside the American Jewish Establishment*, Addison-Wesley, New York, (1996), pp. 327-328.
130. P. S. Mowrer, "The Assimilation of Israel", *The Atlantic Monthly*, Volume 128, Number 1, (July, 1921), pp. 101-110, at 106.
131. "Hylan Takes a Stand on National Issues", *The New York Times*, (27 March 1922), p. 3.

Is there aught of remedy for this neglect of rural life? Let us, at least, yield ourselves to the gratifications of a beautiful dream that there is. In our dream, we have limitless resources, and the people yield themselves with perfect docility to our molding hand. The present educational conventions fade from our minds; and, unhampered by tradition, we work our own good will upon a grateful and responsive rural folk. We shall not try to make these people or any of their children into philosophers or men of learning or of science. We are not to raise up from among them authors, orators, poets, or men of letters. We shall not search for embryo great artists, painters, musicians. Nor will we cherish even the humbler ambition to raise up from among them lawyers, doctors, preachers, politicians, statesmen, of whom we now have ample supply."

The book of *Obadiah* verse 8 teaches the Jews to destroy the intellectual class of non-Jews and deprive the Gentiles of knowledge,

"Shall I not in that day, saith the LORD, even destroy the wise *men* out of Edom, and understanding out of the mount of Esau?"

Through their disproportionate wealth and their ownership of the mass media, as well as through disproportionate representation in colleges and universities, Eastern European Jews corrupted American culture to suit their own ends and to degenerate American Gentile society. Neal Gabler boasted in the film documentary *Hollywood: An Empire of Their Own*, Video Documentary by A&E, directed by Simcha Jacobovici, which originally aired as *Hollywoodism: Jews, Movies and the American Dream*, in 1997,

"They created their own America. An America which is not the real America, it's their own version of the real America. But ultimately this shadow America becomes so popular and so widely disseminated, that its images and its values come to devour the real America. And so the grand irony of all of Hollywood is that Americans come to define themselves by the shadow America that was created by Eastern European Jewish immigrants, who weren't admitted to the precincts of the real America."

The corruption of American culture by Eastern European Jews in the motion picture industry was already apparent in 1921, a few short years after it had begun. In the Nineteenth Century, composer Richard Wagner had criticized the Jewish monopolization and corruption of the opera. In 1921, Ralph Philip Boas, a Jew, criticized the Jewish monopolization of the motion picture and clothing industries—as did THE DEARBORN INDEPENDENT.

German Jews owned sweat shops in Chicago and New York. German Jews exploited Eastern European Jewish labor in these clothing factories. The Eastern European Jews, descendants of the Frankists, took the opportunity to infiltrate American society with Communism and Anarchism by means of the labor unions, which they attempted to subvert—in many instances did subvert. Americans were leery of murderous Jewish Bolshevism, having witnessed the mass murders of millions of Russian Christians. Boas wrote in 1921,

> "And of all non-Saxon groups Jews are the most obvious, because of their temperament, their appearance, their ability, and, above all, their fatal gift of complete absorption in the game of life. They have never acquired the habit of nonchalance. Every Jew has in him the making of a thoroughgoing fanatic. It is his greatness and his doom. It has placed him in the front rank of greatness and it has made him a marked man, the prey of a complex of repressions and of fears. He cannot hide himself if he would; and wherever he is, he must live with the eyes of the world upon him.
>
> Jews are not accustomed to take stock of their own shortcomings. Persecution has saved them the trouble. To be alive at all after twenty centuries is in itself a triumph, which can excuse a few faults. Moreover, Judaism as a religion has been but little given to spiritual introspection. The consciousness of a guilty soul, the dread of eternal punishment, the longing to be one with God, the search for salvation, all the yearning mysticism which, to the Christian, is the very life and essence of religion, means comparatively little to the religious Jew. The Jewish religion is a stately monotheism, with a dignified and noble system of ethics and a theology and code of laws which lie at the basis of modern civilization. But this religion is an intellectual possession—it is not a haven for perturbed spirits, a beacon for the troubled wayfarer, a life-giving draught for parched souls. Jews, when attacked, do not rally to the defense of their religion: they rally to the defense of their good name as a social group. It is but rarely that Jews talk of religion: they take it for granted. But they talk vehemently of their rights as an oppressed people, or of social justice, or of their contributions to civilization. The triumph of prophetic Judaism over the Judaism of the Psalmist explains the shortcomings of Jews in the very points that are made most of by their critics. The greatest Orthodox rabbis are interpreters of the law; the greatest Reform rabbis are prophets of social righteousness. There are few to preach that teaching which Jews most need—personal consecration to righteousness, humility in success, a gentleman's regard for the sensitiveness of others, a willingness to yield one's legal rights before the quality of mercy. And yet it is this very preaching that thoughtful Jews the country over are craving, hardly conscious of what they crave. The time is ripe for the coming of a personality who will interpret in his life and his teaching the spirit that is dimly conscious in the hearts of many Jews.

II

These shortcomings of the Jews explain the concrete criticisms that Americans constantly make, not as conscious anti-Semites, but in all friendliness and good-will. They see that Jews form large settlements in our great cities. Are the cities better for their presence? They see that Jews virtually control certain businesses—for example, the clothing trade, the theatre, and the department store. They ask themselves if these businesses are the better because of Jewish control. Has Jewish domination of the theatre improved theatrical art and morals? Has Jewish domination of the clothing trade shown an example of the progress that can be made toward industrial peace? And these questions are asked, not by foolish theorists, who shrink at the spectacle of Jewish world-domination, not by anti-Semites, who are impervious to ideas of justice and fair play, but by thoughtful and fair-minded Americans, whose memories are long enough to recall a day when Jews were refugees from persecution, craving sanctuary in a land of freedom.

And it is these questions which Jews proud of their heritage and jealous of their good name would gladly avoid answering; for the truth is painful and disillusioning. There is but one answer. Theatres and clothing trade alike are controlled by two passions: a passion for wealth and a passion for power. Thoughtful Jews have no defense for the condition in which the theatre finds itself to-day: the drama gone, driven out by salacious and gaudy spectacle; the moving picture keeping just within the law, seemingly ignorant of any artistic responsibility, and as carefully devised for the extraction of dollars as a window-display of women's finery. It is the bald commercialism of the whole business that is so discouraging—its utter lack of moral and artistic altruism, its cultivation of a background of triviality and immorality. That the American public has allowed itself to be artistically debauched is no excuse for the men who have served up the poisonous fare. They have betrayed their heritage and their race; they have been worse than a wilderness of anti-Semites. For they have created a condition in which their success has furnished a fuel for racial attack that no amount of regulation anti-Semitic propaganda could have furnished; they have made the great refusal. A chance that no theatrical producers in the world have ever had was theirs, and they have, with deliberate cynicism, thrown it way. Their argument that they were merely giving the public what it wanted is worthless, for they have created their public. Nor is their other defense any better. What they have done, it is maintained, they have done, not as Jews, but as other Americans. Yet they remain Jews to themselves and to the world. And they are not as other Americans. They are marked men, heirs of the noble ideals of a race which gave Western civilization religion and morals. And they have betrayed their race for twenty pieces of silver [*Genesis* 37:26-28].

In a lesser degree, the same is true of the clothing trade. Sweating of labor, cutthroat competition, an utter inability to coöperate and compromise, chicanery, pettiness, reaction—all these have characterized this industry. And although, fortunately, some of the great clothing manufacturers have shown a wisely progressive spirit in their relations with their employees, and have set a standard that others would do well to follow, yet it is certainly true that in one of the greatest sections of the clothing-trade, obstinacy, an exaggerated individualism, and stubborn reaction characterize the employers; fanaticism and doctrinaire social theories characterize the employees. The sobering fact for the Jewish apologist is that, in too many cases, when Jews control an industry, they do not improve it: they merely make it more lucrative.

All this is, of course, only to say that Jews, being highly imitative and adaptable, have thoroughly mastered one kind of American business method, the method of driving and selfish efficiency. What the Steel Corporation has done on a large scale, the clothing manufacturers have done on a small scale. Jews have learned well the lesson of American industrial exploitation. But the defense, true as it is, will bear little weight with the public; for the Jews have the misfortune to control enterprises that are constantly before the public. Christian control of steel mills and copper mines may be even worse than Jewish control of clothing shops and motion picture theatres, but the steel mills and the mines are beyond the view of the great American public, while everyone comes in daily contact with the theatre and the clothing shop. Jews in their business life have a fatal obviousness—all the world reads their names on the signs of Fifth Avenue and Broadway; who visits the steel mills of Bethlehem, or the mines of Anaconda?"[132]

Perhaps the examples Jews had set in the motion picture and clothing trades were among the reasons why Americans were reluctant to hand over influential American universities to "Eastern Jews". *The World's Work* published the following article in August of 1922,

"The Jews and the Colleges"

THE ever-increasing importance which the Jewish question is assuming in American life is apparent in the way that it is agitating the colleges. Like every problem affecting Jewish immigration this one is primarily a city problem. It is only the colleges and universities located in or near large cities that feel the necessity of restricting their Jewish students. Again this particular phase of a daily increasing perplexity affects only one element among the Jewish citizenry—and that is the Russian or Polish Jews.

If the public can only get this latter fact clearly in mind the so-called Jewish question will appear in a clearer light. The large Jewish communities which are now found in most American cities are of comparatively recent growth. Jewish immigration to the United States has three well defined phases. At the time of the American Revolution there were only about 2,000 Jews in this country. Practically all of these were Spanish or Portuguese Jews, or their descendants; they had for centuries represented, as they do at the present time, the aristocracy of their race. They lived on the terms of the utmost friendship and respect with their Gentile compatriots; they occupy an important position in Jewish history, for the new American Constitution completely freed and enfranchised them; they were thus the first Jews since

---

132. R. P. Boas, "Jew-Baiting in America", *The Atlantic Monthly*, Volume 127, Number 5, (May, 1921), pp. 658-665, at 662-664.

the fall of Jerusalem that had ever been admitted as citizens of a free state on terms of exact equality with all other citizens.

The second phase of Jewish immigration came from Germany and was part of the general German immigration that began in the 'forties. These German Jews had for centuries lived in an environment which, while cruelly intolerant and discriminating on the social side, had still opened to them most of the economic and educational advantages that go with a superior civilization. These German Jews represented a comparatively small group; they were intelligent and industrious and for the most part prosperous; their habits and tastes were not materially different from those of the people among whom they lived; their children attended the public schools and the higher institutions and mingled, frequently on terms of intimacy, always on terms of good feeling and tolerance, with the offspring of the old established breed. More often than not they were 'unorthodox' in religion; most of them had long since abandoned the dietary practices that cause the Jews to be regarded as a peculiar people. Among them had originated the so-called 'reform' movement in religion; this was fundamentally an attempt to make their religious services lose something of their exotic flavor and correspond somewhat to that of their Christian brethren. The question of the assimilation of the German Jews was hardly ever discussed; their capacity for citizenship was taken for granted and the high position that they frequently attained in the arts, in education, science, and the professions certainly indicated that they had qualities that would be useful in our common American life.

About 1881, however, the systematic persecution of the Jews began in Russia, and from that time dates that enormous influx of Russian Jews which only the recent immigration laws have temporarily checked. The coming of the Russian and Polish Jews—a better term is Eastern Jews—forms the third chapter in the story of Jewish immigration. These Jews were almost as alien to our Spanish and German Jewish population as they were to the native American stock. They came from a country where even the Christian population had for centuries lived in ignorance, uncleanliness, and squalor; their lives had always been an almost hopeless struggle against disease and poverty; to them the old proverb, 'as rich as a Jew' certainly was a cruel misnomer, for as a mass they were extremely poor—as they are still. These representatives of their race presented far greater problems in assimilation than did their predecessors. A greater proportion were orthodox in religion; their racial conciousness had been sharpened by especially atrocious segregation and ill treatment; and as a mass they had had little training in the amenities and delicacies of civilized existence. In their struggles in the new country they developed a competitive zeal that usually made them the conquerors of the occupations in which they specialized. Their competition was especially directed against their own co-religionists. Before they came, the German Jew had been the master of the clothing trades; but the Russian Jew eventually supplanted him; and so it was in other lines.

The second generation of this immigrant body has now reached college age; the Jews have always shown a great aptitude for education, and it is to be expected that they would enter the universities in great numbers. It is only the universities located in large cities that especially feel this pressure. In New York the City College has long been almost exclusively a Jewish institution; New York University is probably seventy-five per cent. Jewish; at one time Columbia had a quota of forty per cent. though the proportion is now believed not to be so large. Yale has a comparatively small number—perhaps 10 per cent.; such places as Dartmouth, Princeton, Williams, and Amherst have practically none; the reason is that the first is located in a comparatively small city, and thus has a smaller Russian Jewish colony to draw upon, while the others are located in the country. The point is that nearly all this Jewish influx comes from the university town itself. Harvard, being near a large urban community, naturally has a larger proportion. The newspaper reports place this at 20 per cent. and President Lowell, in a recent letter, apparently foresees the early day when this will amount to 40.

Such a proportion means more than that Harvard would become, to a great extent, a Jewish institution. It means that its character would be completely changed. Like Yale and Princeton, the Cambridge University is national in scope; it draws its students from all parts of the United States. But the Eastern Jews who are hammering for admission come almost entirely from the Boston community. Most of them live at their own homes and thus do not become part and parcel of the college life. If they number 40 per cent.—and this proportion is likely to increase as time goes on—Harvard will lose its national character to that extent, and be a place given up largely to educating the sons of a particular racial element living in Boston. That is the present function of the City College of New York and New York University, though at the beginning they too were educational institutions of wider scope. There is therefore every reason why the Harvard authorities should deal frankly with this situation."[133]

2.4.1 Jewish Disloyalty

Whereas the prejudice Eastern European Jews faced from Western Jews was principally racism, the "anti-Semitism" the Jews of Eastern Europe faced from Gentiles was primarily political and economic. It resulted from the Jews' harboring loyalty only to the chosen "race" of the "House of Israel", while being openly disloyal to the Nation States in which they resided.

For example, in Poland the Jews segregated themselves into Ghettoes, and sought to take Polish land and turn it into a Jewish nation. In 1911, Israel Zangwill wrote in his article "The Jewish Race",

---

[133]. "The Jews and the Colleges", *The World's Work*, Volume 44, Number 4, (August, 1922), pp. 351-352.

"But if from the Gentile point of view the Jewish problem is an artificial creation, there is a very real Jewish problem from the Jewish point of view—a problem which grows in exact proportion to the diminution of the artificial problem. Orthodox Judaism in the diaspora cannot exist except in a Ghetto, whether imposed from without or evolved from within."[134]

Paul Scott Mowrer wrote in 1921,

"The Ghetto, which the Jews had formed of their own free will, was now imposed on them by force."[135]

In 1923, Burton J. Hendrick wrote in his article, "The Jews in America: III. The 'Menace' of the Polish Jew",

"The orthodox Jew in Poland not only lives, by preference, in crowded ghettoes in the cities, but he dresses in a way—a long gabardine of black cloth reaching to his ankles and a skull cap trimmed with fur—which emphasizes his Jewish particularism."[136]

Burton J. Hendrick also wrote in 1923,

"[Polish Jews] always resented—as they do to-day—the idea that they were Poles or a part of the Polish State; they insisted on being Jews and nothing else. Nor does it seem to be the case that the Jews in Poland were compelled to lead a distinct existence by the Government as a part of an anti-Jewish policy; the Ghetto was their own creation and their own choice; the fact that they were able to enjoy this privilege and many others, was what made their sojourn in Poland so agreeable and so free from the persecutions to which they were subject in other countries."[137]

Jan Drohojowski wrote in 1937,

"Let's nevertheless consider the origins of the 'ghetto'. To many it may seem that Jews have been mercilessly sequestrated in 'ghettos' by cruel Poles or other Christians. The truth is that the 'ghetto' is a purely Jewish arrangement. The 'erub ha-azaroth', a chain or wire joining two, or more,

---

[134]. I. Zangwill, *The Problem of the Jewish Race*, Judean Publishing Company, New York, (1914), p. 7; which was first published as an article, "The Jewish Race", *The Independent*, Volume 71, Number 3271, (10 August 1911), pp. 288-295, at 289.
[135]. P. S. Mowrer, "The Assimilation of Israel", *The Atlantic Monthly*, Volume 128, Number 1, (July, 1921), pp. 101-110, at 104.
[136]. B. J. Hendrick, "The Jews in America: III The 'Menace' of the Polish Jew", *The World's Work*, Volume 44, Number 4, (February, 1923), pp. 366-377, at 368.
[137]. B. J. Hendrick, "Radicalism among the Polish Jews", *The World's Work*, Volume 44, Number 6, (April, 1923), pp. 591-601, at 593.

homes permits the Jew to obviate some prescription regarding the Sabbath. Gradually entire Jewish districts were wired. In such manner Jews separated themselves from Christians."[138]

Adolf Eichmann stated in 1960,

" I would not say I originated the ghetto system. That would be to claim too great a distinction. The father of the ghetto system was the orthodox Jew, who wanted to remain by himself. In 1939, when we marched into Poland, we had found a system of ghettos already in existence, begun and maintained by the Jews. We merely regulated those, sealed them off with walls and barbed wire and included even more Jews than were already dwelling in them. The assimilated Jew was of course very unhappy about being moved to a ghetto. But the Orthodox were pleased with the arrangement, as were the Zionists. The latter found ghettos a wonderful device for accustoming Jews to community living. Dr. Epstein from Berlin once said to me that Jewry was grateful for the chance I gave it to learn community life at the ghetto I founded at Theresienstadt, 40 miles from Prague. He said it made an excellent school for the future in Israel. The assimilated Jews found ghetto life degrading, and non-Jews may have seen an unpleasant element of force in it. But basically most Jews feel well and happy in their ghetto life, which cultivates their peculiar sense of unity."[139]

Polish Jews strongly resented any assertion that they ought to become Poles, and saw themselves only as Jews—Jews who spoke Yiddish, not Polish. Jewish apologists were obliged to recognize that modern anti-Semitism was largely a political reaction by Gentiles to anti-Gentile Jewish racism and Jewish supremacism. Racist Zionist Theodor Herzl believed that religious anti-Semitism was a thing of the past, and that political anti-Semitism is fully justified. In an article entitled, "The Jewish State Idea", in *The New York Times*, 15 August 1897, on page 9, it states,

"Dr. Herzl says that anti-Semitism is economic and social, not religious—and the cure, therefore, is the establishment of the Jewish State. [***] In answer to his critics, Dr. Herzl reasserts his claims, and adds that the resettlement of Palestine by Jews would avoid European complications as to national interests there; that it would come to the aid of shattered Turkish finances by paying a tribute of $500,000 per annum, guaranteeing a loan of $10,000,000, and that this tribute should be increased in proportion to the

---

138. J. Drohojowski, *Brief Outline of the Jewish Problem in Poland*, Polish National Alliance of Brooklyn, U.S.A. (Zjednoczenie Polsko Narodowe), Brooklyn, New York, (1937), p. 22.
139. A. Eichmann, "Eichmann Tells His Own Damning Story", *Life Magazine*, Volume 49, Number 22, (28 November 1960), pp. 19-25, 101-112; at 106; *see also:* "Eichmann's Own Story: Part II", *Life Magazine*, (5 December 1960), pp. 146-161.

increasing population."

Paul Scott Mowrer wrote in 1921,

"This cause [of popular sentiment against the Jews] is neither religious, as is often averred, nor economic, as many believe; it is political. It is based on the observation that the Jews, through innumerable transmutations of time and place, not only have kept their identity as a people, but have opposed a vigorous, if passive, resistance to most attempts at assimilation. The Jew, in short, is regarded as a foreigner, whose 'laws are diverse from all people'; and as such, he is considered to be an enemy to the state.

The underlying reason for Jewish exclusiveness is, perhaps, the law of Moses. The sole object of life, according to the teachings of the rabbis, is the knowledge and the practice of the law, for 'without the law, without Israel to practise it, the world would not be. God would resolve it into chaos. And the world will know happiness only when it submits to the universal empire of the law, that is to say, to the empire of the Jews. In consequence, the Jewish people is the people chosen by God as the depository of his will and his desires.' This strong and narrow spirit, instead of diminishing with the lapse of time, seemed only to increase; until, with the victory of the rabbis over the more liberal Jewish schismatists, in the fourteenth century, the doctors of the synagogue, says Bernard Lazare, 'had reached their end. They had cut off Israel from the community of peoples; they had made of it a being fierce and solitary, rebellious to all law, hostile to all fraternity, closed to all beautiful, noble or generous ideas; they had made of it a nation small and miserable, soured by isolation, stupefied by a narrow education, demoralized and corrupted by an unjustifiable pride.' [***] The Ghetto, which the Jews had formed of their own free will, was now imposed on them by force. [***] But though many Western European Jews have been more or less assimilated during the last hundred years, there are still many others who, though emancipated so far as external restrictions are concerned, have not desired, or have been unable, to shake off the clannishness, the peculiar mentality, inbred by twenty or thirty centuries of almost unbroken tradition; they may not go to synagogue, or even to the reformed tabernacle, but they would be repelled at the idea of marrying outside the race, and they preserve a special and seemingly ineradicable tenderness for their fellow Israelites, of no matter what social stratum, or what geographical subdivision. [***] The restrictive measures of the prevailing governments have merely served to accentuate a distinction ardently desired by the Jews themselves, whose devotion to both the civil and religious aspects of the Jewish Law is here as fervent as it is complete. The net result is that the typical Polish Jew, like the Lithuanian, Bessarabian, and Ukranian Jew, is a being absolutely apart from his Christian neighbors. [***] We are thus, in the end, brought squarely back again to the surmise from which we started, namely, that the Jewish question is, above all, political, and may indeed be reduced to this one inquiry: Is it, or is it not, possible to assimilate the Jews?"[140]

In an article entitled, "Mr. Balfour on Zionism", *The London Times* wrote on 12 February 1919 on page 9, that Arthur James Balfour, who had signed the "Balfour Declaration" and issued it to the Jewish financier Rothschild, stated that the Jews of Eastern Europe were racists and were disloyal to their home States,

"MR. BALFOUR ON ZIONISM.
THE CASE FOR A NATIONAL HOME.

Mr. Balfour, in whose hands has been placed the interests of Palestinian Jewry at the Peace Conference, has written a preface to the History of Zionism, shortly to be published from the pen of M. Sokolow, one of the four leaders of the Zionist Executive Committee.

Mr. Balfour says that convinced by conversations with Dr. Weizmann in January, 1906, that if a home was to be found for the Jewish people, homeless now for nearly 1900 years, it was vain to seek it anywhere but in Palestine. Answering the question why local sentiment is to be more considered in the case of the Jew than (say) in that of the Christian or the Buddhist, Mr. Balfour says:—'The answer is, that the cases are not parallel. The position of the Jews is unique. For them race, religion, and country are interrelated, as they are interrelated in the case of no other race, no other religion, and no other country on earth. By a strange and most unhappy fate it is this people of all others which, retaining to the full its racial self-consciousness, has been severed from its home, has wandered into all lands and has nowhere been able to create for itself an organized social commonwealth. Only Zionism—so at least Zionists believe—can provide some mitigation of this great tragedy.

'Doubtless there are difficulties, doubtless there are objections—great difficulties, very real objections.... Yet no one can reasonably doubt that if, as I believe, Zionism can be developed into a working scheme, the benefit it would bring to the Jewish people, especially perhaps to that section of it which most deserves our pity, would be great and lasting.'

The criticism that the Jews use their gifts to exploit for personal ends a civilization which they have not created, in communities they do little to maintain, Mr. Balfour declares to be false. He admits, however, that in large parts of Europe their loyalty to the State in which they dwell is (to put it mildly) feeble compared with their loyalty to their religion and their race. How, indeed, could it be otherwise? he asks. 'In none of the regions of which I speak have they been given the advantages of equal citizenship; in some they have been given no right of citizenship at all.'

---

140. P. S. Mowrer, "The Assimilation of Israel", *The Atlantic Monthly*, Volume 128, Number 1, (July, 1921), pp. 101-110, at 103-105, 108-109.

'It seems evident that Zionism will mitigate the lot and elevate the status of no negligible fraction of the Jewish race. Those who go to Palestine will not be like those who now migrate to London or New York... . They will go in order to join a civil community which completely harmonizes with their historical and religious sentiments; a community bound to the land it inhabits by something deeper even than custom; a community whose members will suffer from no divided loyalty nor any temptation to hate the laws under which they are forced to live. To them the material gain should be great; but surely the spiritual gain will be greater still.'

Mr. Balfour goes on to consider the position of those, though Jews by descent, and often by religion, who desire wholly to identify themselves with the life of the country wherein they have made their home, many of them distinguished in art, medicine, politics, and law. 'Many of this class,' he says, 'look with a certain measure of suspicion and even dislike upon the Zionist movement. They fear that it will adversely affect their position in the country of their adoption. The great majority of them have no desire to settle in Palestine. Even supposing a Zionist community were established, they would not join it... .

'I cannot share these fears. I do not deny that, in some countries where legal equality is firmly established, Jews may still be regarded with a certain measure of prejudice. But this prejudice, where it exists, is not due to Zionism, nor will Zionism embitter it. The tendency should surely be the other way. Everything which assimilates the national and international status of the Jews to that of other races ought to mitigate what remains of ancient antipathies; and evidently this assimilation would be promoted by giving them that which all other nations possess—a local habitation and a national home."

While Balfour and other segregationist racists tried to lay the blame for "anti-Semitism" on the Czar and the Gentile governments of the East, those governments often tried to welcome the Jews to assimilate and become genuine and loyal citizens. Racist Jews did not want to assimilate and it was largely the racist Jews who created—insisted upon, the Jewish Ghettoes of the East.

When the Czar tried to integrate the Jews into society and combat racist Zionism in 1903, the racist Zionist Jews attacked him and his Government and incited strikes and a bankers' boycott of the nation, which crippled Russia's economy. The racist Zionist Jews fomented a revolution against the Czar on a massive scale in the period of 1903-1905, and the Jewish bankers made the people of Russia starve. Jewish bankers also created the Russo-Japanese war in this period and financed Japan and Russian Revolutionary Jews against Russia, while concurrently blocking Russia's access to international finance. The Jewish bankers did this, not to free the Jews from segregation, but rather to ensure that the Jews remained segregated and form a disloyal and subversive Jewish nation within the Gentile nations of the East. The Jewish bankers did this, not to free the workers of Russia from their chains, but rather to starve and enslave them, and to turn them against the Czar who was trying to save them.

While, due to the lies spread in the Jewish press, the striking workers blamed the Czar for their pain, their dire situation was caused by Jewish bankers who deliberately bankrupted the country. Jewish Communists deliberately tore down society in order to herd the hurting masses toward the cliff of revolutionary suicide. Though the press around the world blamed the Czar for the woes of the Russian people, the Czar tried to save his people from this foreign influence of Jewish bankers, which ruined the Russian People. Though the press, under the influence of Jewish financiers, told the world that the Czar was segregating the Jews and starving the people of Russia, the Czar was in fact trying to integrate the Jews into Russian society and rescue the Russian economy from the Jewish bankers who were deliberately burying it. A bit of truth did, however, filter out in the press.

The London Times reported on 2 September 1903 on page 3,

"THE RUSSIAN GOVERNMENT AND ZIONISM.

(FROM OUR RUSSIAN CORRESPONDENTS.)

A secret circular against Zionism issued by the Russian Minister of the Interior to the Governors, Prefects, and other authorities is published by the Jewish Labour League. It begins with an explanation of the motives for the change in the Government policy towards Zionism which M. de Plehve hinted at in his letter to Dr. Herzl. The Zionists have, it is alleged, departed from their original purpose of creating a Jewish State in Palestine, and now endeavour to develop and strengthen the Jewish national idea, which encourages racial differences. This is inimical to the assimilation of the Jews with the other subjects of the Tsar and contrary, therefore, to the Russian Imperial idea. The circular then prescribes to Governors and others to take the following measures:—

(1) To prohibit the action of the 'Mahids,' or travelling agitators, who make speeches in the synagogues and at public meetings; (2) not to allow public meetings or assemblies of any kind; (3) to forbid conferences of delegates and members of the Zionist organizations; (4) to stop the collection of money for the Jewish National Fund and the circulation of shares issued abroad in connexion with that fund; (5) to compel the Zionist leaders to sign a document not to collect any more funds, to transfer all the funds which are at present in their hands to the Odessa Society for Helping Jewish Farmers and Artisans in Syria and Palestine, and to confiscate all the shares of the Jewish National Fund now in circulation in Russia; (6) to keep a close watch over schools, libraries for adults, and other institutions in which old Hebrew is taught, and which tend to keep the Jews as a race apart; (7) to report as to the Zionist inclinations of all candidates for the position of Rabbi and other offices."

On 11 September 1903, on page 3, *The London Times* reported on the anti-racist, integrationist policies of the Czar, which racist segregationist Jews loathed,

"M. DE PLEHVE AND ZIONISM.

The *Jewish World* of to-day publishes the text of the secret circular to which allusion was made in a despatch from our Russian Correspondents in *The Times* of September 2:—

Strictly confidential.
Ministry of the Interior, Special Police Department.
To the Governors, City Prefects, and Chiefs of Police.

According to information at the disposal of the Police Department, regarding the so-called Zionist societies, they originally set themselves the task of furthering the emigration of Jews to Palestine in order to establish there an independent Jewish State. Now the realization of this idea is being put into the distant future and activity directed to the development and strengthening of the national Jewish idea by the endeavour to form an inner organization of Jews in their present place of domicile.

This tendency, which is hostile to the assimilation of the Jews with the other races, and which widens the national gulf between the former and the latter, is against the fundamental principles of the State, and cannot, therefore, be tolerated. Consequently, I consider it necessary to make the following decision in regard to the Zionist organization.

You will please let me have immediately detailed information on the Zionist groups and gatherings in your district, as well as on their significance from Government and national points of view. But as I regard it as urgent to take measures for the checking and stopping the Zionist organization, which had at first been permitted, and to hinder its further development in that harmful tendency, I consider it my duty, even before a definite decision can be come to on the whole question, to give you the following instructions:—

1. The propaganda of the Zionist idea in public places, as well as in assemblies bearing a public character, is to be forbidden. In this respect it is necessary to stop the activity of the special agitators, the so-called Maggidim, who travel about preaching in synagogues and at general meetings in order to make their audiences, particularly those from the lower classes, become adherents of Zionism.

2. In the same way, so far as they extend their activity to public meetings and gatherings, all existing Zionist organizations, which are spread all over Russia, including Siberia, the kingdom of Poland and Russian Central Asia, must be suppressed and prohibited.

3. Congresses and conferences of members of Zionist organizations, no matter the purpose for which they be held, are always to be prohibited.

4. All collections not authorized by the Government for the shares and coupons of the London Jewish Colonial Trust, whose entrance into Russia was permitted according to No. 92, section I. of the Code of Laws for 1902; the collections for the Jewish National Fund; as also the general collections in some towns, among the general body of the Jews, all are, at the first information obtained, to be at once suppressed. The persons standing at the

head of the Zionist organizations have to bind themselves in writing to withdraw from the management and not to institute any collection. The moneys in their possession are, as collections not authorized by the Government, to be handed over to a Jewish benevolent institution, such as, for instance, the Odessa Society for Assisting Jewish Agriculturalists and Artisans in Palestine and Syria. Shares and coupons of the Jewish Colonial Trust, and the stamps of the Jewish National Fund, are liable to confiscation, and the persons who have concerned themselves in their sale have to bind themselves in writing to stop their activity. The latter is the more harmful, as the persons contributing to the Zionist funds are mostly recruited from those who are least able to afford it.

5. The lectures delivered in the Jewish Chedarim, libraries, reading-rooms, and Saturday schools are to be constantly watched.

6. At the elections of Rabbis, assistant Rabbis, and communal officials it is necessary to be informed as to the measure of their participation in the Zionist organizations.

(Signed) PLEHVE.
LOPUKHIN."

Racist Zionist Jews combated the Czar's progressive anti-racist and integrationist policies. Jews bankrupted and eventually overthrew the Russian Government—mass murdering tens of millions of Gentiles. Far from protecting Jews from racism directed against Jews, racist Jews cheered Hitler's racist policies, financed Hitler and anti-Semitic propaganda, and then put the Nazi Party into power—mass murdering tens of millions of Gentiles, in order to ensure that the Jews become and remain segregated and form a racist apartheid "Jewish State".

Racist Jews were determined to not let holy Jewish blood mix with Slavic blood which they considered sub-human. Racist Jews were determined to ruin Russia in order to prevent the desecration of divine Jewish blood. Racist Zionist Israel Zangwill wrote in his book, *The Problem of the Jewish Race*, Judaen Publishing Company, New York, (1914), pp. 20-21; which was first published as an article, "The Jewish Race", *The Independent*, Volume 71, Number 3271, (10 August 1911), pp. 288-295, at 295,

"Moreover, while as already pointed out the Jewish upper classes are, if anything, inferior to the classes into which they are absorbed, the marked superiority of the Jewish masses to their environment, especially in Russia, would render *their* absorption a tragic degeneration. But if dissolution would bring degeneracy and emancipation dissolution, the only issue from this delimma is the creation of a Jewish State or at least a Jewish land of refuge upon a basis of local autonomy to which in the course of the centuries all that was truly Jewish would drift."

Racist Jews blamed the ruin of the Russian people on those who tried to prevent it. The racist and intolerant Jews, who deliberately caused the famine, unemployment and slaughter, pretended that they were the innocent victims of racism and religious intolerance. Racist Jews even promoted anti-Semitism in order to keep the holy blood of Jews segregated from the Slavic "cattle". The Zionists caused two World Wars and the genocide of the Russian people by the Bolsheviks, which cost the Russians many tens of millions of innocent lives, in order to fulfill the Zionists' dreams of a "World Ghetto"[141] for Jews in Palestine.

In 1922, Henry Morgenthau, a highly influential American Jew, reported on a Commission to Poland ordered by the Zionist President of the United States Woodrow Wilson,[142] which Commission Morgenthau had led in 1919, and which revealed to Morgenthau, among other things, the duplicitous nature of the Zionist Jews of Poland,

> "'Mr. Dmowski,' I said, 'I understand that you are an anti-Semite, and so I want to know how you feel toward our Commission.'
> Instantly he relaxed his severity. He replied in an almost propitiating manner:
> 'My anti-Semitism isn't religious: it is political. And it is not political outside of Poland. It is entirely a matter of Polish party-politics. It is only from that point of view that I regard it or your Mission. Against a non-Polish Jew I have no prejudice, political or otherwise. I'll be glad to give you any information that I possess.'
> He then sketched, with vigor, the arguments against Jewish Nationalism and touched on the Socialist activities of one section of the Polish Jews. He also said: 'There never was a pogrom in Poland. Lithuanian Jews, fleeing Russian persecution in 1908, spoke Russian obtrusively and banded together to employ only Jewish lawyers and doctors; they started boycotting; the Poles' boycott was a necessary retaliation. On the other hand, the Posen Jews speak either German or Yiddish, which is based on German: we want the Polish language in Poland.
> [***]
> 'Pogroms?' Pilsudski had thundered when I first called on him. It was in the Czar's Summer Palace near Warsaw that he was living, and he received me in the 'library' where there was not a book to be seen. 'There have been no pogroms in Poland! Nothing but unavoidable accidents.'
> I asked the difference.
> 'A pogrom,' he explained, 'is a massacre ordered by the Government, or not prevented by it when prevention is possible. Among us no wholesale killings of Jews have been permitted. Our trouble isn't religious; it is economic. Our petty dealers are Jews. Many of them have been war-profiteers, some have had dealings with the Germans or the Bolsheviki, or both, and this has created a prejudice against Jews in general.'"[143]

---

141. T. Herzl, English translation by H. Zohn, R. Patai, Editor, *The Complete Diaries of Theodor Herzl*, Volume 1, Herzl Press, New York, (1960), p. 172.
142. "Mr. Wilson and Zionism", *The London Times*, (7 September 1918), p. 5.
143. H. Morgenthau, "The Jews in Poland", *The World's Work*, Volume 43, Number 5,

In 1921, Henry Morgenthau, one of the most prominent Jews in American history, clarified the fact that Zionist Jews were out to fulfill Jewish Messianic prophesies, which would make the Jews the exclusive rulers over the entire Earth,

"Zionism is a surrender, not a solution. It is a retrogression into the blackest error, and not progress toward the light. I will go further, and say that it is a betrayal; it is an eastern European proposal, fathered in this country by American Jews, which, if it were to succeed, would cost the Jews of America most that they have gained of liberty, equality, and fraternity. [***] Zionism is based upon a literal acceptance of the promises made to the Jews by their prophets in the Old Testament, that Zion should be restored to them, and that they should resume their once glorious place as a peculiar people, singled out by God for His especial favor, exercising dominion over their neighbors in His name, and enjoying all the freedom and blessings of a race under the unique protection of the Almighty. Of course, the prophets meant these things symbolically, and were dealing only with the spiritual life. They did not mean earthly power, or materialistic blessings. But most Jews accepted them in the physical sense; and they fed upon this glowing dream of earthly grandeur as a relief from the sordid realities of the daily life which they were compelled to lead."[144]

In its article "Jews", the *Great Soviet Encyclopedia: A Translation of the Third Edition*, Volume 9, Macmillan, New York, (1975), pp. 292-293, at 293, wrote,

"After World War II, chauvinist tendencies and Zionist ideology, with its antiscientific assertion of the 'messianic' role of the Jews and the idea of the 'chosen people,' were artificially revived among Jews in the developed capitalist countries. Zionism has become an ideology of militant chauvinism and anticommunism, acting in the interests of international imperialism."

In its article "Judaism", the *Great Soviet Encyclopedia: A Translation of the Third Edition*, Volume 11, Macmillan, New York, (1976), pp. 311-313, at 312, wrote,

"Attempting to win over the masses of working Jews and to divert them from the world revolutionary labor and national liberation movements as well as to justify Israel's expansionist policies, Zionism began to use the tenets of Judaism for its political aims (for example, messianism, which proposes the creation of a new, 'ideal' Israel, with Jerusalem as its center,

---

(April, 1922), pp. 617-630, at 618, 622, 626-627.
144. H. Morgenthau, "Zionism a Surrender, Not a Solution", *The World's Work*, Volume 42, Number 3, (July, 1921), pp. i-viii, at i-ii.

that would include the whole of Palestine). Since the second quarter of the 20th century Zionism has found support among the most reactionary Jews, especially in the USA. In its chauvinist and annexationist policy Zionism makes use of Judaic dogma that the Jews are god's chosen people and employs Judaism to substantiate the concept of a 'worldwide Jewish nation' and other reactionary positions."

See also: N. S. Alent'eva, Editor, *Tseli i metody voinstvuiushchego sionizma*, Izd-vo polit. lit-ry, Moskva, (1971). Н. С. Алентьева, Редактор, Цели и методы воинствующего сионизма, Издательство Политической Литературы, Москва, (1971).

2.4.2 In Answer to the "Jewish Question"

Burton J. Hendrick, Associate Editor, published a series of articles in *The World's Work* in 1922-1923, in which he launched a two-pronged attack, one against Henry Ford's alleged anti-Semitism, the other against the segregationist tribalism of "Polish Jews"—the Jews of Eastern Europe who were migrating by the millions through Germany to England and eventually to the United States. Hendrick extolled the virtues of the Sephardic and German Jews who had emigrated to America long before, but obviously sought to curb the influx of Russian Jews into the United States. Hendrick's articles are particularly noteworthy, because they evince the common view in Germany, England and America; that Eastern Jews were too often the dregs of society. Russian Jews were commonly seen as prostitutes, liquor and tobacco peddlers—the promoters and exploiters of vice, gangsters (such as Meyer Lansky, a Polish Jew from Grodno, born Majer Suchowlinski; and "Bugsy" Siegel, born Benjamin Hymen Siegelbaum, who was popular among the powerful Jews of Hollywood—organized crime has always been, and continues to be run behind the scenes by Jews, many of whom are Israelis and Russian Jews, who deal in drugs, weapons and the white slave trade in women and children), revolutionary assassins, shyster lawyers, corrupt stock traders, corrupt politicians who sought to destroy America, and other despicable sorts. On the other hand, while acknowledging the stereotypes that were already pervasive in 1902, Dr. Maurice Fishberg wrote more enthusiastically about the Russian Jew in "The Russian Jew in America", *The American Monthly Review of Reviews*, Volume 26, Number 3, (September, 1902), pp. 315-318; however, this journal was created by William T. Stead to promote the views of Cecil Rhodes, who was himself a Rothschild agent.[145]

The strongest prejudice Eastern Jews faced came not from Gentiles, but from their Western Jewish co-religionists who knew them best. Western Jews were often as intolerant and tribalistic as were their Eastern co-religionists. Ironically, both groups suffered from the intolerance they had passed on to the

---

145. G. E. Griffin, *The Creature from Jekyll Island: A Second Look at the Federal Reserve*, Fourth Edition, American Media, Westlake Village, California, (2002), p. 208.

Gentiles in the forms of Christianity and Islam, and from the Gentiles' reaction to Jewish tribalism and criminal behavior.

The North American Review, Volume 60, Number 127, (April, 1845), pp. 329-368, published an article, "The Modern Jews", which revealed at pages 329-330, and 351, that the Jews were trying to catch up after lagging behind the Gentiles in the Enlightenment, and that some Jews believed that they bore the prophesied burden of telling Gentiles how they ought to think and to learn, as well as how to run their governments. Note the important, though spurious, linkage of Jewish persecution with the Messianic aspirations of some Western Jews (especially the Rothschilds and their agent Montefiore). These incompressibly wealthy Jewish racists also bought the services of merciless Christians and Moslems, who had been corrupted and cajoled behind the scenes by Western Zionist Jews (especially the Rothschilds and their agent Montefiore) and instructed to persecute Jews in order to force them into accepting segregation and ultimately Zionism—most anti-Semitism was artificially manufactured by Jewish leadership,

> "A NEW and rapidly increasing interest in the affairs of the Jewish people has of late years pervaded Protestant Christendom. Among the Jews themselves, too, our day reveals new elements of life, struggling to break the stupor of centuries. Some strange changes are taking place, also, in the external condition of this people. In one country, we behold revived against them a persecuting popish inquisition; in another, an imperial edict is even now sending them, by hundreds of thousands, into exile; in a third,—a Protestant country, too,— the long established policy of excluding them from political privileges altogether has withstood a bold onset from the liberal spirit of the age, and triumphed. Our own land has recently witnessed the singular spectacle of Jews dictating to a Christian people, how the children of that people should be educated; and forbidding to teach, or even name, Jesus Christ in the public schools. Meanwhile, the Protestant church, especially in Great Britain, is putting forth fresh energies, in widely extended missionary enterprises, to win Israel to the acknowledgment of her Messiah, still looked for, though long since come,—perseveringly rejected, yet the object of her fondest hopes. [***] The rank and power which many European Jews have acquired by their learning, or more frequently by their wealth, have been at times an important safeguard to their poor, despised countrymen. None can estimate the influence, in this respect, of the Rothschilds, who, a few years ago, were five in number, with houses at London, Frankfort, Paris, Vienna, and Berlin; guiding the commercial, and sometimes almost the political, destinies of Europe; 'holding in their hands the purse-strings of the civilized world.' One of the brothers was presented to the pope in 1838; and his brethren in Rome profited by his presence to obtain permission to work at their trades. The pope not only granted this request, but also distributed alms among the poor Jews. Sir Moses Montefiori, a princely Israelite of London, was one of the deputation to the Turkish Sultan to obtain relief for the persecuted Jews of Damascus and

Rhodes, and was the chief agent in procuring the firman already mentioned. He profited by this occasion to visit Palestine, and manifested a lively interest in the condition of his brethren in that land. A Jewish banker of Antwerp, M. Cohen, has lately received a knighthood of the order of Isabella from Spain!"

The expulsion of the Jews from Spain and the Inquisition were a means by which racist Jews prevented the assimilation of Sephardic Jews into Catholic Spanish society and Moorish Islamic society. They were a means to maintain the "purity" of the "Jewish race" and were the product of Jewish racism, not Catholic intolerance. *The North American Review* wrote in 1845 (note that crypto-Jews, for example the Marranos of Spain and the Dönmeh in Turkey, were often the most observant members of their feigned religions—the most deceptive and subversive members of their societies, just as the crypto-Jews Reinhard Heydrich, Joseph Goebbels and Julius Streicher were the most vitriolic anti-Semites in Nazi Germany and deliberately brought about the downfall of Germany),

"No estimate can be formed of the number of Jews residing in Roman Catholic countries, particularly in Spain and Portugal, who conceal their religion under a Christian garb; probably, there are several hundred thousand of them. [***] Ferdinand and Isabella, after vanquishing the Moors, commanded all the Jews of Spain either to embrace Christianity, or to leave the kingdom within four months. Eight hundred thousand, according to the Spanish accounts,—according to the Jews, a million,—preferred exile, and suffered inconceivably in their emigration. Some of them took refuge in Portugal, whence, however, with all other Jews, they were soon expelled. Hundreds of thousands in both countries submitted to baptism in preference to exile; but in secret they still practised the rites of Judaism; some carrying dissimulation so far as even to take orders in the Roman Catholic church, and to become judges of the Inquisition, which, it is well known, was originally established in Spain about this time, principally to deal with relapsing Jews and Moors, who had preferred an outward profession of Christianity to banishment, and who were called 'New Christians.' In Spain, the Jews have never since been openly tolerated. To Portugal they were readmitted by John the Sixth about the year 1817, because some Jews had imported large cargoes of corn during a scarcity; and, at the request of the pope, they were allowed the same privileges that were accorded to them in the Roman States. Previously, in that kingdom, the name of Jew was so odious, that a law was passed, giving impunity to any one so called, who should slay the offender on the spot; and there, as well as in Spain, the descendants of the 'New Christians,' who still are Jews at heart, maintain the deception; though in Portugal, where some degree of liberty of conscience has for a few years been enjoyed, these will probably, it is said, soon return to the synagogue. Most of the avowed Jews in that country, at present, are recent immigrants. No longer ago than 1827, a person was put to death in Spain for the heresy of Judaism. The dissemblers there, to make the deception complete, often affect unusual Christian zeal. If a Spanish dwelling superabounds with religious ornaments and utensils, there is good reason for believing the family to be dissembling Jews."[146]

Eastern Jewish emigrants to America sought to continue the noble ancient Jewish tradition of higher education, which had given the Western Jews great advantages in the world. The 1845 article in *The North American Review* continued (note the racism of Sephardic Jews directed against Ashkenazi Jews):

"The rank and power which many European Jews have acquired by their learning, or more frequently by their wealth, have been at times an important safeguard to their poor, despised countrymen. None can estimate the influence, in this respect, of the Rothschilds, who, a few years ago, were five in number, with houses at London, Frankfort, Paris, Vienna, and Berlin; guiding the commercial, and sometimes almost the political, destinies of Europe; 'holding in their hands the purse-strings of the civilized world.' One of the brothers was presented to the pope in 1838; and his brethren in Rome profited by his presence to obtain permission to work at their trades. The pope not only granted this request, but also distributed alms among the poor Jews. Sir Moses Montefiori, a princely Israelite of London, was one of the deputation to the Turkish Sultan to obtain relief for the persecuted Jews of Damascus and Rhodes, and was the chief agent in procuring the firman already mentioned. He profited by this occasion to visit Palestine, and manifested a lively interest in the condition of his brethren in that land. A Jewish banker of Antwerp, M. Cohen, has lately received a knighthood of the order of Isabella from Spain!

The Jews have nowhere preserved faithful genealogical records, but almost always have abundant traditions of their descent, which, of course, are unworthy of credit. Yet supposing that the twelve tribes are now generally amalgamated, some portions of the mass, taken separately, must be less mixed than others. There are, no doubt, among them, though the distinction cannot certainly be traced, not a few pure descendants of some tribes; and none were so likely to keep themselves distinct as the tribe of Judah, claiming, as they did, preëminence. The Spanish and Portuguese Jews have always asserted a superiority in this respect; some said, that they were of the united tribes of Judah and Benjamin, including the Levites; others, that they were of pure descent from Judah; and others, still more arrogantly, that they were of David's royal line [which would make them the self-anointed bearers of the royal Messianic line—CJB]. Since they probably came from Judea about the time of the destruction of Jerusalem, they may undoubtedly be considered among the purest representatives of

---

[146]. "The Modern Jews", *The North American Review*, Volume 60, Number 127, (April, 1845), pp. 329-368, at 336, 350.

the two tribes. The German and Polish Jews, who were reinforced from the East, in the tenth century and subsequently, are of more heterogeneous elements. The latter are denominated Ashkenazim, from Ashkenaz, grandson of Japhet; [*Footnote:* Genesis, x. 3.] the former, Sephardim, from Sepharad, [*Footnote:* Obadiah, 20.] a name which the modern Jews have given to Spain. These are found interspersed with each other in most parts of the world; but in general, it may be said, that the Sephardim belong to the different countries, European, Asiatic, and African, upon the Mediterranean sea. Thus, the forefathers of most of the present native Jews in Constantinople and Palestine came, as exiles, from Spain and Portugal, at the end of the fifteenth and the beginning of the sixteenth century. They have everywhere separate synagogues, and refuse intermarriage with the Ashkenazim. If any of their number marries one of the inferior race, excommunication immediately follows. Early in the present century, the daughter of a Portuguese Jewish physician, at Berlin, married a German Jew, and her family went into mourning for her, as for one dead. In this country, the same distinctions and pretensions are found, gradually wearing away, however, under the combined influences of Jewish neology and American democracy. 'The Hebrew Portuguese Congregation' of Philadelphia has already been mentioned in another connection; this title itself indicates the still existing distinction. The Sephardim are generally more polished than the Ashkenazim; and in Europe, for the most part, are superior to them also in moral and religious principle. Along the shores of the Mediterranean, they have a dialect of their own, originally Spanish, but now modified by Hebrew words, phrases, and idioms, and called Judæo-Spanish. The Jews of Russia and Poland are represented as the worst to be found in any country; some would make them out to be little better than hordes of robbers; this, however, is an exaggeration. Bad as they may be, it is believed they are superior in morals to their Gentile neighbours: 'He lives like a Christian,' is with them an accusation of the grossest immorality."[147]

Herbert N. Casson wrote in 1906, in his article, "The Jew in America",

"The Russian Jew, who was the last to discover America, but who will soon outnumber all the rest, has little education when he arrives. But he is hungrier for knowledge than for money. Scholarship—that is what he worships. He will live five in a room to let little Jacob go to college. And the young Russian Jew will at any time prefer an Idea to a meal. On several occasions, in the North End of Boston and the East Side of New York, I have heard boys of nineteen discussing the poetry of Heine, the music of Mendelssohn, the philosophy of Spinoza, the revolutionism of Marx, as though they had no personal problem to solve in the slum and the sweat-

---

[147]. "The Modern Jews", *The North American Review*, Volume 60, Number 127, (April, 1845), pp. 329-368, at 351-353.

shop."[148]

The otherwise virtuous love of education often became a destructive force in the hands of tribalistic and racist Jews, who were obsessed with self-glorification and clannishly demanded obedience to their Jewish heroes of the arts and sciences. In so doing, these racist Jews stifled progress and discouraged reasonable persons from pursuing fields they otherwise would have entered. It was important to racist Jews that they not only accumulate disproportionate wealth, but also that they prevented others from accumulating enough wealth to pose an organized opposition to the Messianic goals of racist Jews. It was important to them to keep Gentiles comparatively poor and uneducated.

Note that Marx, Spinoza, Mendelssohn and Heine were not only second rate philosophers and artists, but that each was Jewish and a hero to these young Jews, who would impose their hero worship on all of humanity and who would dogmatically and vociferously resist any challenges to their adolescent cults of personality—apparently exclusively Jewish personalities. Seemingly, in their minds one would have to be an anti-Semite not to recognize the vast superiority of their mediocre heroes, who were largely plagiarists.

---

[148]. H. N. Casson, "The Jew in America", *Munsey's Magazine*, Volume 34, Number 4, (January, 1906), pp. 381-395, at 386.

3 ROTHSCHILD, *REX IVDÆORVM*

The banking family known as the "House of Rothschild" desired to become the "King of the Jews". According to Jewish myth, the King of the Jews will bring all Gentile nations, cultures and religions to ruins through world wars. The King of the Jews, whom the Jews call "Messiah", will then rule the world from Jerusalem. According to Jewish myth, the remnant of the Gentile peoples ("Esau") left after the wars to come, will be enslaved, welcoming their enslavement as a joyful opportunity to obey their divine Jewish masters ("Jacob" and "Joseph"). Then the Gentile peoples will be exterminated. The process is well underway and is accelerating. The Rothschilds eventually succeeded in their Messianic goal to found a racist "Jewish State". The Balfour Declaration was written directly to Lord Rothschild, who no doubt took the title literally.

> "15 For the day of the LORD *is* near upon all the heathen: as thou hast done, it shall be done unto thee: thy reward shall return upon thine own head. 16 For as ye have drunk upon my holy mountain, so shall all the heathen drink continually, yea, they shall drink, and they shall swallow down, and they shall be as though they had not been. 17 ¶ But upon mount Zion shall be deliverance, and there shall be holiness; and the house of Jacob shall possess their possessions. 18 And the house of Jacob shall be a fire, and the house of Joseph a flame, and the house of Esau for stubble, and they shall kindle in them, and devour them; and there shall not be *any* remaining of the house of Esau; for the LORD hath spoken *it*."—OBADIAH 15-18

> "8 And the remnant of Jacob shall be among the Gentiles in the midst of many people as a lion among the beasts of the forest, as a young lion among the flocks of sheep: who, if he go through, both treadeth down, and teareth in pieces, and none can deliver. 9 Thine hand shall be lifted up upon thine adversaries, and all thine enemies shall be cut off."—MICAH 5:8-9

> "In European capitals there are Hebrew bankers who dictate certain international relations because they hold the purse-strings of governments; and every European country owes much to the men of great genius that the race has contributed to the arts and to statecraft."—*The World's Work*[149]

3.1 Introduction

Throughout history, the world has faced the radical tendency of many Jews to destructive polarized extremes, which undermined the sovereignty and the cultures of other peoples and led those peoples into wars and revolutions, which fulfilled Jewish Messianic prophecies of Jewish supremacy in the world. Casson wrote, in admiration,

> "Whenever the country has been split in two by a political question, there have been Jews on both sides. Judah P. Benjamin, cabinet officer in the

---

[149]. "The Jews in the United States", *The World's Work*, Volume 11, Number 3, (January, 1906), pp. 7030-7031.

Confederate government, supported the gray as stubbornly as Joseph Seligman did the blue. And in the largest sense we may say that international capital marches under the banner of Rothschild, and international labor under the flag of Karl Marx—Jews both, and irreconcilable."[150]

The Rothschilds and Karl Marx worked together to undermine Gentile nations and gather wealth and power unto the Jews, as was prophesied in *Deuteronomy, Isaiah, Obadiah, Zechariah*, and other Jewish religious literature. The Rothschilds were a highly religious Jewish family and Marx came from a rabbinical family, originally named "Marx Levi". Like Moses Mendelssohn,[151] Karl Marx was a devout Talmudist, which made him devoutly anti-Christian and devoutly anti-Gentile.[152] In hopes that the Gentiles could be persuaded that it was in their best interests to surrender to Jewish world rule, the Rothschilds deliberately caused perpetual wars,[153] which made the Gentile peoples clamor for peace. The Rothschilds then sponsored the myth that the only means to end the wars they themselves had caused, was to eliminate the Gentile nations.

Since Rothschild considered himself to be the messiah of the Jews, he felt duty bound to fulfil Jewish messianic prophecies which demand the extermination of the Goyim, the theft of all the wealth of the World, and the theft of Jerusalem and Palestine.

Anka Muhlstein wrote in her book *Baron James: The Rise of the French Rothschilds*,

> "The success of the five brothers, who seemed to share Europe among themselves without ever dividing their forces, fired the imagination. The best writers of the time gave credence to the notion of the all-powerful bankers. 'The Jewish banker says: '[God] has given me the royalty of wealth and the understanding of opulence, which is the scepter of society.... . 'A Jew now reigns over the Pope and Christianity. He pays monarchs and buys nations,' stated Alfred de Vigny.[12] Jewish polemists joined in and declared: 'There is only one power in Europe, and that is Rothschild... and speculation is his sword... . Rothschild needed states in order to become Rothschild... . However, today he no longer needs the state, but the state needs him.'[13] Heine put it more humorously, recalling olden times when the King would have had the teeth of M. de Rothschild pulled out had he refused a loan. Today, 'Rothschild, Baron and Knight of the Order of Isabella the Catholic,

---

150. H. N. Casson, "The Jew in America", *Munsey's Magazine*, Volume 34, Number 4, (January, 1906), pp. 381-395, at 386.

151. "The Modern Jews", *The North American Review*, Volume 60, Number 127, (April, 1845), pp. 329-368, at 361-365.

152. P. S. Mowrer, "The Assimilation of Israel", *The Atlantic Monthly*, Volume 128, Number 1, (July, 1921), pp. 101-110, at 107.

153. G. E. Griffin, "The Rothschild Formula", *The Creature from Jekyll Island: A Second Look at the Federal Reserve*, Chapter 11, Fourth Edition, American Media, Westlake Village, California, (2002), pp. 217-234.

may stroll over to the Tuileries whenever it strikes his fancy, without fear of losing even one tooth to a cash-hungry King.'[14] Metternich's spies took the same line and in their reports feigned surprise at the presence in James' office of General Rumigny, whom Louis-Philippe had put in charge of everything related to the Bourse. Rothschild, wrote a certain Klindworth to Count Apponyi, the Austrian Ambassador, 'places in each ministry, in every department at all levels, his own creatures who feed him the greatest variety of information.'[15]

While the European press devoted long articles to the Rothschild triumphs, American newspapers, often influenced by a group of German Jews to whom the image of an all-powerful Court Jew was familiar, portrayed Rothschild as fabulous and fantastical: 'The Rothschilds govern a Christian world. Not a cabinet moves without their advice. They stretch their hand, with equal ease, from Petersburg to Vienna, from Vienna to Paris, from Paris to London, from London to Washington. Baron Rothschild, the head of the house, is the true king of Judah, the prince of the captivity, the Messiah so long looked for by this extraordinary people... . The lion of the tribe of Judah, Baron Rothschild, possesses more real force than David—more wisdom than Solomon.'[16] Even the wildest rumors came to seem credible. In 1830 an American weekly reported on its front page that the Rothschilds had purchased Jerusalem: 'We see nothing improbable that in the pecuniary distress of the sultan, he should sell some parts of his dominions to preserve the rest; or that the Rothschilds should purchase the old capital of their nation. They are wealthy beyond desire, perhaps even avarice; and so situated, it is quite reasonable to suppose that they may seek something else to gratify their ambition, that shall produce most important effects. If secured in the possession, which may be brought about by money, they might instantly, as it were, gather a large nation together, soon to become capable of defending itself, and having a wonderful influence over the commerce and condition of the east—rendering Judah again the place of deposit of a large portion of the wealth of the 'ancient world.' To the sultan the country is of no great value; but, in the hands of the Jews, directed by such men as the Rothschilds, what might it not become, and in a short period of time?'[17]"[154]

Marxist Jews preached that the only means to attain peace was to abolish the nations and establish a world government run by them; for, after all, with no nations left but Israel, how could there be any war? This was the method that Jewish leadership used to undermine the sovereignty of the nations in fulfilment of Jewish Messianic prophecy.[155] They did not always openly depend upon

---

154. A. Muhlstein, *Baron James: The Rise of the French Rothschilds*, The Vendome Press, New York, Paris, (1982), pp. 104-106.
155. "Salluste", "Henri Heine et Karl Marx. Les Origines Secrètes du Bolchevisme", *La Revue de Paris*, Volume 35, Number 11, (1 June 1928), pp. 567-589; **and** "Henri Heine et Karl Marx II. Les Origines Secrètes du Bolchevisme", *La Revue de Paris*, Volume 35,

Communism, *per se*, but also upon such bodies as the League of Nations, the United Nations, the European Union, etc.; which, like Communism itself, were conspicuously over represented by Jewish leadership.

Many Jews have interpreted the Old Testament to predict that the when the Messiah arrives, the Jews will horde all the gold, silver and jewels of the world and keep this treasure in Jerusalem. Judaism teaches that the Garden of Eden contained all the jewels of the world, and many Jews believe that these will all fall into Jewish hands in Jerusalem in the "end times". The Jewish book of *Proverbs* 1:13-14, states,

"13 We shall find all precious substance, we shall fill our houses *with* spoil: 14 Cast in thy lot among us; let us all have one purse:"

Ezekiel 28:13 states,

"Thou hast been in Eden the garden of God; every precious stone *was* thy covering, the sardius, topaz, and the diamond, the beryl, the onyx, and the jasper, the sapphire, the emerald, and the carbuncle, and gold: the workmanship of thy tabrets and of thy pipes was prepared in thee in the day that thou wast created."

Zechariah 14 states,

"Behold, the day of the LORD cometh, and thy spoil shall be divided in the midst of thee. 2 For I will gather all nations against Jerusalem to battle; and the city shall be taken, and the houses rifled, and the women ravished; and half of the city shall go forth into captivity, and the residue of the people shall not be cut off from the city. 3 Then shall the LORD go forth, and fight against those nations, as when he fought in the day of battle. 4 And his feet shall stand in that day upon the mount of Olives, which is before Jerusalem on the east, and the mount of Olives shall cleave in the midst thereof toward the east and toward the west, *and there shall be* a very great valley; and half of the mountain shall remove toward the north, and half of it toward the south. 5 And ye shall flee *to* the valley of the mountains; for the valley of

---

Number 12, (15 June 1928), pp. 900-923; **and** "Henri Heine et Karl Marx III. Les Origines Secrètes du Bolchevisme", *La Revue de Paris*, Volume 35, Number 13, (1 July 1928), pp. 153-175; **and** "Henri Heine et Karl Marx IV. Les Origines Secrètes du Bolchevisme", *La Revue de Paris*, Volume 35, Number 14, (15 July 1928), pp. 426-445. ***See also, Rabbi Liber's Response:*** "Judaïsm et Socialisme", *La Revue de Paris*, Volume 35, Number 15, (1 August 1928), pp. 607-628; ***To which "Salluste" Replied:*** "Autour d'une Polémique: Marxism et Judaïsm", *La Revue de Paris*, Volume 35, Number 16, (15 August 1928), pp. 795-834. ***See also:*** "Salluste", *Les Origines Secrètes du Bolchevisme: Henri Heine et Karl Marx*, Jules Tallandier, Paris, (1930). ***See also:*** D. Fahey, *The Mystical Body of Christ in the Modern World*, Browne and Nolan Limited, London, (1935). ***See also:*** R. H. Williams, *The Ultimate World Order—As Pictured in "The Jewish Utopia"*, CPA Book Publisher, Boring, Oregon, (1957?).

the mountains shall reach unto Azal: yea, ye shall flee, like as ye fled from before the earthquake in the days of Uzziah king of Judah: and the LORD my God shall come, *and* all the saints with thee. 6 And it shall come to pass in that day, *that* the light shall not be clear, *nor* dark: 7 But it shall be one day which shall be known to the LORD, not day, nor night: but it shall come to pass, *that* at evening time it shall be light. 8 And it shall be in that day, *that* living waters shall go out from Jerusalem; half of them toward the former sea, and half of them toward the hinder sea: in summer and in winter shall it be. 9 And the LORD shall be king over all the earth: in that day shall there be one LORD, and his name one. 10 All the land shall be turned as a plain from Geba to Rimmon south of Jerusalem: and it shall be lifted up, and inhabited in her place, from Benjamin's gate unto the place of the first gate, unto the corner gate, and *from* the tower of Hananeel unto the king's winepresses. 11 And *men* shall dwell in it, and there shall be no more utter destruction; but Jerusalem shall be safely inhabited. 12 ¶And this shall be the plague where*with* the LORD will smite all the people that have fought against Jerusalem; Their flesh *shall* consume away while they stand upon their feet, and their eyes shall consume away in their holes, and their tongue shall consume away in their mouth. 13 And it shall come to pass in that day, *that* a great tumult from the LORD shall be among them; and they shall lay hold every one on the hand of his neighbour, and his hand shall rise up against the hand of his neighbour. 14 And Judah also shall fight at Jerusalem; and the wealth of all the heathen round about shall be gathered together, gold, and silver, and apparel, in great abundance. 15 And so shall be the plague of the horse, of the mule, of the camel, and of the ass, and of all the beasts that shall be in these tents, as this plague. 16 And it shall come to pass, *that* every one that is left of all the nations which came against Jerusalem shall even go up from year to year to worship the King, the LORD of hosts, and to keep the feast of tabernacles. 17 And it shall be, *that* whoso will not come up of *all* the families of the earth unto Jerusalem to worship the King, the LORD of hosts, even upon them shall be no rain. 18 And if the family of Egypt go not up, and come not, *that* have no *rain*; there shall be the plague, where*with* the LORD will smite the heathen that come not up to keep the feast of tabernacles. 19 This shall be the punishment of Egypt, and the punishment of all nations that come not up to keep the feast of tabernacles. 20 In that day shall there be upon the bells of the horses, HOLINESS UNTO THE LORD; and the pots in the LORD's house shall be like the bowl's before the altar. 21 Yea, every pot in Jerusalem and in Judah shall be holiness unto the LORD of hosts: and all they that sacrifice shall come and take of them, and seethe therein: and in that day there shall be no more the Canaanite in the house of the LORD of hosts."

Joshua 6:19 states,

"But all the silver, and gold, and vessels of brass and iron, *are* consecrated unto the LORD: they shall come into the treasury of the LORD."

Haggai 2:7-8 states,

"7 And I will shake all nations, and the desire of all nations shall come: and I will fill this house *with* glory, saith the LORD of hosts. 8 The silver *is* mine, and the gold *is* mine, saith the LORD of hosts."

Joel 3:5 states,

"Because ye have taken my silver and my gold, and have carried into your temples my goodly pleasant things:"

In 1932, Michael Higger divulged the intentions of Cabalistic Jews in his book *The Jewish Utopia*,

"All the treasures and natural resources of the world will eventually come in possession of the righteous. This would be in keeping with the prophecy of Isaiah: 'And her gain and her hire shall be holiness to the Lord; it shall not be treasured nor laid up; for her gain shall be for them that dwell before the Lord, to eat their fill and for stately clothing.[*Isaiah* 23:18]'[20] Similarly, the treasures of gold, silver, precious stones, pearls, and valuable vessels that have been lost in the seas and oceans in the course of centuries will be raised up and turned over to the righteous.[21] Joseph hid three treasuries in Egypt: One was discovered by Korah, one by Antoninus, and one is reserved for the righteous in the ideal world.[22] [***] Gold will be of secondary importance in the new social and economic order. Eventually, all the friction, jealousy, quarrels, and misunderstandings that exist under the present system, will not be known in the ideal Messianic era.[319] The city of Jerusalem will possess most of the gold and precious stones of the world. That ideal city will be practically full of those metals and stones, so that the people of the world will realize the vanity and absurdity of wasting their lives in accumulating those imaginary valuables.[320]"[156]

The Jewish Encyclopedia reveals the designs of Jews on all the wealth of the world, and the Jewish desire to ruin all nations save Israel,
"With regard to the text 'This is the law when a man dieth in a tent' (Num. xix. 14), they held that only Israelites are *men*, quoting the prophet, 'Ye my flock, the flock of my pasture, are men' (Ezek. xxxiv. 31); Gentiles they classed not as men but as barbarians (B. M. 108b [*see also: Baba Mezia* 114*b*]). [***] The barbarian Gentiles who could not be prevailed upon to observe law and order were not to be benefited by the Jewish civil laws, framed to regulate a stable and orderly society, and based on reciprocity.

---

156. M. Higger, *The Jewish Utopia*, Lord Baltimore Press, Baltimore, (1932), pp. 12-13, 57.

The passage in Moses' farewell address: 'The Lord came from Sinai, and rose up from Seir unto them; he shined forth from Mount Paran' (Deut. xxxiii. 2), indicates that the Almighty offered the Torah to the Gentile nations also, but, since they refused to accept it, He withdrew His 'shining' legal protection from them, and transferred their property rights to Israel, who observed His Law. A passage of Habakkuk is quoted as confirming this claim: 'God came from Teman, and the Holy One from Mount Paran.... He stood, and measured the earth; he beheld, and drove asunder [רתיו = 'let loose,' 'outlawed'] the nations' (Hab. iii. 3-6); the Talmud adds that He had observed how the Gentile nations steadfastly refused to obey the seven moral Nachian precepts, and hence had decided to outlaw them (B. K. 38a [*see also: Baba Kamma* 113*a-b*])."[157]

Indeed, the Talmud "grants" the Jews all of the wealth and property of the Gentiles, at *Baba Kamma* 38*a*,

"WHERE AN OX BELONGING TO AN ISRAELITE HAS GORED AN OX BELONGING TO A CANAANITE THERE IS NO LIABILITY etc. But I might here assert that you are on the horns of a dilemma. If the implication of *'his neighbour'* has to be insisted upon, then in the case of an ox of a Canaanite goring an ox of an Israelite, should there also not be exemption? If [on the other hand] the implication of *'his neighbour'* has not to be insisted upon, why then even in the case of an ox of an Israelite goring an ox of a Canaanite, should there not be liability? — R Abbahu thereupon said: The Writ says, *He stood and measured the earth; he beheld and drove asunder the nations,*[2] [which may be taken to imply that] God beheld the seven commandments[3] which were accepted by all the descendants of Noah, but since they did not observe them, He rose up and declared them to be outside the protection of the civil law of Israel [with reference to damage done to cattle by cattle].[4] R. Johanan even said that the same could be inferred from this [verse], *He shined forth from Mount Paran,*[5] [implying that] from Paran[6] He exposed their money to Israel. The same has been taught as follows: If the ox of an Israelite gores an ox of a Canaanite there is no liability,[7] but if an ox of a Canaanite gores an ox of an Israelite whether the ox [that did the damage] was *Tam* or whether it had already been *Mu'ad*, the payment is to be in full, as it is said: *He stood and measured the earth, he beheld and drove asunder the nations,*[2] and again, *He shined forth from Mount Paran.*[5] Why this further citation? — [Otherwise] you might perhaps think that the verse *'He stood and measured the earth'* refers exclusively to statements [on other subjects] made by R. Mattena and by R. Joseph; come therefore and hear: *'He shined forth from Mount Paran,'* implying that from Paran[1] he exposed their money to Israel."[158]

---

[157]. "Gentile", *The Jewish Encyclopedia*, Funk and Wagnalls Company, New York, (1903), pp. 615-626, at 619-620.
[158]. Rabbi Dr. I. Epstein, Editor, *The Babylonian Talmud: Seder Nezikin: Baba Kamma*,

Genesis 27:29 states,

"Let people serve thee, and nations bow down to thee: be lord over thy brethren, and let thy mother's sons bow down to thee: cursed be every one that curseth thee, and blessed be he that blesseth thee."

According to the Masoretic Text, which is the version of the Old Testament that most accurately reflects of the views of Jews, *Deuteronomy* 6:10-11 and 11:24-25 (*see also: Joshua* 1:2-5; 24:13) state,

"6:10 And it shall be, when the LORD thy God shall bring thee into the land which He swore unto thy fathers, to Abraham, to Isaac, and to Jacob, to give thee—great and goodly cities, which thou didst not build, 6:11 and houses full of all good things, which thou didst not fill, and cisterns hewn out, which thou the didst not hew, vineyards and olive-trees, which thou didst not plant, and thou shalt eat and be satisfied— [***] 11:24 Every place whereon the sole of your foot shall tread shall be yours: from the wilderness, and Lebanon, from the river, the river Euphrates, even unto the hinder sea shall be your border. 11:25 There shall no man be able to stand against you: the LORD your God shall lay the fear of you and the dread of you upon all the land that ye shall tread upon, as He hath spoken unto you. [version of the Jewish Publication Society]"

Isaiah 2:1-4; 40:15-17, 22-24; 54:1-4; 60:5, 8-12, 16-17; and 61:5-6 state,

"2:1 The word that Isaiah the son of Amoz saw concerning Judah and Jerusalem. 2:2 And it shall come to pass in the end of days, that the mountain of the LORD'S house shall be established as the top of the mountains, and shall be exalted above the hills; and all nations shall flow unto it. 2:3 And many peoples shall go and say: 'Come ye, and let us go up to the mountain of the LORD, to the house of the God of Jacob; and He will teach us of His ways, and we will walk in His paths.' For out of Zion shall go forth the law, and the word of the LORD from Jerusalem. 2:4 And He shall judge between the nations, and shall decide for many peoples; and they shall beat their swords into plowshares, and their spears into pruninghooks; nation shall not lift up sword against nation, neither shall they learn war any more. [***] 40:15 Behold, the nations are as a drop of a bucket, and are counted as the small dust of the balance; behold the isles are as a mote in weight. 40:16 And Lebanon is not sufficient fuel, nor the beasts thereof sufficient for burnt-offerings. 40:17 All the nations are as nothing before Him; they are accounted by Him as things of nought, and vanity. [***] 40:22 It is He that sitteth above the circle of the earth, and the inhabitants thereof are as

---

Volume 23, The Soncino Press, London, (1935), pp. 213-216, at 213-214.

grasshoppers; that stretcheth out the heavens as a curtain, and spreadeth them out as a tent to dwell in; 40:23 That bringeth princes to nothing; He maketh the judges of the earth as a thing of nought. 40:24 Scarce are they planted, scarce are they sown, scarce hath their stock taken root in the earth; when He bloweth upon them, they wither, and the whirlwind taketh them away as stubble. [***] 54:1 Sing, O barren, thou that didst not bear, break forth into singing, and cry aloud, thou that didst not travail; for more are the children of the desolate than the children of the married wife, saith the LORD. 54:2 Enlarge the place of thy tent, and let them stretch forth the curtains of thy habitations, spare not; lengthen thy cords, and strengthen thy stakes. 54:3 For thou shalt spread abroad on the right hand and on the left; and thy seed shall possess the nations, and make the desolate cities to be inhabited. 54:4 Fear not, for thou shalt not be ashamed. Neither be thou confounded, for thou shalt not be put to shame; for thou shalt forget the shame of thy youth, and the reproach of thy widowhood shalt thou remember no more. [***] 60:5 Then thou shalt see and be radiant, and thy heart shall throb and be enlarged; because the abundance of the sea shall be turned unto thee, the wealth of the nations shall come unto thee. [***] 60:8 Who are these that fly as a cloud, and as the doves to their cotes? 60:9 Surely the isles shall wait for Me, and the ships of Tarshish first, to bring thy sons from far, their silver and their gold with them, for the name of the LORD thy God, and for the Holy One of Israel, because He hath glorified thee. 60:10 And aliens shall build up thy walls, and their kings shall minister unto thee; for in My wrath I smote thee, but in My favour have I had compassion on thee. 60:11 Thy gates also shall be open continually, day and night, they shall not be shut; that men may bring unto thee the wealth of the nations, and their kings in procession. 60:12 For that nation and kingdom that will not serve thee shall perish; yea, those nations shall be utterly wasted. [***] 60:16 Thou shalt also suck the milk of the nations, and shalt suck the breast of kings; and thou shalt know that I the LORD am thy Saviour, and I, the Mighty One of Jacob, thy Redeemer. 60:17 For brass I will bring gold, and for iron I will bring silver, and for wood brass, and for stones iron; I will also make thy officers peace, and righteousness thy magistrates. [***] 61:5 And strangers shall stand and feed your flocks, and aliens shall be your plowmen and your vinedressers. 61:6 But ye shall be named the priests of the LORD, men shall call you the ministers of our God; ye shall eat the wealth of the nations, and in their splendour shall ye revel. [version of the Jewish Publication Society]"

Obadiah states,

"1 The vision of Obadiah. Thus saith the Lord GOD concerning Edom: We have heard a message from the LORD, and an ambassador is sent among the nations: 'Arise ye, and let us rise up against her in battle.' 2 Behold, I make thee small among the nations; thou art greatly despised. 3 The pride of thy heart hath beguiled thee, O thou that dwellest in the clefts of the rock, thy habitation on high; that sayest in thy heart: 'Who shall bring me down to the ground?' 4 Though thou make thy nest as high as the eagle, and though thou set it among the stars, I will bring thee down from thence, saith the LORD. 5 If thieves came to thee, if robbers by night—how art thou cut off!—would they not steal till they had enough? If grape-gatherers came to thee, would they not leave some gleaning grapes? 6 How is Esau searched out! How are his hidden places sought out! 7 All the men of thy confederacy have conducted thee to the border; the men that were at peace with thee have beguiled thee, and prevailed against thee; they that eat thy bread lay a snare under thee, in whom there is no discernment. 8 Shall I not in that day, saith the LORD, destroy the wise men out of Edom, and discernment out of the mount of Esau? 9 And thy mighty men, O Teman, shall be dismayed, to the end that every one may be cut off from the mount of Esau by slaughter. 10 For the violence done to thy brother Jacob shame shall cover thee, and thou shalt be cut off for ever. 11 In the day that thou didst stand aloof, in the day that strangers carried away his substance, and foreigners entered into his gates, and cast lots upon Jerusalem, even thou wast as one of them. 12 But thou shouldest not have gazed on the day of thy brother in the day of his disaster, neither shouldest thou have rejoiced over the children of Judah in the day of their destruction; neither shouldest thou have spoken proudly in the day of distress. 13 Thou shouldest not have entered into the gate of My people in the day of their calamity; yea, thou shouldest not have gazed on their affliction in the day of their calamity, nor have laid hands on their substance in the day of their calamity. 14 Neither shouldest thou have stood in the crossway, to cut off those of his that escape; neither shouldest thou have delivered up those of his that did remain in the day of distress. 15 For the day of the LORD is near upon all the nations; as thou hast done, it shall be done unto thee; thy dealing shall return upon thine own head. 16 For as ye have drunk upon My holy mountain, so shall all the nations drink continually, yea, they shall drink, and swallow down, and shall be as though they had not been. 17 But in mount Zion there shall be those that escape, and it shall be holy; and the house of Jacob shall possess their possessions. 18 And the house of Jacob shall be a fire, and the house of Joseph a flame, and the house of Esau for stubble, and they shall kindle in them, and devour them; and there shall not be any remaining of the house of Esau; for the LORD hath spoken. 19 And they of the South shall possess the mount of Esau, and they of the Lowland the Philistines; and they shall possess the field of Ephraim, and the field of Samaria; and Benjamin shall possess Gilead. 20 And the captivity of this host of the children of Israel, that are among the Canaanites, even unto Zarephath, and the captivity of Jerusalem, that is in Sepharad, shall possess the cities of the South. 21 And saviours shall come up on mount Zion to judge the mount of Esau; and the kingdom shall be the LORD'S. [version of the Jewish Publication Society]"

Micah 5:7-8 (*Micah* 5:8-9 in the KJV) states:

"7 And the remnant of Jacob shall be among the nations, in the midst of many peoples, as a lion among the beasts of the forest, as a young lion among the flocks of sheep, who, if he go through, treadeth down and teareth in pieces, and there is none to deliver. 8 Let Thy hand be lifted up above Thine adversaries, and let all Thine enemies be cut off. [version of the Jewish Publication Society]"

Zechariah 8:20-23; and 14:9 state,

"8:20 Thus saith the LORD of hosts: It shall yet come to pass, that there shall come peoples, and the inhabitants of many cities; 8:21 and the inhabitants of one city shall go to another, saying: Let us go speedily to entreat the favour of the LORD, and to seek the LORD of hosts; I will go also. 8:22 Yea, many peoples and mighty nations shall come to seek the LORD of hosts in Jerusalem, and to entreat the favour of the LORD. 8:23 Thus saith the LORD of hosts: In those days it shall come to pass, that ten men shall take hold, out of all the languages of the nations, shall even take hold of the skirt of him that is a Jew, saying: 'We will go with you, for we have heard that God is with you.' [***] 14:9 And the LORD shall be King over all the earth; in that day shall the LORD be One, and His name one. [version of the Jewish Publication Society]"

3.2 Jewish Messianic Supremacism

In order to understand why so many viewed racist Jews like Albert Einstein, Karl Marx, and the Rothschilds, as a threat to humanity; it is helpful to understand that Judaism prophesies the violent destruction of Gentile humanity. The same racist Jewish forces who were promoting the racist Jew Albert Einstein to the public, were destroying the nations and religions of Europe in their pursuit of the fulfillment of Old Testament prophecy.

Many have written exposés on the Jewish-Messianic nature of Communism, among them Denis Fahey, who stated, *inter alia*,

"As there is only one world and one Divine Plan for that world, the Messias to whom the Jews look forward must be purely natural. The unity and peace of the coming Messianic era, must, accordingly, be brought about by the subjection of all nations to the Jewish nation. Thus they dream of establishing, on the purely natural level, the union which God is striving to bring about on the supernatural level of the Mystical Body, respectful of national characteristics and of the diversity of national vocations in Christ. The Jews are, therefore, opposed to the whole order of the world, built on the Divinity of Jesus, and their influence in every sphere, in Freemasonry and in Communist movements, in Finance, in the Press and in the Film-world, will favour the naturalistic aims of Masonry and of revolutionary societies while at the same time impelling them in the direction of a world-state in which the Jewish race will he supreme. Accordingly, when we read, in the sermon broadcast by Chief Rabbi Julian Weill (Radio-Paris, March 27th, 1931): 'The Jewish Passover... is turned to the future and affirms with a definite and joyous conviction the liberation to come and the Messianic Passover of the peoples of the world,' we know what that means for those who believe in our Lord's Divinity. We know, too, That this Jewish view of the world may be expressed in another fashion, for it presents another aspect to the Gentile peoples who are being 'liberated.' The *Pilori*, a newspaper published at Geneva, puts that other point of view as follows:—

'Of course, all cannot grasp that it is international high finance, dominated by the Jews and supported by Freemasonry, that started the world-war, brought about the revolutions in Russia and Spain, and now throws the economic life of peoples into confusion. Lengthy reflection is required in order to see that a hundred Jewish bankers... are engaged in liquidating the remaining stocks of the ancient Christian civilization of Europe.'[*Footnote:* Issue of September 25th, 1931.]
[***]
[*Footnote:* 'When people talk about the Jewish religion, they think only of the Bible, of the religion of Moses. This is an illusion... . According to the *Univers Israélite* 'For two thousand years... the Talmud has been the religious code of Israel'... A work of hatred and impiety, the Talmud definitely confirmed the apostasy of modern Jewry... It is a systematic deformation of the Bible... . The pride of race with the idea of universal domination is therein exalted to the height of folly... . For the Talmudist, the Jewish race alone constitutes humanity. The non-Jews are not men. They are of a purely animal nature.' (*L'Histoire et les Histoires dans la Bible*, by Mgr. Landrieux, Bishop of Dijon, pp. 101, 102, 99.) For texts of Talmud, cf. *Les Sources de l'Impérialisme Juif*, pp. 21-40, by Mgr. Jouin.] [***]
[*Footnote:* Mrs. Webster even says that 'it is in the Cabala, still more than in the Talmud, that the Judaic dream of world-domination recurs with the greatest persistence.' (*Secret Societies and Subversive Movements*, p. 370.)][***] The official head of the Anti-God Association of the U.S.S.R. is the Jew, Yaroslawsky, whose real name is Goublemann.[*Footnote: R. I. S. S.*, January 1st, 1933, p. 18. Cf. Appendix I, 'Jewish Power.'] [***]
[*Footnote:* 'The deification of humanity by the Freemasons of the Grand Orient finds its counterpart in the deification of Israel by the modern Jew.' (Mrs. Webster in *Secret Societies and Subversive Movements*, p. 374.)]
[***] A few words must suffice here, but they will be enough to show that many of the Gentile instruments, who figure as leaders, are really the dupes of Jewish capitalism. [***] The proletariat class, which produces the material goods on which human society lives, is a Messianic class destined by its rule to bring about a new era for the world. This Messianic vocation of the proletariat, according to Marx, found an answering echo in the Messianic expectations of the Russian people.[*Footnote:* Cf. *The Russian Revolution*, by N. Berdyaev, pp. 74, 75.] But both the proletariat in general and the Russian people in particular are only means for the realization of the Messianic dreams of Marx's own people. Masters of production through

finance, they will shape the destinies of the world-God or collectivity-God. [***] It would be too long to recount the whole story of the growth of the Communist movement in Europe. The plan of the revolution is always substantially the same. The reins of government of some great nation must be captured and then that nation must be made a sort of battering-ram, in order to impose the revolutionary ideal on the neighbouring peoples. The France of 1789 and its people were used as revolutionary ammunition, to be hurled at Europe. If Marx had succeeded through his agents in the Paris Commune of 1871, France would have had the fate which was reserved to the Russia of 1917. In Russia the vast sums invested in Communism by Jewish capitalists bore fruit and the sovereign thought of the Hegelian philosopher of Berlin has passed from the passive state to the free state, with the results we know. The ideas of God, our Lord Jesus Christ, the native land, the family, and the personality of the child, are all being swept away in the name of 'progress,' while the financiers laugh at their poor dupes. The Russian revolutionary Bakunin, who knew Marx well and who used to describe him and his following as the 'German-Jew Company,' complained in his day of the contempt of Marx and Engels for the poor. Marx spoke of the poor and destitute workers as the 'ragged proletariat' (*Lumpenproletariat*). [***] If we now turn to Mrs. Webster's *The Surrender of an Empire* (pp. 74-79), we get some additional information about the rise of Bolshevism. It seems that the real name of the individual mentioned above in Section III, under the designation of Parvus, is Israel Lazarevitch Helphand and that he is a Jew of the province of Minsk, in White Russia. In the second half of the eighties he took part in revolutionary work in Odessa. In 1886 he went abroad and finally, after many wanderings, went to Copenhagen, where he amassed a large fortune as the chief agent for the supply of German coal to Denmark, working through the Danish Social Democratic Party. Dr. Ziv, in his *Life of Trotsky*, relates that when he was in America in 1916 he said to Trotsky: 'How is Parvus? ' to which Trotsky replied laconically: 'Completing his twelfth million.' It is this Jewish multi-millionaire who, after Karl Marx, was the great inspirer of Lenin. It was through the intervention of Parvus that Lenin was sent back to Russia by the Germans. Lenin was dispatched from Switzerland to Russia in a locked train and was provided with no less than £2,500,000 by the German Imperial Bank. It was not, therefore, as a needy revolutionary, setting forth on a precarious mission, his soul lit with pure zeal for the cause of the workers, that Lenin journeyed into Russia, but as a well-tried agent, versed in all the tricks of intrigue and the art of propaganda and backed by the powerful organization of international finance. The people accompanying him were predominantly aliens: out of a list of 165 names published, 23 are Russian, 3 Georgian, 4 American, 1 German and 128 Jewish.'[*Footnote:* An illuminating sketch of Lenin's career is to be found in an article by Salluste in *La Revue de Paris* (December 15, 1927). Lenin, according to this able writer, was, at the same time, a paid agent of the Russian secret police and of the Jewish financiers engaged in furthering the Marxist conspiracy. He profited by his position as police agent to prepare the triumph of the schemes of the financiers.] The English accuse the Germans of having sent Lenin to Russia. We have seen the influences at the back of that action. On the other hand, the Germans accuse the English of having sent Trotsky back, for Trotsky was set free from arrest by order of the British Government (he had been arrested at Halifax), when he was needed by Jacob Schiff and the others, as we saw above. The truth is that Jewish financial influences were working behind the Governments of both peoples for their own ends. 'Russia' is not a triumph for the workers; but seems to be a gigantic investment of Jewish capitalists for their own ends. Amid the welter of details about 'Russia,' the great fact must not be lost sight of, that the men who seized power and retain it, as the taskmasters of the rationed and ticketed people of Russia, were put there by a certain number of Jewish capitalists. The Russian middle-class and the nobles, the natural leaders of the people, were exterminated, while the manual workers, who were too uneducated to see through the plans of the investors, were extolled to the skies. [***] Of course, Muscovite propaganda, when attacking God and the hierarchical order of human society, will not inform the people who are urged on to the class-war and revolution that a new and savage feudalism or rather slavery will be the result. The members of the Bolshevik party are the new supreme class, and against the party and its members no rights exist, for there is no such thing as a right in the correct sense. [***] One question, however, always returns: 'What about the Jewish international financiers who financed Lenin and Trotsky in 1917?'[2] That their control over the figure-heads of the Communist party, like Stalin, exists is certain. In her book, *Trois ans chez Tsars rouges* (p. 96), Madame Éise Despreaux speaks of the appearance of anti-Semitism in the Communist party and continues:

> 'It is its preponderance amongst the Communists which has brought about the success of Stalin, in 1926 and 1927. Nevertheless, if the Georgian dictator maintains his position, it is at the price of a manifest capitulation in face of the higher power of international finance. The part played by this power in the destinies of the U.S.S.R. is undeniable. Of course, the exact nature of the part is difficult to prove, on account of its secret character. The influence of this power has, however, been exercised recently in favour of the Jews, without whom the Russians would find it difficult to manage commercially and economically.'[1]

[***]

It is to the influence of international finance that the relative stability of the Russian revolution is due. Just as greater skill in carrying out successful revolutions has been acquired by experience since 1789, so also progress has been made in the art of maintaining the figure-heads in power, in spite of the discontent of the majority of the people and the unceasing struggle against the laws of nature.[2] [***] Again, Marxian Communism is a neo-Messianic movement, based on Jewish rejection of the Messias Who has come, and the workers are merely the tools by which Israel hopes to exercise world domination. [***] The complete triumph of the so-called Christian Workers' Republic can have no other result than the extermination of all those who believe in the Divinity of Christ the King. 'No man can serve two masters' (Matthew vi. 24). Of course anyone, Bishop, priest or layman, who stands up for the integral rights of Christ the King will be got rid of, ostensibly as an enemy of the republic and a counter-revolutionary. And be it noted that ideas work themselves out in act, or rather men are spurred on to draw the final conclusions from the ideas they hold. Marxian republicans cannot stop halfway and compromise with Catholicism. They must seek to exterminate its adherents and educate a new generation which will worship only matter, machinery and—Satan. [***] A few extracts from Waldemar Gurian's able work from which we have already quoted will confirm these statements:—'[***] This produces an oppression of unparalleled magnitude. All intellectual life that does not serve Bolshevik aims must be annihilated; intellectual freedom and independence must yield to the dogmas of the Bolshevik creed; religion must disappear, and scientific research be exclusively directed to results which are in harmony with the doctrines of dialectical materialism and above all serve the Bolshevik rule. [***]'"[159]

3.3 The "Eastern Question" and the World Wars

In an article entitled "Modern Jewish Worship", the *New York Evangelist*, Volume 12, Number 40, (2 October 1841), p. 1, wrote,

"Through all their wanderings, they have followed the direction of Moses, to be *lenders* and not *borrowers*. The sovereigns of Europe and Asia, and the republics of America, are their debtors to an immense amount. The Rothschilds are Jews; and they have wealth enough to purchase all Palestine if they choose; a large part of Jerusalem is in fact mortgaged to them. The oppressions of the Turkish government, and the incursions of hostile tribes, have hitherto rendered Syria an unsafe residence; but the Sultan has erected it into an independent power, and issued orders throughout his empire, that the Jews shall be as perfectly protected in their religious and civil rights, as any other class of his subjects; moreover, the present controversy between European nations and the East seems likely to result in placing Syria under the protection of Christian nations. It is reported that Prince Metternich, Premier of Austria, has determined, if possible, to constitute a Christian kingdom out of Palestine, of which Jerusalem is to be the seat of government."

The Rothschilds, and their agent, Karl Marx, saw to it that Gentile nations

---

[159]. D. Fahey, *The Mystical Body of Christ in the Modern World*, Browne and Nolan Limited, London, (1935), pp. 74-77, 82, 84, 86-87, 92-93, 98-102.

and peoples did not advance peacefully and prosperously to the highest achievements they could otherwise have attained without the influence of these corrosive forces. The results of Rothschild and Marx agendas have been the same—tax the Gentiles into comparative poverty, financially, intellectually and even genetically; primarily through wars and revolutions, and through control of the monarchies, press, politics, education and the professions. For centuries, Jewish bankers agitated the nations and artificially created the "Eastern Question" in an effort—which was ultimately successful—to provoke world wars, which would net them Palestine and obstruct the progress of Gentile nations. This was already apparent to many in 1820—after Napoleon Bonaparte had devastated Europe in order to emancipate the Jews and "restore" them to Palestine.

The Atheneum; or, Spirit of the English Magazines, Volume 2, Number 10, (15 August 1820), pages 398ff. stated,

"RUSSIA AND TURKEY"

There is a madness of thrones, and it is the madness of perpetual desire—the madness of avarice and accumulation. No extent of dominion can satisfy it; the utter worthlessness of the object cannot restrain it; desart is added to desart, marsh to marsh, a sickly and beggared population is gathered to the crowd that are already perishing in the midst of their uncultured fields;—yet the passion is still keen, and thousands of lives are sacrificed, years of desperate hazard are encountered, and wealth, that might have transformed the wilderness into a garden, is flung away, for the possession of some leagues of territory, fit only to make the grave of its invaders. Austria, at this hour the mistress of a prodigious empire, one half of which is forest, heath, or mountain, unpeopled, or only peopled by barbarians—Austria, the mistress of Croatia, the Bannat, and Transylvania, is longing for Albania, a country of barren mountain and swampy valley, with a population of robbers. Russia, with a territory almost the third of the old world, stretching from the Black Sea to the Pole, and from Finland to the wall of China, is longing for the fatal marshes of Wallachia and Moldavia; for the desarts of Romelia, and the sovereignty of the fiercest race of barbarians on earth, alien by their creed, alien by their habits, and cursing the ground that has been defiled by the tread of a Russian. With two capitals already hostile to each other, she is struggling for a third, incurably and furiously hostile to both. With an extent of dominion that no single sceptre can adequately rule, and which a few years will see either torn asunder by the violence of rebellion, or falling in pieces by the natural changes of overgrown territory, she is at this hour marshalling her utmost strength, and laying up debility for many a year, in the frantic eagerness to add the Turkish empire to the Muscovite, the Siberian, and the Tartar.

And in this tremendous chase of power, what is to be trampled under the foot of the furious and guilty pursuer! The heart sickens at the reckless waste of life and the means of life, the myriads that must perish in the field, the more miserable myriads that must perish of disease, famine, and the elements let loose upon their naked heads; the still deeper wretchedness of those lonely and deserted multitudes, whose havoc makes no display in bulletins and gazettes, but whose history is registered where the eternal eye of justice and vengeance alone reads—the innumerable host of the widow and the orphan. Yet this weight of calamity is let fall upon mankind at the word of a single individual:—often the most worthless of human beings, an empty, gaudy, ignorant slave of alternate indolence and sensuality; trained by the habitual life of foreign courts to the perpetual indulgence of personal excess, and differing from the contemptible race generated by the habits of foreign life, only by his being the more open dupe of sycophancy, the more prominent object of public alarm, and the more unbridled example of every profligacy that can debase the individual, or demoralize the nation.

Europe is again threatened with universal hostilities by the passion of the Czar to be master of Constantinople.—The nominal cause of the war with Turkey is the removal of the hospodars of Wallachia and Moldavia by the Porte. A treaty in 1801 had established that those governors of the provinces should be removed only at the end of every seven years; a period fixed by the customary cunning of the Russian cabinet, as one in which the hospodars, thus rendered secure from the bow-string, might connect themselves more effectually with Russia. The hospodars were Greeks, and their national prejudices allied them to their new protectors; they were like all the Greeks of Fanar—ambitious, corrupt, and crafty; and the gold of Russia was the virtual sceptre of the hospodariates."

It necessary to interject some explanatory comments, before proceeding with the rest of the above article "Russia and Turkey". Jewish bankers orchestrated an alliance of Greek and Russian Orthodox Christians to diminish or utterly destroy Turkish influence, especially in Greek and Slavic regions, which confrontation benefitted the Jews by opening up Palestine—which was a part of the Turkish Empire—to Jewish colonization, and setting up the groundwork for the world wars, which would lead to peace conferences that would establish a Jewish state and a world government run by Jews.

An article entitled, "The Modern Jews", *The North American Review*, Volume 60, Number 127, (April, 1845), pp. 329-368, at 337-339, wrote,

"Since the last conquest of Constantinople, Turkish policy has inclined to tolerate the Jews; and the consequence has been a great increase of their numbers in that city. They are often bankers for the grandees, and custom, acquiring the force of law, has established them as collectors of the customs and purveyors for the seraglio. Their taxes are not greater than those paid by other races in a similar condition. 'The Jews,' says Judge Noah, 'are at this day the most influential persons connected with the commerce and monetary affairs of Turkey, and enjoy important privileges; but hitherto they have had no protecting influence.'¹⁶⁰ [***] In Syria, the Jews are in a state of real servitude, and no change of masters has bettered their condition. Mohammedans and Christians alike hate and maltreat them; and this hatred is heartily returned, as the latter find, whenever any circumstance gives their enemies a temporary advantage. When the Turkish succeeded the Egyptian troops in Damascus, a few years ago, they were stirred up by the Jews to persecute the Christians of every sect. When the Greeks rose against the Turks in 1822, the Jews eagerly joined against the Christians, especially in Constantinople; while the Greeks, in revenge, murdered all the Jews on whom they could lay their hands."

3.3.1 Dönmeh Crypto-Jews, The Turkish Empire and Palestine

Jewish bankers, including Camondo, Allatini, Modiano and Maître Salem,¹⁶¹ had long overseen Turkish finances. The Jewish bankers oversaw and governed the "Greek" and "Armenian" control of Turkish finances,¹⁶² and eventually bankrupted the Turkish Empire and destroyed the Egyptian economy.

Major Osman Bey wrote in the 1870's,

> "Ilahmi Pascha, the son of the Viceroy of Egypt, had inherited a fabulous fortune, amounting to not less than 150,000,000 francs. The Jew Oppenheim, in Alexandria, became his banker, and administered the affairs of the young Ilahmi so masterly that three years of his administration sufficed to make the Prince a bankrupt."¹⁶³

The Jewish bankers feared that the Egyptians would oppose the formation of a Jewish kingdom in Palestine, even if the Sultan of Turkey and the lands of Palestine could be bought by Rothschild. In an article entitled "Modern Jewish Worship", the *New York Evangelist*, Volume 12, Number 40, (2 October 1841), p. 1, wrote,

> "Through all their wanderings, they have followed the direction of Moses, to be *lenders* and not *borrowers*. The sovereigns of Europe and Asia, and the republics of America, are their debtors to an immense amount. The Rothschilds are Jews; and they have wealth enough to purchase all Palestine if they choose; a large part of Jerusalem is in fact mortgaged to them. The oppressions of the Turkish government, and the incursions of hostile tribes, have hitherto rendered Syria an unsafe residence; but the Sultan has erected

---

160. The article cites: "*Lecture on the Restoration of the Jews*. By M. M. NOAH. Delivered October 28th, 1844, in the Tabernacle, New York City."
161. "Hopes of an Understanding", *The London Times*, (24 February 1912), p. 7.
162. M. M. Noah, *Discourse on the Restoration of the Jews: Delivered at the Tabernacle, Oct. 28 and Dec. 2., 1844*, Harper, New York, (1845), p. 38.
163. Major Osman Bey, English translation by F. W. Mathias, *The Conquest of the World by the Jews, an Historical and Ethnical Essay*, Part 12, St. Louis, (1878), pp. 30-31.

it into an independent power, and issued orders throughout his empire, that the Jews shall be as perfectly protected in their religious and civil rights, as any other class of his subjects; moreover, the present controversy between European nations and the East seems likely to result in placing Syria under the protection of Christian nations. It is reported that Prince Metternich, Premier of Austria, has determined, if possible, to constitute a Christian kingdom out of Palestine, of which Jerusalem is to be the seat of government."

Agitated by Jews and crypto-Jews, who hated Christians, the Sultan retaliated against innocent Armenians who were blamed for allegedly stealing the wealth of the Kingdom—wealth which had been stolen by Jewish financiers. These attacks on innocent Armenians benefitted the Jewish financiers by weakening an ancient Christian enemy in the region, one associated with the mythical exile of the lost ten northern tribes of Israelites and one associated with the Christians in Jerusalem and elsewhere in Palestine, which Christians then outnumbered the Jews in Palestine. It also deflected attention away from the crimes of the Jewish financiers. Furthermore, these attacks left the Sultan dependent on Jewish influence in the mass media to safeguard the image of the Empire from exposure of the atrocities the Turks committed against Armenians due to the instigation of Jews and crypto-Jews. The Jews led the Christians and Moslems to devour one another.

When crypto-Jewish "Young Turks"[164] finally succeeded in overthrowing the bankrupt Sultan, the crypto-Jews mass murdered the Armenians in a genocide of some 1.5 million lives lost—far worse atrocities than had ever been committed under the Sultan, which genocide benefitted the Jews in that it diminished Christian influence in the region of Palestine. The Zionist Jews also hoped that the atrocities could be used as wartime propaganda to inspire hatred of the Turks and of the Germans in America and elsewhere; and would draw the British and French into the region—a goal Cabalistic Jews had lusted after for centuries.

An article entitled, "The Turkish Situation by One Born in Turkey", *The American Monthly Review of Reviews*, Volume 25, Number 2, (February, 1902), pp. 182-191, at 186-188 states:

"Turkish treasury accounts have always been kept by Greeks and Armenians. If a Turk owns land, some Christian keeps its rent-roll. If he has a business, Christian clerks manage it, If he owns mines or works the richer placer of official extortion, some Christian engineer or scribe manages and manipulates his accounts. Such prosperity as there was through the twenty years of Abdul Hamid's reign, which seemed prosperous, went to

---

164. I. Zangwill, *The Problem of the Jewish Race*, Judaen Publishing Company, New York, (1914), pp. 9, 11; which was first published as an article, "The Jewish Race", *The Independent*, Volume 71, Number 3271, (10 August 1911), pp. 288-295, at 290-291. J. Prinz, *The Secret Jews*, Random House, New York, (1973), pp. 111-112.

Christians."

The Zionists deliberately bankrupted Turkey, which owned Palestine, so that they could blackmail the Sultan into surrendering the territory to the Jews. Soon after the Young Turk revolutionaries gained power under their *Dönmeh* crypto-Jewish leadership,[165] the Zionist bankers largely had their way. The Zionists scripted Young Turks to betray the interests of the Turkish Empire and the Moslem faith, and favor the interests of Zionist Jews. *The London Times* reported on 12 March 1909 on page 4,

> "A TURKISH DEPUTY ON ZIONISM.—The *Jewish Chronicle* of to-day states:—Dr. Riza Tewfik, a member of the Chamber of Deputies and one of the foremost leaders of the Young Turk party, delivered a lecture on the Jewish question recently in Constantinople, under the auspices of the Society of Young Jews. At the close of the lecture, Dr. Riza Tewfik invited questions, and in reply to the inquiry, whether a good Ottoman could be a Zionist, he replied, 'Certainly, I myself am a Zionist. Zionism is fundamentally nothing more than the expression of the solidarity which characterizes the Jewish people. What is the aim of Zionism? A humanitarian one: to find a more friendly fatherland for unfortunate co-religionists, where they can live as free men in the enjoyment of their rights. The methods of Zionism are exclusively peaceful. Palestine is your land more than it is ours; we only became rulers of the country many centuries later than you. A service would be rendered to our common fatherland by undertaking the colonization of that uncultivated land, Palestine. Your nation has incomparable qualifications for trade; your fellow-Jews are sober and industrious. They would restore this desolate land. They would devote all their energies to the service of our dear fatherland, and I assure you that my co-operation will never fail you in order to attain this aim.'"

The London Times reported on the Turks' suspicion of cryto-Jewish and Zionist Jewish financial influence on the Empire, on 3 March 1911, on page 5,

"THE TURKISH CHAMBER AND ZIONISM.
(FROM OUR CORRESPONDENT.)
CONSTANTINOPLE, MARCH 1.

In to-day's debate on the Budget in the Chamber Ismail Hakki, Deputy for Gumuldjina, made a long criticism of Djavid Bey's financial policy, at the close of which, after expressly declaring his confidence in the loyalty of the great majority of the Ottoman Jews, he hinted that the Minister had

---

[165]. I. Zangwill, *The Problem of the Jewish Race*, Judaen Publishing Company, New York, (1914), pp. 9, 11; which was first published as an article, "The Jewish Race", *The Independent*, Volume 71, Number 3271, (10 August 1911), pp. 288-295, at 290-291. J. Prinz, *The Secret Jews*, Random House, New York, (1973), pp. 111-112.

shown undue preference to Jewish capitalists and their agents, some of whom he accused of favouring Zionism. He also drew the attention of the House to the growth of Zionist propaganda in Turkey and to the efforts of the foreign Jewish agents on behalf of that cause.

The leader of the 'People's Party' then treated the House to something of an anticlimax, naming Sir Ernest Cassel and other unlikely persons as presumable Zionists. The Grand Vizier explained that Sir Ernest Cassel was a member of the Anglican Church, and was an intimate friend of the late King, and therefore a 'true and loyal friend of the Ottoman Empire.'

Talaat Bey, answering the statement of Ismail Hakki, said that proposals had been made to him and to Djavid Bey by the Jewish General Colonization Society, which they had been unable to accept. He admitted Zionist activity, but said that the law preventing Jewish immigration into Palestine remained in force.

Ismail Hakki Bey Babnzadeh has been appointed Minister of Public Instruction.

The monopolies which the Government intend to create, as announced by Djavid Bey in his recent Budget speech, do not include petroleum. I understand that the Government proposes, subject to the consent of the interested Powers, to establish an Excise duty on petroleum instead of creating a monopoly."

Zionist activity in Turkey became so noxious that it threatened to lead to anti-Semitism in the Turkish Empire, which Turkey had not known. Note that before the Zionists stabbed Germany in the back in favor of England, the German Government and the Zionists had worked together and the German Government was very good to Jews, and to Zionists in particular. *The London Times* stated on 14 April 1911 on page 3,

"THE YOUNG TURKS AND ZIONISM.

HOSTILITY TO THE MOVEMENT.
(FROM OUR OWN CORRESPONDENT.)
CONSTANTINOPLE, April 9.

A curious incident, the news of which has just reached the capital from Salonika, reveals in unmistakable fashion the rapid growth of Turkish hostility to the Zionist movement. A well-known Zionist propagandist, Santo-Semo Effendi, having obtained the permission of the Committee of Union and Progress to use its Club at Salonika for the purpose of a lecture on immigration into Mesopotamia, a large number of Jewish and Turkish members of the Committee promised to be present on this occasion.

They kept their promise, but when the lecturer, after discussing various schemes for the colonization of Mesopotamia, delivered a violent attack on Great Britain, accusing her of opposing German commercial schemes in Mesopotamia simply with a view to the eventual economic and political conquest of Irak, many of the Turks present hooted the lecturer and the meeting was for a time so disturbed that several of the leading Jews present withdrew. Quiet was soon restored, but on the following day the Turkish *Rumeli*, which is now the organ of the Salonika Committee and is believed especially to reflect the views of its military members, published a violent attack on Zionism, which it described as being simply and solely a cloak for German designs and notably for schemes for the economic conquest and exploitation of Mesopotamia. These views certainly appear now to prevail among many Turks both withing and without the Committee organization, who profess to find evidence of German support of Zionism in the strongly Germanophile and Anglophobe tendencies of the principal Zionist organs published in Turkey, and the fact that some of the chief Zionist propagandists here are German subjects. However this may be, it is to be hoped that the anti-Zionist feeling, which has become very marked of late, may not degenerate into Anti-Semitism from which Turkey has till now been free."

At various times, duplicitous Zionist Jews used the French, Russians, Germans, and English against the Turks, leading each nation to believe it was in its own best interests to war with the Turks and install a Jewish nation in the region. The Jewish Zionists were loyal to no nation but themselves. France, Russia, Germany and England each suffered for the loyalty they showed to Zionist Jews—as did the Turkish Empire, which had also been very good to Jews. The Zionists even used themselves as bait to create a war between the Germans and the British over Mesopotamia—and Palestine, and to drive a wedge between the Germans and the Turks on the eve of the First World War.

These facts were becoming increasingly obvious to the Turks, such that the Zionists felt obliged to protest loudly against such accusations. The Zionists even went so far as to blame the Turks for the Zionists' continued intrigues in Turkey, on the sophistical and false premise that they were obliged to continue to intrigue in Turkey so as to dispel the alleged myth that they were intriguing in Turkey. The fact that the Zionists played both sides of the struggles the Zionists themselves had fomented is further revealed in their denials of the facts—the Zionists were primarily Russian Jews operating around the world—disloyal Russian Jews who wanted to bring England, Germany, Russia and Turkey into war. *The London Times* reported on 9 May 1911 on page 7,

"ZIONISM AND TURKEY.
(FROM A CORRESPONDENT.)
COLOGNE, May 4.

The International Council of the Zionist Organization, which has just concluded a two days' Conference at the Central Office, conducted most of its proceedings in private, as they were devoted to a discussion of the Zionist situation in the Ottoman Empire. It was announced that the following resolution had been adopted:—

The International Council, having carefully considered the Zionist situation in Turkey and the reports which it has received from there, declares that the charges recently brought against Zionism are based upon a deficient knowledge of the real character of the movement, and upon an incorrect conception of its aims and endeavors. It is firmly convinced that Zionist aspirations are in complete accord with the interests of the Ottoman Empire, and considers it its duty to continue its efforts in Turkey so that the real import and aims of the Zionist movement may be rightly understood.

In connexion with the Conference, meetings of the Jewish National Fund, the Anglo-Palestine Company, and the Anglo-Levantine Banking Company—which are all Zionist institutions—also took place."

In yet another of the countless instances where Zionists have played both sides of an issue with mutually exclusive and contradictory arguments, a Zionist leader named Wolffsohn attacked the *London Times'* reporting on the basis that the Jews had no desire to take over Palestine. The Zionists later would reverse this stance and go so far as to claim that the Balfour Declaration of 1917 was their deed to the land—this in spite of the fact that England had no right to issue the Declaration and it did not give Palestine to the Jews for the formation of State, but merely looked favorably on the idea of Jews living under a Palestinian Government. It had perhaps escaped Wolffsohn's memory that Theodor Herzl's book was titled, "The Jewish State", which would lead a reasonable person to believe that the political Zionists sought to form a State, no matter what lies the political Zionists told the world public as a means to regulate public opinion, and no matter what public political expressions they were forced to accept. History has put the lie to Wolffsohn's sophistry. The brazen dishonesty of the Zionists is apparent, given the events of the First World War, which contradict Wolffsohn's deceitful reassurances.

On 10 May 1911, on page 8, *The London Times* published the following Letter to the Editor,

"THE YOUNG TURKS AND ZIONISM.

TO THE EDITOR OF THE TIMES.

Sir,—I shall feel much obliged if you will allow me to make a few observations upon the article of your Constantinople Correspondent on the 'Young Turks and Zionism,' which appeared in your issue of April 14, and regret that my recent absence from Cologne has prevented me from writing to you before. I particularly regret this inevitable delay, as several statements in the article are quite incorrect, and as they have not yet been challenged or rectified in your columns, I fear they may have found acceptance in certain quarters. Knowing, however, that you are far from desiring that any injustice should be done through any article in your paper to the cause that I represent, I feel sure that you will grant hospitality to few notes of correction and explanation.

While fully admitting the evident desire of your Correspondent to present an objective and impartial account of Zionism in the Ottoman Empire, I regret that his limited knowledge of our movement and the sources from which he appears to have derived it made it impossible for him to realize that desire. The cardinal defect of his article consists in the assumption that Zionism is a scheme for the foundation of a Jewish State in Palestine. This assumption is wrong. His comments upon our movement and his account of the views upon it in Turkish circles are mainly dependent upon this assumption. As his premiss is incorrect, his conclusions are of interest only in so far as they represent the state of mind shared by others in Turkey who have likewise been misled as to our aims and intentions.

The object of Zionism is clearly defined in its programme adopted at our first Congress at Basel in 1897, and hence known as the Basel Programme. This programme is 'To create a publicly recognized and legally secured home for the Jewish people in Palestine.' The aim thus formulated is essentially different from the aspiration to found a State, and those who attribute to us such an aspiration misrepresent us in a very serious degree, as they are likely, however unwittingly, to cause difficulties being put in our way. It is because this erroneous notion has secured a strong hold upon the minds of many people that disparaging remarks were made upon Zionism in the Turkish Chamber several weeks ago. The misinterpretation of our position is all the more strange and inexcusable as I expressly declared at the ninth Zionist Congress at Hamburg in December, 1909, that our work is guided and governed by the deepest respect for the Constitution and by the fullest recognition of the sovereignty of the Porte. We are simply desirous of making Palestine once again the national home of the Jewish people; and, to achieve that end, we are working for the economic and intellectual regeneration of the Holy Land in full conformity with the law.

Our object is so peaceful and our aims are calculated so highly to benefit the interests of the Ottoman Empire that we are painfully surprised that our movement should arouse any distrust in authoritative circles in Turkey. This circumstance can be ascribed only to the prevalence of various fantastic legends that have been put into circulation by our opponents, who, I regret to say, include many Jews. The latest of these legends is that Zionist activity is being conducted in the specific interests of Germany. This story is utterly without foundation in substance or fact, as we have no relations of any kind that can be construed as specially favouring the economic interests of Germany. The *data* advanced in support of the story are also incorrect. The *Jeune Turc* cited by your Correspondent is a purely Turkish paper, which, it is quite true, has more than once advocated a Jewish immigration into the Ottoman Empire in the interests of the Empire itself, but there is not the least ground for deducing from this that we are even in the least responsible for the policy of the paper. It is therefore immaterial to us whether the proprietor, Herr Hochberg, is a German Jew, or, as I have just been informed on excellent authority, a Russian Jew. Dr. V. Jacobson, who is one of the leading Zionists in Constantinople and manager of an English company—the Anglo-Levantine Banking Company—is also a Russian subject.

Finally, I wish to point out that the Zionist Organization has absolutely no connexion with the General Jewish Colonizing Organization of Berlin. Hence the activity of this organization, or rather of its representative, Dr. Nossig, does not form a 'new phase'—or, indeed, any 'phase'—of Zionism, and the conclusions derived from this activity cannot be used as an argument against our movement.

I feel sure that when those who are interested in Zionism will have purged their minds of the various fantastic fables that have been put into circulation to damage it, they will realize its peaceful intentions and beneficent aims. Our organization has already given a powerful impetus to commercial and industrial life in Palestine during the few years it has been active in the country, mainly through our companies which carry on their operations there. These companies—the Anglo-Palestine Company (Limited), the Jewish National Fund (Limited), and the Palestine Land Development Company (Limited)—have all been registered in London as English companies. The part they are playing in the economic amelioration of Palestine is but an earnest of the great work that Zionism is destined to do, and which, with the good will of the Ottoman Government, it will accomplish.

Yours obediently,
D. WOLFFSOHN,
President of the Zionist Organization.
Cologne, May 1."

3.3.2 The World Wars—A Jewish Antidote to Jewish Assimilation

The racist Zionists failed in their attempts to buy Palestine and populate it with Jewish colonists, because the vast majority of Jews did not want to go to Palestine. The Zionists caused the First World War in order to break up the Turkish Empire and weaken the Moslem nations, which they feared would unite to fight against the formation of a "Jewish State". *The London Times'* Vienna Correspondent published a Letter to the Editor, which was published un der the heading "Jews and the Situation in Albania" in *The London Times* on 27 July 1911, on page 5, and which stated, *inter alia*,

"4. 'The principle of the State resting on homogeneous nationality, and forcible nationalization of different elements,' which Dr. Gaster declares the Young Turks to have learned in 'Western Europe,' might have been better studied by them in Eastern Prussia, Russia, and Hungary. In neither of these countries has its application been attended by such success as to justify its adoption by the far less homogeneous Turkey. In Hungary, where its application receives, unfortunately, the support of the overwhelmingly Jewish Press and of many Jewish Freemasons, professors, and politicians, striking analogies to the blunders of the Salonika Committee might be found."

The Zionists knew that the First World War would end with a peace conference, where the breakup of the Empires and the formation of small, ethnically segregated nations would be discussed. That deliberately manufactured opportunity would give the Zionists a chance to petition for the creation of the "Jewish State". However, since the majority of Jews were happily assimilating into Gentile societies and had no desire to move to Palestine, the Zionists' plans, which were otherwise largely successful, ultimately failed.

The Zionists then felt they had the right to manufacture the Second World War and the Holocaust in order to change the Jews' collective mind by means of force. They did not care at all what most Jews wanted for themselves and the racist Zionists were willing to mass murder millions of Jews in the hopes that the "remnant" would be persuaded to emigrate to the "Holy Land" at war's end. Racist political Zionist Israel Zangwill predicted in 1923 that Zionism would lead to an unprecedented world-wide conflagration.[166] He knew whereof he spoke. The Zionists Lloyd George and "Mentor" also realized at the end of the First World War that there would be second.[167]

In 1906, Leo Tolstoy recognized that the Zionists were leading the world, and especially the Jews, towards disaster. On 9 December 1906, on page SM2, *The New York Times* published a translation of Tolstoy's ominous warnings, which were translated by Herman Bernstein—note the name,

"ZIONISM

An Argument against the Ambition for Separate National Existence. A Plea for Devotion to the Idea of Common Humanity.

By COUNT LEO TOLSTOY

(Translated from the Russian by Herman Bernstein.)

THIS movement has always interested me, not because it offers to the Jews a way out of their painful condition—it offers if them no way out of it—it has interested me because of the example of the enormous influence to which people, who have suffered a great deal and have experienced all the vanity of a certain project, will occasionally submit. Before our eyes an old, wise, and well-experienced people, which had gone through one of the most terrible maladies of mankind, is now falling back into the same malady. There is an awakening of the thirst for imperialism and an evil desire to govern and to play an important part. Again they want to provide themselves with all this show of outward nationalism, with armies—with banners awl inscriptions.

The leaders, without realizing it themselves, have fallen into the terrible

---

166. "Mr. Zangwill on Zionism", *The London Times*, (16 October 1923), p. 11. I. Zangwill, "Is Political Zionism Dead? Yes", *The Nation*, Volume 118, Number 3062, (12 March 1924), pp. 276-278.

167. "Peace, War—and Bolshevism", *The Jewish Chronicle*, (4 April 1919), p. 7. "1918 Peace Views of Lloyd George", *The New York Times*, (26 March 1922), Editorial Section, p. 33.

sin of separating themselves from others, and they are eminently battering this sin into the consciousness of the people to whom they represent the matter not at all as it really is.

They are forever repeating that Zionism is a progressive movement of the national spirit which is eager to throw off at last the chains of captivity and to give the nation an opportunity to live a free and independent life on the sacred mounts where their great past is buried. I have been told of a Jewish preacher who in one of the synagogues of Tula struck himself on the chest and, sobbing, called the people to Palestine, saying: 'There we will see the rock on which Jacob had rested, and we will walk along the same path that Abraham bad trodden. This awakens our feelings!'

But the horror of it all is that this movement is neither progressive nor national, nor does it awaken any feelings.

Jacob's rock and Abraham's path are such distant things that they cannot stir a people and make them take up the wanderer's staff. A nation is Dot an archaeologist, and to break new ground it will not go in a horde of ten millions from the places where they have lived for many centuries, and where they feel more at home than amid the rocks of Jacob and the paths of Abraham. This can be seen on those that go to America, and tortured with homesickness, exhausted, they return and kiss the ground of their native land, the black soil of the same Russia they still love, notwithstanding that the terrible oppressors are shamelessly trying to make of the life of the Jews here a hell of suffering.

If their memory of the sacred places of Palestine were really so strong and their eagerness to live there had been inherent in the Jewish people, they had numerous occasions during these 1,800 years to return there and to live once more in those ancient places.

But the people consciously never wanted it, even as they do not want it now. And that is why I do not regard Zionism as a national movement The real Jewish spirit is against a separate territory of their own. It does not want the old toy of empire, and it has renounced it once for all. I cannot think without emotion of the beautiful saying about a certain Jewish sage of the times of the destruction of the Temple. He had rendered a great service to Vespasian, and Vespasian told him to ask for anything he pleased, and he would grant his request. It would seem that that was an excellent opportunity to ask him to raise the siege and restore the freedom to his land. But the sage said:

'Allow me to go with my pupils to the town of Yamnia and to establish there a school for the study of the Thorah.'

This answer seemed strange to the Roman, who had become brutalized in wars and slaughters.

But it was a conscious, powerful, and beautiful answer of the entire nation.

The sage understood correctly the secret of the people's spirit and asked for something which seemed insignificant. This voluntary fate of the sage—this substitution of the spiritual for the corrupt—is the grandest moment in the history of Judaism, something which has not as yet been sufficiently appreciated, and of which even the Jews have not entirely availed themselves.

And this nation feels it and resists it with all its powers, unwilling to rush into the old adventure which is foreign to its soul.

It is not the land, but the Book, that has become its fatherland. And this is one of the grandest spectacles in history, the noblest calling man can only hope for. Absorbed by this Book, the Jewish people did not notice how centuries had passed over their heads, how nations had appeared and then been wiped off the face of the earth, how new lands bad been discovered and steam power invented, while the black, heavy smoke of the factory chimneys had overcast the clear sky, hiding it from the people who walked in darkness under a dense network of wires along which a mute hut cruel power carried tidings, one more cruel than the other, one more bloody than the other—such tidings as the world had never heard before.

This roaring noise of civilization which is rushing like a waterfall toward the precipice, which kindles in men only wretched desires for worthless comforts, had not reached the ears of the great Wanderer who was absorbed reading the great Book. And the foam of the gushing waterfall is striving to besprinkle the holy pages and to cover them with rusty stairs of mockery and unbelief.

And the leaders of Zionism are helping on the work of this foam, majestically ignoring the religious question and putting forth only immigration and politics, politics and immigration.

'Let us first come together from all sides of the globe,' they say, 'and then we shall also work out a religion.'

This is just as unnatural and unwise as it is not national, especially with regard to the Jews. One recalls the splendid chapter of Deuteronomy, where, after the thundering words of cursings and blessings, the young spirit of the new-born nation utters words of profound significance: 'And it shall come to pass, when all these things are come upon thee, the blessing and the curse, which I have set before thee, and thou shalt call them to mind among all the nations, whither the Lord thy God hath driven thee, And shalt return unto the Lord thy God, and shalt obey His voice—thou and thy children, with all thine heart, and with all thy soul; That then the Lord thy God will turn thy captivity, and have compassion upon thee and gather thee from all the nations, whither the Lord thy God hath scattered thee. And will bring thee into the land which thy fathers possessed. * *'

This is the hope of the people. First turn to God, and then God Himself will do His own work and will give the land to the people and will grant them more favors than He had granted their fathers.

The leaders of Zionism reason differently. They seem to have changed roles with God. They want to gather the Jews from among all nations into the land of their fathers, and there God would take care that the people should turn their hearth to Him.

And God says to them:

'Try to do My work.'
And He turns away from them.

And thus childish colonial banks are started, toy congresses are held, with small and large committees, which, authorized by nobody, are carrying on unnecessary negotiations concerning childish charters and the Sultan's favors. The people see all the vanity of these projects and also turn away from this movement. It isn't God's work—there is too much of the human, the invented, too much of the medical prescription in this work.

That is why, I hear, there are some rabbis who curse this work, condemning Zionism as a doctrine that is foreign to the people and that threatens them with great misfortune. And, indeed, although this view is held by the orthodox rabbis, who usually occupy a dark position on religious questions, yet in this case the orthodox Jews stand upon firm ground, and their opposition is entirely legitimate.

There is no progressive spirit in this movement, which is cut out according to European fashion—it has not even the character of progress of which they speak so eloquently at their congresses. And this is the most amazing feature of it all. If the leaders of Zionism, generally sensitive and sensible men, but far from their people, were unable to create a healthy national movement, they are not to be blamed. They are eager to do something, but they cannot. But if all these people, with their quick understanding of everything that is progressive and striking, did not understand what really moves the higher life of Europe and what constitutes the power of the summits of the European minds, they cannot be excused under any circumstances. Believing that the strength of Europe lies in its imperialism—that is, in its gun power, with all the horrors of militarism—they have decided to array their old man also in the armor of a warrior and give him a rifle in his hands. They felt like creating a new Juden-Staat. The best minds in Europe, and also in America, all those that think truthfully and sincerely, are agitated to the very depths of their souls at the madness and horror of this abyss whither savage mankind, so called civilized, is drifting head foremost.

All that is right, sensible, and not enslaved by fear or money is striving with all its powers to undeceive the people and to remind them that the strength of mankind does not at all lie in the cannon power of imperialism, and that the future of mankind is not in the passion to separate themselves and to live in small States. Those that are truly progressive see the happiness of mankind in just the reverse, in broad union and in the complete absence of cannon and mortars and those groups which are now held together only by the power of mortars, thus ruining the life of the people. All the rational work of the rational portion of mankind is against such imperialism. And they, the leaders of Zionism, want to give life to this antiquity and call such a wild aspiration—progress.

This is a great sin. It borders on blasphemy against the most sacred things that we have in life now.

We need no new Governments; we need loving people who see in their love the mission of life and love of God.

What is it that tempted them, what is it that they like so much in this nationalist, which is in reality a military, movement among the European little nations which the leaders of Zionism are apparently trying to imitate with all their might? Is it the toy freedom of Servia, where the word of the Austrian Ambassador is of greater importance than the orders of the King, and where all their freedom comes to nothing but endless slaughter and intrigues among the parties, and finally to the ruination of the peasants and the exhaustion of the land, which is overburdened with taxes in order to maintain the great number of officials and soldiers, who could be mowed down by two or three volleys from a small battery? Do they like this? Or do they like the seeming freedom of Bulgaria, which is also torn asunder by riots on account of their temporary little Czars, and which will soon be swallowed up by some other power? Or do they like Roumania, Macedonia, Montenegro, Crete, Greece—which of these does Zionism like? I say nothing of Italy, France, England, Germany, and some of the countries still nearer to us, where the cry also goes up to Heaven from the tortured people who are becoming savage and impoverished, thanks to militarism and organization.

The healthy seed of immigration which is striving to break up the congestion of the Jews and to bring them back to long-forgotten agriculture—this undoubtedly a pure and beautiful movement, which the Zionists now claim as their own—does not at all belong to the Zionists. The tendency toward colonization existed before; Zionism has boldly usurped it and given it an unnatural and unnecessary political coloring, and has thus completely checked the return of the Jews to agriculture. The vision of a Jewish State was started, and this has only complicated the simple and clear desire of the people to leave the cities and take up the only proper, healthy, living, and honest work of God—the tilling of the soil."

Racist Zionist Theodor Herzl spoke at the first Zionist Congress of 1897 and disclosed the machinations of the Zionists and their centuries' old desire to destroy the Turkish Empire and bankrupt the Sultan. Herzl had a covert plan to have Turks mass murder Armenians, which would cause an outrage around the world, so as to leave the Turkish Empire at the mercy of the Jewish controlled press, which Herzl pledged would cover up the atrocities if the Sultan would agree to give the Zionists Palestine.[168] *The New York Times* reported on 31 August 1897 on page 7,

"ZIONIST CONGRESS IN BASEL.

---

[168]. "The Turkish Situation by One Born in Turkey", *The American Monthly Review of Reviews*, Volume 25, Number 2, (February, 1902), pp. 182-191, at 186-188. "Zionism", *Encyclopædia Britannica*, Eleventh Edition, (1911).

**The Delegates Adopt Dr. Herzl's Programme
for Re-establishing the Jews in Palestine.**

BASEL, Switzerland, Aug., 30.—At to-day's session of the Zionist Congress the delegates present unanimously adopted, with great enthusiasm, the programme for re-establishing the Hebrews in Palestine, with publicly recognized rights.

A dispatch was sent to the Sultan of Turkey, thanking his Majesty for the privileges enjoyed by the Hebrews in his empire.

The Zionist Congress opened at Basel yesterday with 200 delegates in attendance from various parts of Europe. Dr. Theodor Herzl, the so-called 'New Moses' and originator of the scheme to purchase Palestine and resettle the Hebrews there, was elected President and Dr. Max Nordau was elected Vice President of the Congress.

Dr. Herzl has only recently come into prominence. He seeks to float a limited-liability company in London for the purpose of acquiring Palestine from the Sultan of Turkey and thoroughly organizing it for resettlement by the Hebrews. He has, it is said, already won converts to the Zionistic movement in all parts of the world.

When asked to outline his plans, Dr. Herzl said:

'We shall first send out an exploring expedition, equipped with all the modern resources of science, which will thoroughly overhaul the land from one end to the other before it is colonized, and establish telephonic and telegraphic communication with the base as it advances. The old methods of colonization will not do here.

'See here,' continued Dr. Herzl, showing a good-sized book, 'this is one of the four books which contain the records of the movement—the logbooks of the Mayflower,' he added, with a smile. That one watchword, the 'Jewish State,' has been sufficient to rouse the Jews to a state of enthusiasm in the remotest corners of the earth, though there are those forming the so-called philanthropic party who predict that the watchword will provoke reprisals from Turkey. Inquiries in Constantinople and Palestine show that nothing is further from the truth.

'My plan is simple enough. We must obtain the sovereignty over Palestine—our never-to-be-forgotten, historical home. At the head of the movement will be two great and powerful agents—the Society of Jews and the Jewish Company. The first named will be a political organization, and spread the Jewish propaganda. The latter will be a limited-liability company, under English laws, having its headquarters In London and a capital of, say, a milliard of marks. Its task will be to discharge all the financial obligations of the retiring Jews and regulate the economic conditions in the new country. At first we shall send only unskilled labor—that is, the very poorest, who will make the land arable. They will lay out streets, build bridges and railroads, regulate rivers, and lay down telegraphs according to plans prepared at headquarters. Their work will bring trade, their trade the market, and the markets will cause new settlers to flock to the country. Every one will go there voluntarily, at his or her own risk, but ever under the watchful eye and protection of the organization.

'I think we shall find Palestine at our disposal sooner than we expected. Last year I went to Constantinople and had two long conferences with the Grand Vizier, to whom I pointed out that the key to the preservation of Turkey lay in the solution of the Jewish question.

'The Jews, in exchange for Palestine, would regulate the Sultan's finances and prevent disintegration, while for Europe we should form a new outpost against Asiatic barbarism and a guard of honor to hold intact the sacred shrines of the Christians.

'We can afford to play a waiting game, and either take over Palestine from the European Congress called together to divide the spoils of disintegrated Turkey, or look out for another land, such as Argentina, and say: 'Your Zion Is there.'

'It is to confer over this point that the congress was arranged for at Basel.

'I am sure that the Jews are even better colonists than Englishmen. There are already colonies of Jews in Palestine, and I have on my table excellent Bordeaux, Sauterne, and cognac grown in that country. It is well known that in Galicia and the Balkans the Jews perform the roughest kind of manual labor. There the wealth they bring is not their money, but themselves.'"

Racist Zionist leader Theodor Herzl, and his Jewish financier predecessors, collaborators and successors, promoted anti-Semitism as a means to force reluctant Jews to Palestine against their will—as will be shown later on in this text. An article entitled, "The Jewish State Idea", in *The New York Times*, 15 August 1897, on page 9, evinces the Zionists' designs for a world war centered on the "Eastern Question" which world war the Zionists had been fomenting for centuries; and the article further evinces the fact that the Zionists knew that anti-Semitism was a means to drive Jews to Palestine—as will be confirmed later in this text by citation to Herzl and other Zionists,

"The question of colonization was agitated so early as 1840 by the late Sir Moses Montefiore, but it was not until 1878 that the first colony was planted at Pethach-Thikvah. This was an utter failure, due to the poor selection of colonists, who soon returned to Jerusalem. But in 1880, under the stress of Roumanian oppression, immigrants founded the villages and settlements of Sichron-Ja'akob and Rosh-Pinah. The Russian persecutions brought about the founding of Rishou-l'Zion and the re-establishment of Pethach-Thikvah in 1882. [***] With the bursting of the storm of Russian hate came perilous times for the Palestinian colonists. Their friends in Russia, who had promised their aid, had all they could do to care for themselves, and Palestine was overrun with poverty-stricken Russian exiles. [***] As to the question of the advisability of establishing a Jewish State there, it is natural that opinions vary most widely. Holman Hunt, R. A., the famous English artist, who has lived in Syria, wrote not long since: 'Palestine will soon become a direful field of contention to the infernally armed forces of the European powers, so that it is calculated to provoke a curse to the world of the most appalling character. Russia and Greece will contend for the interests of the Greek Church, France and Italy for the Latin, Prussia and Germany for the German political interests. In addition to the above named contenders for Palestine, there would be England. The only remedy is a Jewish State. Both in Europe and America there are many Jews who oppose the founding of this State on the ground that it could be only a small, weak State, existing by sufferance. It is also urged that Israel's mission is no longer political, but purely and simply religious, and that the establishment of the State would do incalculable harm, and could do no good."

In 1844, an early American Zionist Jew, Mordecai Manuel Noah, revealed the Jewish method to force Jews into segregation and migration to Palestine—oppression,

"I am right in this interpretation, and that this is the land which is beyond the rivers of Ethiopia, what a glorious privilege is reserved for the free people of the United States: the only country which has given civil and religious rights to the Jews equal with all other sects; the only country which has not persecuted them, selected and pointedly distinguished in prophecy as *the* nation which, at a proper time, shall present to the Lord his chosen and trodden-down people, and pave the way for their restoration to Zion. But will they go, I am asked, when the day of redemption arrives? All will go who feel the oppressor's yoke. *We* may repose where we are free and happy, but those who, bowed to the earth by oppression, would gladly exchange a condition of vassalage for the hope of freedom: that hope the Jews never can surrender; they cannot stand up against the prediction of our prophets, against the promises of God; they cease to be a nation, a people, a sect, when they do so."[169]

Prominent and influential racist Zionist Israel Zangwill wrote in 1914, shortly before the First World War began, in his booklet *The Problem of the Jewish Race*, Judaen Publishing Company, New York, (1914), pp. 9-10, 21; which was first published as an article, "The Jewish Race", *The Independent*, Volume 71, Number 3271, (10 August 1911), pp. 288-295, at 290, 295,

"Rabbinic opportunism, while on the one hand keeping alive the hope that these realities, however gross, would come back in God's good time, went so far in the other direction as to lay it down that the law of the land was the law of the Jews. Everything in short—in this transitional period between the

---

[169]. M. M. Noah, *Discourse on the Restoration of the Jews: Delivered at the Tabernacle, Oct. 28 and Dec. 2., 1844*, Harper, New York, (1845), p. 50.

ancient glory and the Messianic era to come—was sacrificed to the ideal of mere survival. The mediaeval teacher Maimonides laid it down that to preserve life even Judaism might be abandoned in all but its holiest minimum. Thus—under the standing menace of massacre and spoliation—arose Crypto-Jews or Marranos, who, frequently at the risk of the stake or sword, carried on their Judaism in secret. Catholics in Spain and Portugal, Protestants in England, they were in Egypt or Turkey Mohammedans. Indeed the *Dönmeh* still flourish in Salonika and provide the Young Turks with statesmen, the Balearic Islands still shelter the *Chuetas*, and only half a century ago persecution produced the *Yedil-al-Islam* in Central Asia. Russia must be full of Greek Christians who have remained Jewish at heart. Last year a number of Russian Jews, shut out from a university career, and seeking the lesser apostacy, became Mohammendans, only to find that for them the Trinity was the sole avenue to educational and social salvation. Where existence could be achieved legally, yet not without social inferiority, a minor form of Crypto-Judaism was begotten, which prevails to-day in most lands of Jewish emancipation, among its symptoms being change of names, accentuated local patriotism, accentuated abstention from Jewish affairs, and even anti-Semitism mimetically absorbed from the environment. Indeed, Marranoism, both in its major and minor forms, may be regarded as an exemplification of the Darwinian theory of protective coloring. The pervasive assimilating force acts even upon the most faithful, undermining more subtly than persecution the life-conceptions so tenaciously perpetuated. [***] A host of political rivalries, perilous to the world's peace, center around Palestine, while in the still more dangerous quarter of Mesopotamia, a co-operation of England and Germany in making a home under the Turkish flag for the Jew in his original birthplace would reduce Anglo-German friction, foster world-peace and establish in the heart of the Old World a bridge of civilization between the East and the West and a symbol of hope for the future of mankind."

Israel Zangwill had a close relationship with the Rothschilds, who had offered to sponsor his education.[170] In the 1800's, Jewish bankers prompted what would become "German" leadership to oppose this racialist Pan-Slavic push to conquer Eastern Europe, with a Pan-Germanic movement based a racialist principles. Jewish bankers led all of these elements, including the Turkish, British and French, into perpetual war for expanded territory, so as to destroy Europe and replace it with a world government run by them, and in order to open up the way for the Jews to enter Palestine *en masse*. Jewish bankers led the Czar to destroy Russia with wars, and eventually bankrupted her by closing off Russia's access to funds, while heavily funding Japan's economy in their war with Russia, as well as funding revolutionary elements against the Czar. Hitler

---

170. "How Zangwill Fought His Way", *Current Literature*, Volume 27, Number 2, (February, 1900), p. 107.

was an agent of the Jewish bankers, and he likewise saw to it that Europe, Germany included, was consumed by perpetual and expanding war, which killed off millions of the best Germans and Slavs. After Hitler's reign, the Jewish bankers succeeded in taking Palestine from the indigenous population and in expanding the Soviet Empire across Eastern Europe—and very nearly all of Europe and America.

The article "Russia and Turkey", *The Atheneum; or, Spirit of the English Magazines*, Volume 2, Number 10, (15 August 1820), page 398ff. quoted above, continued as follows:

> "The determination of Russia to seize upon the European dominions of the Sultan, was at length practically exhibited by the march of her troops, under Wittgenstein, to the Danube. The Turks, after some affairs of posts, retreated before the powerful army which now rushed down from Podolia and Moscow on their scattered parties; and the three sieges of Shumla, Silistria, and Varna, were immediately and rashly undertaken.
>
> The result of the campaign undoubtedly disappointed, to a great extent, the expectations formed of the Russian arms. The Turks were often the assailants even upon level ground, and were not unfrequently left masters of the field. Some of their incursions into Wallachia put the Russian corps into such imminent hazard, that they were saved only by an instant retreat—large convoys were intercepted by the Turkish cavalry, and the campaign was speedily discovered to be only the beginning of a dubious and protracted struggle. The assaults on the Turkish posts were generally repulsed with heavy loss; and, of the three great sieges, but one offered the slightest hope of success. Shumla, the grand object of the campaign, was early found to be totally impracticable. Silistria was nearly despaired of, and finally was abandoned by a disorderly and ruinous flight. Varna alone gave way, after a long succession of attacks; and, from the singular circumstances of its surrender, is still said to have been bought from the Governor, Yusuf Pacha, a Greek renegade.
>
> The campaign was urged into the depths of winter, and the weather was remarkably inclement; the Turks were elated by success, and their attacks kept the enemy perpetually on the alert; the walls of the great towns would not give way; the villages were burnt, and could give shelter no longer; and, as the general result, the Russian army were ordered to retreat from the Danube. The retreat was a second march from Moscow. Everything was lost, buried, or taken. The horses of the cavalry and artillery were totally destroyed, the greater part of the artillery was hidden in the ground, or captured, and the flying army, naked, dismantled, and undisciplined, was rejoiced to find itself once more in the provinces from which it had poured forth a few months before, to plant its standards on the seraglio.
>
> Russia, beaten as she has been, has yet showed that she is too strong for the Turk; she has mastered Varna, a situation of high importance to her further movements, and she has been able to baffle every exertion to wrest it out of her hands. She has seized some minor fortresses, and in every

instance she has been equally able to repel the efforts of the enemy. She has also conquered a city between the Balkan and Constantinople, which, if she shall pass the mountains, will be a place of arms for her troops, and a formidable obstacle on the flank of the Turkish army. The system of the Russian discipline, finance, and influence over the population of the North, is so immeasurably superior to the broken and disorderly polity of the Turk, that if the war be a work of time, victory must fall to the Czar. On the other hand we must remember the daring and sagacious spirit of the Sultan, the fierce bravery of his people, the power of the most warlike superstition on earth, the national abhorrence of the Muscovite, and even the new intrepidity of recent success. A still more powerful element of defence remains, the jealousy or prudence of the great European kingdoms. The possession of Constantinople, by the masters of Moscow and St. Petersburg, would shake the whole European system, by giving, for the time, at least, an exorbitant influence to Russia. England would see in it the threatened conquest of India: France, the complete supremacy of the Levant, and the exposure of her own shores to a Russian fleet on the first hostilities. Spain, though fallen in the scale, must still resist a measure which would lay open her immense sea-line from Barcelona to Cadiz. Austria, alone, might look upon it with some complacency, if she were bribed by the possession of Albania, or the prospect of planting her banners in the Morea. But the aggrandizement of Austria would be resisted by Prussia, and then the whole continent must hear the Russian trumpets as a summons to prepare for universal war.

The possession of Constantinople would be, not merely the mastery of the emporium of Asiatic trade, nor of a great fortress from which Asia and the East of Europe might be awed; but it would be an immediate and tremendous instrument of European disturbance by its perpetual transmission of the whole naval strength of Russia into the centre of Europe. The Russian fleet is unimportant, while it is liable to be locked up for half the year in the ice of the North; or while, to reach the Mediterranean, it must make the circuit of Europe. But if the passage of the Dardanelles were once her own, there is no limit for the force which she might form in the Black Sea, and pour down direct into Levant. There can be no doubt, that with this occasion for the employment of a naval force, Russia would throw a vast portion of her strength into a naval shape; and that while the Circassian forests furnished a tree, or the plains, from the Ukraine to Archangel, supplied hemp and tar, fleet upon fleet would be created in the dock-yards of the Crimea, and be poured down in overwhelming numbers into the Mediterranean.

Thus it is impossible that the Czar shall attack Constantinople without involving the world in war, and in that war England must be a principal. The premier's opinion has been distinctly stated on this subject, and so far as we can rely on the fluctuating wisdom of cabinets, it coincides with that of France and Prussia. To arrange more systematically the resistance to the ruin of Turkey, the Duke of Wellington is said to be on the eve of an extensive

European tour, in which he will ascertain the dependence to be placed upon the courts, and discover how far the Czar may have learned moderation from his last campaign. But the world is in a feverish state: ambition is reviving; conspiracy is gathering on the Continent, and the first hour that sees the Russian superiority in the field decisive, will see the great sovereignties remonstrating, arming, and finally rushing, as to a new crusade, but with the sword unsheathed, nor for the fall, but for the defence of the turban!

That this will be the ultimate consequence we have no doubt. But the time may not be immediate. We are inclined to think that the French war has not yet been sufficiently forgotten by the states of central Europe to suffer them to run the hazards of collision without the most anxious efforts for its avoidance. There is a general deficiency of money. All the great powers are actually, at this hour, living on *loans*. There is no power in Europe whose revenue is enough for its expenditure. Even in England we are borrowing. Our three millions of exchequer bills, issued in the fifteenth year of peace, shows us how little the finance system has sustained our expectations. A war, even for a year, would double our expenditure. On the continent, Rothschild is the true monarch. Every state is in his books, and what must be the confusion, the beggary, and the ultimate bankruptcy of hostilities. The fall of every throne must follow the bankruptcy of every exchequer, and the whole social system be broken up amid revolutionary havoc and individual misery. We believe that the four great powers are so fully convinced of the evil of this tremendous hazard, that they are struggling in every shape of diplomacy to avert the continuance of a war between Turkey and Russia. If they succeed, peace will, in all probability, continue for a few years more; if they fail, Europe must instantly arm, and a scene of warfare be roused, to which there has been no equal since the fall of the Roman Empire."

3.4 Rothschild Warmongering

As anti-Communist Myron Fagan argued, the Rothschilds had hoped that the Napoleonic Wars would have made the world so weary of war that the nations would have eagerly surrendered their sovereignty to the Rothschilds' Jewish world-government at the Congress of Vienna of 1814-1815. Jewish bankers were behind these wars, in which they financed all sides to destroy each and shatter the empires which stood in the way of the Rothschilds' establishing a Jewish kingdom in Palestine from which to rule the world—in agreement with Jewish Messianic myth.

Much of the monarchy of Europe had been infiltrated by Jews and crypto-Jews either through intermarriage and disingenuous Christian conversion, or through finance. Many of these rulers intentionally bankrupted the nations over which they ruled. These nations were then subverted by revolutions and dictatorships under the leadership of Jews, or the agents of Jews. The largest revolutionary movement came in 1848, and it was organized, led and financed by Jews—as Disraeli had noted, in 1844, four years before it happened.

One hundred years after the article "Russia and Turkey" appeared, and shortly after the Zionists had had their First World War, it was again apparent to many that a group of radical Jews sought to rule the world and focused their attention on the "Eastern Question" and the development of a Second World War, which would pit Japan and Germany against America and Great Britain. On 19 June 1920, John Clayton wrote in the *Chicago Daily Tribune* on the front page,

"TROTZKY LEADS RADICAL CREW TO WORLD RULE

Bolshevism Only a Tool for His Scheme
BY JOHN CLAYTON.
(Chicago Tribune Foreign News Service.)
(By Special Cable.)
(Copyright: 1920: By the Tribune Company.)

PARIS, June 18.—For the last two years army intelligence officers, members of the various secret service organizations of the entente, have been bringing in reports of a world revolutionary movement other than Bolshevism. At first these reports confused the two, but latterly the lines they have taken have begun to be more and more clear.

Bolshevism aims for the overthrow of existing society and the establishment of an international brotherhood of men who work with their hands as rulers of the world. The second movement aims for the establishment of a new racial domination of the world. So far as the British, French and our own department's inquiry have been able to trace, the moving spirits in the second scheme are Jewish radicals.

Use Local Hatreds.

Within the ranks of communism is a group of this party, but it does not stop there. To its leaders, communism is only an incident. They are ready to use the Islamic revolt, hatred by the central empires for England, Japan's designs on India, and commercial rivalry between America and Japan.

As any movement of world revolution must be, this is primarily anti-Anglo-Saxon. It sees its greatest task in the destruction of the British empire and the growing commercial power of America. The brains of this organization are in Berlin.

Trotzky at Head.

The directing spirit which issues the orders to all minor chiefs and finds money for the work of preparing the revolt is in the German capital. Its executive head is none other than Trotzky, for it is on the far frontiers of India, Afghanistan, and Persia that the first test of strength will come. The organization expert of the present Russian state is recognized, even among the members of his own political party, as a man of boundless ambition, and his dream of an empire of the east is like that of Napoleon.

The organization of the world Jewish-radical movement has been perfected in almost every land. In the states of England, France, Germany, Poland, Russia, and the east it has its groups. It is behind the Islamic revolt with all the propaganda skill and financial aid at its command because it hopes to control the shaping of the new eastern empire to its own ends. Sympathy with the eastern nationals probably is one of the chief causes for the victory of the pro-nationals in the bolshevik party, which threw communism solidly behind the nationalist aspirations of England's colonies.

Out to Grab Trade Routes.

The aims of the Jewish-radical party have nothing of altruism behind them beyond liberation of their own race. Except for this their aims are purely commercial. They want actual control of the rich trade routes and production centers of the east, those foundations of the British empire which always have been the cornerstone of its national supremacy.

They are striking for the same ends as Germany when she entered the war of 1914 to establish Mittel Europa and so give the Germans control of the Bagdad railway. They believe Europe is tired of conflict and that England is too weak to put down a concerted rebellion in part of her eastern possessions. Therein lies the hope of success. They are staking brains and money against an empire.

'Westward the course of empire makes its way,' but even it swings backward to the old battleground where for countless ages peoples have fought. Nations have risen and crumbled around control of eastern commerce."[171]

3.4.1 Inter-Jewish Racism

German Jews generally disliked *Ostjuden*—the Jews of Eastern Europe. Peter Grose wrote in his book, *Israel in the Mind of America*,

> "It was I. M. Wise, typically, who broke the silence of the established Jews as they saw what was happening to the good name of their faith. From the fresh air of Cincinnati, Wise observed the noisy, smelly scenes in the eastern seaports and was revolted. 'It is next to an impossibility to associate or identify ourselves with that half-civilized orthodoxy which constitutes the bulk of the [Jewish] population in those cities,' he stormed. 'We are Americans and they are not. We are Israelites of the nineteenth century and a free country, and they gnaw the dead bones of past centuries.' Wise was never a man to mince words. 'The good reputation of Judaism must naturally suffer materially, which must without fail lower our social status.' The prosperous 'Uptown' Jew of New York found identification with the unsavory 'Downtown' Jew dangerous in the extreme. It was in the Uptown salons of the German-Jewish aristocracy that the word 'kike' first appeared,

---

171. *See also:* "World Mischief", *The Chicago Tribune*, (21 June 1920), p. 8.

to deride the uncultured and unclean immigrants. Yet the emotional dilemma was acute, for the Uptown Jew was not without a sense of obligation and guilt."[172]

Albert Einstein was the most prominent and vocal advocate of Eastern Jewish emigration to Germany, England and America; which was unusual given that Einstein was a German Jew, and most German Jews opposed the immigration of Eastern Jews into Germany, England and America. The conclusion many drew was that Einstein was a willing stooge exploited by Eastern European Jewish Zionists, who used him to promote their interests. In exchange, they gave Einstein fame and protection from criticism. Note that the Zionist Nazis first attacked assimilatory German Jewry, and then went after the Orthodox Jewry of Eastern Europe who opposed Zionism on religious grounds, while privileging the Zionist Jewry of Eastern Europe. Zionist Jews used their agents the Nazis to punish assimilatory and anti-Zionist Jewry and to degrade and deplete the population of adversarial Jews. Zionist Jews, Albert Einstein chief among them, had long been attacking assimilatory German Jews. "Mentor" wrote in *The Jewish Chronicle* on 11 April 1919 on page 9 in an article entitled "From My Note Book",

> "On the other hand, there are anti-Zionists who wish to see tradition perish from Judaism so that it may be left a religion only, and who recognise in Zionism the strongest possible counter-force. These have their spiritual home in Germany, the cradle of de-traditioned Judaism."

In 1922, Burton J. Hendrick wrote, among other things,
> *"The wave of anti-Semitism, which has been sweeping over the world since the ending of the World War, has apparently reached the United States. An antagonism which Americans had believed was peculiarly European, is gaining a disquieting foothold in this country. The one prejudice which would seem to have no decent cause for existence in the free air of America is one that is based upon race and religion. Yet the most conservative American universities are openly setting up bars against the unlimited admittance of Jewish students; the most desirable clubs are becoming more rigid in their inhospitable attitude towards Jewish members; a weekly newspaper, financed by one of the richest men in America, has filled its pages for three years with a virulent campaign against this element in our population; secret organizations have been established for the purpose of 'fighting' the so-called 'Jewish predominance' in American life; Congress has passed and the President has signed an immigration law chiefly intended—it is just as well to he frank about the matter—to restrict the entrance of Jews from eastern Europe. It is*

---

[172]. P. Grose, *Israel in the Mind of America*, Alfred A. Knopf, New York, (1983), pp. 31-32.

an impressive fact that these manifestations of a less cordial attitude toward the Jews find their counterpart in another country which, in modern times, has been friendly to them—that is, England itself. That anti-Semitism should prevail in Russia, Germany, France, indeed in the whole continent of Europe, is not surprising; but its development in the Anglo-Saxon countries is something entirely new. Yet such conservative organs as the London Morning Post *and London* Spectator *are picturing the activities of English Jews as one of the most disrupting and dangerous influences in British life.* [***] This Jewish community—and similar Sephardic colonies were established in most important American cities, such as Boston, Philadelphia, Baltimore, and Charleston—had since led a career of exclusiveness and hauteur that is typically Spanish. As in Spain centuries ago these Israelites constantly associated with the best in the intellectual and social life of the old grandees, so to-day the New York Mendozas, Cardozos, Acostas, Pintos, and Cordobas—for they all still retain their old Spanish names—find their most congenial associates among cultivated Gentiles. They have always looked down upon their Russian co-religionists, and even upon the Germans, as inferior breeds. No anti-Semite among the native American stock has ever regarded the poor Polish immigrant with greater aversion. There was a time when a Spanish Jew or Jewess who married a German or Russian co-religionist would be promptly disowned; the hostility to such alliances was much stronger than it has ever been between Protestant and Catholic. The Sephardim have always had their own graveyards in which German and Russian Jews have not found rest. Part of this feeling has been due to ancestral pride; part had a more rational basis, for it is incontestable that, from most points of view, the Spanish Jews are superior to other representatives of Israel. There are only a few of them; they are nearly all rich or at least prosperous; they are merchants, bankers, and land owners; they are not pawnbrokers or peddlers or rag-pickers; and they have a distinct talent for public life. It is no accident that the most distinguished Jewish statesman of Great Britain, Disraeli, was a descendant of Spanish Jews and that the greatest public man of American Jewry, Judah P. Benjamin, Secretary of State of the Southern Confederacy and probably the most adroit brain in the Secession movement, belonged to the same branch of the race. It is also significant that the Jew who has reached the most powerful position of any member of his race in recent American life, Mr. Bernard Baruch, also traces his origin to the Jews of Spain.

So long as the Jewish population was limited chiefly to Spanish Jews America had nothing that remotely resembled a Jewish 'problem.' Before the American Revolution practically the whole Jewish population of this country consisted of these Sephardim. They played an honorable part in the Revolution and lived on terms of friendship and respect with the other racial elements. There were only about 2,000 of them in the whole United States at that time. Just how many there are now is not known; that their number is steadily decreasing is apparent and here again the explanation has a great importance; the Spanish Jews are becoming fewer through inter-marriage not with other branches of the race, but with Gentiles. In England it is said that the Spanish Jews have practically disappeared, and, here again, through inter-marriage with Christians. I have instanced above three Sephardic Jews who have reached high public station in Great Britain and the United States: Disraeli, Benjamin, and Baruch. All three of these men married Christians. The tendency that was so common five and six hundred years ago in Spain, when cardinals and kings acknowledged a mixture of Jewish blood, is similarly apparent in the England and America of the present time.

Neither did the second phase of Jewish immigration create anything that could be called a 'problem.' This was the much larger influx of German Jews, which began soon after the Battle of Waterloo, reached a considerable. proportion in the 'forties and 'fifties and fell off appreciably in the late 'seventies. These dates indicate that German Jewish immigration had about the same rise and fall as German immigration in general, and it is a fact that it was not a distinct movement but was merely part of the general flow of German immigrants to this country. German Jews came here for the same reason that other Germans came; in part the motive was economic, the desire to get a better chance at life, and in part the motive was political. German Jews participated extensively in the German liberal movement of '48; when it failed they emigrated in large numbers, precisely as did their Christian associates; the two most distinguished of these political refugees were Carl Schurz, a Gentile, and Abraham Jacobi, a Jew. But racially and culturally the German Jew seemed an entirely different person from his Spanish predecessor. He belonged to the second and northern division of Israel, the type which the Jewish writers designate as the Askenazim. Physically he was probably inferior to the Sephardim. His features were inclined to be coarser, his lips thicker, his hair more woolly in its texture, his head round rather than long; his physical type was not invariably brunette, for blond hair and blue eyes were not uncommon. These points, however, can be pushed too far; the women were not infrequently exceedingly beautiful, and the most famous of American Jewesses belonged to the Germanic branch. This was Rebecca Gratz, a Jewess distinguished for her beauty and piety, and for her friendships with eminent Americans. There is a tradition that Henry Clay was an unsuccessful suitor, and one of her most distinguished friends was Washington Irving. This later association had important literary consequences; Irving was likewise a close friend of Sir Walter Scott, whom he used frequently to visit at Abbotsford; it is said that his description of Miss Gratz, of her loveliness of person, the fineness of her character, her devotion to her religion and her race—a devotion that had prevented her from marrying, most of the men with whom she associated having been Christians—so fired the romantic imagination of Scott that he put her in the novel that he was then writing. In this way it happened that Scott's most famous woman character, his Rebecca of 'Ivanhoe,' was drawn from Rebecca Gratz of Philadelphia.

In the main, however, the German Jew was inferior, in manners, intelligence, and social adaptability, to the Spanish type. In numbers he was much greater; from 1815 to about 1880, when German Jewish immigration, on a large scale, came to an end—in this following the course of German immigration in general, of which, as already said, it was merely one phase—probably not far from 200,000 German Jews arrived, though scientific statistics are not available. With them arrived those characteristically Jewish figures—the rag picker, the itinerant peddler, the pawnbroker, the petty tradesman. These German Jews were not workers; for the most part they were middlemen. Many of the best known Jewish families of the United States founded their fortunes in these humble occupations. The Seligmans, who established one of the most important Jewish-American banking houses, were originally peddlers and clothing merchants; so was Solomon Loeb, who founded the great banking house of Kuhn, Loeb & Company; and Benjamin Altman, who died the owner of the most distinguished department store in New York and the possessor of one of the greatest collections of paintings ever assembled by an American—a collection which, with fine public spirit, he willed to the Metropolitan Museum of Art—is said to have started his business career with a pack on his back. Mr. Oscar S. Straus, ex-Ambassador to Turkey, has recently given, in his very interesting memoirs, a charming picture of a German Jewish family attempting to establish itself economically in its new environment. Mr. Straus' father was an itinerant peddler in the South; he drove a wagon from plantation to plantation, disposing of a miscellaneous cargo of 'Yankee notions.' Such a peddler was a welcome figure in Southern life preceding the Civil War; his coming was an annual event that was eagerly anticipated; he usually became the guest of one of the planters in the community in which he set up his temporary emporium, taking his meals at the family table; his host would never accept pay for this entertainment, but the Jewish merchant, as an acknowledgment of the hospitality, invariably made a parting gift to the wife or daughter—not uncommonly an unusually fine piece of dress goods. It may well be imagined that the arrival of an exotic figure of this kind, with his conversation of great cities and his reminiscences of European life, gave a welcome and bazaar-like color to the somewhat monotonous life of a Southern plantation; and this scene also is typical of the entirely kindly relations that prevailed sixty years ago between the native population and the Jewish immigrant.

The great point to be kept in mind is that these German Jews did not congregate in vast colonies in the great seaboard cities. [***] Perhaps the public feeling now and then was a little contemptuous; the Jewish sharpness in trading created a veritable literature of Jewish anecdotes; but the American attitude was always good natured; the idea that this race was a 'menace' to American institutions never occurred to the most harebrained of contemporary thinkers. In certain respects the German Jew displayed a greater tendency to "assimilation" than did his Spanish predecessor. The change in the ritual of the synagogue, for which the German Jew was responsible, is most significant from this point of view. Fundamentally this represented an attempt to Occidentalize somewhat the Jewish services—to make them more like the proceedings in Christian churches. Meetings were held Sunday instead of Saturday; English sermons were introduced; organs and choirs became regular features of the programme; the men removed their hats and the women appeared in bonnets instead of shawls. The German Jews greatly shocked their more conservative Spanish co-religionists by the extent to which they ignored the dietary laws; ham and bacon not infrequently appeared upon their breakfast tables; and oysters, lobsters and other forbidden creatures tempted the Jewish appetite as irresistibly as the Gentile. Jewish children formed a small minority in every public school and high school; a still smaller contingent appeared in all the colleges—thirty and forty years ago Yale, Harvard, and Princeton usually had four or five in every graduating class; now and then a German Jew was elected to one of the most exclusive city clubs—though here, it must be admitted, progress was more difficult. It would be absurd to deny that a certain prejudice existed against the Jews, even in the days when the Spanish and German elements constituted almost exclusively American Israel, but it was not intense or bitter, and never reached the proportions of a public issue. Occasionally the desire of Jews to be exempted from the provisions of Sunday laws—on the ground, that, as orthodox Hebrews, they kept their establishments closed on Saturdays—caused a ripple of dissatisfaction; the refusal of summer hotels to admit them led to several law suits of sensational character; but, in the main, the Gentile population showed little alarm about their progress, and anti-Semitism was a word whose significance few Americans remotely understood.

    The facts to be kept in mind are that the Jewish population before 1880 consisted almost exclusively of Spanish and German Jews, or their descendants; that they were comparatively few in number; that they were bankers or tradesmen, large and small; that they did not form a compact mass of wretchedness in large cities; that, in education, manners, and social opportunities their past did not compare unfavorably with that of the other immigrating races, It is the year 1881 that marks the beginning of the American Jewish 'problem' as that word is commonly understood. Then began the influx, on an enormous scale, of an entirely different type of Judaism from the staid Spanish and the energetic German of the previous generations. It is customary to speak of Israel as a scattered people, as a race that is constantly seeking a home among other nations, as one that really possesses no settled abode of its own. In a sense that is true; but in its larger aspects it is not true at all. For the Jews, as a mass, have inhabited the same territory for at least a thousand years. At the present time there are perhaps 9,000,000 Jews in Europe. Comparatively small numbers are found in all countries—perhaps 100,000 in France, 240,000 in the United Kingdom—despite the ribald accusation that Scotland is no place for the Jews, the record discloses about 27,000 north of the Tweed—15,000 in Belgium, 8,000 in Greece and so on. These are merely the fringes of European Israel; of the 9,000,000 Jews living in Europe, not far from 7,000,000 are congregated as a mass in one rather restricted area.* This territory comprises western Russia, eastern Prussia and northern Austria. One hundred and fifty years ago not a square mile of this region belonged to the three countries named; all of it was part of the ancient Kingdom of Poland. Until the partitions of Poland, in the Eighteenth Century, neither Russia, Prussia, nor Austria had any large number of Jews; their present Jewish populations, that is, are an inheritance from that unholy piece of statecraft. There is thus a certain inaccuracy in referring to Russian and Austrian and Polish Jews; in reality they are all Polish Jews. For some reason which is not perfectly understood the great majority of all the Jews in the world found their way into Poland in the Middle Ages and in that country their descendants have remained until the present time. Here, then, is the present Jewish home—or at least here it was in 1881, but there is one country now which also has a very large Jewish population. That is the United States. In forty years, that is, American Jews have grown in numbers from 200,000 to 3,000,000. And the significant fact is that this growth represents a type of Jew that was hardly known to this country in 1881. Almost all of our American Jews have come from those provinces of Poland which were until recently parts of Russia, Prussia, and Austria. The transplantation of millions of Jews from their mediaeval home in Central Europe—a transplantation which was perhaps not at first deliberate and conscious, but which is becoming increasingly so—forms not only the most startling migration in the history of Israel, but gives the United States its great 'Jewish problem.' Unless the influx is artificially dammed there is not the slightest question that, in less than a generation, this great mass of central European Jews will have been moved to this country America will fulfil the rôle which Poland filled in the Middle Ages as the great home of the Jewish race.

It would have been strange if this eastern European Jew did not present such dissimilarities to the type of Jew which had already been domesticated here as to seem almost to belong to an entirely different race. His history had been a deplorable one. Possibly his remote ancestors may have resembled the Spanish Jew or the Jew from Bavaria and the Rhineland, but centuries of separation, in the era when means of communication were all but unknown, had produced a type that had little in common except a common religion. The Polish Jew had lived for centuries among Slavs and physically he had taken on so many Slavic characteristics that there is little doubt that in his veins there flows a considerable amount of Slavic blood— just as in the Spanish Jews there flows a considerable mixture of Spanish blood. The brunette type—the Jew of coal-black eyes and raven hair—is perhaps the most commonly met among the Polish Jews, but there was a considerable proportion of blonds—Jews and Jewesses with the fair hair and the blue and gray eyes that unquestionably indicate a considerable racial mixture with the Slav. Even that feature which is so dear to the cartoonist, the hooked nose, is infrequently found among the so-called Russian Jews; their nose is more commonly retroussé or pug. The hair is not always kinky or curly, but more commonly straight—again a Slavic characteristic. While physically the Eastern Jew frequently resembled the peoples among which he had lived for centuries, and so presented traits which greatly contrasted with his co-religionists already established in this country, mentally and spiritually he is something entirely different.

The thing that marked him most conspicuously was his religious orthodoxy. The long unkempt beards, the trailing hair, the little curls about the ears—these carefully preserved stigmata of traditional Israel were merely the outward signs of lives that were lived strictly according to the teachings of rabbinical law. It is perhaps not strange that the Jewish communities already established in this country regarded these strange apparitions as peoples alien to themselves, and, that, although they sympathized with their sufferings and gladly assisted in establishing them in their new environment, they refused to regard them as social equals, abhorred the idea of intermarriage, called them 'Polaks' and 'hinter Berliners,'—and practised against them, indeed, many of the discriminations which all Jews have for generations suffered at the hands of their Gentile compatriots. [***] These expulsions and these massacres had another purpose—and one which was chiefly interesting to the United States. When the Jews protested against these proceedings to Count Ignatieff, the author of the May laws, he made this laconic answer: 'The Western borders are open to you Jews.' Up to this time Russia had had vigorous laws prohibiting emigration; but now she began to relax these laws. One privilege was extended to the Jews that was withheld from all other denizens of the Czar's dominion: they were not only permitted but invited to leave the country. Such was the original impetus of the movement that, in forty years, increased the Jewish population of the United States from 200,000 to 3,000,000."[173]

Sephardic and German Jews had long opposed the emigration of Russian Jews into the United States. They considered them to be racially and socially inferior and an embarrassment to the modern faith of "Reformed Judaism". As is always the case, the worst enemy of the Jews was Jewish racism and Jewish religious intolerance. Burton J. Hendrick wrote in his article, "The Jews in America: III. The 'Menace' of the Polish Jew",

"From the standpoint both of the citizen and business man, no more abrupt change could be imagined than that which the Eastern Jew made when he transplanted himself from the old cities of Poland to the Atlantic seaboard of the United States. This Jew had never been a citizen, and had never developed the slightest sense of citizenship, as that word is understood. For thousands of years he had merely been the member of a tribe, governed by tribal laws and tribal chiefs. With the Jews from western Europe who had preceded him to America, in much smaller numbers, the Polish or Eastern

---

173. B. J. Hendrick, "The Jews in America: I How They Came to This Country", *The World's Work*, Volume 44, Number 2, (December, 1922), pp. 144-161.

Jew had little in common except a common religion. I have made this point before, but it cannot be made too frequently or too emphatically, for it is the fundamental fact in the existing Jewish problem. [***] As candidates for assimilation these Jews, as they land at Ellis Island, are about as promising as a similarly inflowing stream of Hindus or Syrian Druses. This may seem an extreme statement, but a glance at the Jews of eastern Europe, especially Poland, makes it clear that it is not. For these Eastern Jews have never been Europeanized. For ages they have lived, in Poland, in Russia, in Galicia, in Hungary, in Rumania, not as a nation or part of a nation, but essentially as a tribe. With them the Jewish religion has been the all-important consideration, far more important than nationality; the right to practise their faith, to observe their Sabbath and religious holidays, to limit their diet to the most rigid teachings of the Talmud, has been valued much higher than the mere right to enjoy political equality. A Jew of the old breed in America takes pride in calling himself an American and resents any imputation that he is not; a Jew in Germany, as the Great War showed, is almost fanatical in his assertion of his Germanism; but a Jew in Poland just as vehemently resents being called a Pole. 'I am not a Pole; 1 am a Jew,' he retorts. After a sojourn of 800 or 1,000 years in Poland he does not speak the Polish language; his dialect is a form of middle low German which was spoken in certain parts of Germany in the Middle Ages and which is still spoken in a few remote areas. The orthodox Jew in Poland not only lives, by preference, in crowded ghettoes in the cities, but he dresses in a way—a long gabardine of black cloth reaching to his ankles and a skull cap trimmed with fur—which emphasizes his Jewish particularism. His long beard and the ringlets about his ears are also part of his religion. He treats his womankind in a way that suggests his Asiatic origin. 'Thank God I am not a dog, a woman, or a Christian,' is the prayer of thanksgiving with which he begins his day. [...]"

This prayer, which Jewish men recite each morning, appears in the Talmud, *Menachos* 43*b*, and in the *Tosefta Berakhot* 6:18, and is still widely used:

"6:18 A. R. Judah says, 'A man must recite three benedictions every day:
(1) 'Praised [be Thou, O Lord...] who did not make me a gentile';
(2) 'Praised [be Thou, O Lord...] who did not make me a boor';
(2) 'Praised [be Thou, O Lord...] who did not make me a woman.';
    B. 'A gentile—as Scripture states, *All the nations are nothing before him, they are accounted by him as less than nothing and emptiness* (Isa. 40:17).
    C. 'A boor—for *'A boor does not fear sin'* [M. Abot 2:5].
    D. 'A woman—for women are not obligated [to perform all] the commandments.'"[174]

---

174. J. Neusner and R. S. Sarason, Editors, "Berakhot 6:18", *The Tosefta: Translated from the Hebrew*, Volume 1, Ktav Publishing House Inc., Hoboken, New Jersey, (1986), pp. 40-41, at 40.

Menachos 43*b* states:

> "A MAN IS OBLIGED TO RECITE THREE specific BLESSINGS EVERY DAY, [***] —and THEY ARE THE FOLLOWING: [***] — (1) *Blessed are You, Hashem, our God, King of the Universe, WHO HAS MADE ME A JEW;* [***] —(2) ... *WHO HAS NOT MADE ME A WOMAN;*[42] AND [***] —(3) ... *WHO HAS NOT MADE ME A BOOR.* [*Footnote:* Nowadays, this blessing is recited in the form of: [***] *Who has not made me a gentile*"[175]

Time Magazine wrote in the issue of 3 March 1923,

> "'Thank God I am not a dog, a woman, or a Christian,' is the prayer with which the orthodox Jew in Poland begins his day."

Evelyn Kaye wrote in her book, *The Hole in the Sheet: A Modern Woman Looks at Orthodox and Hasidic Judaism*, L. Stuart Inc., Secaucus, New Jersey, (1987), p. 89:

> "During the prayers which a Jewish man recites every morning are a series of blessings, which include: 'Thank you, Lord, for not making me a non-Jew, for not making me a slave, for not making me a woman.'"[176]

The prayer takes on somewhat different forms in different traditions, though it always expresses a Jew's gratitude to God for not being born a Goy. Burton J. Hendrick continued in his article, "The Jews in America: III. The 'Menace' of the Polish Jew",

> "[...]Just as Japanese women blacken their teeth and Chinese women bind their feet, so the orthodox Polish Jewesses, after marriage, shave their heads. These are merely the outward indications of an Orientalism that controls all phases of Jewish life. For centuries the orthodox Jews existed in Poland under an order that was tribal and patriarchal—never national. They were not subject to the laws and the civil and criminal administration of the country but they were ruled, in all departments of life, by their own rabbis, who administered the law as it is laid down in the Old Testament and the Talmud. They even counted time, not according to the Christian, but according to the Jewish Calendar. The British Commission sent to investigate the condition of the Jews in Poland were astonished to find, in

---

175. Rabbi E. Herzka and Rabbi M. Weiner, Elucidators, "Tractate Menachos 43*b*", *Talmud Bavli: The Schottenstein Edition*, Volume 59, Mesorah Publications, Ltd., Brooklyn, New York, (2002), 43b⁵.
176. E. Kaye, *The Hole in the Sheet: A Modern Woman Looks at Orthodox and Hasidic Judaism*, L. Stuart Inc., Secaucus, New Jersey, (1987), p. 89.

interrogating witnesses, that few knew the day of the week, the month, or the year; the reason is that they all reckoned time according to the orthodox Jewish calendar. That this exclusiveness is not necessarily enforced upon an unwilling people is evident from the fact that the Jews of Poland demanded of the Versailles Peace Conference—and successfully—the right to be regarded as a 'minority' people in a resurrected Poland. This means that the Jews intend to maintain themselves in Poland as a separate people, with the right to a certain number of seats in every municipal council and the national parliament, with important powers of legislation and taxation, with their own law courts, the privilege of using their own language, and other important advantages which they are to enjoy not as Poles but as Jews. Thus the organization of the Eastern Jews in Europe, in its political and social aspects, is primitive, tribal, Oriental; and their economic status represented just about the same stage of progress. Though the population did contain a considerable number of handicraftsmen, especially in the tailoring trades, for the most part the Polish Jews were middlemen—hucksters, hawkers, peddlers, small tradesmen, petty bankers, and the like. The Polish masses were agriculturists, and the Jews, who were for the most part city dwellers, acted as middlemen in the distribution of their products. They would travel into the surrounding country, chaffer with the peasants for their vegetables, and sell them in the city. Poland of course was not an industrial state; factories were few; there was thus no opportunity, had the Jew really had the inclination, for training in industrial life. They were the small shopkeepers in the town; they hawked their wares up and down the streets; such occupations, however, could not furnish support for the entire Jewish population, the result being that the great masses lived under conditions of appalling poverty and social degradation. That they were uncleanly in their habits was perhaps the inevitable consequence of the over-crowded conditions under which they existed, for their poverty was so great that a great population struggled from hand to mouth, never knowing whence their daily bread was to come. Such was the exotic mass that the steamships began dumping on the Atlantic seaboard forty years ago, and which has been attempting since to adjust itself to the economic conditions of the United States. [***] The three-per-cent. restriction on immigration therefore represents statesmanlike wisdom of the highest kind, and all attempts to break this down should be vigorously resisted."[177]

The Judaification of American institutions would only have been a bad thing if it resulted in a degeneration of those institutions and served to reduce what would have otherwise been the participation and productive talents of Gentiles in the progress of humanity; or if it led to subversive political movements and worked against the interests of Americans at large. So the question arises, "What

---

[177]. B. J. Hendrick, "The Jews in America: III The 'Menace' of the Polish Jew", *The World's Work*, Volume 44, Number 4, (February, 1923), pp. 366-377, at 366-368, 377.

were the effects?"

One of the effects, which no doubt had many benefits, was to tend to secularize these institutions, many, if not most, of which had a Christian foundation. This resulted from Jewish tribalism, Jewish secularism, and the schism which existed between Christian and Jew which vanished on the neutral ground of secularism. This is not to say that there was no such push towards secularism among the Gentile community of professionals and scientists, as well. On the downside, the massive influx of *Ostjuden* lent a kosher talmudic flavor to both the content of the curriculum and the atmosphere of the universities—and more broadly to professional and scientific debate—which was unpalatable to many Gentiles and Jews alike, and which discouraged Gentile participation. Debates increasingly became festivals of *ad hominem* attack, where racist Jews would subvert open scientific debate and substitute in its place personal insult, smear campaign, the self-glorifying hero worship of Jews made famous by the Jewish press, and the dogma (often plagiarized and corrupted Metaphysical nonsense) their feted Jewish leaders promoted. One sees a similar shift toward adolescent behavior in the modern media, which has increasingly come under the influence of Zionists, and which tends to discourage reasonable Gentiles and Jews from becoming involved in the political process. The deleterious political effects of Eastern Jewish emigration, were, among other things, the unnecessarily involvement of Americans in numerous wars, and will be addressed at length later in the text.

3.4.1.1 Rothschild Power and Influence Leads to Unbearable Jewish Arrogance

The tribalistic intolerance of some racist Jews in the press and at the universities did enormous harm to the reputation of Jews in general after emancipation, as did the tribalistic attacks many Jews in the press made on Catholics during the *Kulturkampf*, which ultimately resulted in the anti-Jewish spirit in France of the Dreyfus Affair. The rise in Jewish influence through the Rothschild family at the expense of the Roman Catholic Church was so apparent in the 1870's, that some felt a need to defend themselves against a general vilification of Jews based on the Rothschilds' corruption of international politics. *The Chicago Daily Tribune* reported on 28 June 1874 on page 2,

"Disraeli and the Jews.
London Correspondence of the Cincinnati Commercial.

Every now and then there are little intimations of the bitterness with which the Jews regard the desertion of their ancient religion and fraternity by Disraeli. All the glory which his genius and eminence reflect upon them ethnologically is lost again by his condemnation of them religiously, by his example,—that is, allowing himself to be spoken of at May anniversaries as a 'converted Jew.' Disraeli is so plainly a Jew in physiognomy that his look has unconsciously reminded the public again and again of the debt they owe to the intellectual distinction of the race. A very clever Jewish writer of

London,—Mr. Levy,—recently wrote a very remarkable article showing to what a large extent European nations are at present under the influence of Jews (as Castelar, Gambetta, the Rothschilds, etc.), and contrasted the fact with the decay of Roman Catholic power over the politics of Europe—the implication being that the historic position of the two, Jews and Romanists, might one of these days be reversed. The clever writer of the article might have given it more point by reference to certain facts in the career of the late Sir David Salomons, who, above all others of his race who have lived in England, deserves to be remembered as the true representative of his people. Through his influence Parliament altered the declaration, 'On the faith of a true Christian,' which he refused to make, thereby annulling his election to the office of Alderman twice. He then obtained very civic distinction, and in 1855-'56 became Lord Mayor of London. His first work after being raised to this distinction was to secure two things which relieved the Roman Catholics of special grievances. He put down the before boisterous and general observance of Guy Fawkes Day, which was always the occasion of insults to the Catholics, and he caused so much of the inscription on the monument near Billingsgate, which attributed the great fire of London to the Catholics, to be erased. Pope wrote of that column, which—
   Towering to the skies,
   Like a tall bully lifts its head and lies.
But that it no longer slanders the Catholics is due to the determination of a Jew. Baron Lionel de Rothschild was the first Jew elected to the House of Commons, but he had omitted the declaration, 'On the true faith of a Christian,' and withdrew. In 1851, Sir David Salomons was elected to Parliament by the borough of Greenwich. He also refused the declaration, and was requested to withdraw. He did so, but not until he had made a wise and temperate speech to the House which made it feel ashamed of the disabilities imposed on Jews. The late Lord Westbury took the matter up, and after a time the 'Jewish Disabilities bill' was passed. From that time Sir David, who, meanwhile, was created a Baronet of the United Kingdom, sat in Parliament, where he was considered the highest authority on finance, a subject on which he wrote several valuable books. He was one of the founders of the London and Westminster Bank, and was its Chairman until the day of his death. It is a notable circumstance that the Catholic organs of London should have attacked the Jews generally because of the loan the Rothschilds are said to have made to the Italian Government, saying that they were as ready to crucify Christ, when the first acts of the first Jews who got into power in London were the abolition of the two things which most annoyed them. When he was before the people for election as Sheriff, they were curious to know whether some of his views might not impair his official work. Some one asked him what he would do in case a reprieve for a criminal came on Friday night—riding being then prohibited to Jews— and he promptly responded, 'I would order my carriage and go at once.' Some propositions have been made lately that the large and increasing body of Theists should graft themselves on to the ancient Jewish stem; but there

is in England no society of Jews who have dispensed with the old formulas and usages—paschal, sabbatarian, etc.,—which would, of course, render such amalgamation impossible. However, amenities have been passing between the Theists and the Jews, and not a few of the latter are now found attending the religious services of Mr. Voysey and other rationalists."

It should be noted that the seemingly altruistic actions of David Salomon towards Catholics had an ulterior motive. Jews were traditionally staunchly anti-Catholic, but they saw an opportunity to benefit themselves by the emancipation of Catholics. This freedom for Catholics in England would set the precedent for religious tolerance for Jews in England—which is ironic given that it was Cabalist Jews who created Protestantism, Puritanism and Theism as a means to destroy Catholicism and convert it into Judaism. *The North American Review* wrote in 1845,

> "Strange to say, in England the Jews still suffer under grievous civil disabilities. In 1290, Edward the First banished all in his kingdom, and seized on their property. The exclusion was so rigid and complete, that no traces of them in that country occur again till the period of the Commonwealth. Cromwell made an unsuccessful movement in their behalf; and in his time they began to return in small numbers. In the reigns of Charles the Second and James the Second, some privileges were granted them; which, however, were withdrawn after the Revolution of 1688. In 1753, a bill was passed in parliament, not without virulent opposition, permitting Jews, who had been residents of Great Britain or Ireland three years, to be naturalized; but so odious did the law prove to the nation at large, that the ministry who had encouraged the enactment shrunk from its support, and it was repealed at the very next session. From the pulpit generally, by the mercantile corporations, and by a bigoted populace, it was vehemently opposed. Dean Tucker, who, almost alone among the clergy, wrote decidedly in favor of the naturalization of the Jews, was very roughly treated, and, by the people of Bristol, burnt in effigy in full canonicals, with his obnoxious writings. In May, 1830, on the back of the Roman Catholic emancipation act, another effort was made in parliament to emancipate the Jews; but it was opposed by the ministry, and failed. In short, the decree of Edward the First has never been formally abrogated; and though several acts of parliament have recognized, and thus legalized, their presence in the kingdom, England, with all her boasting of Roman Catholic and negro emancipation, still treats native-born Jews as foreigners, admitting them to few privileges but those of alien residents and traders. To a single inch of the soil they cannot obtain a title."[178]

---

[178]. "The Modern Jews", *The North American Review*, Volume 60, Number 127, (April, 1845), pp. 329-368, at 346.

Alas, the Catholics had to wait their turn in line to enter the British Parliament behind the Catholic hating Jews. The above article *Disraeli and the Jews* gave the false impression that the Jews helped the Catholics gain full emancipation. Alas, the Catholics had to wait their turn in line to enter the British Parliament behind the Catholic hating Jews, who in fact stood in their way. Though the Jews sought to ride on the backs of the Catholics when Catholic initiatives forwarded Jewish interests, they also denied the Catholics rights when able to do so, even to their own detriment, as proven by the following article in *The New York (Daily) Times* of 13 June 1854, on page 4,

"The Jews in Parliament

Lord JOHN RUSSELL displays consistency in connection with the Jewish Disabilities bill very unusual in so fickle and procrastinating a Minister. On May the 25th, the second reading of the bill was moved, when a singular and unexpected debate took place. The great champion of Israeliteism in England, the eulogist of it in the House of Commons, the glorifier of it in his novels, the steadfast, eloquent defender of the Hebrew race—Mr. D'ISRAELI—voted against the bill which contemplated the admission of Jews to a seat in the British Senate. The reasons which induced this singular opposition of his own opinions, were stated by him at length, in a speech distinguished by his usual earnestness and happy sarcasm. The bill framed by Lord JOHN RUSSELL was intended to remove certain obnoxious clauses from the Parliamentary oath, and substitute other forms, which should be unobjectionable to Hebrew or Roman Catholic. The words, 'on the faith of a Christian,' were to be abolished for the Jews, while the anti-Jacobite and anti-Papal clauses were to be cancelled in favor of the Roman Catholics. This was, no doubt, looked on by Lord JOHN RUSSELL as a skillful combination, by which his bill would secure a double support. The Irish members, from association or actual interest, would vote for a bill abolishing an oath by which every Roman Catholic entering the House was required to declare himself no traitor, while those members who on other occasions passed the previous Jewish bills through the Lower House, only in order to have them assassinated in the House of Lords, would doubtless sweep this through with an overwhelming majority. Lord JOHN, however, miscalculated. His double blossomed liberality was nipped in the bud, and the bill was rejected by a small majority. This is curious and significant. Heretofore the House of Commons has passed the Jew bills triumphantly, while the Upper House butchered them one after another with dogged determination. The moment, however, that license to the Jews was coupled with license to the Roman Catholics, that instant the Lower House was alarmed, and not daring to trust the insidious bill even to the tender mercies of the Peers, strangled it in its infancy on the spot.

It was a singular sight to see the leader of the despised Hebrew race, disdainfully rejecting constitutional rights for his party, because the same hand held out the same gifts to the Roman Catholics. Mr. D'ISRAELI displayed, however, considerable astuteness in this opposition. It is his policy to disconnect the Jewish question from all possible odium, whether religious or political. At some future, and more favorable period, he calculates that the untiring animosity of the House of Lords will be exhausted, and Baron ROTHSCHILD, and Mr. SOLOMONS will be legally elected members of the House. To mix up the Jewish interests, therefore, with those of the Roman Catholics, and let the bill appear before the Lords with this duplicated offensiveness, would be to sustain another defeat, and strengthen still more the enmity entertained against the original bill, Mr. D'ISRAELI, therefore, used his influence to have the bill crushed before it could go before the Lords, and so bides his time until he can present the claims of the Jews at a more favorable opportunity, and unclogged by the weight of an obnoxious addition.

That the Jew will eventually conquer opposition, and enter the British Parliament, no one can have a shadow of a doubt who looks at all closely into his claims and the course of events in England. There is nothing in the Jewish character to alarm the supporters of the Established Church. The Hebrew makes no proselytes, for the Jewish faith is as much a matter of race and blood as it is of religion. It is physical as well as moral, and like the Poet, the Jew must be born, not made. There is little fear then of the Jews using political influence to subvert the established religion, and as to the disbelief in Christianity, they are surely as well qualified to hold a place in the English Senate as Lord BOLINGBROKE was, who made no concealment of his Atheism. They represent large commercial and political interests. They sway the destinies of many nations, and the issue of more than one great European question is dependent on their word. The basis of this power to be sure is a monetary one, but English gold is represented in the Commons by Mr. BARING; why not Jewish capital by Baron ROTHSCHILD?

That the House of Commons holds these opinions is evident by the alacrity with which they passed every Jew bill up to the last; but they seem to hesitate about extending the favor to the Roman Catholics. The Jew is virtually obliged to recant his faith, if he wishes to enter the House at present, which is of course a complete barrier, while the Roman Catholic is merely forced to forswear any designed or secret allegiance to the House of STUART, and promise to keep his fingers out of the pie of the Established Church.

After the anti-papal feeling exhibited by the House of Commons on this question, it is probable that the Jewish Disabilities bill will take another shape when it next appears before the Senate. Meantime, Mr. D'ISRAELI and his sister will write novels in which it is proved that the world owes everything, from the mariner's compass up to the steam engine, to the Hebrew race. Baron ROTHSCHILD will be again elected, perhaps, and again martyred for his faith, until that incapable congregation of old women in the House of Lords think fit to admit a little of the liberal light of the nineteenth century into their chamber, darkened with worm-eaten prejudices."

Note that the Jews saw themselves as Jews, not as British, and did what they believed was "good for the Jews" and would never consider doing what was good for humanity other than as a means of somehow profiting the Jews. Note that the Jews saw themselves as international lords over Europe, not as loyal citizens of any given nation in Europe. The Rothschild family, who held government posts in several different and often hostile nations, were loyal to none but themselves and the Jews whom the ruled.

Wealthy Jews prevented the emancipation of the Jews of England, though as Jews so often do, they scapegoated Christians for their heinous actions. The famous American Zionist Jew, Mordecai Manuel Noah, noted in 1818, that it was wealthy Jews who had opposed the emancipation of English Jews,

> "Great Britain,[7] by an act of parliament, passed in the year 1753, granted to the Jews the rights of citizens; the clamours of the people, and, indeed, the discontent of a large portion of the Jews themselves, caused this honourable law to be revoked; and from the organization of the government, there exists no hope at present of its revival. [***] [7] *Great Britain*. The act of Parliament alluded to produced a considerable clamour among the wealthy Jews, who were fearful that giving rights to a vast body of their nation who were ignorant of their value, would have a tendency to create a stronger current of prejudice against the nation. This objection, joined to the indignation of an ignorant populace, induced its repeal. The Jews in England have almost forgot the cruelties formerly practised towards their nation during several reigns; particularly the third Henry, and John; however, these things had better be forgotten, and are compensated by the liberality of the present times."[179]

Einstein claimed that anti-Semites were correct to be believe that Jews exercised undue influence in Germany. Einstein wrote in the *Jüdische Rundschau*, on 21 June 1921, on pages 351-352,

> "This phenomenon [*i. e.* Anti-Semitism] in Germany is due to several causes. Partly it originates in the fact that the Jews there exercise an influence over the intellectual life of the German people altogether out of proportion to their number. While, in my opinion, the economic position of the German Jews is very much overrated, the influence of Jews on the Press, in literature, and in science in Germany is very marked, as must be apparent to even the most superficial observer. This accounts for the fact that there are many anti-Semites there who are not really anti-Semitic in the sense of being Jew-haters, and who are honest in their arguments. They regard Jews as of a nationality different from the German, and therefore are alarmed at the increasing Jewish influence on their national entity. [***] But in

---

179. M. M. Noah, *Discourse Delivered at the Consecration of the Synagogue of [K. K. She'erit Yisra'el] in the City of New-York on Friday, the 10th of Nisan, 5578, Corresponding with the 17th of April, 1818*, Printed by C.S. Van Winkle, New-York, (1818), pp. 13, 38.

Germany the judgement of my theory depended on the party politics of the Press[.]¹⁸⁰

Einstein also stated,

"The way I see it, the fact of the Jews' racial peculiarity will necessarily influence their social relations with non-Jews. The conclusions which—in my opinion—the Jews should draw is to become more aware of their peculiarity in their social way of life and to recognize their own cultural contributions. First of all, they would have to show a certain noble reservedness and not be so eager to mix socially—of which others want little or nothing. On the other hand, anti-Semitism in Germany also has consequences that, from a Jewish point of view, should be welcomed. I believe German Jewry owes its continued existence to anti-Semitism."¹⁸¹

Nazi Zionist Joseph Goebbels, sounding very much like political Zionist Albert Einstein, was quoted in *The New York Times*, on 29 September 1933, on page 10,

"It must be remembered the Jews of Germany were exercising at that time a decisive influence on the whole intellectual life; that they were absolute and unlimited masters of the press, literature, the theatre and the motion pictures, and in large cities such as Berlin, 75 percent of the members of the medical and legal professions were Jews; that they made public opinion, exercised a decisive influence on the Stock Exchange and were the rulers of Parliament and its parties."

Max Born knew that a Albert Einstein and his sycophantic Jewish promoter Alexander Moszkowski would be used as examples to justify a Dühring-style general vilification of Jews—which could also hurt the sales of Born's book and spoil his efforts to profit from the Einstein name in the desperate times which followed the First World War. Eugen Karl Dühring, who wrote important historical treatises on Physics which are on a par with those of Ernst Mach, including an analysis of space-time theories and the underlying principles of what was to become the general theory of relativity, promoted racial anti-Semitism to modern Germany and inspired Theodor Herzl's racist political Zionist movement.¹⁸² Dühring was a Socialist who combated Lasalle, Marx and

---

180. A. Einstein, "Jewish Nationalism and Anti-Semitism", *The Jewish Chronicle*, (17 June 1921), p. 16.
181. A. Einstein, A. Engel translator, "How I became a Zionist", *The Collected Papers of Albert Einstein*, Volume 7, Document 57, Princeton University Press, (2002), pp. 234-235, at 235.
182. M. Samuel, "Diaries of Theodor Herzl", in: M. W. Weisgal, *Theodor Herzl: A Memorial*, The New Palestine, New York, (1929), pp. 125-180, at 129. T. Herzl, English translation by H. Zohn, R. Patai, Editor, *The Complete Diaries of Theodor Herzl*, Volume

Engels over the future of Socialism in Germany. The Socialists Dühring, Lasalle and Marx each used the tactic of Jew-baiting for political gain. Engels, in at least one instance, spoke out against it.[183]

Shrill cries of "anti-Semite!" and "dirty Jew!" increasingly filled the air in both political and scientific debates, and were most often the product of those Jewish minds who wanted to deflect interest from the facts, and who wanted to keep Jews segregated from non-Jews. Anti-Semitism was a favorite tool of racist Jews to manipulate both Jews and Gentiles, and it was racist Jews who deliberately caused most of the anti-Semitic persecutions of Jews throughout history, either by posing as anti-Semites, or hiring or otherwise recruiting Gentiles to pose as anti-Semites. As fantastic as it sounds, this is easily proven, and will be proven later in the text.

The context of the polemic battles between these Socialists is given in the endnote,[184] which reprints an important and quite readable history of the Socialist

---

1, Herzl Press, New York, (1960), pp. 4, 111.

183. P. W. Massing, *Rehearsal for Destruction: A Study of Political Anti-Semitism in Imperial Germany*, Howard Fertig, New York, (1967), pp. 311-312.

184. R. H. Fife, Jr., *The German Empire Between Two Wars: A Study of the Political and Social Development of the Nation Between 1871 and 1914*, Macmillan, New York, (1916), pp. 177-199 and 359-388:
"CHAPTER IX
THE PROLETARIAN IN POLITICS

IF we were obliged to cover with one word the development of Germany in the four decades between the two great wars, that word would certainly be "socialism." It is not merely that in philosophy, literature and art the welfare of the masses is the leading motif running through the eighties and nineties until it became lost after 1900 in the swelling music of national ambition. In the field of political economy also socialistic ideas marked the age. They began by conquering the professorial chairs in the universities in the seventies, where such "socialists of the chair" as Adolf Wagner of the university of Berlin set their stamp on the generation of political economists which followed the war with France, and they found expression in the compulsory insurance measures and similar legislation of the following decade. Such ideas were indeed nothing new in Germany since the sixteenth century, when cities such as Augsburg and Strasburg were models of a hard and fast organization, in which capital played a small part and the workers formed the commonwealth on the principle of a closed shop, where communal undertakings largely supplanted private enterprise and every detail of life, including the details of food and dress, was fixed by law. The paternalism of the petty despotisms which preceded German unity had disciplined the Germans to live under efficient supervision, and the ideals of the Manchester school of British economists did not take lasting hold on German economic life.

Socialism then grew in Germany on well-prepared soil. State ownership of railroad and telegraph had come naturally soon after the coming of these utilities, and municipal control of many forms of enterprise descended as a tradition from the later middle ages. That the individual should look to the government to provide for his welfare and that state and communal funds should supplant private capital in many undertakings had long been the case when Bismarck undertook his compulsory insurance policy in the eighties. This program was, as we have seen, an effort to strike the ground from beneath the Social Democrats by removing some of the causes of proletarian dissatisfaction. Here and there Bismarck's successors went further on the road, with such measures as the purchase of the *Hercynia* potash mine (cf. page 166). That they did not go still further in this and other fields of state socialism was due in large measure to the existence of the Social Democratic party. This Ishmael in Germany's political life by its very advocacy of measures made them impossible for the government.

What is it that has made the Socialist unfitted to be an ally and unwelcome as a coworker with nearly all other parties? What is there in the advocacy by the Social Democrats of any reform that has caused not only the East Elbian *Junker* and the Westphalian manufacturer, but even the National Liberal physician and shopkeeper to look askance at it? The answer is to be found both in the doctrinaire character of the party and in the violence of Socialist editors and orators. Karl Lamprecht has shown that all German political parties are antiquated in that all cling to formulas and doctrines that have outlived their applicability to present-day affairs. In this sense the Social Democratic party is the most antiquated and the least opportunist. In this has lain its strength as a class party and its weakness in electoral and parliamentary strategy. Beginning with the removal of the coercive laws in 1890, it cast at all national elections the largest vote of any party, and after 1903 held under its discipline nearly one-third of all the electors to the national parliament, more than all the other Liberal fractions combined. Nevertheless it exercised less influence on legislation than any other of the major groups in the empire. To understand the reason for this one must glance at the development of socialism as a political force.

When in 1867 Friedrich Liebknecht and August Bebel were elected to the first *Reichstag* of the new-born North German Confederation, they found ready at hand both the gospel of socialism in the works of Karl Marx and the needed fighting force in the German Workingmen's Party (*Allgemeiner Deutscher Arbeiter-Verein*), which had been founded four years earlier by Ferdinand Lassalle. Two years later at the famous Eisenach Convention Liebknecht and Bebel called the Social Democratic Workingmen's Party into existence, on a platform built of Marx' theory of the destructive rule of capital and his call to the workingmen of all lands to unite, and finally in 1875 the followers of Lassalle forsook their nationalistic ideals and were won over to the internationalism of the Marxists. Immediately the triumphal march of the Social Democrats began, a march which has continued with few halts since. Aided by the hardships brought on by the financial crises of the seventies, the Marxian theories of the misery caused by the capitalistic state and the exploitation of the working class through the capitalistic organization of society found eager acceptance in all quarters of industrial Germany. Already in 1876 there were twenty-four papers and journals published in the interest of the party with nearly one hundred thousand subscribers: by the next year the number of party periodicals had increased to forty-one, and that year the party cast nearly half a million votes and elected twelve members to the national legislature. From that time the Social Democracy kept pace closely with the forward movement of industrial Germany. Wherever factories sprang up and workmen came to live together, the theories of Marx took root. The workingmen were organized into Socialist unions, which became at once fighting units in the industries and the elections; with the capacity for organization so characteristic of an industrial age and of German society in particular, the Social Democracy was solidified by the establishment of central bureaus under the control of secretaries. These latter quickly developed into a class of experienced leaders, at once clever agitators in the industries and skillful strategists in political campaigns.

Bismarck watched the rise of the party and its often unscrupulous means of agitation with growing distrust. He put no confidence in the alleged peaceful program of socialism: for him the party bore nothing but red revolution on its banners. In 1878 two attempts were made on the life of Emperor William which were unjustly ascribed to the effect of socialist agitation; and the Chancellor took advantage of the popular outcry to dissolve the Liberal *Reichstag* and appeal to the electors on an anti-socialist program. The result was the enactment of rigid laws forbidding Socialist propaganda. The following ten years, 1880 to 1890, were for the party a period of almost subterranean existence. Clubs were suppressed, newspapers and journals confiscated, many of the leaders, Liebknecht and Bebel among them, went to prison. In spite of prosecution and imprisonment, however, the propaganda went straight ahead. Political clubs were reorganized as singing societies and bowling clubs and the party organization was perpetuated by these and by the trade unions, which continued to spread like a vast network throughout industrial Germany. During the ten years of the anti-socialist laws the total vote of the party increased, a larger number of deputies was chosen to the *Reichstag*, and more important still, the inner organization and solidity of the party gained tremendously under persecution. This was shown immediately on the expiration of the anti-socialist laws in 1890. In that year the party cast nearly one and one-half million votes in the national elections, and became thereby the strongest party in the empire. In 1898 the Social Democratic vote had risen to two millions, in 1907 to three and one quarter millions, in 1912 to more than four and one-quarter millions, more than one-third of all votes cast in the imperial elections of that year.

  The great Chancellor was, however, too far-seeing a statesman to think that the mere forbidding of socialist propaganda would stop the growth of socialism, which to his mind was only revolution in disguise. He set out, as we have seen, to cut the ground from beneath the feet of the proletarian agitators by a system of legislation which should ban from the empire the direst poverty by insuring to the working class compensation in case of injury and care in sickness and old age. These needs, which were outlined in an imperial message of 1881, formed the basis of debate and experiment through the following eight years and were finally met in the various compulsory insurance measures which, so to speak, set their stamp upon Germany's internal politics in the eighties. In the Workingmen's Compensation or Accident Insurance Act of 1884, the burden of insurance was laid entirely upon the employer; the cost of the Sick Insurance Act of 1883 fell upon both employer and employee; for carrying out the provisions of the Old Age Pension Act of 1889, the empire joined with both capital and labor in providing for the veterans of labor. By this legislation, which though several times amended in minor parts, has remained essentially the same, Germany took a long step in the direction of state socialism and assumed the first place among nations in the protection of its army of labor. Both Radical and Socialist have found much to criticise in the laws, and the amendments which reformers suggested should long ago have received attention at the hands of the government; nevertheless, with all of their imperfections, the compulsory insurance acts have been a guiding star for the social legislation of other lands and one of the brightest decorations on the bosom of modern Germania. They are no less a superb monument to the liberal view and modern spirit of Bismarck in social legislation.

  But they did not win over the Socialists. The representatives of the fourth estate accepted the socialistic laws of the eighties not as a gift from the hands of benevolent capital, but as a right conceded through the fear of the rising strength of the proletariat. There is evidence that the old Chancellor had wearied of the struggle to win the working classes to a national and patriotic spirit and that at the expiration of the anti-socialist laws in 1890 he was preparing a stroke against the constitution, which by the abolition of manhood suffrage should undo the work of 1866 and exclude the non-propertied classes from a share in government (cf. page 127). However, young Emperor William thought otherwise, and with the fall of Bismarck, legislation against the Social Democracy was dropped and the Emperor sought to accomplish by conciliation what suppressive laws had failed to do. He summoned an international congress in Berlin to consider measures for the further welfare of the working classes, and outlined for adoption various propositions, such as a complete Sunday holiday, which had been advocated in the Socialist platform. But the effort to win the workingmen to fealty to monarch and Fatherland by kindness broke against the hard class consciousness of the fourth estate. No royal enticements could prevail against the teachings of Marx, ably and speciously interpreted by Socialist speakers, no words of the sovereign could make progress against the class feeling which had been bred in the industrial proletariat for two decades in trade union, tavern debating club and Socialist journal. From that day on the crown and indeed all of the upper classes and a large part of the middle classes in Germany parted company with the proletariat. Henceforth every representative of the existing organization of society from the sovereign to the Rhenish crockery dealer denounced the Social Democrats as enemies of the Fatherland. But whether ridiculed as a "transitory phase" or threatened with a holy war of extermination by "all lovers of God and Fatherland," the Socialist forces marched on in ever increasing numbers, a solid phalanx of industrial workers, soaked with the doctrines of Marx and Engel and ably led by labor secretary and editor.

In his opposition to the monarchy and the entire capitalistic state, the Social Democrat included of course the army, under feudal and capitalistic leadership. Nowhere, however, has the German military spirit found better expression than in the organization and discipline of the Social Democratic party. Who could watch the orderly, shoulder to shoulder march of tens of thousands of workingmen through the streets of Berlin on the occasion of the burial of a leader or on the anniversary of the "victims of March," the revolutionists who fell in the street fighting of March 1848, without seeing in imagination these same men clad in the blue and red or khaki of active soldiers? And who could see the eyes-to-the-front, fingers-on-the-trouser-seam carriage with which the individual workman follows his leader in strike or electoral campaign without recalling the Prussian military discipline? In August 1911 at Treptow, a suburb of Berlin, a mighty Socialist demonstration was made against the threatened war with France and England over the Morocco affair. A vast crowd of men and women, estimated at eighty thousand, gathered on a Sunday afternoon about a tribune to hear their leaders denounce war as a diabolical game at which the capitalist must win and the proletarian lose. Only a few of the mighty audience could hear a word of the orators, but all stood at respectful attention in the intense heat until the speeches were over and then at a given signal waved their arms in a mighty storm wave, voting affirmatively on a resolution which protested in the name of labor against the threatened war. And throughout the day not one case of disorder, scarcely even a chance hard word at an over-officious policeman, among the tens of thousands of workingmen and working women who spent the hot Sunday journeying back and forth from their homes in almost all parts of Greater Berlin!

The same iron discipline that has taught moulder and stoker and street paver that he owes it to his class to suppress even a natural outburst of resentment, because it may give the representatives of feudalism and capitalism an advantage, holds sway over leader and editor. The annual party convention, the *Parteitag*, is the court of last resort, before which even those highest in the councils of the party must appear and justify their actions. Prominent Socialists, including some of the leading parliamentarians of the party and the editors of such journals as *Vorwärts* and the *Sozialistische Monatshefte*, have been

movement in Germany in the Nineteenth and early Twentieth Centuries found in Robert Herndon Fife, Jr.'s book, *The German Empire Between Two Wars*, which was published in 1916. Fife also analyzed contemporary German newspapers, and provides the modern reader with an understanding of the background which gives context as to why Einstein was often viewed as a Socialist and Communist agitator. Fife also documents the unabashed political partisanship of the contemporary newspapers in Germany. According to Fife, Socialists tended to be rigidly dogmatic and vicious to those with whom they disagreed. They tended to be very intolerant of dissent and/or mere disagreement.

Einstein had many Socialist friends in the press and publishing business. Most of them were ethnically-biased Jews, who were prone to make personal attacks against Einstein's critics through their journals and newspapers. These pro-Einstein Socialists often called Einstein's critics "anti-Semitic" without grounds. Socialists in the Dühring camp were in turn vicious to Einstein and to Jews in general.

Communists were also rigidly dogmatic[185] and murderous to their critics. Communists are notorious for manufacturing patently false historical revisionism and for suppressing the truth, which false revisionism favors their equally notorious penchant for creating cults of personality around megalomaniacal and genocidal dictators like Lenin (born Ulyanov), Trotsky (born Bronstein) and Stalin (born Djugashvili). Socialists and Communists created personality cults around Marx and Lasalle and used anti-Semitism for political gain, as did the German Jews Karl (born Mordecai) Marx (whose family name was originally Marx Levi) and Ferdinand Lasalle (born Lasal), who promoted anti-Jewish hatred as a means to promote crypto-Jewish Socialists and Jewish Communists into power.[186] The Communist German-Jewish agitator

---

called upon to defend the orthodoxy of their faith, and prominent leaders have been unceremoniously thrust out of the party. It became an accepted canon that when a man found that his position, reached after scientific inquiry, was no longer that of the party, and when he could not persuade the party to accept his position, he was by that very fact no longer a Social Democrat. This tyranny of the majority was due not merely to a democratic intolerance of strong individualities, it proceeded also from the extreme doctrinarianism of the party.

    This doctrinarianism is the very bone of the Social Democracy. No orthodox theologian of years agone ever clung to the verbal inspiration of Holy Writ with greater zeal than Socialist orator and editor and private soldier have held to every jot and tittle of the Erfurt Platform. This declaration of faith was adopted in 1891, soon after the expiration of the anti-socialist laws, and has had no official revision since. It could not be expected, however, that the Marxian theories, as enunciated in that instr

185. F. S. Meyer, *The Moulding of Communists: The Training of the Communist Cadre*, Harcourt, Brace and Co., New York, (1961). ***See also:*** W. Chambers, *Witness*, Random House, New York, (1952). ***See also:*** D. A. Hyde, *Dedication and Leadership Techniques*, Mission Secretariat, Washington, (1963); **and** *Dedication and Leadership: Learning from the Communists*, University of Notre Dame Press, (1966).

186. E. Bernstein, "Jews and German Social Democracy", *Die Tukunft* (New York),

Ferdinand Lassalle wrote to Marx on 24 June 1852,

> "... Party struggles lend a party strength and vitality; the greatest proof of a party's weakness is its diffuseness and the blurring of clear demarcations; a party becomes stronger by purging itself... "[187]

3.4.1.2 Jewish Intolerance and Mass Murder of Gentiles

Russian-Jewish anarchist Emma Goldman, who was accused of inciting the assassination of U. S. President William McKinley in 1901, stated in 1920 that "we" always knew that Marxism would inevitably lead to tyranny. John Clayton reported in *The Chicago Tribune* on 18 June 1920 on the front page,

"RUSSIAN SOVIET 'ROTTEN,' EMMA GOLDMAN SAYS

**U. S. Flag on Bureau;
Longs for Home.**
BY JOHN CLAYTON
(Chicago Tribune Foreign News Service.)
(By Special Cable.)
(Copyright: 1920: By the Tribune Company.)

PARIS, June 17:—On the bureau of Emma Goldman's room in Hotel Astoria at Petrograd draped over a corner of the picture of her niece is the American flag. Emma Goldman, deported from America as an anarchist, makes no apologies for this flag.

The communist leaders living at the hotel josh her a little about it, but Emma says:

'That's the flag of my niece's country. I'm going back there some day, for I love America as I love no other land.'

Emma: 'Bolshevism is Rotten.'

Emma Goldman is sick of bolshevik Russia. When I called on her in

---

Volume 26, (March, 1921), pp. 145ff.; English translation in: P. W. Massing, *Rehearsal for Destruction: A Study of Political Anti-Semitism in Imperial Germany*, Howard Fertig, New York, (1967), pp. 322-330. ***See also:*** H. Hirsch, "The Ugly Marx: Analysis of an 'Outspoken Anti-Semite'", *Philosophical Forum*, Volume 8, (1978), pp.150-162. ***See also:*** P. L. Rose, *Revolutionary Antisemitism in Germany from Kant to Wagner*, Princeton University Press, (1990), pp. 296-305. ***See also:*** R. Grooms, "The Racism of Marx and Engels", *The Barnes Review*, Volume 2, Number 10, (October, 1996), pp. 3-8.

187. Quoted in V. I. Lenin, "What is to be Done?", *V. I. Lenin: Collected Works*, Volume 5, Foreign Languages Publishing House, Moscow, (1961), pp. 347-530, at 347; **and** *What is to be Done? Burning Questions of Our Movement*, International Publishers, New York, (1969), p. 5. Note that the title of Lenin's work, "What is to be Done?", was a repetition of revolutionary Nikolai Gavrilovich Chernyshevsky's work of 1873, *What is to be Done?*

Petrograd she asked: 'What do you think of it? You have been here six weeks. How do you feel about it?'

'It is rotten,' I replied. 'It's so rotten I'm sick with it.'

'You're right, it is rotten,' she said. 'But it is what we should have expected. We always knew the Marxian theory was impossible, a breeder of tyranny. We blinded ourselves to its faults in America because we believed it might accomplish something.

'I've been here four months now, and I've seen what it has accomplished. There is no health in it. The state of socialism or state of capitalism—call it what you will—has done for Russia what it will do for every country. It has taken away even the little freedom the man has under individual capitalism and has made him entirely subject to the whims of a bureaucracy which excuses its tyranny on the ground it all is done for the welfare of the workers.'

More Freedom in United States.

'Where did you find the greater degree of freedom, Miss Goldman?' I asked. 'In the United States or in communist Russia?'

'Any form of government is bad enough,' she replied, 'but between this and individual capitalism, the choice lies with the latter. At least the individual has a chance to express his individuality.'

Of all the deportees who entered Russia with Miss Goldman, only one or two have accepted the doctrines of communism. Miss Goldman, Berkman, and Novikov, the leaders of the group, refused to work with the government in any way except purely humanitarian labor.

Expects to Go to Jail.

'We are studying conditions in Russia,' said Miss Goldman at another time. 'We want to make a trip through the country districts and talk with the peasants. Then we will be ready to speak. We probably will go to jail when we start criticising, but that doesn't matter. We've been in jail before. We cannot be true to our principles and not speak.'

Miss Goldman and Novikov refused places in the reviewing stand at the May day procession, nor will they accept places at any government meeting.

Emma: 'Hit Hard.'

I spent much of my week in Petrograd with them. When I was ready to leave she said to me: 'Be careful what you write, if you want to return to Russia. If you don't, then hit hard. You may be called an agent of the capitalistic class by the people in America who don't understand.

'If you are, tell them we have been here four months and now we know. We have investigated the factories, homes, and institutions as no newspaper man can be permitted to investigate them, and we've found them bad. I know from my conversation with you you have gotten at the heart of the matter. It's up to you to tell the American people, and tell them straight.'

And that is what I intend to do. Emma Goldman has found, as I did, that the best cure for bolshevism is a trip to bolshevik Russia. She told me to hit out straight from the shoulder. Well, as an American, I'll let that little flag on Emma's bureau hit for me."

Jewish leaders sponsored Marxism, Bolshevism and the Russian Revolution. After news arrived in the West of the Bolshevik mass murders of millions of Christians, Jewish leaders made a great show of denouncing Bolshevism in the West, especially after the First World War ended. They feared retaliation against all Jews for the crimes committed by Jewish Bolsheviks in the East.

Russian and Polish Jews committed genocide against the Russian People as an act of revenge and mass murdered millions of innocent Christians. This was part of a series of vengeful acts which Jewish bankers had been carrying out against the Russians at least since the 1870's, which vengeful acts resulted in Pogroms in the 1880's—a series of vengeful acts which Jews continue to this day. It was the Jews who began the cycle of violence and death, by their refusal to assimilate into Russian society, while taking from that society a disproportionate share of its wealth—which they continue to do to this day. *The Chicago Daily Tribune* wrote on 21 July 1878, on page 13,

"BEACONSFIELD'S LUCK.

Bismarck's Hand Disclosed in the
Workings of the Congress
at Berlin.

How the Jew Bankers Revenged
Themselves for Insults to Their
Race.

Correspondence New York Graphic.
LONDON, July 6.—All hail, Beaconsfield!

He is the hero of the hour. He is looked upon by all loyal Englishmen as the pivot on which has turned all the deliberations of the Berlin Congress. But is this the correct view?

Not at all. England's triumphs at Berlin are simply incidents in the 'streak of luck' which has marked the career of this great political adventurer.

I am enabled to furnish the *Graphic* with the first true account of the recent moves on the chess-board of European politics.

The result of the Congress may be briefly stated as the complete humiliation of Russia. True, she receives Batoum, with conditions that render the concession practically valueless. True, she regains her little strip of Bessarabia that had been given to Roumania, and she is permitted to retain Kars. But it is her rivals who have secured the material advantages at the Congress, and, worse than all, it is England, her special rival, who has been made the chief recipient of the fruits of Russia's expenditure of blood and treasure.

It is now certain—it will be published in the journals and confirmed in Parliament ere this letter is 1,000 miles on its way to you—that England is to have Cyprus as her own, and is to acquire a protectorate of the whole of Asiatic Turkey, with practically illimitable possibilities of the extension of trade in the Levant and down the Valley of the Euphrates. Egypt is virtually hers; the Suez Canal is absolutely in her control.

Russia has acquired neither facilities for the extension of her trade nor territory; and she has lost all the prestige acquired by the war.

What does this mean?

The answer to this question involves three names—Rothschild, Bismarck, Andrassy.

First, as to Rothschild. The sympathy of the Hebrews all over the world has been with Turkey and against Russia. Russia, in the nineteenth century, has oppressed and persecuted the Jews with the most bitter and malignant cruelty. The hatred of the Greek Church for the Jews to-day is as intense as was that of some of the bigoted Catholics in the Middle Ages for that long suffering and persecuted race. The success of the Russian arms against Turkey filled the Jews with indignation and alarm. The Turks in their rule in Europe and in Asia have been tolerant alike to Christian and to Jew; it may be said they have been forced to award this tolerance; but it was not in violation of their faith nor of the will of their great Prophet, for to this day there exists the authenticated manuscript of the famous decree of Mohammed, in which he commands the faithful to abstain from persecuting and to treat charity and kindness the Jews and Christians dwelling under their rule. But, against the personal wishes of the Czar, the blind and bitter hatred of the Russians for the Jews continually manifests itself, and their persecution of the chosen people has never ceased.

Russia was forced to make great pecuniary sacrifices to keep her armies in the field; she taxed her monetary resources to the utmost; and when the San Stefano treaty had been negotiated and the question of war or peace hung trembling in the balance, she found to her dismay that if she ventured upon a war with England she must reckon with a potent foe, of whose existence she had hitherto been disdainful, if not ignorant.

This foe was the most powerful element in Continental Europe.

All bankers are not Jews. But the Hebrew element among the money-lenders and money-masters of Europe is so widespread and so powerful that it was easy for it to effect combinations by which Russia was shut out from the privilege of borrowing money to continue to renew her march of conquest.

She tried to borrow in England—no money! She sought to effect a loan in Paris—no money! She intrigued through her most skillful agents in all the minor Bourses of Europe—not a rouble could she obtain. And now, as you will probably learn in a few days, she is in such desperate financial straits that, as a last resort, she is about to call upon her patriotic subjects—if she has any—to put their hands in their pockets and lend her their own money,—if they have any, which is doubtful.

Yes! In the very hour of Russia's military triumph, when, flushed with her dearly-bought victories, and with the Sultan willing to prostrate himself as a vassal at her feet, the despised and persecuted Israelite was able to say to the Czar: 'Thus far and no farther!'

It was not England who forced Russia to appear before the Berlin Congress, and submit to a revision of her extorted treaty with Turkey.

Russia was forced into this humiliation by the Jew bankers of the world.

Once in the Congress, Gortschakoff and Schouvaloff found to their dismay and horror that they were contending single-handed against all Europe.

Bismarck proved to be the arch enemy of Russia in the Congress, the master-spirit who formed the combination to humiliate her by the Treaty of Berlin after her victories more than she had been humiliated by the Treaty of Paris after her defeats.

Now for a State secret, hinted at in various ways, but which has never come to light in any official form, and the details of which cannot be fully known until after Kaiser William and Prince Bismarck are dead.

Bismarck, with true statesmanlike prescience, detests Russia. Russia is a military power of incalculable possibilities, capable, perhaps, in time, of overrunning and conquering all Europe. A war that would increase the military prestige or augment the territorial domain of Russia, Bismarck regarded with alarm and indignation.

Why, then, did he not put an end to the Russian and Turkish war?

The answer is—Kaiser William.

The German Emperor is swayed by his personal affections and his dynastic prejudices. The old gentleman never had much political sense. He supposed his personal honor was pledged to Russia. The Czar had not interfered with Prussia in her wars with Austria and France. He, then, should not interfere in Russia's contest with Turkey. Bismarck had been quite willing to have an amicable understanding with Russia as regarded Austria and France; but he had no intention of permitting Russia to gain a military and territorial predominance that might overshadow Germany.

Thus it was Bismarck who formed the combination that robbed Russia of the fruits of her great victories.

How did he effect this? Here comes in the third name—Andrassy.

The Prime Minister of Hungary, be it remembered, is a Hungarian statesman. Blood with him, also, is thicker than water. He remembers that, when Hungary had German-Austria at her feet in 1848, Russia sent 60,000 troops to the aid of Austria, turned the tide of victory, and crushed out forever the hopes of Hungary for independent neutrality. The hated Slav was thus used to overcome the legitimate and patriotic aspirations of Hungary.

I state upon the best authority that, in the conferences held in the beginning of the late war by Bismarck and Andrassy, the scheme was concocted which culminated in the yet unsigned Treaty of Berlin. It was in these conferences determined that Russia should be despoiled of the fruits of her victories. One of the results is seen in the virtual annexation of Bosnia and Herzegovina by Austria, and the great strengthening of that Power thereby.

Here, then, is the key to the mysteries of the Congress of Berlin. Rothschild, the representative of the Jews, closing the Bourses Europe against Russia; Bismarck, intent on the purpose of curbing and manacling the giant of the North in the interests of Western civilization; Andrassy paving off Russia for the injuries inflicted on Hungary in 1848, and turning her victories into Dead Sea fruit,—pleasant to the sight, but turning to ashes upon the lips.

But how about Disraeli—Beaconsfield? Is he not the real hero of this great dama? Not at all.

True, again, blood with him is thicker than water; and undoubtedly he placed himself in relation with the Jewish money-kings to effect the humiliation of Russia. True, he withdrew the timid and hesitating Lord Derby at the right moment, and put the courageous Marquis of Salisbury in his place. But the cession of Cyprus to England, and investing her with protectorate of Asiatic Turkey, was really the work of Bismarck.

Cyprus should have been given to France. The trade of the Levant properly belongs to her and to Italy more than to England. But Bismarck, in view of the prejudices of his own people,—not that he shares these prejudices, for he is a true statesman, but merely out of deference to these narrow hatreds and dislikes,—was compelled to permit England to take what really belongs to France, and by doing this he has crowned with a new chaplet the brow of that strange personage, the novelist and the political adventurer who is now Premier of England, who will certainly become a Duke, and who is possibly destined—as gossip will have it—to still further honor, to wear the Royal robes of Prince Consort and to occupy the long vacant bed of 'Albert the Good.'"

Despite their public protests of the atrocities Eastern Jews committed against Russian Christians, Western Jewish leaders believed that they had a duty to perpetuate Bolshevism in Russia and with it the mass murder of Russian Christians, lest the freed Russian Gentiles take revenge on the Jews—Jews who had mass murdered their people.[188] A similar debate took place earlier when the

---

188. "Hope Strong Man Will Rule Russia", *The New York Times*, (9 November 1917), pp. 1-2. *See also:* "Jews Against Bolsheviki", *The New York Times* reported on (19 November 1917), p. 2. *See also:* C. Weizmann, *The Letters and Papers of Chaim Weizmann*, Volume 1, Series B, August 1898-July 1931, Transaction Books, Rutgers University, (1983), pp. 241-242. *See also:* "Bolshevism and the Jews", *The Jewish Chronicle*, (28 March 1919), p. 11. *See also:* X, "Flight from Bolshevism", *The London Times*, (14 October 1919), p. 14; **and** "The Horrors of Bolshevism", *The London Times*, (14 November 1919), pp. 13-14. *See also:* I. Cohen, "Jews and Bolshevism", *The London Times*, (21 November 1919), p. 8; **and** "Jews and Bolshevism", *The London Times*, (25 November 1919), p. 8; **and** "Jews and Bolshevism: The Mosaic Law in Politics: Racial Temperament", *The London Times*, (27 November 1919), p. 15; **and** "Jews and

Jewish Young Turks mass murdered Christians. Well intentioned persons in the West pleaded with Western Jews to repudiate the actions of the Jews and crypto-Jews who were behind the Young Turks.[189] That element of Jewish leadership which received the most attention in the press was consistent only in its public dishonesty. More sensible Jewish leaders were often largely ignored by the press, or, when they could no longer be ignored, ridiculed.

In addition to the pure blood lust Jewish bankers had expressed for centuries—the blood lust of Judaism itself—those Jewish leaders who brought about the Russian Revolution must also have concluded that it would be to their advantage to weaken Russian society and culture, so as to minimize any retaliatory actions taken against Jews at some future date. They had their agents pillage the land and execute its best citizens, which, in addition to minimizing any risk of any backlash against Jews, fulfilled the Jewish prophecies that Jews

---

Bolshevism: A Further Rejoinder", *The London Times*, (1 December 1919), p. 10. *See also:* Philojudaeus, "Jews and Bolshevism: The Group Round Lenin", *The London Times*, (22 November 1919), p. 8. *See also:* Janus, "Jews and Bolshevism: Revolutionary Elements", *The London Times*, (26 November 1919), p. 8. *See also:* Judaeus, *The London Times*, (26 November 1919), p. 8; **and** "Jews and Bolshevism: A Reply to 'Verax.'", *The London Times*, (28 November 1919), p. 8. Verax, "Jews and Bolshevism: The Mosaic Law in Politics: Racial Temperament", *The London Times*, (27 November 1919), p. 15; **and** "Bolshevism and the Jews: A Larger Issue: The Danger in Russia", *The London Times*, (2 December 1919), p. 10. *See also:* J. H. Hertz, Chief Rabbi, "Jews and Bolshevism: The Chief Rabbi's Reply", *The London Times*, (29 November 1919), p. 8. *See also:* Pro-Denikin, "A Witness from Russia", *The London Times*, (29 November 1919), p. 8. *See also:* An English-Born Jew, *The London Times*, (1 December 1919), p. 10. *See also:* Ivan Ivanovich, "The Jews and Bolshevism", *The London Times*, (6 December 1919), p. 10. **"Epatism" defined in *The London Times:*** "Epatism", *The London Times*, (10 December 1919), p. 15. *See also:* I. Zangwill, "Is Political Zionism Dead?", *The Nation*, Volume 118, Number 3062, (12 March 1924), pp. 276-278, at 276. **189**. "Jews and the Situation in Albanian", *The London Times*, (11 July 1911), p. 5. *See also:* M. Gaster, "Jews and the Situation in Albanian", *The London Times*, (12 July 1911), p. 5. *See also:* M. A. Syriotis, "The Jews and the Young Turks", *The London Times*, (19 July 1911), p. 5. *See also:* "Jews and the Situation in Albanian", *The London Times*, (27 July 1911), p. 5. *See also:* M. Gaster, "Jews and the Situation in Albanian", *The London Times*, (1 August 1911), p. 11. *See also:* G. F. Abbott, "To the Editor of the Times", *The London Times*, (1 August 1911), p. 11. *See also:* H. C. Woods, "The Adana Massacres", *The London Times*, (3 August 1911), p. 4. *See also:* The Israelite Community of Salonika, "Jews in Turkey", *The London Times*, (4 August 1911), p. 11. *See also:* "Jews and the Situation in Albanian", *The London Times*, (8 August 1911), p. 5. *See also:* "Jews and the Situation in Albanian", *The London Times*, (9 August 1911), p. 3. *See also:* M. Gaster, "To the Editor of the Times", *The London Times*, (9 August 1911), p. 3. *See also:* "The Jews and the Young Turks", *The London Times*, (9 August 1911), p. 9. *See also:* H. C. Woods, "The Jews and the Situation in Albanian", *The London Times*, (11 August 1911), p. 3. *See also:* S. Schiff, "To the Editor of the Times", *The London Times*, (11 August 1911), p. 3. *See also:* A Citizen of London, "To the Editor of the Times", *The London Times*, (11 August 1911), p. 3. *See also:* "The Jews and Albania", *The London Times*, (19 August 1911), p. 3. *See also:* M. A. Syriotis, "The Jews and Young Turks", *The London Times*, (25 August 1911), p. 3.

should destroy other nations and take their wealth, then rule the world, a world which would suffer only supplicant and stupid Gentiles to survive.

When this cultureless Soviet society led to better relations between Jews and Gentiles and to the assimilation of Jews into Gentile Soviet society, Zionist leaders feared that the Jews were losing their unique identity. These Jewish leaders once again promoted anti-Semitism to prevent the assimilation of Jews into Soviet society. They also advocated the segregation of Jews. Jewish leadership intentionally caused great harm and prolonged suffering to both Russian Gentiles and Russian Jews, as will be shown later in this text—their deliberate mass murder and general inhumanity is truly shocking.

It bears repeating that on 19 June 1920, John Clayton published an article in *The Chicago Tribune* on the front page, which alleged that an international Jewish organization sought Jewish supremacy over the world, largely through the destruction of the British Empire,

"TROTZKY LEADS RADICA L CREW TO WORLD RULE

Bolshevism Only a
Tool for His Scheme
BY JOHN CLAYTON.
(Chicago Tribune Foreign News Service.)
(By Special Cable.)
(Copyright: 1920: By the Tribune Company.)

PARIS, June 18.—For the last two years army intelligence officers, members of the various secret service organizations of the entente, have been bringing in reports of a world revolutionary movement other than Bolshevism. At first these reports confused the two, but latterly the lines they have taken have begun to be more and more clear.

Bolshevism aims for the overthrow of existing society and the establishment of an international brotherhood of men who work with their hands as rulers of the world. The second movement aims for the establishment of a new racial domination of the world. So far as the British, French and our own department's inquiry have been able to trace, the moving spirits in the second scheme are Jewish radicals.

Use Local Hatreds.

Within the ranks of communism is a group of this party, but it does not stop there. To its leaders, communism is only an incident. They are ready to use the Islamic revolt, hatred by the central empires for England, Japan's designs on India, and commercial rivalry between America and Japan.

As any movement of world revolution must be, this is primarily anti-Anglo-Saxon. It sees its greatest task in the destruction of the British empire and the growing commercial power of America. The brains of this organization are in Berlin.

Trotzky at Head.

The directing spirit which issues the orders to all minor chiefs and finds money for the work of preparing the revolt is in the German capital. Its executive head is none other than Trotzky, for it is on the far frontiers of India, Afghanistan, and Persia that the first test of strength will come. The organization expert of the present Russian state is recognized, even among the members of his own political party, as a man of boundless ambition, and his dream of an empire of the east is like that of Napoleon.

The organization of the world Jewish-radical movement has been perfected in almost every land. In the states of England, France, Germany, Poland, Russia, and the east it has its groups. It is behind the Islamic revolt with all the propaganda skill and financial aid at its command because it hopes to control the shaping of the new eastern empire to its own ends. Sympathy with the eastern nationals probably is one of the chief causes for the victory of the pro-nationals in the bolshevik party, which threw communism solidly behind the nationalist aspirations of England's colonies.

Out to Grab Trade Routes.

The aims of the Jewish-radical party have nothing of altruism behind them beyond liberation of their own race. Except for this their aims are purely commercial. They want actual control of the rich trade routes and production centers of the east, those foundations of the British empire which always have been the cornerstone of its national supremacy.

They are striking for the same ends as Germany when she entered the war of 1914 to establish Mittel Europa and so give the Germans control of the Bagdad railway. They believe Europe is tired of conflict and that England is too weak to put down a concerted rebellion in part of her eastern possessions. Therein lies the hope of success. They are staking brains and money against an empire.

'Westward the course of empire makes its way,' but even it swings backward to the old battleground where for countless ages peoples have fought. Nations have risen and crumbled around control of eastern commerce."[190]

The man behind Joseph Stalin's genocide of the Slavs and anti-Semitism was an alleged "self-hating Jew",[191] Lazar Moiseyevich Kaganovich. Kaganovich caused the deaths of tens of millions of innocents, including many Jews. American Communists, many, if not most of whom were ethnic Jews, largely turned a blind eye to these atrocities in their attempts to sponsor the cult of personality of Joseph Stalin and bring Communism to America and the rest of the world. After the creation of the State of Israel, the Communists used anti-Semitism as a means to try to force Jews towards Israel. The Jewish Communists also tried to take over Moslem nations in the hopes that they could ruin the

---

190. *See also:* "World Mischief", *The Chicago Tribune*, (21 June 1920), p. 8.
191. S. Kahan, "Preface", *The Wolf of the Kremlin*, William Morrow and Company, Inc., New York, (1987).

Moslem religion, culture, and Moslem governments—and to create the illusion that Israel was strategically important to the United States—and to artificially make the Moslem nations enemies of the United States. Communists lured Moslems toward self-destruction by pretending to be the enemies of Zionism, though they ultimately hoped to instill Communist régimes led by Jews in the nations surrounding Israel, and thereby secure the hegemony of the Jews in the Mideast. Some believe the Saudi Royal family descends from Jews, and if the current President of Iran is not an agent of Israel, he could not be doing a better job of serving the Zionists' perceived self-interests.

Adolf Hitler used the same principles as Lasalle to make himself a dictator, to mass murder his perceived political rivals in the *SA* and to justify the *Gleichschaltung* and the *Ermächtigungsgesetz* laws in Nazi Germany, which forbade dissent of any kind. Lenin iterated his infamous doctrine of "Democratic Centralism" in 1901-1902 in his famous article "What is to be Done?",[192] which doctrine prohibited dissent, or even discussion, on issues of Party dogma (note that the title of Lenin's work, "What is to be Done?", was a repetition of revolutionary Nikolai Gavrilovich Chernyshevsky's work of 1873, *What is to be Done?*). Communist Party dogma covered all aspects of life, including science. Lenin employed this principle of "Democratic Centralism" to make himself a dictator, as did Joseph Stalin. Lenin censored the press and prohibited the publication even of revolutionary literature by such notables as Maxim Gorky, which dared to advocate democracy and freedom of thought. In 1948, Communists used terror tactics to close down the play "Thieves' Paradise" by outspoken Jewish anti-Communist Myron Fagan.[193] The Communists largely destroyed Fagan's career and his life.

The Jewish Bolshevist Leon Trotsky (born Lev Davidovitch Bronstein) tried to justify dictatorship, terrorism ("Red Terror") and murder in his book: *The Defence of Terrorism (Terrorism and Communism) a Reply to Karl Kautsky*, Labour Pub. Co. and G. Allen & Unwin, London, (1921); republished as: *Dictatorship vs. Democracy (Terrorism and Communism) a Reply to Karl Kautsky*, Workers party of America, New York City, (1922). The Jewish publicity which promoted Einstein as a sort of law-giver Moses, with whom no one could disagree because his laws supposedly came from God, was immediately criticized as the intrusion of totalitarian Bolshevism into science, by Charles Lane Poor in November of 1919.[194]

In 1843, Karl Marx reviewed Bruno Bauer's anti-Semitic works "On the Jewish Question".[195] Marx's anti-Semitic responses were published in the

---

192. V. I. Lenin, "What is to be Done?", *V. I. Lenin: Collected Works*, Volume 5, Foreign Languages Publishing House, Moscow, (1961), pp. 347-530,; **and** *What is to be Done? Burning Questions of Our Movement*, International Publishers, New York, (1969).
193. "Communists Closed Play", *The New York Times*, (6 February 1948), p.29.
194. "Jazz in Scientific World", *The New York Times*, (16 November 1919), p. X8.
195. B. Bauer, *Die Judenfrage*, Friedrich Otto, Braunschweig, 1843; English translation, H. Lederer, *The Jewish Problem*, Hebrew Union College-Jewish Institute of Religion, Cincinnati, (1958); **and** "Die Fähigkeit der heutigen Juden und Christen, frei zu werden",

Deutsch-Französische Jahrbücher in 1844 at a critical time in the struggle of Jews to obtain political freedom and equality. Karl Marx, like Bauer, denounced Jews as anti-social segregationists, who worshiped and accumulated gold, and despised art and science. Marx concluded,

> "The *social* emancipation of the Jews is the *emancipation of society from world Jewry* [or: *Judaism.*]."

> "Die gesellschaftliche Emancipation des Juden ist die Emancipation der Gesellschaft vom Judenthum."[196]

Many leading Jews desperately sought to keep Jews segregated from Gentiles and used anti-Semitism as a means to accomplish this end. Their racism stems from their religion.

Marx and his Jewish friend the racist Zionist Moses Hess were two early Socialists, who defamed Jews in order to promote themselves and their political agenda. Hess later became the founding father of a racist theory of National Socialist Zionism, which eventually morphed into the Nazi Party.[197] Marx and Hess were followed by an unbroken line of Socialist anti-Semites, that eventually perpetrated the Holocaust in a Socialist totalitarian regime led by a dictator—the NSDAP (National Socialist German Worker's Party) led by Adolf Hitler. Hitler, himself, was a former Bolshevik reputedly of Jewish descent.[198]

Judaism is absolutely intolerant of dissent or disagreement, promotes dictatorship though its Messianic myths, and promotes a rigid belief system

---

Einundzwanzig Bogen aus der Schweiz, Herausgegeben von Georg Herwegh, Zürich und Winterthur, (1843), pp. 56-71 *See also:* B. Bauer, "Die Judenfrage", *Deutsche Jahbücher für Wissenschaft und Kunst*, Numbers 274-282, (17-26 November 1842), pp. 1093-1126; **and** "Neueste Schriften über die Judenfrage", *Allgemeine Literatur-Zeitung*, Volume 1, (December, 1843), pp. 1-17; **and** "Neueste Schriften über die Judenfrage", *Allgemeine Literatur-Zeitung*, Volume 4, (March, 1844), pp. 10-19.

196. K. Marx, "Zur Judenfrage. 1) Bruno Bauer: Die Judenfrage. Braunschweig 1843. — 2) Bruno Bauer: Die Fähigkeit der heutigen Juden und Christen frei zu werden. Ein und zwanzig Bogen aus der Schweiz. Herausgegeben von Georg Herwegh. Zürich und Winterthur. 1843. S. 56-71", *Deutsch-Französische Jahrbücher*, Herausgegeben von Arnold Ruge und Karl Marx, 1ste und 2te Lieferung, Paris, (1844), pp. 182-214, at 214.

197. M. Hess, *Rom und Jerusalem: die letzte Nationalitätsfrage*, Eduard Wengler, Leipzig, (1862); English: *Rome and Jerusalem: A Study in Jewish Nationalism*, Bloch, New York, (1918).

198. H. Koehler, *Inside the Gestapo: Hitler's Shadow Over the World*, Pallas Pub. Co. Ltd., London, (1940). *See aslo:* H. Frank, *Im Angesicht des Galgens; Deutung Hitlers und seiner Zeit auf Grund eigener Erlebnisse und Erkenntnisse. Geschrieben im Nürnberger Justizgefängnis*, F. A. Beck, München-Gräfelfing, (1953), pp. 330-331. *See aslo:* D. Bronder, *Bevor Hitler kam: Eine historische Studie*, Hans Pfeiffer Verlag, Hannover, (1964), p. 204 (p. 211 in the 1974 edition). *See aslo:* H. Kardel, *Adolf Hitler, Begründer Israels*, Verlag Marva, Genf, (1974); English translation *Adolf Hitler: Founder of Israel*, Modjeskis' Society Dedicated to Preservation of Cultures, San Diego, (1997).

centered on the illusion of absolute law. Communism (and its absurd bastard child, the National Socialist German Worker's Party) was merely a temporary means of achieving the goals of Judaic Messianic myth. Those goals include the destruction of Gentile peoples, their "racial" distinctions, their independence and liberty, their religions and nations, even their very lives. This is succinctly proven in Robert H. Williams' booklet, *The Ultimate World Order—As Pictured in "The Jewish Utopia"*, CPA Book Publisher, Boring, Oregon, (1957?). Williams proves that the "New World Order" is in fact the "Jewish World Order" of Judaic prophecy. The ancient and medieval Jewish myths which call for the destruction of Gentiles will be quoted, and their implications explored, further on in this text.

Maurice Samuel wrote in his collection of contemporary Jewish clichés, which he styled *You Gentiles*,

"IF anything, you must learn (and are learning) to dislike and fear the modern 'assimilated' Jew more than you did the old Jew, for he is more dangerous to you At least the old Jew kept apart from you, easily recognizable as an individual, as the bearer of the dreaded Jewish world-idea: you were afraid of him and loathed him. But to a large extent he was insulated. But the Jew assimilates, acquires your languages, cultivates a certain intimacy, penetrates into your life, begins to handle your instruments, you are aware that his nature, once confined safely to his own life, now threatens yours. You are aware of a new and more than concerting character at work in the world you have built and are building up, a character which crosses your intentions and thwarts your personality. The Jew, whose lack of contact with your world had made him ineffective, becomes effective. The vial is uncorked, the genius is out. His enmity to your way of life was tacit before. To-day it is manifest and active. He cannot help himself: he cannot be different from himself: no more can you. It is futile to tell him: 'Hands off!' He is not his own master, but the servant of his life-will. [***] It is to this Jew that liberals among you will point to refute my thesis. And it is precisely this Jew who best illustrates its truth. The unbelieving and radical Jew is as different from the radical gentile as the orthodox Jew from the reactionary gentile. The cosmopolitanism of the radical Jew springs from his feeling (shared by the orthodox Jew) that there is no difference between gentile and gentile. You are all pretty much alike: then why this fussing and fretting and fighting? The Jew is *not* a cosmopolitan in your sense. He is not one who feels keenly the difference between national and nation, and overrides it. For him, as for the orthodox Jew, a single temper runs through all of you, whatever your national divisions. The radical Jew (like the orthodox Jew) is a cosmopolitan in a sense which must be irritating to you: for he does not even understand why you make such a fuss about that most obvious of facts—that you are all alike. The Jew is altogether too much of a cosmopolitan—even for your internationalists. [***] Philosophies do not remold natures. What your radicals want is another form of the Game, with other rules. Their discontent joins hands with Jewish discontent. But it is not the same kind of discontent. A little distance down the road the ways part for ever. The Jewish radical will turn from your social movement: he will discover his mistake. He will discover that nothing can bridge the gulf between you and us. He will discover that the spiritual satisfaction which he thought he would find in social revolution is not to be purchased from you. I believe the movement has already started, the gradual secession of the Jewish radicals, their realization that your radicalism is of the same essential stuff as your conservatism. The disillusionment has set in. A century of partial tolerance gave us Jews access to your world. In that period the great attempt was made, by advance guards of reconciliation, to bring our two worlds together. It was a century of failure. Our Jewish radicals are beginning to understand it dimly. We Jews, we, the destroyers, will remain the destroyers for ever. *Nothing* that you will do will meet our needs and demands. We will for ever destroy because we need a world of our own, a God-world, which it is not in your nature to build. Beyond all temporary alliances with this or that faction lies the ultimate split in nature and destiny, the enmity between the Game and God. But those of us who fail to understand that truth will always be found in alliance with your rebellious factions, until disillusionment comes. The wretched fate which scattered us through your midst has thrust this unwelcome role upon us. [***] You are bound to find 'spiritual value' in science because you do not want ultimate spiritual value—only the spiritual value of immediate lyric enjoyment. You who worship gods instead of God must naturally worship science. Science is merely idol-worship: for eikons instruments, for incantations formulæ: the palpable, the material, the enjoyable. Science is not a serious pursuit: your grave professors of chemistry, astronomy, physics, your Nobel prize winners are but bald or bearded schoolboys playing mental football for their own delight and the delight of spectators. Science, then, is an art, though its technique is of so peculiar a nature as to divide it from all the other arts: but we most easily recognize it as an art because the true scientist takes an artistic delight in science. And because your science is not serious, we Jews have never achieved in it any peculiar preëminence. We have our few exceptions: we can master as well as you the system and the scheme, but we lack the spiritual urge, the driving joy, the illusion that this is the all in all. We know nothing of science for science's sake—as we know nothing of art for art's sake. We only know of art for God's sake. If there is art or beauty in our supreme production, the Bible, it is not because we sought either. The type of the artist is alien to us, and just as alien is the delight of the artist. The artist is one who seeks beauty, goes out of his way to find her. But the Hebrew prophet, who wrought so beautifully, did not go out of his way to find God. God pursued him and caught him; hunted him out and tortured him so that he cried out. Until this day we have no artists in your sense: such art as we have created has been the byproduct of a fierce moral purpose. Art and science—this is your gentile world, a lovely and ingenious world. Kaleidoscopic, graceful, bewilderingly seductive, a world, at its best, of lovely apparitions, banners, struggles, triumphs, gallantries, noble gestures and conventions. But not our world, not for us Jews. For such Field-of-the-Cloth-of-Gold delights we lack imagination and inventiveness. We are not touched with this vigor of productive playfulness. Under duress we take part in the ringing mêlée, and give an indifferently good account of ourselves. But we have not the heart for this world of yours."[199]

Note that Samuel repeats the ancient accusation that Jews lack imagination for the arts and sciences, and that art and science are irreligious. The enduring existence of this theory is one reason why Jews so vigorously hyped Albert Einstein as if he were a great scientist. They hoped to add a "Jewish Newton" to the list of greats who have revolutionized science, because no Jew had yet made a breakthrough discovery on the level of a Copernicus, Galileo or a Newton; and Jews were roundly criticized, by Jew and Gentile alike, as if parasites instead of contributors. It terribly irked the Jews that they had not produced a Galileo, a Mozart, nor a Rembrandt. What they could not accomplish in fact, some Jews accomplished through plagiarism and hype. Other Jews justified their insecurities with the sour grapes of their religious beliefs. They asserted that the Jews were the chosen people of God—chosen to obey supreme law, not to artistically create new laws and images.

Note further Samuel's subtle argument that Jewish segregation is better for Gentiles than Jewish assimilation, because assimilated Jews become radicals and revolutionaries who will ultimately fulfill the "Jewish mission" to destroy Gentile nations, cultures, religions and peoples; and will Judaize the world. This was part of an ongoing Zionist campaign against Gentile nations and assimilatory Jews, which employed the carrot and the stick method of persuading Gentiles to segregate Jews and prevent Jewish assimilation. Racist Jews loathed assimilation and told Gentiles that they had to chose between a segregated "Jewish State", or a subjugated world under Jewish tyranny. This will be discussed in detail further on in this text in section "7.6 The Carrot and the Stick". These Jewish propagandists failed to mention that the formation of a Jewish State heralded the extermination of the Gentiles in Jewish Messianic prophecy.

3.4.2 The Messiah Myth

Jewish leaders have, for thousands of years, corrupted international politics and culture in order to fulfill their Messianic prophecies of Jewish world domination. The Rothschilds and other Jewish financiers have used their great wealth to destroy nations and religions through wars, Communism, and control of the mass media and government. Jewish financiers brought about the calamitous events of the Twentieth Century, the mass murder of tens, if not hundreds, of millions

---

199. M. Samuel, *You Gentiles*, Harcourt, Brace & Co., New York, (1924), pp. 144-145, 150-151, 154-155, 174-176.

of human beings, in order to: force assimilating Jews back to the racist segregationist prophecies of Judaism; to force the establishment of a Jewish State which will eventually extend from the Nile to the Euphrates; to force the destruction of all other nations and their peoples, who will be killed off or enslaved and ruled by Jews; to force the destruction of all other religions; to force the destruction of the Dome of the Rock and Al Aqsa Mosque to be replaced with a Jewish Temple; and such petty and spiteful acts which fulfill prophecy as the destruction of the orchards and farms of the Palestinians, etc. Both the "Proclamation of Independence"[200] of the racist "Jewish State" and the "Law of Return 5710-1950"[201] are segregationist instruments which assert the same racist doctrines of *"Blut und Boden"* as Nazism.

On 28 December 1960, racist Zionist David Ben-Gurion, who was the first Prime Minister of the undemocratic and racist "Jewish" State of Israel, revealed that the allegedly *political* motivations of the Zionists, were in fact *religious*; and that, though the declaration of independence of Israel claimed that the state was founded as a result of the Holocaust, the formation of the state was in fact the fulfilment of an ancient religious Messianic plan of the Jews to rule the world, which the "Jewish People" had themselves fulfilled because God had failed to give them the promised Messiah. Note that racist Zionist Jews deliberately caused both World Wars and the Holocaust in order fulfill the "apocalyptic goals" of their genocidal religious mythologies, as will be proven throughout much of this text. Note also that Ben-Gurion's Hitler-like cry for Jews to tribalistically unite in blind loyalty to one another and to segregate, or face extinction through assimilation. This warning should be heeded by American and Russian Jews, for they will face the same fate at the hands of racist Zionist Jews in the coming Third World War, as the assimilatory Jews of Europe faced in the Second World War. Racist Zionist Jews directed the exact same threats at the Jews of Europe from the 1880's through the 1930's, and then they put Adolf Hitler into power in order to herd up the Jews of Europe and march them out—or into their graves. Note still further the fanatical arrogance of racist, religious Jews, who believe that they have the sole God-given right to govern the fate of humanity and determine the religion and "redemption" of others. According to racist Jews and their Messianic mythologies, all laws worldwide must emanate from Jerusalem, and no individual has the right of free choice and no nation the right of self-determination (*Exodus* 34:11-17. *Psalm* 2; 72. *Isaiah* 2:1-4; 9:6-7; 11:4, 9-10; 42:1; 61:6. *Jeremiah* 3:17. *Joel* 3:16-17. *Micah* 4:2-3. *Zechariah* 8:20-23; 14:9). Judaism differs from Christianity, in that Jews believe that their Heaven is on Earth and that their rewards are found on Earth. If evil actions bring them earthly success, then they believe that God will judge those actions as good. Racist Zionist Jews believe it is righteous to fulfill God's plan by human political action. They are not concerned by judgements in an afterlife, nor do they aspire to attain rewards in Heaven. They want everything here and now, and view

---

200. Provisional Government of Israel, "The Declaration of the Establishment of the State of Israel", *Official Gazette*, Number 1, Tel Aviv, (14 May 1948), p. 1.
201. *Sefer Ha-Chukkim*, Number 51, (5 July 1950), p. 159.

immortality not as an individual achievement, but as the survival of the "Jewish People". Ben-Gurion stated,

> "But through all these changes there was a continuity, a basic nucleus that did not change, and this nucleus is the Messianic vision of redemption, the vision of redemption for the Jewish nation and for all mankind.
>
> This vision is also intimately intertwined with our ancient homeland and our cultural heritage, and it has close and organic bonds with the apocalyptic goals: the goal of international peace and human fraternity cherished by the prophets of Israel and the best men of all nations.
>
> The Jewish faith and the Messianic hope enabled the Jews to overcome the sufferings, restrictions and humiliations that they underwent in most countries and in most generations. Their ability withstand external pressure, undismayed by tortures and persecution, were examples of great moral heroism, but this was only a passive heroism. This was an inner heroism, accompanied by a submission to fate and a feeling of helplessness and impotence in practice. The salvation which they expected and desired was to be brought about by supernatural forces from above.
>
> The emancipation, the Haskalah and the revolutionary developments in the nineteenth century; the movements for national liberation and unity that arose among the enslaved and divided peoples of Europe (Italy, Germany, Poland, the Balkan States), the awakening of the working class to struggle for a new social regime; the mass migration from Europe to countries across the seas; the new Hebrew literature which inspired the Hebrew reader with the spirit of the Bible in its early glory—all these gave a new direction to the aspiration for redemption, a natural, active, deliberate and planned direction.
>
> ### Active Faith in Ability
>
> There awoke the active faith in the ability and power of the Jew to change his fate with his own hands, and to advance his redemption through natural means. This faith became the common property of the best sons of the people, both among the religious (like Rabbi Alkalai, Rabbi Kalisher, etc.) and among the non-religious. And from the deepest wellsprings of the people there arose the latent but powerful will, the pioneering will, which is not discouraged by difficulties, obstacles and dangers from fulfilling its historical mission. [***] I regard the unity of the Jewish people as a primary condition for its survival—and the survival of Israel as well—and as I have said elsewhere, I am a Jew first, and an Israeli afterwards. [***] In our Proclamation of Independence, we declared that 'the State of Israel will be open for Jewish immigration and the ingathering of the exiles,' and in 1950 we enacted in the Knesset the Law of the Return, which is one of our basic laws, characteristic of the mission and the unique character of the Jewish State that we have established.
>
> This law lays down the national principle through which and for which the state was established, namely that it is a natural and historic right of every Jew, wherever he may dwell, to return and settle in Israel.

It is not the state that grants the Jews of the Diaspora the right of return; it is inherent in every Jew. This right preceded the revival of the State of Israel; it was this right that built the state. [***] This was the Messianic vision, the vision of national redemption and revival, which in the last seventy years was given the name of Zionism but was real and live before the term was coined, and it lived in the hearts of thousands and tens of thousands of Jews who settled in Israel after it was coined, but never described themselves as 'Zionists,' and the term has remained strange to them to this day. [***] On the other hand, the Messianic vision of redemption for the Jewish people and all mankind is not something that has been created by European Jewry in recent times; it is the soul of prophetic Jewry, in all its forms and metamorphoses until this day, and it is the secret of the open and hidden devotion of world Jewry to the State of Israel.

While before the rise of the state, the Messianic vision was reinforced by the pressure of Jewish distress in the Diaspora, in our days it is strengthened by the attractive force of the state itself, as it is today and as it ought to be, namely by the reality of the state and by its historic mission in the realization of the Messianic vision.

This vision is not the outcome of any local or temporary conditions; it was created by the prophetic concept of the universe, the destiny of man on earth and the millennial era. It does not recognize idols of gold and silver; it does not accept the robbery of the poor, the oppression of peoples, the lifting up of swords by nation against nation or the study of war; it foretells the coming of the Redeemer whose loins are girt with righteousness; it looks forward to the day when the nations will cease to do evil.

2 Forms of Redemption

This Messianic vision depends on the redemption of Israel, which will assume two forms: The ingathering of the exiles and the creation of a model nation, as Isaiah, the son Amotz, prophesied:

'Fear not, for I have redeemed thee. From the East I will bring thy seed and from the West I will gather thee. I will say to the North: Give, and to the South: Hold not back, bring my sons from far and my daughters from the end of the earth' (43:5-6). And he also said: 'And I will hold thee by the hand, and I will form thee, and I will make thee a covenant of the people, a light to the nations' (42:6).

These are no empty figures of speech—in our own day we are seeing the first signs of their realization. [***] This really the most important aspect of the picture, for our very survival—which involves the survival of Jewry in the world—depends on it. [***] [T]he Judaism of the Jews of the United States and similar countries is losing all meaning, and only a blind man can fail to see the danger of extinction, which is spreading without being noticed. [***] A large part of the laws cannot be observed in the Diaspora, and since the day when the Jewish state was established and the gates of Israel were flung open to every Jew who wanted to come, every religious Jew has daily violated the precepts of Judaism and the Torah of Israel by remaining in the Diaspora. Whoever dwells outside the land of Israel is considered to have no God, the sages said.

Every Jew who is concerned for the future of the Jewish people, and who holds the name of Jew dear above every other, must realize that without Jewish education for the younger generation, to imbue him with a more profound Jewish consciousness and deepen his roots in Israel's history and the unity of the people, Jewry in the Diaspora is on the road to assimilation and extinction.

Those who are devoted to Judaism must see the dagger facing Diaspora Jewry courageously and with open eyes. In several totalitarian and Moslem countries, Judaism is in danger of death by strangulation; in the free and prosperous countries it faces the kiss of death, a slow and imperceptible decline into the abyss of assimilation."[202]

Ben-Gurion, *de facto* "King of the Jews", or Messiah, wrote in his *Memoirs*,

"Jews are activists, that is they have a Messianic spirit. They are not missionaries since they don't seek to convert others to their ways. But they are merciless with themselves. The Bible has imparted to them that divine discontent leading at its best to initiatives such as the pioneering life, at its worst to persecution by their fellow men. It has never allowed them as a people to enjoy for long comfortable mediocrity. Certainly in Israel today we are Messianic. The Jews feel themselves to have a mission here; they have a sense of mission. Restoration of sovereignty is tied to a concept of redemption. This had determined Jewish survival and it is the core of Jewish religious, moral and national consciousness. It explains the immigration to Israel of hundreds of thousands of Jews who never heard of Zionist doctrine but who, nevertheless, were moved to leave the lands wherein they dwelt to contribute with their own effort to the revival of the Hebrew nation in its historic home. [***] The Jewish people are not easily overwhelmed. They have their Messianic tradition which binds them together and gives their existence purpose. More than one sea of eastern or western culture has attempted to swallow them up but never has succeeded. They have influenced the world far more than the world has influenced them. Israel is far better equipped to resist cultural extinction than were the Jewish exiles during two thousand years. Our evident role here is to give new life to all that is meant by the 'Covenant' of the Jewish people whereby they remain one. That is hardly a role leading to 'drowning' in alien cultures. On the contrary, it represents a revival of our own cultural activity."[203]

It is interesting to note that Adolf Hitler fit in very well with Jewish apocalyptic mythology, especially the prophecies recorded in the *Sefer*

---

[202]. "Text of Ben-Gurion's Address Before the World Zionist Congress in Jerusalem", *The New York Times*, (8 January 1961), pp. 52-53.

[203]. D. Ben-Gurion, *Memoirs*, The World Publishing Company, New York, Cleveland, (1970), pp. 122, 162.

Zerubbabel (*Book of Zerubbabel*), *The Wars of King Messiah* and the writings of Rabbi Simon Ben Yohai. These predicted that an evil pseudo-Messiah named Armilus would emerge as a child born of a statue in Rome, and of Satan. Though this prophecy was probably meant to ridicule Jesus, a contemporary of Hitler who sought to convince himself and others that prophecies were being fulfilled could have argued in retrospect that the birth of Armilus represented the rise of Adolf Hitler as the product of Mussolini's fascism. This monster of Jewish lore would gain power through his charisma and attempt to conquer the world and lead people to believe that he is the Biblical Messiah destined to lead a thousand-year Empire, the Messianic Era—one might say in this context: *Ein tausendjähriges Reich*. Adolf Hitler's crypto-Jewish propagandists did in fact promote Hitler to the German People as if he were the Messiah, who would lead Germany through a period of tribulations into the 1,000 year Messianic Era (*Revelation* 20:1-7), the thousand-year German Empire.

The *Encyclopaedia Judaica* writes in its article "Zerubbabel, Book of",

"The victory of the Messiah and his mother over Armilus represents that of Judaism over the Roman Empire and the Christian Church."[204]

This victory heralds the "restoration" of the Jews to Palestine and the enslavement, then extermination of the Gentiles after "the times of the Gentiles" has expired (*Luke* 21:24. See also: *Matthew* 24. *Romans* 9; 11).

According to the Jewish prophecies, the Jews would oppose the pseudo-Messiah, and he would be defeated by Messiah Son of Joseph, and then the Jews would be restored to Palestine—as happened in the case of Hitler and Joseph Stalin, though by human design, Jewish design. The name "Stalin" is a pseudonym. Joseph "Stalin" was born Joseph Djugashvili. "Stalin" means "steel" in Russian. He was said to rule with an iron fist, one might even say, with an iron scepter (*Numbers* 24:17-20. *Psalm* 2:9). While the names are coincidental legacies, they may have been seen and exploited as fortuitous by Cabalistic Jews, who tend to be highly superstitious, and who practice such occult beliefs as numerology.

In any event, it is a fact that Joseph Stalin's government, like that of Adolf Hitler, was rotten with genocidal Jews and crypto-Jews, who committed genocide against the Slavs, Georgians, Germans, and other peoples under their control. They insisted upon the segregation of the Jews at all costs, including the mass murder of Jews, terrorism against Jews committed by Jews, who disguised themselves as non-Jews, and who blamed non-Jews for the atrocities they themselves committed so as to artificially cause enmity between Jews and the rest of the world. They sought the diminution of the genetic stock of other peoples, and the improvement of the genetic stock of the Jews through vicious natural and artificial selection, and perhaps sought the injection of fresh blood

---

204. "Zerubbabel, Book of", *Encyclopaedia Judaica*, Volume 16 Ur-Z, Macmillan, Jerusalem, (1971), col. 1002.

into the "tribe" from kidnaped children after the war.

They sought a world government led by Jews, that would blend other "races" into one amorphous whole, without a unique heritage, and without a religion, in keeping with Jewish Messianic myth. While racist Jews commonly blame Jewish segregation on non-Jews, it has commonly been the case that the Jews themselves have sought to segregate from the non-Jews. It was the Jews who created the segregated Ghettoes of Poland before the Nazis rose to power, as Adolf Eichmann and others have noted.[205] Intrinsic Jewish racism even caused the Jews to segregate among Jews, with the Sephardim refusing to integrate with the Askenazim, and with each forming racist subgroups. In 1845, *The North American Review* wrote, and note that the Jews were very much involved in slavery, the secession of the Confederacy which began in South Carolina, and the KKK,

> "The first great fact which strikes the observer of this people, in their present state, is their dispersion throughout the world, while they are still a separate race, excepting where, at the confines of their channel, they mingle enough with the surrounding waters to manifest that tendency to amalgamation, which characterizes all human kind, and in them is overborne only by some mysterious power opposing the diffusive force of the natural current. The narrative of their dispersion is necessarily involved at many points in great obscurity, which Jewish superstition and fondness for traditionary lore have served in no small degree to thicken. The agricultural life of the early Hebrews, as well as all the Mosaic institutions, opposed their mingling freely with other nations [***] The first who settled in the United States are said to have been Spaniards and Portuguese, who fled from the inquisition to the Dutch colony of New Amsterdam. To South Carolina the Jews came long before the Revolution, being German, English, and Portuguese emigrants; and they are now more numerous there than in any other Southern State. To Georgia a few came over in 1733, soon after General Oglethorpe. In Virginia we find them before the year 1780. The Jews of this country are as mixed a people as those among whom they dwell, and much less disposed than the latter to forget petty differences, real or imaginary, in family or caste, among themselves; and therefore not so rapidly assuming a

---

205. P. S. Mowrer, "The Assimilation of Israel", *The Atlantic Monthly*, Volume 128, Number 1, (July, 1921), pp. 101-110, at 104. *See also:* B. J. Hendrick, "The Jews in America: III The 'Menace' of the Polish Jew", *The World's Work*, Volume 44, Number 4, (February, 1923), pp. 366-377, at 368; **and** "Radicalism among the Polish Jews", *The World's Work*, Volume 44, Number 6, (April, 1923), pp. 591-601, at 593. *See also:* J. Drohojowski, *Brief Outline of the Jewish Problem in Poland*, Polish National Alliance of Brooklyn, U.S.A. (Zjednoczenie Polsko Narodowe), Brooklyn, New York, (1937), p. 22. *See also:* A. Eichmann, "Eichmann Tells His Own Damning Story", *Life Magazine*, Volume 49, Number 22, (28 November 1960), pp. 19-25, 101-112; at 106; *see also:* "Eichmann's Own Story: Part II", *Life Magazine*, (5 December 1960), pp. 146-161.

homogeneous aspect."[206]

The Hitler and Stalin régimes, as do the American régime, and the emerging Chinese régime, fit the mythological prophecies of *Daniel* 7, which religious Jews employ as a political guide, and which state, *inter alia*,

> "3 And four great beasts came up from the sea, diverse one from another. 4 The first *was* like a lion, and had eagle's wings: I beheld till the wings thereof were plucked, and it was lifted up from the earth, and made stand upon the feet as a man, and a man's heart was given to it. 5 And behold another beast, a second, like to a bear, and it raised up itself on one side, and *it had* three ribs in the mouth of it between the teeth of it: and they said thus unto it, Arise, devour much flesh. 6 After this I beheld, and lo another, like a leopard, which had upon the back of it four wings of a fowl; the beast had also four heads; and dominion *was* given to it. 7 After this I saw in the night visions, and behold a fourth beast, dreadful and terrible, and strong exceedingly; and it had great iron teeth: it devoured and brake in pieces, and stamped the residue with the feet of it: and it *was* diverse from all the beasts that *were* before it; and it had ten horns."

The myth of Zerrubbabel is noteworthy today for another reason. It calls on the Jews to use a Christian empire to clear the way for the Jewish Messiah. The Zionists, who have long believed that politics can play the rôle of Messiah, and the evil pseudo-Messiah the Christians call the "anti-Christ". The Zionists are currently using the United States of America to smash Islam and spread a corrupted form of Christianity, which will condition the peoples of the world to accept Jewish Messianic myth and monotheism. The Zionists are using America as the "anti-Christ" to make way for the Jewish Messiah, who will then crush America. The *Encyclopaedia Judaica* writes of the myth of Zerrubabel in its article "Messiah",

> "Only after such unity is achieved by a Christian 'messiah' can the Jewish Messiah appear and overcome the enemy."[207]

In describing another pervasive Jewish Messianic myth, the *Encyclopaedia Judaica* writes in its article "Messianic Movements",

> "[T]he Messiah is to take the crown from the head of the alien sovereign by his virtue of appearance alone and redeem and avenge the Jews by miraculous means."[208]

---

[206]. "The Modern Jews", *The North American Review*, Volume 60, Number 127, (April, 1845), pp. 329-368, at 331-332.
[207]. "Messiah", *Encyclopaedia Judaica*, Volume 11 LEK-MIL, Macmillan, Jerusalem, (1971), cols. 1407-1417, at 1413.
[208]. "Messianic Movements", *Encyclopaedia Judaica*, Volume 11 LEK-MIL,

Racists Jews are settled upon the idea that they can fool the foolish by using modern science to accomplish things their future subjects will be conditioned to believe are "miraculous". For example, the use of biological agents to kill off populations. Recall that the Zionists declared HIV/AIDS to be a scourge of God upon the homosexuals. This misuse of Science was already discussed, in a way, in the writings of Maimonides and other Jewish scholars, and was an ancient and Medieval theme taken from the story of "Atlantis" found in Plato's writings. One also wonders what smoke and mirror illusions the racist Jews will use to promote their Messiah, as if he descended from the heavens and carries with him supernatural powers.

The racist Jews would have an easy time deceiving Gentiles who are deliberately raised in ignorance. The Bolsheviks tried very hard to keep the Peoples of the Soviet Union from discovering the true nature of life in the West and Jewish organizations are now imposing Soviet style restrictions on the Peoples of the West. The American news media keeps the American People in ignorance of world events and disproportionately focuses attention on Israel and does so with an heavily pro-Israeli bias. Many of those same Americans who criticized the Soviets for submitting to such autocratic and oppressive tactics sheepishly laud those who are oppressing them today in America.

The genocidal Zionists justify their inhuman actions as manifestations of the Messianic myth of *hevlei Mashiah*, or "the birth pangs of the Messiah".[209] They believe it is alright to mass murder fellow Jews and the rest of humanity, because it will supposedly hasten the Messianic Era, in which the Jewish "remnant", or "the Elect" will enslave the rest of humanity and then exterminate it. In Biblical prophecy, the "remnant" are a minority in the Jewish community, who embrace genocidal Judaism while other Jews have abandoned it; and to Dispensationalist Christians, the "remnant" will be those Jews who convert to Christianity and rule the world from Zion, *see: Isaiah* 1:9; 6:9-13; 10:20-22; 11:11-12; 17:6; 37:31-33; 41:9; 42; 43; 44; 45:4; 59:20-21. *Ezekiel* 20:38; 25:14; 37. *Daniel* 12:1, 10. *Amos* 9:8-10. *Joel* 2:32. *Obadiah* 1:18. *Micah* 5:8. *Matthew* 24. *Romans* 9:27-28; 11:1-5, 17, 26-27.

Racist Jews have succeeded in creating the "Jewish State" through these means—through the Holocaust. To this day, the Zionists justify their genocide of the Palestinians as *hevlei Mashiah*, and ask their fellow Jews—especially those who dominate the mass media—to conceal the Jewish genocide of the Palestinians, and to call those who object to it, "anti-Semites". Preterist Christians, in contrast to Dispensationalist Christians, believe that the prophecies of the Old Testament have already been fulfilled and do not wish to make

---

Encyclopaedia Judaica, Jerusalem, The Macmillan Company, New York, (1971), cols. 1417-1427, at 1419.

209. "Messianic Movements", *Encyclopaedia Judaica*, Volume 11 LEK-MIL, Encyclopaedia Judaica, Jerusalem, The Macmillan Company, New York, (1971), cols. 1417-1427, at 1418. G. Scholem, *Kabbalah*, New American Library, New York, (1974), p. 284. Compare to: *Matthew* 24:7-8. *Mark* 13:7-8.

themselves the slaves of Jewish tyrants. Since the Jews' Messianic myth will never be fulfilled, they will forever trouble the world and justify their villainy as *hevlei Mashiah*.

David Ben-Gurion admitted in 1956 that the Jews had stolen the Palestinians' land,

> "I don't understand your optimism,' Ben Gurion declared. 'Why should the Arabs make peace? If I was an Arab leader I would never make terms with Israel. That is natural: we have taken their country. Sure, God promised it to us, but what does that matter to them? Our God is not theirs. We come from Israel, it's true, but two thousand years ago, and what is that to them? There has been antisemitism, the Nazis, Hitler, Auschwitz, but was that their fault? They only see one thing: we have come here and stolen their country. Why should they accept that? They may perhaps forget in one or two generations' time, but for the moment there is no chance. So it's simple: we have to stay strong and maintain a powerful army. Our whole policy is there. Otherwise the Arabs will wipe us out.'"[210]

When Black leader Stokely Carmichael stated essentially the same thing at a lecture in George Washington University in 1970, pro-Israel supporters jeered at him.[211] When Iranian President Mahmoud Ahmadinejad stated essentially the same thing on 14 December 2005, Zionists called him "anti-Semitic" and made his statements a *casus belli* for annihilating Iran. President Ahmadinejad stated,

> "Today, they have created a myth in the name of Holocaust and consider it to be above God, religion and the prophets, [***] If you committed this big crime, then why should the oppressed Palestinian nation pay the price? This is our proposal: If you committed the crime, then give a part of your own land in Europe, the United States, Canada or Alaska to them so that the Jews can establish their country."[212]

The Zionists have been in a quandary for over half a century on how to justify the theft of Palestine from its native population. The Zionists put the Nazis into power in order to chase the reluctant Jews of Europe into Palestine. When their efforts failed in the late 1930's, they caused the Second World War and blamed it on the Jews, so as to provoke the Germans into humiliating and murdering Jews, which indescribably painful experience the Zionists hoped would then inspire the Jews to flee to Palestine—though it did not. The Zionists then caused problems for the Jews of Hungary, Romania, Russia, Iraq, Egypt,

---

210. D. Ben-Gurion, quoted in: N. Goldmann, *The Jewish Paradox*, Grosset & Dunlap, New York, (1978), p. 99.
211. "Carmichael, in Washington, Terms Arab Struggle Just", *The New York Times*, (10 April 1970), p. 22.
212. S. Delacourt, "Iranian President Denies Holocaust, Sparks Outrage", *The Toronto Star*, (15 December 2005), p. A1.

etc. to force them to Palestine against their own wishes, with marginal success. They doubtless plan to create more problems for the Jews of America and Russia so as to increase the population of Israel.

In *The Washington Post* on 11 July 2003 on page A1, Rebecca Dana and Peter Carlson quoted excerpts from the diary of Harry "S" Truman, President of the United States of America:

> "'He'd no business, whatever to call me,' Truman wrote. 'The Jews have no sense of proportion nor do they have any judgement [sic] on world affairs. Henry brought a thousand Jews to New York on a supposedly temporary basis and they stayed.'
>
> Truman then went into a rant about Jews: 'The Jews, I find, are very, very selfish. They care not how many Estonians, Latvians, Finns, Poles, Yugoslavs or Greeks get murdered or mistreated as D[isplaced] P[ersons] as long as the Jews get special treatment. Yet when they have power, physical, financial or political neither Hitler nor Stalin has anything on them for cruelty or mistreatment to the under dog. Put an underdog on top and it makes no difference whether his name is Russian, Jewish, Negro, Management, Labor, Mormon, Baptist he goes haywire. I've found very, very few who remember their past condition when prosperity comes.'"

In his Forward to Israel Shahak's book *Jewish History, Jewish Religion: The Weight of Three Thousand Years*, Pluto Press, London, (1997/2002), pp. vi-vii, at vi, Gore Vidal wrote of Truman,

> "Sometime in the late 1950s, that world-class gossip and occasional historian, John F. Kennedy, told me how, in 1948, Harry S. Truman had been pretty much abandoned by everyone when he came to run for president. Then an American Zionist brought him two million dollars in cash, in a suitcase, aboard his whistle-stop campaign train. 'That's why our recognition of Israel was rushed through so fast.' As neither Jack nor I was an antisemite (unlike his father and my grandfather) we took this to be just another funny story about Truman and the serene corruption of American politics."

After the Second World War ended, Zionist racists like Albert Einstein callously demanded Palestine on a *quid pro quo* basis for the human sacrifice of millions of Jews, which the Zionists had wrought.[213] But where was the logic in this? If the Europeans had murdered six million Jews, as the Zionists claimed, why should the Palestinians pay with their lives and property for the crimes of the European Nazis? In typical fashion, the Zionists exhibited their infamous dishonesty and argued both sides of the same issue as opposing and mutually

---

213. A. Einstein, "Unpublished Preface to a Blackbook", *Out of My Later Years*, Philosophical Library, New York, (1950), pp. 258-259, at 259.

exclusive arguments suited their needs. David Ben-Gurion wrote in his *Memoirs* of 1970,

> "I have called the Arab attitude towards Israel irrational. Nevertheless, the Arab world has levelled several concrete accusations against us and it might be well to answer these here.
>
> They have said, for instance, that the Moslem portion of the globe is paying for Nazism in Europe, that without the holocaust we would never have come here as a mass and never have founded a State. And, complain the Arab propagandists, it isn't fair that this part of the world should pay for the persecutions carried out in Europe.
>
> I have already gone exhaustively into the reasons for our being here, reasons that I as a pioneer of 1906 can affirm have nothing to do with the Nazis! I think that Hitler did much to retard, not advance, our nationhood. In the middle thirties, it looked as though we were soon to achieve a Jewish State. But with war in Europe looming ever closer, thanks to the Nazis, Britain cracked down on Jewish nationalist aspirations with the famous White Paper of 1939. Ripe as we were for nationhood at that time, we had the greatest difficulty in helping even a fraction of European Jewry escape the gas chambers. Certainly Israel's population contains no massive element of direct victims of Nazism or their descendants. We just were unable to save the majority of these people. And those who did escape from Germany and the other countries didn't always come here as we weren't equipped to get them in their hundreds of thousands past the British embargo on immigration or offer them a true nation once they got here.
>
> I would agree, however, that the advent of Nazism and its consequences in Europe did have one direct effect on Israel. It indicated to us all, to every Jew, the potential danger of being without a homeland. Nazism proved that Jews could live for five hundred years in peace with their neighbours, that they could all but assimilate in national society save for a few traditions and separate religious practices. They could believe themselves integral citizens of states professing freedom of belief and granting full rights to all inhabitants. Such was the situation prevailing in Germany, France, Italy, Holland, Denmark, Norway. Yet one raving maniac could blame the world's troubles on a group constituting less than six per cent of Europe's population and the holocaust was at hand!
>
> So, many a Jew realized that to be fully Jewish and fully a human being, and fully safe as both, one had to have a country of one's own where it was possible to live and work for something belonging to a personal cultural heritage. In this sense, Nazism did bring many Jews to Israel, from everywhere on earth. Not as victims of persecution but as believers in the positive good of a Jewish national home.
>
> I have said that personally I was never a victim of anti-Jewish persecution. I have, however, seen and marked the 'outsider' status of the Jews in even the most enlightened countries, as opposed to their full

participation in our society here."[214]

Ben-Gurion lied when he implied that he had tried to help the Jews of Europe escape death in the Holocaust. The Zionists delighted in the suffering of the Jews of Europe and were the instigators of it. David Ben-Gurion stated,

"The First World War brought us the Balfour Declaration. The Second ought to bring us the Jewish State."[215]

Michael Bar-Zohar wrote in his book *Ben-Gurion: The Armed Prophet*,

"The danger soon became a reality. Many were unable to distinguish between the British Government and the British people, and when war broke out, the extremists adopted radical methods. Supporters of Abraham Stern, who dreamed of a Kingdom of Israel extending from the Nile to the Euphrates, fired the first shots against the British. They even committed the unpardonable crime of recommending an alliance with Nazi Germany, against Britain. When the British shot Stern, his gang avenged him by bomb attacks. These men were few in number and represented a very small part of the *Yishuv*, but their terrorist activities began a new, violent phase in the struggle against the British, a phase which was to lead to open warfare between various factions and groups in Palestine, when Jew fought against Jew and disaster almost came to the Zionist cause."[216]

David Ben-Gurion stated,

"If I knew that it would be possible to save all the children in Germany by bringing them over to England, and only half of them by transporting them to *Eretz Yisrael*, then I would opt for the second alternative. For we must weigh not only the life of these children, but also the history of the People of Israel."[217]

In 1944, while the Nazis were massacring innocent and helpless Slavs, Jews, Gypsies, etc., Zionist David Ben-Gurion stated,

"One Degania [resident of the first communal settlement of Zionists in

---

214. D. Ben-Gurion, *Memoirs*, The World Publishing Company, New York, Cleveland, (1970), pp. 163-164.
215. M. Bar-Zohar, *Ben-Gurion: The Armed Prophet*, Prentice-Hall, Englewood Cliffs, New Jersey, (1967), p. 69.
216. M. Bar-Zohar, *Ben-Gurion: The Armed Prophet*, Prentice-Hall, Englewood Cliffs, New Jersey, (1967), p. 68.
217. D. Ben-Gurion, quoted in: Y. Gelber, "Zionist Policy and the Fate of European Jewry (1939-1942), *Yad Vashem Studies*, Volume 13, Martyrs' and Heroes Remembrance Authority, Jerusalem, (1979), pp. 169-210, at 199.

Palestine] is worth more than all the 'Yevsektzias' [Jewish Bolsheviks who sought to secularize Jews] and assimilationists in the world."[218]

and boasted,

"This people was the first to prophesy about 'the end of days,' the first to see the vision of a new human society. [***] Our small and land-poor Jewish people, therefore, lived in constant tension between the power and influence of the neighboring great empires and its own seemingly insignificant culture—a culture poor in material wealth and tangible monuments, but rich and great in its human and moral concepts and in its vision of a universal 'end of days.'"[219]

Christopher Sykes wrote,

"[...]Zionist leaders were determined at the very outset of the Nazi disaster to reap political advantage from the tragedy."[220]

David Ben-Gurion stated in 1932,

"What Zionist propaganda for years and years could not do, disaster has done overnight. Palestine is today the fiery question for the Jews of East and West, and the New World as well."[221]

Ben-Gurion also stated,

"The disaster facing European Jewry is not directly my business."[222]

and,

"It is the job of Zionism not to save the remnant of Israel in Europe but

---

[218]. D. Ben-Gurion, *Ba-Maarachah*, Volume 3, Tel-Aviv, (1948), pp. 200-211, English translation in A. Hertzberg, *The Zionist Idea*, Harper Torchbooks, New York, (1959), pp. 606-619, at 616.

[219]. D. Ben-Gurion, *Ba-Maarachah*, Volume 3, Tel-Aviv, (1948), pp. 200-211, English translation in A. Hertzberg, *The Zionist Idea*, Harper Torchbooks, New York, (1959), pp. 606-619, at 607-608.

[220]. K. Polkehn, "The Secret Contacts: Zionism and Nazi Germany, 1933-1941", *Journal of Palestine Studies*, Volume 5, Number 3/4, (Spring-Summer, 1976), pp. 54-82, at 58; citing C. Sykes, *Crossroads to Israel*, London, (1965); *Kreuzwege nach Israel; die Vorgeschichte des jüdischen Staates*, C. H. Beck, München, (1967), p. 151.

[221].C. Weizmann, "The Key to Immigration", *Rebirth and Destiny of Israel*, Philosophical Library, New York, (1954), p. 41.

[222]. T. Segev, *The Seventh Million: The Israelis and the Holocaust*, Hill and Wang, New York, (1993), p. 98.

rather to save the land of Israel for the Jewish people and the yishuv."[223]

In the 1937, David Ben-Gurion stated that the Zionist Jews want to take not just Palestine, but all of southern Syria and southern Lebanon, as well as Jordan and the Sinai, from their rightful inhabitants—they want the land of the Covenant from the Nile to the Euphrates.[224] Ben-Gurion stated in 1936,

> "The acceptance of partition does not commit us to renounce Transjordan; one does not demand from anybody to give up his vision. We shall accept a state in the boundaries fixed today, but the boundaries of Zionist aspirations are the concern of the Jewish people and no external factor will be able to limit them."[225]

Ben-Gurion stated to the General Staff,

> "I proposed that, as soon as we received the equipment on the ship, we should prepare to go over to the offensive with the aim of smashing Lebanon, Transjordan and Syria. [***] The weak point in the Arab coalition is Lebanon [for] the Moslem regime is artificial and easy to undermine. A Christian state should be established, with its southern border on the Litani River. We will make an alliance with it. When we smash the [Arab] Legion's strength and bomb Amman, we will eliminate Transjordan, too, and then Syria will fall. If Egypt still dares to fight on, we shall bomb Port Said, Alexandria, and Cairo. [***] And in this fashion, we will end the war and settle our forefathers' accounts with Egypt, Assyria, and Aram."[226]

In her book *Israel's Sacred Terrorism*, Livia Rokach reproduced an excerpt from a 26 May 1955 entry in Moshe Sheratt's personal diary, which recounts his impressions of Moshe Dayan's plans to provoke the Arabs to respond by first attacking them, then stealing their land when they sought to defend themselves,

> "The conclusions from Dayan's words are clear: This State has no international obligations, no economic problems, the question of peace is

---

[223]. T. Segev, *The Seventh Million: The Israelis and the Holocaust*, Hill and Wang, New York, (1993), p. 129.

[224]. D. Ben-Gurion, "On Ways of Our Policy", *Report of the Congress of the World Council of Poaley Zion*, (Zurich, July 27-August 1937), Tel-Aviv, (1938), pp. 206-207. *Cf.* I. Shahak, "The 'Historical Right' and the Other Holocaust", *Journal of Palestine Studies*, Volume 10, Number 3, (Spring, 1981), pp. 27-34, at 30. N. Chomsky, *Fateful Triangle: The United States, Israel, and the Palestinians*, Second Revised Edition, South End Press, Cambridge, Massachusetts, (1999), p. 161.

[225]. N. Chomsky, *Fateful Triangle: The United States, Israel, and the Palestinians*, Second Updated Edition, South End Press, Cambridge, Massachusetts, (1999), p. 161.

[226]. D. Ben-Gurion, quoted in: M. Bar-Zohar, *Ben-Gurion: A Biography*, Delacorte Press, New York, (1978), p. 166.

nonexistent.... It must calculate its steps narrow-mindedly and live on its sword. It must see the sword as the main, if not the only, instrument with which to keep its morale high and to retain its moral tension. Toward this end it may, no—it must—invent dangers, and to do this it must adopt the method of provocation-and-revenge.... And above all—let us hope for a new war with the Arab countries, so that we may finally get rid of our troubles and acquire our space. (Such a slip of the tongue: Ben Gurion himself said that it would be worth while to pay an Arab a million pounds to start a war.) (26 May 1955, 1021)"[227]

Menachem Begin stated in 1948,

"The partition of the Homeland is illegal. It will never be recognized. The signature of institutions and individuals of the partition agreement is invalid. It will not bind the Jewish people. Jerusalem was and will forever by our capital. Eretz Israel [the Land of Israel] will be restored to the people of Israel. All of it. And forever."[228]

As Ben-Gurion and many other leading Jewish figures have declared, Jews set about to fulfill the Messianic prophecies themselves, without God's intervention and without any concern for the rights, or the lives, of others. The Zionists were not reacting to the Holocaust when they took away the Palestinians' homes by force. Rather, they created the Holocaust as a means to achieve Jewish prophecy and force the Jews out of Europe, then the Zionists continued their Nazi practices in Palestine. The Zionists were not justified in taking the Palestinians' land because of the Holocaust. Rather, they were themselves responsible for the rise of the Nazis, and in no event did anything the Nazis did give the Jews the right to maim, murder, terrorize or displace the Palestinians. It is important to note that Nazism was but one phase of the Zionists' plan to terrorize humanity and that the Zionists' terror tactics were widely used during the formation of the "Jewish State" and have continued throughout Israel's existence. The Zionists will eventually cause a Third World War to bring on the apocalypse that they believe will hasten the Messianic Era and the miraculous creation of a new Earth with only "righteous" Jews to populate it (*Isaiah* 11:4; 42:1; 65; 66. *Jeremiah* 33:15-16). Racist cabalistic Jews believe that they are duty bound to destroy the living environment of the earth and ruin the genetics of the human species so as to provoke God to obliterate this earth and "create new heavens and a new earth"—the so-called "New World Order" or "Jewish Utopia". These racist cabalistic Jews are taught that they will have new and improved bodies in this new world and need not worry about the genetic damage they are intentionally causing to human beings across the earth.

---

227. L. Rokach, *Israel's Sacred Terrorism*, Third Edition, AAUG Press, Belmont Massachusetts, p. 41.
228. N. Chomsky, *Fateful Triangle: The United States, Israel, and the Palestinians*, Second Updated Edition, South End Press, Cambridge, Massachusetts, (1999), p. 161.

They believe that only Jews will be left alive and that they will not only be restored, but improved upon. The books of *Isaiah* chapters 65 and 66 and *Ezekiel* chapters 36 through 38 are the primary sources of these concepts, which were more fully developed in subsequent Jewish literature including the apocalyptic apocryphal Jewish books of *Enoch* and others. Note that the "elect", the "chosen" are exclusively the Jews.

The *Zohar*, I, 28*a-b*, states,

> "At that time every Israelite will find his twin-soul, as it is written, 'I shall give to you a new heart, and a new spirit I shall place within you' (Ezek. XXXVI, 26), and again, 'And your sons and your daughters shall prophesy' (Joel III, 1); these are [28b] the new souls with which the Israelites are to be endowed, according to the dictum, 'the son of David will not come until all the souls to be enclosed in bodies have been exhausted', and then the new ones shall come."[229]

Isaiah 65 states,

> "1 I am sought of *them that* asked not *for me;* I am found of *them that* sought me not: I said, Behold me, behold me, unto a nation *that* was not called by my name. 2 I have spread out my hands all the day unto a rebellious people, which walketh *in* a way *that was* not good, after their own thoughts; 3 A people that provoketh me to anger continually to my face; that sacrificeth in gardens, and burneth incense upon altars of brick; 4 Which remain among the graves, and lodge in the monuments, which eat swine's flesh, and broth of abominable *things is in* their vessels; 5 Which say, Stand by thyself, come not near to me; for I am holier than thou. These *are* a smoke in my nose, a fire that burneth all the day. 6 Behold, *it is* written before me: I will not keep silence, but will recompense, even recompense into their bosom, 7 Your iniquities, and the iniquities of your fathers together, saith the LORD, which have burned incense upon the mountains, and blasphemed me upon the hills: therefore will I measure their former work into their bosom. 8 Thus saith the LORD, As the new wine is found in the cluster, and *one* saith, Destroy it not; for a blessing *is* in it: so will I do for my servants' sakes, that *I* may not destroy them all. 9 And I will bring forth a seed out of Jacob, and out of Judah an inheritor of my mountains: and mine elect shall inherit it, and my servants shall dwell there. 10 And Sharon shall be a fold of flocks, and the valley of Achor a place for the herds to lie down in, for my people that have sought me. 11¶ But ye *are* they that forsake the LORD, that forget my holy mountain, that prepare a table for *that* troop, and that furnish the drink offering unto *that* number. 12 Therefore will I number you to the sword, and ye shall all bow down to the slaughter: because when I called, ye did not

---

229. H. Sperling and M. Simon, *The Zohar*, Volume 1, The Soncino Press, New York, (1933), p. 108.

answer; when I spake, ye did not hear; but did evil before mine eyes, and did choose *that* wherein I delighted not. 13 Therefore thus saith the Lord GOD, Behold, my servants shall eat, but ye shall be hungry: behold, my servants shall drink, but ye shall be thirsty: behold, my servants shall rejoice, but ye shall be ashamed: 14 Behold, my servants shall sing for joy of heart, but ye shall cry for sorrow of heart, and shall howl for vexation of spirit. 15 And ye shall leave your name for a curse unto my chosen: for the Lord GOD shall slay thee, and call his servants by another name: 16 That he who blesseth himself in the earth shall bless himself in the God of truth; and he that sweareth in the earth shall swear by the God of truth; because the former troubles are forgotten, and because they are hid from mine eyes. 17¶ For, behold, I create new heavens and a new earth: and the former shall not be remembered, nor come into mind. 18 But be ye glad and rejoice for ever *in that* which I create: for, behold, I create Jerusalem a rejoicing, and her people a joy. 19 And I will rejoice in Jerusalem, and joy in my people: and the voice of weeping shall be no more heard in her, nor the voice of crying. 20 There shall be no more thence an infant of days, nor an old man that hath not filled his days: for the child shall die an hundred years old; but the sinner *being* an hundred years old shall be accursed. 21 And they shall build houses, and inhabit *them;* and they shall plant vineyards, and eat the fruit of them. 22 They shall not build, and another inhabit; they shall not plant, and another eat: for as the days of a tree *are* the days of my people, and mine elect shall long enjoy the work of their hands. 23 They shall not labour in vain, nor bring forth for trouble; for they *are* the seed of the blessed of the LORD, and their offspring with them. 24 And it shall come to pass, that before they call, I will answer; and while they are yet speaking, I will hear. 25 The wolf and the lamb shall feed together, and the lion shall eat straw like the bullock: and dust *shall be* the serpent's meat. They shall not hurt nor destroy in all my holy mountain, saith the LORD."

Isaiah 66:22-24 states,

"22 For as the new heavens and the new earth, which I *will* make, *shall* remain before me, saith the LORD, so shall your seed and your name remain. 23 And it shall come to pass, *that* from one new moon to another, and from one sabbath to another, shall all flesh come to worship before me, saith the LORD. 24 And they shall go forth, and look upon the carcases of the men that have transgressed against me: for their worm shall not die, neither shall their fire be quenched; and they shall be an abhorring unto all flesh."

Ezekiel 36:24-38 states,

"24 For I will take you from among the heathen, and gather you out of all countries, and will bring you into your own land. 25 ¶Then will I sprinkle clean water upon you, and ye shall be clean: from all your filthiness, and from all your idols, will I cleanse you. 26 A new heart also will I give you, and a new spirit will I put within you: and I will take away the stony heart out of your flesh, and I will give you an heart of flesh. 27 And I will put my spirit within you, and cause you to walk in my statutes, and ye shall keep my judgments, and do *them*. 28 And ye shall dwell in the land that I gave to your fathers; and ye shall be my people, and I will be your God. 29 I will also save you from all your uncleannesses: and I will call for the corn, and will increase it, and lay no famine upon you. 30 And I will multiply the fruit of the tree, and the increase of the field, that ye shall receive no more reproach of famine among the heathen. 31 Then shall ye remember your own evil ways, and your doings that *were* not good, and shall lothe yourselves in your own sight for your iniquities and for your abominations. 32 Not for your sakes do I *this*, saith the Lord GOD, be it known unto you: be ashamed and confounded for your own ways, O house of Israel. 33 Thus saith the Lord GOD; In the day that I shall have cleansed you from all your iniquities I will also cause *you* to dwell in the cities, and the wastes shall be builded. 34 And the desolate land shall be tilled, whereas it lay desolate in the sight of all that passed by. 35 And they shall say, This land that was desolate is become like the garden of Eden; and the waste and desolate and ruined cities *are become* fenced, *and* are inhabited. 36 Then the heathen that are left round about you shall know that I the LORD build the *ruined* places, *and* plant that that was desolate: I the LORD have spoken *it*, and I will do *it*. 37 Thus saith the Lord GOD; I will yet *for* this be inquired of by the house of Israel, to do *it* for them; I will increase them *with* men like a flock. 38 As the holy flock, as the flock of Jerusalem in her solemn feasts; so shall the waste cities be filled *with* flocks of men: and they shall know that I *am* the LORD."

Ezekiel 37 states:

"1 The hand of the LORD was upon me, and carried me out in the spirit of the LORD, and set me down in the midst of the valley which *was* full *of* bones, 2 And caused me to pass by them round about: and, behold, *there were* very many in the open valley; and, lo, *they were* very dry. 3 And he said unto me, Son of man, can these bones live? And I answered, O Lord GOD, thou knowest. 4 Again he said unto me, Prophesy upon these bones, and say unto them, O ye dry bones, hear the word of the LORD. 5 Thus saith the Lord GOD unto these bones; Behold, I *will* cause breath to enter into you, and ye shall live: 6 And I will lay sinews upon you, and will bring up flesh upon you, and cover you with skin, and put breath in you, and ye shall live; and ye shall know that I *am* the LORD. 7 So I prophesied as I was commanded: and as I prophesied, there was a noise, and behold a shaking, and the bones came together, bone to his bone. 8 And when I beheld, lo, the sinews and the flesh came up upon them, and the skin covered them above: but *there was* no breath in them. 9 Then said he unto me, Prophesy unto the wind, prophesy, son of man, and say to the wind, Thus saith the Lord GOD;

Come from the four winds, O breath, and breathe upon these slain, that they may live. 10 So I prophesied as he commanded me, and the breath came into them, and they lived, and stood up upon their feet, an exceeding great army. 11 Then he said unto me, Son of man, these bones *are* the whole house of Israel: behold, they say, Our bones are dried, and our hope is lost: we are cut off for our parts. 12 Therefore prophesy and say unto them, Thus saith the Lord GOD; Behold, O my people, I *will* open your graves, and cause you to come up out of your graves, and bring you into the land of Israel. 13 And ye shall know that I *am* the LORD, when I have opened your graves, O my people, and brought you up out of your graves, 14 And shall put my spirit in you, and ye shall live, and I shall place you in your own land: then shall ye know that I the LORD have spoken *it*, and performed *it*, saith the LORD. 15¶ The word of the LORD came again unto me, saying, 16 Moreover, thou son of man, take thee one stick, and write upon it, For Judah, and for the children of Israel his companions: then take another stick, and write upon it, For Joseph, the stick of Ephraim, and *for* all the house of Israel his companions: 17 And join them one to another into one stick; and they shall become one in thine hand. 18 And when the children of thy people shall speak unto thee, saying, Wilt thou not shew us what thou meanest by these? 19 Say unto them, Thus saith the Lord GOD; Behold, I *will* take the stick of Joseph, which *is* in the hand of Ephraim, and the tribes of Israel his fellows, and will put them with him, *even* with the stick of Judah, and make them one stick, and they shall be one in mine hand. 20 And the sticks whereon thou writest shall be in thine hand before their eyes. 21 And say unto them, Thus saith the Lord GOD; Behold, I *will* take the children of Israel from among the heathen, whither they be gone, and will gather them on every side, and bring them into their own land: 22 And I will make them one nation in the land upon the mountains of Israel; and one king shall be king to them all: and they shall be no more two nations, neither shall they be divided into two kingdoms any more at all: 23 Neither shall they defile themselves any more with their idols, nor with their detestable things, nor with any of their transgressions: but I will save them out of all their dwellingplaces, wherein they have sinned, and will cleanse them: so shall they be my people, and I will be their God. 24 And David my servant *shall be* king over them; and they all shall have one shepherd: they shall also walk in my judgments, and observe my statutes, and do them. 25 And they shall dwell in the land that I have given unto Jacob my servant, wherein your fathers have dwelt; and they shall dwell therein, *even* they, and their children, and their children's children for ever: and my servant David *shall be* their prince for ever. 26 Moreover I will make a covenant of peace with them; it shall be an everlasting covenant with them: and I will place them, and multiply them, and will set my sanctuary in the midst of them for evermore. 27 My tabernacle also shall be with them: yea, I will be their God, and they shall be my people. 28 And the heathen shall know that I the LORD do sanctify Israel, when my sanctuary shall be in the midst of them for evermore."

Ezekiel 38 states:

"1 And the word of the LORD came unto me, saying, 2 Son of man, set thy face against Gog, the land of Magog, the chief prince of Meshech and Tubal, and prophesy against him, 3 And say, Thus saith the Lord GOD; Behold I *am* against thee, O Gog, the chief prince of Meshech and Tubal: 4 And I will turn thee back, and put hooks into thy jaws, and I will bring thee forth, and all thine army, horses and horsemen, all of them clothed with all sorts *of armour, even* a great company *with* bucklers and shields, all of them handling swords: 5 Persia, Ethiopia, and Libya with them; all of them *with* shield and helmet: 6 Gomer, and all his bands; the house of Togarmah of the north quarters, and all his bands: *and* many people with thee. 7 Be thou prepared, and prepare for thyself, thou, and all thy company that are assembled unto thee, and be thou a guard unto them. 8 After many days thou shalt be visited: in the latter years thou shalt come into the land *that is* brought back from the sword, *and is* gathered out of many people, against the mountains of Israel, which have been always waste: but it is brought forth out of the nations, and they shall dwell safely all of them. 9 Thou shalt ascend and come like a storm, thou shalt be like a cloud to cover the land, thou, and all thy bands, and many people with thee. 10¶ Thus saith the Lord GOD; It shall also come to pass, *that* at the same time shall things come into thy mind, and thou shalt think an evil thought: 11 And thou shalt say, I will go up to the land of unwalled villages; I will go *to* them that are at rest, that dwell safely, all of them dwelling without walls, and having neither bars nor gates, 12 To take a spoil, and to take a prey; to turn thine hand upon the desolate places *that are now* inhabited, and upon the people *that are* gathered out of the nations, which have gotten cattle and goods, that dwell in the midst of the land. 13 Sheba, and Dedan, and the merchants of Tarshish, with all the young lions thereof, shall say unto thee, Art thou come to take a spoil? hast thou gathered thy company to take a prey? to carry away silver and gold, to take away cattle and goods, to take a great spoil? 14 Therefore, son of man, prophesy and say unto Gog, Thus saith the Lord GOD; In that day when my people of Israel dwelleth safely, shalt thou not know *it?* 15 And thou shalt come from thy place out of the north parts, thou, and many people with thee, all of them riding upon horses, a great company, and a mighty army: 16 And thou shalt come up against my people of Israel, as a cloud to cover the land; it shall be in the latter days, and I will bring thee against my land, that the heathen may know me, when I shall be sanctified in thee, O Gog, before their eyes. 17¶ Thus saith the Lord GOD; *Art* thou he of whom I have spoken in old time by my servants the prophets of Israel, which prophesied in those days *many* years that *I* would bring thee against them? 18 And it shall come to pass at the same time when Gog shall come against the land of Israel, saith the Lord GOD, *that* my fury shall come up in my face. 19 For in my jealousy *and* in the fire of my wrath have I spoken, Surely in that day there shall be a great shaking in the land of Israel;

20 So that the fishes of the sea, and the fowls of the heaven, and the beasts of the field, and all creeping things that creep upon the earth, and all the men that are upon the face of the earth, shall shake at my presence, and the mountains shall be thrown down, and the steep places shall fall, and every wall shall fall to the ground. 21 And I will call *for* a sword against him throughout all my mountains, saith the Lord GOD: every man's sword shall be against his brother. 22 And I will plead against him with pestilence and with blood; and I will rain upon him, and upon his bands, and upon the many people that *are* with him, an overflowing rain, and great hailstones, fire, and brimstone. 23 Thus will I magnify myself, and sanctify myself; and I will be known in the eyes of many nations, and they shall know that I *am* the LORD."

Ezekiel 39 states:

"1 Therefore, thou son of man, prophesy against Gog, and say, Thus saith the Lord GOD; Behold, I *am* against thee, O Gog, the chief prince of Meshech and Tubal: 2 And I will turn thee back, and leave but the sixth part of thee, and will cause thee to come up from the north parts, and will bring thee upon the mountains of Israel: 3 And I will smite thy bow out of thy left hand, and will cause thine arrows to fall out of thy right hand. 4 Thou shalt fall upon the mountains of Israel, thou, and all thy bands, and the people that *is* with thee: I will give thee unto the ravenous birds of every sort, and *to* the beasts of the field to be devoured. 5 Thou shalt fall upon the open field: for I have spoken *it,* saith the Lord GOD. 6 And I will send a fire on Magog, and among them that dwell carelessly in the isles: and they shall know that I *am* the LORD. 7 So will I make my holy name known in the midst of my people Israel; and I will not let *them* pollute my holy name any more: and the heathen shall know that I *am* the LORD, the Holy One in Israel. 8 Behold, it is come, and it is done, saith the Lord GOD; this *is* the day whereof I have spoken. 9 And they that dwell in the cities of Israel shall go forth, and shall set on fire and burn the weapons, both the shields and the bucklers, the bows and the arrows, and the handstaves, and the spears, and they shall burn them with fire seven years: 10 So that they shall take no wood out of the field, neither cut down *any* out of the forests; for they shall burn the weapons with fire: and they shall spoil those that spoiled them, and rob those that robbed them, saith the Lord GOD. 11 And it shall come to pass in that day, *that* I will give unto Gog a place there of graves in Israel, the valley of the passengers *on* the east of the sea: and it *shall* stop the *noses of the* passengers: and there shall they bury Gog and all his multitude: and they shall call *it* The valley of Hamon-gog. 12 And seven months shall the house of Israel be burying of them, that *they* may cleanse the land. 13 Yea, all the people of the land shall bury *them;* and it shall be to them a renown the day that I shall be glorified, saith the Lord GOD. 14 And they shall sever out men of continual employment, passing through the land to bury with the passengers those that remain upon the face of the earth, to cleanse it: after the end of seven months shall they search. 15 And the passengers *that* pass through the land, when *any* seeth a man's bone, then shall he set up a sign by it, till the buriers have buried it in the valley of Hamon-gog. 16 And also the name of the city *shall be* Hamonah. Thus shall they cleanse the land. 17 And, thou son of man, thus saith the Lord GOD; Speak unto every feathered fowl, and to every beast of the field, Assemble yourselves, and come; gather yourselves on every side to my sacrifice that I do sacrifice for you, *even* a great sacrifice upon the mountains of Israel, that ye may eat flesh, and drink blood. 18 Ye shall eat the flesh of the mighty, and drink the blood of the princes of the earth, of rams, of lambs, and of goats, of bullocks, all of them fatlings of Bashan. 19 And ye shall eat fat till *ye* be full, and drink blood till *ye* be drunken, of my sacrifice which I have sacrificed for you. 20 Thus ye shall be filled at my table *with* horses and chariots, with mighty *men,* and *with* all men of war, saith the Lord GOD. 21 And I will set my glory among the heathen, and all the heathen shall see my judgment that I have executed, and my hand that I have laid upon them. 22 So the house of Israel shall know that I *am* the LORD their God from that day and forward. 23 And the heathen shall know that the house of Israel went into captivity for their iniquity: because they trespassed against me, therefore hid I my face from them, and gave them into the hand of their enemies: so fell they all by the sword. 24 According to their uncleanness and according to their transgressions have I done unto them, and hid my face from them. 25¶ Therefore thus saith the Lord GOD; Now will I bring again the captivity of Jacob, and have mercy upon the whole house of Israel, and will be jealous for my holy name; 26 After that they have borne their shame, and all their trespasses whereby they have trespassed against me, when they dwelt safely in their land, and none made *them* afraid. 27 When I have brought them again from the people, and gathered them out of their enemies' lands, and am sanctified in them in the sight of many nations; 28 Then shall they know that I *am* the LORD their God, which cause them to be led into captivity among the heathen: but I have gathered them unto their own land, and have left none of them any more there. 29 Neither will I hide my face any more from them: for I have poured out my spirit upon the house of Israel, saith the Lord GOD."

Christians who believe that these prophecies are miraculously being fulfilled in modern times are admonished to realize that what has happened in recent centuries is not the product of divine intervention, but rather the result of the deliberate actions of racist Cabalistic Jews meant to destroy Christians. *Revelation* 20:7-8 clearly states that the battle against Gog and Magog is post-Millennial. It is not the work of God, but rather the deliberate destruction is wrought by ill-intentioned racist Jewish leadership who intend to exterminate the Christians. Jesus warned against obeying racist Jewish leadership and in Christianity the covenant with God has passed from the Jews to all Peoples (*Matthew* 12:30; 21:43-45. *Romans* 4; 9; 11:7-8. *Galatians* 3:16, 28-29; 4. and *Hebrews* 8:6-10).

In a "Letter to the Editor", signed by Isidore Abramowitz, Hannah Arendt, Abraham Brick, Rabbi Jessurun Cardozo, Albert Einstein, Herman Eisen, M. D., Hayim Fineman, M. Gallen, M. D., H. H. Harris, Zelig S. Harris, Sidney Hook, Fred Karush, Bruria Kaufman, Irma L. Lindheim, Nachman Majsel, Seymour Melman, Myer D. Mendelson, M. D., Harry M. Orlinsky, Samuel Pitlick, Fritz Rohrlich, Louis P. Rocker, Ruth Sager, Itzhak Sankowsky, I. J. Schoenberg, Samuel Shuman, M. Znger, Irma Wolpe, Stefan Wolpe; dated "New York. Dec. 2, 1948."; published as: "New Palestine Party; Visit of Menachen Begin and Aims of Political Movement Discussed", *The New York Times*, (4 December 1948), p. 12; it states, *inter alia*,

> "Among the most disturbing political phenomena of our time is the emergence in the newly created state of Israel of the 'Freedom Party' (Tnuat Haherut), a political party closely akin in its organization, methods, political philosophy and social appeal to the Nazi and Fascist parties. It was formed out of the membership and following of the former Irgun Zvai Leumi, a terrorist, right-wing, chauvinist organization in Palestine. The current visit of Menachen Begin, leader of this party, to the United States is obviously calculated to give the impression of American support for his party in the coming Israeli elections, and to cement political ties with conservative Zionist elements in the United States. Several Americans of national repute have lent their names to welcome his visit. It is inconceivable that those who oppose fascism throughout the world, if correctly informed as to Mr. Begin's political record and perspectives, could add their names and support to the movement he represents. [***] The public avowals of Begin's party are no guide whatever to its actual character. Today they speak of freedom, democracy and anti-imperialism, whereas until recently they openly preached the doctrine of the Fascist state. It Is in its actions that the terrorist party betrays its real character; from its past actions we can judge what it may be expected to do in the future. [***] The Deir Yassin incident exemplifies the character and actions of the Freedom Party. Within the Jewish community they have preached an admixture of ultranationalism, religious mysticism, and racial superiority. Like other Fascist parties they have been used to break strikes, and have themselves pressed for the destruction of free trade unions. In their stead they have proposed corporate unions on the Italian Fascist model. [***] This is the unmistakable stamp of a Fascist party for whom terrorism (against Jews, Arabs, and British alike), and misrepresentation are means, and a 'Leader State' is the goal. In the light of the foregoing considerations, it is imperative that the truth about Mr. Begin and his movement be made known in this country. It is all the more tragic that the top leadership of American Zionism has refused to campaign against Begin's efforts, or even to expose to its own constituents the dangers to Israel from support to Begin."

While the mass media in America has traditionally covered up the fascistic nature of the Israeli Government and its leaders, certainly not all Israelis have approved of the territorial and political ambitions of leading Zionists murderers like David Ben-Gurion and Menachem Begin. Anthony Lewis quoted Avraham Burg in an article titled, "Hope Against Hope" in *The New York Times*, Section 4, on 17 April 1983 on page 19,

> "'When we established Israel,' [Avraham Burg] said, 'it was based on the feeling that we needed a new basis for Jewish continuity, Jewish existence. Now, for many, the state has become the end of existence instead of the means. It has become the Messiah.
>
> 'That is dangerous because in Judaism there is no Messiah now. You walk toward it. It is your ideal. If you achieve it, it's a false Messiah. And our history knows many false Messiahs who endangered Jewish existence. I'm afraid that if the Jewish state becomes such a false Messiah, such a substitute for our ideals, the day will come when we will recognize that and there will be a mortal crisis. I am against it totally.
>
> Judaism is not territories. It is more than a piece of the land.'"

Pious and compassionate Jews must realize that the racist and genocidal Jewish Messianic myths guiding the actions of the leading Zionists like Ben-Gurion and Begin remain troubling today, because they predict an apocalyptic war between the "Messiah Son of Joseph" (in a secular view, the State of Israel) and the King of Persia (President of Iran), which, after a nine month period of tribulations for Israel and the death of the Messiah Son of Joseph, will result in the ascendence of the "Messiah Son of David" (in a secular view, the State of Greater Israel extending from the Nile to the Euphrates), and the subjugation, then extermination, of the Gentile peoples of the Earth.

The Lubavitcher Jews have announced that they are prepared to anoint the Messiah and that it will happen soon. They are broadly disseminating propaganda to condition the world to accept this event.

Karl Marx took advantage of Gentile prejudice against pious Jews to bring about the ruin of Gentile nations, in fulfilment of Jewish Messianic prophecy. Pious Jews hated science, art and Gentiles—refused even to eat at the same table with Gentiles—as Shakespeare's Shylock in *The Merchant of Venice* noted.[230] Pious Jews felt a loyalty only to God, to the Law and to each other. To a pious Jew, Greek science was a product of human reason and an affront to the Law, which had supposedly been given to the Jews, and only to the Jews, by God. Art depicted graven images and idols, and the Gentiles were individualistic in the pejorative sense and the Jews considered them to be soulless and cruel animals. For a pious Jew, immortality was meant for the Jews as a "race", and they did not accept the Christian belief in the immortality of the individual soul. In order to achieve their "racial" immortality, the Jews had to remain segregated, and this

---

230. W. Shakespeare, *The Merchant of Venice*, Act 1, Scene 3, "Yes, to smell pork; to eat of the habitation which your prophet, the Nazarite, conjured the devil into! I will buy with you, sell with you, talk with you, walk with you, and so following; but I will not eat with you, drink with you, nor pray with you."

meant that they ultimately had to kill off the Gentiles. The God of the Old Testament is a creator God, and the creations of mankind, such as science and art, were considered to be an affront to this God's authority. After the emancipation movement, begun by the French Revolution and advanced by Napoleon, came into full swing, several Jewish movements tried to reconcile the Enlightenment, and the insights of science, with the antagonism of Judaism to human creations and the obvious falsehoods expressed in the religion (see, for example, *Babylonian Talmud,* Tractate Menahoth, folio 99*b*, which discourages the study of "Greek science"). These organizations created Marxism as a stumbling stone for the Gentiles to trip over. Marx took this opportunity to defame his fellow Jews in order to promote himself and use the Gentiles' own prejudices to destroy them.

Many newly emancipated secular Jews embraced art and science and excelled at them. They found themselves hated by many pious Jews, and some returned that hatred and ridicule. This was a painful dilemma for secular Jews, because all of their traditions taught them to find security in community, and their quest for individuality often resulted in alienation from both the Jewish and Gentile communities. This struggle between secular and pious Jews continued through the Twentieth Century and is depicted in Chaim Grade's story "My War with Hersh Rasseyner", *Commentary*, Volume 16, Number 5, (November, 1953), pp. 428-441; and yet more poignantly in the 1991 film based on this story, *The Quarrel* directed by Eli Cohen.

3.5 Jewish Dogmatism and Control of the Press Stifles Debate

If Robert Herndon Fife, Jr.'s book, *The German Empire Between Two Wars: A Study of the Political and Social Development of the Nation Between 1871 and 1914*, Macmillan, New York, (1916), at pages 177-199 and 359-388, bore a political bias, it appears to have been a pro-Socialist bias tending toward Marxist Socialism, though certainly not anti-Semitism. His book is dated in its relevance to Einstein by two factors: the founding of the Weimar Republic, and the interjection of politics into scientific matters practiced by Einstein and his advocates, as well as his opponents. In matters related to Einstein, the normally responsible scientific reporting of the German press surrendered ground to their typically irresponsible political reporting.

Just as a terrible propaganda machine had evolved in Germany, which apparatus of propaganda truly became a monster during the war, Lord Northcliffe and many others had established numerous propaganda outlets in Great Britain and America to promote Allied interests, often with outrageous lies.[231] After the war, these highly advanced propaganda factories consolidated

---

231. H. E. Barnes, *The Genesis of the World War: An Introduction to the Problem of War Guilt*, A.A. Knopf, New York, London, (1927), pp. 590-653; **and** *In Quest of Truth and Justice: De-bunking the War Guilt Myth*, National Historical Society, Chicago, (1928),

to promote Einstein to the world. They successfully brought him undeserved fame and defamed and largely silenced his critics. Their vitriolic and racist attacks on Einstein's critics, coupled together with organized campaigns to destroy the careers of any scientists who would speak out against the theory of relativity, had the desired chilling effect on the effort to expose Einstein to the public as an irrational plagiarist.

Sir Gilbert Parker, who was in charge of British propaganda in America, revealed the organized power of the highly developed art of propaganda at the time, in *Harper's Magazine* in March of 1918. Parker discussed many of the corrupt tactics that were put to use soon afterwards to promote Einstein and to attack his critics and suppress dissent against Einstein, against Einstein's self-promotion and against Einstein's irrationality,

> "Perhaps here I may be permitted to say a few words concerning my own work since the beginning of the war. It is in a way a story by itself, but I feel justified in writing one or two paragraphs about it. Practically since the day war broke out between England and the Central Powers I became responsible for American publicity. I need hardly say that the scope of my department was very extensive and its activities widely ranged. Among the activities was a weekly report to the British Cabinet on the state of American opinion, and constant touch with the permanent correspondents of American newspapers in England. I also frequently arranged for important public men in England to act for us by interviews in American newspapers; and among these distinguished people were Mr. Lloyd George (the present Prime Minister), Viscount Grey, Mr. Balfour, Mr. Bonar Law, the Archbishop of Canterbury, Sir Edward Carson, Lord Robert Cecil, Mr. Walter Runciman, (the Lord Chancellor), Mr. Austen Chamberlain, Lord Cromer, Will Crooks, Lord Curzon, Lord Gladstone, Lord Haldane, Mr. Henry James, Mr. John

---

pp. 30-34, 98-105. *See also:* A. Ponsonby, *Falsehood in War-Time, Containing an Assortment of Lies Circulated Throughout the Nations During the Great War*, G. Allen & Unwin, ltd. London, E. P. Dutton, New York, (1928). A. J. Dawe, Letter to the Editor, "The Crime Of Louvain. Vivid Account By An Eye-Witness. *See also:* A Ruthless Holocaust. The Real Horrors Of War", *The London Times*, (3 September 1914), p. 4. *See aslo:* J. Bryce, *Report of the Committee on alleged German outrages appointed by His Britannic Majesty's Government and presided over by the Right Hon. Viscount Bryce. Evidence and Documents laid before the Committee on alleged German outrages: (appendix to the Report).*, Printed Under the Authority of His Majesty's Stationery Office, London, (1915); **French** *Rapport de la Commission d'Enquête sur les Atrocités Allemandes*, Darling & Son, London, (1915); **Italian** *Relazione della Commissione d'Inchiesta sulle Atrocità Tedesche*, Vincenzo Bartelli, Perugia, (1915), **Portugese** *Relatorio da Commissão sobre as Barbaridades Attribuidas aos Allemães, nomeada pelo Governo de Sua Magestade Britannica presidida pelo Visconde Bryce*, Thomas Nelson & Sons, Paris, Edimburgo, (1915); **Spanish** *Informe Acerca de los Atentados Atribuidos á los Alemanes, Emitido por la Comisión Nombrada por el Gobierno de su Majéstad Británica y Presidida por el muy Honorable Vizconde Bryce*, Thomas Nelson & Sons, Paris, Edimburgo, (1915).

Redmond, Mr. Selfridge, Mr. Zangwill, Mrs. Humphry Ward, and fully a hundred others.

Among other things, we supplied three hundred and sixty newspapers in the smaller States of the United States with an English newspaper, which gives a weekly review and comment of the affairs of the war. We established connection with the man in the street through cinema pictures of the Army and Navy, as well as through interviews, articles, pamphlet etc.; and by letters in reply to individual American critics, which were printed in the chief newspaper of the State in which they lived, and were copied in newspapers of other and neighboring States. We advised and stimulated many people to write articles; we utilized the friendly services and assistance of confidential friends; we had reports from important Americans constantly, and established association, by personal correspondence, with influential and eminent people of every profession in the United States, beginning with university and college presidents, professors and scientific men, and running through all the ranges of the population. We asked our friends and correspondents to arrange for speeches, debates, and lectures by American citizens, but we did not encourage Britishers to go to America and preach the doctrine of entrance into the war. Besides an immense private correspondence with individuals, we had our documents and literature sent to great numbers of public libraries, Y. M. C. A. societies, universities, colleges, historical societies, clubs, and newspapers.

It is hardly necessary to say that the work was one of extreme difficulty and delicacy, but I was fortunate in having a wide acquaintance in the United States and in knowing that a great many people had read my books and were not prejudiced against me. I believed that the American people could not be driven, preached to, or chivied into the war, and that when they did enter it would be the result of their own judgment and not the result of exhortation, eloquence, or fanatical pressure of Britishers. I believed that the United States would enter the war in her own time, and I say this, with a convinced mind, that, on the whole, it was best that the American commonwealth did not enter the war until that month in 1917 when Germany played her last card of defiance and indirect attack. Perhaps the safest situation that could be imagined actually did arise. The Democratic party in America, which probably would not have supported a Republican President had he declared war, were practically forced by the logic of circumstances to support President Wilson when be declared war, because he had blocked up every avenue of attack."[232]

After the war ended, both the media of the Allies and that of the Central Powers were applied to making Einstein a celebrity and the fine art of controlling public opinion, which had become so refined during the war, was applied to the

---

[232]. G. Parker, "The United States and the War", *Harper's Magazine*, Volume 136, Number 814, (March, 1918), pp. 521-531, at 521-522.

task of making Einstein famous. The methods learned and employed in wartime were also used to suppress and quash open debate on important scientific and ethical questions related to Einstein's plagiarism, the fatal flaws in the theory of relativity and the misrepresentation of the physical evidence used to justify the theory.

Many were struck by the speed with which Einstein became famous. No scientist had ever become so famous so quickly. Many were skeptical and suspicious that something unseemly was taking place.

In his book, Alexander Moszkowski recounts Albert Einstein's assuredness as to the results of the eclipse observations that made Einstein famous—*before the photographs of the eclipse had been taken*, an assurance that worried Max Planck and struck Heinrich Zangger as odd.[233] Einstein was absolutely confident that the results of the eclipse observations would confirm "his" prediction. Einstein's apparent knowledge of the results before they were obtained leads one to believe that the published conclusions of the eclipse observations, no matter what the evidence actually showed or was capable of showing, was a foregone conclusion arrived at in collusion, not through experimentation and observation. Moszkowski wrote,

> "In no sense did Einstein himself entertain a possibility of doubt.
>
> On repeated occasions before May 1919 I had opportunities of questioning him on this point. There was no shadow of a scruple, no ominous fears clouded his anticipations. Yet great things were at stake.
>
> Observation was to show 'the correctness of Einstein's world system' by a fact clearly intelligible to the whole world, one depending on a very sensitive test of less than two seconds of arc.
>
> 'But, Professor,' said I, on various occasions, 'what if it turns out to be more or less? These things are dependent on apparatus that may be faulty, or on unforeseen imperfections of observation.' A smile was Einstein's only answer, and this smile expressed his unshakeable faith in the instruments and the observers to whom this duty was to be entrusted.
>
> Moreover, it is to be remarked that no great lengths of time were available for comfortable experimentation in taking this photographic record. For the greatest possible duration of a total eclipse of the sun viewed at a definite place amounts to less than eight minutes, so that there was no room for mishaps in this short space of time, nor must any intervening cloud appear. The kindly co-operation of the heavens was indispensable—and was not refused. The sun, in this case the darkened sun, brought this fact to light.
>
> Two English expeditions had been equipped for the special occasion of the eclipse—one to proceed to Sobral and the other to the Island of Principe, off Portuguese Africa; they were sent officially with equipment provided in

---

233. Letter from M. Planck to A. Einstein of 4 October 1919, *The Collected Papers of Albert Einstein*, Volume 9, Document 121, Princeton University Press, (2004). Letter from H. Zangger to A. Einstein of 22 October 1919, *The Collected Papers of Albert Einstein*, Volume 9, Document 148, Princeton University Press, (2004).

the main by the time-honoured Royal Society. Considering the times, it was regarded as the first symptom of the revival of international science, a praiseworthy undertaking. A huge apparatus was set into motion for a purely scientific object with not the slightest relation to any purpose useful in practical life. It was a highly technical investigation whose real significance could be grasped by only very few minds. Yet interest was excited in circles reaching far beyond that of the professional scientist. As the solar eclipse approached, the consciousness of amateurs became stirred with indefinite ideas of cosmic phenomena. And just as the navigator gazes at the Polar Star, so men directed their attention to the constellation of Einstein, which was not yet depicted in stellar maps, but, from which something uncomprehended, but undoubtedly very important, was to blaze forth.

In June it was announced that the star photographs had been successful in most cases, yet for weeks, nay for months, we had to exercise patience. For the photographs, although they required little time to be taken, took much longer to develop and, above all, to be measured; in view of the order of smallness of the distances to be compared, this was a difficult and troublesome task, for the points of light on the plate did not answer immediately with Yes or No, but only after mechanical devices of extreme delicacy had been carefully applied.

At the end of September they proclaimed their message. It was in the affirmative, and this Yes out of far-distant transcendental regions called forth a resounding echo in the world of everyday life. Genuinely and truly the $1\frac{7}{10}$ seconds of arc had come out, correct to the decimal point. These points representing ciphers, as it were, had chanted of the harmony of the spheres in their Pythagorean tongue. The transmission of this message seemed to be accompanied by the echoing words of Goethe's 'Ariel':

> 'With a crash the Light draws near!
> Pealing rays and trumpet-blazes,—
> Eye is blinded, ear amazes.'

Never before had anything like this happened. A wave of amazement swept over the continents. Thousands of people who had never in their lives troubled about vibrations of light and gravitation were seized by this wave and carried on high, immersed in the wish for knowledge although incapable of grasping it. This much all understood, that from the quiet study of a scholar an illuminating gospel for exploring the universe had been irradiated.

During that time no name was quoted so often as that of this man. Everything sank away in face of this universal theme which had taken possession of humanity. The converse of educated people circled about this pole, could not escape from it, continually reverted to the same theme when pressed aside by necessity or accident. Newspapers entered on a chase for contributors who could furnish them with short or long, technical or non-technical, notices about Einstein's theory. In all nooks and corners social evenings of instruction sprang up, and wandering universities appeared with errant professors that led people out the three-dimensional misery of daily life into the more hospitable Elysian fields of four-dimensionality. Women lost sight of domestic worries and discussed co-ordinate systems, the principle of simultaneity, and negatively-charged electrons. All contemporary questions had gained a fixed centre from which threads could be spun to each. Relativity had become the sovereign password. In spite of some grotesque results that followed on this state of affairs it could not fail to be recognized that we were watching symptoms of mental hunger not less imperative in its demands than bodily hunger, and it was no longer to be appeased by the former books by writers on popular science and by misguided idealists.

And whilst leaders of the people, statesmen, and ministers made vain efforts to steer in the fog, to arrive at results serviceable to the nation, the multitude found what was expedient for it, what was uplifting, what sounded like the distant hammering of reconstruction. Here was a man who had stretched his hands towards the stars; to forget earthly pains one had but to immerse oneself in his doctrine. It was the first time for ages that a chord vibrated through the world invoking all eyes towards something which, like music or religion, lay outside political or material interests.

The mere thought that a living Copernicus was moving in our midst elevated our feelings. Whoever paid him homage had a sensation of soaring above Space and Time, and this homage was a happy augury in an epoch so bare of brightness as the present.

As already remarked, there was no lack of rare fruits among the newspaper articles, and a chronicler would doubtless have been able to make an attractive album of them. I brought Einstein several foreign papers with large illustrations which must certainly have cost the authors and publishers much effort and money. Among others there were full-page beautifully coloured pictures intended to give the reader an idea of the paths pursued by the rays from the stars during the total eclipse of the sun. These afforded Einstein much amusement, namely, *e contrario*, for from the physical point of view these pages contained utter nonsense. They showed the exact opposite of the actual course of the rays inasmuch as the author of the diagrams had turned the convex side of the deflected ray towards the sun. He had not even a vague idea of the character of the deflection, for his rays proceeded in a straight line through the universe until they reached the sun, where they underwent a sudden change of direction reminiscent of a stork's legs. The din of journalistic homage was not unmixed with scattered voices of dissent, even of hostility. Einstein combated these not only without anger but with a certain satisfaction. For indeed the series of unbroken ovations became discomfiting, and his feelings took up arms against what seemed to be developing into a star-artist cult. It was like a breath of fresh air when some column of a chance newspaper was devoted to a polemic against his theory, no matter how unfounded or unreasoned it may have been, merely because a dissonant tone broke the unceasing chorus of praise. On one occasion he even said of a shrill disputant, 'The man is quite right!' And these words were uttered in the most natural manner possible. One must know him personally if one is to understand these excesses of toleration. So did Socrates defend his opponents."[234]

Albert Einstein marveled at the spurious evidence which had made him a cult figure. Moszkowski informs us that,

"A copy of this photograph had been sent to Einstein from England, and he told me of it with evident pleasure. He continually reverted to the delightful little picture of the heavens, quite fascinated by the thing itself, without the slightest manifestation of a personal interest in his own success. Indeed, I may go further and am certainly not mistaken in saying his new mechanics did not even enter his head, nor the verification of it by the plate; on the contrary, he displayed that disposition of the mind which in the case of genius as well as in that of children shows itself as *naïveté*. The prettiness of the photograph charmed him, and the thought that the heavens had been drawn up as for parade to be a model for it."[235]

We know that Eddington was biased, and that photographs taken in 1918 failed to show any displacement—though it is difficult to believe that any photographs taken in that era were accurate enough to measure such things. The Annual Meeting of the British Association for the Advancement of Science, Bournemouth, 1919, in its "Transactions of Section A", Friday, September 12, pages 156-157, reported:

"1. *Photographs taken at Principe during the Total Eclipse of the Sun, May 29th.* By Professor A. S. EDDINGTON, *F.R.S.*, *and* E. T. COTTINGHAM, *followed by a Discussion on Relativity, opened by* Professor EDDINGTON, *F.R.S.*

Professor Eddington gave an account of the observations which had been made at Principe during the solar eclipse. The main object in view was to observe the displacement (if any) of stars, the light from which passed through the gravitational field of the sun. To establish the existence of such an effect and the determination of its magnitude gives, as is well known, a crucial test of the theory of gravitation enunciated by Einstein. Professor Eddington explained that the observations had been partially vitiated by the presence of clouds, but the plates already measured indicated the existence of a deflection intermediate between the two theoretically possible values 0 87" and 1 75" He hoped that when the measurements were completed the latter figure would prove to be verified. Incidentally Professor Eddington pointed out that the presence of clouds had resulted in a solar prominence

---

[234]. A. Moszkowski, *Einstein: The Searcher*, E. P. Dutton, New York, (1921), pp. 12-15.

[235]. A. Moszkowski, *Einstein: The Searcher*, E. P. Dutton, New York, (1921), p. 19.

being photographed and its history followed in some detail; some very striking photographs were shown.

Following on this account Professor Eddington opened the discussion on relativity, and referred again to the bending of the wave front of light to be expected from Einstein's new law when the light passes near a heavy body. It should be possible to test experimentally this law, which demands that the speed of light varies as $1 - 2\Omega$ where Ω is the gravitational potential. He showed that whether Einstein's solution of the problem be correct or not, it has at any rate given a new orientation to our ideas of space and time. Sir Oliver Lodge regarded the relativity theory of 1905 as a supplement to Newtonian dynamics by the adoption of the factor $\left(1 - \frac{v^2}{c^2}\right)$ and its powers necessitated by experimental results; but he did not consider this dependence of mass and length on velocity as entailing any revolutionary changes of our ideas of space and time, or as rendering necessary the further complexities of 1915. He compared the difficulties involved with the case of measuring temperature, defined in terms of a perfect gas, and made with gases which only approximate to this ideal state. Dr. Silberstein pointed out that Einstein's theory of gravitation predicts three verifiable phenomena, *i.e.*, a shift of spectral lines, the bending of light round the sun and the secular motion of the perihelion of a planet. In the neighbourhood of a radially symmetric mass, such as our sun, the line element *ds* is given by:—

$$ds^2 = \left(1 - \frac{2M}{c^2 r}\right) c^2 dt^2 - (1 - \frac{2M}{c^2 r})(dx^2 + dy^2 + dz^2)$$

The coefficient $c^2 dt^2$ gives by itself a lengthening of the period of oscillation for a terrestrial observer in the ratio $\left(1 + \frac{M}{c^2 r}\right) : 1$, demanding a shift of spectral lines of about 01Å.U. Secondly, the path of rays of light is obtained by putting *ds* = *0* and the first and second coefficients give jointly a bending which, for rays almost grazing the sun, is **1 75"**. Thirdly, Keplerian motion is predicted with a progressively moving perihelion which in the case of Mercury turns out to be **43"** per century. He drew attention to the fact that St. John's results in 1917 showed no shift of the spectral lines, a fact which in itself would overthrow the theory in question. Father Cortie pointed out that Campbell's photographs, taken in 1918 and measured by Curtis, gave no trace of any displacement of the images of 43 stars distributed irregularly round the sun."

Regarding this meeting and the evidence against general relativity which was known to Freundlich and Einstein, *see also: Nature*, Volume 103, (1919), p. 394; and *The Observatory*, Volume 42, (1919), pp. 298-299, 361-366; and the letter from E. Freundlich to A. Einstein of 15 September 1919, *The Collected Papers of Albert Einstein*, Volume 9, Document 105, Princeton University Press, (2004); as well as Einstein's response to Freundlich on 19 September 1919, *ibid.* Document 106.

On 9 October 1919, Albert Einstein reported in *Die Naturwissenschaften* (J. Springer), Volume 7, Number 42, (17 October 1919), p. 776,

> "Zuschriften an die Herausgeber.
> Prüfung der allgemeinen Relativitätstheorie.
> Nach einem von Prof. *Lorentz* an den Unterzeichneten gerichteten Telegramm hat die zur Beobachtung der Sonnenfinsternis am 29. Mai ausgesandte englische Expedition unter *Eddington* die von der allgemeinen Relativitätstheorie geforderte Ablenkung des Lichtes am Rande der Sonnenscheibe beobachtet. Der bisher provisorisch ermittelte Wert liegt zwischen **0,9** und **1,8** Bogensekunden. Die Theorie fordert **1,7**.
> Berlin, den 9. Oktober 1919.   *A. Einstein.*"

Lorentz followed his telegram with a letter of 7 October 1919. Einstein delighted in Lorentz' news and forwarded the information to numerous friends and family.[236]

Vossische Zeitung began actively promoting Albert Einstein at least as early as 26 April 1914.[237] On 23 July 1918, *Vossische Zeitung* reported,

> **"Das Weltbild des Physikers.**
> Professor Einstein über die Motive des Forschens.
> Anläßlich des 60. Geburtstages von Max Planck, dem Schöpfer der Quantentheorie, veranstaltete die Deutsche Physikalische Gesellschaft eine besondere Sitzung, in der Plancks Verdienste um die Wissenschaft in Ansprachen hervorragender Physiker gewürdigt wurden. Diese Ansprachen liegen jetzt gedruckt vor. (C. F. Müllersche Hofbuchhandlung, Karlsruhe). Der Frankfurter Physiker M. von Laue schildert Plancks thermodynamische Arbeiten, der Münchener Physiker A. Sommerfeld zeigte, wie Planck zur Entdeckung der Quanten kam, Einstein, der Physiker der Berliner Akademie, untersuchte die Motive des Forschens und kommt dabei auf das Weltbild des theoretischen Physikers zu sprechen. Dieses stellt die höchsten Anforderungen an die Straffheit und Exaktheit der Darstellung der Zusammenhänge, wie sie nur die Benutzung der mathematischen Sprache verleiht. Aber dafür muß sich der Physiker stofflich um so mehr bescheiden, indem er sich damit begnügen muß, die allereinfachsten Vorgänge abzubilden, die unserem Erleben zugänglich gemacht werden können, während alle komplexen Vorgänge nicht mit jener subtilen Genauigkeit und Konsequenz, wie sie der theoretische Physiker fordert, durch den menschlichen Geist nachkonstruiert werden können. Höchste Reinheit, Klarheit und Sicherheit auf Kosten der Vollständigkeit. „Was kann es aber

---

236. *The Collected Papers of Albert Einstein*, Volume 9, Documents 110, 112, 113, 117, 121,124, 127, 149, 151, 164, 165, etc., Princeton University Press, (2004). R. W. Clark, *Einstein: The Life and Times*, World Publishing, New York, (1971), pp. 230-231.

237. A. Einstein, "Vom Relativitäts-Prinzip", *Vossische Zeitung*, Morning Edition, (26 April 1914), pp. 1-2; reproduced in *The Collected Papers of Albert Einstein*, Volume 6, Document 1, Princeton University Press, (1996), pp. 3-5.

für einen Reiz haben, einen so kleinen Ausschnitt der Natur genau zu erfassen, alles Feinere und Komplexe aber scheu und mutlos beiseite zu lassen? Verdient das Ergebnis einer so resignierten Bemühung den stolzen Namen „Weltbild"? Ich glaube, der stolze Name ist wohlverdient, denn die allgemeinsten Gesetze, auf welche das Gedankengebäude dr theoretischen Physik gegründet ist, erheben den Anspruch, für jegliches Naturgeschehen gültig zu sein. Aus ihnen sollte sich auf dem Wege reiner gedanklicher Deduktion die Abbildung, d. h. Theorie eines jeden Naturprozesses einschließlich der Lebensvorgänge finden lassen, wenn jener Prozeß der Deduktion nicht weit über die Leistungsfähigkeit menschlichen Denkens hinausginge. Höchste Aufgabe des Physikers ist also das Aufsuchen jener allgemeinsten elementaren Gesetze, aus denen durch reine Deduktion das Weltbild zu gewinnen ist. Zu diesen elementaren Gesetzen führt kein logischer Weg, sondern nur die auf Einfühlung in die Erfahrung sich stützende Intuition ... Die Entwicklung hat gezeigt, daß von denkbaren theoretischen Konstruktionen eine einzige jeweilen sich als unbedingt allen anderen überlegen erweist. Keiner, der sich in den Gegenstand wirklich vertieft hat, wird leugnen, daß die Welt der Wahrnehmungen das theoretische System praktisch eindeutig bestimmt, trotzdem kein logischer Weg von den Wahrnehmungen zu den Grundsätzen der Theorie führt. Mit Staunen sieht der Forscher das scheinbare Chaos in eine sublime Ordnung gefügt, die nicht auf das Walten des eigenen Geistes, sondern auf die Beschaffenheit der Erfahrungswelt zurückzuführen ist; dies ist es, was Leibniz so glücklich als „prästabilierte Harmonie" bezeichnete."[238]

On 15 April 1919, *Vossische Zeitung*, evening edition, reported,

"Grundgedanken der Relativitätstheorie.
Professor Einstein am Vortragstisch.
Nicht nur in der Politik, auch in der Wissenschaft wird der Fortschritt aus der Not geboren, so begann Professor Einstein, das an Jahren jüngste Mitglied unserer Akademie, der Mitschöpfer der modernen Relativitätstheorie, seine Betrachtungen über diese Theorie. Da der Redner bei der überaus zahlreichen Zuhörerschaft, die sich in der Aula der Viktoria-Luisen-Schule auf Einladung des sozialistischen Studentenvereins zusammengefunden hatte, weder auf besonders mathematische, noch physikalische Vorkenntnisse rechnen konnte, so verzichtete er fast völlig auf das anscheinend unentbehrliche mathematische Rüstzeug. Auch die grundlegenden physikalischen Experimente konnten nur kurz in ihren

---

[238]. The published lecture was: A. Einstein, "Motive des Forschens", *Zu Max Plancks sechzigstem Geburtstag. Ansprachen, gehalten am 26. April 1918 in der Deutschen Physikalischen Gesellschaft von E. Warburg, M. v. Laue, A. Sommerfeld und A. Einstein*, C. F. Hofbuchhandlung, Karlsruhe, (1918), pp. 29-32; reprinted in *The Collected Papers of Albert Einstein*, Volume 7, Document 7, Princeton University Press, (2002), pp. 54-59.

entscheidenden Endergebnissen herangezogen werden.

In seinen Betrachtungen geht Einstein von der Relativität der Bewegung aus, wie sie Galilei und Newton gelehrt haben. Er zeigt, daß wir eine absolut gleichförmige Translationsbewegung in keiner Weise definieren können. Zwei sich gleichförmig gegeneinander bewegende Bezugssysteme (Koordinaten-Systeme) sind mechanisch vollkommen äquivalent. Es sind Aussagen von vollkommen gleichem Inhalt, wenn wie einmal das eine System als ruhend und das andere als bewegt ansprechen oder umgekehrt. Es kommt gar nicht darauf an, welches Bezugssystem das ruhende, welches das bewegte ist. Dieses Relativitätsprinzip der Mechanik läßt sich aber nicht ohne weiteres auf die Vorgänge beim Licht, oder allgemeiner, auf die elektrodynamischen Erscheinungen anwenden. Dem widerspricht anscheinend der Fizeausche Versuch. In einer mit gleichförmiger Geschwindigkeit strömenden Flüssigkeit möge sich Licht in Richtung der Strömung fortpflanzen. Nach dem Relativitätsprinzip Galileis müßte ein im Strom treibender Beobachter die gleiche Fortpflanzungsgeschwindigkeit wahrnehmen, wie wenn die Flüssigkeit ruhte. Der außenstehende Beobachter müßte also die Fortpflanzungsgeschwindigkeit des Lichts um die volle Geschwindigkeit der Flüssigkeit vermehrt finden. Das ist aber nicht der Fall. Auch im luftleeren Raum pflanzt sich der Lichtstrahl mit derselben Geschwindigkeit fort. Michelson hat versucht festzustellen, ob die Bewegung der Erde einen Einfluß auf die Lichtgeschwindigkeit hat, aber sowohl seine Experimente, wie die seiner Nachfolger verliefen so, als ob das Relativitätsprinzip der Mechanik auch in der Optik gilt, während das nach dem Fizeauschen Versuch nicht der Fall war. Wie läßt sich dieser Widerspruch lösen? Er liegt, wie Einstein weiter ausführt, [??? *three words illegible*] Voraussetzungen unserer Ueberlegung. Wenn der nicht mitbewegte Beobachter einen Einfluß der Bewegung für den mitbewegten Beobachter festzustellen meint, den dieser selbst nicht wahrnimmt, so liegt das daran, daß beide Beobachter mit verschiedenem Maße messen, daß es verschiedene Dinge sind, die sie als identisch bezeichnen, gleiche Zeitintervalle und gleiche Längen ansprechen. Was gleichzeitig in bezug auf das eine Bezugssystem ist, ist nicht gleichzeitig auf ein anderes Bezugssystem, ebenso ist der Begriff der Länge ebenfalls relativ. Bewegte starke Körper und bewegte Uhren verhalten sich anders als ruhende. Der bewegte Körper verkürzt sich. Eine Uhr, die vom nichtbewegten System aus beurteilt wird, läuft langsamer. Der bewegte Beobachter beurteilt mit seinen Instrumenten die bewegte Welt anders, als der unbewegte Beobachter.

In der knappen Zeit von 1½ Stunden ist es unmöglich, die ganze Gedankenarbeit auch nur in kurzen Umrissen zu schildern, die zur heutigen Relativitätstheorie geführt hat. Aber man erhält doch einen Einblick, wie die Physiker die gedanklichen und physikalischen Schwierigkeiten zu beseitigen versuchen. Wir sehen, wie das moderne Relativitätsprinzip dazu zwingt, die Beziehungen zwischen wägbarer Masse und Energie neu zu gestalten, wie nach dem Relativitätsprinzip jede Energiezunahme auch eine

Massenzunahme zur Folge hat. Tatsächlich haben die neueren Untersuchungen über die Elektronen diese Forderung bestätigt. Auch die Perihelbewegung des Merkur bestätigt die Relativitätstheorie, auch die Aberration des Lichts der Fixsterne dient zu ihrer Stütze. Ende dieses Monats soll ein neuer experimenteller Beweis für sie geführt werden. In Brasilien will man die Sonnenfinsternis daraufhin beobachten, ob eine Ablenkung der Sonnenstrahlen entsprechend dem modernen Relativitätsprinzip stattfinden. K. J."

On 13 May 1919, *Vossische Zeitung* reported,

"**Sonnenfinsternis und Relativitätstheorie.** Die am 29. Mai stattfindende Sonnenfinsternis, deren Totalitätszone sich in einem nach Süden offenen Bogen von Arequipa an der Westküste von Südamerika bis etwa nach Mikindani, an der Ostküste von Afrika erstreckt, gewinnt dadurch eine ganz besondere Bedeutung, daß sie durch ihre lange Totalitätsdauer für die Prüfung der Einsteinschen Theorie besonders geeignet ist. Zu ihrer Beobachtung haben, wie die ,,Naturwissenschaften'' nach englischen Quellen berichten, die Engländer zwei Unternehmungen ausgerüstet. Die eine unter Crommelin geht nach Sobral in Brasilien (etwa 130 Kilometer landeinwärts von der Küste), die zweite unter Eddington auf die portugiesische Isla do Principe (etwa 180 Kilometer von der afrikanischen Küste). Abgesehen von der langen Totalitätsdauer ist diese Sonnenfinsternis durch das reiche Feld an Sternen rings um die Sonne bemerkenswert, und es ist die Aufmerksamkeit auf die dadurch gegebene, überaus günstige Gelegenheit gelenkt worden, die Einsteinsche Relativitätstheorie zu prüfen. Nach diesen muß ein Strahl, der von eiem Stern aus tangential zur Sonne verläuft, 1,74" abgelenkt werden und die Ablenkung für andere Sterne umgekehrt proportional ihrem Abstande vom Mittelpunkte der Sonne sein. Fällt die Entscheidung für Einstein, so würde das zusammen mit seinem Erfolge in der Erklärung der Bewegung des Merkurperihels, genügen, um seine Lehre als das wirkliche System des Universums anzunehmen. Auch ihre endgültige Widerlegung aber würde von Nutzen sein, da sie die Verschwendung weiterer Kraft auf ihre Ausarbeitung verhindern würde, obwohl diese Theorie, wie die ,,Nature'' bemerkt, als scharfsinniges System idealer Geometrie noch immer unsere Bewunderung verdienen würde."

On 21 July 1919, *Vossische Zeitung* reported,

"**Die Sonnenfinsternis am 29. Mai.** Wie die englische Zeitschrift ,,Nature'' vom 5. Juni meldet, hat die englische Expedition, die in Sobral in Brasilien arbeitete, günstiges Wetter gehabt. Die gestellten Aufgaben ließen sich befriedigend durchführen. Alle zu erwartenden Sterne sind auf den photographischen Platten herausgekommen. Auch die nach Eddington an der Küste Westafrikas gesandte Expedition ist mit ihren Erfolgen zufrieden. Beide Expeditionen sollten, wie schon gemeldet, die dicht bei der Sonne stehenden Sterne photographisch aufnehmen, um die Einsteinsche Theorie zu prüfen. Die Aufnahmen während der Sonnenfinsternis dienen zum Vergleich mit Aufnahmen derselben Himmelsgegend bei Nacht, um eine etwaige Verschiebung zu entdecken, die man auf die Anwesenheit der Sonne in diesem Feld als Ursache zurückführen kann."

On 15 October 1919, *Vossische Zeitung* reported,

"**Sonnenfinsternis und Relativitätstheorie.** Nach einer Mitteilung des neuesten Heftes der „Naturwissenschaften" hat die zur Beobachtung der Sonnenfinsternis am 29. Mai ausgesandte englische Expedition die von der allgemeinen Relativitätstheorie geforderte Ablenkung des Lichtes am Rande der Sonnenscheibe beobachtet. Der bisher provisorisch ermittelte Wert (die Durchrechnung der Beobachtungsresultate ist noch nicht beendet) liegt zwischen **0,9** und **1,8** Bogensekunden, die Theorie fordert **1,7**.

Eine der wichtigsten Folgerungen der Einsteinschen Theorie ist die Abhängigkeit der Lichtgeschwindigkeit von dem sogenannten Gravitationspotential, und die sich dadurch ergebende Krümmung eines Lichtstrahles bei seinem Durchgang durch ein Gravitationsfeld. Die Theorie ergibt für einen dicht an der Sonne vorbeigehenden Lichtstrahl, der z. B. von einem Fixstern herkommt, eine Krümmung seiner Bahn. Infolge der Krümmung muß man den Stern gegen seinen wahren Ort am Himmel um einen Betrag verschoben sehen, der am Sonnenrande **1,7** Sekunden beträgt und proportional dem Abstande vom Sonnenmittelpunkte abnimmt. Da aber die photographische Aufnahme des an der Sonne vorbeigehenden von einem Fixstern herkommenden Lichtes nur dann möglich ist, wenn das alles überstrahlende Licht der Sonne am Eintritt in unsere Atmosphäre gehindert wird, so kommen nur die seltenen totalen Finsternisse für diese Beobachtung und die Lösung der Aufgabe in Betracht. Die Sonnenfinsternis am 29. Mai dieses Jahres, während der die Engländer auf zwei Beobachtungsstationen im Hinblick auf dieses Problem photographische Aufnahmen gemacht haben, hat das erforderliche Material zur Entscheidung geliefert."

On 18 November 1919, *Vossische Zeitung* reported,

"Einstein und Newton.

Die Ergebnisse der Sonnenfinsternis vom Mai 1919.

Wie erinnerlich hatte England eine Expedition ausgesandt mit der Aufgabe, die Erscheinungen der Sonnenfinsternis vom 29. Mai d. J. photographisch festzuhalten. Als geeigneter Ort hierfür war Sobral in Nord-Brasilien bezeichnet worden. Es wurde damals telegraphisch gemeldet, daß die Abordnung ihre Aufgabe voll erfüllen konnte. Inzwischen sind die Mitglieder der Expedition nach England zurückgekehrt und haben der britischen Astronomischen Gesellschaft Bericht erstattet.

Professor C. Davidson von der Greenwich-Sternwarte sprach sich des näheren einem „Times"-Redakteur gegenüber über diese Ergebnisse aus. Davidson bestätigte, daß die im Augenblick der totalen Verfinsterung der Sonnenscheibe an Kappa 1 und Kappa 2, nahe dem Sternbild der Hyaden, angestellten Beobachtungen die vollständige Richtigkeit der Ablenkung der Lichtstrahlen durch die Schwerkraft der Sonne ergeben haben. Auf den vom Professor Rewall von der Universität Cambridge erhobenen Einwand, daß diese Ablenkung durch eine noch unbekannte Sonnen-Atmosphäre von ungeahnter Ausdehnung und noch unbekannter Kraft verursacht sein könnte, erwidert Professor Davidson: „Das ist nicht möglich, denn um eine derartige Ablenkung hervorzurufen, müßte eine Atmosphäre vorhanden sein, die jeder bisherigen Theorie und Beobachtung widerspricht. Ueberdies sind Kometen beobachtet worden, die in einem, den Sonnenraum fast streifenden Abstande von der Sonne ihre Bahn ohne jede Störung verfolgt haben." Davidson trennt sich demnach nicht von der Anschauung, daß die Entdeckung einer Lichtquelle, die sowohl Gewicht als Körper besitzt, einen Fortschritt für die Auffassung bedeutet, daß außhalb des drei-dimensionalen Raumes, wie wir ihn heute kennen, noch besondere Bedingungen vorhanden sind. Professor Einsteins Theorie, so bemerkte Davidson, verlangt u. a. eine Verschiebung der Spektrallien nach dem Rot hin. Diese Forderung hat auch Dr. St. John auf Mount Wilson in Amerika nachgeprüft, doch bisher ohne jeden Erfolg. Nichtsdestoweniger sind gewisse Abweichungen in dem Verhalten der Spektrallien vorhanden, für die, nach Meinung einer großen Zahl von Gelehrten, eine befriedigende Erklärung gefunden werden könnte. Was aber jene in Brasilien gemachte hauptsächlichste Entdeckung anbelangt, so pflichtet Professor Davidson voll der Meinung bei, daß das Newtonsche Prinzip umgeworfen worden sei und daß Professor Einstein wenigstens bezüglich zweier seiner drei Voraussagen recht hat. Seine Vermutung bezüglich des Spektrums, versicherte der Greenwicher Professor, bleibt noch den Beweis schuldig. Betreffs der Lichtablenkung aber haben die in Brasilien vorgenommenen Beobachtungen ergeben, daß an Stelle einer Ablenkung von **0,87** Bogensekunden am Sonnenrande, wie man sie nach dem Newtonschen Gesetze allenfalls hätte erwarten können, diese Ablenkung **1,75** betrug, wie sie nach Einsteins Theorie auch sein sollte."

Vossische Zeitung continued to promote the eclipse observations and Einstein on 8 December 1919, 27 January 1920, 7 February 1920, and 24 February 1920. On 30 November 1919, Erwin Freundlich, a Jewish man who considered himself to have been Einstein's friend, though Einstein had ridiculed him behind his back,[239] and a man who had a personal interest in the promotion of the eclipse observations, published an article in the morning edition of *Vossische Zeitung*, which promoted Einstein. Freundlich had been the brains

---

239. *See:* Letter from A. Einstein to A. Sommerfeld of 2 February 1916, *The Collected Papers of Albert Einstein*, Volume 8, Document 186, Princeton University Press, (1998).

behind Einstein's plagiarism of the general theory of relativity from Marcel Grossmann and David Hilbert, though Einstein took all of the credit.

Freundlich was trying to advance his career and increase his salary and his success depended on the acceptance of the general theory of relativity by German astronomers. Times were hard in Europe after the First World War. Einstein's friends desperately needed money and believed they could not succeed without promoting Einstein. Einstein's friends often complained to him that they needed money and asked for his help in furthering their careers. Freundlich sought to profit from a book he had published on relativity theory, and from its translation into English—as did Einstein's acquaintance Moritz Schlick—and they had Einstein intervene with the publishers to increase their profits.[240] Freundlich was corrupt through and through, as were Einstein and Schlick.

Freundlich's article is notable for many things. "Einstein's" theory was not initially popular—in fact it was very unpopular in the scientific world. Freundlich was keenly aware that his own institution would not back him due to the lack of support for relativity theory. The majority of physicists and astronomers opposed the general theory of relativity. He also knew that there was strong evidence against the general theory of relativity.[241] Einstein wrote to Freundlich on 19 September 1919,

> "You are entirely right that getting you a position in Potsdam should not be attempted *for the present*. The Gen. Th. of Rel. must win acceptance among astronomers beforehand."[242]

Einstein and his friends knew that they needed a public following and the acceptance of astronomers in order to be successful in setting aside the "old" ideas—in order to forward their careers. Knowing that they had plagiarized it,

---

240. E. Freundlich, *Die Grundlagen der Einsteinschen Gravitationstheorie*, J. Springer, Berlin, (1916); English translation: *The Foundations of Einstein's Theory of Gravitation*, Cambridge University Press, (1920). *See also:* M. Schlick, *Raum und Zeit in der gegenwärtigen Physik. Zur Einführung in das Verstandnis der allgemeinen Relativitätstheorie*, Springer, Berlin, (1917); English translation: *Space and Time in Contemporary Physics: An Introduction to the Theory of Relativity and Gravitation*, Oxford University Press, New York, (1920). *See also: The Collected Papers of Albert Einstein*, Volume 9, Documents 105, 119, 222, 228, 234, 240, 249, 275, 285, 392, 393, Princeton University Press, (2004).
241. Letter from E. Freundlich to A. Einstein of 15 September 1919, *The Collected Papers of Albert Einstein*, Volume 9, Document 105, Princeton University Press, (2004).
242. Letter from A. Einstein to E. Freundlich of 19 September 1919, English translation by A. Hentschel, *The Collected Papers of Albert Einstein*, Volume 9, Document 106, Princeton University Press, (2004), pp. 89-90, at 89. Freundlich's fortunes changed after Einstein began to spread word of Lorentz's news that the English confirmed that a deflection of light at the limb of the sun had been measured. *See: The Collected Papers of Albert Einstein*, Volume 9, Documents 119, 168 and 194, Princeton University Press, (2004).

they nevertheless speciously promoted the theory of relativity as a completely new approach, one which was unique to Einstein and one which he allegedly thought up in his head without any empirical inspiration. They did this in part to deceive the public and make a hero out of Einstein. They also were forced to do this, because Einstein had plagiarized the works and failed to reference his sources.

Note that Freundlich lauds Einstein; but the names of Poincaré, Mach, Bateman, Hilbert, Gerber, Maxwell, FitzGerald, Larmor, Cohn, Lorentz, Minkowski, etc. are conspicuously missing from his piece; such that one must conclude that it was not the ideas which were considered significant, because they were not considered significant under the pens of Einstein's predecessors, but it was instead the promotion of Albert Einstein as a hero that was foremost on Freundlich's mind. Freundlich was also able to blackmail Einstein as a means to promote himself, Freundlich, because Freundlich could have exposed Einstein as a plagiarist and a fraud at any time.

Furthermore, it would have been impossible to have advertised Einstein the way Einstein's friends sought to advertize him, and to still have named a just handful of Einstein's predecessors—the historical facts and the circus promoter's fancy simply did not agree. For example, the perihelion motion of the planet Mercury was taken as proof that Einstein was correct and the implication was that Einstein had predicted a previously unknown effect with a non-Newtonian theory of gravity premised on the belief that gravity propagates at light speed. In fact, the perihelion motion of Mercury was observed long before Einstein was born. The equations Einstein used to describe it in 1915 were first published by Paul Gerber in 1898. Gerber believed that gravity propagates at light speed and attempted to prove it with Mercury as an empirical example. Einstein and Freundlich were aware of these facts and deliberately lied to the public.

Einstein, himself, admitted that the hype promoting him was unfounded,

"'There has been a false opinion widely spread among the general public,' [Einstein] said, 'that the theory of relativity is to be taken as differing radically from the previous developments in physics from the time of Galileo and Newton—that it is violently opposed to their deductions. The contrary is true. Without the discoveries of every one of the great men of physics, those who laid down preceding laws, relativity would have been impossible to conceive and there would have been no basis for it. Psychologically, it is impossible to come to such a theory at once without the work which must be done before. The four men who laid the foundations of physics on which I have been able to construct my theory are Galileo, Newton, Maxwell, and Lorenz.'"[243]

---

243. A. Einstein quoted in "Einstein, Too, Is Puzzled; It's at Public Interest", *The Chicago Tribune*, (4 April 1921), p. 6.

When Einstein critic Ernst Gehrcke made similar statements, Einstein called him "anti-Semitic".

Speaking anecdotally, it amazes your author how relativists praise specific ideas, when they attribute them to Einstein, but when it is proven to them that another person wrote the same thing before Einstein, these same relativists call these same ideas "insignificant" and "obvious". They then change the subject to another idea they wrongfully attribute to Einstein, and the pattern repeats itself, until they feel forced to change the subject from mathematical formalism to Metaphysics, or vice versa, then to the combination of mathematical formalism with Metaphysics in the theory of relativity which they mistakenly attribute to Einstein, and when even this is proven unoriginal, they either circle back to the start as if their views had not been refuted, or they launch a personal attack, or they change the subject to racial, nationalistic or humanitarian politics and issues. There appears to be a deep need for the hero not to be toppled—especially among racist and ethnically biased Jews, and it is the childish and fawning love of this hero, "Einstein", not his mythologies, which is at the core of the Einstein legend. The theories may be debunked, diminished or demeaned, but the love of the man cannot be shaken among his devout and blind followers—no matter what the facts tell us about him.

So powerful was the initial propaganda of self-interested liars like Alexander Moszkowski, Erwin Freundlich, Max Born, and the others, so vulnerable and gullible are his admirers, that nothing can shake off their religious fervor for the man. They are eager to excuse his sadistic mistreatment of his family and friends, his career of plagiarism, his irrationality, his racism, his misogyny, and his nationalistic segregationist bigotry. Nothing can make them fall out of love with their shaggy-haired comic book hero. What is worse for them is the fact that Einstein has been so shamelessly overrated for so long, that for them to admit to the truth is to admit to their past gullibility, or deliberate dishonesty and, often, racist bias.

Similar hero worship had attended the cults which arose around Aristotle, Spinoza, Copernicus, Des Cartes, Newton, and, in the time of Einstein, Leonardo da Vinci. Einstein and his promoters knew their history and knew how to manufacture a "star-artist cult" around Einstein, which they could then use to promote a theory with no practical implications (believed by them at the time), which would make Einstein a powerful political force in the international arena, who could then do great good—in their eyes, by creating a race war between Jews and Gentiles.

R. S. Shankland stated,

> "About publicity Einstein told me that he had been *given* a publicity value which he did not *earn*. Since he had it he would use it if it would do good; otherwise not."[244]

---

[244]. R. S. Shankland, "Conversations with Albert Einstein", *American Journal of Physics*, Volume 31, Number 1, (January, 1963), pp. 47-57, at 56. Also see Einstein's letters to Zangger of late December, 1919, and of January, 1920, in which he discusses

Albert Einstein stated on 27 April 1948,

"In the course of my long life I have received from my fellow-men far more recognition than I deserve, and I confess that my sense of shame has always outweighed my pleasure therein."[245]

Albert Einstein told Peter A. Bucky,

"Peter, I fully realize that many people listen to me not because they agree with me or because they like me particularly, but because I am Einstein. If a man has this rare capacity to have such esteem with his fellow men, then it is his obligation and duty to use this power to do good for his fellow men."[246]

Einstein "had been *given* a publicity value which he did not *earn*" so that he could promote political Zionism among Jews. Political Zionism is a racist movement among Jews meant to segregate Jews in Palestine in order to end the assimilation of Jews into other cultures and "races". In 1919, most Jews opposed this racist movement and the Zionists needed a famous spokesman to help overcome this resistance to Zionism among Jews.

Albert Einstein confided to his old friend and confidant Michele Besso, on 12 December 1919, that he planned to attend a Zionist conference dedicated to founding a Hebrew university in Palestine. Einstein wrote,

"The reason I am going to attend is not that I think I am especially well qualified, but because my name, in high favor since the English solar eclipse expeditions, can be of benefit to the cause by encouraging the lukewarm kinsmen."[247]

In his book *The Jewish State*, Theodor Herzl laid emphasis on the need of celebrity and publicity to promote Zionism. The same is true of his diary. In 1897, Theodor Herzl told the First Zionist Congress,

"We Zionists wish to urge self-help on the people; thereby no exaggerated and unsound hopes will be awakened. On this ground, also, publicity in

---

the cult surrounding him.
245. A. Einstein, "On Receiving the One World Award", *Out of My Later Years*, Philosophical Library, New York, (1950); here quoted from: *Ideas and Opinions*, Crown, New York, (1954), pp. 146-147.
246. P. A. Bucky, Einstein, and A. G. Weakland, *The Private Albert Einstein*, Andrews and McMeel, Kansas City, (1992), p. 32, *see also:* pp. 110, 116-117.
247. Letter from A. Einstein to M. Besso of 12 December 1919, English translation by A. Hentschel, *The Collected Papers of Albert Einstein*, Volume 9, Document 207, Princeton University Press, (2004), pp. 178-179, at 178.

dealing with this point is of the highest value. [***] The confidence of the State, which is necessary for a settlement of large masses of Jews, can only be gained by publicity and by loyal action."[248]

Paul Ehrenfest wrote to Albert Einstein on 9 December 1919,

"I hear, for ex., that your accomplishments are being used to make propaganda, with the 'Jewish Newton, who is simultaneously an ardent Zionist' (I personally haven't *read* this yet, but only *heard* it mentioned). [***] But I cannot go along with the propagandistic fuss with its *inevitable* untruths, precisely *because* Judaism is at stake and *because* I feel myself so thoroughly a Jew."[249]

Most people probably think that we today are the most politically sophisticated generation of all times, having the benefit of the recorded history of all other times to guide us. I do not think we today are, in general, nearly as politically sophisticated as the Europeans of the early Twentieth Century. The reasons for this are many, and I suspect include the overspecialization of today's students, which does not give them a broad enough knowledge of many fields of study to gain the insights needed to absorb the fuller meanings of what they are told, and they too often lack the willingness and ability to judge all aspects of the information presented to them as if facts. Many too often succumb to the opinions of others based on their credentials alone and are reluctant to rely upon logic and research, and instead submit to authority. Physicist Ernst Gehrcke noted that this was already becoming a problem in the 1920's, and Sociologist Max Weber's concerns over the bureaucratic control of human behavior have since been justified. Another problem is the fact that the internationalization and attendant standardization of thought has diminished competition in the arena of ideas and replaced it with cult figures who dominate the debate, not through talent, but through relentless commercial promotion.

At any rate, Einstein's friends were very sophisticated politically. Einstein was himself manipulative. Einstein had a good teacher in his mother on how to manipulate people and circumstances. His friends in the scientific community, and in the press, came to his aid in a most corrupt fashion whenever he needed their help. It appears odd that these scientists were determined to promote Einstein as if a revolutionary figure in the popular press, when they knew that he was not, until one realizes that they were his friends and had selfish interests in promoting and perpetuating the cult of Einstein for personal profit.

Article after article appeared in the popular press aggrandizing and sanctifying the man, but nothing was written about how "his" theory allegedly

---

[248]. "The Zionist Congress: Full Report of the Proceedings", *The Jewish Chronicle*, (3 September 1897), pp. 10-15, at 11.
[249]. Letter from P. Ehrenfest to A. Einstein of 9 December 1919, English translation by A. Hentschel, *The Collected Papers of Albert Einstein*, Volume 9, Document 203, Princeton University Press, (2004), pp. 173-175, at 174.

changed everyday life so as to make it deserving of the abundant *news* coverage that it received—all of which is why Reuterdahl dubbed Einstein the "Barnum of the scientific world". While others made important discoveries that benefitted humanity in unprecedented ways, it was Einstein who was aggressively promoted in the press. The wealthy internationalist Richard Fleischer wrote to Einstein on 21 December 1919 offering grant money for research into any practical applications the theory of relativity might have, with the goal of promoting international cooperation in the sciences. The best Einstein could offer was a self-serving experiment on spectral lines by Grebe and Bachem meant to eliminate the doubts cast on the general theory of relativity by the experiments of St. John and others.[250] This had no practical implications to the man on the street.

The astronomer W. J. S. Lockyer was quoted in *The New York Times* on page 17, 10 November 1919,

> "The discoveries, while very important, did not, however, affect anything on this earth. They do not personally concern ordinary human beings; only astronomers are affected."

The New York Times later reported on 25 November 1919, page 17,

> "The effects on practical astronomy of the verification of Einstein's theory were not very great. It was chiefly in the field of philosophical thought that the change would be felt."

Einstein was quoted in *The Chicago Tribune* on 4 April 1921 on page 6,

> "Whatever the value of relativity, it will not necessarily change the conceptions of the man in the street, said Prof. Einstein. 'The practical man does not need to worry about it,' he said."

Erwin Freundlich, in his article which follows, does not acknowledge the fact that the empirical basis of the theory was known before the theory was developed and applied to it, and that the alleged experimental confirmations and predictions were known beforehand, or were corrupted and misrepresented to fit the theory. Freundlich, as a scientist, must have known that his declarations were, at best, incorrect and premature.

The fundamental belief of science is that of generalization. A non-Newtonian theory of gravitation which describes the known motion of the perihelion of Mercury automatically leads to a non-Newtonian prediction of the deflection of a ray of light grazing the sun, and a shift in the spectral lines, and vice versa. The inductive analysis of one of these known problems leads to

---

250. *The Collected Papers of Albert Einstein*, Volume 9, Documents 227, 238, and 283, Princeton University Press, (2004).

generalizations which deduce the solution to the other, such that there was no great insight in clarifying the known problems with known solutions, which is to say that geometrical laws circularly defined to describe one motion ought to describe all of Nature, if Nature is truly uniform, *cæteris paribus*.

A key facet (and specious *fecit*) of the modern propaganda promoting Einstein is the myth that he had thought up the physical problems in his head and derived their solutions by himself with original thought experiments. The solutions and approaches, contrary to Moszkowski and Freundlich's self-serving propaganda, were developed before Einstein by Voigt, FitzGerald, Lorentz, Larmor, Poincaré, Poisson, Gerber, Cohn, Minkowski, Bateman, Varičak, Grossmann, Hilbert, Schwarzschild, and many others; and the physical problems were known through the research of Soldner, Leverrier, Michelson and *Freundlich*, among many others, before Einstein.

Freundlich, of course, knew most of this, though he failed to disclose these facts to the public. Freundlich himself worked on the eclipse idea and Eddington expressed regret that Freundlich was not the first to experimentally test the theory, though he was "first in the field"—a comment which caught Einstein's attention.[251] As is proven by a letter from Max Born to David Hilbert dated 23

---

251. *The Collected Papers of Albert Einstein*, Volume 9, Documents 186, 187 and 216, Princeton University Press, (2004). ***See also:*** *The Collected Papers of Albert Einstein*, Volume 5, Documents 492 and 506, Princeton University Press, (1993). ***See also:*** Letter from A. Einstein to P. Ehrenfest of 19 August 1914, *The Collected Papers of Albert Einstein*, Volume 8, Part A, Document 34, Princeton University Press, (1998), pp. 56-57, *especially* p. 57, note 4. ***See also:*** E. Freundlich, "Über einen Versuch, die von A. Einstein vermutete Ablenkung des Lichtes in Gravitationsfeldern zu prüfen", *Astronomische Nachrichten*, Volume 193, (1913), cols. 369-372; **and** "Zur Frage der konstanz der Lichtgeschwindigkeit", *Physikalische Zeitschrift*, Volume 14, (1913), pp. 835-838; **and** "Über die Verschiebung der Sonnenlinien nach dem roten Ende auf Grund der Hypothesen von Einstein und Nordström", *Physikalische Zeitschrift*, Volume 15, (1914), pp. 369-371; **and** "Über die Verschiebung der Sonnenlinien nach dem roten Ende des Spektrums auf Grund der Äquivalenzhypothese von Einstein", *Astronomische Nachrichten*, Volume 198, (1914), cols. 265-270; **and** *Astronomische Nachrichten*, Volume 199, (1915), cols. 363-365; **and** "Über die Gravitationsverschiebung der Spektrallinien bei Fixsternen", *Physikalische Zeitschrift*, Volume 16, (1915), pp. 115-117; **and** *Beobachtungs-Ergebnisse der Königlichen Sternwarte zu Berlin*, Number 15, (1915), p. 77; **and** "Über die Erklärung der Anomalien im Planeten-System durch die Gravitationswirkung interplanetarer Massen", *Astronomische Nachrichten*, Volume 201, (1915), cols. 49-56; **and** "Über die Gravitationsverschiebung der Spektrallinien bei Fixsternen", *Astronomische Nachrichten*, Volume 202, (1915), cols. 17-24; **and** "Über die Gravitationsverschiebung der Spektrallinien bei Fixsternen", *Astronomische Nachrichten*, Volume 202, (1916), cols. 17-24; **and** *Astronomische Nachrichten*, Volume 202, (1916), col. 147; **and** "Die Grundlagen der Einsteinschen Gravitationstheorie", *Die Naturwissenschaften*, Volume 4, (1916), pp. 363-372, 386-392; **and** *Die Grundlagen der Einsteinschen Gravitationstheorie*, Multiple Revised and Enlarged Editions; **and** "Über die singulären Stellen der Lösungen des n-Körper-Problems", *Sitzungsberichte der Königlichen Preussischen Akademie der Wissenschaften zu Berlin*, (1918), pp. 168-188; **and** "Zur Prüfung der allgemeine Relativitätstheorie", *Die Naturwissenschaften*, Volume

November 1915,[252] Erwin Freundlich knew that David Hilbert had first derived and discovered the generally covariant field equations of gravitation of the general theory of relativity, which Freundlich and Einstein plagiarized from Hilbert on 25 November 1915—Freundlich likely being the true primary author of the subsequent paper on the field equations of gravitation attributed to Einstein.[253]

Fruendlich, Born and Moszkowski were but a few of Einstein's many dishonest friends. Max Planck and Max von Laue were well aware that Poincaré had anticipated Einstein, which we know because they cited Poincaré's work in their early works on Poincaré's principle of relativity. In 1905 and 1906, Paul Ehrenfest considered Lorentz' 1904 paper[254] on special relativity and Poincaré's 1905 Rendiconti paper[255] on space-time to be the most significant work (both

---

7, (1919), pp. 629-636, 696; **and** "Über die Gravitationsverschiebung der Spektrallienien bei Fixsternen. II. Mitteilung", *Physikalische Zeitschrift*, Volume 20, (1919), pp. 561-570.

252. Letter from M. Born to D. Hilbert of 23 November 1915, Niedersächische Staats- und Universitätsbibliothek Göttingen, Cod. Ms. D. Hilbert 40 A: Nr. 11; the relevant part of which is reproduced in D. Wuensch, *„zwei wirkliche Kerle"*: *Neues zur Entdeckung der Gravitationsgleichungen der Allgemeinen Relativitätstheorie durch Albert Einstein und David Hilbert*, Termessos, Göttingen, (2005), pp. 73-74.

253. A. Einstein, "Die Feldgleichungen der Gravitation", *Sitzungsberichte der Königlich Preussischen Akademie der Wissenschaften zu Berlin der physikalisch-mathematischen Classe*, (1915), pp. 844-847.

254. H. A. Lorentz, "Electromagnetische Verschijnselen in een Stelsel dat Zich met Willekeurige Snelheid, Kleiner dan die van Het Licht, Beweegt", *Koninklijke Akademie van Wetenschappen te Amsterdam, Wis- en Natuurkundige Afdeeling, Verslagen van de Gewone Vergaderingen*, Volume 12, (23 April 1904), pp. 986-1009; translated into English, "Electromagnetic Phenomena in a System Moving with any Velocity Smaller than that of Light", *Proceedings of the Royal Academy of Sciences at Amsterdam* (*Noninklijke Nederlandse Akademie van Wetenschappen te Amsterdam*), 6, (May 27, 1904), pp. 809-831; reprinted *Collected Papers*, Volume 5, pp. 172-197; a redacted and shortened version appears in *The Principle of Relativity*, Dover, New York, (1952), pp. 11-34; a German translation from the English, "Elektromagnetische Erscheinung in einem System, das sich mit beliebiger, die des Lichtes nicht erreichender Geschwindigkeit bewegt," appears in *Das Relativitätsprinzip: eine Sammlung von Abhandlungen*, B. G. Teubner, Leipzig, (1913), pp. 6-26.

255. H. Poincaré, "Sur la Dynamique de l'Électron", *Rendiconti del Circolo matimatico di Palermo*, Volume 21, (1906, submitted July 23rd, 1905), pp. 129-176; reprinted in H. Poincaré, *La Mécanique Nouvelle: Conférence, Mémoire et Note sur la Théorie de la Relativité / Introduction de Édouard Guillaume*, Gauthier-Villars, Paris, (1924), pp. 18-76; reprinted *Œuvres*, Volume IX, pp. 494-550; redacted English translation by H. M. Schwartz with modern notation, "Poincaré's Rendiconti Paper on Relativity", *American Journal of Physics*, Volume 39, (November, 1971), pp. 1287-1294; Volume 40, (June, 1972), pp. 862-872; Volume 40, (September, 1972), pp. 1282-1287; English translation by G. Pontecorvo with extensive commentary by A. A. Logunov with modern notation, *On the Articles by Henri Poincaré ON THE DYNAMICS OF THE ELECTRON*, Publishing Department of the Joint Institute for Nuclear Research, Dubna, (1995), pp. 15-78.

historically and scientifically) on the subject of the principle of relativity. Ehrenfest and his wife Tatiana attended David Hilbert's 1905 Göttingen seminars on electron theory, which described Lorentz' and Poincaré's work on special relativity. In 1911 in a long and well-referenced paper[256] written in consultation with Lorentz on the principle of relativity, space-time and the perihelion motion of Mercury; Willem de Sitter extensively cited Poincaré, but did not mention Einstein, and de Sitter knew that Lorentz and Poincaré had created the theory of relativity before Einstein. Minkowski, at times, took credit for many of Poincaré's insights, and falsely credited Einstein with Poincaré's ideas on time in Minkowski's most famous lecture "Space and Time" of 28 September 1908 delivered in Cologne. David Hilbert must have been aware of these facts—we know that Minkowski was, because he acknowledged Poincaré's work in earlier statements. Arnold Sommerfeld, whom Einstein characterized as deceitful,[257] was aware of this, and, according to Lewis Pyenson,

> "Sommerfeld was unable to resist rewriting Minkowski's judgment of Einstein's formulation of the principle of relativity. [***] Sommerfeld also suppressed Minkowski's conclusion, where Einstein was portrayed as the clarifier, but by no means as the principal expositor, of the principle of relativity."[258]

Lorentz and Sommerfeld failed to include any of Poincaré's work in their famous collection of papers *Das Relativitätsprinzip* of 1913, though they included Einstein's papers and Minkowski's lecture "Space and Time". No scientist would today dare to try to lay claim to all that preceded her the way that Einstein and his friends did, even if she assembled specific known empirical facts and predictions with known theory the way that Einstein and his friends did—often with mistakes and contradictions.

Note Feundlich's overblown title and bear in mind that it was written soon after Germany's defeat in the First World War. Freundlich wrote in the 30 November 1919 morning edition of *Vossische Zeitung*:

"Albert Einstein.
Zum Siege seiner Relativitätstheorie.

Von
Erwin Freundlich, Neubabelsberg.

---

[256]. W. de Sitter, "On the Bearing of the Principle of Relativity on Gravitational Astronomy", *Monthly Notices of the Royal Astronomical Society*, Volume 71, (March, 1911), pp. 388-415.
[257]. Letter from A. Einstein to H. and M. Born of 27 January 1920, *The Collected Papers of Albert Einstein*, Volume 9, Document 284, Princeton University Press, (2004).
[258]. L. Pyenson, *The Young Einstein: The Advent of Relativity*, Adam Hilger, Boston, (1985), p. 82.

In Deutschland hat ein wissenschaftliches Ereignis von außerordentlicher Bedeutung noch nicht den Widerhall gefunden, den es seiner Bedeutung nach verdient. Anläßlich der Sonnenfinsternis am 29. Mai dieses Jahres haben englische Astronomen eine wichtige Voraussage der Einsteinschen Relativitätstheorie, nämlich die Ablenkung eines Lichtstrahles im Gravitationsfelde der Sonne, bestätigt gefunden und damit eine Erkenntnis sichergestellt, die von ausschlaggebender Bedeutung für unsere Auffassung von Raum, Zeit und Materie in der Physik ist. Es ist keine Uebertreibung, wenn wir dieses Ereignis als einen Wendepunkt in der Geschichte der Naturwissenschaften feiern, nur zu vergleichen mit Epochen, welche mit den Namen Ptolemäus, Kopernikus, Kepler und Newton verknüpft werden.

Wenn es auch nicht möglich ist, an dieser Stelle die Grundzüge der Einsteinschen Theorie darzulegen, so will ich doch versuchen, die große Linie in der Entwicklung der Physik bis zur Einsteinschen Relativitätstheorie aufzuzeichnen, um die volle Würdigung seiner genialen Leistungen zu ermöglichen.

Das Weltbild, welches sich das Altertum gebildet hatte, ist durch den Umstand gekennzeichnet, daß in den Mittelpunkt der Welt der Mensch, d. h. die Erde, gesetzt wurde, um welche alle Himmelskörper in Kreisen sich bewegen sollten, Gäbe es keine Planeten, so wäre die Durchführung dieser Auffassung nicht auf solche Schwierigkeiten gestoßen. Da tat Kopernikus um 1543 den ersten großen Schritt. Er entthronte die Erde und erhob die Sonne zum Mittelpunkt der Welt. Diese Tat stellt wohl den entscheidendsten Fortschritt in der Gestaltung unseres Weltbildes dar; doch hafteten ihr zu Anfang noch mannigfache Schwächen an, bis Kepler seine bekannten Gesetze aufstellte.

Was die Entwicklung bis dahin charakterisiert, ist der Umstand, daß man sich noch nicht bemühte, durch Aufstellung allgemeiner Prinzipien zu einer einheitlichen Auffassung der mannigfachen auch auf der Erde beobachteten Bewegungserscheinungen fortzuschreiten. Den Beginn mit einer so vertieften Naturbeschreibung machte Galilei, als er den Begriff der Trägheit schuf und den Grundsatz aufstellte: Jeder bewegte Körper behält infolge seiner Trägheit eine einmal gewonnene Geschwindigkeit bei, es sei denn, daß eine bremsende Kraft sie allmählich verringert. Als Galilei seine Bewegungsgesetze aufstellte, stand ihm vielleicht eine einheitliche Erfassung aller Bewegungsvorgänge, auch der der Himmelskörper, als fernes Ziel vor Augen. Zu diesem führte uns aber erst Newton hin. Er verschmolz die Fallerscheinungen auf der Erde mit den Bewegungsvorgängen der Planeten und Monde, indem er neben dem Begriff der Trägheit den der Schwere eines Körpers schuf und sein mathematisch außerordentlich einfaches Gravitationsgesetz aufstellte. Auf seinen Aufsätzen baut sich die „klassische" Mechanik auf, die in einer Kette unerhörter Erfolge alle Bewegungsvorgänge im Sonnensystem mit einer solchen Genauigkeit zu verfolgen erlaubte, daß viele glaubten, hier sei man zu einer ganz endgültigen Theorie der Bewegungserscheinungen gelangt, die in ihren Fundamenten niemals erschüttert werden können. Und doch nagte schon damals der Wurm an den Wurzeln des hochgeschossenen und weit verästelten Baumes; und niemand verspürte vielleicht tiefer die angeborenen Schmächen der Theorie als ihr Schöpfer, Newton, selbst.

Die Newtonsche Mechanik arbeitet nämlich mit verschiedenen Grundbegriffen, über deren physikalische Bedeutung und Beziehung zueinander man nie so recht ins Reine kam. Z. B., obwohl wir ausschließlich die Bewegungen von Körpern relativ zueinander wahrnehmen, tritt doch in der Newtonschen Mechanik der leere Raum als ein physikalisches Ding auf, welches für das Auftreten der Zentrifugalkräfte, die wir auf rotierenden Körpern feststellen, verantwortlich gemacht wird. Schon Newton empfand das physikalisch Unbefriedigende einer solchen Auffassung. Oder, um noch ein Beispiel anzuführen: in die Newtonsche Mechanik werden zwei von einander unabhängige Grundattribute eines jeden Körpers, nämlich seine Schwere und seine Trägheit, eingeführt. Als man an die Messung der Beträge dieser beiden Größen heranging, entdeckte man das anscheinend mit aller [??? *Three to five words illegible on my photocopy.*], daß die träge und schwere Masse aller Körper stets absolut gleich sind. Sollte diese Uebereinstimmung ein reiner Zufall sein? Oder ist nicht vielmehr zu vermuten, daß eine Theorie wie die Newtonsche, in welcher dieses Grundgesetz für alle Materie keine tiefere Begründung findet, in ihren Grundlagen verfehlt ist?

Schließlich stieß man sogar auf eine zahlenmäßige Abweichung zwischen Theorie und Beobachtung, nämlich beim Planeten Merkur, die sich im Rahmen der Newtonschen Theorie nicht beheben ließ. Ihre sonstigen Erfolge waren jedoch so groß, daß man lange Zeit nicht glauben konnte und wollte, daß sie in ihren Grundlugen einen Todeskeim trage. Den Anstoß zu ihrem Zusammenbruch erfuhr sie auch nicht von innen heraus, sondern von seiten der Elektrodynamik. Als nämlich diese dazu überging, die elektrischen Vorgänge bei bewegten Körpern zu studieren, geriet man in eine äußerst mißliche Lage. Es zeigte sich nämlich, daß uns die bestehende Physik nicht die erforderlichen Hilfsmittel zur befriedigenden Beschreibung solcher Erscheinungen an die Hand gab. Nachdem man sich einige Zeit vergeblich abgemüht hatte, den fühlbaren Mangel befriedigend zu beheben, trat Albert Einstein im Jahre 1905, damals noch ein junger, 26jähriger, unbekannter Physiker, hervor und zeigte, daß in den ganz prinzipiellen, tiefliegenden Schwächen der Newtonschen Theorie der Grund der Schwierigkeiten zu suchen sei. Und nun begann er in einer Folge groß angelegter Arbeiten, die in den letzten Jahren einen gewissen Abschluß gefunden haben, ein ganz neues Gebäude der theoretischen Physik von so unerhörter Kühnheit aufzuführen, daß er sicherlich nicht so schnell Mitarbeiter und Anhänger gefunden hätte, wenn nicht folgende drei Momente jeden objektiv Forschenden gewonnen hätten. Erstens, die grundsätzlichen begrifflichen Schwierigkeiten der Newtonschen Theorie, von denen wir schon einige andeuteten, waren unbestritten vorhanden.

Dadurch, daß Einstein seine Theorie frei von diesen Schwächen begründete, kam er einem lang empfundenen Bedürfnis entgegen. Zweitens, schon die ersten Ansätze im Anschluß an die Probleme der Elektrodynamik lieferten eine so befriedigende Darstellung aller Beobachtungen, daß man an der Fruchtbarkeit seiner neuen Gesichtspunkte nicht zweifeln konnte. Drittens, in mutiger Verfolgung der letzten Folgerungen seiner allgemein durchgeführten Ideen hat Einstein neue Erscheinungen vorausgesagt, die sich bisher alle fast restlos haben bestätigen lassen. Wer es weiß, wie furchtlos, ohne sich gewissermaßen durch Geschwindigkeit seiner Ansätze einen Rückweg zu sichern, Einstein seine Theorie begründet und aufgebaut hat, der vermag die Bedeutung dieser praktischen Erfolge zu würdigen.

Zum Ausgangspunkt seiner Reform wählte Einstein das Relativitätsprinzip der Newtonschen Mechanik. Dieses Prinzip fordert, daß in den Bewegungsvorgängen z. B. auf der Erde, deren, in jedem Augenblick mit genügender Annäherung, geradlinig gleichförmiger Bewegungszustand mit bemerkbarwird. Diese durch die Erfahrung gesicherte Tatsache äußert sich mathematisch in den Formeln der Mechanik darin, daß die Bewegungsgleichungen ihre Gestalt bewahren, ganz gleich, auf welches System die Raum-Zeit-Messungen, die den Vorgang festzulegen und zu verfolgen erlauben, bezogen werden, solange man sich auf geradlinig gleichförmig gegeneinander bewegte Systeme beschränkte. Transformationsformeln, welche den Uebergang von den Raum-Zeit-Messung eines solchen Systems zu denen in einem anderen bewerkstelligen sollten, hatte man abgeleitet und lebte in der falschen Vorstellung besangen, diese Formeln seien die einzigen, die diesem Zweck dienen könnten. Da zeigte Einstein als erster, daß, wenn man den Uebergang von einem System zu einem anderen [*about seven words are illegible on my photocopy*: perhaps Bewegungssystem und insbesondere eine neu gewonnene Erfahrung,] nämlich die besondere Bedeutung der Lichtgeschwindigkeit in der Natur in Rücksicht zieht, man gezwungen ist, andere Transformationsformeln als die bisher üblichen zu verwenden, und ein neues Relativitätsprinzip formulieren muß. Diese neue Erkenntnis war von geradezu revolutionärem Charakter. Denn einmal folgte aus den neuen Formeln, daß wir unsere Anschauungen über das Wesen der Raum-Zeit-Messungen von Grund auf ändern müssen, da nach ihnen die Länge eines Gegenstandes, der Zeitpunkt eines Ereignis ihren absoluten, d. h. unabhängig vom Bewegungszustand des Beobachters geltenden Wert verlieren. Sodann aber zeigte sich, daß die Gleichungen der Newtonischen Theorie dem neuen Relativitätsprinzip entsprechend umgestaltet werden mußten. Dafür behob aber Einsteins Neugestaltung des Relativitätsprinzips für geradlinig gleichförmig bewegte Systeme mit einem Schlage alle Schwierigkeiten, auf die die Elektrodynamik gestoßen war. Dies war die erste Etappe auf seinem Wege zur Neubegründung der Physik.

Bis hierher folgten ihm bald viele, sobald man die Richtigkeit und Ueberlegenheit seines Standpunktes erkannt hatte. Und während schon fleißige Hände und Köpfe an die Aufgabe gingen, die Gleichungen der

Newtonschen Mechanik dem neuen sogenannten „speziellen" Relativitätsprinzip anzupassen, da war Einstein, in voller Klarheit über die begrenzte Leistungsfähigkeit der bis dahin gewonnenen Erkenntnisse, in seinen Gedanken seinen Mitarbeitern einen großen Schritt voraus. Er war sich darüber im klaren, daß der Boden für die Neubegründung der Mechanik noch nicht erreicht war. Mit der Erkenntnis der Relativität der der beschleunigten Bewegung ein tieferes Erfassen der Gravigeschwindigkeit war wohl eine Schwäche der bestehenden Theorie aufgedeckt, aber vielleicht keineswegs ihre fundamentalste. Ein Anpassen der Mechanik an die spezielle Relativitätstheorie wäre ein Stehenbleiben auf halbem Wege gewesen.

Einstein übersah sofort, daß eine Reform der Newtonschen Mechanik nur in einer radikalen Umgestaltung derselben in eine solche bestehen konnte, welche ausschließlich Aussagen über Relativbewegungen enthielt und den Begriff des absoluten Raumes ausschaltete. Er erkannte auch sofort, daß eine Berücksichtigung der beschleunigten Bewegungen ein tieferes Erfassen der Gravitationserscheinungen erforderte. Und hier tritt besonders eindringlich eine Besonderheit der Einsteinschen Forschungsart zutage, die, trotz des ausgesprochen philosophischen Grundzuges seines Wesens, ihn als reinen Naturforscher kennzeichnen. Zwei alte Erfahrungstatsachen, die wir alle in der Schule gelernt haben, an denen wir aber alle mehr oder minder gedankenlos vorübergegangen sind, nämlich die Gleichheit der trägen und schweren Masse aller Körper und die völlige Unabhängigkeit der Fallbeschleunigung von der physikalischen und chemischen Beschaffenheit des fallenden Körpers, diese gewannen durch Einstein erst Leben und tieferen Sinn. Er erkannte, daß diese zwei Tatsachen uns im wesentlichen alle erforderliche Erkenntnisse liefern, um eine Mechanik der Relativbewegungen der Massen und eine Theorie der Gravitationserscheinungen aufzubauen. Allerdings hatte die letzte Säule unserer Anschauungen über das Wesen von Raum-Zeit in der Physik zu fallen.

Durch die „spezielle" Relativitätstheorie war der absolute Charakter der Raum-Zeit-Messungen zwar beseitigt worden. Doch behielt immerhin jedes System das Recht, eine Messungen nach den Formeln der euklidischen Geometrie auszuwerten. Bei der Ausgestaltung der allgemeinen Relativitätstheorie kam aber die schon im Jahre 1854 von dem genialen Mathematiker Bernhard Riemann ausgesprochene Erkenntnis zutage, daß die Erforschung der geometrischen Verhältnisse in der materiellen Welt ein Grundproblem der Physik sei und nicht eine rein mathematische Angelegenheit. Ganz unabhängig gelangte Einstein zu derselben Einsicht, fand aber zugleich als erster eine Lösung für diese tiefliegende Problemstellung. Er zeigte, daß die Erforschung der geometrischen Zusammenhangsverhältnisse der physikalischen Welt gleichbedeutend ist mit der Erforschung ihrer Gravitationsverhältnisse.

Auf Fundamente von solcher Tiefe und Breite baute Einstein seine neue Mechanik auf; immer, trotz aller Abstraktheit der Gedankengänge und trotz der schwierigen neuen mathematischen Hilfsmittel, die er heranzog, immer bestrebt, durch beobachtbare Folgerungen seiner Ansätze ihre Ueberlegenheit über die früherer Theorie zu erweisen. Er schuf neue Bewegungsgesetze für die Planeten und zeigte, daß sie nicht nur dasselbe leisten wie diejenigen der Newtonschen Mechanik, sondern darüber hinausgehend, sofort die beim Merkur beobachtete und oben erwähnte Bewegungsanomalie restlos deutete. Seine Theorie ergab, daß die Eigenschaft der Schwere und Trägheit, bisher von uns als spezifisches Merkmal der Materie aufgefaßt, auch jeglicher Energie, also Licht, Wärmestrahlung usw. zukommt. Daraus zog er sofort die für die neue Auffassung entscheidende Folgerung, daß ein in unmittelbarer Nähe an der Sonne vorübergehender Lichtstrahl eines Sternes abgelenkt werden müsse. Zwei englische Expeditionen, die am 29. Mai dieses Jahres speziell zur Prüfung dieser Folgerung der Einsteinschen Theorie ausgerüstet worden waren, haben seine Voraussage vollauf bestätigt gefunden. Auch eine dritte Folgerung seiner allgemeinen Relativitätstheorie, ein Einfluß der Schwere auf die Lage der Spektrallinien ist, wenn auch nicht sicher erwiesen, doch schon heute in hohem Grade wahrscheinlich gemacht.

So hat die beispiellose Gestaltungskraft eines Mannes in 15 Jahren die Physik auf eine ganz neue Grundlage gestellt, so daß wir am Beginn einer ganz neue Epoche der Naturbeschreibung stehen, geknüpft an den Namen Einstein, so wie frühere an die Namen Ptolemäus, Kopernikus und Newton geknüpft werden. Er hat die Physik vor ganz neue Probleme gestellt, die Mathematik vor die Aufgabe, die neuen mathematischen Hilfsmittel auszubauen, die benötigt werden, da seine Theorie die bisher üblichen Formeln der euklidischen Geometrie verläßt; die Philosophie vor die Notwendigkeit, unsere Anschauungen über Raum — Zeit — Materie einer gründlichen Revision zu unterziehen, und die Astronomie vor die Ehrenpflicht, die Prüfung der letzten Konsequenzen der neuen Theorie an der Erfahrung durchzuführen."

The *Berliner Illustrirte Zeitung*, Volume 28, Number 50, (14 December 1919), printed a large portrait of Einstein on the cover with the following caption,

"Eine neue Größe der Weltgeschichte: Albert Einstein,
dessen Forschungen eine völlige Umwälzung unserer Naturbetrachtung bedeutet und den Erkenntnissen eines Kopernikus, Kepler und Newton gleichwertig sind."

Einstein's acquaintance Max Born wrote in the *Frankfurter Zeitung und Handelsblatt* (which Zionist Theodor Herzl called a "Jewish paper"[259]), first

---

259. Racist political Zionist Theodor Herzl wrote on 12 June 1895,

"Jewish papers! I will induce the publishers of the biggest Jewish papers (*Neue Freie Presse, Berliner Tageblatt, Frankfurter Zeitung,* etc.) to publish editions over there, as

morning edition, on 23 November 1919 (*see also: Frankfurter Zeitung und Handelsblatt*, first morning edition of 30 September 1919, for an article on the eclipse expeditions):

"Raum, Zeit und Schwerkraft.
Von Professor Dr. M. Born.

Am 29. Mai dieses Jahres fand eine Sonnenfinsternis statt, die einen schmalen Streifen der südlichen Erdhälfte einige Minuten verdunkelte, in Europa aber unsichtbar blieb. Mit diesem unscheinbaren Ereignis ist einer der größten Siege verknüpft, die der Menschengeist der Natur abgetrotzt hat, kein Triumph dröhnender Technik, sondern des reinen Erkennens: die Bestätigung der Einsteinschen Theorie der Gravitation und der allgemeinen Relativität.

Zur Beobachtung der Finsternis war eine englische Expedition unter dem Astronomen Eddington ausgeschickt worden; ihre Aufgabe war nicht die Aufzeichnung und Messung jener glänzenden Erscheinungen, die jede totale Verfinsterung so eindrucksvoll machen, wie Protuberanzen, Corona, Fackeln, sondern die Messung der Stellung einiger Fixsterne, die während der Finsternis in unmittelbarer Nähe des Sonnenrandes standen und nur während der Verdeckung der alles überstrahlenden Sonne durch den Mond dem Auge und der photographischen Platte zugänglich waren.

Der Zweck dieser höchst mühseligen, schwierigen Messung war die Prüfung, ob diese Sterne die von der Einsteinschen Theorie geforderte scheinbare Verschiebung zeigten. Der beschränkte Raum gestattet nicht, die Entwicklung dieser Theorie hier darzustellen. Nur soviel sei gesagt, daß es zuerst Erfahrungen bei optischen und elektrischen Präzisionsmessungen waren, die sich mit Hilfe der überkommenen Vorstellungen von Raum, Zeit, Bewegung nicht deuten ließen, und die Einstein veranlaßten, eine Revision dieser Grundbegriffe vorzunehmen.

Der Hauptinhalt seiner Lehre ist folgender: Man denke sich einen Beobachter, der sich mit seiner Umgebung geradlinig und gleichförmig durch den Raum bewegt; dies ist tatsächlich unsere Situation auf der im Weltenraum dahineilenden Erde, wenn man von der schwachen Krümmung der Erdbahn absieht. Richtet der Beobachter seinen Blick auf andere Körper,

---

the *New York Herald* does in Paris."—T. Herzl, English translation by H. Zohn, R. Patai, Editor, *The Complete Diaries of Theodor Herzl*, Volume 1, Herzl Press, New York, (1960), p. 84.

THE DEARBORN INDEPENDENT, praised the *New York Herald*. "When Editors Were Independent of the Jews", THE DEARBORN INDEPENDENT, (5 February 1921). *See also:* T. Herzl, English translation by H. Zohn, R. Patai, Editor, *The Complete Diaries of Theodor Herzl*, Volumes 1 and 2, Herzl Press, New York, (1960), pp. 37, 97, 170, 455, 457, 480. *See also:* A. Elon, *Herzl*, Holt, Rinehart and Winston, New York, (1975), pp. 167-168. *See also: The Collected Papers of Albert Einstein*, Volume 7, Document 35, Princeton University Press, (2002), pp. 296-297, note 8.

die an seiner Bewegung nicht teilnehmen (etwa auf entfernte Gestirne), so wird er an der allmählichen Verschiebung dieser Körper merken, daß sein Standpunkt sich gegen sie bewegt. Die Frage ist nun aber, ob er seine Ortsveränderung auch feststellen kann, wenn er nicht fremde Körper beobachtet, sondern sich auf Messungen in seinem Laboratorium mit seinen mechanischen, elektrischen, optischen Apparaten beschränkt. Die klassische Mechanik gibt darauf die Antwort, daß ihm seine Bewegung verborgen bleiben muß; denn die mechanischen Gesetze in gleichförmig und geradlinig bewegten Systemen von Körpern stimmen vollständig mit denen überein, die im Falle der Ruhe dieser Körper gelten, daher funktionieren alle mechanischen Apparate, wie Pendel, Wage usw. genau so, als wenn sie sich auf ruhender Grundlage befänden. Versagt also die mechanische Apparatur, so wird der Beobachter die elektrische, magnetische und optische zum Nachweise seiner Bewegung heranholen. Hier könnte man zunächst ein positives Ergebnis erwarten, denn als Träger der elektromagnetischen und optischen Erscheinungen gilt der Weltäther, und wenn das ganze Laboratorium des Beobachters auf der Erde mit der gewaltigen Geschwindigkeit dieses Planeten von etwa 30 Kilometern in der Sekunde durch den Aether rast, so müßte ein heftiger Aetherwind durch das Laboratorium wehen, entsprechend dem Gegenwinde, den der Automobilfahrer bei schneller Fahrt spürt. Der Aetherwind würde mancherlei Wirkungen ausüben, z. B. Lichtwellen verwehen, ihre Richtung und Geschwindigkeit ändern; man hat nun mit den schärfsten Meßmethoden versucht, diese Wirkungen nachzuweisen, aber immer vergebens: Der Aetherwind existiert nicht, die Lichtwellen laufen auf der bewegten Erde gerade so, als wenn sie ruhte, und von allen elektrischen und magnetischen Vorgängen gilt dasselbe. Das heißt aber nichts anderes als daß auch mit elektromagnetischen und optischen Messungen die Feststellung einer absoluten gleichförmigen und geradlinigen Bewegung durch den Raum nicht möglich ist. Feststellbar sind nur relative Bewegungen eines Körpers gegen den anderen.

Diese Tatsache ist aber, wie das Bild des Aetherwindes zeigt, mit der gewöhnlichen Auffassung von Raum, Zeit, Bewegung vollständig unbegreiflich. Einstein faßte nun den kühnen Gedanken, zugleich mit der Vorstellung des absoluten Raumes auch die der absoluten Zeit, als einer physikalisch meßbaren Größe, auszugeben. Auf diese Weise gelang es tatsächlich, alle elektromagnetischen und optischen Erfahrungen ebenso gut wie die mechanischen mit der Relativität in Einklang zu bringen.

Diese erste Einsteinsche Relativitätstheorie vom Jahre 1906 heute die „spezielle" genannt, war noch ziemlich harmlos zwar brachte sie außer der Auflösung der überlieferten Begriff von Raum und Zeit noch zahlreiche umstürzende Gedanken wie den, daß die Masse keine konstante Eigenschaft der Materie sondern von ihrer Geschwindigkeit und ihrem Energieinhalt abhängig sei, aber es bedurfte nur weniger Jahre, um so ziemlich alle Physiker zu Relativisten zu machen. Denn diese spezielle Relativitätstheorie hatte eine große Anzahl von Konsequenzen, die sich durch Versuche prüfen ließen, und nachdem ein Experiment nach dem anderen zu ihren Gunsten entschied mußten selbst hartnäckige Verfechter des Absoluten die Waffen strecken.

Die Beschränkung auf gradlinige und gleichförmige Relativbewegung ist für den auf Allgemeinheit der Erkenntnis gerichteten Geist zweifellos ein Stein des Anstoßes. Aber primitive Erfahrungen scheinen dafür einzustehen, daß diese Beschränkung wesentlich ist. Hierher gehören die bekannten Erscheinungen, auf Grund deren man die Rotation der Erde durch irdische, nicht astronomische Messungen nachweist; z. B. die Drehung der Schwingungsebene des Foucauldschen Pendels oder die Zentrifugalkraft, durch die eine scheinbare Aenderung der Schwerkraft mit der geographischen Breite und der Abplattung der Erde an den Polen erzeugt wird. Nach der klassischen Mechanik sind das alles Erscheinungen, die auf die Widerstande der Massen gegen Geschwindigkeitsänderungen (Beschleunigungen), der sogenannten Massenträgheit, beruhen. Durch die Rotation der Erde werden solche Trägheitswiderstände hervorgerufen; obwohl die Mechanik behauptet, die gleichförmige und geradlinige Bewegungen gegen den absoluten Raum nicht feststellbar sind, hält sie daran fest, daß ungleichförmige oder nicht gradlinige Bewegungen, z. B. Rotationen, gegen den leeren Raum bestimmte physikalische Wirkungen hervorbringen. Auch wenn die Erde allein im Weltraume schwebte, müßten die Menschen ihre Drehung etwa mit dem Foucauldschen Pendel oder durch Beobachtung der Abplattung der Erdkugel feststellen können, also eine Drehung gegen den leeren Raum, gegen das Nichts. Vor Einstein haben nur wenige Denker an diesem Unding des im leeren Raume bewegten Körpers Anstoß genommen, so vor allem Ernst Mach, der Physiker und Philosoph, der ausdrücklich eine Revision der mechanischen Grundgesetze zur Beseitigung jeder absoluten Bewegung forderte. Aber erst Einstein besaß die Kraft der Abstraktion, die zu einer solchen Leistung notwendig war. Der Schlüssel für die Lösung war die Entdeckung des Zusammenhangs zwischen dem Raum-Zeit-Problem und dem Problem der Gravitation oder allgemeinen Schwerkraft. Eine sehr sicher begründete, aber wenig beachtete Erfahrung besagt, daß alle Körper (im luftleeren Raume) gleich schnell fallen. Man denke sich einen Beobachter in einem allseits geschlossenen Kasten mit allerlei Gegenständen untergebracht, und dieser Kasten falle herab, dann wird der Beobachter, da alle Dinge im Kasten gleich schnell fallen, feststellen können, daß die Dinge ihre Schwere verlieren. Hier erkennt man die Brücke zwischen der Bewegungslehre und der Gravitation. Der Widerstand, den die Masse der Körper einer Beschleunigung entgegensetzt, und die Anziehung einer schweren Masse durch die andere werden zwei Erscheinungsformen desselben Grundgesetzes. Nun ist die Massenanziehung offenbar eine relative Wirkung zweier Körper; somit muß auch der Beschleunigungswiderstand relativiert werden, auch er ist nur vorhanden, wenn andere Körper zugegen sind, nicht aber im leeren Raume. Die zum Nachweis der Rotation der Erde gebrauchten Erscheinungen der Massenträgheit, z. B. die Abplattung der Erde, sind nach Einstein

Wirkungen fremder Massen, nämlich des Systems aller Himmelskörper, vor allem des Heeres der Fixsterne, und sie würden verschwinden, wenn die Erde allein im Weltenraume schwebte. Das Argument der im leeren Raume allein rotierenden Erde ist für Einsteins Wirklichkeitssinn nichtig; für ihn ist nur reell, was feststellbar ist, also relative Oerter, relative Zeiten, relative Bewegungen. Aber der Weg, der ihn von dieser subjektiven Ueberzeugung bis zur objektiven Behauptung der allgemeinen Relativität aller Bewegungsvorgänge, aller physischen Vorgänge führte, war ein Anstieg auf steilsten Hängen, über Hindernisse, die jeden andern abgeschreckt hätten.

Nur einer vor ihm hatte ähnliche Pfade eingeschlagen, der Mathematiker Bernhard Riemann, doch war seine Zeit (Mitte des 19. Jahrhunderts) noch nicht reif, die Summe der Erfahrungen zu beschränkt. Die Durchführung der allgemeinen Relativität erfordert nämlich nicht mehr und nicht weniger als den Verzicht auf die allgemeine Gültigkeit der Euklidischen Geometrie, die seit 2000 Jahren als der Grundstein allen Wissens gilt, und ihre Ersetzung durch die von Riemann zuerst entworfene allgemeine Raumlehre. In dieser gibt es weder gerade Linien noch ebene Flächen, die nach Euklid wie ein starres Gerüst den Raum durchziehen. Am besten kann man sich eine Vorstellung von dieser Riemannschen Geometrie machen, wenn man an die Geometrie auf einer krummen, komplizierten Oberfläche, etwa einer Alpenlandschaft, denkt; auch da kann man keine geraden Linien von beträchtlicher Länge auf dem Erdboden ziehen, und ein Feldmesser, der nur mit der Meßkette, ohne optische Visierinstrumente ausgerüstet wäre, hätte eine heillose Mühe: und doch würde er die Aufgabe bewältigen. Er würde, von irgend einem Netz von Fixpunkten ausgehend die kürzesten Wege zwischen irgend zwei Punkten mit der Meßkette festzustellen suchten, dann die Krümmungseigenschaften der Berge, Täler und Sättel ausmessen und so allmählich eine Aufzeichnung des Geländes herstellen, die von dem zugrunde gelegten Netze von Fixpunkten unabhängig ist und nur die tatsächlichen Beziehungen der Oertlichkeiten enthält. In ganz ähnlicher Lage ist der Mensch im Raume, wenn man diesen nicht von vornherein als Euklidisch voraussetzt, sondern ihn ohne Voreingenommenheit mit der Meßkette ausmißt.

Das ist der Standpunkt Riemanns, den Einstein, durch Einbeziehung der Zeit auf das physikalische Geschehen übertragen hat; zur Meßkette muß dann noch eine Uhr treten. Gestützt auf ein beliebiges Gerüst physischer Fixpunkte sucht man durch Messung die den Dingen eigentümlichen Raumgesetze zu ergründen, die in unserm Bilde den Krümmungsverhältnissen der Erdoberfläche analog sind. Die Einsteinsche Theorie führt dann zu der Vorstellung, daß der Raum nur da „ungekrümmt", „Euklidisch" ist, wo keine merkbaren Massen sind; in der Nähe der Massen aber zeigt er Abweichungen von den Euklidischen Gesetzen „Krümmungen", und auf diesen beruhen die Krümmungen der Bahnen bewegter Körper, die in der klassischen Mechanik als Wirkungen der Schwerkraft angesehen werden.

Man sieht, wie diese auf dem Boden der Erfahrung gewachsene Theorie hinübergreift über die Grenzen der Naturwissenschaft und die Philosophie zur Stellungnahme herausfordert.

Für die tatsächliche Gültigkeit der Einsteinschen Theorie konnten bislang nur wenige Tatsachen der Astronomie angeführtwerden. Die klassische Himmelsmechanik Newtons ist nämlich vom Standpunkte der neuen Theorie nur näherungsweise richtig und muß in der Nachbarschaft großer, gravitierender Massen, wie der Sonne, in bestimmter Weise korrigiert werden; in der Tat konnte Einstein auf diesem Wege eine bisher unerklärte Abweichung des sonnennächsten Planeten Merkur von seiner Newtonschen Bahn quantitativ genau erklären. Außerdem fordert die Einsteinsche Theorie gewisse Verschiebungen der Spektrallinien des Lichtes der Sonne und der Fixsterne; auch diese Erscheinung ist heute sicher nachgewiesen. Endlich sollen Lichtstrahlen, die nahe an der Sonne vorbeistreichen, von dieser abgelenkt werden; dies zu prüfen, war die Aufgabe der englischen Finsternis-Expedition. Nach einer Mitteilung in der Zeitschrift „die Naturwissenschaften" [*Footnote:* 7. Jahrg., Heft 42 vom 17. Oktbr. 1919. S. 775.]) hat nun Einstein ein Telegramm des holländischen Physikers Lorentz bekommen, wonach die von Einstein vorhergesagte Ablenkung der Lichtstrahlen im vollen Betrage (1,7 Bogensekunden) wirklich vorhanden ist.

Ist es nun aber nötig, das ganze Gebäude der tausendjährigen Geometrie einzureißen, um diese winzige, unauffällige Erscheinung zu erklären?

Sicherlich wird der, der nichts anderes als diese eine Uebereinstimmung kennt, ein solches Beginnen töricht nennen; gibt es doch genug physikalische Kräfte, die man ersinnen könnte, um die Lichtstrahlablenkung durch die Sonne zu erklären. Aber wer das ganze System der allgemeinen Relativitätstheorie gründlich durchdacht hat, der ist hinreichend vorbereitet, um an sie zu glauben, sobald ein schlagendes Experiment den Einklang des Gedachten mit dem Wirklichen beweist. Darum kann man dem Vorsichtigen, Ungläubigen nur sagen: geh hin und studiere, die Mühe lohnt; du wirst eine geistige Befreiung erleben, vergleichbar der, die Kopernikus der Menschheit bereitet hat.

Es steht wohl außer Zweifel, daß die physikalischen Wissenschaften sich in Zukunft streng relativistisch einstellen werden. Für die Philosophie aber bedeutet die Einsteinsche Lehre den Sturz der räumlichen und zeitlichen Kategorien von der Höhe des *a priori* in die Niederungen der „platten Empirie". Die Behauptung Kants, daß die Urteile über Raum und Zeit synthetische Urteile *a priori* seien, stützt sich auf die zu seiner Zeit geltende Ansicht, daß man an der Wahrheit der geometrischen Erkenntnisse in der überkommenen Form Euklids nicht zweifeln dürfe, daß es vielmehr die Aufgabe der Philosophie sei, die „Möglichkeit" einer solchen Erkenntnis nachzuweisen, die Gründe für sie aufzusuchen. Da nun die Möglichkeit solcher objektiven und vollkommen genauen Urteile weder auf reiner Logik (analytisch Urteile *a priori*) noch auf Erfahrung (synthetische Urteile *a posteriori*) beruhen konnten, so entstand die Vorstellung einer besonderen Erkenntnisquelle, die „synthetische Urteile *a priori*" ermöglichen soll. Raum und Zeit sind nach Kant „Formen der Anschauung" und ihre Gesetze *a priori* gültig. Inzwischen hat die Entwicklung der Geometrie die Sonderstellung der Euklidischen Geometrie durch die Entdeckung von logisch widerspruchsfreien „nicht-Euklidischen" Geometrien durchbrochen, sodann hat die Physik die allgemeinste Form dieser übergeordneten Geometrien, die Riemannsche, ihrer Darstellung der Wirklichkeit zu Grunde gelegt. Natürlich bleibt davon die logische Sicherheit des Euklidischen Systems von Sätzen unangetastet; aber daß die Axiome Euklids, aus denen diese Sätze folgen, die adäquate Darstellung der räumlichen Beziehungen der Dinge sind, das leugnet die heutige Physik. Damit ist die Grundlage der kantischen Lehre von der Unantastbarkeit der geometrischen Wahrheiten durchbrochen. Die Empirie hat sie verworfen und sich allgemeinere Grundlagen geschaffen. Ob die „Formen der Anschauung" Kants als Ausdruck gewisser psychologischer Eigenschaften des menschlichen Geistes eine Daseinsberechtigung haben, das zu prüfen ist nicht Sache des Physikers. Allerdings steht die Exaktheit der geometrischen Sätze zu der Verschwommenheit aller psychologischen in krassem Widerspruche.

Wer diese Entwicklung miterlebt hat, der wird sich des Zweifels am apriorischen Charakter auch anderer Kategorien des Denkens nicht erwehren können. Einstein selbst steht in seinen philosophischen Ueberzeugungen den größten unter den exakten Naturforschern nahe, einem Gauß, einem Riemann, einem Helmholtz, die sich alle trotz Kant zum Empirismus bekannten und unmittelbar an Hume anknüpften.

Die relativistischen Ideen sind zuerst in deutscher Sprache gedacht und aufgezeichnet worden; das *Experimentum Crucis* haben englische Forscher durchgeführt. Ein so kostspieliges Unternehmen wie die Finsternis-Expedition zu rein Theoretischen Zwecken beweist eine starke Teilnahme der Oeffentlichkeit an wissenschaftlichen Problemen. Großen Anteil daran hat die berühmte, im besten Sinne populäre Zeitschrift „The Nature", die unter Mitwirkung der ersten Gelehrten erscheint und ungeheuer verbreitet ist. Auch wir besitzen ähnliche, nach denselben Grundsätzen geleitet Wochenschriften, vor allem die schon genannten „Naturwissenschaften"; doch spielen sie noch nicht die gleiche Rolle im Geistesleben der Nation wie in England. Erst wenn die Kenntnis der wissenschaftlichen Probleme das Interesse an ihnen geweckt hat, kann die Opferwilligkeit entstehen, die für ideelle Ziele Mühe und Geld nicht scheut."

It is interesting to observe how Einstein's followers like Max Born, Robert Daniel Carmichael[260] and Moritz Schlick[261] tried to justify Einstein's many fallacies of *Petitio Principii*. These fatal fallacies were obvious to Einstein's

---

260. R. D. Carmichael, *The Theory of Relativity*, Mathematical Monographs No. 12, John Wiley & Sons, Inc., New York, Chapman & Hall, Limited, London, (1920).
261. M. Schlick, *Space and Time in Contemporary Physics*, Oxford University Press, New York, (1920).

critics Robert Drill (whom Born had attacked),[262] and more significantly Franz Kleinschrod[263] and Hugo Dingler.[264]

Albert Einstein, Karl Marx and Sigmund Freud were each plagiarists promoted by the "Jewish Press" and each lacked the ability to form rational theories which proceeded from fundamental principles to logical conclusions. They and their defenders argued in circles—redundancies, and stagnated science with their irrational dogmas. When relativity critics pointed out the fatal flaws in the theory of relativity, they were told that the theory was irrefutable and that it was the finest example of logical perfection in the history of science. Redundancies are not theories and it is irrational to state conclusions as premises, which is what Einstein did in order to mask his plagiarism.

Nobel Prize laureate Friedrich August Hayek encountered the same type of irrational devotees defending the irrational dogmas of Marx and Freud, as those who defended the similarly irrational dogmas of Einstein. Hayek stated,

> "The two chief subjects of discussion among students of the University of Vienna in the years immediately after the war were Marxism and psychoanalysis, as they were to become much later in the West. I made a conscientious effort to study both the doctrines but found them the more unsatisfactory the more I studied them. It seemed to me then and has so appeared ever since that their doctrines were thoroughly unscientific because they so defined their terms that their statements were necessarily true and unrefutable, and therefore said nothing about the world. It was in the struggle with these views that I developed views on the philosophy of science rather similar to, but of course much less clearly formulated than, those which Karl Popper formed from much the same experiences; and it was only natural that I read his views when he published *The Logic of Scientific Discovery* in 1935, some years before I made his acquaintance. [***] Karl Popper is four or five years my junior, so we did not belong to the same academic generation. But our environment in which we formed our ideas was very much the same. It was very largely dominated by discussion, on the one hand, with Marxists and, on the other hand, with

---

[262]. R. Drill, "Die Kultur der Haeckel-Zeit", *Frankfurter Zeitung*, (18 August 1919); **and** "Nachwort", *Frankfurter Zeitung*, (2 September 1919); **and** "Ordnung und Chaos. Ein Beitrag zum Gesetz von der Erhaltung der Kraft. I-II", *Frankfurter Zeitung*, (30 November 1919 / 2 December 1919).

[263]. F. Kleinschrod, "Das Lebensproblem und das Positivitätsprinzip in Zeit und Raum und das Einsteinsche Relativitätsprinzip in Raum und Zeit", *Frankfurter Zeitgemäße Broschuren*, Volume 40, Number 1-3, Breer & Thiemann, Hamm, Westphalen, (October-December, 1920), pp. 17, 47.

[264]. H. Dingler, *Die Grundlagen der Physik; synthetische Prinzipien der mathematischen Naturphilosophie*, Second Edition, Walter de Gruyter & Co., Berlin, (1923); **and** *Physik und Hypothese Versuch einer induktiven Wissenschaftslehre nebst einer kritischen Analyse der Fundamente der Relativitätstheorie*, Walter de Gruyter & Co., Berlin, Leipzig, (1921); **and** "Kritische Bemerkungen zu den Grundlagen der Relativitätstheorie", *Physikalische Zeitschrift*, Volume 21, (1920), pp. 668-669.

Freudians. Both these groups had one very irritating attribute: They insisted that their theories were, in principle, irrefutable. I remember particularly one occasion when I suddenly began to see how ridiculous it all was when I was arguing with Freudians, and they explained, 'Oh, well, this is due to the death instinct.' And I said, 'But this can't be due to the death instinct.' 'Oh, then this is due to the life instinct.' Naturally, if you have these two alternatives available to explain something, there's no way of checking whether the theory is true or not. And that led me, already, to the understanding of what became Popper's main systematic point: that the test of empirical science was that it could be refuted, and that any system which claimed that it was irrefutable was by definition not scientific. I was not a trained philosopher; I didn't elaborate this. It was sufficient for me to have recognized this, but when I found this thing explicitly argued and justified in Popper, I just accepted the Popperian philosophy for spelling out what I had always felt."[265]

Max Born's condescending tone when addressing Einstein's critics is perhaps reflective of his insecurity surrounding his overblown claims. His strikingly incomplete and nationalistically biased history is one example of his duplicitous character. Note that Poincaré's name is conspicuously absent from Born's article.

Max Born was educated at the Göttingen Academy and this was typical of their attitude toward their mathematical and national rival Henri Poincaré, as Jules Leveugle has shown.[266] Hilbert and Minkowski, both of Göttingen, lectured Born in 1905 on the works of Hertz, Voigt, FitzGerald, Larmor, Lorentz, and Poincaré,[267] the real founders of the theory of relativity; and Born would later acknowledge their contributions—after Einstein had died. While Einstein lived, and after Whittaker had completed the second volume of his *A History of the Theories of Aether and Electricity*, which disputed Einstein's priority for the theory of relativity based upon the facts and primary sources, Born felt obliged to write to Einstein to emphatically deny that he had helped his very good friend Sir Edmund Whittaker to write it. Born then later endorsed Whittaker's and G.

---

265. F. A. Hayek, edited by S. Kresge and L. Wenar, *Hayek on Hayek: An Autobiographical Dialogue*, University of Chicago Press, (1994), pp. 48-49, 50-51.

266. J. Leveugle, *La Relativité, Poincaré et Einstein, Planck, Hilbert: Histoire véridique de la Théorie de la Relativité*, L'Harmattan, Paris, (2004).

267. M. Born, *My Life: Recollections of a Nobel Laureate*, Charles Scribner's Sons, New York, (1975), pp. 98, 130; **and** *The Born-Einstein Letters*, Walker and Company, New York, (1971), p. 1; **and** "Physics and Relativity", *Physics in my Generation*, second revised edition, Springer, New York, (1969), p. 101. *See also:* J. Leveugle, "Hilbert et Poincaré", *Poincaré et la Relativité : Question sur la Science*, Chapter 10, (2002), ISBN: 2-9518876-1-2, pp.147-230; **and** *La Relativité, Poincaré et Einstein, Planck, Hilbert: Histoire véridique de la Théorie de la Relativité*, L'Harmattan, Paris, (2004). *See also:* L. Pyenson, *The Young Einstein and the Advent of Relativity*, Bristol, Adam Hilger, (1985), pp. 103-104. *See also:* C. Reid, *Hilbert*, Springer Verlag, Berlin, Heidelberg, New York, (1970), pp. 100, 105.

H. Keswani's view that Lorentz and Poincaré published the special theory of relativity before Einstein, in a letter Born wrote to Prof. Keswani. Born's early papers on what he, like many others, sometimes called the "Lorentz-Einstein principle of relativity",[268] did not emphasize the work of Einstein, but instead emphasized the work of Lorentz and Minkowski, to the exclusion of Poincaré.[269] Minkowski, like Born, was Jewish and many thought that Lorentz was also Jewish. It should be noted that Felix Klein was an important figure at Göttingen and that Arnold Sommerfeld kept close ties to the Göttingen community.

Note also that David Hilbert's name is not to be found in Born's article. Born, who at one time was Hilbert's lecture assistant, knew that Einstein had plagiarized the generally covariant field equations of gravitation of the general theory of relativity from Hilbert. Max Born wrote to David Hilbert on 23 November 1915,[270] two days before Einstein submitted a paper which plagiarized Hilbert's equations. Max Born knew that Hilbert had the equations before Einstein, and that Einstein and Freundlich copied them from Hilbert.

Einstein's sycophantic friends, Moszkowski, Freundlich, Born, Planck, Laue,[271] etc. had a vested interest in the Einstein image and they desired to make fortunes from it. Moszkowski, Laue and Born were especially greedy. This explains Nobel Prize laureate Max von Laue's disingenuous attempts (which are reprinted later in this text) to change the subject from Einstein's sophistry, self-promotion, plagiarism, and the evidence against the general theory of relativity; to racially charged personal attacks on Einstein's critics Paul Weyland, Ernst Gehrcke and Philipp Lenard, which vicious attacks shocked Nobel Prize laureate Lenard, who had been completely objective in his criticisms of relativity theory and had treated Einstein with great respect.[272]

Lenard was assistant to Heinrich Hertz, who was half-Jewish, and Lenard posthumously edited Hertz' works. Lenard was perhaps himself of Jewish

---

268. *See, for example:* M. Born, "Zur Kinematik des starren Körpers im System des Relativitätsprinzips", *Nachrichten von der Königlichen Gesellschaft der Wissenschaften und der Georg-Augusts-Universität zu Göttingen*, (1910), pp. 161-179, at 161.

269. *See, for example:* M. Born, "Eine Ableitung der Grundgleichungen für die elektromagnetischen Vorgänge in bewegten Körpern vom Standpunkte der Elektronentheorie. Aus dem Nachlaß von Hermann Minkowski", *Mathematische Annalen*, Volume 68, (1910), pp. 526-551; **and** "Zur Kinematik des starren Körpers im System des Relativitätsprinzips", *Nachrichten von der Königlichen Gesellschaft der Wissenschaften und der Georg-Augusts-Universität zu Göttingen*, (1910), pp. 161-179.

270. Letter from M. Born to D. Hilbert of 23 November 1915, Niedersächsische Staats- und Universitätsbibliothek Göttingen, Cod. Ms. D. Hilbert 40 A: Nr. 11; the relevant part of which is reproduced in D. Wuensch, „zwei wirkliche Kerle'': Neues zur Entdeckung der Gravitationsgleichungen der Allgemeinen Relativitätstheorie durch Albert Einstein und David Hilbert, Termessos, Göttingen, (2005), pp. 73-74.

271. M. Born, *The Born-Einstein Letters*, Walker and Company, New York, (1971), p. 5.

272. *See:* Letter from P. Lenard to J. Stark 8 September 1920 in A. Kleinert and C. Schönbeck, "Lenard und Einstein. Ihr Briefwechsel und ihr Verhältnis vor der Nauheimer Diskussion von 1920", *Gesnerus*, Volume 35, Number 3/4, (1978), pp. 318-333, at 328-329.

descent,[273] though he later publicly espoused Nazism after Einstein and Einstein's friends had smeared him with lies in the international press and had refused to retract their admitted lies. The financial and egotistical interests of Einstein's friends also explains Planck's corrupt methods at the Bad Nauheim debate, and the deceptive articles by the experts Freundlich and Born which gave credence to the promotional campaign for Einstein in the press, promotions tainted with the foul smell of highly unethical ethnic and political bias.

Born attempted to obstruct Moszkowski's efforts to profiteer off of the Einstein brand Moszkowski had created, by blocking publication of Moszkowski's book *Einstein, the Searcher; His Work Explained from Dialogues with Einstein*. Born feared that the publication of this shameless book would confirm Weyland, Gehrcke and Lenard's accusations that Einstein was a sophistic, plagiarizing, publicity-seeking egomaniac and that many wished to profit from his name. Born, Einstein and others believed that the unprecedented Einstein hype by Einstein's Jewish friend Moszkowski revealed that Jews and Jewish-owned media interests were manufacturing an Einstein legend for the purposes of profit and self-promotion. Hedwig and Max Born wanted to calm this rising storm and protect their financial interests.

The "Magazine Section" of *The Minneapolis Journal* reported on 24 October 1920,

"Dr. Einstein at the present is meeting a wave of opposition in Germany. Professors and scientific men recently have banded together in a campaign against him. They accuse him of fostering a great propaganda with the aid of Jewish funds to put himself on the pedestal of fame. They go so far as to call his work plagiarism and his theories sophistry.

The Tidende of Bergen, Norway, prints in detail the record of a meeting in Germany in which the name Einstein was hooted by the assembly. A writer sent to interview the famous doctor disagreed with the tales of modesty attributed to him and characterized Einstein as a man having a very exalted opinion of himself."

The Literary Digest wrote in April of 1921,

"There are two men in Germany to-day who are traditionally inaccessible to newspaper men, Mr. Tobinkin notes. One is the financier, Hugo Stinnes. The other is Einstein. We are told:

Einstein has been greatly abused by a section of the German press, and he therefore shuns publicity."[274]

---

[273]. D. Bronder, *Bevor Hitler kam: Eine historische Studie*, Hans Pfeiffer Verlag, Hannover, (1964), p. 204 (p. 211 in the 1974 edition). H. Kardel, *Adolf Hitler, Begründer Israels*, Verlag Marva, Genf, (1974); English translation *Adolf Hitler: Founder of Israel*, Modjeskis' Society Dedicated to Preservation of Cultures, San Diego, (1997), pp. 4, 73.
[274]. "Personal-Glimpses: Einstein Finds the World Narrow", The Literary Digest, (16

Einstein confirmed that Moszkowski wanted to profit from the Einstein brand Moszkowski had created, and that Einstein approved of the profiteering, while attempting to quash legitimate criticism of the theory of relativity by the world-famous physicists Philipp Lenard and Willy Wien. Einstein wrote to Max Born in 1920,

> "However, I still prefer [Moszkowski] to Lenard and Wien. The latter two squabble because of a passion for squabbling, while the former does it only to earn money (which is, after all, better and more reasonable)."[275]

Einstein interceded on behalf of Erwin Freundlich and Moritz Schlick in an effort to help them profiteer from the Einstein brand on 27 January 1920.[276]

Max Born was peddling a book of his own, *Einstein's Theory of Relativity*.[277] Born, who was eager to prevent the public disclosure of the truths carelessly revealed in Einstein's conversations, wrote many desperate letters to Einstein trying to prevent the publication of Moszkowski's book and stated, *inter alia*,

> "It seems that you are less excited about it than your friends. My wife has already written to you saying what I think about this affair. (She is already regretting that she, too, has tried to turn your name into gold by sending me to America; women, poor creatures, carry the whole burden of existence, and grasp at any relief.) You will have to shake off [Moszkowski], otherwise Weyland will win all along the line, and Lenard and Gehrcke will triumph. [***] Forgive the officiousness of my letter, but it concerns everything dear to me (and Planck and Laue, etc.) You do not understand this, in these matters you are a little child. We all love you, and you must obey judicious people (not your wife). Should you prefer to have nothing further to do with the whole business, give me *written* authority. If necessary, I will go to Berlin, or even to the North Pole."[278]

Bear in mind that Einstein was, at that time, just a friend of these men and not the awe inspiring superhero of science they made him out to be through their deceptive self-aggrandizing promotion. They knew that they were lying to the

---

April 1921), pp. 33-34.

275. M. Born, *The Born-Einstein Letters*, Walker and Company, New York, (1971), p. 41.

276. Letter from A. Einstein to Cambridge University Press of 27 January 1920, *The Collected Papers of Albert Einstein*, Volume 9, Document 285, Princeton University Press, (2004).

277. M. Born, *Die Relativitätstheorie Einsteins und ihre physikalischen Grundlagen: gemeinverständlich dargestellt*, J. Springer, Berlin, (1920).

278. M. Born, *The Born-Einstein Letters*, Walker and Company, New York, (1971), pp. 39-40.

public, and they constructed the modern myth from their lies and misrepresentations. Born later changed his opinion of Moszkowski's book when he read it many decades later, seemingly having come to believe in his own mythologies. However, Max Born conceded in 1962 in the preface of the revised edition of his book *Einstein's Theory of Relativity* (first edition 1920), that the chief cause of interest in the eclipse expeditions, which made Einstein famous in 1919, was deliberate sensationalism—and he was himself a very active participant in that campaign to promote Einstein,

> "This text was originally an elaboration of a series of lectures given at Frankfurt am Main to a large audience when a wave of popular interest in the theory of relativity and in Einstein's personality had spread around the world, following the first confirmation by a British solar-eclipse expedition of Einstein's prediction that a beam of light should be bent by the gravitational action of the sun. Though sensationalism was probably the main cause of this interest, there was also a considerable and genuine desire to understand."[279]

Born states that the first edition of his book of 1920 resulted from a series of lectures given to large audiences. Born's lectures, which were promoted in the *Frankfurter Zeitung*,[280] might have been polemic, as well as promotional. Born states in his book,

> "There are opponents of the principle of relativity, simple minds who, when they have become acquainted with this difficulty in determining the length of a rod, indignantly exclaim, 'Of course, everything can be derived if we use false clocks; here we see to what absurdities blind faith in the magic power of mathematical formulae leads us,' and then condemn the theory of relativity at one stroke."[281]

Born did indeed profiteer from the Einstein name,

> "At that time a wave of interest in Einstein and his theory of relativity was sweeping the world. He had predicted the deflection, by the sun, of light coming from a star. Several expeditions, amongst them a British one under Eddington, had been sent out to tropical regions where a total eclipse of the sun was visible and the deflection could be observed. Now after laborious measurements and tedious calculations the conclusion was arrived at that Einstein was right, and this was published under sensational headlines in all

---

[279]. M. Born, "Preface", *Einstein's Theory of Relativity*, Revised and Enlarged Edition, Dover, New York, (1962/1965).
[280]. "Die Einsteinsche Relativitätstheorie" *Frankfurter Zeitung*, Number 46, (18 January 1920), p. 2; Number 61, (23 January 1920), p. 2; Number 82, (31 January 1920), p. 2.
[281]. M. Born, *Einstein's Theory of Relativity*, Revised and Enlarged Edition, Dover, New York, (1962/1965), p. 246.

the newspapers. It caused a tremendous stir in the civilized world, as I have already described in another chapter. There was an Einstein craze, everybody wanted to learn what it was all about, and he became the victim of a publicity racket. I used this for my own purposes. I announced a series of three lectures in the biggest lecture-hall of the University of Einstein's theory of relativity and charged an entrance fee for my Department. It was a colossal success, the hall was crowded and a considerable sum collected. My friends in the Frankfurt business world told me that I would have done even better if I had sent out private invitations to a lecture in the most expensive hotel, in evening dress and with cocktails, and had asked for an assistance fund. But that was not in my line.

The money thus earned helped us for some months, but as inflation got worse, it evaporated quickly and new means had to be found. One day I met a friend of the Ehrenberg family who told me that he had been engaged for years to an American girl from whom he had been separated by the war, and now he was going to New York to be married. I said jokingly: 'If you find a German-American who is still interested in the old country, tell him I need dollars for important experiments in my Department.' I had quite forgotten this remark when a few weeks later a postcard arrived, signed by this man: 'I am happily married and have found your man. Write to Henry Goldman, 998 Fifth Avenue, New York.' At first I took it for another joke, but on reflection I decided that an attempt should be made. With Hedi's help a nice letter was composed and despatched, and soon a most charming reply arrived and a cheque for some hundreds of dollars which helped us out of all our difficulties."[282]

Felix Ehrenhaft also sought to profiteer from the Einstein name and wrote to Einstein on 6 December 1919 requesting that he lecture for the Chemical-Physical Society of Vienna, stating, "[...]I would expect extraordinary profit[... .]"[283]

3.5.1 Advertising Einstein in the English Speaking World

Ernst Gehrcke's *Die Massensuggestion der Relativitätstheorie: Kulturhistorisch-psychologische Dokumente*, Hermann Meusser, (1924), is a valuable reference for newspaper and journal articles promoting Einstein as well as criticisms of Einstein up until 1924. I am only able to reproduce some of the articles cited in Gehrcke's important work and add a few others I have found.

---

282. M. Born, *My Life: Recollections of a Nobel Laureate*, Charles Scribner's Sons, New York, (1975), pp. 195-196.
283. Letter from F. Ehrenhaft to A. Einstein of 6 December 1919, English translation by A. Hentschel, *The Collected Papers of Albert Einstein*, Volume 9, Document 196, Princeton University Press, (2004), pp. 166-167, at 166. Einstein rejected the offer *ibid*. Document 211; and expressed reservations about Ehrenhaft's personality, *ibid*. Documents 269 and 270.

The London Times wrote on 7 November 1919,

"REVOLUTION IN SCIENCE.

NEW THEORY OF THE UNIVERSE.

NEWTONIAN IDEAS OVERTHROWN.

Yesterday afternoon in the rooms of the Royal Society, at a joint session of the Royal and Astronomical Societies, the results obtained by British observers of the total solar eclipse of May 29 were discussed.

The greatest possible interest had been aroused in scientific circles by the hope that rival theories of a fundamental physical problem would be put to the test, and there was a very large attendance of astronomers and physicists. It was generally accepted that the observations were decisive in the verifying of the prediction of the famous physicist, Einstein, stated by the President of the Royal Society as being the most remarkable scientific event since the discovery of the predicted existence of the planet Neptune. But there was difference of opinion as to whether science had to face merely a new and unexplained fact, or to reckon with a theory that would completely revolutionize the accepted fundamentals of physics.

SIR FRANK DYSON, the Astronomer Royal, described the work of the expeditions sent respectively to Sobral in North Brazil and the island of Principe, off the West Coast of Africa. At each of these places, if the weather were propitious on the day of the eclipse, it would be possible to take during totality a set of photographs of the obscured sun and of a number of bright stars which happened to be in its immediate vicinity. The desired object was to ascertain whether the light from these stars, as it passed the sun, came as directly towards us as if the sun were not there, or if there was a deflection due to its presence, and if the latter proved to be the case, what the amount of the deflection was. If deflection did occur, the stars would appear on the photographic plates at a measurable distance from their theoretical positions. He explained in detail the apparatus that had been employed, the corrections that had to be made for various disturbing factors, and the methods by which comparison between the theoretical and the observed positions had been made. He convinced the meeting that the results were definite and conclusive. Deflection did take place, and the measurements showed that the theoretical degree predicted by Einstein, as opposed to half that degree, the amount that would follow from the principles of Newton. It is interesting to recall that Sir Oliver Lodge, speaking at the Royal Institution last February, had also ventured on a prediction. He doubted if deflection would be observed, but was confident that if it did take place, it would follow the law of Newton and not that of Einstein.

DR. CROMMELIN and PROFESSOR EDDINGTON, two of the actual observers, followed the Astronomer Royal, and gave interesting accounts of their work, in every way confirming the general conclusions that had been enunciated.

'MOMENTOUS PRONOUNCEMENT.'

So far the matter was clear, but when the discussion began, it was plain that the scientific interest centred more in the theoretical bearings of the results than in the results themselves. Even the President of the Royal Society, in stating that they had just listened to 'one of the most momentous, if not the most momentous, pronouncements of human thought,' had to confess that no one had yet succeeded in stating in clear language what the theory of Einstein really was. It was accepted, however, that Einstein, on the basis of his theory, had made three predictions. The first, as to the motion of the planet Mercury, had been verified. The second, as to the existence and the degree of deflection of light as it passed the sphere of influence of the sun, had now been verified. As to the third, which depended on spectroscopic observations, there was still uncertainty. But he was confident that the Einstein theory must now be reckoned with, and that our conceptions of the fabric of the universe must be fundamentally altered.

At this stage Sir Oliver Lodge, whose contribution to the discussion had been eagerly expected, left the meeting.

Subsequent speakers joined in congratulating the observers, and agreed in accepting their results. More than one, however, including Professor Newall, of Cambridge, hesitated as to the full extent of the inferences that had been drawn and suggested that the phenomena might be due to an unknown solar atmosphere further in its extent than had been supposed and with unknown properties. No speaker succeeded in giving a clear non-mathematical statement of the theoretical question.

SPACE 'WARPED.'

Put in the most general way it may be described as follows: the Newtonian principles assume that space is invariable, that, for instance, the three angles of a triangle always equal, and must equal, two right angles. But these principles really rest on the observation that the angles of a triangle do equal two right angles, and that a circle is really circular. But there are certain physical facts that seem to throw doubt on the universality of these observations, and suggest that space may acquire a twist or warp in certain circumstances, as, for instance, under the influence of gravitation, a dislocation in itself slight and applying to the instruments of measurement as well as to the things measured. The Einstein doctrine is that the qualities of space, hitherto believed absolute, are relative to their circumstances. He drew the inference from his theory that in certain cases actual measurement of light would show the effects of the warping in a degree that could be predicted and calculated. His predictions in two of three cases have now been verified, but the question remains open as to whether the verifications prove the theory from which the predictions were deduced."

The London Times wrote on 8 November 1919,

"THE REVOLUTION IN SCIENCE.

EINSTEIN v. NEWTON.

VIEWS OF EMINENT PHYSICISTS.

Wide interest in popular as well as in scientific circles has been created by the discussion which took place at the rooms of the Royal Society on Thursday afternoon on the results of the British expedition to Brazil to observe the eclipse of the sun on May 29. (These were referred to in an interview with Sir Frank Dyson, the Astronomer Royal, which appeared in *The Times* of September 9.) The subject was a lively topic of conversation in the House of Commons yesterday, and Sir Joseph Larmor, F. R. S., M. P. for Cambridge University, on arriving at a lecture before the Royal Astronomical Society last evening, said he had been besieged by inquiries as to whether Newton had been cast down and Cambridge 'done in.'

Mr. C. Davidson, of Greenwich Observatory, one of the astronomers who took the photographs of the sun's eclipse at Sobral, in Northern Brazil, last May, in conversation with a representative of *The Times* last night, said he agreed that the observations taken of $Kappa^1$ and $Kappa^2$, near the constellation of Hyades, at the moment of totality, were conclusive of the deflection of their rays by the gravitation of the sun. In reply to the suggestion made by Professor Newall, of Cambridge, that the deflection might be due to an unknown solar atmosphere further in its extent than had been supposed and with unknown properties, Mr. Davidson said:—'That does not seem possible, because to produce such a deflection there would have to be an atmosphere of a kind unknown to theory and observation. Moreover, comets have been known to pass within grazing distance of the sun without any apparent retardation in their motion.'

Mr. Davidson was also prepared not to dissent from the view that the discovery of light possessing weight as well as mass might mark progress towards a conception of conditions outside three-dimensional space as we at present know it. 'Professor Einstein's theory', he remarked, 'demanded a good deal more of the dimensions existing in space than can be at present mathematically proved. It requires the curvature of space, variable time, and the displacement of the spectral lines towards the red. The latter has been very carefully tested by Dr. St. John at Mount Wilson in the United States, but so far without success. Nevertheless, there are some anomalies in the behaviour of the spectral lines which a good many scientific people believe may have compensations to explain them.'

On the main discovery, however, Mr. Davidson fully endorsed the opinion that the Newtonian principle had been upset, and that Professor Einstein had been right in at least two of his three predictions. 'His surmise with regard to the spectrum,' Mr. Davidson said, 'remains to be demonstrated. As to the phenomena of light, the Brazil observations have established that instead of a deflection of ·87 of a second of arc at the sun's limit which would have been expected by the application of Newton's law, it was **1·75**, which accords with Professor Einstein's theory. Our observations also proved that the outstanding discrepancy in the perihelion of Mercury can now also be accounted for.'

THE ETHER OF SPACE.

SIR OLIVER LODGE'S CAUTION.
TO THE EDITOR OF THE TIMES.

Sir,—To avoid misunderstanding, permit me to explain that my having to leave the meeting, reported in your issue of to-day (Friday), was due to a long-standing engagement and a 6 o'clock train.

The eclipse result is a great triumph for Einstein; the quantitative agreement is too close to allow much room for doubt, and from every point of view the whole thing is of intense interest.

I have more to say about it, and your excellent report gives a good idea of the general position; but I must deprecate the notion that last February I ventured on anything so serious as a prediction concerning the probable result.

I was rash enough to express a hope for a result equal to half Einstein's value. But the double-valued result can be assimilated and specified in various ways, one of which is the ponderability of light coupled with a definite effect of motion on the Newtonian constant of gravitation, an effect which the behaviour of Mercury and other planets has already rendered probable; while another is the vaguer suggestion that one of the two etherial constants, responsible for the velocity of light, is affected by a gravitational field, so as to cause a kind of refraction.

In any case, I would issue a caution against a strengthening of great and complicated generalizations concerning space and time on the strength of the splendid result: I trust that it may be accounted for, with reasonable simplicity, in terms of the ether of space.

Meanwhile I heartily congratulate Professor Einstein, and also the skilled and painstaking observers who have so admirably verified his striking and original prediction.

Yours faithfully,
OLIVER LODGE.
Llwynarthan, Castleton, Cardiff, Nov. 7.

DR. ALBERT EINSTEIN.

Dr. Albert Einstein, whose astronomical discoveries were described at the meeting of the Royal Society on Thursday as the most remarkable since the discovery of Neptune, and as propounding a new philosophy of the universe, is a Swiss Jew, 45 years of age. He was for some time Professor in Mathematical Physics at the Polytechnic at Zurich, and then Professor at Prague. Afterwards he was nominated a member of the Kaiser Wilhelm Academy for Research in Berlin, with a salary of 18,000 marks (£900) per annum, and no duties, so that he should be able to devote himself entirely to research work.

During the war, as a man of liberal tendencies, he was one of the signatories of the protest against the German manifesto of the men of science who declared themselves in favour of Germany's part in the war, and at the time of the Armistice he signed an appeal in favour of the German revolution. He is an ardent Zionist and keenly interested in the proposed Hebrew University at Jerusalem, and has offered to cooperate in the work there."

Note that *The London Times*, which had been one of the Director of British War Propaganda Lord Northcliffe's wartime propaganda organs, wanted to stress that Einstein opposed "Germany's part in the war". It also emphasized the claim that Newtonian theory had been overthrown. This drew harsh criticism from the nationalistic British, who took great pride in Isaac Newton. *The New York Times* emphasized the idea that Einstein's theory was incomprehensible to all but twelve persons in the world.[284] This myth aided Einstein, in that it allowed him to avoid criticism by claiming that anyone who criticized the theory of relativity did not understand it. The myth also enthralled a gullible public, which found the notion of incomprehensibility intriguing, and felt no need to try to judge the merits of the theory for themselves. In the introduction to the abridged version of the collection of some of Einstein's statements entitled *The World As I See It*, it says, among other things,

> "Einstein, therefore, is great in the public eye partly because he has made revolutionary discoveries which cannot be translated into the common tongue. We stand in proper awe of a man whose thoughts move on heights far beyond our range, whose achievements can be measured only by the few who are able to follow his reasoning and challenge his conclusions."[285]

---

[284]. J. Crelinsten, "Einstein, Relativity, and the Press", *The Physics Teacher*, (February, 1980), pp. 115-122; **and** "Physicists Receive Relativity: Revolution and Reaction", *The Physics Teacher*, (March, 1980), pp. 187-193. On the reaction of the British to the idea that Newton had been defeated, see A. F. Lindemann's letter to A. Einstein of 23 November 1919, *The Collected Papers of Albert Einstein*, Volume 9, Document 174, Princeton University Press, (2004).

[285]. "Introduction to the Abridged Edition", in A. Einstein, *The World As I See It*,

The New York Times wrote on 9 November 1919 on page 6,

"ECLIPSE SHOWED GRAVITY VARIATION

Diversion of Light Rays
Accepted as Affecting
Newton's Principles.

HAILED AS EPOCHMAKING

British Scientist Calls the Discovery
One of the Greatest of
Human Achievements.

Copyright 1919, by The New York Times Company.
Special Cable to THE NEW YORK TIMES.
LONDON, Nov. 8.—What Sir Joseph Thomson, President of the Royal Society, declared was 'one of the greatest—perhaps the greatest—of achievements in the history of human thought' was discussed at a joint meeting of the Royal Society and the Royal Astronomical Society in London yesterday, when the results of the British observations of the total solar eclipse of May 29 were made known.

There was a large attendance of astronomers and physicists, and it was generally accepted that the observations were decisive in verifying the prediction of Dr. Einstein, Professor of Physics in the University of Prague, that rays of light from stars, passing close to the sun on their way to the earth, would suffer twice the deflection for which the principles enunciated by Sir Isaac Newton accounted. But there was a difference of opinion as to whether science had to face merely a new and unexplained fact or to reckon with a theory that would completely revolutionize the accepted fundamentals of physics.

The discussion was opened by the Astronomer Royal, Sir Frank Dyson, who described the work of the expeditions sent respectively to Sobral, in Northern Brazil, and the Island of Principe, off the west coast of Africa. At each of these places, if the weather were propitious on the day of the eclipse, it would be possible to take during the totality a set of photographs of the obscured sun and a number of bright stars which happened to be in its immediate vicinity.

The desired object was to ascertain whether the light from these stars as it passed by the sun came as directly toward the earth as if the sun were not there, or if there was a deflection due to its presence. And if the

---

translated by A. Harris, Citadel, New York, (1993), p. vii.

deflection did occur the stars would appear on the photographic plates at measurable distances from their theoretical positions. Sir Frank explained in detail the apparatus that had been employed, the corrections that had to be made for various disturbing factors, and the methods by which comparison between the theoretical and observed positions had been made. He convinced the meeting that the results were definite and conclusive, that deflection did take place, and that the measurements showed that the extent of deflection was in close accord with the theoretical degree predicted by Dr. Einstein, as opposed to half of that degree, the amount that would follow if the principles of Newton were correct.

Dr. Crommelin, one of the observers at Sobral, who spoke next, said that eight exposures of twenty-eight seconds each were made during the totality of the eclipse. Seven of these plates showed seven stars in each. One showed no stars, owing to the presence of a thin cloud, but gave well-defined images of the inner corona of the sun and of great prominence. Seven exposures of the same star field were made for comparison between July 14 and July 18 in the morning sky, the sun being then 45 degrees or more away from it. The results reduced to the sun's limb were 2.08 seconds and 1.94 seconds respectively. The combined result was 1.98 seconds, with a probable error of about 6 per cent. This was a strong confirmation of Einstein's theory, which gave a shift at the limb of 1.7 seconds. The evidence in favor of the gravitational bending of light was overwhelming, and there was a decidedly stronger case for the Einstein shift than for the Newtonian one.

Though the results were fairly conclusive, Dr. Crommelin said the question of the revision of Newton's law of gravitation was one of such fundamental importance that consideration was already being given to the next total eclipse in September, 1922, visible in the Maldive Islands and Australia.

Two of the consequences of Einstein's theory, he continued, namely, the motion of Mercury's perihelion and the bending of light by gravitation, might now be looked on as established, 'at least with great probability.' There was, however, a third predicted consequence, which was a shift of the lines in the spectrum toward the red in a strong gravitational field. The effect in the solar spectrum would amount to one-twentieth of the Angstrom unit, the same as that due to a motion of one-half kilometer per second away from the sun. Dr. St. John had looked for this effect without success. If this failure were taken as final it would mean that parts of Einstein's theory would need revision, but the parts already verified would remain.

The effects on practical astronomy, Dr. Crommelin said, of the verification of Einstein's theory were not very great. It was chiefly in the field of philosophical thought that the change would be felt. Space would no longer be looked on as extending indefinitely in all directions. Euclidian straight lines could not exist in Einstein's space. They would all be curved, and if they traveled far enough they would regain their starting point.

Sir Joseph Thomson, summing up the discussion, said:

'These are not isolated results that have been obtained. It is not the discovery of an outlying island, but of a whole continent of new scientific ideas of the greatest importance to some of the most fundamental questions connected with physics. It is the greatest discovery in connection with gravitation since Newton enunciated that principle.'"

On page 17, 10 November 1919, *The New York Times* reported:

"LIGHTS ALL ASKEW IN THE HEAVENS

Men of Science More or Less
Agog Over Results of Eclipse
Observations.

EINSTEIN THEORY TRIUMPHS

Stars Not Where They Seemed
or Were Calculated to be,
but Nobody Need Worry.

A BOOK FOR 12 WISE MEN

No More in All the World Could
Comprehend It, Said Einstein When
His Daring Publishers Accepted It.

Special Cable to THE NEW YORK TIMES.

LONDON, Nov. 9.—Efforts made to put in words intelligible to the nonscientific public the Einstein theory of light proved by the eclipse expedition so far have not been very successful. The new theory was discussed at a recent meeting of the Royal Society and Royal Astronomical Society, Sir Joseph Thomson, President of the Royal Society, declares it is not possible to put Einstein's theory into really intelligible words, yet at the same time Thomson adds:

'The results of the eclipse expedition demonstrating that the rays of light from the stars are bent or deflected from their normal course by other aerial bodies acting upon them and consequently the inference that light has weight form a most important contribution to the laws of gravity given us since Newton laid down his principles.'

Thomson states that the difference between theories of Newton and those of Einstein are infinitesimal in a popular sense, and as they are purely mathematical and can only be expressed in strictly scientific terms it is useless to endeavor to detail them for the man in the street.

'What is easily understandable,' he continued, 'is that Einstein predicted the deflection of the starlight when it passed the sun, and the recent eclipse has proved a demonstration of the correctness of the prediction.

'His second theory as to the anomalous motion of the planet Mercury has also been verified, but his third prediction, which dealt with certain sun lines, is still indefinite.'

Asked if recent discoveries meant a reversal of the laws of gravity as defined by Newton, Sir Joseph said they held good for ordinary purposes, but in highly mathematical problems the new conceptions of Einstein, whereby space became warped or curled under certain circumstances, would have to be taken into account.

Vastly different conceptions which are involved in this discovery and the necessity for taking Einstein's theory more into account were voiced by a member of the expedition, who pointed out that it meant, among other things, that two lines normally known as parallel do meet eventually, that a circle is not really circular, that three angles of a triangle do not necessarily make the sum total of two right angles.

'Enough has been said to show the importance of Einstein's theory, even if it cannot be expressed clearly in words,' laughed this astronomer.

Dr. W. J. S. Lockyer, another astronomer, said:

'The discoveries, while very important, did not, however, affect anything on this earth. They did not personally concern ordinary human beings; only astronomers are affected. It has hitherto been understood that light traveled in a straight line. Now we find it travels in a curve. It therefore follows that any object, such as a star, is not necessarily in the direction in which it appears to be astronomically.

'This is very important, of course. For one thing, a star may be a considerable distance further away than we have hitherto counted it. This will not affect navigation, but it means corrections will have to be made.'

One of the speakers at the Royal Society's meeting suggested that Euclid was knocked out. Schoolboys should not rejoice prematurely, for it is pointed out that Euclid laid down the axiom that parallel straight lines, if produced ever so far, would not meet. He said nothing about light lines.

Some cynics suggest that the Einstein theory is only a scientific version of the well-known phenomenon that a coin in a basin of water is not on the spot where it seems to be and ask what is new in the refraction of light.

Albert Einstein is a Swiss citizen, about 50 years of age. After occupying a position as Professor of Mathematical Physics at the Zurich Polytechnic School and afterward at Prague University, he was elected a member of Emperor William's Scientific Academy in Berlin at the outbreak of the war. Dr. Einstein protested against the German professor's manifesto approving of Germany's participation in the war, and at its conclusion he welcomed the revolution. He has been living in Berlin for about six years.

When he offered his last important work to the publishers he warned them there were not more than twelve persons in the whole world who would understand it, but the publishers took the risk."

On 11 November 1919, on page 17, *The New York Times* reported:

"ACCEPTS EINSTEIN GRAVITATION THEORY

Prof. Currier of Brown University
Calls Eclipse Demonstration
Great Achievement.

SOME SCIENTISTS CAUTIOUS

They Want Full Reports from the
Observers Before Forming Their
Final Conclusions.

Special to The New York Times.
PROVIDENCE, Nov. 10.—The two expeditions which went out from the Royal Observatory at Greenwich, England, in connection with the total solar eclipse of May 29, accomplished one of the greatest scientific achievements of modern times, Clinton H. Currier, Professor of Astronomy at Brown University, declared tonight in commenting on the results recently announced at the joint meeting of the Royal Society and the Royal Astronomical Society in London.

As the result of the observations made by these scientists in Sobral, Brazil, and on the island of Principe, in the Gulf of Guinea, Professor Currier said, the Einstein relativity theory had apparently been confirmed.

Professor Currier pointed out that, according to Newton's theory, gravitation would not affect the direction of a ray of light. With the development of the electro-magnetic theory of light, however, it was asserted that gravitation would bend a ray of light as if it were a material projective moving at the same rate.

'It was not until 1915,' he said, 'that the four-dimensional theory of the universe, with time as a fourth dimension, was definitely conceived. This was contained in Einstein's famous relativity theory.

'According to Einstein, a ray of light is deflected by gravitation, the amount of deflection being twice that predicted by the electro-magnetic theory. The only way yet devised to test these theories is by means of stars near the sun at the time of a total eclipse of the sun. At such a time, a ray of light from a distant star passing close to the sun would be bent, according to these theories, causing the star to appear displaced from the position it normally occupied.'

This apparent displacement, according to recent dispatches from London, was observed by the scientists last May.

Special to The New York Times.

POUGHKEEPSIE, Nov. 10.—Miss Caroline Ellen Furness, Ph. D., Professor of Astronomy and Director of the Observatory at Vassar, says:

'Einstein's theory is one of the most difficult parts of mathematical physics. As yet I have not followed strictly its application to astronomy. Its results are remarkable and are such that they must be accepted. Since it was made from a study of photographs taken May 29, 1919, it ought to be easily verified by study of photographs of previous eclipses. At the time of every eclipse photographs are taken to see if there are any planets between Mercury and the sun. It ought to be possible to use these for this purpose.

'This phenomenon means that light does not travel in straight lines; that a ray from a star passing near another body of matter is slightly deflected from its original course.

'Ordinarily the positions of the stars are not affected by their nearness to the sun. They cannot be seen when near the sun except at an eclipse. The course of a star may be deflected many times, according to the new theory, and the true positions of stars will be confused for a while,' Professor Edna Carter of the Department of Physics says:

'This is the first positive proof for Einstein's theory of gravitation. It is of great importance. Einstein claimed that light was constant only when in uniform gravitation, and that when it came in the field of the sun it was deflected somewhat. His theory affects the theory of gravitation with relation to generalized relativity. The proof for Einstein's new theory seems indisputable.'

Special to The New York Times.

HANOVER, N. H., Nov. 10.—John M. Poor, Professor of Astronomy at Dartmouth College, said concerning the Einstein theory:

'If, as reported in the daily papers, Einstein's theory has received confirmation as a result of observations of photographs made at the time of the recent eclipse, it represents another approximation to the ultimate truth which the scientist is continually seeking. The Newtonian mechanics will need modification. That will be a matter which for the present, at least, will concern the student in mathematics and pure science. But what the ultimate effect will be on practical life cannot now be foretold.'

Astronomers and physicists and other scientific men in New York are much interested in the news from London that British observations of the total solar eclipse of May 29 bore out the theories of Dr. Einstein, Professor of Physics in the University of Prague, which, in effect, would bring about a revision of Newton's law of gravitation. They are reluctant to express an opinion on the deductions from the observations until they have full information. However, they regard the discovery as important; but one prominent physicist said that he would not regard it as being of such importance as to revolutionize the accepted fundamentals of physics.

Another said that he did not doubt the correctness of the observations, but that he would not be willing to accept the conclusions until it had been more definitely shown that the bending of light from stars passing close to the sun on its way to the earth was not due to the refraction of light gases surrounding the sun. He said that the theory was probably all right, but pointed out that it was one very hard of proof."

Numerous other articles appeared in the period from 1919 through 1921 and those interested in these articles are encouraged to reference the *New York Times* and *London Times* indices, as well as Gehrcke's *Die Massensuggestion der Relativitätstheorie*.

3.5.2 Reaction to the Unprecedented Einstein Promotion

Sir Oliver Lodge was one of Einstein's many critics. *The New York Times* published some of Sir Oliver Lodge's comments on 25 November 1919,

"A NEW PHYSICS, BASED ON EINSTEIN

Sir Oliver Lodge Says It Will
Prevail, and Mathematicians
Will Have a Terrible Time.

SPACE OF FOUR DIMENSIONS

In Which Gravity Ceases to be
a Force and Becomes
a Quality.

ATTEMPT TO MEASURE IT

Its Radius Put at 16,000,000 Light-
Years, or 80 Times the Distance to
Farthest Star Cluster Known.

Copyright. 1919, by the New York Times Company.
Special Cable to THE NEW YORK TIMES.

LONDON, Nov. 24.—To a small and distinguished gathering at Lord Glenconner's residence tonight, Sir Oliver Lodge explained the theory of Einstein, whose predictions were recently partially confirmed by the solar expedition and given to the world by the Astronomer Royal.

So complicated has this revolutionary theory proved that even some of the most learned have been confounded. Sir Oliver gave the foundation of the theory in this way:

'So long as matter is stationary with matter, its motion with respect to the ether produces no sort of optical effect, though this effect has been sought by observers in the last half century. Hence Einstein said 'let us assume that it is impossible to observe motion through the ether, but that the compensation will always be complete and let us work out a physics on that hypothesis. We do not know,' he said, 'whether the earth is moving a thousand miles a second or only an inch an hour. All our attempts to measure such ideas of motion are frustrated by some compensations influences which are embedded in the ether.'

'So in 1905 Einstein virtually said: 'We must assume that we shall never be able to get anything about the motion of matter through the ether, and we can only make deductions from the relativity of other motions of matter.' '

Hence the new physics, declared Sir Oliver, required four co-ordinates, not merely length, breadth, and thickness, but time. Gravity, too, ceases to become a force but becomes a quality in a fourth dimensioned space.

'The death knell of ether has been sounded,' he said, 'and there come strangely varying properties out of emptiness. Einstein's theory is not dynamical. Euclid becomes incorrect when applied to existing realities. Either there is boundary to space or there is not, but personally I cannot conceive either, though we must assume that one of these theories is right. To my mind, the great achievement of Einstein is his discovery of gravity in its relation to other forces.'

Sir Oliver concluded with the prediction that the new physics would dominate all other physics, and that the next generation of mathematical professors would have a terrible time of it, at which there was laughter.

'For university courses and for all purposes of scholastic instruction,' he said, 'we shall have the Galilean and Newtonian dynamics, but they would reign as a 'limited monarchy,' and, sooner or later, the Einstein physics would influence the intelligent man.'

Replying to Dr. Schuster, who voiced the thanks of the company, Sir Oliver said that the younger scientists of today were pursuing Einstein's path with brilliant success.

'Some day,' he remarked, 'I think that perhaps gravitation will give up its secret, but I must leave all the 'transcendental' methods to the young men.'

More Details Made Known.

The observations confirmatory of 'the Einstein theory' were made during the total eclipse of the sun on May 29 last, by two British expeditions, one sent to Principe on the west coast of Africa, the other to Sobral, in North Brazil. The results of these observations were communicated to a joint meeting of the Royal Society and the Royal Astronomical Society in London on Nov. 6. Perhaps the clearest and fullest account was supplied by Dr. A. C. Crommelin of the Royal Observatory at Greenwich, who was one of the observers with the Sobral expedition.

Dr. Crommelin said that the purpose of the expeditions was to test whether the light of the stars that are nearly in a line with the sun is bent by its attraction, and if so, whether the amount of bending is that indicated by the Newtonian law of gravitation (viz., seven-eighths of a second at the sun's limb), or the amount indicated by the new Einstein theory, which postulates a bending just twice as great. The fact that the new theory explained the anomalous motion of the major axis of Mercury's orbit impressed astronomers with a sense of its truth, and they took advantage of the recent eclipse to test it further. Two cameras were employed by the party at Sobral.

The first had a lens of 4 inches aperture and 20 feet focus; this camera and its coelostat were lent by the Royal Irish Academy. It was with this instrument that the best results were obtained. Eight exposures of 28 seconds each were made during the totality of the eclipse; seven of these plates showed seven stars each; one (the sixth exposure) showed no stars, owing to the presence of thin cloud, but gave well-defined images of the inner corona of the sun and of a great prominence. Seven exposures of the same star field were made for comparison between July 14 and 18, in the morning sky, the sun being then 45 degrees or more away from it.

The results, reduced to the sun's limb, were 2.08 and 1.94 seconds respectively. The combined result was 1.98 seconds, with a probable error of about 6 per cent. This was a strong confirmation of Einstein's theory, which gave a shift at the limb of 1.75 *seconds*. The results from the individual stars were consistent, and incidentally they confirmed the theoretical law that the shift ought to vary inversely as the distance from the sun's centre. If the shift were due to refraction produced by a gaseous envelope round the sun, it would vary according to a less simple law. The second camera used at Sobral was the object-glass of the Greenwich astrographic equatorial, of aperture 13 inches (which was reduced to 8 inches, as it was found to improve the definition), and focal length 11¼ feet, mounted in a steel tube, and supplied with light from a 13-inch coelostat. The focus was obtained by photographs of Arcturus. Unfortunately the images secured were not good, evidently owing to the coelostat mirror not being flat, for the quality of the object-glass was known to be very good.

Observations at Principe were much interfered with by clouds; however, five stars were recorded on some plates. No comparison plates of the field could be taken here; the observers did not arrive early enough to obtain them before the eclipse, and it was impossible to wait long enough to obtain them after it. The plan adopted was to photograph a check field near Arcturus. Both this field and the eclipse field had been photographed with the same object-glass at Oxford (without using the coelostat) and the Oxford plates enabled the eclipse field to be connected with the check one.

The shift at the sun's limb came out 1.60 seconds, with a probable error of about 0.30 second. It could be seen that the mean of this result and that of the four inch at Sobral exactly agreed with the value predicted by Einstein. The evidence in favor of gravitational bending of light was overwhelming, and there was a decidedly stronger case for the Einstein shift than for the Newtonian one. Though the results were fairly conclusive, the question of the revision Newton's law of gravitation was one of such fundamental importance that consideration was already being given to the next total eclipse, in September, 1922, visible in the Maldive Islands and Australia.

Two of the consequences of Einstein's theory, viz. the motion of Mercury's perihelion and the bending of light by gravitation, might now be looked on as established (at least with great probability). There was, however, a third predicted consequence, which was the shift of the lines in the spectrum toward the red in a strong gravitational field. The effect in the solar spectrum would amount of 1-20 of an Angstrom unit, the same as that due to a motion of ½ kilometre per second away from us. Dr. St. John had looked for this effect without success. If this failure were taken as final it would mean that parts of Einstein's theory would need revision, but the parts already verified would remain.

The effects on practical astronomy of the verification of Einstein's theory were not very great. It was chiefly in the field of philosophical thought that the change would be felt. Space would no longer be looked on as extending indefinitely in all directions; if they went far enough they would re-enter the same ground. Euclidian straight lines could not exist in Einstein's space. They were all curved, and if they traveled far enough they would regain the starting point. Mr. de Sitter had attempted to find the radius of space. He gave reasons for putting it at about 1,000,000,000 times the distance from the earth to the sun, or about 16,000,000 light-years. This was eighty times the distance assigned by Dr. Shapley to the most distant stellar cluster known. The fourth dimension had been the subject of vague speculation for a long time, but they seemed at last to have been brought face to face with it."

The New York Times published numerous articles which mentioned Sir

Oliver Lodge. Lodge was a vocal critic of Einstein's work.[286] *The New York Times* published the following on 26 November 1919, on page 12,

"Bad Times for the Learned.

It must indeed have been 'a small and distinguished gathering' that Sir OLIVER LODGE addressed in London, this week, if they were helped toward an understanding of the Einstein theory when he presented, as its foundation, the statement that 'so long as matter is stationary with matter, its motion with respect to the ether produces no sort of optical effect.'

So darkling and so seemingly irrelevant to anything in particular is that statement that one refrains with difficulty from suspecting a cable operator of having edited the dispatch. By no means all of it, however, was incomprehensible, even to the wayfaring man, and some of it even he could enjoy. Nothing could have been simpler, or pleasanter, for instance, that Sir OLIVER'S admission of his personal inability to conceive of space either as having a boundary or as not having one, though obviously it either is or is not unlimited. Some of us cannot see how anybody can conceive space otherwise than as going on and on, forever and forever. At least to do so is vastly easier than to elude the natural question, What except more space can there be beyond the place where space ends, if it does end? If Sir OLIVER can, he is lucky, or queer, or something.

Thoroughly human was his prophecy that as a result of the Einstein discoveries 'terrible times' are coming for the mathematicians—at any rate the tone of satisfaction in which he said it was thoroughly human. Mathematicians have caused so many other people to have terrible times so often and so long that it's only fair for them to have their own troubles at last. Not one woman in a hundred will give them any sympathy, whatever their suffering may be, and innumerable boys and girls will simply gloat if the mathematicians are forced to admit the wrongness of their haughty pronouncements. Their infallibility had been admitted long enough, and those of us who always thought there were errors in the multiplication table, especially where it deals with sevens, eights, and nines, at last are to be brilliantly vindicated."

On 15 December 1919, *The New York Times* wrote on page 14:

"Obviously a Rash Prophecy.

---

[286]. O. Lodge, "The New Theory of Gravity", *Nineteenth Century*, (December, 1919); and "The Ether Versus Relativity", *Fortnightly Review*, (January, 1920). *Cf.* "A New Physics Based on Einstein", *The New York Times*, (25 November 1919), p. 17. *The London Times*, (8 November 1919), p. 12, col. *d*; (25 November 1919), p.16, col. *d*; (29 November 1919), p. 9, col. *d*; (13 December 1919), p. 13, col. *a*. Lodge also published an article in *Nature*, Volume 106, (17 February 1921). *Confer:* J. Crelinsten, "Physicists Receive Relativity: Revolution and Reaction", *The Physics Teacher*, (March, 1980), pp. 187-193.

As it was before the Royal Society that Sir OLIVER LODGE last week discussed atomic energies and the possibilities they offer, it is to be presumed that he spoke with some care. Yet, when he prophesied that within a century the power now derived from burning 1,000 tons of coal would be obtained by setting free the force latent in two ounces of some unnamed substance, one cannot help remembering that Sir OLIVER has two personalities—that he is an eminent scientist and a credulous listener to 'mediums.'

That the atoms, instead of being mere ultimate divisions of dead matter, are alive with force nobody now doubts, but it seems hardly scientific to emphasize as Sir OLIVER did the astonishing velocity at which move the missiles which some atoms shoot out without at the same time calling attention to the size of the missiles. He knows, of course, the formulae relating to speed, mass, and momentum, and that to get any appreciable amount of 'work' done by the radium particles he described it would seem that they would have to move far more rapidly than they do. And a way to harness them is hardly imaginable, as yet.

Curved Space Before Einstein.
To the Editor of The New York Times:

In so far as concerns Einstein's 'new theory' that space is curved, which carries with it implications necessarily overturning current scientific dicta that parallel lines can never meet, that astronomical parallaxes cannot be relied upon for giving approximate distances of faraway stars, it may be interesting to note that Einstein is a late investigator in this field of speculative research.

For instance, Professor A. E. Dolbear in his 'Matter, Ether and Motion' (edition of 1892, page 57) says:

'We are assured that, for all we know, and therefore for all we can reason from, space itself may be curved so that if one were to start in what we call a straight line, in any direction, and travel in it on and on he would find himself after a long time coming to his starting point from the opposite direction; that what one would see if his sight were prolonged in any direction would be the back of his own head much magnified. * * * If the space we live in and the geometric relations are only practically true upon a small scale; if we may have a kind of space of four or more dimensions, whether we can now conceive of it or not, then should one understand that spaces and distances and velocities, and all computations formed upon them, though practically true, for all our experience, must not be pushed up into statements that shall embrace all things in the heavens as well as on the earth.'

It will appear from the above that one of our own foremost American physicists, one who is credited as having antedated Marconi in all the theoretical possibilities of wireless telegraphy, had covered, nearly three decades ago, all the essentials of what is now being attributed as a 'new theory' of the universe to Dr. Einstein.

GEO: H. HADLEY.
Fairfield, Conn. Dec. 12, 1919."

Sir Oliver Lodge believed in the utility of atomic energy. Contrary to popular modern myth, Albert Einstein opposed the idea of atomic energy. It turns out that Lodge was right and Einstein was wrong; but, amazingly, it is Einstein, and not his predecessors, who is today considered the father of atomic energy, which is an idea Einstein had found silly. The modern association of Einstein and the formula E with atomic energy and the atomic bomb probably originally stems not from Einstein, but from Pflüger and Moszkowski, as will be shown further on in this text.

Charles Lane Poor was another outspoken critic of Einstein and of the disingenuous promotion of the man. *The New York Times* wrote on 16 November 1919:

"JAZZ IN SCIENTIFIC WORLD

Prof. Charles Lane Poor of Columbia
Explains Prof. Einstein's Astronomical Theories.

WHEN is space curved?
When do parallel lines meet?
When are the three angles of a triangle not equal to two right angles?

Why, when Bolshevism enters the world of science, of course!

It is thus that Charles Lane Poor, Professor of Celestial Mechanics at Columbia University, explains the extraordinary cable announcements from London about Professor Albert Einstein's theories, which some suppose to have been verified by observations of the recent total eclipse of the sun. These observations were assumed to show that the rays of stars were deflected as they passed the sun, which led to the Q. E. D. that they were subject to the attraction of the sun, that is to gravitation: and from this premise it was easy to jump to the conclusion that Sir Isaac Newton's theory had been knocked to smithereens.

Well, Sir Isaac, after he saw the apple fall in his gardens at Woolsthorpe, and evolved therefrom his theory of gravitation, couldn't prove it for a long time. He made his calculations from a wrong estimate of the radius of the earth; and it was not until years later, when another scientist had corrected the figure for the radius, that he was able to give the gravitational principle to a shocked and incredulous world. Once the incredulity had evaporated in the light of proof, and the theory had become an established fact, it still was not immune from mistaken attack, as Professor Poor points out.

'For some years past,' Professor Poor said the other day, after reading the cable dispatches about the Einstein theory, 'the entire world has been in a state of unrest, mental as well as physical. It may well be that the physical aspects of the unrest, the war, the strikes, the Bolshevist uprisings, are in reality the visible objects of some underlying, deep mental disturbance, worldwide in character. This mental unrest is evidenced by the widespread intent in social problems, by the desire, on the part of many, to throw aside the well-tested authors of Governments in favor of radical and untried experiments.

'This same spirit of unrest has invaded science, and today there is just as great a conflict in the realm of scientific thoughts as there is in the realm of political and social life. There are many who would have us throw aside the well-tested theories upon which have been built the entire structure of modern scientific and mechanical development in favor of psychological speculations and fantastic dreams about the universe.

'Whenever a new observation is made which apparently does not directly fit into the old-time theories these modern disciples of scientific unrest rush into some weird explanation involving psychological speculations as to the constitution of matter or our fundamental concepts of mathematics.

'The eclipse observations reported to have been made on May 29 last are a case in point. If these observations are as reported (and such seems unquestionably to be the case), then these explanations, under present accepted theories, may be difficult, but such observations certainly do not warrant the acceptance of the speculations of Einstein.

'It may be that history is merely repeating itself. When Newton's theory of universal gravitation was given to the world in 1685 it was received with incredulity, especially among scientists on the Continent of Europe. Observations were adduced which these scientists asserted proved the fallacy of the Newtonian ideas. One by one these observations were shown to be in harmony with the law, to be direct consequences of it.

'Nearly one hundred years later (1770) Euler, one of the greatest mathematicians of the age, who had devoted a lifetime to developing and perfecting the Newtonian theory, in discussing the observed motion of the moon, wrote:

"There is not one of its equations about which any uncertainty prevails, and it now appears to be established by indisputable evidence that the secular inequality in the moon's motion *cannot* be produced by the forces of gravitation.'

'The essay in which this statement was made appeared during a time of profound mental and political unrest, such as now pervades the world. It won the prize of the Paris Academy of Sciences. To explain this peculiar motion of the moon, the greatest scientists of that age adopted theories involving a resisting medium in space, or introduced a time element into gravitation. Yet only a few years later Laplace found a full and complete explanation in certain intricate relationships between the motion of the moon and the varying shape of the earth's orbit, which had been overlooked by Euler and his followers, and found that this motion was a direct result of the forces of gravitation.

'Now, the so-called Einstein theories, or rather speculations, are such as completely overthrow not only the law of gravitation, but the fundamental conceptions on which all geometry and physics rest. And to sustain such a complete overturning of the entire basis on which scientific thought has been built, two—just two—observed facts are quoted; the motion of the perihelion of Mercury and certain displacements of stars when photographed near the sun.

'There is no need to go outside the law of gravitation to explain the motion of Mercury's perihelion. The explanation may well be in some term of the most complicated formulas which the mathematicians have overlooked or in some distribution of matter near the sun which the astronomer has hitherto failed to properly note. As a matter of fact, in order to make their equations usable, the mathematical observer assumes that the sun is a perfect sphere and that the space between the sun and the planets is empty. Yet both these assumptions are known to be false; the well-known sun spots and the many photographs of its corona prove the sun to be not perfectly spherical and to be surrounded by an irregular and changeable mass of matter. The real trouble is that the mathematicians have not yet been able to introduce the effects of these into their equations and to deduce their possible effects upon the motion of Mercury.

'The displacement of the stars noted in the recent eclipse photographs may be a phenomenon analogous to the refraction of light. All rays of light, when they pass from one medium to another, from air to glass, for example, are bent or refracted. Upon this principle are based the ordinary eyeglass, or the telescope. When the rays from the stars enter the earth's atmosphere they are bent and travel in curved paths. Now, the sun is surrounded by an envelope of gases of irregular shape and of varying densities, an envelope which certainly extends to the orbit of the earth, and probably, millions of miles beyond. Would it not be in accord with all known laws of optics if the rays of light from distant stars were bent and refracted when passing through such an envelope?

'The fact that such a bending effect has now been measured is of great scientific importance, and the results may change some of the hitherto accepted ideas as to the density and distribution of matter near the sun, but I fail to see how such an observation can prove the existence of a fourth dimension, or can overthrow the fundamental concepts of geometry.

'I have read various articles on the fourth dimension, the relativity theory of Einstein and other psychological speculation on the constitution of the universe; and after reading them I feel as Senator Brandegee felt after a celebrated dinner in Washington. 'I feel,' he said, 'as if I had been wandering with Alice in Wonderland and had tea with the Mad Hatter.'"

3.5.3 The Berlin Philharmonic—The Response in Germany

It was often difficult for scientists in Germany to publish their works in opposition to relativity theory or their condemnation of Einstein's plagiarism. Paul Weyland and Hermann Fricke organized a group of scientists to stand up against the suppression of dissent. They called themselves the *Arbeitsgemeinschaft deutscher Naturforscher zur Erhaltung reiner Wissenschaft*. Their plan was to publish the facts surrounding the promotion of Einstein and the theory of relativity and to hold public meetings exposing Einstein as a fraud and the theory of relativity as a "mass suggestion" imposed on the world public by the press. Einstein knew well the power "of coercive manipulation of public opinion"[287]. Einstein wrote to Lorentz on 21 September 1919 in the context of his, Einstein the Zionist's, hatred of the German People's loyalty to their nation,

"Those on the outside have no conception of how difficult it is to escape mass suggestion."[288]

The first meeting of the *Arbeitsgemeinschaft deutscher Naturforscher zur Erhaltung reiner Wissenschaft* was held in the Berlin Philharmonic on 24 August 1920. Einstein attended the meeting with his stepdaughter Ilse,[289] who was a reluctant member of Albert Einstein's "small harem".[290] Young Ilse Einstein wrote to Georg Nicolai about Albert Einstein's sexual advances toward her,

"I have never wished nor felt the least desire to be close to [Albert Einstein] physically. This is otherwise in his case—recently at least.—He himself even admitted to me once how difficult it is for him to keep himself in check."[291]

At the meeting, Paul Weyland and Ernst Gehrcke publicly exposed Einstein as a sophist and a plagiarist and discredited the evidence taken to support the theory of relativity. After the meeting, Einstein was convinced that all of German science knew he was a fraud. Panicked, Einstein wanted to run away from

---

287. Letter from A. Einstein to H. A. Lorentz of 21 September 1919, English translation by A. Hentschel, *The Collected Papers of Albert Einstein*, Volume 9, Document 108, Princeton University Press, (2004), pp. 92-93, at 93.
288. Letter from A. Einstein to H. A. Lorentz of 21 September 1919, English translation by A. Hentschel, *The Collected Papers of Albert Einstein*, Volume 9, Document 108, Princeton University Press, (2004), pp. 92-93, at 93.
289. D. K. Buchwald, *et al.*, Editors, *The Collected Papers of Albert Einstein*, Volume 7, Princeton University Press, (2002), p. 106.
290. A. Einstein quoted in M. Born, *The Born-Einstein Letters*, Walker and Company, New York, (1971), p. 8.
291. Ilse Einstein to Georg Nicolai, English translation by A. M. Hentschel, *The Collected Papers of Albert Einstein*, Volume 8, Document 545, Princeton University Press, (1998), p. 565. *See also:* D. Overbye, *Einstein in Love: A Scientific Romance*, Viking, New York, (2000), pp. 343, 404, note 22. *See also:* A. Einstein to Ilse Einstein, *The Collected Papers of Albert Einstein*, Volume 8, Document 536, Princeton University Press, (1998).

Germany without another word. A few days later, Einstein learned that his friends and friendly newspapers had instigated a smear campaign against Einstein's critics. Learning that there were others dishonest enough to defend him, and knowing that he would not have to defend himself, but instead would be defended by more competent persons than himself, Einstein decided to join in the fray with an article he published in the *Berliner Tageblatt*. He threw an undignified fit, which juvenile rant found a ready outlet in a pro-Einstein "Jewish newspaper".

Hendrik Antoon Lorentz and Paul Ehrenfest had been trying to persuade Albert Einstein to move to Leyden. Einstein refused because he knew that Lorentz would quickly discover that Einstein had no talent for original thought. Ehrenfest realized this and wrote to Einstein on 2 September 1919 to reassure him that they were not interested in Einstein's work, but merely wanted to use his name,

"No one here expects any accomplishments, all simply want you nearby."[292]

Soon after the press began to promote Einstein as if he were a new Newton, Albert Einstein wrote to Lorentz (whose work Einstein had plagiarized in 1905) about Lorentz' offer to join him in Leyden, or at least to spend a couple of weeks a year in Leyden. The press claimed that Einstein was the greatest and most original thinker the world had ever seen. Einstein wrote to Lorentz on 19 January 1920,

"Nevertheless, unlike you, nature has not bestowed me with the ability to deliver lectures and dispense original ideas virtually effortlessly as meets your refined and versatile mind. [***] This awareness of my limitations pervades me all the more keenly in recent times since I see that my faculties are being quite particularly overrated after a few consequences of the general theory stood the test."[293]

Pacifist Lorentz was very interested in the success of the eclipse observations as an opportunity for *rapprochement*, as were Einstein's supporters Arthur S. Eddington,[294] and Robert W. Lawson and Hans Thirring, who were apparently friends.[295] Thirring, like Einstein, never doubted the results of the

---

[292]. Letter from P. Ehrenfest to A. Einstein of 2 September 1919, English translation by A. Hentschel, *The Collected Papers of Albert Einstein*, Volume 9, Document 98, Princeton University Press, (2004), pp. 81-82, at 82.
[293]. Letter from A. Einstein to H. A. Lorentz of 19 January 1920, English translation by A. Hentschel, *The Collected Papers of Albert Einstein*, Volume 9, Document 265, Princeton University Press, (2004), p. 220.
[294]. Letter from A. S. Eddington to A. Einstein, *The Collected Papers of Albert Einstein*, Volume 9, Document 271, Princeton University Press, (2004), pp. 224-225, at 224.
[295]. *The Collected Papers of Albert Einstein*, Volume 9, Documents 146 and 177, Princeton University Press, (2004).

eclipse expeditions. Bertrand Russell, Georg Friedrich Nicolai and Romain Rolland were also Socialist Pacifists, who supported Einstein. Russell profited from a popular book he published on the theory of relativity, which helped to promote the theory, Einstein, and Russell.[296] As so often asserted by the researchers themselves, the eclipse observations were a publicity stunt to advertise a *rapprochement* between British and German science.

When this stunt was exposed, Einstein, in cooperation with a few pro-Einstein newspapers, tried to change the subject to anti-Semitism from Einstein's plagiarism, Einstein's misrepresentations of the scientific evidence, and the exposure of the contradictions in Einstein's theories. Certain papers made it quite clear to all, that anyone who criticized Einstein would be viciously smeared as if anti-Semitic, no matter what the nature of their complaint might be, and whether or not they had made any anti-Semitic statements—even Nobel Prize winning physicists were smeared around the world. There was no to be no fair hearing for Einstein's many critics. There views would not be made known to the public through the major press outlets of the world. This, of course, had a chilling effect on the debate, and when the press had effectively silenced all but a few of Einstein's many critics, the press disseminated the lie that no scientists of renown had ever disagreed with Einstein.

Einstein was right to run from his critics. He had been exposed as a plagiarist and a fraud. However, the proven threat of public smears undoubtedly quieted many who opposed Einstein and the theory of relativity, which group constituted the majority of scientists at the time. The pro-Einstein papers were especially vicious to Paul Weyland, probably because he had dared to accuse them of what they were doing—of shamelessly hyping Einstein, of misrepresenting the facts, and of making false accusations of anti-Semitism in a cowardly attempt to change the subject.

After an exchange of newspaper articles between Max von Laue and his opponents, and after the pro-Einstein press misrepresented the events at and surrounding the meeting in the Berlin Philharmonic, Paul Weyland printed his Philharmonic speech and reprinted several newspaper articles in the second volume of works published by the press of the *Arbeitsgemeinschaft deutscher Naturforscher zur Erhaltung reiner Wissenschaften*. The anti-Einstein press (Einstein used the term "pan-German press"[297]) and Weyland were generally fair to the extent that they allowed both sides of the argument to be heard. Such was not, and is not, the case with the pro-Einstein press.

Paul Weyland's brochure:

---

296. B. Russell, *The A B C of Relativity*, Harper & Brothers, New York, London, (1925).
297. Letter from A. Einstein to H. Delbrück of 26 January 1920, *The Collected Papers of Albert Einstein*, Princeton University Press, (2004), p. 235.

> Schriften aus dem Verlage der Arbeitsgemeinschaft deutscher
> Naturforscher zur Erhaltung reiner Wissenschaft e. V.
> Heft 2.

Betrachtungen

über

Einsteins Relativitätstheorie

und die Art ihrer Einführung

von

Paul Weyland

Vortrag gehalten am 24. August 1920 im großen Saal der
Philharmonie
zu Berlin

Berlin 1920

Verlegt bei der
Arbeitsgemeinschaft deutscher Naturforscher zur Erhaltung reiner
Wissenschaft e. V. Berlin N 113.

Als sich die Arbeitsgemeinschaft deutscher Naturforscher zur Erhaltung reiner Wissenschaft gründete, um als eins ihrer Hauptziele die Auswüchse der Allgemeinen Relativitätstheorie einerseits und die Art ihrer Propaganda andererseits zu bekämpfen, waren sich die Gründer von vornherein darin klar, daß es hier nicht glatt gehen würde. Der Umstand, daß Herr Einstein zufälligerweise jüdischen Glaubens sei und seine Gegner, die sich z. T. in der genannten Arbeitsgemeinschaft zusammenfanden, auch Christen aufweisen, ließ die Vermutung begründet erscheinen, daß, wenn sachliche, von den Rednern der Arbeitsgemeinschaft angeführte Gegengründe nicht sachlich erwidert werden können, diese zu schimpfen anfangen und dann mit dem Rettungsanker, dem Vorwurf des Antisemitimus kommen.

Diese Vermutung, die allerdings erst für die eigentlichen, späteren Vorträge erwartet wurden, hat sich überraschender Weise schon beim ersten Abend bestätigt — ein Umstand, der deutlich beweist, wie schwach man sich auf der Gegenseite fühlt.

Es ist nicht meine Absicht gewesen, meine, ausdrücklich als die Vorträge einleitenden Bemerkungen und Begrüßungsworte an das Auditorium, im Druck erscheinen zu lassen. Ich glaubte meiner polemischen Taktik dadurch Genüge getan zu haben, daß ich einige Artikel in die Tagespresse lenkte. Im übrigen war es — und ist es noch heute — mein Standpunkt, daß nur die Widerlegung des Themas selbst nötig und erwünscht sei. Ich bin eines besseren belehrt worden. Ein Teil jener Presse, die ich als „gewisse" Presse bezeichne, beginnt, sich deutlich abzuheben und durch entstellte Berichte den Wert einer Aktion in den Augen der Öffentlichkeit herabzusetzen, für die sie bestimmt sind. Ich durchbreche deshalb in diesem Falle mein Prinzip nur unbedingt wissenschaftlich zu sein, indem ich mich mit der Technik der Einsteinschen Regie befasse. Immerhin trösten mich die in dieser Schrift angeführten Tatsachen: Der genaue Nachweis der Methode, wie die Einsteinleute arbeiten, ist vielleicht kein wissenschaftlicher Gewinn, aber doch wohl [*4*] Mittel zum Zweck, uns solchem Gewinn näherzubringen. Denn bisher ist es m. E. noch nicht belegt worden, wie systematisch und skrupellos man dort zu Werke geht.

Der Leser möge nun ja nicht glauben, daß ich die „kritischen" Glanzleistungen des „Berliner Tageblatt", der „Vossischen Zeitung", des „Vorwärts" oder des „8-Uhr-Abentblattes" für ernst nehme, daß ich ihnen die Ehre eines Abdruckes zolle. Mein Zweck ist ein anderer. Da, wie gesagt, vermutet wurde, daß die Gegenpartei alles aufbieten wird, um der Aktion zu schaden, so haben wir zunächst auf sachliche Einwände gewartet. Diese sind ausgeblieben Man schimpft. Man kommt mit dem schwarzen Mann, dem Antisemitismus. Was hat der schon bei schiefen Situationen helfen können! Ich will dem interessierten Publikum nun Gelegenheit geben, selbst zu urteilen, wer „Zur Sache" zu rufen ist. Jene Skandalmacher, die um jeden Preis stören wollten, oder ich in meinem Vortrag, der alles, was er behauptete, ausgiebig bewies. Daß ich speziell nicht sprach, habe ich gleich in den ersten einleitenden Worten betont und auf die spezielle Behandlung an einem späteren Termin hingewiesen.

Ich übergebe deshalb meinen Vortrag der Öffentlichkeit in der Hoffnung, daß er dem edlen Zweck, dem die Vortragsreihe dienen soll, ein weiterer Baustein sei. Mit dem Erkennen der Einsteinschen Methode ist schon ein gewichtiger Schritt zum Erkennen der wahren Sachlage gedient. Daß aber die Gegenpartei derartig schnell die Flinte ins Korn wirft und in unsachgemäßes Schimpfen verfällt, hat sich selbst der kühnste Optimist auf unserer Seite nicht träumen lassen. Mein Vortrag ist genau wörtlich nach dem Konzept abgedruckt. Wo es mir wichtig erschien, habe ich Ergänzungen gemacht, diese aber als Fußnoten angebracht. Vorher jedoch die Abdrucke der klassischen Beispiele objektiver Berichterstattung: Zunächst das Tageblatt vom 25. August 1920, Morgenausgabe. (Nr. 398, Ausgabe A Nr. 210):

Die Relativitäts-Theorie.

Von Dr. V. Engelhardt (Berlin-Friedenau)

Gestern begann die „Arbeitsgemeinschaft deutscher Naturforscher", über deren Zusammensetzung uns Näheres nicht bekannt ist, in der Philharmonie eine Reihe von Vorträgen, die sich gegen Einsteins „Relativitäts-Theorie" richten sollen. Obwohl diese Art öffentlicher Polemik gegen einen [*5*] Forscher von der Bedeutung Einsteins uns wenig angemessen erscheint, werden wir über den Eindruck des ersten Abends sachlich berichten. Damit aber die Leser zunächst auch wissen, worum es sich eigentlich handelt, sei in den folgenden Zeilen der Versuch gemacht, über den Sinn der Relativitäts-Theorie einiges in populärer Form zu sagen. Daß ein Problem von dieser Tiefe in dem begrenzten Raum einer Tageszeitung auch nicht annähernd erschöpft werden kann, wird jedem Nachdenklichen klar sein. Die Redaktion.

Es folgt nun ein Einstein-Artikel.

Erst bekommt also das Publikum schnell eine Einstein-Spritze. Die „sachliche" Entgegnung sieht folgendermaßen aus: (Berliner Tageblatt, Nr. 399. Ausgabe B Nr. 189, Mittwoch, 25. August 1920, abends).

Die Offensive gegen Einstein.

E. V. Nachdem die Gegner Einsteins und seiner Relativitätstheorie sich in einer „Arbeitsgemeinschaft deutscher Naturforscher" organisiert hatten, erfolgte gestern abend in der Philharmonie der erste Vorstoß. Die beruhigende Erklärung des einen Forschers und Gelehrten, daß entsprechende Maßnahmen getroffen seien, um Skandalmacher an die Luft zu setzen, mußte den rein wissenschaftlich interessierten Besucher, der gekommen war, einer gelehrten Auseinandersetzung, einer streng sachlichen Beweisführung zu lauschen, etwas eigenartig berühren. Immerhin scheint die Erkenntnis, daß Stuhlbeine als Gegenargumente nur bedingten Wert haben, auch in dieser Arbeitsgemeinschaft deutscher Naturforscher vorhanden zu sein. Obwohl Professor Einstein, in einer Loge sitzend, eine bequeme Zielscheibe bot, wurde er doch nur mit solchen kleine Invektiven wie „Reklamesucht", „wissenschaftlicher Dadaismus", „Plagiat" usw. bombardiert.

Auf die bibelfesten Naturforscher, die einst so wild gegen Darwin vom Leder zogen, sind die gesinnungstüchtigen Naturforscher gefolgt, die jetzt dem wahrscheinlich höchst prinzipienlosen Relativitätsprinzip zuliebe wollen. Gesinnung ist etwas sehr Schönes, aber es wirkt immer ein wenig komisch, sie in der Mathematik verwendet zu sehen; sie hat die

Eigentümlichkeit, den aufgestellten Lehrsatz nur mangelhaft zu beweisen. Das ehrlichste im wissenschaftlichen Kampf bleibt doch immer das *argumentum in rem*. Die *argumenta in personam* sind außerdem ein zweischneidiges Schwert, und als einzige Gesinnung des Angreifers entpuppte sich schon öfter der Neid. Und wenn Namen von so glänzender Unbekanntheit sich erheben, so haben sie doch unbedingt nötig, sich mit Beweisen zu legitimieren.

Daß Herr Paul Weyland mit seiner Volksversammlungsrede die sogenannte „Einsteinsche Relativitätstheorie" zu Fall gebracht hätte, kann auch der stärkste Mann der Wissenschaft, ja selbst Herr Weyland nicht behaupten. Er wandte sich auch lediglich gegen die Person Einsteins und „seine Reklamepresse", [*6*] und verfehlte dabei nicht, für die eigene Presse gebührend Reklame zu machen. Sein Ton war nicht überzeugend, bisweilen aber peinlich. Wenn man dem Gegner unlautere Propaganda seiner Idee vorwirft, sollte man diese Idee nicht mit unlauterer Propaganda bekämpfen. Und wenn man dem anderen die Suggestion der Massen nicht verzeihen kann, so sollte man selber nicht auf die Gasse laufen.

Vornehmer und wissenschaftlicher war der Vortrag von Professor Gehrcke, und sein Spott auf die „junggeschüttelten Organismen" und andere „Experimente" der Relativität der Bewegung und der Relativierung von Zeit und Raum wäre vielleicht sehr treffend gewesen, wenn er in den Bildern nicht so stark aufgetragen hätte. Was er über die Beweise der Rotverschiebung der Spektrallinien und über die Perihelbewegung des Merkur vorbrachte, wird hoffentlich Professor Einstein zu wissenschaftlichen Entgegnungen reizen.

Von gleichem sachlichen Geist zeugt der Bericht der „Vossischen Zeitung", die schon leise zum Rettungsanker des Antisemitismus schielt:

Der Kampf gegen Einstein.

Der Feldzug gegen die Einsteinsche Relativitätstheorie oder wohl mehr gegen Einstein selbst wurde gestern Abend in der Philharmonie ziemlich temperamentvoll eröffnet. Eine zahlreiche Zuhörerschaft hatte sich eingefunden, darunter namhafte Mitglieder der Gelehrtenwelt, auch Prof. Einstein sah man in einer Loge, an seiner Seite die Tochter und nicht weit von ihm Prof. Nernst. Der angegriffene Forscher folgte mit gelassener Ruhe, mitunter sogar leise lächelnd, den Ausführungen der Redner oben auf der Bühne.

Mit schwerem Geschütz rückte Herr Paul Weyland, der die Kampagne eröffnete, an. Er wandte sich gegen die „sogenannte Einsteinsche Relativitätstheorie", die „Einsteinschen Fiktionen", ohne auch nur mit einem Worte zu erklären, worin diese eigentlich beständen. Daneben machte er wacker Reklame für Schriften, die im Vorraum käuflich seien; um deren Absatz zu befördern, wurde sogar bald eine einviertelstündige Pause eingelegt. Daneben wurden Physiker, die für Einstein eintraten, gehörig verdächtigt, dieser selber beschuldigt, daß er und seine Freunde die

Tagespresse und sogar die Fachpresse zu Reklamezwecken für die Relativitätstheorie eingespannt hätten. Da man immer noch nicht erfuhr, worum es sich eigentlich handelte, erscholl wiederholt der Ruf: „zur Sache!" Herr Paul Weyland erwiderte auf diese freundliche Aufforderung: „Es sind entsprechende Maßnahmen getroffen, um Skandalmacher an die Luft zu setzen." Nach etlichen Ausfällen gegen die Professorenklique, wobei der Redner bei Schopenhauer fleißige Anleihe machte, wurde über die geistige Verflachung unseres Volkes geklagt, selbst der Dadaismus wurde herangezogen und Herrn Einstein und seinen Anhängern wissenschaftlicher Dadaismus vorgeworfen. [*7*] Daneben klang ganz schwach eine antisemitische Note an und zum Schlusse Herrn Einstein ohne weiteres vorgeworfen, daß seine Formeln über die Perihelbewegung des Merkur einfach von Gerber abgeschrieben worden sei.

Eine ganz andere Tonart schlug der nächste Redner, Prof. Gehrcke, ein. Er bemühte sich, völlig sachlich seinen gegnerischen Standpunkt gegen die Relativitätstheorie klarzulegen. Diese sei eine geistige Strömung; ob gesund oder verhängnisvoll ist eine andere Frage. Er geht kurz auf die Relativität der Bewegung ein, bemüht sich sodann, zu zeigen, wie Einstein seine Relativitätstheorie mehrfach geändert habe; was er als Schwankungen bei Einstein bezeichnete, würden vielleicht andere als eine Entwicklung auffassen. Dann geht Gehrcke auf die Relativierung von Zeit und Raum ein. Nicht ohne Humor sucht er die Einsteinschen „Organismen", die sich der relativierten Zeit anpassen müssen, zu verspotten. Die Relativierung der Zeit führe, so meinte der Kritiker, zur Relativierung des Seins und damit zum physikalischen Solipsismus. Wie stehe es nun mit den Folgerungen, die Einstein aus seiner Theorie gezogen hatte? Es seien freilich nur winzige Effekte zu erwarten, aber die Rotverschiebung der Spektrallinien hat sich nicht feststellen lassen. Die Perihelbewegung des Merkur sei auch auf andere Weise zu erklären, ebensowenig seien die Ergebnisse der letzten Sonnenfinsternis-Beobachtung ein zwingender Beweis für den Einstein-Effekt. Zum Schluß meint Gehrcke, daß auch die Gedanken der Relativitätstheorie, nämlich die Idee der Union von Zeit und Raum von einem ungarischen Philosophen schon im Jahre 1901 ausgesprochen sei. Die heutigen Vorträge können noch keine abschließende Antwort über die Einsteinsche Relativitätstheorie gehen. Im übrigen möge sich jeder selbst ein Urteil bilden, die Grundlagen dazu werden die späteren Abende, die dieser Theorie gewidmet sind, liefern. K. J.

Der freundliche Leser wolle sich an Hand meines Vortrages genau überzeugen, wo ich bei Schopenhauer Anleihe machte und ob zum Thema geredet wurde oder nicht.

Seiner Tendenz entsprechend besitzt der Vorwärts das größte Maß an Unverfrorenheit, der die Veranstaltung sogar für Vorgänge verantwortlich macht, die sich auf der Straße abspielen. Jedes Kind weiß, daß man in dieser herrlichen Republik nicht in seinem Haufe kommandieren kann, daß also auch bei Veranstaltungen, Theatern usw. Zeitungs- und sonstige Verkäufer in dan Pausen bis in die Säle dringen. Daß Zigarettenverkäufer, „Freiheits"-Zeitungshändler ebenfalls da Publikum belästigen, hat der wackere Vorwärtsmann natürlich nicht gesehen. Es entfließt folgender Erguß dem Gehege seines Schreibtisches:

Der Kampf um Einstein. Gestern Abend entbrannte in der Philharmonie der Kampf um Einstein. Die Arbeitsgemeinschaft deutscher Naturforscher zur [*8*] Erhaltung reiner Wissenschaft hatte geladen. Der Anfang war häßlich und hatte mit Wissenschaft nichts zu tun, weder mit „reiner" noch mit „unreiner". Am Tore wurden Hakenkreuze verkauft — solche, die man die Rockklappe stecken konnte. Der erste Vortrag des Herrn Weyland paßte zu diesem Empfang. Er versprach eine wissenschaftliche Bekämpfung der Relativitätstheorie und mußte fortwährend zur Sache gerufen werden. Die höchst „sachliche" Entgegnung des Vortragenden war die Versicherung, daß man auf solche Zwischenrufe gefaßt sei und Vorsorge getroffen hätte, unliebsame Störenfriede an die Luft zu setzen. Jedenfalls auch eine Methode, um wissenschaftliche Fragen glatt zu lösen!

Doch genug von diesem Schmutz, der schließlich in persönlichen Angriffen das höchste leistete. Der nachfolgende Redner, Prof. Gehrcke, ein in der physikalischen Welt anerkannter Forscher, hatte nach dieser ihm scheinbar unerwarteten Einleitung sichtlich mit Befangenheit zu kämpfen. Bald aber festigte sich seine Stimme und er brachte in wohltuend ruhiger Weise seine Bedenken gegen die Relativitätstheorie vor. Die Widersprüche dieser Theorie sind nach Gehrcke nur zu lösen, wenn wir uns auf den Standpunkt eines „physikalischen Solipsismus" stellen und behaupten, daß jeder Mensch in seiner eigenen Welt lebt, die mit der des anderen gar nichts zu tun hat. Die Schwierigkeiten, welche die Relativitätstheorie unserem Denken bereitet, liegen wohl darin, daß wir immer und immer wieder unser gefühlsmäßiges „Zeiterlebnis" mit dem exakt definiertem „Zeitmaß" Einsteins verwechseln. Die Einwendungen Gehrckes gegen die Relativitätstheorie gingen ebenfalls von dieser „erlebten" Zeit aus, die mit dem physikalisch definierten Zeitmaßnichts zu tun hat — und können darum nicht stichhaltig genannt werden. Über den Ausfall der experimentellen Prüfung der Theorie wurde etwas einseitig berichtet. Die Akten sind hier noch nicht geschlossen. Den Stimmen gegen Einstein stehen ebenso gewichtige für Einstein gegenüber. Erst die Zeit wird lehren, ob Einsteins Theorie die experimentelle Fenerprobe wirklich besteht.

Am entzückendsten und sachlichsten äußert sich das „8-Uhr-Abendblatt", das Blatt der Dezimeter großen Überschriften, anerkannter Sachlichkeit, pp.:

Ein Einstein-„Kenner".
Der Kampf gegen die Relativitätstheorie.

Ein Herr Weyland, dessen Verdienste um die Wissenschaft weitesten

Kreisen bisher verborgen geblieben sind, versprach gestern in der Philharmonie einen Vortrag über „Einsteins Relativitätstheorie eine Massensuggestion". Als der Vorleser aber immer wieder von einer „gewissen Presse", die für Einstein Reklame machte, sprach, aus dieser „gewissen Presse" ihm passende Artikelstellen zitierte und dann aber selbst für einige „geschäftliche Mitteilungen" Gehör [*9*] verlangte, wurde der Vorleser aus der Mitte des Saales lebhaft „Zur Sache!" gerufen. Aber Herr Weyland hatte darauf nur zu erwidern, daß dafür gesorgt sei, Skandalmacher an die frische Luft zu befördern. Diejenigen, die wirklich Eintrittsgeld gezahlt hatten und nicht als persönliche Leibgarde des Herrn Vorlesers erschienen waren, hatten — so dünkt uns — doch einen Anspruch darauf, zu verlangen, daß gehalten werde, was in den Ankündigungen versprochen worden war. Tatsächlich sah man im Auditorium neben einigen wenigen ausgesprochenen Gelehrtenköpfen — Einstein selbst saß in der Nähe von Nernst in einer Loge — eine Anzahl junger handfester Burschen, deren ganzes Gehaben deutlich zeigte, in welchem Zusammenhang sie mit der Einsteinschen Lehre stünden. Schon beim Betreten des Saales wurden ja die berüchtigten antisemitischen Hetzbroschüren und blätter laut angepriesen. — Der Vorleser gedachte nicht mit einer Silbe der Genialität Einsteins, die von seinen wissenschaftlich geschulten Gegnern ohne weiteres anerkannt wird. Dafür erwähnte er aber die „sogenannte Einsteinsche Relativitätstheorie", die einen Umsturz in den Massen hervorgerufen habe, und prompt sagte eine hinter mir sitzende biedere Frau zu ihrem Mann: „Nu siehste, ick habe dir doch jesagt, daß er een Bolschewist ist." Der Mann nickte resigniert. Als der Vorleser dann, ohne es zu beweisen, von der „gewisse Presse" sprach, die vollkommen im Dienste Einsteins stünde, und man im Saal „Verleumdung Beweise!" rief, war es das biedere Ehepaar, das Herrn Weyland am begeisterten Beifall klatschte! Wollte man Herrn Weylands Ausführungen für ernst nehmen, dann müßte man folgerichtig die Universitätsfakultäten und Akademien, die Einstein mit Ehrenprofessuren und anderen akademischen Würden auszeichneten, für Reklameorganisationen von Stümpern und Idioten halten. Als der Vorleser schließlich eine Brücke zwischen Einsteins Lehren und dem Dadaismus zu schaffen sich anschickte, brachte ihm dies aus meiner Umgebung Kosenamen ein, die ich aus Höflichkeit hier lieber nicht wiedergeben möchte. Sie sind auch recht unparlamentarisch. Nach dieser vielversprechenden und verheißungsvollen Ouvertüre glaubte ich der Fortsetzung dieser eigenartigen Veranstaltung nicht weiter beiwohnen zu müssen. Diese taten desgleichen: ergriffen mit der einen Hand ihren Hut, mit der andern die — Flucht. K. M.

Hoffentlich nimmt der glänzende Vertreter einwandfreier Berichterstattung am 2. September Veranlassung, alsdann mit der anderen Hand sitzen zu bleiben, and jenem 2. September, wo speziell begonnen wird, Einsteins Theorie zu zergliedern.

Inzwischen erscheint — zur Verwendung für diese Broschüre nicht mehr geeignet — im Berliner Tageblatt (Nr. 402 Ausgabe A Nr. 212) vom

Freitag, den 27. August, Morgen-Ausgabe, Einsteins Antwort. Hier sei nur soviel bemerkt, daß Herr Einstein sachlich ebenfalls nichts [*10*] hervorbringt und ganz offen hinter dem Antisemitismus Schutz sucht. Es ist also soweit gekommen, eine sachliche Erklärung von ihm nicht zu erlangen. Er fertigt seine Gegner als kleine Geister ab, hat aber doch soviel Respekt vor ihnen, daß er schleunigst ins Ausland geht, statt sie mit seinen „erdrückenden" Beweisen zu schlagen. Nicht einmal den ersten der speziellen Vorträge hat Herr Einstein abgewartet! Die ersten allgemeinen Ausführungen genügten vollständig, ihn zum Rückzug zu veranlassen!

Ich lasse meinen Vortrag folgen:

Meine sehr verehrten Damen und Herren!

Ich habe die Ehre und das Vergnügen, Sie heute mit einigen einleitenden Worten zu einer Reihe von Darlegungen zu begrüßen, die sich mit der sogenannten Einsteinschen Relativitätstheorie befassen. Es handelt sich darum, kritisch zu untersuchen, ob die Einsteinschen Fiktionen eine konkrete Stütze durch die Wissenschaft, insbesondere die Naturwissenschaft erfahren kann, oder philosophische Punkte zu ihrer Bestätigung anzuführen hat.

Meine Damen und Herren! Es übersteigt den Rahmen der uns heute zugemessenen Zeit, daß ich Ihnen in diesem erten Vortrag eine gründliche Kritik der Einsteinschen Relativitätstheorie vom speziellen Standpunkt aus gebe. Diese Darstellung wird später mathematisch erfolgen. Ich habe mich heute lediglich damit zu befassen, zu untersuchen, wie es kam, daß die Allgemeine Relativitätstheorie seit geraumer Zeit die Massen in Aufruhr versetzen konnte. Ehe ich mich jedoch dieser einleitenden Aufgabe entledige, möchte ich einige geschäftliche Bemerkungen vorneweg schicken. Es wird mir soeben mitgeteilt, daß die Druckerei den heutigen Vortrag des Herrn Professor Dr. Gehrcke fertiggestellt hat und eine gewisse Anzahl Exemplare noch heute hierher senden wird. Ich werde diese Bücher im Foyer aufstellen lassen, wo selbst diese nach dem Vortrage käuflich zu haben sind. Ebendort wird eine Schrift des Heidelberger Physikers P. Lenard ausgelegt, die ich allen denen, die sich über den Wert der Einsteinschen Relativitätstheorie in wirklich sachlicher Weise informieren wollen, recht empfehlen möchte. Das Buch erfreut sich nach meinem Dafürhalten neben strenger Wissenschaftlichkeit ungemeiner Eindringlichkeit und Gemeinverständlichkeit.

Meine Damen und Herren! Wohl selten ist in der Naturwissenschaft [*11*] mit einem derartigen Aufwand von Reklame ein wissenschaftliches System aufgestellt worden, wie bei dem Allgemeinen Relativitätsprinzip, daß sich bei näherem Zusehen als höchst beweisbedürftig entpuppte. Dieses System, das unter Heranziehung aller möglichen Philosopheme, mit Mathematik verbrämt, teils in reiner Abstraktion, teils in konkreten Abstrusitäten als Relativismus oder allgemeine Relativitätstheorie bezeichnet wird, wollen wir uns im Verlaufe der vorliegenden Vortragsreihe unter der Führung von Spezialforschern etwas näher ansehen.

Es handelt sich um ein System, welches beansprucht, die alleinige Wahrheit zu bringen über alle Vorgänge des Naturgeschehens. Es soll uns die tiefste Wahrheit über das, was in der Erfahrungswelt geschieht, enthüllt werden. Wie begründet nun aber der Erfinder der Relativitätstheorie diese, seine Absicht. Er sagt: „Es ist mein Hauptziel, meine Theorie so zu entwickeln, daß jeder psychologische Natürlichkeit des eingeschlagenen Weges empfindet." Statt uns mit Tatschen zu kommen, statt Beweise zu bringen, wird uns „die psychologische Natürlichkeit der Theorie", „empfindend" nahegelegt, an anderen Stellen „die Schönheit der Theorie", in noch anderem Falle „die Kühnheit der Theorie" angepriesen. Meine Damen und Herren! Kühnheit des Gedankens ist sehr wohl eine Notwendigkeit des erfolgreichen Forschers, nur hat diese Kühnheit sich selbst Grenzen zu ziehen, die im menschlichen Taktgefühl und in wissenschaftlicher Einsicht begründet sind. Treffender kann sich niemand über diesen Punkt äußern als P. Lenard [*Footnote:* P. Lenard, Über Relativitätsprinzip, Äther, Gravitation. Verlag von S. Hirzel, Leipzig 1920. Preis M. 6.—] in seiner kleinen Schrift. Ich möchte Ihnen diese Stelle hier nicht vorenthalten. Lenard sagt zu diesen Punkt auf Seite 1 folgendes:

„Den Tatsachen kühn voraneilen wollen — Hypothesen machen — gehört dabei dennoch immer zu den schönsten, auch nützlichsten Vorrechten des Naturforschers. Aber er darf auch hierbei nicht rücksichtslos verfahren, sondern muß jeden Augenblick bereit sein, vor Tatsachen sich zu beugen, und er muß nie vergessen, daß er wirklich nur Zufall ist; wenn eine seiner Hypothesen dauernd die Probe an der Wirklichkeit besteht und also einen Fund bedeutet, und daß er also, will er gewissenhaft sein, nur zögernd das, was ursprünglich Hypothese, Dichtung des Geistes war, als Wahrheit auszugehen oder anzuerkennen wird bereit [*12*] sein dürfen. Je „kühner" ein Naturforscher sich gezeigt hat, desto mehr Stellen finden sich im allgemeinen in seinen Veröffentlichungen, die nicht dauernd standhalten; man kann dies mit Beispielen aus alter und neuer Zeit (besonders leicht aus letzterer) belegen. Deshalb verdient die Kühnheit des Naturforschers auch lange nicht die Hochschätzung wie die des Kriegers; denn letzterer setzt mit seiner Kühnheit sein Leben ein, während ersterer meist bequeme Nachsicht und Vergessenheit für seine Fehlschläge findet. Manchmal scheint die Naturforschern zugeschriebene „Kühnheit" wirklich nur darin zu bestehen, daß ziemlich skrupellos zu Ungunsten der Gediegenheit der Wissenschaftliteratur von vornherein auf eigene Schadlosigkeit gerechnet wird. Deutsche Eigenschaft ist diese Kühnheit nicht."

Meine Damen und Herren! Es ist eine ganz auffallende Erscheinung, daß die Einstein-Presse und -Literatur sich mit ganz geringen Ausnahmen in einer derartigen überschwänglichen Lobhudelei gefällt, wie ich sie oben angeführt habe, daß aber diesen Phrasen nicht das geringste Positive entgegensteht. Ich könnte noch stundenlang in der Aufzählung solcher Äußerungen fortfahren — alle aus Einsteins oder seiner Anhänger wissenschaftlichen Veröffentlichungen, aus Arbeiten — die in den Annalen der Physik, in den Sitzungsberichten der Preußischen Akademie und in vielen anderen ernsten wissenschaftlichen Zeitschriften gedruckt worden sind.

Diese Redensarten, die nun schon in der Fachpresse auftraten, werden durch die Veröffentlichungen, welche sich an ein breiteres Publikum wenden, noch erheblich übertroffen. Es soll Einsteins Theorie einen „Wendepunkt des menschlichen Denkens und der menschlichen Kultur" bedeuten. „Die großen Genies der Vergangenheit Kopernikus, Kepler, Newton verblassen gegenüber der alles überstrahlenden Theorie von Einstein!" „Abgrundtiefe eisige Höhen", „höchste Gipfel", „gewaltigste Gedankenarchitektur" — das sind die Beiworte, die dieser Fiktion gezollt werden. „Die wissenschaftliche Welt beugt sich vor der siegenden Kraft, vor dem glänzenden Triumph des menschlichen Geistes der an theoretischer Bedeutung noch die berühmte Errechnung des Planeten Neptun durch Leverrier und Adams in den Schatten stellt. Von überraschender Folgerichtigkeit, physikalisch und philosophisch gleich befriedigend ist der Bau des Alls, den die allgemeine Relativitätstheorie vor uns enthüllt. Überwunden sind alle Schwierigkeiten, die auf Newtonschen Boden erwuchsen, alle Vorzüge jedoch, durch die das moderne Weltbild sich [*13*] über die engen antiken Anschauungen erhob, strahlen im reineren Glanze als zuvor. Die Welt ist durch keine Grenzen eingeengt und doch in sich harmonisch geschlossen, sie ist vor der Gefahr der Verödung gerettet! Von neuem erkennen wir die erlösende Kraft der Relativitätstheorie die dem menschlichen Geist eine Freiheit und ein Kraftbewußtsein schenkt, wie kaum eine andere wissenschaftliche Tat sie je zu geben vermochte!"

Meine Damen und Herren! Was ich Ihnen hier eben erzählte, sind nicht etwa von mir ausgedachte Parodien, sondern wörtliche Zitate aus der Einstein-Presse, die ich Ihnen hundertfältig ergänzen könnte und die in unzähligen Auflagen in einer wahren Massenflut auf die bedauernswerte Öffentlichkeit losgelassen wurde.

Wenn man sich diese Ausprüche vergegenwärtigt, so drängt sich dem kritisch veranlagten Geist unwillkürlich die Frage auf: „Sollte hier nicht etwas vorliegen, was mit ernster wissenschaftlicher Arbeit und Sachlichkeit nichts zu tun hat? Wie will ein heute lebender Mensch imstande sein, eine menschliche Entdeckung oder Erfindung in eine Linie mit den Taten eines Kopernikus, Kepler oder Newton zu setzten, von denen uns heute Jahrhunderte trennen? Wie will der heutige Mensch irgend einer wissenschaftlichen Neuheit heute schon ansehen können, daß sie sich dereinst in Jahrhunderten aus dem Getriebe der Zeit so herausheben wird, wie dies bei den großen Namen der Vergangenheit der Fall ist? Spricht bei solch exaltierten Ausdrücken wie wir sie soeben gehört haben, überhaupt noch der nüchterne wissenschaftliche Verstand, oder sind wir hier in einem Gefühlsrausch hineingeraten, der vor anderen Räuschen nur das voraus hat, daß es sich auf die Wissenschaft bezieht? Solche überschwänglichen Ausdrücke sind jedenfalls in der wissenschaftlichen Welt etwas ungewöhnliches und lassen deutlich eine gesuchte Beeinflussung mit Reklamemitteln vermuten, wo durch strenge Sachlichkeit nichts erreicht werden kann.

Aber nun wird behauptet, der Erfinder der Relativitätstheorie habe mit allen diesen Dingen nichts zu tun. Ihn kümmerte nur der weitere Ausbau seiner Theorie und die reine Wissenschaft in stiller Gelehrtenzurückgezogenheit. Ein Büchlein [*Footnote:* Max Hasse, Das Einsteinsche Relativitätsprinzip, Magdeburg, Selbstverlag des Verfassers.] dem ich einen Teil der Lobeshymnen entnommen habe, schreibt nun in seinem Vorwart: „Der Verfasser nahm sich die Freiheit, die Druckbogen Prof. Dr. A. Einstein [*14*] einzusenden, der ihn mit folgender Antwort erfreute: „Ihre populäre Darstellung scheint mir in der Tat dem Geiste des Nicht-Physikers in glücklicher Weise entgegenzukommen. Ich sende Ihnen die Korrekturbogen mit einigen Randbemerkungen zurück, damit Sie einige kleine Böcke daraus entfernen können."

In einem Zeitungsartikel verwandte ich diese Niedlichkeit und werde von einem hervorragenden Berliner Physiker darauf mit folgenden Worten angegriffen:

In Nr. 171 dieses Blattes ereifert sich Herr Weyland gegen Einsteins allgemeine Relativitätstheorie; gegen die Art ihrer Verbreitung in der größeren Öffentlichkeit sowie gegen ihren Inhalt. Es liegt mir durchaus ferne, alles das decken zu wollen, was kleinere Geister bei der Verbreitung der neuen Lehre durch Ungenauigkeiten, Übertreibungen und Geschmacklosigkeiten gelegentlich gesündigt haben, und die im besonderen herangezogenen Äußerungen von Archenhold und Max Hasse kann ich nicht beurteilen, weil ich sie nicht kenne. Zu einem solchen Angriff auf Einsteins Persönlichkeit, wie ihn Herr Weyland macht, bieten diese Dinge aber doch nicht den mindesten Anlaß.

Demgegenüber möchte ich festellen, daß Herrn Einstein die Mitwirkung der jetzt abgeschüttelten kleineren Geister doch wohl höchst angenehm war, denn sonst hätte er sich nicht zu der soeben verlesenen Antwort veranlaßt gefühlt. Aber einen Menschen, der in seiner Naivität und Unkenntnis des Themas soweit geht, daß er noch ausdrücklich in seinem Vorwort hervorhebt, nicht mehr einen Satz der euklidischen Geometrie beweisen zu können, vor seinen Wagen zu spannen, ist nach meinem Dafürhalten Reklamemache um jeden Preis — oder Unwissenschaftlichkeit. Wenn Herr Einstein gewollt hätte, diesem Geschreibsel ein Ende zu machen, hätte er jahrelang Zeit gehabt. Durch eine einzige Äußerung, durch der mit seinem Kreise vorzüglich in Verbindung stehenden Presse hätte er es erreichen können, daß der ganze Schwall von Verherrlichung und Bewunderung ein Ende findet, das hat Einstein nicht gewollt, sonst hätte er sich dementsprechend geäußert und was noch wichtiger ist, dementsprechend gehandelt. Das ist die systematische Massensuggestion zum Preis und Ruhm eines Einzelnen, der die breite Öffentlichkeit bitter notwendig hat, nachdem ihm sachlich Opposition über Opposition erwächst. Aber auch in wissenschaftlichen Kreisen wird das Äußerste versucht, um Beweise für die Relativitätstheorie an den Haaren herbei zu ziehen. [*15*] Da es um die Frage der Rotverschiebung still geworden ist, [*Footnote:* Wer sich über den neuesten Stand der Rotverschiebung informieren will, dem sei die Schrift von L. C. Glaser, Über Versuche zum Beweise der Relativitätstheorie (Heft 3 der vorliegenden Sammlung) empfohlen.] schaut man nach anderen Objekten aus und findet leider recht dürftige Ausbeute. Da setzt dann nun an gewissen Stellen, wo man die Beziehung und die Macht hat, die Taktik des Totschweigens ein. Einsteins ständige Referenten geben von Forschungsberichten auf anderem Standpunkt stehender Gelehrten in ihren Referaten entweder gar keine oder durch einschränkende Bemerkungen entstelle Berichte, z. B. werden solche Forschungsergebnisse gegenüber den Einsteinschen „Axiomen" stets als unbewiesene offene Fragen behandelt. [*Footnote:* Unter einem Referat versteht man gemeinhin die Wiedergabe der Meinung eines Autors, ohne daran einschränkende Kritiken zu knüpfen. Die „Physikalischen Berichte", deren Redaktion durchaus unter Einsteinschen steht, wendet diese nicht übliche Praxis der indirekten Stimmungsmache an, wo es absolut nicht zu vermeiden ist, über gegenteilige Ansichten zu referieren.] So wird eine Arbeit von Sir Oliver J. Lodge mit folgenden Worten abgefertigt: „ Es wird in dieser ganz kurzen Notiz versucht, das Wesen der Ablenkung eines Lichtstrahles, nach der allgemeinen Relativitätstheorie eine Folge der Schwere der Energie, auf Grund früherer Anschauungen plausibel zu machen.

Weiter heißt es (Physik. Ber. 1920, Heft 15, S. 947) J. v. Kries: Über die zwingende und eindeutige Bestimmbarkeit des physikalischen Weltbildes. Die „Naturwissenschaften", 8, 237-44, 1920: Kries wirft die Frage auf, ob das Weltbild der modernen Physik zwingend und eindeutig genannt werden kann, und vertritt die Anschauung, daß diese Forderung für das Weltbild der Relativitätstheorie nicht durchgeführt ist, diese also nur als eine mögliche Erscheinungsform unter vielen anderen erscheint. Für den Physiker, dem die Relativitätstheorie heute als der befreiende Weg aus den Dunkelkammern der bisher klassischen Wissenschaften erscheint, muß diese Auffassung befremdend anmuten usw.

Einen anderen, noch instruktiven Fall finden Sie in der letzten Nummer der Naturwissenschaften. [*Footnote:* Die Naturwissenschaften 1920, Heft 34, Seite 667-673. Der Bericht der englischen Sonnenfinsternisexpedition über die Ablenkung des Lichtes im Gravitationsfeld der Sonne. Von Erwin Freundlich.] In dieser Zeitschrift, die nicht nur [*16*] von Fachleuten gelesen wird, sitzen die Eistein-Leute besonders fest. Von dort aus wird quasi als deren Hauptquartier Stimmung für ihn gemacht.

Es werden in einem langen Artikel die Untersuchungen der englischen Sonnenfinsternisexpedition, die nach Brasilien gesandt wurde, Herz und Nieren geprüft, ob sich etwas für das Relativitätsprinzip günstiges herauspressen ließe. Dabei kann der Referent — natürlich ein Freund Einsteins — nun nicht umhin, sich den Schein der Objektivität zu geben. Er zitiert ausdrücklich die Bedenken der Expeditionsleiter gegen eine

Annahme einer Bestätigung im Sinne des allgemeinen Relativitätsprinzipes, wo es heißt:

Die Aufnamen mit dem 8zölligen astrographischen Objektiv, die ebenfalls in Brasilien gewonnen wurden, liefern zwar auch einen Hinweis für die vermutete Lichtablenkung, aber die Sternbilder auf den Patten sind nach den Angaben der englischen Beobachter so unscharf und diffus, daß die aus ihnen abgeleiteten Resultate nur ein geringes Gewicht haben. Anscheinend hatte sich der Coelostatenspiegel infolge der Sonnenstrahlen stark verworfen und die Abbildungen verdorben. Es ergibt sich für den Wert von a am Sonnenrand der Wert **0",93 Nimmt man aber an**, daß der Skalenwert auf den Finsternisplatten in Wahrheit nicht weiter verändert war, als er es nach dem Einfluß der Refraktion und Aberration sein mußte — **eine sehr wahrscheinlich richtige Annahme**, denn die Unschärfe der Bilder rührte wohl kaum von einer reellen Änderung der Fokusierung des Objektivs her —, so resultiert für a der Wert **1",52** am Sonnenrand.

Und was macht der Einstein-Mann aus dieser deutlichen Einschränkung?
Er leitet daraus folgendes ab:

„Zusammenfassend kann man sagen:
„Die Sonnenfinsternisplatten in Sobral wie in Principe offenbaren unzweideutig eine systematische Verlagerung der Sternbilder, wie sie zutage treten müßte, wenn das Licht im Gravitationsfelde der Sonne abgelenkt würde. Diese Ablenkung verläuft dem Betrage nach durchaus [*17*] so, wie sie von der Relativitätstheorie vorausgesagt worden war."
[*Footnote:* Die Frage der Refraktion, die, wenn ein Effekt in Frage kommt, sowie der sogen. Eberhard-Effekt, der jedem Astrophysiker bekannt ist, wird hier nicht berührt. Falls Opponenten hier die Beobachtungen auf Principe für sich in Anspruch nehmen, verweise ich auf Heft 3 dieser Sammlung: Dr.-Ing. L. C. Glaser: Über Versuche zum Beweise der Relativitätstheorie, wo dieser Einwand vornherein widerlegt wird.]

Gegenüber solchen Unglaublichkeiten versagt einem Menschen normaler Denkungsweise das Ausdrucksvermögen. Ein Kaufmann hat dafür den treffenden Ausdruck: Bilanzverschleierung.

An diesen kleinen Beispielen, die sich, wie die oben angeführte Lobhudelei in beliebigem Maße fortsetzen lassen, können Sie ersehen, daß auch hier die Macht des Einsteinschen Armes wirkt und die Beeinflussung in diesem Falle der wissenschaftlichen Welt genau so versucht und durchgeführt wird, wie der breiten Öffentlichkeit gegenüber. Wo es absolut nicht geht, die berühmte Konjugation, über die sich bereits Schopenhauer in seiner Abhandlung über die Universitätsphilosophie in so satyrischer Weise ausgelassen hat, anzuwenden, nämlich nach der Formel: ich schweige tot, du schweigst tot, er schweigt tot — wir schweigen tot, ihr schweigt tot, sie schweigen tot außer Kraft zu setzen, da beginnt die indirekte Methode, nämlich Forschern, die sich durch räumliche Entfernung oder sonst wie nicht gleich zur Sache äußern können, den Wert ihrer Abhandlungen durch einschränkendes oder kritisches Referat herabzusetzen.

Warum hat nun Einstein Veranlassung, mit seinen Hypothesen die breiten Massen und die Wissenschaft zu beeinflussen zu versuchen? Wohl nur deshalb, weil ihm in wissenschaftlichen Kreisen dauernd Gegner erwachsen — Tatsachen, die man gern verschweigt und, wenn sie gedruckt werden sollen, gern unterbindet durch die Beziehungen, die man hat. Noch ein in den letzten Tagen erscheinenes Buch eines gewissen Harry Schmidt (Verlag Hartung, Hamburg) erkühnt sich, alle Gegengründe gegen Einsteins Theorie, ohne die Spur eines Gegenbeweises anzutreten, abzuweisen, unglaubliche Unrichtigkeiten und Unsachlichkeiten in das Publikum zu werfen und, was das Unverschämteste an dieser Arbeit ist, Beweise als gesichert anzugeben, wo das [*18*] Gegenteil einwandfrei feststeht. [*Footnote:* Das Schmidt'sche Buch werde ich an anderer Stelle behandeln.] Aber nicht nur in der Literatur, sondern auch in öffentlichen Vorträgen wird die Massensuggestion im Einsteinschen Sinne emsig betrieben, ohne daß die interessierte Öffentlichkeit den wahren Stand der exakten Naturforschung zu hören bekommt. So hielt kürzlich ein Berliner Popularastronom im Blüthner-Saal einen Propagandavortrag, [*Footnote:* Während der Pause nahm Herr Archenhold Veranlassung, mich im Künstlerzimmer aufzusuchen und sich erregt über meinen Angriff auszusprechen. Herr Archenhold erklärte, daß er den Vortrag aus eigener Iniative hielte, Einstein ebenso gut und schlecht kenne, wie mich. Ferner machte Herr Archenhold Bemerkungen darüber, daß er an der Treptower Sternwarte mit seinem Herzen hängt und genau so arm einst aus ihr herausgehe, wie er hineingekommen ist. Diese zum Thema nicht gehörige Bemerkung möchte ich dahin berichtigen, daß es mir erstens nie eingefallen ist, gegen die verdienstvolle und ehrwürdige Persönlichkeit des Herrn Archenhold auch nur in irgend einer Form vorzugehen, Was Herr Archenhold auf seinem Gebiet — nämlich für die Popularisierung der Astronomie — geschaffen hat, bin ich der letzte, nicht anzuerkennen. Ich verwahre mich aber sachlich mit Entschiedenheit dagegen, daß er seine große Popularität dazu benutzt, die Einsteinsche Relativitätstheorie zu interpretieren, die er, wie sein Vortrag bewies, in ihren Prinzipien und Konsequenzen nicht erkannt hat. Und wenn er sie erkannt hätte, wäre es verdammte Pflicht und Schuldigkeit des ernsten Forschers gewesen, sich über die Qualität des referierten Gebietes zu überzeugen, ehe er es kritiklos dem bedauernswerten Publikum vorsetzte. Herr Archenhold trug aber nur Einstein-Literatur vor. Der Arbeiten von Hale, Silberstein, St. John, Evershed, Davidson, Eddington u. a. Forscher, die gewichtiges Material gegen Einstein anführen, gedachte er keines Wortes. Selbst wenn hier, was ich im Interesse des Herrn Archenhold annehme, Gutgläubigkeit vorliegt, so ist doch diese Gutgläubigkeit im vorliegenden Falle unbedingt verwerflich. Meine kritische Bemerkung war in diesem Falle also sachlich durchaus gerechtfertigt. Gerade Herr Archenhold hat sich durch die Eigenart seiner Position doppelt vorzusehen, unfertige Wahrheiten zu behandeln, denn er spricht vor einer Gemeinde die ihm unbedingt glaubt.] den er nebenbei bemerkt vom Einsteinschen Standpunkte aus betrachtet, schlecht genug interpretierte. Auch hierbei wurde das Publikum in mehr als fragwürdiger und unsachlicher Weise über den Wert der Einsteinschen Relativitätstheorie unterrichtet und bewiesene Gegengründe nach bewährter Methode einfach totgeschwiegen.

[*19*]

Meine Damen und Herren! Es liegt mir heute ob, zu ergründen und nachzuweisen, wie es kam, daß diese sogenannte Hypothese, die sich bei näherer Prüfung als glatte Fiktion herausstellte, die Welt dauernd in Atem halten konnte. Wissenschaftlich genommen, ist dieses leicht erklärlich. Durch die Verbrämung verschiedener wissenschaftlicher Disziplinen mit einander ist es dem Spezialforscher nicht möglich gewesen, sich in ein ihm fremdes Gebiet, schnell genug hinein zu finden. Gründliche Forscherarbeit und Prüfung erfordert eben Zeit.

Aber noch ein anderer Grund spricht hier ein wichtiges Wort mit. Wohl nicht zum geringsten Teile hat diese Erscheinung ihre Ursache in der mehr oder minder geistigen Verflachung, in die uns die gegenwärtige Zeit versenkt hat. Wir haben erst kürzlich erleben können, mit welchem Aufwand von Reklame heutzutage Wissenschaft gemacht wird. Es ist leider soweit gekommen, daß die Wissenschaft nicht mehr Selbstzweck ist, sondern Mittel zum Zweck, gewissen Personen mit dem Glorienschein wissenschaftlicher Päpstlicher zu umgeben. Sie alle, meine Damen und Herren haben es mit eigenen Augen gesehen und mit eigenen Ohren gehört, in welchem Tiefstand sich die geistigen ethischen, und moralischen Qualitäten derer bewegten, die uns die gegenwärtigen Zustände brachten. Das schlimmste Übel war eine gewisse Presse, die die neben einer bereits bestehenden wie Pilze aus der Erde schoß, die alle moralischen und sittlichen Werte im deutschen Volke erstickte, um aus dem geschaffenen Trümmerhaufen für sich brauchbares herauszuscharren. Um diese Presse gruppierten sich Abenteurer jeder Art, nicht nur in der Politik, sondern auch in Kunst und Wissenschaft. Genau wie die Herren Dadaisten mangels jeden Erfahrungsgedankens in ihrer Kunst- und Weltanschauung, Aufbau, Entwicklung und Reife vermissen lassen und dieses unreife Zeug durch einen Teil der alten, hauptsächlich aber die neue Literatur propagieren lassen, weil sie geistig nicht imstande waren, sich selbst durchzusetzen, genau so vollzieht sich in der Einstein'schen Relativitätstheorie als ein völliges Analogon das Hineinwerfen der Relativitätstheorie in die Massen. Auch hier liegt bewußte Ablehnung erfahrungsmäßiger Kenntnisse und Erkenntnisse vor. Wir stehen bei der Betrachtung der Einsteinschen Ideen genau vor demselben Gedankenchaos der Dadaisten, die wohl etwas wollen und wünschen, es aber nicht begreiflich machen und beweisen können.

[*20*]

Meine Damen und Herren! Niemand wird sich wundern, wenn gegen diesen wissenschaftlichen Dadaismus eine Bewegung entstanden ist, mit dem Ziele, die Öffentlichkeit aufzuklären, was denn eigentlich an der Einsteinschen Relativitätstheorie ist, und was man vor allen Dingen unter Fortschritten der Wissenschaft zu verstehen hat. Es sollen in einer Reihe von Vorträgen andere Gesichtspunkte und Anregungen zur Geltung kommen, als sie bisher in allzu einseitiger und aufdringlicher Weise der Öffentlichkeit geboten worden sind. Zu Einzelheiten wissenschaftlicher Art mich zu äußern bin ich heute noch nicht an der Reihe. Den Herren, die schon lange in der Bewegung stehen und die Einsteinschen Phantasmen unentwegt bekämpften, gebührt der Vortritt. Ehe ich jedoch schließe, noch eine kurze Bemerkung. Ich bin in der Tagespresse, wie ich schon vorhin erwähnte, von einem hervorragenden deutschen Physiker angegriffen worden. [*Footnote*: Ich habe im Anhang dieses Heftes die Polemik abgedruckt, um sie besser bekannt zu geben.] Mir wurde u. a. entgegengehalten, daß ich annehme, die Ergebnisse mancher Forscher hinsichtlich der Prüfung der Relativitätstheorie könnten durch Voreingenommenheit beeinflußt sein. Dem gegenüber stelle ich fest, daß alle für Einstein sprechenden Gründe in Deutschland besonders aufgebauscht und die gegenteiligen Beweisgründe in angeführter Manier totgeschwiegen wurden. Ferner wird mir meine Behauptung vorgeworfen, Herr Einstein habe eine Formel von Gerber abgeschrieben. Hierzu stelle ich fest, daß das peinlich jahrelange Schweigen von Herrn Einstein über diesen nicht nur von mir, sondern auch von einer ganzen Reihe von Fachgenossen und unvoreingenommenen Beurteilern erhobenen Vorwurf als sehr eigentümlich empfunden wird. Ich stelle fest, daß es doch allgemein üblich ist, sich zu Vorwürfen solcher Art und Schwere selbst und zwar sofort zu äußern.

[*21*]

Abdruck aus: ,,Tägliche Rundschau'', Freitag, 6. August, Abendausgabe.

Einsteins Relativitätstheorie—eine wissenschaftliche Massensuggestion.

Von Paul Weyland.

Wir leben in einem Zeitalter des Amerikanismus. Die Geschäftswut Englands ist in Dollarika zur Potenz erhoben, führte dort auf allen Gebieten des wirtschaftlichen und geistigen Lebens zu Rekordleistungen, die rein technischer, zivilisatorischer Art waren, hinter denen kulturelle Bestrebungen zurückstehen mußten. Die Rekordjägerei endigte im Bluff, und wir stehen vor der traurigen Tatsache, daß auch diese Bluffmacherei vor der reinen Wissenschaft nicht Halt machte, so daß die Sache neben der Person verschwand.

Ich erinnere an den bekanntesten Fall dieser Art, an den Entdeckerstreit Cook-Peary, der in der Öffentlichkeit am besten bekannt wurde. In Deutschland erlebte man, nach dem der Amerikanismus hier Eingang fand, gegenüber diesen Reisenbluffs bislang nur Sensationchen, die aber so lebhaft von dem Geist Zeugnis ablegten, der gewisse wissenschaftliche

Kreise auch unseres Vaterlandes ergriffen hat. Ich erinnere an Friedmanns Tuberkulin, an die Herstellung von Mehl aus Stroh usw., um an diesen Beispielen zu zeigen, daß man es in gewissen Kreisen nicht mehr für nötig hält, die Bestätigung eines Laboratoriumversuches in der Praxis abzuwarten, sondern mit Hilfe einer gefügigen Presse sich mit seiner halbfertigen Sache dem Publikum vorstellt, den werten Namen nebst Photographie in alle Windrichtungen hinausbläst, um einige Zeit später, wenn — wie fast stets — die Hinfälligkeit der Entdeckung durch ernste Forscher beweisen wird, beharrlich zu schweigen. Davon aber erfährt das Publikum natürlich nichts, und die Masse schwört blindlings auf die „großen" Namen.

Mittlerweile hat sich Deutschland — endlich — neben solchen Sensatiönchen auch eine richtige Sensation geleistet. Herr Albertus Magnus ist neu erstanden, guckte in die ernsten Arbeiten stiller Denker wie Riemann, Minkowsky, Lorentz, Mach, Gerber, Palagyi u. a. m., räusperte sich und sprach ein großes Wort gelassen aus. [*Footnote:* Um endlose Wiederholungen zu vermeiden, wird das Relativitätsprinzip beim Leser als bekannt vorausgesetzt.] Die Wissenschaft staunte. Die Öffentlichkeit war starr. Alles [*22*] brach zusammen. Herr Einstein spielte mit der Welt Fangball. Er brauchte nur zu denken, und flugs relativierte sich alles Geschehen und Werden.

Einsteins Methode war nun so bewußt abstrakt, daß es dem Fachmann ernstliche Schwierigkeiten bereitete, sich hindurchzuarbeiten. Zunächst verquickte er mehrere wissenschaftliche Disziplinen miteinander, ja er errichtete für seine Zwecke ein ganz neues mathematisches Gebäude, so daß der nachprüfende Naturforscher vor lauter Nebensachen zunächst gar nicht an den Kern der Sache heran kam, weil diese Nebensächlichkeiten, die erst geprüft werden mußten, ja den Aufbau seines Theorems bedeuteten. Dieses Drum und Dran ist von Forschern wie P. Lenard, Gehrcke, Kraus u. a. geprüft worden, es stellte sich heraus, daß nicht einmal das Skelett einer kritischen Betrachtung standhielt. Was soll da aber erst aus dem Hauptteil werden?

So bemängelt z. B. P. Lenard mit unbedingtem Recht, daß bei Einstein der einfachsten Logik Hohn gesprochen wird. Ich zittiere Lenard wörtlich: [*Footnote:* P. Lenard, Über Relativitätsprinzip, Äther, Gravitation. Verlag S. Hirzel, Leipzig 1920.]

„Man lasse den bekannten Eisenbahn eine deutlich ungleichförmige Bewegung machen. Während hier durch Trägheitswirkung im Zuge alles in Trümmer geht, während draußen alles unbeschädigt bleibt, so wird, meine ich, kein gesunder Verstand einen anderen Schluß ziehen wollen als den, daß es eben der Zug war, der mit Ruck seine Bewegung geändert hat und nicht die Umgebung. Das verallgemeinerte Relativitätsprinzip verlangt es, seinem einfachen Sinne nach, auch in diesem Falle, zuzugeben, daß es möglicherweise doch die Umgebung sei, welche die Geschwindigkeitsänderung erfahren habe und daß dann das ganze Unglück im Zuge nur Folge dieses Rucks der Außenwelt sei, vermittelt durch eine

„Gravitationswirkung" der Außenwelt auf das Innere des Zuges. Für die naheliegende Frage, warum denn der Kirchturm neben dem Zuge nicht umgefallen sei, wenn er mit der Umgebung den Ruck gemacht habe — warum solche Folgen des Rucks so einseitig nur im Zuge sich zeigen, während dennoch kein einseitiger Schluß auf den Sitz der Bewegungsänderung möglich sein sollte — hat das Prinzip anscheinend keine den einfachen Verstand befriedigende Antwort."

Hier hat Lenard mit wenigen klar verständlichen, an den Verstand gerichteten Worten den mathematischen Unfug getroffen, der sich aus dem Theorem entwickelte. Was nützt alle hochgelahrte Mathematik, aller verwickelter Formelkram, wenn er — verkehrt aufgebaut wird? Zu obigem Einwand, den Lenard bereits 1918 in dem Jahrbuch für Radioaktivität und Elektronik erhob, hat sich Einstein bis heute nicht geäußert. Mit diesem Einwand oder seiner Widerlegung fällt und steht aber das ganze Prinzip.

Doch sehen wir weiter zu. Einsteins Theorie verlangt, daß infolge der Gravitationswirkung der Sonne ihr Gravitationsfeld passierende Lichtstrahlen [*23*] eine Verzögerung, eine zeitliche Abbremsung erfahren müssen. Die Theorie berechnet eine Verschiebung nach dem roten Teil des Spektrums um 0.01 Angström-Einheiten, d. h. den zehntausendmillionsten Teil eines Millimeters, eine fast unvorstellbare Kleinheit, die aber mit unseren feinen Gitterspektrographen sehr gut zu messen ist. St. Juhn hat („Astrophysik. Journ." 46, S. 249, „Nature" 100, S. 433) an 43 Linien in der Sonnenmitte I 0.00 A.—E., also ein negatives Resultat erzielt, für die Sonnenkorona + 0.0018 A.—E. Ferner hat Schwarzschild (Berl. Ber. 1914 S. 1201) ein ebenfalls negatives Ergebnis festgestellt. Auch andere Forscher von Ruf haben diese Einsteinsche Hauptbedingung nicht bestätigt gefunden. Grebe und Bachem, ausgesprochene „Relativisten", glauben nun, die gefundenen Werte + 0.0018 für Einstein deuten zu können und ziehen mit einer Kompensationserklärung vom Leder. Einem jungen Forscher, Glaser, ist es aber gelungen, den Nachweis zu führen, daß das Grebe und Bachemsche Ergebnis lediglich auf Beobachtung mit einem fehlerhaften Rowlandschen Gitter zurückzuführen ist. Das Material hierüber wird dem Naturforschertag in Nauheim im September vorgelegt werden. Mit der Verschiebung der Spektrallinien nach Rot ist es also auch nichts. Bleibt somit nur noch die berühmte Ablenkung der Perihelbewegung des Merkur um 41 Sekunden übrig.

Es ist auch hier wieder das Verdienst von Prof. Gehrcke (Berlin), der festgestellt hat, daß Einstein für seine Zwecke eine äußerst schwer zugängliche Arbeit von Gerber benutzte, die bereits vor achtzehn Jahren erschien. Hier gestattete er sich die Abschrift einer Formel, verwendete diese für sich und ließ den wahren Entdecker unerwähnt. Prof. Gehrcke sorgte flugs für zugänglichen Neudruck der seltenen Gerberschen Arbeit, und jedermann kann heute feststellen, wer der Autor dieser Erklärung der Perihel-Abweichung des Merkur ist und ob es nötig ist, dafür ein Relativitätsprinzip zu erfinden.

Unzählige andere Beispiele können noch angeführt werden. Diese wenigen mögen hier genügen. Ein großer Teil deutscher Forscher, der sich zuerst zu Einstein bekannte, sieht den Irrtum ein. Mancher hat schon widerrufen in der richtigen Erkenntnis, daß es ruhmvoller ist, einen Irrtum ehrlich zu bekennen, als in ihm hartnäckig zu verharren. Diese Forscher stellen sich ein ehrenvolles Zeugnis aus, daß sie der Sache, der Wahrheit die Ehre geben und die Person zurückstellen. Noch einige taktische Bemerkungen seien angeführt.

Da, wie gesagt, Einstein eine gewisse Presse, eine gewisse Gemeinde hat, so wird von dieser immer wieder die Oeffentlichkeit im Einsteinschen sinne beeinflußt. So hielt z. B. vor vierzehn Tagen Herr Archenhold im Blüthner-Saal einen Vortrag über dieses Thema. Kundige haben den Kopf geschüttelt, daß Herr Archenhold gar nichts von den Gegengründen erwähnte, sondern sie stillschweigend überging, dagegen die unbedingt strittige Ablenkung des Lichtes um **1.7"** im Gravitationsfeld der Sonne postulierte. Herrn Archenhold sei erwidert, daß solche Stellungnahme vor einem Publikum, das in der großen Mehrzahl seine Ausführungen nicht beurteilen [*24*] konnte, entschieden zu verurteilen ist daß Parteinahme wohl politisch gerechtfertigt, wissenschaftlich aber verwerflich ist. Es dürfte Herrn Archenhold als Fachmann und „Sonnenforscher" wohl nicht unbekannt sein, daß die Sonne eine Atmosphäre besitzt und daß diese für die Ablenkung des Lichtstrahles mit mindestens demselben Recht in Frage kommt wie die sehr hypothetische Wirkung des Gravitationsfeldes, wie das schon Lindemann 1918 festgestellt hat. Daß Einstein den Aether durch ein Dekret abschaffte, ihn aber durch einen anderen Begriff mit gleichen Funktionen wieder einführte, sei hier nur, um mit Einstein selbst zu reden, der „Drolligkeit" halber erwähnt.

Schließlich sie noch der unzulässigen Art der Propaganda kurz gedacht, die Einstein zum ersten Male in die deutsche Universität einführte. Welcher Mittel sich Einstein zur Verbreitung seiner Ideen bedient, ist an dem Wust von Referaten zu erkennen, von denen die meisten ihn nicht einmal verstehen. Der entzückendste Witz dieser Art ist eine Schrift von Max Hasse, A. Einsteins Relativitätslehre (Magdeburg 1920, Selbstverlag des Verfassers), wo es im Vorwort heißt: „Der Verfasser gesteht freimütig ein, nicht mehr einen Lehrsatz euklidischer Geometrie beweisen zu können — die Zeit hat früher Gelerntes verwischt." Und solch ein Mensch wagt es, über die tollste mathematische Abstraktion, die es je gegeben, zu berichten! Und was sagt Einstein dazu? Es heißt nämlich im Vorwort weiter: „Der Verfasser nahm sich die Freiheit, die Druckbogen Prof. Dr. A. Einstein einzusenden, der ihn mit folgender Antwort erfreute: „Ihre populäre Darstellung scheint mir in der Tat dem Geiste des Nicht-Physikers in glücklicher Weise entgegenzukommen. Ich sende Ihnen die Korrekturbogen mit einigen Randbemerkungen zurück, damit Sie einige kleine Böcke daraus entfernen können."

Das ungefähr kennzeichnet Einsteins Methodik. Wenn aber die deutsche Wissenschaft demnächst geschlossen gegen Einstein auftreten wird und mit ihm zu Gericht geht, dann hat er sich diese Wirkung seiner,

sagen wir ungewöhnlichen Kampfesweise selbst zuzuschreiben.
[*25*]
Abdruck aus: „Tägliche Rundschau", Mittwoch, 11. August, Abendausgabe.

Zur Erörterung über die Relativitätstheorie.
Entgegnung an Herrn Paul Weyland. Von M. v. Laue.

In Nr. 171 dieses Blattes ereifert sich Herr Weyland gegen Einsteins allgemeine Relativitätstheorie; gegen die Art ihrer Verbreitung in der größeren Öffentlichkeit sowie gegen ihren Inhalt. Es liegt mir durchaus ferne, alles das decken zu wollen, was kleinere Geister bei der Verbreitung der neuen Lehre durch Ungenauigkeiten, Übertreibungen und Geschmacklosigkeiten gelegentlich gesündigt haben, und die im besonderen herangezogenen Äußerungen von Archenhold und Max Hasse kann ich nicht beurteilen, weil ich sie nicht kenne. Zu einem solchen Angriff auf Einsteins Persönlichkeit, wie ihn Herr Weyland macht, bieten diese Dinge aber doch nicht den mindesten Anlaß.

Welche Einwände richtet aber Weyland gegen den Inhalt? Daß hier reines Denken eine neue Naturauffassung begründet, scheint ihm, wenn ich recht verstehe, gegen die Begründung der Physik in der Erfahrung zu verstoßen. Ist ihm aber nicht bekannt daß Einstein von einer Tatsache ausgeht, die, längst bekannt, noch in den letzten Jahren durch besonders gute Messungen auf das genaueste festgestellt ist? Daß nämlich alle Körper unter der Wirkung der Schwere gleich rasch fallen? Oder fehlt ihm das Verständnis für die Größe einer Leistung, welche uns bei einer so alten Tatsache endlich etwas zu denken lehrt? Bisher galt es doch stets als der größte dem menschlichen Geiste in einer Naturwissenschaft mögliche Triumph, wenn in Umkehrung des gewöhnlichen Ganges die Theorie der Beobachtung erfolgreich voranschritt.

Nun kann man ja freilich noch bestreiten, daß die Folgerungen aus der Theorie, wie die Rotverschiebung der Spektrallinien und die Lichtablenkung an der Sonne durch die Erfahrung endgültig bestätigt sind. Darüber ist in der Tat das letzte Wort nicht gesprochen. Wenn aber Herr Weyland entgegen den sonstigen Gepflogenheiten in wissenschaftlichen Erörterungen andeutet, es könne Voreingenommenheit die Ergebnisse mancher Forscher beeinflußt haben, so möchten wir ihm mitteilen, daß die Engländer, denen wir die Lichtablenkungsmessungen [*26*] verdanken, vorher durchaus nicht Anhänger des Relativitätsgedankens in Einsteinscher Prägung waren. [*Footnote:* Hätte Herr v. Laue die englische Literatur etwas aufmerksamer verfolgt, so hätte er diese Behauptung sicher nicht aufgestellt. Die Tagespresse, wohl meist der Niederschlag der inspirierten öffentlichen Meinung schreibt z. B. darüber: Westminster-Gazette: 14. August 1920: „Obwohl die Exped. nach Sobral und Principe in Bezug auf die Bestätigung der Theorie erfolgreich waren, wurde der damals erlangte, etwas dürftige Beweis (*somewhat meagre evidence*) in einem gewissen Grade durch das Versagen des astrographischen Fernrohres in Sobral beeinträchtigt. Aus diesem Grunde sollen eben bei der Sonnenfinsternis am 20. IX. 22 neue Prüfungen vorgenommen werden."

Hieraus geht z. B. auch hervor, daß die unter atmosphärischen Beeinträchtigungen behinderte Beobachtung auf Principe nicht für einwandsfrei betrachtet wird. Im Übrigen verweise ich auf die schon erwähnte Arbeit von Glaser in Heft 3 dieser Sammlung.]

Unbestreitbar gibt die allgemeine Relativitätstheorie jene minimalen, aber sicher festgestellten Abweichungen der Merkurbahn von der nach der älteren Theorie der Schwere errechneten Form zahlenmäßig richtig wieder. Man mag dies Zusammentreffen als einen Zufall ohne besondere Beweiskraft abtun. Aber man darf Einsteins Ableitung, welche eine entfernte Folgerung einer großen, aus ganz anderen Gesichtspunkten entsprungenen Theorie darstellt, denn doch nicht in einem Atem nennen mit der Arbeit von Gerber, welche nach einer Fülle von Unklarheiten, Mißverständnissen und Ungenauigkeiten die Perihelbewegung aus einem eigens zu diesem Zweck ersonnenen, sonst zu nichts brauchbaren, aus der Geschichte der Wissenschaft nur zu gut verständlichen mathematischen Ansatz errechnet. Hat sich doch auch der Münchener Astronom H. v. Seeliger, ein entschiedener Gegner der Relativitätstheorie, scharf gegen dies Machwerk gewandt. Wie Herr Weyland hier gegen Einstein den Vorwurf erheben konnte, die Gerbersche Formel „abgeschrieben" zu haben, darüber mag er sich einmal selbst Rechenschaft zu geben versuchen.

Etwas näher wollen wir eingehen auf P. Lenards, von Herrn Weyland angeführten Einwand. Einstein hat in der Tat nie auf ihn geantwortet. Man tritt eben einem verdienten Fachgenossen nicht immer entgegen, wenn ihm einmal eine weniger richtige Äußerung entschlüpft; zumal in einem Falle, in welchem der Sachverhalt so leicht zu durchschauen ist, wie hier. Wie steht es denn? Um den Grundgedanken seiner Lehre klarzumachen, knüpft Einstein an das alltägliche Erlebnis einer Eisenbahnfahrt an. Fährt mein Zug auf idealen, stoßfreien Schienen mit unveränderter Geschwindigkeit immer in derselben Richtung a, so sind es zwei physikalisch gleichwertige Annahmen, ob ich mein Abteil als bewegt und die Umgebung als ruhend bezeichne oder umgekehrt verfahre. Das war die Meinung schon seit jeher. Nun aber sagt Einstein, man könne, [*27*] auch wenn der Zug bremst und alle Körper im Abteil das Streben zeigen, sich gegen dessen vordere Wand zu bewegen, die Auffassung in allen ihren physikalischen Folgerungen vertreten, das Abteil bleibe in Ruhe, während die Umgebung, die mir bisher mit konstanter Geschwindigkeit entgegenkam, jetzt in ihrer Bewegung aufgehalten wird. Nur muß dann in dem Bezugsystem, in welchem mein Abteil dauernd ruht, ein Schwerefeld in der Richtung a neu entstanden sein, welches die Umgebung aufhält. Im Innern des ruhenden Abteils bemerke ich das Feld an der erwähnten Bewegungstendenz der Körper. In der Umgebung ruft es außer der gemeinsamen Geschwindigkeitsverminderung aller Gegenstände keine Wirkungen hervor, eben weil alle Körper gleich schnell fallen. Geschieht doch auch in einem Aufzug, der sich von der Aufhängung gelöst hat, kein Unheil, solange er frei fällt; erst beim

Aufschlagen auf den Erdboden wird das anders. Herr Lenard übersieht, daß infolge des gleich raschen Falls aller Körper das neue Schwerefeld im Außenraum keine Lageänderungen der Gegenstände gegeneinander hervorruft, wohl aber im Innenraum die Dinge gegen die ruhenden Wände des Abteils in Bewegung setzt.

Soviel gegen P. Lenard. Herrn Weyland aber möchte ich zum Schluß einen Rat geben, dessen Befolgung in seinem eigensten Interesse liegen dürfte: sollte er sich nämlich noch einmal gegen Einstein wenden, sich über diesen Mann mit etwas mehr Achtung zu äußern. Die Relativitätstheorie mag man für richtig oder falsch halten, es äußert sich auf jeden Fall in ihr eine Genialität, die auf anderen Gebieten der Physik schon zu den schönsten Ergebnissen geführt und ihm verdientermaßen Weltruhm verschafft hat. Die stolze Wissenschaft ist stolz darauf, ihn zu den Ihrigen zählen zu dürfen!

Wir haben Herrn Weyland, wie üblich, von dieser Entgegnung Kenntnis gegeben und erhalten darauf von ihm folgende Zuschrift:

Raummangel verbietet mir, an dieser Stelle eine Erwiderung zu geben, wie sie eine Persönlichkeit wie Herr v. Laue erfordert. Ich werde mich am 24. August im großen Saal der Philharmonie mit Herrn E. Gehrcke zunächst allgemein zur Sache äußern, späterhin im besonderen. Ich bitte Herrn v. Laue, zu diesem Abend anwesend zu sein. Des weiteren werden Herr Kraus (Prag) und Herr Glaser (Berlin) am 2. September im gleichen Saale zum Thema sprechen.

Hier nur soviel: Ich wende mich nicht gegen eine Theorie, sondern gegen mathematische Fiktionen und maßlose Übertreibungen. Daß die Frage der Rotverschiebung für Herrn v. Laue nunmehr ebenfalls keine absolute Tatsache ist, freut mich. Früher, als keine Kritiker, die es kontrollieren konnten, (ich erinnere an Herrn Freundlichs Märzvortrag), da waren, las man's anders. Ferner ist Herr v. Laue anscheinend über den neuesten Stand der englischen und amerikanischen Forschung nicht ganz im Bilde. Anders kann ich seine Bemerkung nicht verstehen. Näheres im Vortrag. Hinsichtlich der Gerberschen [*28*] Formel verweise ich auf die Arbeiten von E. Gehrcke (Verhandlg. d. Deutschen Physikal. Gesellschaft 1918 S. 165, Ann. d. Physik, 4. Folge, Band 51, 1916, S. 119.) Die Sache ist ja für Herrn Einstein sehr peinlich, aber nicht zu ändern. Es wundert mich nur, daß man die ganze Gerbersche Arbeit verdonnert — Schwächen seien zugegeben, aber: wo sind keine? — und gerade das Ergebnis so schön findet, daß man es, sagen wir, verwendet. Hier hilft kein Drehen und Deuteln. Oder soll ich noch deutlicher werden? Ich erinnere an Palagyi, Mach! Weiß Herr v. Laue nicht, wie sich Herr Einstein hinsichtlich der Verwendung, Machscher Gedanken herausgeredet hat?

Zu dem Einwand gegen Herrn L. Lenard äußere ich mich nicht. Dieser hervorragende Heidelberger Gelehrte wird seinerzeit selbst das Wort gegen Einstein ergreifen. [*Footnote:* Herr Lenard teilt mir seine Antwort brieflich mit, die ich hier wiedergeben möchte: „Herrn v. Laues Äußerungen zu meiner Schrift haben mich stark befremdet, insofern sie mir die Sachlage nicht zu treffen scheinen. 1. Trifft es nicht zu, daß Herr Einstein auf meine Einwände nie geantwortet habe. Vielmehr wird seine Antwort in der soeben erschienenen 2. Auflage meiner Schrift „Über Relativitätsprinzip, Äther, Gravitation" nicht nur genau zitiert, sondern auch besprochen, aber nicht als befriedigend befunden (siehe meine Fußnote auf S. 31) und es wird sogar angegeben, wo Herr Einstein oder einer der Verteidiger der allgemeinen Relativitätstheorie einsetzen müßten, um den Beweis — oder genügenden Hinweis — für die Berechtigung der Verallgemeinerung zu liefern, wobei ich garnicht zweifle, daß es nicht nur mir allein gegenüber lohnend wäre, dies wirklich zu tun, — falls es möglich ist. Es scheint mir hiernach, daß Herr v. Laue die neue Auflage meiner Schrift noch garnicht, die alte aber auch nur unvollkommen kennt, beziehlich überlegt hat. Denn 2. trifft es außerdem auch nicht zu, daß ich das Nichtauftreten von Trägheitswirkungen infolge gleichschnellen Fallens aller Körper bei Wirkung von Gravitation übersehen hätte. Sondern ich finde nur große Schwierigkeiten gegen die Annahme der Einsteinschen Gravitationsfelder und erörtere diese Schwierigkeiten — die sofort auftreten, sobald man einfache Beispielsfälle zu Ende zu überlegen versucht — ausführlich mit dem Resultate, daß eine Einschränkung des verallgemeinerten Relativitätsprinzipes notwendig sei, um es von seinen gegen den Verstand gerichteten Härten zu befreien. — Eine selbst bei Zutreffen der von Herrn Einstein gemachten, experimentell kontrollierbaren Voraussagen irgendwie gesicherte Allgemeingiltigkeit des Relativitätsprinzips kann bisher nicht behauptet werden, womit aber auch jede Betonung einer philosophischen auf die Grundauffassung des Naturgeschehens gerichteten Bedeutung zunächst wegfallen sollte. Gerade weil solche Betonung zu oft zu auffallend vor die Allgemeinheit gebraucht worden ist, schien es und scheint es nun eben nötig, neben den Vorzügen auch die der gegenwärtigen Erfahrung entsprechenden Grenzen des Relativitätsprinzips, oder die Übertreibungen, die man sich mit demselben gestattet hat, hervorzuheben. Wer hierüber im Einzelnen orientiert sein will, wie es meiner Auffassung nach dem wirklichen Stand der Kenntnis entspricht, muß für jetzt auf die erwähnte 2. Auflage meiner Schrift verwiesen werden.] Herr v. Laues Einwand werde ich ihm übermitteln.

[*29*]

Für den mir erteilten Rat danke ich bestens. Ich bin mit anderen Herren so frei, über die Relativitätstheorie meine besondere Meinung zu haben. Die Beweise werden in einer Vortragsreihe, an der erste Physiker und Astronomen teilnehmen, dargelegt werden.

P. Weyland

Tägliche Rundschau Nr. 180.
Zur Erörterung über die Relativitätstheorie.
Entgegnung an Herrn Professor Dr. M. v. Laue.
Von Dr.-Ing. L. C. Glaser (Berlin).

In Nummer 175 dieses Blattes sagt M. v. Laue, daß man Einsteins Erklärung für die Abweichung der Perihelbewegung der Planetenbahnen, insonderheit des Merkurs, nicht in einem Atem mit der Arbeit von Gerber nennen darf, welcher nach seiner Meinung nach einer Fülle von Unklarheiten, Mißverständnissen und Ungenauigkeiten die Perihelbewegung aus einem eigens zu diesem Zweck ersonnenen, sonst zu nichts brauchbaren, aus der Geschichte der Wissenschaft nur zu gut verständlichen, mathematischen Ansatz errechnet. Man ist, wie von P. Lenard bereits schon bemerkt ist, mit der Arbeit des verstorbenen Oberlehrers Paul Gerber besonders scharf ins Gericht gegangen. Im Hinblick darauf, daß M. v. Laue sich schützend vor Einstein stellt, ist es Pflicht der Menschlichkeit, das Ergebnis dieser Arbeit des verstorbenen Oberlehrers Paul Gerber gegen die Bezeichnung „Machwerk" in Schutz zu nehmen. Die Ereiferung M. v. Laues über die Arbeit von Gerber ist unverständlich, zumal diese Arbeit im Auslande auf Grund des Wiederabdruckes in den „Annalen für Physik" von Herrn L. Silberstein, der ja bekanntlich gegen die allgemeine Relativitätstheorie Einsteins eine durchaus ablehnende Stellung einnimmt, gelegentlich einer Arbeit „über die Perihelbewegung des [*30*] Merkurs, abgeleitet nach der klassischen Theorie der Relativität" in den „Monthly Notices" der *Roy. Astr. Soc.* 1917, 503-610, als Gerbers Formel aufgeführt und anerkannt wird. Daß nun den Anhängern der Relativitätstheorie das Bestehen der Gerberschen Formel, über deren Ansatz man im einzelnen denken kann, wie man will, recht unbequem ist, ist ja sehr leicht verständlich, zumal die Forderungen und sogenannten Bestätigungen der Einsteinschen Relativitätstheorie im ganzen äußerst beweisbedürftig sind. Da die Arbeit Gerbers der Geschichte angehört, das Einsteinsche Ergebnis vorwegnimmt, aber gern totgeschwiegen wird, ist es besonders erfreulich, festzustellen, daß diese bereits Aufnahme in der zweiten Auflage des Lehrbuches der Physik von Riecke, herausgeben von Lecher, gefunden hat.

Tägliche Rundschau Nr. 175, Abendausgabe.
Zur Erörterung über die Relativitätstheorie.
Von M. v. Laue.

Auf meinem Aufsatz in Nr. 176 dieses Blattes hin haben mich verschiedene Fachgenossen auf Einsteins „Dialog über die Einwände gegen die Relativitätstheorie". [*Footnote*: Diese Arbeit war mir bekannt. Als Einwand habe ich sie nicht gelten lassen. Herr Lenard ist lt. seinem Briefe genau derselben Ansicht.] (Naturwissenschaften, 6. Jahrgang, Seite 6-697, 1918) aufmerksam gemacht, in welchem Einstein selbst zu dem Lenardschen Einwand Stellung nimmt. Was dort steht, deckt sich zwar nicht mit dem, was ich neulich an dieser Stelle — übrigens als die Ansicht sehr vieler — darüber sagte, doch besteht auch kein Widerspruch; ich gebe diesen Hinweis hiermit weiter.

Ein wenig ausführlicher aber möchte ich in Hinblick auf Herrn Glasers

Entgegnung in Nr. 178 auf die Gerbersche Erklärung der Perihelbewegung beim Merkur eingehen. Zwar kann man eine sozusagen philosophische Kritik dieser Arbeit und ihrer Schlußformel nur einem fachmännischen Publikum verständlich machen, so daß ich hier darauf verzichten muß. Aber ich möchte doch einmal fragen, was diese Arbeit denn eigentlich leistet.

Eine Tatsache physikalisch erklären, heißt doch, sie in Beziehung zu anderen physikalischen Tatsachen setzen. Darin bin ich hoffentlich mit den Gegnern der Relativitätstheorie einig. Mit welcher anderen Tatsache setzt nun Gerber die Perihelbewegung in Beziehung? Die Überschrift seiner Veröffentlichung könnte die Antwort nahelegen: Mit der (zwar nie unmittelbar beobachteten, [*31*] aber doch sehr wahrscheinlichen) Ausbreitung der Schwere mit endlicher, und zwar mit Lichtgeschwindigkeit. [*Footnote:* Diese sehr interessante Einschränkung eines der wichtigsten Einstein'schen Postulate werde ich an anderer spezieller Stelle entsprechend würdigen. Daß v. Laue das Einsteinsche Postulat von der Lichtgeschwindigkeit als äußerste Grenze aller Geschwindigkeiten so einschräkend behandelt, ist aus der Feder diese bedeutendsten Relativisten von außerordentlicher Wichtigkeit.] Aber diese Antwort wäre nicht richtig. Unmittelbar nach dem Wiederabdruck in den Annalen der Physik habe ich an derselben Stelle (Band 53, Seite 214) darauf hingewiesen, daß Gerbers Formeln die Schwere als eine unvermittelte Fernwirkung hinstellen. Einen Widerspruch gegen diesen Nachweis habe ich bisher weder öffentlich noch privatim vernommen. Und welche andere Tatsache ließe sich hier erwähnen? Ich wüßte keine.

Nun lege wir einmal denselben Maßstab an Einsteins Erklärung. Sie bringt die Perihelbewegung in Zusammenhang mit der Äquivalenz der trägen und der schweren Masse, die der Versuch mit einer seltenen Schärfe bewiesen hat; natürlich auch mit der Lichtablenkung und der Verschiebung der Spektrallinien an der Sonne — doch diese Tatsachen sind ja noch bestritten. Sicher aber ist, daß die allgemeine Relativitätstheorie die beschränkte (ich vermeide gern das Fremdwort „spezielle") als fast stets brauchbare Näherung einschließt. Sie setzt damit die Perihelbewegung in Beziehung zu allen den berühmten Versuchen, welche durch Beobachtung auf der Erde deren Bewegung um die Sonne vergeblich nachzuweisen suchten; ferner zu den vielen sicher festgestellten Tatsachen der Elektrodynamik und Optik der bewegten Körper. Weiter: Die beschränkte Relativitätstheorie steht — ich glaube unbestritten — im Einklang zur gesamten mechanischen Erfahrung, einschließlich der verhältnismäßig neuen Beobachtungen über die Dynamik schnell bewegter Elektronen. Kurz: Einsteins Erklärung reiht die Perihelbewegung in den großen Zusammenhang von Tatsachen ein, den wir als das physikalische Weltbild bezeichnen.

Der Weg, auf dem das erreicht wird, mag manchem nicht gefallen. Dafür habe ich durchaus Verständnis. Aber man soll die relativistische Theorie der Perihelbewegung wirklich nicht auf eine Stufe stellen mit der Gerberschen Erklärung, die, abgesehen davon, was sonst über sie zu sagen wäre, überhaupt keine Erklärung ist."

Ernst Gehrcke addressed Albert Einstein to his face in the Berlin Philharmonic on 24 August 1920. Ernst Gehrcke was the second and last speaker at the event. Gehrcke stated, as recorded in a the published transcript of his talk: *Die Relativitätstheorie. Eine Wissenschaftliche Massensuggestion, gemeinverständlich dargestellt*, Volume 1 of the Press of the Arbeitsgemeinschaft deutscher Naturforscher zur Erhaltung reiner Wissenschaft e. V., Köhler, Berlin, (1920); which was reprinted in Gehrcke's booklet *Kritik der Relativitätstheorie*, Hermann Meusser, Berlin, (1924), pp. 54-68:

"Was ist eigentlich die Einsteinsche Relativitätstheorie? Diese Frage wird heute nicht nur in gelehrten Kreisen erörtert, sondern sie beschäftigt sehr viele, denen akademische und gelehrte Dinge sonst fern liegen. Das Thema der Relativitätstheorie, der Streit über ihre Bedeutung und Richtigkeit ist heute bis in die Tagespresse aller möglichen Richtungen gedrungen. Aber um was es sich eigentlich dreht, das dürfte trotz aller Zeitungsartikel und populären Broschüren, die wie Pilze aus der Erde schießen, nur sehr wenigen klar sein. Dem soll im Folgenden abgeholfen werden.

Es wird dabei zu beachten sein, daß die Relativitätstheorie nicht wie ein deus ex machina plötzlich eines Tages da war, sondern dass sie, wie alle geistigen Strömungen, eine längere Entwicklung gehabt hat und schrittweise und allmählich gewachsen ist. Daß die Relativitätstheorie eine geistige Strömung darstellt, kann niemand bezweifeln, nur darüber wird man verschiedener Meinung sein können, ob diese Strömung eine gesunde, verheißungsvolle ist, ob sie, kurz gesagt, einen Fortschritt darstellt, oder ob das Gegenteil der Fall ist, ob sie ungesund, unfruchtbar und falsch, also kurz gesagt ein Irrlicht der geistigen Entwicklung war. Die Meinungen hierüber sind sehr geteilte. Der Gemeinde der Relativitätsgläubigen steht eine Schar von Zweiflern und Kritikern gegenüber, hüben und drüben haben anerkannte Autoritäten Partei ergriffen, und wie die Dinge liegen, werden nicht allein wissenschaftliche, sondern auch politische und andere Gesichtspunkte in die Debatte hineingetragen. In dieses Chaos der durcheinander wogenden Behauptungen und Interessen soll hier also hineingeleuchtet werden. Nur unter dem Gesichtspunkt der Entwicklung wird es aber möglich sein, das Durcheinander zu verstehen und sich über das Gewirr der Meinungen ein Urteil zu bilden. Wir fragen im Folgenden nicht, was ist die Relativitätstheorie? sondern: wie hat sie sich entwickelt? und beginnen mit demjenigen Punkte, welcher der Relativitätstheorie den Namen gegeben hat, mit dem
Relativitätsprinzip.
Gemäß dem Obigen werden wir nicht fragen, was ist das Relativitätsprinzip? sondern: wie hat sich das Relativitätsprinzip entwickelt? Erst die Darlegung dieser Entwicklung wird uns zu einem Standpunkt gegenüber dem Relativitätsprinzip führen, der von dem augenblicklichen Tagesurteil frei ist.

Das Relativitätsprinzip ist in der Tat kein erst in unsern Tagen aufgestellter Grundsatz, sondern es hat eine lange Geschichte, die bis in das griechische Altertum und möglicherweise noch weiter zurückreicht. Die voltständige Darstellung seines Werdeganges wäre eine umfangreiche, historisch-kritische Studie, die hier nicht auf kurzem Raum gegeben werden kann und hier auch nicht behandelt zu werden braucht. Es wird genügen, wenn wir deutlich machen, daß das Relativitätsprinzip an sehr einfache, alltägliche Erfahrungen, die schon mancher gemacht hat, anknüpft.

Stellen wir uns etwa vor, daß wir in einem Eisenbahnzuge sitzen, der auf dem Bahnhof hält. Auf der andern Seite des Bahnsteigs soll ebenfalls ein Zug stehen. Wir warten ungeduldig auf Abfahrt, endlich geht es los, der Zug setzt sich in Bewegung, und wir sehen durch das Fenster, wie wir am jenseitigen Zuge uns vorbeibewegen. Aber mit einem Mal entdecken wir, daß wir uns geirrt haben: wir halten immer noch auf dem Bahnhof, aber der andere Zug fährt! Dieses unliebsame Erlebnis in seiner Alltäglichkeit und Einfachheit ist geeignet, uns dem Relativitätsprinzip näher zu führen: Wir konnten nicht feststellen, ob wir fahren oder der andere Zug, ob wir in Ruhe blieben oder der andere Zug, das einzige, das wir beobachten konnten, war, daß die beiden Züge relativ zueinander in Bewegung waren. Man nennt dies die Relativität der Bewegungen. Alle Bewegung ist relativ, d. h. bezogen auf irgend etwas, außerhalb des Bewegten Befindliches. Alle Naturkörper in unserer Umgebung, auf der Erde, alle Gestirne am Himmel bewegen sich relativ zueinander. Man drückt sich auch so aus, daß man sagt, der Bewegungsbegriff sei ein Relationsbegriff, d. h. ein Begriff, der ohne Bezugnahme auf etwas, gegenüber welchem das Bewegte sich bewegt, nicht gedacht werden kann. Aber die Relativität der Bewegungen ist noch nicht das Prinzip der Relativität. Hierüber ein anderes, alltägliches Beispiel.

Es soll ein Stück Holz mit einer Säge durchgesägt werden. Das kann auf zweierlei Weisen geschehen: erstens so, daß das Stück Holz festgehalten wird, z. B. indem man es auf einen Sägebock legt und die Säge hin und her bewegt, zweitens so, daß die Säge festgehalten, z. B. zwischen die Knie geklemmt wird, und nun das Stück Holz quer zur Säge hin und her bewegt wird. In beiden Fällen wird das gleiche Ergebnis erzielt: das Holz wird durchgesägt. Ob ich also die Säge bewege und das Holz festhalte, oder umgekehrt die Säge festhalte und das Holz bewege, kommt auf dasselbe hinaus. Die beiden Bewegungsvorgänge: Holz fest, Säge bewegt und: Säge fest, Holz bewegt, sind aber in relativer Hinsicht gleich; es bewegt sich in beiden Fällen das eine in bezug auf das andere in gleicher Weise. Dieser Spezialfall läßt sich sogleich verallgemeinern, wenn man behauptet, daß bei irgend zwei Bewegungsvorgängen, die relativ zueinander gleich sind, immer das gleiche Ergebnis herauskommt. Damit wird ein Satz aufgestellt, der durch Beobachtung nahegelegt ist und den man in seiner Allgemeinheit versuchsweise auf alle Bewegungsvorgänge in der Natur erstreckt. Die Behauptung, wenn sie richtig ist, wird damit zu einem allgemeinen Naturprinzip, und man nennt ein solches Naturprinzip das Relativitätsprinzip.

So weit ist die Sache also gar nicht schwierig, und jedermann, der über Beobachtungen an relativ zueinander bewegten Körpern verfügt oder der Holz gesägt hat, kann begreifen, was man unter dem Relativitätsprinzip versteht. Man wird auch begreifen, daß die Gedankengänge, die zum Relativitätsprinzip geführt haben, nicht erst im 20. Jahrhundert von der Menschheit eingeschlagen wurden, sondern erheblich älteren Datums sind. Sonderlich originell ist also das Prinzip nicht, das der Relativitätstheorie den Namen gegeben hat. Es taucht nun aber sogleich die Frage auf: ist denn das Prinzip überhaupt richtig?

Diese Frage zu beantworten ist viel verwickelter, als begreiflich zu machen, was man unter dem Relativitätsprinzip versteht. In der sogenannten klassischen Mechanik, die von Galilei und Newton begründet ist, wird das Relativitätsprinzip als in aller Strenge gültig angesehen für gewisse Bewegungen von Naturkörpern, nämlich solche, die derartig verlaufen, daß die relativen Bewegungen gradlinig sind und mit gleichbleibender Geschwindigkeit erfolgen, sofern dabei keine andern als rein mechanische Erscheinungen hervortreten.

Ob das Relativitätsprinzip auch über diesen engen Bereich hinaus noch im Rahmen der alten klassischen Mechanik tatsächlich gültig ist, darüber sind sich nicht einmal heute die Gelehrten einig. Namhafte Forscher nehmen an, daß alle Bewegungen in der klassischen Mechanik, in denen die Geschwindigkeiten nicht gleichbleiben, in denen also sogenannte Beschleunigungen auftreten, das Relativitätsprinzip durchbrechen, andere nehmen an, daß das Relativitätsprinzip auch für ungleichförmige Bewegungsvorgänge gültig bleibt, sofern dabei Drehbewegungen (Rotationen) ausgeschlossen werden. Für Drehbewegungen jedenfalls gilt das Relativitätsprinzip der klassischen Mechanik nicht. Wer sich näher für diesen Gegenstand interessiert, mag dies in der Fachliteratur nachlesen. [*Footnote:* Vergl. E. Gehrcke. Verhandlungen der Deutschen Physikalischen Gesellschaft 15 S. 260. 1913.]

Wir werden nun weiter gehen und fragen, ob denn das Relativitätsprinzip auch für solche Naturerscheinungen gilt, welche nicht nur hinsichtlich ihrer Bewegung (z. B. wie zwei relativ zueinander bewegte Eisenbahnzüge) oder mechanisch, wie das Zersägen von Holz, betrachtet werden, sondern ob es auch für elektrische, magnetische, optische und andere Erscheinungen gültig bleibt. Auch hierüber besteht keine Einigkeit unter den Forschern. Besonders trennen sich hier die Parteien nach dem Gesichtspunkt, ob die elektrischen, magnetischen, optischen u. a. Erscheinungen in einem unsichtbaren, untastbaren, unwägbaren, aber doch tatsächlich vorhandenen Medium, genannt Weltäther, vor sich gehen oder nicht. Diejenigen Forscher, welche an den Äther glauben — und zu diesen gehören die bedeutendsten Gelehrten der Vergangenheit und der Gegenwart — müssen das Relativitätsprinzip, wie es oben für wägbare Naturkörper eingeführt wurde, allgemein ablehnen, auch für völlig gradlinige Bewegungen mit völlig gleichförmiger Geschwindigkeit (sogenannte gleichförmige Translationen). Diejenigen aber, welche nicht an den Äther glauben, haben die Freiheit, die Gültigkeit des Relativitätsprinzips in den verschiedensten Erweiterungen probeweise anzunehmen. Welchen Gültigkeitsbereich nehmen nun die Anhänger der sogenannten Relativitätstheorien für das Relativitätsprinzip an?

Auch diese Frage ist nicht einfach zu beantworten, weil die Meinungen sehr geteilte sind. Der Erfinder der Relativitätstheorie, Einstein, hat hierüber im Laufe der Zeit sehr verschiedene Ansichten gehabt und seinen Standpunkt mehrfach gewechselt. Er hat zunächst behauptet [*Footnote:* A. Einstein, Annalen der Physik 17, S. 891, 1905. Vgl. ferner die Zusammenstellung von Gehrcke: Die Naturwissenschaften 1, S. 62, 170, 338, 1913; ebenda 1919, S. 147.], daß das Prinzip auch für optische, elektrische usw. Erscheinungen an wägbaren Körpern gültig sei wobei stillschweigend vorausgesetzt war, daß die oben von der klassischen Mechanik für mechanische Erscheinungen zugelassene Bedingung der geradlinigen, gleichbleibenden Geschwindigkeit (gleichförmiger Translation) zutrifft; dann hat er sich zwei Jahre später merkwürdigerweise dahin geäußert, daß das Relativitätsprinzip nur auf beschleunigungsfreie (relative) Bewegungen angewandt worden sei, und überlegt, ob das Prinzip auch für beschleunigte Bewegungen gelte. Er kommt zu dem Schluß, daß dies so ist und glaubt, das Prinzip auf den speziellen Fall gleichförmiger Beschleunigung erweitern zu dürfen. Später hat Einstein in einer mehrere Monate nach meinen Einwänden erschienenen Schrift das Relativitätsprinzip wieder beschränkt auf gleichförmige Translationen. Ferner hat Einstein das Relativitätsprinzip ganz allgemein erweitern zu können geglaubt, und es auf sämtliche, auch ungleichförmige Translationen, und sogar auf Rotationen ausdehnen wollen. Er nannte die auf diese Ansicht gegründete Theorie „allgemeine Relativitätstheorie''. Schließlich hat Einstein noch einen etwas anderen Standpunkt eingenommen, er hat nämlich das Relativitätsprinzip ersetzt durch ein modifiziertes Prinzip, das sogenannte „Äquivalenzprinzip'' [*Footnote:* A. Einstein, Annalen der Physik, Bd. 35, S. 898, 1911.], und wir stehen vor dem bemerkenswerten Ergebnis, daß dasjenige Prinzip, welches der Relativitätstheorie den Namen gegeben hat, in der neueren Theorie Einsteins einem anderen Prinzip Platz gemacht hat. Einstein hat sich übrigens in der Verteidigung des Relativitätsprinzips nicht glücklich geäußert; dies trifft besonders für seine Polemik mit Lenard [*Footnote:* P. Lenard, Über Relativitätsprinzip, Äther, Gravitation. Verlag von Hirzel, Leipzig 1920. P. Lenard, Über Relativitätsprinzip, Äther, Gravitation. Verlag von Hirzel, Leipzig 1920. Hier findet man viele zugehörige Literaturhinweise.] zu, den er sachlich gar nicht widerlegen kann und an dessen Gegengründen er einfach vorbeiredet.

Es hätten die Schwankungen in der Auffassung Einsteins über eine so grundlegende Frage wie das Relativitätsprinzip eigentlich schon genügen können, um die Fachwelt stutzig zu machen und mit Skepsis gegen die Relativitätstheorie zu erfüllen. Wenn diese Skepsis nicht in dem Maße zutage trat, wie es unter gewöhnlichen Umständen zu erwarten gewesen wäre, so werden hierfür Gründe da sein. Darüber soll später im

Zusammenhang mit anderen Dingen einiges gesagt werden. Hier sei noch folgendes zum Relativitätsprinzip bemerkt:

Das Relativitätsprinzip, das in der Relativitätstheorie eine Rolle spielt, betrifft die Relativität von Bewegungsvorgängen. Sachlich gar nichts zu tun hat mit dieser Relativität der Bewegungen alles das, was in der Presse und auch zuweilen in Fachblättern sonst noch mit dem Wort Relativität gemeint wird. Daß „alles relativ" ist, worunter man sich, je nach dem individuellen Bildungsgrad, das Verschiedenste denken kann, mag auch bei den Anhängern der Relativitätstheorie eine wichtige Rolle, möglicherweise zuweilen nur im Unterbewußtsein, spielen, aber mit der theoretischen Relativitätstheorie als solcher haben derartige Allgemeinheiten sachlich nichts zu schaffen. Als Schlagwort, das auf die Massen wirkt, bei dem jeder glaubt, etwas ihm einigermaßen Bekanntes zu hören und bei dem auch kaum zwei an dasselbe denken, ist aber das „Relative" zur Einführung und zur Empfehlung der Relativitätstheorie vorzüglich geeignet. Das „Äquivalenzprinzip" wird niemals so populär werden können wie das „Relativitätsprinzip". Es liegt eine gewisse Tragik darin, daß die Relativitätstheorie in ihrer allmählichen Entwicklung ihr Hauptschlagwort in den Hintergrund geschoben hat; statt dessen wird, je länger je mehr, der Hauptnachdruck auf ein anderes Gebiet der Relativitätstheorie gelegt: auf die sogenannte

Relativierung von Raum und Zeit.

Die „Relativierung von Raum und Zeit" bildet heute die stolzeste Errungenschaft der Relativitätstheorie, deren Erwähnung die Brust des Relativisten schwellen läßt und durch die die philosophisch-erkenntnistheoretische Umwälzung unserer ganzen Weltauffassung gegeben sein soll. Die Relativierung von Raum und Zeit soll eine geistige Erneuerung und einen Wendepunkt in der menschlichen Denkweise bedeuten, demgegenüber die Taten von Kopernikus, Kepler und Newton verblassen.

Die Relativierung von Raum und Zeit wird in den bekannten Darstellungen der Relativitätstheorie als eine grundgelehrte Sache mathematisch eingekleidet vorgetragen, sodaß vielfach der Nichtmathematiker den Eindruck erhalten hat, er werde nie imstande sein, die Tiefe dieser weltstürzenden Gedanken je zu ermessen und zu begreifen. Und dabei ist kaum ein Gegenstand der ganzen Relativitätstheorie mit so wenig Aufwand an gelehrten Ausdrücken und Formeln klar zu machen, als gerade dieser. Das ist eigentlich von vornherein klar. Denn über Dinge, die so grundlegend sind wie Raum und Zeit, auf denen sich so vieles, Mathematisches und Nichtmathematisches, aufbaut, muß sich der Verstand mit einem Minimum an künstlichem, mathematischen Handwerkszeug klar werden können — wenn er dazu überhaupt imstande ist. Die mathematischen Formeln geben uns ja auch nur Aufschluß darüber, wie groß im einzelnen die errechneten Effekte sind, sie sagen jedoch nichts aus über den ihnen zugrunde liegenden Standpunkt. Aber die Anhänger der Relativitätstheorie sind anderer Meinung. Ihnen ist der mathematische

Aufbau offenbar unlösbar verknüpft mit den allgemeinen, erkenntnistheoretischen Grundauffassungen, vor denen sie staunen. An keiner Stelle liegt aber die Wurzel der Relativitätstheorie klarer, als bei der ihr eigentümlichen Auffassung von Raum and Zeit, und an keinem Punkte wird die Lage für die Zukunft der Relativitätstheorie bedenklicher als beim Raum und bei der Zeit.

Einstein hat, wenn auch nicht seine Grundauffassung, so doch seine Folgerungen hinsichtlich des raumzeitlichen Geschehens durch allgemein verständliche Bilder zu erläutern gesucht. Hier nur eine Probe.

Einstein erörterte gelegentlich eines Vortrages in Zürich [*Footnote:* A. Einstein, Vierteljahrsschrift der Naturforschenden Gesellschaft Zürich 56, S. 11 und folgende.] die Vorgänge, die sich nach seiner Theorie in einer hin and her bewegten Uhr angeblich abspielen sollen. Eine solche hin and herbewegte Uhr soll nach Einstein gegenüber einer ruhenden Uhr nachgehen. Er äußert sich dann, um recht deutlich and populär zu sein, folgendermaßen: „Wenn wir z. B. einen lebenden Organismus in eine Schachtel hineinbrächten und ihn dieselbe Hin- und Herbewegung ausführen ließen wie vorher die Uhr, so könnte man es erreichen, daß dieser Organismus nach einem beliebig langen Fluge beliebig wenig geändert wieder an seinen ursprünglichen Ort zurückkehrt, während ganz entsprechend beschaffene Organismen, welche an dem ursprünglichen Orte ruhend geblieben sind, bereits längst neuen Generationen Platz gemacht haben. Für den bewegten Organismus war die lange Zeit der Reise nur ein Augenblick, falls die Bewegung annähernd mit Lichtgeschwindigkeit erfolgte! Das ist eine unabweisbare Konsequenz der von uns zugrunde gelegten Prinzipien, die die Erfahrung uns aufdrängt."

Also kurz gesagt: Die Zeitfolge aller Ereignisse auf einem Naturkörper soll nach Einsteins Theorie abhängig sein vom Bewegungszustand des Körpers, derart, daß die Bewegung des Naturkörpers alle auf ihm sich abspielenden Vorgänge verlangsamt: es soll hiernach z. B. ein lebender Organismus durch Schütteln, wegen der dadurch bedingten Verzögerung aller an ihm und in ihm sich abspielenden Prozesse, jung erhalten werden können. Diese Geschichte hat Einstein und ebenso seine Anhänger als „unabweisbare Konsequenz" der Relativitätstheorie einem staunenden Publikum erzählt! Sie ist von den Relativisten mannigfach variiert and weiter ausgebaut worden: Von zwei Zwillingen wird der eine gleich nach seiner Geburt auf eine lange Reise geschickt, von welcher er als Schuljunge zurückkehrt; er findet dann seinen Bruder als Greis mit weißen Haaren vor! Solche and ähnliche Betrachtungen sind, um es noch einmal hervorzuheben, nicht etwa Märchen oder Witze, sondern „unabweisbare Konsequenzen" der Relativitätstheorie! Die genannten Konsequenzen muß man mitmachen, wenn man an die Relativitätstheorie glaubt.

Statt auf mathematische Formeln einzugehen, können wir an den genannten Bildern das Wesen der erkenntnistheoretischen Grundlagen der Theorie erfassen. Wir wollen uns fragen: 1. Welche Grundansicht über die Zeit liegt diesen Betrachtungen zugrunde? 2. Was folgt weiter daraus?

Fassen wir jetzt also irgendeine den Folgerungen ins Auge, die den relativistischen Zeitablauf kennzeichnen, z. B. das obige, Einsteinsche Beispiel der gegeneinander bewegten Organismen. Wir wollen tatsächlich annehmen, es wäre experimentell gefunden, daß der bewegte Organismus jünger geblieben ist als der ruhende; über die Unwahrscheinlichkeit und die technischen Schwierigkeiten einer solchen Feststellung wollen wir uns hinwegsetzen. Dann wäre alles, so sonderlich es wäre, immerhin verständlich, wenn Bewegung als solche die Eigenschaft haben würde, eine Verlangsamung aller auf dem bewegten Körper vor sich gehenden chemischen und physikalischen Prozesse hervorzubringen. Gerade die Bewegung als solche, auch genannt „absolute Bewegung", wird aber von Einstein geleugnet, und er muß daher die gegebene Erklärung für das merkwürdige Jungbleiben des bewegten Organismus von sich weisen. Statt dessen nimmt er eine „Relativierung den Zeit" an; das bedeutet, daß der bewegte Organismus nur vom Standpunkt des ruhenden Organismus aus der jüngere ist, daß aber andererseits auch vom Standpunkt des andern Organismus aus der erste Organismus der bewegte und daher der jüngere ist. Nach der Relativitätslehre soll jeder Standpunkt dem andern gleichberechtigt, keiner von dem andern bevorzugt sein. Ein solcher Ausweg führt nun aber zu höchst bedenklichen Folgerungen. Dies ist unschwer einzusehen, wenn wir die beiden Organismen miteinander reden lassen, nachdem die Reise beendet ist und sie beide wieder relativ zueinander ruhen. Der eine Organismus wird z. B. behaupten: ich habe weiße Haare, and Du bist jung geblieben; der andere Organismus wird ebenfalls behaupten: ich habe weiße Haare and Du bist jung geblieben, denn ich bin ja von meinem Standpunkt aus der ruhende, und Du der bewegte! Also die beiden Organismen wenden sich gegenseitig für jung und jeder sich selbst für gealtert erklären!

Die beiden kommen also zueinander in Widerspruch. Man könnte auf den Einfall kommen, daß der Widerspruch beseitigt wäre, wenn in der Unterhaltung der eine immer das Gegenteil von dem hören würde, was der andere sagt, aber auch das rettet nicht aus der Schwierigkeit. Denn wenn die Reise des bewegten Organismus lange genug gedauert hat, ist der ruhende Organismus tot (vgl. oben Einsteins Worte). Dann ist es aber eine „unabweisbare Konsequenz", wenn der jung gebliebene Organismus zum Toten spricht: Nicht Du bist tot, sondern ich! Denn vom Standpunkt des jungen Organismus aus war ja er selbst der ruhende, der andere der bewegte [*Footnote:* Der empirische Einwand, daß ein Toter nicht sprechen kann, steht dem Relativisten nicht zu, der selbst als Begründung für seine Behauptungen über Zeit und Raum nichts anderes anzuführen weiß, als daß sich „a priori" nichts gegen sie einwenden ließe.]! Es ist zu bedauern, daß die Relativitätstheoretiker das Einsteinsche Organismenbeispiel nicht gründlich weiter gedacht haben. Vielleicht wären ihnen dann noch einige Zweifel aufgestiegen, ob die Vertauschbarkeit den Standpunkte, die sie hinsichtlich des zeitlichen Geschehens unter der Bezeichnung „Relativierung der Zeit" eingeführt haben, sich durchführen läßt.

Es ist nur eine einzige Möglichkeit ersichtlich, aus den Widersprüchen, zu denen die „Relativierung den Zeit" führt, herauszukommen, wenn man nämlich dazu übergeht, jedem Standpunkt, Organismus, Beobachter, Subjekt oder „Monade" eine eigene Welt zuzuordnen, die mit den Welten anderer, bewegter Monaden nichts zu tun hat. Der „Relativierung der Zeit" fügt man so eine „Relativierung des Seins" hinzu, d. h. mit anderen Worten: die Eindeutigkeit des Naturgeschehens für alle bewegten Monaden wind aufgehoben. Man kann auch so sagen: es wird der Standpunkt eines physikalischen Solipsismus eingenommen. Es weist kein Anzeichen darauf hin, daß die in den erkenntnistheoretischen Fragen sehr unklaren Relativitätstheoretiker einen solchen Ausweg beabsichtigt oder überhaupt nur erwogen haben. Auch Minkowski, der von seiner eigenen „Verwegenheit mathematischer Kultur" spricht, scheint diese Verwegenheit der Relativierung des Seins, zu der er bei konsequentem Festhalten an dem einmal beschrittenen Wege gedrängt wird, nicht im Auge gehabt zu haben. Wie denn überhaupt die Denkrichtung den Relativitätstheoretiker auf den mathematischen Ausbau and die formalistische Struktur der Theorie gerichtet ist, und nicht in die erkenntnistheoretische Vertiefung und Klarstellung.

Immerhin deuten manche Äußerungen Einsteins, gerade in seinen sogenannten „allgemeinverständlichen" Darlegungen, darauf hin, daß ihm die inneren Schwierigkeiten seiner Lehre nicht ganz fremd waren. Wenn er z. B. gelegentlich behauptet hat, daß der Begriff der Gleichzeitigkeit zweier Ereignisse keinen Sinn habe, so läßt diese zunächst mystische Ausdrucksweise vermuten, daß Einstein gefühlt hat, etwas Besonderes erfinden zu müssen, um innere Widersprüche zu vermeiden. Bei Klarlegung des erkenntnistheoretischen Standpunkts der Relativitätstheorie als eines Solipsismus erscheint allerdings das Sinnlose der Gleichzeitigkeit als eine zulässige Selbstverständlichkeit. Es ist aber keine Kunst, einen Widerspruch dadurch zu vermeiden, daß man implicite den Grundsatz einführt: es bezieht sich die eine Aussage, die einer zweiten Aussage widerspricht, auf eine ganz andere Welt als die zweite. Die Sonderbarkeiten der Relativitätstheorie, ihre angebliche Reform der Erkenntnistheorie mündet immer wieder in den oben gekennzeichneten Standpunkt aus, den man physikalischen Solipsismus nennen kann. Dieser Standpunkt ist der eines Menschen, welcher in die äußerste Enge getrieben ist, der seine Sache bis aufs letzte verficht, und schließlich, um sich zu retten, die Erklärung abgibt: ich habe nicht, denn Du hast auch recht, weil wir beide verschiedenen Welten angehören und deshalb unsere Aussagen gar nicht miteinander vergleichen können! Wenn man den „Zeitbegriff relativiert", so zerstört man die Idee der einen, allgemeinen, objektiven Natur; wenn die eine Monade ihre Eigenzeit, von den Relativisten t genannt, die andere ihre Eigenzeit, t' genannt, hat, so muß auch jede Monade ganz für sich ihre eigene Welt oder Natur haben, und so wenig man den Zeiten t und t' „gleichzeitige" Augenblicke erlaubt, ebensowenig sind auch in den Welten der beiden Monaden ein und dieselben Dinge vorhanden, höchstens können beide Welten miteinander gewisse Ähnlichkeiten aufweisen. Die Relativitätstheorie fühnt also nur zu einem alten, abgelebten, skeptischen Standpunkt. Das ist die „neue Revolution des modernen Denkens", die die Relativitätstheorie enzeugt hat!

Wir werden es uns versagen können, nach dem Obigen noch die Relativierung des Raumes in der Relativitätstheorie näher zu erörtern. Wenn Minkowski von sich sagt, er habe Einsteins „Hinwegschreiten über die Zeit" durch ein „Hinwegschreiten über den Raum" vervollständigt, so hat er damit eine Folgerung gezogen, die ihm nur deshalb bewundernswürdig erschienen ist, weil er selbst sich prinzipiell so unklar war.

Relativitätstheorie und Gravitation.

Die erste Relativitätstheorie Einsteins, welche er später „die spezielle" genannt hat, wurde von ihm ersetzt durch eine zweite „allgemeine" Relativitätstheorie, die die ursprünglichen Mängel der ersten Theorie nicht haben sollte. Nun ist aber das Verhältnis der beiden Theorien zueinander nur in formaler Hinsicht das des Speziellen zum Allgemeinen, während in grundsätzlichen Fragen ein erheblicher, bis zum Widerspruch gesteigerter Unterschied besteht. Die allgemeine Relativitätstheorie ist dadurch gekennzeichnet, daß in ihr die allgemeine Schwere (Gravitation) eine besondere Rolle spielt, ferner ist besonders bezeichnend für sie ein allgemeines Relativitätsprinzip, d. h. die Behauptung den Relativität aller Bewegungen, auch die der Rotationen.

Abgesehen von den mit den „Relativierung von Zeit und Raum" verbundenen, oben erwähnten Schwierigkeiten sind es auch Bedenken mehr empirischer Natur, die die allgemeine Form der Einsteinschen Relativitätstheorie als undurchführbar erscheinen lassen. Ein Beispiel wird dies deutlich machen können. Angenommen, wir setzen uns auf den in manchen Vergnügungsstätten sehr beliebten Apparat, genannt Drehscheibe, oder wir setzen uns auf eins der altmodischen Karussels, so soll es nach der Relativitätstheorie ebensogut möglich sein zu behaupten, daß das Karussel fährt, als daß das Karussel still steht und die ganze Außenwelt sich um das Karussel dreht. Also der Auffassung des gewöhnlichen Menschen: das Karussel fährt: soll die Behauptung des Relativisten gleichwertig sein: die ganze Welt fährt um das stillstehende Karussel im Kreise herum! Hierbei kommt der Relativist nicht nur nur zu der von seinem eigenen, theoretischen Staudpunkt aus störenden Folgerung, daß er den in großen Abständen vom Karussel stehenden Naturkörpern, wie z. B. allen Fixsternen, ungeheure Geschwindigkeiten beilegen muß, welche die auch der Theorie höchst zulässige Geschwindigkeit, die Lichtgeschwindigkeit, erheblich übersteigen, er muß auch noch besondere, seltsame Naturerscheinungen hinzudichten, um den Ablauf der Erscheinungen, wie er sich abspielt, beschreiben zu können. Er muß nämlich annehmen, daß die bei der Rotation der Welt auftretenden Zentrifugalkräfte durch eine Schwerkraft kompensiert werden, welche proportional dem Abstand von der Drehungsachse des Karussels zunimmt und welche im Raume des Karussels selbst ihr Vorzeichen umkehrt. Für ein solches Schwerkraftfeld ist aber keine Veranlassung erkennbar, abgesehen davon, daß sich auch mathematisch überhaupt keine Massenanordnung ersinnen läßt, die ein Schwerefeld erzeugen können, welches den mathematischen Bedingungen des Problems zu genügen vermöchte. In der Tat ist das Vorgehen des Relativisten, der die ganze Welt in Rotation, um ein Karussel versetzt und der zu diesem Zweck ein physikalisch unmögliches Gravitationsfeld voraussetzt, rein fiktiv, physikalisch unzulässig. Der Standpunkt des Relativisten gleicht dem eines Menschen, welchem ein Geldstück gestohlen worden ist und der behauptet: ich kann entweder annehmen daß der Dieb das Geldstück gestohlen hat, oder ich kann annehmen, daß der Dieb die ganze Welt gestohlen hat, nur nicht das Geldstück. Die zweite „Denkmöglichkeit" scheidet aus Gründen der Erfahrung, „a posteriori", aus, und es ist deshalb nicht möglich, hier eine „Relativität" der Standpunkte einzunehmen. Genau so ist es auch mit dem Standpunkt des Relativitätstheoretikers gegenüber der Rotation eines Karussels, er widerspricht aller Erfahrung. Wer sich über diese Seite der Gegnerschaft gegen die Relativitätstheorie näher unterrichten will, dem seien die Schriften von Lenard angelegentlichst empfohlen, besonders die Broschüre: Über Relativitätsprinzip, Äther, Gravitation. Verlag von S. Hirzel, Leipzig 1920, von der ausgehend man auch den Weg zu der übrigen Literatur üben den Gegenstand findet.

Die Grundlage der allgemeinen Relativitätstheorie leidet auch an dem Mangel, keinen innere Grund für die Annahme eines Schwerefeldes für die zur Durchführung der Theorie benötigten Beschleunigungsfelder erkennen zu lassen. Man kann nicht einsehen, warum gerade die Gravitation berufen ist, als Ursache für Beschleunigungen angesehen zu werden, wo doch auch andere Ursachen für Beschleunigungen denkbar sind, wie Kräfte im Äther, Kapillaritätskräfte usw. Durch die Einführung der Gravitation, also einer empirischen, physikalischen Erscheinung in die Grundgleichungen der Relativitätstheorie, wird jedenfalls der Boden der reinen, mathematischen Konstruktion verlassen und ein physikalisches, empirisches Element hineingezogen. Der Relativist kann sich daher nicht mehr in der Rolle des abstrakten Mathematikers allein verhalten, sondern er muß es sich gefallen lassen, daß der Physiker die Theorie als eine empirisch richtig sein sollende objektiv prüft. Fällt diese Prüfung zu ungunsten des Relativisten aus, so muß dieser seine Theorie aufgeben und kann eventuell eine neue ersinnen. Es geht aber nicht an, daß der Relativist deshalb an seiner Theorie festhält, weil er sie mathematisch schön findet. Abgesehen von allen logischen und erkenntnistheoretischen Erwägungen bleibt die Erfahrung der Hauptprüfstein jeder physikalischen Theorie, und so auch der Relativitätstheorie.

Die experimentelle Prüfung der Theorie.

Wer sich im praktischen Leben oder als Naturforscher betätigt, wird dem theoretischen Unterfangen, eine für alle Beobachter gleiche, objektive Natur in ihrer einen Zeit und ihrem einen Raume aufzugeben, wenig Vertrauen entgegenbringen.

Er wird daher auch nicht sonderlich erstaunt sein, wenn sich herausstellt, daß einzelne praktische Folgerungen einer solchen Theorie mit der Erfahrung in Widerspruch geraten. So wenig einerseits die Bestätigung einer Folgerung die Richtigkeit der Theorie beweisen würde, — kann man doch häufig von ganz verschiedenen Grundlagen aus zu derselben, sich als richtig erweisenden Folgerung kommen, ohne damit etwas über die Richtigkeit der Grundlagen sagen zu können, — so sicher beweist andererseits eine als falsch sich herausstellende Folgerung, daß auch die Grundlage, aus der sie abgeleitet war, falsch sein muß. Die Relativitätstheorie hat die Prüfung an der Erfahrung schlecht bestanden. Dies soll im Folgenden kurz dargestellt werden.

Zunächst sei bemerkt, daß alle Folgerungen den Relativitätstheorie immer auf so winzige Effekte führen, daß es nicht einfach ist, die experimentelle Prüfung vorzunehmen. Das war bisher in gewissem Sinne ein Glück für die Theorie, die ja dadurch in die Lage versetzt ist, auf die Schwierigkeit des Experiments, die Ungenauigkeit den Beobachtungen hinzuweisen, wenn sich ein vorausgesagter Effekt nicht findet. Es gibt aber heute Beobachtungen, die so genau sind, daß man diesen Schluß nicht mehr ziehen kann.

In ersten Linie ist hier die sogenannte Rotverschiebung der Spektrallinien zu erwähnen. Eine Spektralinie wird durch gewisse Schwingungen in einem Gase erzeugt, das leuchtet. Auch auf unserer Sonne, welche nach den Ergebnissen der Astronomie und Astrophysik ein sehr hoch erhitzter Gasball ist, werden Spektrallinien beobachtet. Nur soll nach der Relativitätstheorie die Zeitdauer irgend eines Vorgangs vom Schwerkraft-(Gravitations-)felde abhängig sein, also sollten auch die Schwingungsvorgänge aller Spektrallinien auf der Sonne vom Gravitationsfeld der Sonne abhängen. Dieses letztere ist aber erheblich stärker als das Gravitationsfeld der Erde, so daß die Spektrallinien eines Gases auf der Sonne gegenüber den Spektrallinien derselben Gasart auf der Erde einen Unterschied zeigen sollten — behauptet die Relativitätstheorie. Für die Größe dieses Unterschiedes und sein Vorzeichen sind Formeln aufgestellt worden. Sie besagen, daß die Spektrallinien der Sonne eine geringe Verschiebung nach der roten Seite des Spektrums erleiden müssen, im Betrage von 0,01 sogenannten Angström-Einheiten. Die Kleinheit dieses Betrages ist für jeden ersichtlich, wenn man ihn in Millimeter ausdrückt: er beträgt ein Milliardstel eines Millimeters. Dieser kleine Effekt, dessen Bestehen die Relativitätstheorie prophezeit hat und fordert, kann aber heutzutage mit den hochentwickelten Meßeinrichtungen gesucht werden und würde den modernen Instrumenten nicht entgehen, wenn er da wäre. Der Effekt ist sorgfältig gesucht worden, hat sich aber nicht finden lassen:

Zuerst ist die relativistische Rotverschiebung an Stickstofflinien der Sonne auf dem astrophysikalischen Institut in Potsdam gesucht worden; Schwarzschild [*Footnote:* Sitzungsbericht der Berliner Akademie der Wissenschaft 1914, S. 1201-1213.], der verstorbene Direktor des Instituts, hat das Ergebnis im Jahre 1914 veröffentlicht; er findet keine

Rotverschiebung. Dann hat der bekannte amerikanische Astrophysiker St. John nach der Rotverschiebung gesucht und sie ebenfalls nicht gefunden. St. John sagt in seinem Bericht vom Jahre 1917 über das Ergebnis seiner Versuche [*Footnote:* St. John, Carnegie Institution of Washington, Mount Wilson Solar Observatory Communications to the National Academy of Sciences No. 46. Vol. 3, 450-452, July 1917.]: „Das allgemeine Ergebnis der Untersuchung ist, daß innenhalb der Beobachtungsfehler die Messungen kein Anzeichen eines Effektes von der Größenordnung ergeben, die aus dem Relativitätsprinzip abgeleitet wird." Die Beobachtungsfehler St. Johns waren nur ein Bruchteil von dem geforderten, nicht vorhandenen Einstein-Effekt. Hale, der bekannte Sonnenforschen und Direktor der Mount-Wilson-Sternwarte, hat sich für die Richtigkeit St. Johns Beobachtungen ausgesprochen [*Footnote:* Z. B. im Annual Report of the Direktor of the Mount Wilson Solar-Observatory, Yearbook, Nr. 16, S. 200, 1917.]. Diese Untersuchungen auf Mount Wilson, mit den besten Instrumenten unter den günstigsten Arbeits- und Beobachtungsbedingungen, wie sie zurzeit kein anderes astrophysikalisches Institut auf der Erde aufweisen kann, hätten den Einstein-Effekt unzweifelhaft feststellen müssen, wenn er existierte. Demgegenüber will es wenig heißen, wenn neuerdings ein Mitarbeiter von Einstein, Herr Freundlich, mit der Behauptung aufgetreten ist, daß die Amerikaner eine Fehlerquelle in ihren Messungen gehabt haben; die Zusammenstellung und kritische Würdigung dieses gesamten Materials wird in einer demnächst von fachmännischer Seite in Aussicht gestellten Druckschrift von. L. C. Glaser gegeben werden, auf die hier verwiesen sei.

Die Rotverschiebung der Spektrallinien auf der Sonne stellt bisher den Haupteffekt der Relativitätstheorie dar, er ist entschieden die wichtigste, weil am genauesten zu prüfende Folgerung, deren Nichtvorhandensein als eine experimentelle Widerlegung der Relativitätstheorie anzusehen ist — wenn es einer solchen überhaupt noch bedurft hätte. Andere Folgerungen der Relativitätstheorie sind für die Theorie weniger charakteristisch, weil sich sofort verschiedene andere Erklärungsmöglichkeiten darbieten. Da ist z. B. die sogenannte Perihelstörung des Planeten Merkur zu nennen. Nach den Beobachtungen der Astronomen dreht sich die Bahnellipse des Merkur um einen sehr kleinen Betrag von 43 Bogensekunden in 100 Jahren. Auch dies ist eine ungeheuer kleine Größe, aber sie ist dank der Feinheit der astronomischen Beobachtungsmethoden feststellbar. Es sind schon seit vielen Jahren Erklärungen für diese Bahnstörung des Merkur gegeben worden, insbesondere muß hier die Formel des Oberlehrers Gerber vom Jahre 1898 genannt werden [*Footnote:* Die schwer zugängliche Veröffentlichung Gerbers ist in den Annalen der Physik Bd. 52, Seite 415, 1917 im Neuabdruck erschienen.], die dieser aufgestellt hat, als es noch gar keine Relativitätstheorie gab und die völlig mit der aus der Relativitätstheorie von Einstein abgeleiteten Formel übereinstimmt. Hier könnte die Relativitätstheorie nur dann als eine gewisse, und zwar die zuletzt gegebene, Erklärungsmöglichkeit für eine an sich bekannte Sache angesehen werden, wenn sie im übrigen einwandfrei wäre.

Endlich ist noch ein, neuerdings in der Tagespresse mit besonderer Breite behandelter Effekt zu nennen: die Ablenkung der Sternorte in der Nähe der Sonne. Auch hier ist die Sache durchaus nicht so neu, als es auf den ersten Blick den Anschein hat, denn man kennt in der Astronomie schon lange gewisse systematische Abweichungen der Sternorte in Abhängigkeit von der Stellung des Sterns zur Sonne. Diese Erscheinung, die als jährliche Refraktion bezeichnet wird, ist bisher noch nicht erklärt, obschon ein erhebliches Tatsachenmaterial über den Gegenstand vorliegt, das bis in die Mitte des vorigen Jahrhunderts zurückreicht; man kann sich hierüber z. B. aus einer Abhandlung von L. Courvoisier, Beobachtungsergebnisse der Kgl. Sternwarte zu Berlin Nr. 15 vom Jahre 1913 unterrichten. Einstein hat nun ebenfalls eine Abhängigkeit der Sternorte in Abhängigkeit von der Sonne aus seiner Relativitätstheorie gefolgert und es sind Messungen darüber von englischen Expeditionen gelegentlich der Sonnenfinsternis des Jahres 1919 angestellt werden. Die Beurteilung dieser Beobachtungen ist schwierig, da die Originalberichte noch nicht alle gedruckt vorliegen und die Angaben über die in der englischen Akademie in London vorgelegten Mitteilungen der verschiedenen Forscher nicht einheitlich sind. Jedenfalls steht fest, daß die deutsche Fachwelt und Presse bisher in einseitiger, für Einsteins Theorie zu günstiger Weise unterrichtet worden ist. Dies geht z. B. aus Äußerungen des Londoner Astronomen Silberstein hervor, der darauf aufmerksam macht [*Footnote:* Abgedruckt in: Die Naturwissenschaften 8, 390, 1920.], daß das in der physikalischen Gesellschaft in Berlin erstattete Referat in wesentlichen Punkten Irrtümer enthielt, deren Berichtigung das Ergebnis den Messungen zu Ungunsten von Einsteins Theorie verschiebt. Über den Effekt der Sternorte in der Nähe der Sonne läßt sich also zurzeit nichts Sicheres aussagen. Aber er ist für die Theorie gar nicht so wichtig, da er, selbst wenn die von Einstein angegebene Verschiebung der Sternorte um 1¾ Bogensekunden am Sonnenrande tatsächlich sicher beobachtet wäre, noch eine ganze Reihe anderer Erklärungsversuche, die physikalisch viel verständlicher sind als die Deutung durch die Relativitätstheorie, gegeben werden können. Es ist übrigens hier die Kleinheit des Betrages von nur 1¾ Bogensekunden ein erhebliches Hindernis für das Experiment; um von diesem Betrage eine Vorstellung zu geben, sei erwähnt, daß der kleine Winkel 1¾ Bogensekunden diejenige Größe hat, unter der dem Auge eine Kirsche in 2 Kilometer Entfernung erscheint.

Welches Urteil wird man sich über die Relativitätstheorie zu bilden haben?

Das ist die Frage, die nunmehr zu beantworten ist.

Die Einsteinsche Relativitätstheorie nimmt ihren Ursprung aus einer Theorie des holländischen Physikers Lorentz. Die übereinstimmung mit der Lorentzschen Theorie geht so weit, daß die mathematische Form der Einsteinschen Theorie vom Jahre 1905 wesentlich dieselbe ist, wie die von Lorentz, die Gleichungen dieser Einsteinschen Theorie sind die Gleichungen von Lorentz. Neuartig erschien die Deutung der Theorie, die Interpretation der Grundbegriffe Zeit und Raum. Einstein hat mit dieser

Interpretation etwas getan, von dem seine Bewunderer gesagt haben, es stelle alles bisher Dagewesene in den Schatten. Die Interpretation Einsteins war aber gleichfalls weit weniger neu, als es den Anschein hatte. Schon im jahre 1901 hat der ungarische Philosoph Melchior Palágyi in Engelmanns Verlag in Leipzig eine Schrift in deutscher Sprache [*Footnote:* Neue Theorie des Raumes und der Zeit. Von Dr. Melchior Palágyi. Verlag Engelmann, Leipzig 1901.] erscheinen lassen, die wesentliche Gedanken Einsteins und Minkowskis, des begeisterten, mathematischen Anhängers Einsteins, vorwegnahm: so besonders die Idee der „Union zwischen Zeit und Raum", die Auffassung der „Welt" in 4 Koordinaten, von denen die eine, die Zeit, mit der imagären Einheit $\sqrt{-1}$ multipliziert auftritt usw. Den Physikern waren diese Vorgänge — zum Teil heute noch—unbekannt, sie nahmen die Relativitätslehre Einsteins teils kopfschüttelnd, teils abwartend auf. Als aber anerkannte Autoritäten sich begeistert für die Relativitätstheorie einsetzten, trat auch im Publikum Begeisterung auf, und nun nahm die Entwicklung ihren unaufhaltsamen Gang. Bei der Verknüpfung mathematischer, physikalischer und philosophischer Gedanken in der Relativitätstheorie war es den Fachleuten in unserer Zeit des hochgesteigerten, wissenschaftlichen Spezialistentums sehr schwer gemacht, zu einem selbständigen Urteil über die Theorie zu gelangen, zumal Einstein sein Werk mit Geschicklichkeit zu verteidigen wußte und den Physikern ihre Bedenken mit mathematischen und philosophischen, den Mathematikern ihre Bedenken mit physikalischen und philosophischen, den Philosophen ihre Bedenken mit mathematischen und physikalischen Gegengründen zerstreute: jeder Fachmann beugte sich vor der Autorität des Kollegen im andern Fach, jeder glaubte das, was er nach andern Fachautoritäten als für bewiesen halten zu sollen vermeinte. Niemand wollte sich dem Vorwurf aussetzen, er verstände nichts von der Sache! Und so wurde eine Lage geschaffen, ähnlich der von Andersen geschilderten in seinem Märchen „Des Kaisers neue Kleider": hier sieht ein Kaiser mit seinen Ministern und Untertanen dem Weben eines Gewandes zu, das die Eigenart hat, von denjenigen Menschen nicht gesehen zu werden, die dazu nicht klug genug sind, und schließlich stehen alle staunend vor den leeren Webstühlen, weil niemand sich getraut zu bekennen, daß er nichts sieht. So hat auch die Relativitätstheorie die Geister gefesselt, sie ist zur Massensuggestion geworden. Aber eine Massensuggestion ist an sich nichts Verwerfliches, die Ausschaltung des klaren Verstandes braucht durchaus kein Beweis dafür zu sein, daß das Streben der Masse ein törichtes ist. Alles hing bei der Relativitätstheorie davon ab, ob sie in ein erkenntnistheoretisch annehmbares Fahrwasser geleitet werden konnte.

Einstein hat die Schwächen seiner Theorie öfters zu verbessern und den Einwänden auszuweichen gesucht, er hat z. B. das Relativitätsprinzip hin und hergeworfen (s. oben S. 57 ff.), er hat schließlich geglaubt, den sicheren Hafen erreicht zu haben und im Jahre 1915 erklärt [*Footnote:* Sitzungsberichte der Berliner Akademie 1915, S. 847.], daß endlich die Relativitätstheorie als logisches Gebäude abgeschlossen sei. Ein Punkt bei all diesen Wandlungen ist noch besonders wichtig, hervorgehoben zu werden: so wenig neuartig die mathematische Form der ersten Relativitätstheorie Einsteins ist, die mit der älteren Lorentzschen Theorie übereinstimmt, so wenig ist auch die im weiteren Verlauf der Entwicklung durch Einstein vollzogene Veränderung des mathematischen Gewandes der Theorie besonders neuartig gewesen: daß die Relativitätstheorie in die Formeln der nichteuklidischen Geometrie hineinführt, zeigte zuerst der Mathematiker Varicak; daß die mathematische Komplikation der nichteuklidischen Kontinua von den Mathematikern formal bereits seit langem gelöst war, erkennt sogar Einstein an. Inwieweit Einstein die neueste von Weyl u. a. eingeschlagene, relativitätstheoretische Richtung überhaupt noch mitmacht, ist nicht recht klar. Jedenfalls verbreiten Anhänger von Einstein Nachrichten, die für die Weylschen Arbeiten ungünstig lauten.

Wenn es also feststeht, daß Einstein in seiner Relativitätstheorie keine mathematisch ungewöhnlichen Formen entdeckt hat, wenn die philosophisch-erkenntnistheoretische Grundlage des ganzen Gebäudes unbefriedigend ist, wenn endlich die Experimente der Physiker und Astronomen die Theorie night beweisen können, so wird man fragen, was denn überhaupt noch übrig bleibt, um in der Relativitätstheorie ein Werk zu erblicken, das über die Taten von Kopernikus, Kepler und Newton hinausgeht. Diese Frage werden die heutigen Anhänger und Gegner der Theorie, je nach ihrem persönlichen Gefühl, verschieden beantworten. Eine Antwort, die alle befriedigt, wird sich erst erzielen lassen, wenn die Suggestion der Reklame und der Druckerschwärze, mit welcher die „revolutionäre Relativitätstheorie" arbeitet, von allen als solche erkannt ist. Zu dieser Aufklärung beitragen zu helfen mögen die obigen Zeilen dienen."

Gehrcke effectively accused Einstein of plagiarizing the mathematical formalisms of Lorentz, the space-time concepts of Palágyi,[298] and the non-Euclidean Geometry of Varičak.[299] Albert Einstein's first wife Mileva Marić

---

298. Menyhért (Melchior) Palágyi, *Neue Theorie des Raumes und der Zeit*, Engelmanns, Leipzig, (1901); reprinted in *Zur Weltmechanik, Beiträge zur Metaphysik der Physik von Melchior Palágyi, mit einem Geleitwort von Ernst Gehrcke*, J. A. Barth, Leipzig, (1925).
299. V. Varičak, "Primjedbe o jednoj interpretaciji geometrije Lobačevskoga", *Rad Jugoslavenska Akademija Znanosti i Umjetnosti*, Volume 154, (1903), pp. 81-131; **and** "O transformacijama u ravnini Lobačevskoga" *Rad Jugoslavenska Akademija Znanosti i Umjetnosti*, Volume 165, (1906), pp. 50-80; **and** "Opcéna jednadzba pravca u hiperbolnoj ravnini", *Rad Jugoslavenska Akademija Znanosti i Umjetnosti*, Volume 167, (1906), pp. 167-188; **and** "Bemerkung zu einem Punkte in der Festrede L. Schlesingers über Johann Bolyai", *Jahresbericht der Deutschen Mathematiker-Vereinigung*, Volume 16, (1907), pp. 320-321; **and** "Prvi osnivači neeuklidske geometrije", *Rad Jugoslavenska Akademija Znanosti i Umjetnosti*, Volume 169, (1908), pp. 110-194; **and** "Beiträge zur nichteuklidischen Geometrie", *Jahresbericht der Deutschen Mathematiker-Vereinigung*, Volume 17, (1908), pp. 70-83; **and** "Anwendung der Lobatschefskijschen Geometrie in der Relativitätstheorie", *Physikalische Zeitschrift*, Volume 11, (1910), pp. 93-96; **and** "Die Relativtheorie und die Lobatschefskijsche Geometrie", *Physikalische*

would have been able to have read *all* of Varičak's works. She also would have been able to have understood all of Smoluchowski's lectures. She could also read English, making her the likely source of many of the works Albert Einstein plagiarized from English-speaking authors.[300] Gehrcke also accused Albert Einstein of masking his plagiarism and the weaknesses of the theory of relativity with irrational Metaphysics. Gehrcke stood up and declared that, "the Emperor has no clothes!"—an admission Einstein had already privately made to Heinrich Zangger on Christmas Eve of 1919.[301] Gehrcke said that people were often afraid to admit that they did not understand the theory of relativity, and were in stupefied awe of that which they did not understand, not in informed appreciation of the theory. Einstein had made the exact same statements in his private correspondence, but shamelessly called Gehrcke anti-Semitic when he reiterated Einstein's own beliefs.

Einstein's only response came days later in a frantic, inappropriately emotional and irrational "hand-waving" *ad hominem* attack against Lenard, Weyland and Gehrcke. Einstein simply appealed to authority—his hangers-on, and those from whom he had plagiarized the theory of relativity. Einstein's response appeared in the *Berliner Tageblatt* on pages 1 and 2 on 27 August 1920.

Nobel Prize laureate Philipp Lenard had had no involvement in the Berlin Philharmonic lectures. Even Einstein's friends condemned Einstein's flippant, inaccurate and racially-charged response. Sommerfeld wrote to Lenard and pleaded with Lenard to forgive Einstein, who had misrepresented Lenard's involvement in the event. Lenard must have been outraged that Sommerfeld should be the one to write to him, not Einstein, and Lenard must have been outraged that Einstein apologized not only through a proxy, but privately.

Nobel Prize laureate Philipp Lenard demanded a personal public apology from Albert Einstein to be attended with as much publicity as Einstein's (and Max von Laue's) cowardly and unscrupulous personal attacks against Lenard. Einstein's apology was not forthcoming.[302] *After* the Bad Nauheim debate,

---

Zeitschrift, Volume 11, (1910), pp. 287-294; **and** "Die Relexion des Lichtes an bewegten Spiegeln", *Physikalische Zeitschrift*, Volume 11, (1910), pp. 586-587; **and** "Zum Ehrenfestschen Paradoxon", *Physikalische Zeitschrift*, Volume 12, (1911), pp. 169-170; **and** "Интерпретација теорије релативности у геометрији Лобачевскоυа", *Glas, Srpska Kraljevska Akademija*, Volume 83, (1911), pp. 211-255; **and** *Glas, Srpska Kraljevska Akademija*, Volume 88, (1911); **and** "Über die nichteuklidische Interpretation der Relativitätstheorie", *Jahresbericht der Deutschen Mathematiker-Vereinigung*, Volume 21, (1912), pp. 103-127; **and** *Rad Jugoslavenska Akademija Znanosti i Umjetnosti*, (1914), p. 46; (1915), pp. 86, 101; (1916), p. 79; (1918), p. 1; (1919), p. 100.
300. *Cf.* J. Stachel, Ed., *The Collected Papers of Albert Einstein*, Volume 2, Princeton University Press, (1989), p. 110.
301. Letter from A. Einstein to H. Zangger of 24 December 1919, English translation by A. Hentschel, *The Collected Papers of Albert Einstein*, Volume 9, Document 233, Princeton University Press, (2004), pp. 197-198.
302. A. Kleinert and C. Schönbeck, "Lenard und Einstein. Ihr Briefwechsel und ihr Verhältnis vor der Nauheimer Diskussion von 1920", *Gesnerus*, Volume 35, Number 3/4, (1978), pp. 318-333.

where Lenard destroyed Einstein in a debate, Max Planck and Franz Himstedt stated to the press that Einstein had regretted including Lenard in his personal attack, because Lenard had not granted Weyland leave to place his name on the list of speakers at the Berlin Philharmonic lectures. The *Berliner Tageblatt* morning edition 25 September 1920 ran this story. This was obviously not an adequate apology for Einstein's vicious and deceitful smears.[303]

Einstein could not defend himself or his position other than to change the subject to a personal attack against his opponents. He pouted and whined like a spoiled brat in order to avoid the bulk of accusations made against him and the theory of relativity. Instead of arguing the issues, Einstein wanted to wait for others to speak on his behalf in defense of the theory. He was not competent to defend the theory himself. Einstein, who was himself a racist who believed that anti-Semitism was justified and proper and helpful to Jews, hypocritically tried to change the subject to race in order to attack his opponents as if racists. Albert Einstein wrote in the *Berliner Tageblatt*, Morgen Augabe, 27 August 1920, pp. 1-2:

"Meine Antwort
Ueber die anti-relativitätstheoretische G. m. b. H.
Von
Albert Einstein.

Unter dem anspruchsvollen Namen „Arbeitsgemeinschaft deutscher Naturforscher" hat sich eine bunte Gesellschaft zusammengetan, deren vorläufiger Daseinszweck es ist, die Relativitätstheorie und mich als deren Urheber in den Augen der Nichtphysiker herabzusetzen. Neulich haben die Herren Weyland und Gehrke in der Philharmonie einen ersten Vortrag in diesem Sinne gehalten, bei dem ich selber zugegen war. Ich bin mir sehr wohl des Umstandes bewußt, daß die beiden Sprecher einer Antwort aus meiner Feder unwürdig sind; denn ich habe guten Grund zu glauben, daß andere Motive als das Streben nach Wahrheit diesem Unternehmen zugrunde liegen. (Wäre ich Deutschnationaler mit oder ohne Hakenkreuz statt Jude von freiheitlicher, internationaler Gesinnung, so ...) Ich antworte nur deshalb, weil dies von wohlwollender Seite wiederholt gewünscht worden ist, damit meine Auffassung bekannt werde.

Zuerst bemerke ich, daß es heute meines Wissens kaum einen Forscher gibt, der in der theoretischen Physik etwas Erhebliches geleistet hat und nicht zugäbe, daß die ganze Relativitätstheorie in sich logisch aufgebaut und mit den bisher sicher ermittelten Erfahrungstatsachen im Einklang ist. Die bedeutendsten theoretischen Physiker — ich nenne H. A. Lorentz, M. Planck, Sommerfeld, Laue, Born, Larmor, Eddington, Debye, Langevin, Levi-Civita — stehen auf dem Boden der Theorie und haben meist wertvolle

---

[303]. D. K. Buchwald, *et al.* Editors, *The Collected Papers of Albert Einstein*, Volume 7, Princeton University Press, (2002), p. 110.

Beiträge zu derselben geleistet. Als ausgesprochenen Gegner der Relativitätstheorie wüßte ich unter den Physikern von internationaler Bedeutung nur Lenard zu nennen. Ich bewundere Lenard als Meister der Experimentalphysik; in der theoretischen Physik aber hat er noch nichts geleistet, und seine Einwände gegen die allgemeine Relativitätstheorie sind von solcher Oberflächlichkeit, daß ich es bis jetzt nicht für nötig erachtet habe, ausführlich auf dieselben zu antworten. Ich gedenke es nachzuholen.

Es wird mir vorgeworfen, daß ich für die Relativitätstheorie eine geschmacklose Reklame betreibe. Ich kann wohl sagen, daß ich zeitlebens ein Freund des wohlerwogenen, nüchternen Wortes und der knappen Darstellung gewesen bin. Vor hochtönenden Phrasen und Worten bekomme ich eine Gänsehaut, mögen sie von sonst etwas oder von Relativitätstheorie handeln. Ich habe mich oft lustig gemacht über Ergüsse, die nun zugeterletzt mir aufs Konto gesetzt werden. Uebrigens lasse ich den Herren von der G. m. b. H. gerne das Vergnügen.

Nun zu den Vorträgen. Herr Weyland, der gar kein Fachmann zu sein scheint (Arzt? Ingenieur? Politiker? Ich konnt's nicht erfahren), hat gar nichts Sachliches vorgebracht. Er erging sich in plumpen Grobheiten und niedrigen Anschuldigungen. Der zweite Redner, Herr Gehrke, hat teils direkte Unrichtigkeiten vorgebracht, teils hat er durch einseitige Auswahl des Materials und Entstellung beim unwissenden Laien einen falschen Eindruck hervorzurufen versucht. Folgende Beispiele mögen das zeigen:

Herr Gehrke behauptet, daß die Relativitätstheorie zum — Solipsismus führe, eine Behauptung, die jeder Kenner als Witz begrüßen wird. Er stützt sich dabei auf das bekannte Beispiel von den beiden Uhren (oder Zwillingen), deren eine in bezug auf das Inertialsystem eine Rundreise durchmacht, die andere nicht. Er behauptet — trotzdem ihm dies von den besten Kennern der Theorie schon oft mündlich und schriftlich widerlegt worden ist —, die Theorie führe in diesem Falle zu dem wirklich unsinnigen Resultat, daß von zwei nebeneinander ruhenden Uhren jede der anderen gegenüber nachgehe. Ich kann dies nur als einen Versuch absichtlicher Irreführung des Laienpublikums auffassen.

Herr Gehrke spielt ferner auf Herrn Lenards Einwände an, die viele auf Beispiele der Mechanik aus dem alltäglichen Leben beziehen. Diese sind schon hinfällig auf Grund meines allgemeinen Beweises, daß die Aussagen der allgemeinen Relativitätstheorie in erster Näherung mit denen der klassischen Mechanik übereinstimmen.

Was Herr Gehrke über die experimentelle Bestätigung der Theorie gesagt hat, ist mir aber der schlagendste Beweis dafür, daß es ihm nicht um die Enthüllung des wahren Sachverhalts zu tun war.

Herr Gehrke will glauben machen, daß die Perihelbewegung des Merkur auch ohne Relativitätstheorie zu erklären sei. Es gibt da zwei Möglichkeiten. Entweder man erfindet besondere interplanetare Massen, die so groß und so verteilt sind, daß sie eine Perihelbewegung von dem wahrgenommenen Betrage ergeben; dies ist natürlich ein höchst unbefriedigender Ausweg gegenüber dem von der Relativitätstheorie gegebenen, welche die Perihelbewegung des Merkur ohne irgendwelche besondere Annahme liefert. Oder aber man beruft sich auf eine Arbeit von Gerber, der die richtige Formel für die Perihelbewegung des Merkur bereits vor mir angegeben hat. Aber die Fachleute sind nicht nur darüber einig, daß Gerbers Ableitung durch und durch unrichtig ist, sondern die Formel ist als Konsequenz der von Gerber an die Spitze gestellten Annahmen überhaupt nicht zu gewinnen. Herrn Gerbers Arbeit ist daher völlig wertlos, ein mißglückter und irreparabler theoretischer Versuch. Ich konstatiere, daß die allgemeine Relativitätstheorie die erste wirkliche Erklärung für die Perihelbewegung des Merkur geliefert hat. Ich habe die Gerbersche Arbeit ursprünglich schon deshalb nicht erwähnt, weil ich sie nicht kannte, als ich meine Arbeit über die Perihelbewegung des Merkur schrieb; ich hätte aber auch keinen Anlaß gehabt, sie zu erwähnen, wenn ich von ihr Kenntnis gehabt hätte. Der diesbezügliche persönliche Angriff, welchen die Herren Gehrke und Lenard auf Grund dieses Umstandes gegen mich gerichtet haben, ist von den wirklichen Fachlauten allgemein als unfair betrachtet worden; ich hielt es bisher für unter meiner Würde, darüber ein Wort zu verlieren.

Herr Gehrke hat die Zuverlässigkeit der meisterhaft durchgeführten englischen Messungen über die Ablenkung der Lichtstrahlen an der Sonne in seinem Vortrage dadurch in einem schiefen Lichte erscheinen lassen, daß er von den drei unabhängigen Aufnahmegruppen nur eine erwähnte, welche infolge Verzerrung des Heliostatenspiegels fehlerhafte Resultate ergeben mußte. Er hat verschwiegen, daß die englischen Astronomen selbst in ihrem offiziellen Berichte ihre Ergebnisse als eine glänzende Bestätigung der allgemeinen Relativitätstheorie gedeutet haben.

Herr Gehrke hat bezüglich der Frage der Rotverschiebung die Spektrallinien verschwiegen, daß die bisherigen Bestimmungen noch einander widersprechen, und daß eine endgültige Entscheidung dieser Angelegenheit noch aussteht. Er hat nur die Zeugen gegen das Bestehen der von der Relativitätstheorie vorhergesagten Linienverschiebung angeführt, hat aber verschwiegen, daß durch die neuesten Untersuchungen von Grebe und Buchem und von Perot jene früheren Ergebnisse ihre Beweiskraft eingebüßt haben.

Endlich bemerke ich, daß auf meine Anregung hin in Neuheim auf der Naturforscherversammlung eine Diskussion über die Relativitätstheorie veranstaltet wird. Da kann jeder, der sich vor ein wissenschaftliches Forum wagen darf, seine Einwände vorbringen.

Es wird im Auslande, besonders auf meine holländischen und englischen Fachgenossen H. A. Lorentz und Eddington, die sich beide eingehend mit Relativitätstheorie beschäftigt und darüber wiederholt gelesen haben, einen sonderbaren Eindruck machen, wenn sie sehen, daß die Theorie sowie deren Urheber in Deutschland selbst derart verunglimpft wird."

Einstein knew that he had been very publicly exposed as a fraud. He decided to flee Germany. It was obvious to him that all of German science would stand against him for what he had done. Pro-Einstein newspapers came to his rescue and published alarmist nonsense and personal attacks by Einstein's friends. It came as a surprise to Einstein that Laue, Nernst and Rubens would campaign by personal attack in the newspapers to rescue Einstein's reputation.[304]

It was only reluctantly that Einstein then chose to put up any kind of a fight with his undignified rant in the *Berliner Tageblatt*. If his friends had not rescued him, Einstein would have left Germany in total defeat without having spoken a word in his defense. The *Berliner Tageblatt* reported on 27 August 1920, parroting (as opposed to mocking) the nationalistic tone von Laue and Einstein had condemned as "anti-Semitic", and cried out that the sky was falling, and spoke of Einstein as if of a god,

> "Albert Einstein will Berlin verlassen! Die persönlichen Angriffe, die gegen Dr. Albert Einstein in der an dieser Stelle bereits gekennzeichneten Versammlung der „Arbeitsgemeinschaft deutscher Naturforscher" vorgebracht wurden, haben einen Erfolg gehabt, der für Berlin tief Beschämend ist: Albert Einstein, angewidert von den altdeutschen Anrempelungen und den pseudowissenschaftlichen Methoden seiner Gegner will der Reichshauptstadt den Rücken kehren. So also steht es im Jahre 1920 um die geistige Kultur Berlins! Ein deutscher Gelehrter von Weltruf, den die Holländer als Ehrenprofessor nach Leiden berufen, dem die amerikanische Columbia-Universität die Große goldene Medaille verleiht, den schwedische und norwegische Gesellschaften zu ihrem Ehrenmitglied ernennen, dessen Werk über die Relativitätstheorie als eines der ersten deutschen Bücher nach dem Kriege in englischer Sprache erscheint: ein solcher Mann wird aus der Stadt, die sich für das Zentrum deutscher Geistesbildung hält, herausgeekelt. Eine Schande!
>
> Wir können es noch nicht glauben, daß in dieser Angelegenheit, die nicht nur für die Welt der Wissenschaft von Bedeutung ist, das letzte Wort gesprochen sein soll. Die Berliner Universität hat die Pflicht, alles zu tun, um diesen hervorragenden Lehrer und Gelehrten sich und Berlin zu erhalten. Und Albert Einstein, der über niedrigen Anwürfen steht, wird hoffentlich nach ruhigerer Ueberlegung seinen Feinden nicht den Gefallen erweisen, vor ihrem sinnlosen Geschrei den Platz zu räumen. Wer die Ehre deutscher Wissenschaft auch in Zukunft hochhalten will, muß jetzt zu diesem Manne stehen."

The report in the *Berliner Tageblatt*, adopting and improving upon Lenard's tactics, sought to make it appear unpatriotic for Germans to enter into a scientific dispute with Einstein—the archangel of Berlin. Einstein had called the *Berliner Tageblatt* a hypocritical newspaper in the context of Socialism.[305] The *Berliner*

---

[304]. A. Hermann, *Briefwechsel. Sechzig Briefe aus dem goldenen Zeitalter der modernen Physik*, Schwabe & Co., Basel, Stuttgart, (1968), p. 65.
[305]. Letter from A. Einstein to A. Stodola of 31 March 1919, *The Collected Papers of*

Tageblatt turned Einstein's cowardly flight from the exposure of his plagiarism, the self-contradictions in relativity theory, and the uncertain evidence used to promote the man and his theory, into the crucifixion of the Messiah by a cabal of ungodly anti-German nationalistic Germans. More effective—more boldly dishonest—propaganda than that used to promote and sell Einstein to the public is hard to find.

Einstein had made his *ad hominem* attacks against the Berlin Philharmonic gathering with the cooperation of some members of the international press not only in an effort to smear his outspoken critics, but also to threaten anyone who dared side with them. The press orchestrated an overwhelming international defamation against Einstein's critics.

Einstein believed the majority of physicists sided with Lenard and Gehrcke and sought to suppress any public sympathy for their position. After the terrible hype of the 1919 eclipse observations, the press used Einstein and Einstein used the press. Einstein wrote to Sommerfeld in this context,

"It is a bad thing that every utterance of mine is made use of by journalists as a matter of business."[306]

Ad hominem attack and smear campaigns were Einstein's and his followers' preferred method of response to challenges to Einstein's priority and to relativity theory, as even Einstein's advocates were forced to concede in 1931,

"Even individual fanatic scientific advocates of the Einsteinian theory seem to have finally abandoned their tactic of cutting off any discussion about it with the threat that every criticism, even the most moderate and scrupulous ones, must be discredited as an obvious effluence of stupidity and malice. But even if these monstrous products of the 'Einstein frenzy' [*Einstein-Taumel*] now belong to history and are thus eliminated from consideration, thoroughly respectable reasons for a certain discomfort with relativity theory still do remain[.]"[307]

This was a response to the charge of such *ad hominem* attacks made in *Hundert Autoren gegen Einstein* (*100 Authors Against Einstein*),

"It is the aim of this publication to confront the terror of the Einsteinians with an overview of the quality and quantity of the opponents and opposing

---

Albert Einstein, Volume 9, Document 16, Princeton University Press, (2004).
306. A. Einstein quoted in R. W. Clark, *Einstein: The Life and Times*, The World Publishing Company, (1971), p. 261; referencing A. Einstein to A. Sommerfeld, in A. Hermann. *Briefwechsel. 60 Briefe aus dem goldenen Zeitalter der modernen Physik*, Schwabe & Co., Basel, Stuttgart, (1968), p. 69.
307. A. v. Brunn, quoted in: K. Hentschel, Ed., A. Hentschel, Ed. Ass. and Trans., *Physics and National Socialism: An Anthology of Primary Sources*, Birkhäuser, Basel, Boston, Berlin, (1996), p. 11.

arguments."[308]

Ernst Gehrcke decided to fight propaganda with thoroughly documented fact, but initially came up on the losing side. Einstein's persona, as depicted in the corrupt press, was perhaps too endearing to be successfully countered by the facts. The press also largely made it impossible for Einstein's critics to argue their side to the public. Einstein often opted to hide from criticism, as even his advocates were forced to admit,

"Although Einstein himself, by nature a pure scientist, is uninterested in such academic disputes!"[309]

After decades of misrepresentations which promote Einstein as if he were an angelic figure, it is necessary to show that he was not only capable of plagiarism, but that we know for a fact that he committed far worse moral offenses—Einstein's plagiarism is among the least of his many sins. It is also helpful to know Einstein's habits. Einstein clearly plagiarized the special theory of relativity, as well as many important aspects of the general theory of relativity, from Jules Henri Poincaré and Hendrik Antoon Lorentz. In fact, Einstein evinced a career-long pattern of plagiarism and was often accused of appropriating the work of others. He tried to avoid these accusations and never refuted them.[310]

---

308. From the preface of *Hundert Autoren gegen Einstein* translated by: H. Goenner, "The Reaction to Relativity Theory in Germany, III: 'A Hundred Authors against Einstein'", J. Earman, M. Janssen, J. D. Norton, Eds., *The Attraction of Gravitation: New Studies in the History of General Relativity*, Birkhäuser, Boston, Basel, Berlin, (1993), p. 251.
309. A. v. Brunn, quoted in: K. Hentschel, Ed., A. Hentschel, Ed. Ass. and Trans., *Physics and National Socialism: An Anthology of Primary Sources*, Birkhäuser, Basel, Boston, Berlin, (1996), p. 14.
310. C. J. Bjerknes, *Albert Einstein: The Incorrigible Plagiarist*, XTX Inc., Downer Grove, Illinois, (2002). *See also:* P. Langevin, "Le Physicien", *Revue de Métaphysique et de Morale*, Volume 20, Number 5, (September, 1913), pp. 675-718. *See also:* H. A. Lorentz, "Deux mémoires de Henri Poincaré sur la physique mathématique", *Acta Mathematica*, Volume 38, (1921), pp. 293-308; reprinted in *Œuvres de Henri Poincaré*, Volume 9, Gautier-Villars, Paris, (1954), pp. 683-695; and Volume 11, (1956), pp. 247-261. *See also:* W. Pauli, "Relativitätstheorie", *Encyklopädie der mathematischen Wissenschaften mit Einschluss ihrer Anwendungen*, Volume 5, Part 2, Chapter 19, B. G. Teubner, Leipzig, (1921), pp. 539-775; English translation by G. Field, *Theory of Relativity*, Pergamon Press, London, Edinburgh, New York, Toronto, Sydney, Paris, Braunschweig, (1958). *See also:* H. Thirring, "Elektrodynamik bewegter Körper und spezielle Relativitätstheorie", *Handbuch der Physik*, Volume 12 ("Theorien der Elektrizität Elektrostatik"), Springer, Berlin, (1927), pp. 245-348, *especially* 264, 270, 275, 283. *See also:* S. Guggenheimer, *The Einstein Theory Explained and Analyzed*, Macmillan, New York, (1929). *See also:* J. Mackaye, *The Dynamic Universe*, Charles Scribner's Sons, New York, (1931). *See also:* J. Le Roux, "Le Problème de la Relativité d'Après les Idées de Poincaré", *Bulletin de la Société Scientifique de Bretagne*, Volume 14, (1937), pp. 3-10. *See also:* Sir Edmund Whittaker, *A History of the Theories of Aether and Electricity*, Volume II, Philosophical Library Inc., New York, (1954), *especially* pp.

For example, Einstein wrote to Willy Wien in 1916 when Ernst Gehrcke[311]

27-77; and "Albert Einstein", *Biographical Memoirs of Fellows of the Royal Society*, Volume 1, (1955), pp. 37-67. *See also:* G. H. Keswani, "Origin and Concept of Relativity, Parts I, II & III", *The British Journal for the Philosophy of Science*, Volume 15, Number 60, (February, 1965), pp. 286-306; Volume 16, Number 61, (May, 1965), pp.19-32; Volume 16, Number 64, (February, 1966), pp. 273-294; and Volume 17, Number 2, (August, 1966), pp. 149- 152; Volume 17, Number 3, (November, 1966), pp. 234-236. *See also:* G. H. Keswani and C. W. Kilmister, "Intimations of Relativity before Einstein", *The British Journal for the Philosophy of Science*, Volume 34, Number 4, (December, 1983), pp. 343-354. *See also:* G. B. Brown, "What is Wrong with Relativity?", *Bulletin of the Institute of Physics and the Physical Society*, Volume 18, Number 3, (March, 1967), pp. 71-77. *See also:* C. Cuvaj, "Henri Poincaré's Mathematical Contributions to Relativity and the Poincaré Stresses", *American Journal of Physics*, Volume 36, (1968), pp. 1109-1111. *See also:* C. Giannoni, "Einstein and the Lorentz-Poincaré Theory of Relativity", *PSA: Proceedings of the Biennial Meeting of the Philosophy of Science Association*, Volume 1970, (1970), pp. 575-589. JSTOR link: <http://links.jstor.org/sici?sici=0270-8647%281970%291970%3C575%3AEATLTO%3E2.0.CO%3B2-Z>
See also: J. Mehra, *Einstein, Hilbert, and the Theory of Gravitation*, Reidel, Dordrecht, Netherlands, (1974). *See also:* W. Kantor, *Relativistic Propagation of Light*, Coronado Press, Lawrence, Kansas, (1976). *See also:* R. McCormmach, "Editor's Forward", *Historical Studies in the Physical Sciences*, Volume 7, (1976), pp. xi-xxxv. *See also:* H. Ives, D. Turner, J. J. Callahan, R. Hazelett, *The Einstein Myth and the Ives Papers*, Devin-Adair Co., Old Greenwich, Connecticut, (1979). *See also:* J. Leveugle, "Henri Poincaré et la Relativité", *La Jaune et la Rouge*, Volume 494, (April, 1994), pp. 29-51; **and** *La Relativité, Poincaré et Einstein, Planck, Hilbert: Histoire véridique de la Théorie de la Relativité*, L'Harmattan, Paris, (2004). *See also:* A. A. Logunov, *On the Articles by Henri Poincaré ON THE DYNAMICS OF THE ELECTRON*, Publishing Department of the Joint Institute for Nuclear Research, Dubna, (1995); and *The Theory of Gravity*, Nauka, Moscow, (2001); **and** Анри Пуанкаре и ТЕОРИЯ ОТНОСИТЕЛЬНОСТИ, Наука, Москва, (2004). An English translation of this book will soon appear as: *Henri Poincaré and the Theory of Relativity*. *See also:* E. Gianetto, "The Rise of Special Relativity: Henri Poincaré's Works before Einstein", *ATTI DEL XVIII CONGRESSO DI STORIA DELLA FISICA E DELL'ASTRONOMICA*, pp. 172-207; URL: <http://www.brera.unimi.it/Atti-Como-98/Giannetto.pdf>
See also: S. G. Bernatosian, *Vorovstvo i obman v nauke*, Erudit, St. Petersburg, (1998), ISBN: 5749800059. *See also:* U. Bartocci, *Albert Einstein e Olinto De Pretto: La vera storia della formula piu famosa del mondo*, Societa Editrice Andromeda, Bologna, (1999). *See also:* Jean-Paul Auffray, *Einstein et Poincaré: sur les Traces de la Relativité*, Le Pommier, Paris, (1999). Y. Brovko, "Einshteinianstvo—agenturnaya set mirovovo kapitala", Molodaia Gvardiia, Number 8, (1995), pp. 66-74, at 70. Юрий Бровко, "Эйнштейнианство — агентурная сеть Мирового капитала", Молодая гвардия, № 8, (1995), cc. 66-74; **and** Y. Brovko, "Razgrom einshteinianstvo", Priroda i Chelovek. Svet, Number 7, (2002), pp. 8-10. Юрий Бровко, "Разгром эйнштейнианства", Природа и Человек. Свет, № 7, (2002), cc. 8-10. URL: <http://medograd.narod.ru/einstein.html>
311. E. Gehrcke, "Zur Kritik und Geschichte der neueren Gravitationstheorien", *Annalen der Physik*, Volume 51, (1916), pp. 119-124; reprinted *Kritik der Relativitätstheorie*, Hermann Meusser, Berlin, (1924), pp.40-44.

effectively accused Einstein of plagiarizing Paul Gerber's formula for the perihelion motion of Mercury,

"[...]I am not going to respond to Gehrcke's tasteless and superficial attacks, because any informed reader can do this himself."[312]

It was clear that Einstein had an ethical obligation to acknowledge Gerber's priority. Einstein's close friends Friedrich Adler and Michele Besso wrote to him and pointed out that Einstein had repeated Gerber's formula.[313] It was terribly unfair, unethical and unprofessional of Einstein to respond to Gehrcke in the manner in which he did. Einstein had an ethical obligation to acknowledge Gerber's priority and explain why he had repeated his formula without an attribution. Einstein instead ridiculed Gerber and Gehrcke and asserted that he had no obligation to cite Gerber's work.

In another instance where Einstein took the coward's way out, a meeting was arranged to discuss Hans Vaihinger's[314] theory of fictions in 1920. Einstein pledged that he would attend this meeting. Knowing that Einstein would be devoured in a debate over his mathematical fictions, which confused induction with deduction, Wertheimer and Ehrenfest helped Einstein to fabricate an excuse to miss the meeting he had agreed to attend. Einstein was proven a liar.[315] He also hid from many other criticisms, and Einstein refused to answer T. J. J. See's many charges of plagiarism,[316] and refused to debate Arvid Reuterdahl or to answer his many charges of plagiarism.[317] When Robert Drill[318] criticized the

---

312. Letter from A. Einstein to W. Wien of 17 October 1916, translated by A. M. Hentschel, *The Collected Papers of Albert Einstein*, Volume 8, Document 267, Princeton University Press, (1998), p. 255.
313. M. Besso letter to Einstein of 5 December 1916, translated by A. M. Hentschel, *The Collected Papers of Albert Einstein*, Volume 8, Document 283, Princeton University Press, (1998), p. 271. F. Adler letter to Einstein of 23 March 1917, translated by A. M. Hentschel, *The Collected Papers of Albert Einstein*, Volume 8, Document 316, Princeton University Press, (1998), p. 308.
314. H. Vaihinger, *Die Philosophie des Als Ob, System der theoretischen, praktischen und religiosen Fiktionen der Menschheit auf Grund eines idealistichen Positivismus. Mit einem Anhang über Kant und Nietzsche*, Reuther & Reichard, Berlin, (1911); English translation by C. K. Ogden, *The Philosophy of 'As If'*, Harcourt, Brace & Company, Inc., New York, (1925); reprinted Routledge & K. Paul, London, (1965). **See also:** C. K. Ogden, *Bentham's Theory of Fictions*, K. Paul, Trench, Trubner & Co. Ltd., (1932).
315. H. Goenner, "The Reaction to Relativity Theory. I: The Anti-Einstein Campaign in Germany in 1920", *Science in Context*, Volume 6, Number 1, (1993), pp. 107-133, at 111.
316. "Einstein Ignores Capt. See", *The New York Times*, (18 October 1924), p. 17.
317. "Challenges Prof. Einstein: St. Paul Professor Asserts Relativity Theory Was Advanced in 1866", *The New York Times*, (10 April 1921), p. 21. **See also:** "Einstein Charged with Plagiarism", *New York American*, (11 April 1921). **See also:** "Einstein Refuses to Debate Theory", *New York American*, (12 April 1921).
318. R. Drill, "Die Kultur der Haeckel-Zeit", *Frankfurter Zeitung*, (18 August 1919); **and** "Nachwort", *Frankfurter Zeitung*, (2 September 1919); **and** "Ordnung und Chaos. Ein Beitrag zum Gesetz von der Erhaltung der Kraft. I-II", *Frankfurter Zeitung*, (30 November 1919 / 2 December 1919).

theory of relativity, Einstein tried to persuade Max Born and Moritz Schlick to not respond to the critique, but if they did so, to hide from his arguments and merely ridicule Drill with insults.[319] Einstein hid from the French Academy of Sciences.[320] Einstein hid from Cardinal O'Connell.[321] Einstein hid from Cartmel.[322] Einstein hid from Dayton C. Miller's falsification of the special theory of relativity.[323] Miller challenged Einstein in the press over the course of many years. *The New York Times Index* lists several articles in which Miller's and William B. Cartmels' falsifications of the special theory of relativity are discussed.[324] Einstein and Lorentz were very worried by Miller's results and could not find fault with them.[325] Einstein told R. S. Shankland not to perform an experiment which might falsify the special theory of relativity,

> "[Einstein] again said that more experiments were not necessary, and results such as Synge might find would be 'irrelevant.' [Einstein] told me not to do any experiments of this kind."[326]

Einstein knew he was caught at the Arbeitsgemeinschaft deutscher Naturforscher meeting in the Berlin Philharmonic, and wanted to run away from Germany. Einstein desired to hide from the Bad Nauheim debate, in which he had threatened to devour his opponents,[327] then Einstein—after being talked into

---

[319]. *The Collected Papers of Albert Einstein*, Volume 9, Documents 198, 199 and 222, Princeton University Press, (2004).

[320]. *The New York Times*, (4 April 1922), p. 21.

[321]. "Cardinal Doubts Einstein", *The New York Times*, (8 April 1929), p. 4. *See also:* "Einstein Ignores Cardinal", *The New York Times*, (9 April 1929), p. 10. *See also:* "Cardinal Opposes Einstein", *The Chicago Daily Tribune*, (8 April 1929), p. 33. *See also:* "Cardinal Hits at Einstein Theory", *The Minneapolis Journal*, (8 April 1929). *See also:* "Cardinal Gives Further Views on Einstein", *Boston Evening American*, (12 April 1929). *See also:* "Cardinal Warns Against Destructive Theories", *The Pilot* [Roman Catholic Newspaper, Boston], (13 April 1929), pp. 1-2. *See also:* "Vatican Paper Praises Critic of Dr. Einstein", *The Minneapolis Morning Journal*, (24 May 1929).

[322]. *The New York Times*, (24 February 1936), p. 7. *See also:* "Calls Ether Reality; Differs with Einstein; Proof is Submitted", *The Chicago Tribune*, (23 February 1936).

[323]. M. Polanyi, *Personal Knowledge*, University of Chicago Press, (1958), p. 13. *See also:* A. Pais, *Subtle is the Lord*, Oxford University Press, (1982), pp. 113-114. *See also:* W. Broad and N. Wade, *Betrayers of the Truth: Fraud and Deceit in the Halls of Science*, Simon & Schuster, New York, (1982), p. 139.

[324]. *See also:* "Einstein Theory will be Refuted by an American", *The Chicago Tribune*, (24 October 1929), p. 18. *See also:* "Calls Ether Reality; Differs with Einstein; Proof is Submitted", *The Chicago Tribune*, (23 February 1936).

[325]. R. S. Shankland, "Conversations with Albert Einstein", *American Journal of Physics*, Volume 31, Number 1, (January, 1963), pp. 47-57; **and** "Conversations with Albert Einstein. II", *American Journal of Physics*, Volume 41, Number 7, (July, 1973), pp. 895-901.

[326]. R. S. Shankland, "Conversations with Albert Einstein", *American Journal of Physics*, Volume 31, Number 1, (January, 1963), pp. 47-57, at 54.

[327]. A. Einstein quoted in R. W. Clark, *Einstein: The Life and Times*, The World

appearing and after much hype promoting the event which attracted thousand of visitors—then Einstein, when losing the debate, ran away during the lunch break and again wanted to run away from Germany.[328]

Einstein prospered from hype. Einstein never exhibited his legendary genius in public. Instead, Einstein either appeared like a childish madman in public, or rattled off a script he had been told to recite. The press rescued him again and again, while he and they hid from, and suppressed, legitimate criticism. Einstein was unable to defend "his" theories.

3.5.4 Jewish Hypocrisy and Double Standards

Einstein's plagiarism became an international scandal in the early 1920's and many newspapers owned and/or edited by Jews avoided the legitimate criticisms leveled at Einstein and instead resorted to *ad hominem* attacks against his critics calling anyone who dared speak a word against Einstein *ipso facto* an anti-Semite. The intolerance of criticism in the "Jewish liberal press" had long struck many in Germany as hypocritical. During the *Kulturkampf* (the struggle between Catholics and Protestants in the German Empire in the Nineteenth Century) elements of the "Jewish liberal press" in Vienna and in Berlin relentlessly attacked Catholicism, Catholics and the Gospels, but were intolerant of any criticism directed at them, or Judaism. Ernst Lieber, while defending Jews against discriminatory legislation, stated to the Reichstag in 1895,

> "Those of us in particular who bore the brunt of the *Kulturkampf* will never forget how viciously and brutally Jewish pens attacked, dragged into the mud, reviled, ridiculed and insulted all that is sacred to us and that we were called on to defend so strenuously and painstakingly."[329]

Adolf Stoecker brought attention to this fact in an attempt to justify his call for discriminatory legislation against Jews in 1879. Stoecker stated,

> "It is strange indeed that the Jewish liberal press does not have the courage to answer the charges of its attackers. Usually it invents a scandal, even if there is none. It sharpens its poisonous pen by writing about the sermons in our churches and the discussions in our church meetings; but it hushes up the Jewish question and does everything to prevent its readers from hearing even a whisper from these unpleasant voices. It pretends to despise its

---

Publishing Company, (1971), p. 261; referencing A. Einstein to A. Sommerfeld, in A. Hermann. *Briefwechsel. 60 Briefe aus dem goldenen Zeitalter der modernen Physik*, Schwabe & Co., Basel, Stuttgart, (1968), p. 69.
328. A. Einstein, *Neues Wiener Journal*, (29 September 1920). C. Kirsten and H. J. Treder, *Albert Einstein in Berlin 1913-1933*, Akademie Verlag, Berlin, Volume 2, (1979), pp. 139, 205.
329. Quoted in: P. W. Massing, *Rehearsal for Destruction: A Study of Political Anti-Semitism in Imperial Germany*, Howard Fertig, New York, (1967), p. 296.

enemies and to consider them unworthy of an answer. It would be better to learn from the enemy, to recognize one's own defects, and work together toward the social reconciliation which we need so badly. It is in this light that I intend to deal with the Jewish question, in the spirit of Christian love, but also with complete social truthfulness. [***] People who are in the habit of pouring out the most biting criticism of State and Church, men and events, become highly incensed when anyone takes the liberty of directing even so much as a searching glance at Jewry. They themselves hatefully and sneeringly assail any non-Jewish endeavor. But as soon as a mild word of truth is uttered about them and their doings, they put on an act of injured innocence, of outraged tolerance, of being the martyrs of world history. Nevertheless I shall dare to speak up openly and candidly about modern Jewry tonight. And I am quite prepared for the distorted reports that will come back."[330]

Wilhelm Marr also alleged in 1879,
"The *Kulturkampf* breaks out. Since 1848, if we Germans so much as criticized any little thing Jewish, it was enough to have us entirely outlawed from the press. Jewry, on the other hand, not only mixes in our religious controversies and in the *Kulturkampf* against Ultramontanism but has the most to say about it in our press. In their humor magazines, which are anxiously on the lookout for anything that can be satirized as 'Jew baiting,' they pour boiling oil on Ultramontanism. Why, of course. Ultramontanism was Jewry's competitor for world hegemony! While a sense of delicacy is wholly absent among the Jews, it is demanded of us that we handle them like fine glassware or extremely sensitive plants.

Indeed, there were great newspapers in which we Germans could not even get a hearing. Why not? Because in order to criticize Romish fanaticism, it would have been necessary to show that it was the outcome of Old Testament, Jehovah fanaticism. Even the Ultramontanes suppressed hostile representations from their newspapers as soon as Israel was even lightly grazed!!

Just once try to comment upon Jewish rituals and observances. You will see that no pope is more infallible and unassailable than these doctrines. You would be accused of religious hatred. But when Jews hold forth and have the final say on our church-state matters, that is something quite different! While we embroil ourselves in church-state conflicts, Jewry shouts 'Vae Victis! Woe unto the vanquished!'

I and several of my friends tried, at the outbreak of the *Kulturkampf*, to participate and contribute from a higher cultural and historical point of view. But in vain. We were only permitted to speak without theoretical premises or when, out of the blue, we wished to disparage the clericals. None of our

---

[330]. Quoted in: P. W. Massing, *Rehearsal for Destruction: A Study of Political Anti-Semitism in Imperial Germany*, Howard Fertig, New York, (1967), pp. 278-279; *see also:* pp. 304, 314-315.

letters to the editor were ever printed in the Jewish press. Thus has Jewry monopolized the free expression of opinion in the daily press."[331]

Hermann Bielohlawek expressed his outrage at the defamations issuing from the "Viennese Jewish press beasts" against Mayor Karl Lueger, and the "muzzling and terrorism" of the Social Democrats who prevented fair and open debate, before the Vienna City Council in 1902.[332] Long before Stoecker and Marr, Bruno Bauer argued that "the Jews" hypocritically insulated themselves from criticism, while attacking Christians.[333]

The Zionist Jews believed they had to destroy the Catholic Church and the Turkish Empire in order to steal Palestine from its indigenous population. Protestantism and financial exploitation were two of the methods these Jews used to attack the Turks and the Catholics. Moses Hess wrote in his *Rome and Jerusalem*,

> "My friend, Armond L., who traveled for several years through the Danube Principalities, told me that the Jews were moved to tears when he announced to them the end of their suffering, with the words 'The time of the return approaches.' The more fortunate Occidental Jews do not know with what longing the Jewish masses of the East await the final redemption from the two thousand year exile. They know not that the patriotic Jew cannot suppress his cry of anguish at the length of the exile, even in the midst of his festive songs, as, for instance, the patriotic poem which is read on Chanukah, closes with the mournful call:
>
> 'For salvation is delayed for us and there is no end to the days of evil.'
>
> 'They asked me,' continued my friend, 'what are the indications that the end of the exile is approaching?' 'These,' I answered, 'that the Turkish and the papal powers are on the point of collapse.'"[334]

Those elements in the Jewish press of Vienna and Berlin who participated in the *Kulturkampf*, and relentlessly and viciously attacked the Catholics, ultimately incurred the wrath of both Catholic and Protestant, both Frenchman and German, and provoked much of the anti-Semitism that manifested itself the Dreyfus Affair, where Jews were seen as agents of German Protestants attacking the French Catholics, or that German Protestants were the dupes of those Jews

---

331. W. Marr, *Der Sieg des Judenthums über das Germanenthum*, Rudolph Costenoble, Bern, (1879); English translation in: R. S. Levy, *Antisemitism in the Modern World: An Anthology of Texts*, D. C. Heath and Company, Toronto, (1991), pp. 76-93, at 85-86.
332. H. Bielohlawek, "Yes, We Want to Annihilate the Jews!" in R. S. Levy, *Antisemitism in the Modern World: An Anthology of Texts*, D. C. Heath and Company, Toronto, (1991), pp. 115-120.
333. P. L. Rose, *Revolutionary Antisemitism in Germany from Kant to Wagner*, Princeton University Press, (1990), p. 267.
334. M. Hess, *Rome and Jerusalem: A Study in Jewish Nationalism*, Bloch Publishing Company, New York, (1918/1943), pp. 150-151.

out to destroy Catholicism. French Catholics had been under attack since the early days of the French Revolution—French Catholics gave the Pope the majority of his funding and the Revolution sought to destroy French Catholicism and with it all Catholicism. In the 1893, Anatole Leroy-Beaulieu wrote,

> "CHRISTIANS who belong to the educated classes do not share the antiquated popular prejudices against the Jew. Even in Eastern Europe, in Hungary, Roumania, and Russia, the thin stratum of the cultured, 'the intelligent,' as the Russians call them, are well aware that the Jew does not steal children to give them up to the knife of the *schochet* and that the Synagogue needs no Christian blood to celebrate the Hebrew Passover. The Catholics, Protestants, and members of the Greek church have another grievance against the Jews, a less crude and childish one. They accuse them of being the born enemy of what they style ' Christian civilisation.' The very vagueness of this charge makes it one of the most serious brought against Israel.
>
> If it be not true that, in his secret rites, the talmudic Jew takes delight in spilling Christian blood, the Jews, it is asserted, especially the progressive Jews, do what is still worse: they are bent upon destroying Christian faith, morals, and civilisation. Not satisfied with the toleration accorded to them, they endeavour, openly or secretly, to 'de-christianise' Europe and modern society. Thus considered, Judaism is a disintegrating force, both from the moral and the religious, as well as from the economic and the national, point of view; it is a solvent of our old Christian institutions.
>
> In Evangelical Germany, in Orthodox Russia, in Catholic France and Austria, the Jew is denounced as the most zealous destroyer of what one is pleased to call the Christian state and Christian civilisation. In assailing the Jews and Judaism, Christians of every sect assert, with Pastor Stoecker, that their attack on the Jew is only an act of self-defence. There are men who strive to find hidden springs in every historical event, who believe in prolonged designs, mysteriously followed up through centuries; such persons go so far as to look upon the 'princes of Judah' as the eternal instigators of the secular war waged against Christ, the Church, and the Christian spirit. [*Footnote:* Thus, for instance, *Les Juifs nos Maîtres*, by Chabaudy, Paris, 1882.] To them the ancient, chosen people, having rebelled against the Messiah, has become the enemy of the city of God, the foundations of which it is noiselessly sapping, and on the ruins of which it hopes to establish the site of Israel's dominion. The Jews are the originators and the apostles of the great 'Anticrusade' waged in our times against Christian traditions and institutions. In this sense, Antisemitism is, after a fashion, the counterpart of Anticlericalism; it is a second *Kulturkampf*, a *Kulturkampf* that has recoiled against the secret or avowed enemies of Christian civilisation.
>
> Here we have, indeed, one of the real causes of the Antisemitic movement. It may be recognised by the country and the period in which it first appeared. The fact that it originated in the Germany of Bismarck, in the

very heart of the struggle between the new Empire and the Catholic hierarchy, is not due to mere chance. Whilst the liberal German press, partly led by the Jews, was assailing the Church, the besieged party, trying to find the weak spots in the lines of attack, made a sally in the direction of the Synagogue, where the troops commanded by the Jew Lasker were encamped. That was good strategy. Such a digression had been suggested by the composition of the opposing armies. In fact, it is in a fair way of coming to be considered as one of the classical manœuvres of modern clerical campaigns. The Jew, who was apparently to have been the gainer, thus runs the risk of being the victim in the warfare against Christianity. This incident proves that he does not invariably play a safe game when he incites, or takes part in, religious struggles. Imprudent being! He will get nothing but blows for his pains. The shafts hurled by him, or by his people, against the Clericals, are in danger of rebounding against Israel. It is an unfortunate situation for the Jew when the question is put whose eyes can be offended by the harmless shadow of the Cross, whose hands are interested in effacing from our old countries the noble and precious emblems of the religion of our fathers?

'Why,' said a Silesian German to me, 'should you try to prevent us from returning to the Talmud the blows aimed at the Gospel? When an appeal is made to the state against our clergy and our Christian associations, have we not a right to appeal in our turn to the state and the people against the rabbis and the Jewish associations? Let the toleration which the Jews claim for themselves, who are in the minority, be shown to us, who are in the majority. Otherwise they will again have to listen to the cries of 'Hep! hep!' [*Footnote: Hep! Hep!* the traditional cry against the Jews in Germany. Many explications, almost all imaginary, have been given of it. Some have found in it the initials of the three words: *Hierusalem est perdita*.[335] It is, perhaps, according to the hypothesis of Isidore Loeb, nothing more than a corruption of the word *Hebe! heb!* 'Stop! hold him !' still used, in this sense, in Alsace and the Rhenish lands.] from millions of Christians who persist in believing that the best gifts they can make to their children are the New Testament and the Crucifix.' And such language is used not only by believers; I have heard it from the lips of sceptical or indifferent people, who, in the presence of a Jew, all of a sudden remembered that they were Christians.

Anticlericalism has thus been, by the revulsion it has occasioned, one of the main abettors of Antisemitism. In more than one country its effects have been felt by the Jews even more keenly than by the Catholics. To those who denounced the Church as a foreign body, obedient to a foreign master, the Catholics were naturally led to reply with a denunciation of the Jews as intruders of an alien race, without country, or love of country. To those who in Germany, for instance, accused the spiritual subjects of the pope of being

---

335. *Cf.* "The Modern Jews", *The North American Review*, Volume 60, Number 127, (April, 1845), pp. 329-368, at 347.

thorough-going Ultramontanes, rebellious to the Teutonic spirit, the Catholics were, of course, ready to retaliate with an attack on the Semites, as persons obstinately set against the German spirit and *deutsche Kultur*. 'Make front against Rome,' was said one day, in 1879, in the thick of the *Kulturkampf*, by one of the Berlin journals, managed or edited by Jews. This war-cry was answered by another from the *Germania*, the organ of the *Ultramontane Centre:* 'Make front against New Jerusalem.' Thus, from time immemorial, has intolerance bred intolerance: *abyssus, abyssum*.

'The eyes of the German nation are opened at last,' continued the *Germania;* 'it sees that the struggle for civilisation is the struggle against the ascendancy of the Jewish spirit and of Jewish wealth. In every political movement, it is the Jew who plays the most radical and revolutionary part, waging war to the death against all that has remained legitimate, historical and Christian in national life.' [*Footnote: Germania*, September 10, 1879. In Germany and in Austria this has become the habitual theme of a number of newspapers. *Cf.*, in our country, *La France Juive*, of M. Drumont.]

And this awful charge against Israel was not advanced only by the Catholics, who had to face Prince Bismarck and his short-sighted allies, the national Liberals; Protestant Germany echoed the words of Catholic Germany. The Russian priests, uneasy at seeing that the missiles aimed at the Roman hierarchy, flew higher than the mitres of their bishops and reached the Gospel and the Cross, were themselves perhaps the most ardent preachers of the new crusade. [*Footnote:* I could cite, as an example, the speech of Pastor Stoecker in the *Landtag*, March 22, 1880. *Cf.* the writings of Professor von Treitschke.] The *Kreuz-Zeitung* exceeded the *Germania* in zeal; and, outside of Germany, in states where such a movement seemed out of place, Russian writers took it up, in their turn. The *Rous*, edited by the Moscovite Aksakof, formed the Slav component of the cosmopolitan quartette which was composed of the Evangelical *Kreuz-Zeitung*, the Ultramontane *Germania*, and the Roman *Civiltà Cattolica*. Thus, for the Prussian Protestant, for the Austrian and French Catholic, for the Russian Orthodox, the war against Israel was merely a *Kulturkampf*. It meant nothing less than the preservation to modern nations of the benefits of Christian civilisation, by putting an end to what is called the judaising of European society. To one and all, Slav, Latin, German, and Magyar, the Jew, that odious parasite, was the deadly microbe, the infectious bacteria, that poisoned the blood of modern states and societies."[336]

The active involvement of persons guided by Cabalistic Jews,[337] persons

---

336. A. Leroy-Beaulieu, *Israel chez les nations: Les Juifs et l'antisémitisme*, C. Lévy, Paris, (1893); English translation by F. Hellman, *Israel among the Nations: A Study of the Jews and Antisemitism*, G. P. Putnam's Sons, New York, W. Heinemann, London, (1895), pp. 43-48.
337. B. Lazare, *Antisemitism: Its History and Causes*, (1894); *L'Antisémitisme, son Histoire et ses Causes*, L. Chailley, Paris, (1894); *cf.* "Salluste", "Autour d'une

such as Weishaupt, Nicholai, Bahrdt, Voltaire, Diderot, etc., in the destruction of the Catholic religion in the period preceding the French Revolution is covered by John Robison in his book *Proofs of a Conspiracy Against All the Religions and Governments of Europe: Carried on in the Secret Meetings of Free Masons, Illuminati, and Reading Societies,* Printed for William Creech, and T. Cadell, Junior, and W. Davies, Edinburgh, London, (1797); see especially the fourth edition of 1798, to which Robison added a postscript. Robison stated in the introduction to his book,

> "I have been able to trace these attempts, made, through a course of fifty years, under the specious pretext of enlightening the world by the torch of philosophy, and of dispelling the clouds of civil and religious superstition which keep the nations of Europe in darkness and slavery. I have observed these doctrines gradually diffusing and mixing with all the different systems of Free Masonry; till, at last, AN ASSOCIATION HAS BEEN FORMED for the express purpose of ROOTING OUT ALL THE RELIGIOUS ESTABLISHMENTS, AND OVERTURNING ALL THE EXISTING GOVERNMENTS OF EUROPE. I have seen this Association exerting itself zealously and systematically, till it has become almost irresistible: And I have seen that the most active leaders in the French Revolution were members of this Association, and conducted their first movements according to its principles, and by means of its instructions and assistance, *formerly requested and obtained:* And, lastly, I have seen that this Association still exists, still works in secret, and that not only several appearances among ourselves show that its emissaries are endeavoring to propagate their detestable doctrines among us, but that the Association has Lodges in Britain corresponding with the mother Lodge at Munich ever since 1784.
>
> If all this were a matter of mere curiosity, and susceptible of no good use, it would have been better to have kept it to myself, than to disturb my neighbours with the knowledge of a state of things which they cannot amend. But if it shall appear that the minds of my countrymen are misled in the very same manner as were those of our continental neighbours—if I can show that the reasonings which make a very strong impression on some persons in this country are the same which actually produced the dangerous association in Germany; and that they had this unhappy influence solely because they were thought to be sincere, and the expressions of the sentiments of the speakers—if I can show that this was all a cheat, and that the Leaders of this Association disbelieved *every word* that they uttered, and every doctrine that they taught; and that their real intention was to abolish *all* religion, overturn every government, and make the world a general plunder and a wreck—if I can show, that the principles which the Founder and Leaders of this Association held forth as the perfection of human virtue,

---

Polémique: Marxism et Judaïsm", *La Revue de Paris*, Volume 35, Number 16, (15 August 1928), pp. 795-834, at 825. ***See also:*** D. Fahey, *The Mystical Body of Christ in the Modern World*, Browne and Nolan Limited, London, (1935), pp. 76-77.

and the most powerful and efficacious for forming the minds of men, and making them good and happy, had no influence on the Founder and Leaders themselves, and that they were, almost without exception, the most insignificant, worthless, and profligate of men; I cannot but think, that such information will make my countrymen hesitate a little, and receive with caution, and even distrust, addresses and instructions which flatter our self-conceit, and which, by buoying us up with the gay prospect of what is perhaps attainable by a change, may make us discontented with our present condition, and forget that there never was a government on earth where the people of a great and luxurious nation enjoyed so much freedom and security in the possession of every thing that is dear and valuable.

When we see that these boasted principles had not that effect on the leaders which they assert to be their native, certain, and inevitable consequences, we will distrust the fine descriptions of the happiness that should result from such a change. And when we see that the methods which were practised by this Association for the express purpose of breaking all the bands of society, were employed solely in order that the leaders might rule the world with uncontrollable power, while all the rest, even of the associated, will be degraded in their own estimation, corrupted in their principles, and employed as mere tools of the ambition of their *unknown superiors;* surely a free-born Briton will not hesitate to reject at once; and without any farther examination, a plan so big with mischief, so disgraceful to its underling adherents, and so uncertain in its issue.

These hopes have induced me to lay before the public a short abstract of the information which I think I have received. It will be short, but I hope sufficient for establishing the fact, *that this detestable Association exists, and its emissaries are busy among ourselves.*"[338]

Like Robison and many others including Pope Leo XIII,[339] George Goyau argued that Freemasonry sought to establish a world government—a Jewish Messianic goal.[340] Denis Fahey argued that the Jews, armed with their Cabalistic and Talmudic doctrines and symbolism, were the guiding force behind

---

338. J. Robison, *Proofs of a Conspiracy Against All the Religions and Governments of Europe, Carried on in the Secret Meetings of Free Masons, Illuminati, and Reading Societies, Collected from Good Authorities*, Fourth Edition, Printed and Sold by George Forman, New York, (1798), pp. 6-8.
339. D. Fahey, *The Mystical Body of Christ in the Modern World*, Browne and Nolan Limited, London, (1935), pp. 69, 79-81, 103, 111-112.
340. G. Goyau, *L'Idée de Patrie et l'Humanitarisme: Essai d'Histoire Française, 1866-1901*, Perrin et Cie., Paris, (1902). *See also:* Abbé Barruel *Mémoires pour Servir a l'Histoire du Jacobinisme*, De l'Imprimerie Françoise, Chez P. Le Boussonier, Londres, (1797-1798); English translation by R. Clifford: *Memoirs Illustrating the History of Jacobism*, Printed for the translator by T. Burton and co., London, (1798). *See also:* P. Benoit, *La Franc-Maçonnerie*, Société Générale de Librairie Catholique, Palmé, Paris, (1886). *See also:* E. Cahill, *Freemasonry and the Anti-Christian Movement*, M. H. Gill & Son, Dublin, (1929).

Freemasonry.[341] Freemasonry became truly tigerish under the direction of Adam Weishaupt.

It is interesting to note that the charges made against Jews of corrosive materialism and of leading revolutionary movements to overthrow European civilization and with it Christendom were also made by the racist Zionist Jews, including Moses Hess, Benjamin Disraeli, Bernard Lazare and Theodor Herzl. The Cabalistic Frankist Jews and their progeny wanted to Judaize Europe and destroy Christendom. There are many instances in the Hebrew Bible where Jews are told to destroy all other religions and that Judaism will become the only religion on Earth (*Exodus* 34:11-17. *Psalm* 2; 72. *Isaiah* 2:1-4; 9:6-7; 11:4, 9-10; 42:1; 61:6. *Jeremiah* 3:17. *Micah* 4:2-3. *Zechariah* 8:20-23; 14:9).

Zionist Jews feared that should they openly anoint a Messiah, Christians would attack them for worshiping the anti-Christ. In order to hide their true intentions, Zionist Jews veiled their religious movement in a shroud of secular mythology. Christians, Moslems, and Jews in general were concerned that should a Jewish State be formed in Palestine, that Jewish State would be obliged to obey Mosaic Law and chase out the Christians and Moslems.

Knowledgeable Christians, Moslems and Orthodox Jews, opposed the establishment of a Jewish State in Palestine on the grounds that it would lead to:

1.) The reconstruction of the Temple in Jerusalem, which necessitates the destruction of the Dome of the Rock and Al Aqsa Mosque.

2.) The reestablishment of the Priests of Aaron and the Levitical ministration, together with the re-institution of the animal sacrifices of lambs, bulls and goats as David had done and as Jews are required to do for atonement (*Leviticus* 17:11. I *Samuel* 7:9. II *Samuel* 6; 24:22-25. *Ezra* 3. *Jeremiah* 33:18. *Daniel* 9:27; 11:31; 12:11. I *Peter* 2:5. *Sanhedrin* 20*b*).

3.) The need to honor the year of release, the *Shemmitah* and perhaps even the Jubilee, as well as proscriptions against usury to fellow Jews (*Exodus* 23:10-11. *Leviticus* 25. *Deuteronomy* 15; 23:20; 31:10-13. II *Chronicles* 36:20-21. *Jeremiah* 33:15).

4.) The anointment of the Jewish Messiah (*Sanhedrin* 20*b*), who will signify the "anti-Christ" to Christians.

5.) The need to exterminate the Amalekites (*Exodus* 17:14-16. *Numbers* 24:17-20. *Deuteronomy* 25:17-19. I *Samuel* 15:1-35. *Malachi* 1:1-14. *Sanhedrin* 20*b*. P188L *Dvarim* 25:19).

These were probably among the many reasons why Herzl laid emphasis on

---

341. D. Fahey, *The Mystical Body of Christ in the Modern World*, Browne and Nolan Limited, London, (1935), pp. 76-77.

his assertion that political Zionism was an atheistical movement, so as not to worry Christians and Moslems that he was the anti-Christ and would destroy Moslem Mosques.

The truth is that Cabalistic Jews were always the motive force behind the "political Zionist" movement. Now that the Jews have stolen Palestine from its native inhabitants; and it should be noted that neither Judahites nor Israelites were ever a native population in Palestine; Judaism requires the Israelis to anoint a king, which is to say anoint the Jewish messiah, to build a temple, and to exterminate the "Amalekites".

In 1915, that meant killing off the Armenians, and Zionist Jews set about that task with a Cabalistic Jewish fervor that took 1,500,000 innocent Christian lives. Today, the racist and genocidal Jews of Israel view the Palestinians, Lebanese, and Moslems in general, as the "Amalekites". Religious Jews believe that there may be a decision as to whether to first exterminate the Moslems, or first destroy the Dome of the Rock and Al Aqsa Mosque. They are aggressively pursuing both "religious duties", and have decided to let fate decide the order. We bear witness to the brutal racist Jewish attempt to genocide these innocent Peoples in Israel's latest unprovoked attack on Lebanon. This is all according to an ancient Jewish plan, as revealed in, among many other texts, the Jewish Talmud in the book of *Sanhedrin*, folio 20*b*.

A. Kisch wrote a letter to Editor of *The Jewish Chronicle* which was published on 1 December 1911 on pages 20-21, in which Kisch tried to reassure Christians who were leery of Zionist motives that the Zionists did not want a state and that their Zionism did not herald the appearance of the Jewish Messiah, a. k. a. the anti-Christ,

> "Like the Professor, Mr. Chamberlain contends that religious Jews feel the attraction towards Zion so overpowering a force that should it at any time involve a course of action opposed to the interests of the British Empire, those interests were, he considered, in danger of being disregarded to the peril of the State. Having regard to the recognised ability of the Hebrew race he thinks this supposed possibility a serious matter, but he did not show why the possession of political rights by naturalized foreigners coming from other nations was not open to the like objection. It, therefore, seems clear that his attitude is based on prejudice, not on reason. It is but fair to recognise that he confessed to some ignorance of the Jewish position, and it is only such ignorance that can excuse his attitude. Thinking that he might be under some misapprehension about the meaning and aims of the movement known as Zionism, I rose with the intention of reassuring him that it makes no pretension to herald the approach of the Messiah, or the formation of an independent Jewish State."

Many Zionists pushed this false message in 1911, at a time when they were trying to convince the Turks that they had no reason to fear the Zionists, who had been out to destroy the Turkish Empire for centuries (recall D. Wolffsohn's letter to the Editor of the *The London Times* published on 10 May 1911, on page

8, entitled, "The Young Turks and Zionism"). The Zionists have since founded their "Jewish State" and the Lubavitchers are trying to condition the world to accept the appearance of the Jewish Messiah, whom they are about to anoint. It is good lesson not to trust the assurances of Zionists.

Other Zionists were less guarded in their public statements. In the 8 December 1911 edition of *The Jewish Chronicle*, a letter to the Editor by B. Felz appeared on page 38,

"TO THE EDITOR OF THE 'JEWISH CHRONICLE.'

SIR,—It is not in the least surprising that Mr. Chesterton's lecture to the West End Jewish Literary Society should have proved so unpalatable to the members of that body in general and to your correspondent, Mr. Kisch, in particular.

There are quite a number of ladies and gentlemen with a weathercock cast of mind—the sort of person who though he has never read a single one of M. Bergson's books, can never say anything just now without mentioning his name—who, at prize distributions of Sabbath classes, boys' and girls' clubs, and other functions of the kind, makes it a constant burden of all his speeches, that Jews besides being good Jews should always be good Englishmen. This is the message that the West is repeatedly flashing to the East. When, therefore, a gentleman of Mr. Chesterton's logical cast of mind comes along and very flatly tells them that good Jews cannot be patriotic Englishmen, it is not unnatural that the ladies and gentlemen in question should kick. The patriotism of the Jew is simply a cloak he assumes to please the Englishman and so when Mr. Chesterton is shrewd enough to detect the Jew beneath the Englishman's clothing, the masqueraders become exceedingly angry. They had hoped to placate the Englishman by saying that they loved him and agreed with him. Judge then of their dismay when he turns round and says: I can only accept your love when you hate me and differ from me. The Jew is suspect and he knows it; and in the hope that the suspicion will be drowned in the noise, he becomes most vulgarly loud in his profession of patriotism. This atmosphere of suspicion in which the Jew lives from the moment of his birth, makes him so horribly fidgety, that when he meets a Gentile, the fact that he is a Jew is either the very first or the very last thing he wants to tell him. The Jew never takes the fact that he is one as a matter of course, which shows that he is never sure of himself, since it is only the things we are sure of and easy about that we take as matters of course.

Mr. Kisch seems to think that because some thirty years ago, two eminent men had a quarrel about the question whether good Jews could be patriotic Englishmen that, therefore, the matter has been disposed of at once and for all. To the Jews of this generation, the question is more acute and insistent than ever. We Jews of the younger generation are simply being coerced and intimidated, not through the compulsion of physical force but through the more subtle and insidious compulsion of a tyrannous public opinion, into a profession of patriotism, which, in the nature of things, must always be viewed with distrust and suspicion. I think it can be laid down as a general law, that the more Jews become Englishmen the less they become Jews. That does not imply any moral censure; it is simply a statement of fact, and Jews who pretend that they can at once be patriotic Englishmen and good Jews are simply living lies.

Yours obediently,
B. FELZ."

While Christians were more easily duped, contemporary Orthodox Jews, who were close to the Zionists, remained very worried about the Zionists' intentions. Rabbi Isaac M. Wise was quoted in *The New York Times* on 5 September 1897 on page 14,

"A Jewish State in Palestine and Impossibility.
Rabbi Isaac M. Wise in American Israelite.
Sept. 2.

Dr. Herzl does not profess to be a religious Jew. With most of his followers he maintains only to be a Jew by nationality or race. He has not the least intention to benefit Judaism. He is a politician, loyal and patriotic, no doubt, as so many politicians profess to be, and works to set up a Judenstadt [sic], not a religious congregation at all. Religion is at present out of the question altogether. Some zealous Zionists want the return to Palestine as a revival of Judaism, and hold also to Dr. Herzl's project. Romantic zealots cannot possibly do without a number of contradictions in their creed, religious or political. The establishing of a Jewish State in Palestine is an impossibility in itself, and with the state laws of Moses unimaginable. The years of release and Shemittah (Sabbatical years) can not well be re-introduced, but the genuine Zionists must do it, as they have proved a few years ago. The sacrificial cult with the Aaronitic priesthood and the Levitical ministration, so much the zealous Zionist must admit, cannot be restored, nor can it in Palestine be abolished according to the dogmas of the strict Zionists and the whole orthodoxy. You would not stone to death the Sabbath breaker, the adulterer, the blasphemer, the false prophet. But in the Judenstadt [sic] in Palestine the laws of Moses would be in force and you cannot get over it as orthodox Jews. The contradictions between Dr. Herzl and the orthodox Zionists are as numerous as they are in every rationalistic Lover of Zion. None can leap over two thousand years of history and commence anew where all things were left then."

For Christians, Christ was the ultimate sacrifice (*Isaiah* 53:5-7). Christ foretold the destruction of the Temple in 70 A. D., which ended Jewish animal sacrifices—at least until very recently. The fear that Christians would stand in the way of the formation of a Jewish State and the anointment of a Jewish Messiah gave the Jews an enormous incentive to destroy Christianity and Christians. Zionist Jews also felt obligated to destroy Islam, for they could not rebuild the Temple without destroying the Dome of the Rock and the Al Aqsa

Mosque, which would inflame the Moslem world against them. This is one reason the Israelis seek to destroy the Moslem nations of the world today, and in so doing render their armies impotent and unable to oppose, with military force, the destruction of Moslem holy sites and building of the Jewish Temple. It should be noted that most religious Jews today follow the ancient tradition of Rabbi Yochanan ben Zakkai (*Berakoth* 55a. *Midrash Avot de Rabbi Natan*), who felt that kindness and obedience to God, not animal sacrifices, atoned for sins (I *Samuel* 15:22. *Isaiah* 1:11-15. *Hosea* 6:6. *Amos* 5:22-24. *Micah* 6:6-8).

Moses Mendelssohn was on a "Jewish Mission" to supplant the religions of the world with a modern (and in some senses a more ancient) reformed Judaism, which would make political revolution the Jewish Messiah.[342] However, this mission was the same old Jewish mission of subjugating the world to the Jewish faith, a Jewish world-government, and world-wide obedience to Jewish dogma. Like Karl Marx,[343] Mendelssohn was a strict Talmudist. Like the Frankist revolutionaries, Mendelssohn attacked Rabbinical Judaism in order to promote himself and gain inroads into Gentile society. Mendelssohn set the stage for Communism, political Zionism and the Jewish revolutions which spread like wildfire around the globe. Like the Frankists, Mendelssohnians sought to keep their true objectives hidden behind a veil of "modernism". Like the Frankists, the Marxists are known for loose morals and sexual incontinence, and engaged in orgies and other deviant behavior associated with the Frankists.[344] Einstein also engaged in this deviant Frankist lifestyle. Frankist Jews believed that they could hasten the coming of the Messianic Era by making the majority of Jews infidels or Christians, thereby angering God, who would slaughter masses of Jews and give the "remnant" of "righteous Jews" both Israel and command over the Gentiles.[345] In 1845, shortly before the revolutions of in Europe of 1848, which largely accomplished the emancipation of the Jews across Western Europe, *The North American Review* wrote,

> "We might confidently look for reformers under such a system as Rabbinism; and, even without the name of reformation, for wide departures from the Talmud, either towards the 'old paths,' or to infidelity. The man who in modern times exerted the most commanding influence on Judaism

---

342. A. Leroy-Beaulieu, *Israel chez les nations: Les Juifs et l'antisémitisme*, C. Lévy, Paris, (1893); English translation by F. Hellman, "The Jew is the product of His Tradition and His Law", *Israel among the Nations: A Study of the Jews and Antisemitism*, Chapter 6, G. P. Putnam's Sons, New York, W. Heinemann, London, (1895), pp. 123-147.
343. P. S. Mowrer, "The Assimilation of Israel", *The Atlantic Monthly*, Volume 128, Number 1, (July, 1921), pp. 101-110, at 107.
344. *See, as but one example*: "Soviet 'Orgies'", *The London Times*, (31 August 1922), p. 7.
345. *Exodus* 34:11-17. *Psalm* 2; 72. *Isaiah* 1:9; 2:1-4; 6:9-13; 9:6-7; 10:20-22; 11:4, 9-12; 17:6; 37:31-33; 41:9; 42; 43; 44; 61:6. *Jeremiah* 3:17; 33:15-16. *Ezekiel* 20:38; 25:14. *Daniel* 12:1, 10. *Amos* 9:8-10. *Obadiah* 1:18. *Micah* 4:2-3; 5:8. *Zechariah* 8:20-23; 14:9. *Romans* 9:27-28; 11:1-5.

was Moses Mendelssohn. He was born at Dessau, in 1729, was carefully educated in the Bible and Talmud, but was thrown upon Hebrew charity in Berlin, at the age of thirteen. Following the bent of his own genius, and stimulated by various associations, he left the dreary paths of tradition, to pursue the intricate but flowery ways of Gentile philosophy. He even improved the German language, in which he wrote with great taste. The influence of his works and his example was soon manifest. An enthusiasm for German literature and science was awakened among the Jewish people, when they beheld their kinsman ranking with the first scholars of the age. 'Parents wished to see their children like Mendelssohn. Rashi and Kimchi, the Shulchan, Aruch, and Josaphoth, were laid on the shelf. Schiller and Wieland, Wolff and Kant, were the favorite books of the holy nation.' Mendelsshon was very strict in Talmudical observances, and did not in his works directly oppose them; yet he certainly intended to undermine Rabbinism, and covertly labored to obliterate superstitions and prejudices, and to render his religion consistent with free intercourse between Jew and Gentile, and with the palpable benefits of modern progress in letters and refinement in manners. After all, he was probably at best but a deist; and he certainly lacked that directness, candor, and earnestness of purpose, which true-hearted reformers have usually manifested. Christians must deny to Judaism that vitality which is essential to its maintenance upon the true basis even of a pure pre-Messianic creed. As a system, though not indeed strictly in each individual, it must ever oscillate between Rabbinism, or the like, and rationalism,—finding no stable, middle, spiritual ground.

Mendelssohn died in 1786; but others arose to carry out his innovations. A Jewish literary and philosophical society was formed at Königsberg, in 1783, which supported the first Jewish periodical ever published,—a journal devoted to the cause of reform. The 'new light' rapidly spread; and now Mendelssohnism, in different varieties, inclined more or less to the Talmud, or to infidelity, is the religion of a great majority of the Jews in all Europe west of Poland, into which country itself, especially Austrian Poland, the revolution has in some degree extended. The 'Jews of the New Temple,' or ' Rational ' or 'Reformed Jews,' as they are called, where their numbers have not secured peaceable ascendency, have generally seceded from the Talmudists; who, on their own part, where the so-called reformation has made good progress, adhere to the Talmud scarcely even in name.

The creed of the new sect has never appeared in an authoritative shape, but may be gathered from their writings and practices. The believers in it agree, that the Jews are no longer a chosen people, in the sense hitherto commonly received. They reject the Talmud, professing to receive the Hebrew Scriptures as the true basis of religious belief, and as a divine revelation; though after explaining away their inspiration, and the miracles recorded in them, on rationalistic principles. Regarding the Mosaic institutions as never abrogated, they consider, however, that most of their requirements are applicable only to a state of national establishment in Palestine; and therefore hold, that, until the unknown period of the

Messiah's advent, and Israel's restoration, such laws only are to be observed as are necessary to preserve the essence of religion, or useful to form pious ecclesiastical communities, and which do not interfere with Gentile governments, with any of the existing relations of life, or with intellectual culture. The synagogue service has been remodelled; and the modern languages have been generally substituted for the Hebrew. A weekly lecture has taken the place of the semi-annual sermons of the Rabbinists. Contrary to the precept of the Talmud, instrumental music is introduced into public worship. 'The question of organ or no organ,' says a late journal devoted to the Jews, 'divides Judaism on both sides of the Atlantic.'

Before long, the latitudinarian views of the leaders in this movement clearly discovered themselves; and there was a temporary reaction in favor of Rabbinism, to which the more devout among their converts receded. Yet the new system has signally prevailed and flourished. It is in France, perhaps, that the Jews have thrown off most completely the trammels of Judaism,—indeed, of all religion. They now style themselves *French Israelites*, or *Israelitish Frenchmen*, according to the doctrine of Napoleon's Sanhedrim; and seem anxious to amalgamate themselves more and more with the nation at large. Most of their leaders are infidels, undisguisedly aiming to obliterate all the common notions about a Messiah, as utterly superstitious; referring the prophecies of his advent—which they still nominally treat as prophecies—to the political emancipation of the Jews in the various lands of their sojourn. 'The Regeneration,' a journal published at Paris by some of their most learned and influential men, has represented the French Revolution as the coming of the Messiah, bringing, first, judgment, then, liberty and peace. The grand rabbi of Metz, a few years ago, in addressing the Jews of his district, spoke thus:—

'God has permitted different religions, according to the different necessities of men, in the same way as he has created different plants, different animals, and men of different characters, genius, constitutions, physiognomies, and colors. Consequently, all religions are salutary for those who are born in these religions; consequently, we must respect all religions. All men, without distinction of religion, will be partakers of eternal beatitude, provided they have practised virtue in this life.'

On the 12th of June last, a voluntary Jewish synod met at Brunswick, composed of twenty-five eminent rabbins, from various parts of the continent. It was the first of a proposed succession of annual synods, to deliberate on Jewish affairs. They sat eight days, passed various resolutions proposing important changes, and declared their concurrence in all the decisions of Napoleon's Sanhedrim. The Jews of England, though visibly influenced by residence in so enlightened a kingdom, were all nominally Rabbinists, until, within the last four or five years, a reforming party seceded in London whence their principles and denomination—' British Jews—have since gradually spread. Even among those who remained, great difference of opinion prevails as to Talmudical observances. Both there and in this country, the Portuguese Jews seem most active in the work of revolution.

The tide of Jewish emigration to the United States is rapidly swelling; and as it comes from many lands, it exhibits a variety of hue. But the voluntary emigrant is ever and characteristically a lover of change; and here the Talmud has little sway, and that rapidly declining. Mr. Leeser represents the Bible alone as the basis of the Jewish faith and in the whole article already referred to, does not so much as mention the Talmud. He edits, at Philadelphia, 'The Occident and American Jewish Advocate,' the first Jewish periodical established in this country. Soon after its establishment, 'The Israelite,' a weekly German paper, devoted to the same cause, and also published in Philadelphia, was announced; whether this still survives, we know not. Mr. Leeser expects a literal Messiah, —not God, or a son of God, but a mere man, eminently endowed, like Moses, to accomplish all that is foretold of him. He protests against some of the decisions of the late Brunswick synod, particularly the one reaffirming the *dictum* of the French Sanhedrim, that Jews might intermarry with Gentiles. He has long had in his congregation a Sabbath school, or a school for religious instruction, held, not on the seventh day, but on the Christian Sabbath, which Christian observance makes necessarily a day of convenient leisure for the purpose.

Among the stricter Jews, all over the world, the expectation of Messiah's advent is becoming more and more anxious. They not unfrequently talk, though without serious purpose, of embracing Christianity, should he not appear within a certain time. Migration to the Holy Land is visibly increasing. Multitudes from all parts of the world would hasten thither, could they become possessors of the dear soil, and enjoy reasonable protection. Mr. Noah proposes, that Christian societies and governments interested in the welfare of the Jews should exert their influence to procure these advantages for them in their native land of promise. The suggestion deserves notice.

Of modern efforts for the conversion of Israel to Christianity we can speak but briefly. The chief extraordinary obstacles which have hitherto opposed such efforts have been, a bigotry which treated the bare thought of investigating Christianity as a heinous sin, and which was ever prepared to stifle free inquiry by persecution; the character of Talmudical education, which disqualified the pupil for independent judgment; and accumulated prejudices against a religion too often exemplified only by profligate persecutors. But all these obstacles are gradually sinking away; nor does growing infidelity appear so formidable as the superstition and fanaticism which have given place to it. Moreover, the spirit of inquiry, and the dissensions kindled by the progress of the revolution which Mendelssohn commenced, are favorable to Christian effort. We shall speak only of what Protestants have done."[346]

---

[346]. "The Modern Jews", *The North American Review*, Volume 60, Number 127, (April, 1845), pp. 329-368, at 361-365.

Protestantism, Puritanism and the *Kulturkampf* were instigated by Cabalistic Jews seeking to create a schism in the Church of Rome in order to end its hegemony and desires on Jerusalem, as well as to lead Christianity back to its Jewish roots and then destroy it. Jews were largely left alone when the Christians began to fight each other at the instigation of Jews. When searching for the true forces behind the Reformation and the French Revolution, one should ask, *qui bono?* or who benefits? The *North American Review* wrote in 1845,

"The darkest pages of history are those which exhibit Christianity, so called, as a persecuting religion. Before the epoch of the Reformation, bigotry, clothed with ecclesiastical power, was generally leagued with political tyranny and popular malice to oppress and destroy the Jews. To attempt to convert them to the Christian faith without violence was considered by most Roman Catholics as a wholly chimerical scheme, and the undoubted fact of their rejection by God, even more than the dreaded anathemas of the Church, seemed to place them beyond the pale of human sympathies. Better prospects than at any period of their dispersion brightened before them with the dawn of the Reformation. The principles of that mighty change extended to all the interests of humanity, temporal as well as eternal; and planted the seeds both of religious and political regeneration. The hearts of the Reformers were moved with compassion towards the ancient people of God; and they advocated milder plans than those which had usually been adopted, to bring them over to the Christian faith. They discountenanced and condemned the system of wholesale plunder, from which, under the garb of zeal for the Catholic church, princes and prelates had for ages drawn a bloody revenue. But a period of lethargy among Christians in regard both to the civil and religious state of this people—a period of returning gloom—soon succeeded; and the French Revolution, itself one of the mighty effects of a reformation which necessarily emancipated human error and passion, at the same time with truth and reason, brought the first blessings of permanent civil freedom to any of the Jews of Europe."[347]

Paul Scott Mowrer wrote in 1921,

"But the religious wars had now fairly begun, and in the heat of the struggle between Catholic and Protestant, the Jews, greatly to their good, were well-nigh forgotten. For them, the worst was over."[348]

In 1914, Edward Alsworth Ross, a Professor of Sociology at the University of Wisconsin, wrote in his book, *The Old World in the New: The Significance of Past and Present Immigration to the American People*, The Century Co., New

---

[347]. "The Modern Jews", *The North American Review*, Volume 60, Number 127, (April, 1845), pp. 329-368, at 342-343.
[348]. P. S. Mowrer, "The Assimilation of Israel", *The Atlantic Monthly*, Volume 128, Number 1, (July, 1921), pp. 101-110, at 104.

York, (1914), pages 160 and 163,

> "The good will of a Southern gentleman takes set forms such as courtesy and attentions, while the kindly Jew is ready with any form of help that may be needed. So the South looked askance at the Jews as 'no gentlemen.' Nor have the Irish with their strong personal loyalty or hostility liked the Jews. On the other hand the Yankees have for the Jews a cousinly feeling. Puritanism was a kind of Hebraism and throve most in the parts of England where, centuries before, the Jews had been thickest. With his rationalism, his shrewdness, his inquisitiveness and acquisitiveness, the Yankee can meet the Jew on his own ground."

The *Kulturkampf* followed the anti-Catholic English and French Revolutions, which had emancipated the Jews of many nations. The Old Testament led Jews to believe that Jews would rule the world through their Messiah, who would dwell with the Lord in Jerusalem, which city would serve as the sacred and the profane capital of the world. *Deuteronomy* 18:14-18:

> "14 For these nations, which thou shalt possess, hearkened unto observers of times, and unto diviners: but *as for* thee, the LORD thy God hath not suffered thee so *to do*. 15 The LORD thy God will raise up unto thee a Prophet from the midst of thee, of thy brethren, like unto me; unto him ye shall hearken; 16 According to all that thou desiredst of the LORD thy God in Horeb in the day of the assembly, saying, Let me not hear again the voice of the LORD my God, neither let me see this great fire any more, that I die not. 17 And the LORD said unto me, They have well *spoken that* which they have spoken. 18 I will raise them up a Prophet from among their brethren, like unto thee, and will put my words in his mouth; and he shall speak unto them all that I shall command him. 19 And it shall come to pass, *that* whosoever will not hearken unto my words which he shall speak in my name, I will require *it* of him. 20 But the prophet, which shall presume to speak a word in my name, which I have not commanded him to speak, or that shall speak in the name of other gods, even that prophet shall die. 21 And if thou say in thine heart, How shall we know the word which the LORD hath not spoken? 22 When a prophet speaketh in the name of the LORD, if the thing follow not, nor come to pass, that *is* the thing which the LORD hath not spoken, *but* the prophet hath spoken it presumptuously: thou shalt not be afraid of him."

Psalm 72:1-20,

> "Give the king thy judgments, O God, and thy righteous'ness unto the king's son. 2 He shall judge thy people with righteousness, and thy poor with judgment. 3 The mountains shall bring peace to the people, and the little hills, by righteousness. 4 He shall judge the poor of the people, he shall save the children of the needy, and shall break in pieces the oppressor. 5 They

shall fear thee as long as the sun and moon endure, throughout all generations. 6 He shall come down like rain upon the mown grass: as showers that water the earth. 7 In his days shall the righteous flourish; and abundance of peace so long as the moon endureth. 8 He shall have dominion also from sea to sea, and from the river unto the ends of the earth. 9 They that dwell in the wilderness shall bow before him; and his enemies shall lick the dust. 10 The kings of Tarshish and *of* the isles shall bring presents: the kings of Sheba and Seba shall offer gifts. 11 Yea, all kings shall fall down before him: all nations shall serve him. 12 For he shall deliver the needy when he crieth; the poor also, and *him* that hath no helper. 13 He shall spare the poor and needy, and shall save the souls of the needy. 14 He shall redeem their soul from deceit and violence: and precious shall their blood be in his sight. 15 And he shall live, and to him shall be given of the gold of Sheba: prayer also shall be made for him continually; *and* daily shall he be praised. 16 There shall be an handful of corn in the earth upon the top of the mountains; the fruit thereof shall shake like Lebanon: and *they* of the city shall flourish like grass of the earth. 17 His name shall endure for ever: his name shall be continued as long as the sun: and *men* shall be blessed in him: all nations shall call him blessed. 18 Blessed *be* the LORD God, the God of Israel, who only doeth wondrous *things*. 19 And blessed *be* his glorious name for ever: and let the whole earth be filled *with* his glory; Amen, and Amen. 20 The prayers of David the son of Jesse are ended."

Isaiah 9:6-7,

"6 For unto us a child is born, unto us a son is given: and the government shall be upon his shoulder: and his name shall be called Wonderful, Counsellor, The mighty God, The everlasting Father, The Prince of Peace. 7 Of the increase of *his* government and peace *there shall be* no end, upon the throne of David, and upon his kingdom, to order it, and to establish it with judgment and with justice from henceforth even for ever. The zeal of the LORD of hosts will perform this."

Jeremiah 3:17,

"At that time they shall call Jerusalem the throne of the LORD; and all the nations shall be gathered unto it, to the name of the LORD, to Jerusalem: neither shall they walk any more after the imagination of their evil heart."

Zechariah 8:20-23; 14:9

"20 Thus saith the LORD of hosts; *It shall* yet *come to pass*, that there shall come people, and the inhabitants of many cities: 21 And the inhabitants of one *city* shall go to another, saying, Let us go speedily to pray before the LORD, and to seek the LORD of hosts: I will go also. 22 Yea, many people and strong nations shall come to seek the LORD of hosts in Jerusalem, and to pray before the LORD. 23 Thus saith the LORD of hosts; In those days *it shall come to pass*, that ten men shall take hold out of all languages of the nations, even shall take hold of the skirt of him that is a Jew, saying, We will go with you: for we have heard *that* God *is* with you. [***] 14:9 And the LORD shall be King over all the earth: in that day there shall be one LORD, and his name one. "

In 1862, racist Zionist Moses Hess expressed the motives of the Jews who participated in the *Kulturkampf* as a means to destroy Catholicism in order to end German anti-Semitism and as revenge for Catholic persecutions (as opposed to their other motivations of ending Catholic hegemony and Catholic designs on Jerusalem, which they believed were Jewish provinces),

> "FROM the time that Innocent III evolved the diabolical plan to destroy the moral stamina of the Jews, the bearers of Spanish culture to the world of Christendom, by forcing them to wear a badge of shame on their garments, until the audacious kidnapping of a Jewish child from the house of his parents, which occurred under the government of Cardinal Antonelli, Papal Rome symbolizes to the Jews an inexhaustible well of poison. It is only with the drying-up of this source that Christian German Anti-Semitism will die from lack of nourishment.
>
> With the disappearance of the hostility of Christianity to culture, there ceases also its animosity to Judaism; with the liberation of the Eternal City on the banks of the Tiber, begins the liberation of the Eternal City on the slopes of Moriah; the renaissance of Italy heralds the rise of Judah. The orphaned children of Jerusalem will also participate in the great regeneration of nations, in their awakening from the lethargy of the Middle Ages, with its terrible nightmares.
>
> Springtime in the life of nations began with the French Revolution. The year 1789 marks the Spring equinox in the life of historical peoples. Resurrection of nations becomes a natural phenomenon at a time when Greece and Rome are being regenerated. Poland breathes the air of liberty anew and Hungary is preparing itself for the final struggle of liberation. Simultaneously, there is a movement of unrest among the other subjected nations, which will ultimately culminate in the rise of all the peoples oppressed both by Asiatic barbarism and European civilization against their masters, and, in the name of a higher right, they will challenge the right of the master nations to rule.
>
> Among the nations believed to be dead and which, when they become conscious of their historic mission, will struggle for their national rights, is also Israel— the nation which for two thousand years has defied the storms of time, and in spite of having been tossed by the currents of history to every part of the globe, has always cast yearning glances toward Jerusalem and is still directing its gaze thither. Fortified by its racial instinct and by its cultural and historical mission to unite all humanity in the name of the Eternal Creator, this people has conserved its nationality, in the form of its

religion and united both inseparably with the memories of its ancestral land. No modern people, struggling for its own fatherland , can deny the right of the Jewish people to its former land, without at the same time undermining the justice of its own strivings. [***] No nation can be indifferent to the fact that in the coming European struggle for liberty it may have another people as its friend or foe. [***] The general history of social and political life, as well as the national movement of modern nations, will be drawn upon, so as to throw light upon the undischarged function of Judaism. These sources will be utilized, furthermore, to demonstrate that the present political situation demands the establishment of Jewish colonies at the Suez Canal and on the banks of the Jordan. And, finally, these illustrations will be employed to point out the hitherto neglected fact, that behind the problems of nationality and freedom there is a still deeper problem which cannot be solved by mere phrases, namely, the race question, which is as old as history itself and which must be solved before attempting the solution of the political and social problems."[349]

These revolutions; and the wars fought over the "Eastern Question"—the battles between the Holy Roman, Russian, Turkish, Hungarian, French, German and British Empires; favored Zionism, as did the national unifications of Italy and Germany; though the Papacy remained sovereign in Rome, to the dismay of the Zionists. Jewish enmity towards Christianity continues to this day, most especially in Israel, as Israel Shahak has proven.[350]

The Rothschilds used their incalculable wealth in an attempt to act as Messiah and destroy the economies of Egypt, Russia, and Turkey, so as to leave these nations no choice but to sell Palestine to the Rothschild family. They created wars throughout the world to generate profits for themselves, and to liberate Jews; as well as to open the gates to Palestine. However, they could not persuade large numbers of Jews to emigrate to Palestine, until Jewish financiers put Adolf Hitler into power in order to scare the Jews into emigration.

3.6 The Messiah Rothschilds' War on the Gentiles— and the Jews

It is an ancient trick of the loan shark, and the extortionist criminal, to run a victim into debt, then force the victim to obtain a loan secured by property the loan shark wishes to own, and then to ensure that the victim has no means to repay the loan, such that the loan shark becomes the inevitable owner of said property. Shakespeare told such a tale of a Jewish Shylock in his *Merchant of Venice*. An article appeared in *The Religious Intelligencer*, Volume 9, Number

---

[349]. M. Hess, *Rome and Jerusalem: A Study in Jewish Nationalism*, Bloch, New York, (1918/1943), pp. 35-37, 40.
[350]. I. Shahak, *Jewish History, Jewish Religion: The Weight of Three Thousand Years*, Pluto Press, London, Boulder, Colorado, (1994).

26, (27 November 1824), page 411, which stated,

"PROPOSED RESTORATION OF THE JEWS.

—

The Gazette of Spires, assures its readers, that the house of Rothschilds [an immensely rich Jewish banking house in London] has recently received proposals from the Turkish government, for a loan to a considerable amount, and an offer of the entire of Palestine as a security for the payment. In consequence, adds the paper, a confidential agent has been despatched by that house to Constantinople, to examine into the validity of the pledge offered by the Turkish Cabinet.

The N. Y. Advocate says, that the Jews will be restored to their former country, and possess it in full sovereignty cannot be doubted.

Our country must be an asylum to the ancient people of God. Here they must reside; here, in calm retirement, study laws, governments, sciences, become familiarly known to their brethren of other religious denominations; cultivate the useful arts; acquire a knowledge of legislation, and become liberal and free. So, that appreciating the blessings of just and salutary laws, they be prepared to possess permanently their ancient land, and govern righteously."

Baron Rothschild wanted to beat Jesus Christ to the second coming, by becoming the first Jewish Messiah to wreck the Gentile nations and restore the Jews to Palestine. He tried to justify the theft of Palestine from its indigenous population with the same argument Zionists employed after the Holocaust—that the Jews need a nation in order to be safe from Gentiles—again, note the incentive that Jewish financiers had to create the Holocaust in order to "justify" the theft of the Palestinians' land. However, the vast majority of Jews did not want a Jewish nation. Most Jews did not believe Palestine would be a sanctuary, and certainly did not want to live in Palestine. It was the Zionists who perpetrated the Holocaust in order to force the reluctant Jews into moving to an undemocratic, segregated and racist "Jewish" State. Bear in mind that the word "Holocaust" means burnt sacrifice, and the slain and humiliated Jews of Europe were such a sacrifice to the ambitions of the Zionists.

It is important to note that the sophistical premise for the creation of the "Jewish State" of Israel was asserted more than one hundred years before the Holocaust began, and the Holocaust was created in order to justify the formation of an apartheid and racist "Jewish" State. Jews who want to be safe from further persecution should investigate and prosecute the Zionists and disassemble the State of Israel. The ultimate source of their suffering was, is, and will continue to be the racist Zionists.

The Episcopal Watchman, Volume 3, Number 38, (5 December 1829), p. 304; published the following article:

"ROTHSCHILD AND JERUSALEM.—Without vouching for its authenticity, we copy below, from the London Court Journal, an account of a project which it is said that the great banker Rothschild entertains of purchasing the sovereignty of Jerusalem, and the territory of ancient Palestine. If any credit is to be attached to this statement, the sublime Porte will not find the difficulty which the London journalists anticipate, in complying with the pecuniary demands of Russia. Whether, however, this letter from Smyrna is entitled to any belief or not, it is quite certain that there have been some curious notions propagated of late among the Israelites in Great Britain, and we have seen it mentioned that a number of enthusiastic men—Irving, Cunningham, Drummond, &c. have openly maintained that the Jews will ere long be restored to Palestine, where it is prophesied that Christ will re-appear, in person, and establish a political kingdom. Mr. Wolff, the Christian missionary, is said to have embraced this doctrine, and the following paragraph which has found its way into the newspapers, is alleged to be an extract of a letter from him, dated in Jerusalem in April last.—*N. Y. Eve. Post.*

'I proclaimed for two months to the Jews the great truth—first, that Jesus of Nazareth came the first time to the earth despised and rejected of men to die for poor sinners; and secondly, that he will come again with glory and majesty, and glorious in his apparel, and travelling in the greatness of his strength, he will come the SON OF MAN, in the year 1847, in the clouds of Heaven, and gather all the tribes of Israel, and govern in person as man and God, in the literal city of Jerusalem, with his saints, and be adored in the Temple, which will be rebuilt, and thus he shall govern 1000 years; and I, Joseph Wolf, shall see with my own eyes, Abraham, Isaac, and Jacob, in their bodies, in their glorified bodies! and I shall see thee, Elijah, and thee, Isaiah, and thee, Jeremiah, and thee, David, whose songs have guided me to *Jesus of Nazareth*. I shall see you all here at Jerusalem, where I am now writing these lines! There were the topics upon which I spoke, not only with Jews, but likewise with some Mussulmans.'

The following is the extract of a letter, published in the Court Journal on the subject of the purchase of Jerusalem by Baron Rothschild:—

King Rothschild.—The following curious extract is from a private letter from Smyrna. We give it without note or comment.

The confidence of the children of Israel in the words of the Prophet has not been in vain: the temple of Solomon will be restored in all its splendor. Baron Rothschild, who was accused in having gone to Rome to abjure the faith of his fathers, has merely passed through that city on his way to Constantinople, where he is about to negotiate a loan with the Porte. It is stated, on good authority, that Baron Rothschild has engaged to furnish to the Sultan the enormous sum of 350,000,000 piastres, at three installments, without interest, on condition of the Sultan's engaging, for himself and his successors, to yield to Baron Rothschild for ever, the sovereignty of Jerusalem, and the territory of ancient Palestine, which was occupied by the twelve tribes. The Baron's intention is, to grant to the rich Israelites who are scattered about in different parts of the world, portions of that fine country, where he proposes to establish seigniories, and to give them, as far as possible, their ancient and sacred laws.

Thus the descendants of the Hebrews will at length have a country, and every friend of humanity must rejoice at the happy event. The poor Jews will cease to be the victims of oppression and injustice. Glory to the great Baron Rothschild, who makes so noble a use of his ingots.

A little army being judged necessary for the restored kingdom, measures have been taken for recruiting out of the wrecks of the Jewish battalion raised in Holland by Louis Bonaparte. All the Israelites who were employed in the various departments of the Dutch Administration, are to obtain superior posts under the Government of Jerusalem, and the expenses of their journey are to be paid them in advance."

The New-Yorker, Volume 9, Number 13, Whole Number 221, (13 June 1840), pp. 196-197; wrote of Rothschild's desire to be King of the Jews, and by the implications of Jewish prophecy, King of the World—and by the implications of Christian prophecy, the anti-Christ:

"RESTORATION OF THE JEWS.—On more than one occasion we have called attention to the signs, of one kind or another, by which the exiles of Israel are beginning to express their impatience for the accomplishment of the prophecies that point to their restoration; and the changes, physical and moral, which are gradually breaking down the barriers to the final fulfilment of the promise. These are curious and worth attention; and more significant in their aggregation, and with reference to the character of the people in question, than those of our readers who have looked at them hastily and separately, may have been prepared to suspect. The Malta letters brings accounts from Syria, in which some curious particulars are given of Sir Moses Montefiore's proceedings, during his late visit to the Holy Land. We remember rumors, which had currency some years ago, of the Jewish capitalist's (Rothschild's) design to employ his wealth in the purchase of Jerusalem, as the seat of a kingdom, and bring back the tribes under his own guidance and sovereignty. If the scheme, amid its sublimity, savored sufficiently of the romantic to make the rumor suspicious, the positive acts of Sir Moses, at least, exhibit an anxiety to gather together the wanderers in the neighborhood of their ancient home and future hopes; that they may await events on the ground where they can best be made available to the fulfilment of the promise. During his pilgrimage he sought his way to the hearts of his countrymen, by giving a *talaris* (we believe about fifteen piastres) to every Israelite; and having instituted strict inquiries respecting the various biblical antiquities on his way, and ascertained the amount of duty which the sacred places and villages paid to the Egyptian Government to be about 64,000 purses (a purse being equal to fifteen talaris,) he proposed to the Viceroy of Egypt, that he (Sir Moses) should pay this revenue out of his own pocket, as the price of that prince's permission to him to colonize all those places with the Children of Israel. The offer has been, it is said, accepted, subject to the condition that the colony shall be considered national, and not under European protection. Athenæum."

The *Scientific American* wrote in 1846 of the man who would be King of the Jews, Rothschild, and revealed that orthodox religious fanaticism and a racist desire to keep the Jews segregated from the *Goyim* were the main motives of Messiah Rothschild,

> "THE ISRAELITES IN GERMANY are in great commotion. At Berlin and Frankfort two-thirds of them have separated from the synagogues, to form new societies, and it is thought that their example will be generally followed. The new school are supported by the government; they celebrate the Sabbath of the Christians, and worship with chants, the music of the organ, and sermons. Sir Moses Montefiore, backed by the Rothschilds, is about establishing a Jewish colony in Palestine, and has obtained an ukase from the Emperor Nicholas, authorising the emigration thither of ten thousand Russian Jews."[351]

On 2 October 1866, on page 2, *The Chicago Tribune* reported that Rothschild wanted to rule the Jews and fulfill Messianic prophecy,

"REGENERATION OF THE HOLY LAND.

An important society has been formed in Europe called the 'International Society of the Orient,' to prevent the grave complications arising out of the Eastern question, and to regenerate the East by infusing therein the spirit of Western civilization. To accomplish this great result the Society, which enrolls among its members such men as Napoleon, the Rothschilds and Montefiore, propose to favor the development of agriculture, industry, commerce and public works in the East, especially in Palestine; to obtain from the Turkish Government certain privileges and monopolies, chief of which is the gradual concession and advancement of the lands of Palestine; to distribute at cash prices such of those lands as the company receives, and to effect the colonization of the more fertile villages of the Holy Land.

The Society, after having established its commercial bureau at Constantinople and other cities of the Turkish Empire, will construct a port at Joppa, and a good road or railroad from that city to Jerusalem. Upon the north of this road the Society expect land to be conceded by Turkey, which they will sell to Israelitish families. These in their turn will create new colonies, aided by their Oriental co-religionists, and it is expected special committees will send thither Jews of Morocco, Poland, Moldavia, Wallachia, from the East, and from Africa. The Society claim that this plan will reconstruct the Holy places of Jerusalem in a Christian manner; put an end to the constant conflict between the great powers in reference to them;

---

351. *Scientific American*, Volume 1, Number 42, (9 July 1846), p. 3.

transform the ancient Jerusalem into a new and great city; create European colonies which will become in time the centres whence occidental civilization will spread in Turkey and penetrate to the remote Orient.

The Society is being rapidly formed, with the strongest influences, financial and political, at its back. The Rothschilds, Sir Moses Montefiore, and other great capitalists among the Jews, are actively in sympathy with the undertaking. The plan has also the favor of more than one crowned head in Europe, among them Napoleon, of whose especial theory of nationalities it is a development. Several prominent noblemen of England, and the leading names of the Faubourg St. Germain, are also among its friends."

Mayer Anselm (Bauer), the founder of the Rothschild destiny, was a highly religious Jew and his father urged him to become a rabbi.[352] Mayer aimed higher and sought to become the Messiah, himself, a goal which he passed on to his descendants. On 8 April 1878, *The Chicago Tribune* reported, among other things, in an article "The House of Rothschild" on page 2,

"There is a popular idea that the Rothschilds dream of yet restoring the Temple and the City of Jerusalem. If so, events may even now be working to meet their views. They are all earnest in the faith of their fathers, and proved their Jewish convictions by breaking off all relations with the Roman Government after the abduction of the little Moriara."

The Rothschild's used prominent figures in the "Gentile" community, either "Shabbas Goys" or crypto-Jews, to spread the myth that the Jews were morally and intellectually superior to Gentiles, but were kind enough to condescend to lead the Gentiles. Meanwhile, the Rothschilds accumulated the wealth of the Gentile nations while deliberately destroying their culture, their countries and their genetics. Many have alleged that there is a clear pattern in history, where one can observe that for two thousand years, Jews had preached liberalism to Gentiles as a means to remove barriers against Jewish access to immigration, then government, commerce, higher education and the media. Once in control of those organs of society, Jews have historically instituted the most tyrannical and illiberal of régimes. In a society in which the majority act morally, socially responsibly, and largely independently; a corrupt minority which acts immorally or amorally, considering only their perceived self-interests, and which works collusively—tribalistically to accumulate the wealth of nation and corrupt its media, government and universities, such a deceitful minority can easily overwhelm a society. When the success of Jewish tribalism led to Jewish assimilation, the Rothschilds promoted anti-Semitism as a means to segregate Jews from Gentiles and force the Jews to emigrate to another region, taking with them the wealth of the nation they had overwhelmed, and in some instances

---

352. "Rothschild", *The Encyclopædia Britannica*, Volume 21, Ninth Edition, Charles Scribner's Sons, New York, (1886), p. 3.

brought to ruins.

In 1883, Ernest Renan gave a philo-Semitic lecture. He was introduced by "Baron" Alphonse de Rothschild. *The Chicago Daily Tribune* reported on 25 June 1883 on page 7,

"THE FUTURE OF JUDAISM.

M. Renan Delivers a Panegyric of the Jews and Predicts a Realization of the Religion of Isaiah.

At a recent meeting of the society of Jewish Studies in Paris M. Ernest Renan, presented by the Baron Alphonse de Rothschild, delivered a remarkable lecture on the subject of the original identity and gradual separation of Judaism and Christianity. M. Renan predicted a great future for the Society of Jewish Studies, one clause of whose studies permits Gentiles to form part of the society. Doubtless Jewish studies belonged of right to the Jews; but they belonged also to humanity. Researches relative to the Israelite past interest all the world. All beliefs find in Jewish books the secret of their formation. The Bible has become the intellectual and moral nutriment of civilized humanity. The Jews have this incomparable privilege, that their book has become a book of the whole world—a privilege of universality which they share with the Greeks, a race which has imposed its literature on all centuries and all countries. M. Renan thanked the members of the Society of Jewish Studies for having admitted the Gentiles, like good Samaritans, to work along with them in a work that interests us all equally. Proceeding then to speak of the subject of his life's study, the origins of Christianity, M. Renan said that those origins ought to be placed at least 750 years before Christ, at the epoch of the great prophets, who created an entirely new idea of religion, and under whose influence was definitively accomplished the passage from primitive religion full of unwholesome superstitions to pure religion. After the captivity, in the sixth century B. C., the dream of the prophet of Israel is a worship that might suit all humanity, a worship consisting in the pure ideal of morality and virtue— in short, the reign of justice. This idea constitutes the great originality of the prophets; and the true founders of Christianity, according to M. Renan, were these great prophets, who announced pure religion, freed from all coarse material practices and observances, and residing in the disposition of the mind and heart—a religion, consequently, which can and ought to be common to all, an ideal religion, consisting in the proclamation of the kingdom of God upon the earth and in the hope of an era of justice for poor humanity.

M. Renan next proceeded to show that the first Christian generation is essentially Jewish. The epistles of St. James and St. Jude, representing the spirit of the first church, are altogether Jewish; St. Paul never thought of separating himself from the Jewish Church. The Apocalypse of St. John, composed about A. D. 68 or 69, is a Jewish book and the author is a passionate Jewish patriot. After the capture of Jerusalem comes the composition of the synoptical Gosples. Here there is a division, and yet Luke, the least Jewish of the evangelists, insists upon the fact that Jesus observed all the ceremonies of the law. Toward 75 or 80 A. D. many books were written inspired by Jewish patriotism, such the book of Judith, the Apocalypses of Ezra and Baruch, and even the book of Tobias. There is nothing more Jewish than the book of Judith, for instance, and yet these books are lost among the Jews and preserved only among the Christians, so true is it that the bond between the church and synagog was not yet broken when they appeared. In the epistles and Gospels attributed to St. John and written about A. D. 125, the case is altogether different. In them Judaism is treated as an enemy, and they contain symptoms of the approach of the systems that will lead the Christians to deny their Jewish origin, such gnosticism, for instance, which represents Christianity as being a reaction against Judaism and utterly opposed to it, while Marcion goes still further, and declares Judaism to be a bad religion which Jesus came to abolish.

M. Renan remarked the singularity of such an error having been able to manifest itself only a century after the death of Christ, but insisted on the fact that in the Christian church gnosticism was like a lateral stream to a river. In the second century the orthodox church always considered itself bound in the most intimate manner to the synagog. In the third century the schism becomes more pronounced under the influence of the school of Alexandria. Clement and Origen speak with much injustice of Judaism, and the separation becomes complete when, under Constantine, Christianity becomes a state religion and official, while Judaism remains free. And yet Chrysostom was obliged to rebuke his congregation for going to the synagog. Nevertheless, the separation really grows more and more profound; we enter the middle ages; the barbarians arrive, and then begins that deplorable ingratitude of humanity, become Christian toward Judaism. The crusades give the signal for the massacres of the Jews, while scholastic philosophy largely contributed to embitter the hostility against them.

Reviewing rapidly the condition of the Jews in France in the Middle Ages and subsequently, M. Renan arrived at 'a more consoling epoch, that eighteenth century which proclaimed at length the rights of reason, the rights of man, the true theory of human society—that is to say, the State without official dogma, the State neutral in the midst of metaphysical and theological opinions. It is from that day that equality of rights began for the Jews. It was the revolution that proclaimed the equality of the Jews with the other citizens of the State. The revolution found here the true solution with a sentiment of absolute justice, and everybody will come around to this opinion.' In point of fact, continued M. Renan, the Jews had themselves prepared this solution; they had prepared it by their past, by their prophets, the great religious creators of Israel. The founders of the movement were Isaiah and his successors, then the Essenians, these poetical ascetics who announced an ideal of peace, of right, and of fraternity. Christianity, too, has powerfully contributed to the progress of civilization, but Christianity was only the continuation of the Jewish prophets, and the glory of Christianity and the glory of Judaism are one. And now that these great things are accomplished, let us say with assurance, continued the speaker, that Judaism, which has done so much service in the past will serve in future. It will serve the true cause—the cause of liberalism of the modern spirit."

The cause of Jewish "Liberalism" created the tyranny of the French, Russian, Chinese, Cambodian, Israeli, etc., Terrors. The cause of Jewish Liberalism slaughtered countless Europeans and Americans in the Nineteenth Century, and many millions more human beings in the Twentieth Century. It brought the world to world wars and to genocide. It is interesting to note, however, that when the Jews began to convert the Northern Europeans and the British to Judaism, which is to say, when the Jews began the Protestant cults, the racist Jewish concept of the "Elect" found in *Isaiah* 65 and in the *Book of Enoch* and in the Jewish myth of the "chosen"—in contradiction to the "Universal" or Catholic Church—as well as the Jewish practices of wealth accumulation and sober studies, led the Puritans and Protestants to surpass their philosophical masters. This benefitted the Jews by spreading monotheism around the world and opening up markets and trade routes, but some Jews ultimately sought to eliminate the threat of Gentile world domination by reintroducing Jewish "Liberalism" in the form of Communism, which taught the Gentiles to self-destruct by degrading the practice of wealth accumulation and by degrading the Nationalistic pride inherent in the mythology of the "Elect" (*Isaiah* 65); both of which had worked so well for the Jews for thousands of years. They hoped that this Jewish Liberalism, imposed on the Gentiles, though not on the Jews, would have the same destructive effects on Gentile empires in the modern world, that it had on the Roman Empire in the ancient world.

One need only take a cursory look at the immensely destructive antisocial behavior of the Rothschilds to see that they were not a friendly guiding spirit to the Gentile nations. They caused the stock markets to crash in the "Black Fridays" of Wall Street in New York, as well as other financial calamities, in 1869, 1873, 1879, 1893, 1907, and 1929; in Prussia in the 1870's; in the "Black Friday" of Vienna in 1873; and in London after the battle of Waterloo—an event that began the large scale emigration of German Jews to America, which increased after the Jewish-led revolutions of 1848. While tragic for the nations and for the world at large, these crashes netted the Rothschilds and their agents immense profits—profits made by destruction, not production—profits made without labor. The Rothschilds also deliberately caused wars and revolutions towards the same ends.

The Jewish bankers caused wars to make the peoples of the world clamor for world government, which they alleged could secure peace. Wars also made the Jewish bankers enormous profits and weakened the nations. The Jewish bankers deliberately caused chaos after the revolutions they instigated, in order to make peoples clamor for dictatorships, which the Jewish bankers argued would restore order—dictatorships the Jewish bankers covertly controlled—dictatorships which brought on wars and enabled the Jewish bankers to rob the wealth of the nation and ruin the people. The Jewish bankers deliberately caused depressions in America to make the people clamor for banking reforms which would enable the Jewish bankers to install a privately held central bank in control of the money supply. Depressions also made for wonderful buying opportunities for Jewish bankers.

On 2 June 1873, *The Chicago Daily Tribune* reported on the front page in an article entitled "Vienna's Black Friday",

"Reading off the names of brokers and firms that failed to meet their engagements was like the call of the death-roll in the Reign of Terror. Many of the lighter stocks were swept out of the market. Austrian loans, railroad shares of the best companies, dropped 5, 10, 20, even 50 per cent. On Friday afternoon it seemed impossible to raise a loan on any security. The bears had things their own way. The branch house of Rothschild was accused of 'bearing' without mercy, and two of the firm narrowly escaped being lynched."

Wherever a corrupt cabal controlled the disproportionate wealth the Rothschilds controlled, there was no chance for any individual, or even any government, or even any coalition of governments, to compete with them on a level playing field. The Rothschilds enjoyed a rigged system in which they could steal the wealth of nations at will, and could demand that nations engage in wars, win wars, and even lose wars, or face utter annihilation and death by starvation. Their fortunes eclipsed the wealth of any nation on Earth. Their fortunes eclipsed the wealth of many nations combined.

The Chicago Tribune made a point of pointing out that the Rothschilds had been war profiteers from the beginning of their financial empire, which was built in part on elicit profits gained by spreading the false rumor that the British had lost at Waterloo in order to buy shares at reduced prices, only to sell the next day at inflated prices, which netted the Rothschilds $5,000,000 in one day, while throwing the British Nation into turmoil. The *Tribune* proved that the Rothschilds profited from the havoc they caused in the United States during the Civil War through the American representative of the Rothschild family,[353] Auguste Belmont—a crypto-Jew whose real name was August Schoenberg—the name "Schoenberg" becomes "Belmont" when translated into French, which sounds more *gentil* and Gentile.[354] It had always been a Jewish plan to destroy Gentile nations by pitting brother against brother and fomenting wars and revolutions (*Judges* 7:22. *Haggai* 2:22). While Schoenberg financed the South, the Seligmans (a. k. a. the "American Rothschilds")[355] financed the North, and the country fought its bloodiest and most profitable war to date—against itself.

---

[353]. G. E. Griffin, *The Creature from Jekyll Island: A Second Look at the Federal Reserve*, Fourth Edition, American Media, Westlake Village, California, (2002), p. 208.

[354]. B. J. Hendrick, "The Jews in America: II Do the Jews Dominate American Finance?", *The World's Work*, Volume 44, Number 3, (January, 1923), pp. 266-286, at 267, 277-278.

[355]. B. J. Hendrick, "The Jews in America: II Do the Jews Dominate American Finance?", *The World's Work*, Volume 44, Number 3, (January, 1923), pp. 266-286, at 272, 278.

The Rothschilds desired to divide America up between France and Great Britain.[356] The North would join with Canada and return to the British Empire. The South would go to Mexico, which would in turn serve as a colony of France. Bismarck described this plan, and his description is detailed in Maj.-Gen., Count Cherep-Spiridovich, *The Secret World Government, Or, "The Hidden Hand": the Unrevealed in History: 100 Historical "Mysteries" Explained*, The Anti-Bolshevist Publishing Association, New York, (1926).

The Rothschilds would then have a profitable division between Latin and French Catholics in the South, and Anglo-Saxon Protestants in the North. The Rothschilds could then use the model they had so successfully employed in Europe to create perpetual wars[357] between the North and South which would earn the Rothschilds immense profits, place both Empires further in the Rothschilds' debt, and destroy the competitive threat that American finance posed. Bismarck, who had close contacts with Jewish finance, stated,

> "The division of the United States into federations of equal force was decided long before the Civil War by the high financial powers of Europe. These bankers were afraid that the United States, if they remained in one block and was one nation, would attain economic and financial independence, which would upset their financial domination over Europe and the world. Of course, in the 'inner circle' of Finance, the voice of the Rothschilds prevailed. They saw an opportunity for prodigious booty if they could substitute two feeble democracies burdened with debt to the financiers, ... in place of a vigorous Republic sufficient unto herself. Therefore, they sent their emissaries into the field to exploit the question of slavery and to drive a wedge between the two parts of the Union... . The rupture between the North and the South became inevitable; the masters of European finance employed all their forces to bring it about and to turn it to their advantage."[358]

The Attorney General, then Secretary of War, then Secretary of State of the Confederacy—"the brains of the Confederacy"[359]—was a Jew named Judah Philip Benjamin, who was a close and enduring friend of Jefferson Davis.[360]

---

356. G. E. Griffin, *The Creature from Jekyll Island: A Second Look at the Federal Reserve*, Fourth Edition, American Media, Westlake Village, California, (2002), pp. 222-224. R. McNair Wilson, *Monarchy or Money Power*, Eyre and Spottiswoode Ltd., London, (1933), pp. 81-83.
357. G. E. Griffin, "The Rothschild Formula", *The Creature from Jekyll Island: A Second Look at the Federal Reserve*, Chapter 11, Fourth Edition, American Media, Westlake Village, California, (2002), pp. 217-234.
358. G. E. Griffin, *The Creature from Jekyll Island: A Second Look at the Federal Reserve*, Fourth Edition, American Media, Westlake Village, California, (2002), p. 374. Griffin cites C. Siem, *La Vieille France*, Number 216, (17-24 March 1921), pp. 13-16.
359. "Benjamin, Judah Philip", *The Universal Jewish Encyclopedia*, Volume 2, The Universal Jewish Encyclopedia, Inc., New York, (1940), pp. 181-184, at 182.
360. "Benjamin, Judah Philip", *The Universal Jewish Encyclopedia*, Volume 2, The

President Lincoln was assassinated by a Jewish actor named John Wilkes Booth—some say because Lincoln dared to oppose the desires of the Rothschilds to control American banking.[361] Before Belmont (Schoenberg) helped the Rothschilds to foment the Civil War, the Bohemian Jew Isaac Phillips[362] represented the Rothschilds' interests in America, as did the Jewish Grandmaster of Masons of the State of Virginia, Solomon Jacobs.[363] Later, John Pierpont Morgan, John Davison Rockefeller and "Colonel" Edward Mandell House served as the Rothschilds' agents in America.[364] Though their plan to divide America between North and South largely failed, after the Civil War the Rothschilds and their agents drew a steady profit from the American financial system. An article entitled "Review of the Stock and Money Market for 1879", *The Bankers' Magazine and Statistical Register*, Volume 14, Number 8, (February, 1880), p. 635; reported,

> "The great event of the year was, of course, the resumption of coin payments on the first day of January. It occurred without a jar or ripple and would have been unobserved if the public had not been constantly reminded of it by the newspapers. The parity of paper and coin having been restored several weeks previously, no demand was made for coin. All anxiety on the subject was over in a day, and it was instinctively felt that an era of prosperity was ushered in. The sales of four per cents., under the offer for popular subscriptions, became so large that from January 1 to January 18, both inclusive, calls were issued for the redemption of $90,000,000 of outstanding bonds at a higher rate of interest. On the 21st of January, the Treasury made an arrangement with a syndicate consisting of the following banking firms in London, viz.: Messrs. Rothschild, J. S. Morgan & Co., Morton, Rose & Co., and J. and W. Seligman & Co., for the exclusive sale in Europe of the United States four per cents, They took $10,000,000 on that day, with the option, provided they took $5,000,000 more monthly until July 1, of then having the entire balance (if any) of the loan, which, however, was to remain open until July 1 to popular subscription. The arrangement

---

Universal Jewish Encyclopedia, Inc., New York, (1940), pp. 181-184. B. J. Hendrick, "The Jews in America: I How They Came to This Country", *The World's Work*, Volume 44, Number 2, (December, 1922), pp. 144-161, at 153.

361. Refer to the articles in *The Vancouver Sun* on 2 May 1934 and 4? and 5? May 1934 relating to Gerald Grattan McGeer's speech before the Canadian House of Commons, and the *Vancouver Daily Province* of 2 May 1934. ***See also:*** G. G. McGeer, *The Conquest of Poverty; or, Money, humanity and Christianity*, Garden City Press, Gardenvale, Quebec, (1935).

362. "Phillips, Isaac", *The Universal Jewish Encyclopedia*, Volume 8, The Universal Jewish Encyclopedia, Inc., New York, (1942), p. 492.

363. H. R. London, *Portraits of Jews by Gilbert Stuart and other Early American Artists*, Charles E. Tuttle Company, Rutland, Vermont, (1969), pp. 60-61, 70-71, 187.

364. G. E. Griffin, *The Creature from Jekyll Island: A Second Look at the Federal Reserve*, Fourth Edition, American Media, Westlake Village, California, (2002), pp. 209, 213, 457-459.

with this syndicate was regarded as settling the question of the ability of the Government to obtain all the money it might desire at four-per-cent. interest, The success of resumption, the large and continuous popular subscriptions to the four-per-cent. loan, and the syndicate arrangement of January 21, naturally caused a very buoyant feeling and a general upward tendency in the prices of bonds and shares dealt in at the Stock Exchange."

On 29 March 1861, at the beginning of the Civil War, *The Chicago Tribune* reported on page 2, that Baron Rothschild had arrived in New Orleans,

"Arrival of Baron Rothschild at New Orleans.
The New Orleans *Picayune* of the 22d says:
Among the arrivals in this city yesterday by the steamship Cahawba, from Havana, was Baron Rothschild, of the distinguished family of that name in Paris, who is a guest of the St. Charles. Baron R. has been spending some weeks in Havana, where he was the object of many attentions on the part of the Captain General and other distinguished gentlemen of that city."

The Rothschilds had been working toward a "race war" between Latin Catholics and Anglo-Saxon Protestants centered in Mexico and spreading to the United States, Canada, France, Great Britain, Austria and North Germany, at least since the time of the Civil War. The Rothschilds sought to weaken the United States by dividing it up. They funded both sides of the Civil War. McClellan needlessly prolonged the war, by refusing to attack and pursue the Confederates. The Rothschilds did not desire to end slavery, rather they desired to enslave Mexico and America, and to return the Americas to a colonial status and to embroil the Americas in perpetual war for the sake of Rothschild profits. On 10 June 1862, on page 3, *The Chicago Tribune* reported,

"FRANCE AND MEXICO.

THE SECRET HISTORY OF THE EXPEDITION.

THE ACTUAL ATTITUDE OF THE FRENCH GOVERNMENT.

New Mutterings of Intervention.

[New York Times Correspondent.]

PARIS, May 23, 1862.
The Mexican affair has assumed all at once at Paris a most serious aspect. Never before has the Emperor been attacked by the liberal press with such violence, or rather, with such an outspoken energy, as within the last few days, on this unfortunate Mexican expedition. It is the all-absorbing topic of the moment, and I cannot do better than to give you an *apercu* of

the situation, as we understand it here.

It so happens that, so far as regards the Press, the three papers which have thus far defended the cause of the rebellion in the United States, are exactly those which sustain the Almonte-Maximilian programme for Mexico; while the rest of the journals, with the exception of the Catholics, defend the cause of the Union in the United States, and combat the monarchical programme in Mexico. This striking concurrence in the division of views on the two subjects, indicates, beyond any question, that for the French there is an important connection between the two. It is this connection which gives the question its gravity.

For a long time the Emperor has dreamed of two things:

First—The acquisition of Sonora, with its gold and silver mines.

Second—The reconstruction of the Latin race, and the pitting of this race and Catholicism against the Anglo Saxon race and Protestantism.

The two governments of France and England, and no doubt of Spain also, did not believe till lately that there was any possibility of the suppression of the rebellion in the United States and the reconstruction of the Union. When, therefore, the treaty of London, of last year, in regard to the expedition to Mexico, was drawn up, it was drawn up with an almost complete indifference as to what the United States might think or do about it, and there is now every reason to believe that each of the contracting parties had ulterior views, which were not only concealed from the world, but from each other. The treaty was therefore drawn up in a loose and vague manner, so as to admit of deviations at will, so that each might seize upon whatever advantages offered themselves. And here I ought to recall, for its historical value, an observation made by Mr. Dayton nine months ago, and put upon record at the time in this correspondence, to the effect that, although the French government was full of kind and frank expressions towards the United States in connection with this Mexican expedition, yet that there seemed to be a vagueness and a confusion in their own understanding of the objects and the details of the expedition which foreboded no good to the future relations between France and the United States.

At the time of the arrival of the Soledad Convention at Paris there had been nothing done toward changing the belief of the French Government that a final dissolution of the Union was inevitable, and Napoleon is known at that time to have given Gen. Lorencez hasty and imperative orders to hurry on to the City of Mexico, without regard to consequences. Why? Because, the Government papers here now say, it was recognized as impossible to gain the objects of the expedition without displacing Jaurez from power and establishing in his stead a stable government, capable of offering, besides indemnity for the present, security for the future. And here is where the English and Spaniards deserted Napoleon, and where the great majority of Napoleon's own subjects also deserted him. They divided on the question of an interference in the internal affairs of Mexico, after having obtained satisfaction for the first objects of the expedition. It came out all at once that Napoleon had been serious in his secret transactions with Almonte at Paris, and that the plan of erecting a throne for an Austrian Prince was not an illusion. Knowing the mind of the Mexican people, the Allies and the Liberals of Paris naturally and legitimately jumped to the conclusion that the Emperor was bent on a conquest of the country, for that was the only condition on which he could maintain a foreign Prince in power, and that sooner or later it would terminate with an acquisition of territory and a war with the United States.

The news of the breaking up of the alliance at Orizaba arrived in Europe with that of the capture of New Orleans, and it is hard to tell which event caused most consternation at the Palace. For the first time the fact that the Southern Confederacy might possible prove a failure, penetrated the short vision of the French Government; and now we believe that under the influence of these two events, the French Government has modified its intentions, and that it has sent to Mexico orders not to push matters to the extreme point at first designed.

The opposition press here has said to the Emperor: Your Mexican expedition, under the present aspect of the case, (that is to say, as an agent of the monarchial party,) is either an aberration or a scheme for the ransom of Venetia. If it be the first, comment is unnecessary—there is but one course to follow: withdraw as quickly as possible after securing what Mexico owes us; if it be the ransom of Venetia that is intended, permit us to suggest that a war with Austria in the quadrilateral will cost us infinitely less in time; men, money, and especially in honor, than a war with the United States.

The opposition press also points out with telling effect on the public mind the analogy which exists between the entrance of the allies into France in 1815, bringing with them the exiles who were selling their country in order to gain power for a minority. For whatever may be the faults of Juarez, he is fighting for his native country against the foreigner, which constitutes his patriotism—quite another thing to that of Almonte, Miramon and company.

As we understand the question then, to-day, Napoleon, at the moment he heard of the treaty of Soledad, gave to Gen. Lorencez instructions which conveyed with them the perspective of a monarchy, a more or less permanent occupation, an acquisition of territory, and a strengthening of the Latin race in America. But the late Union victories have changed the programme, and by this time we have every reason to believe Gen. Lorencez has received a modification to his previous orders. But how far this modification extends no one knows or pretends even to conjecture. That the Emperor will renounce the monarchical programme is, however, generally believed, but whether, when his troops arrive at the capital, they will treat with Juarez or insist on putting Almonte into the Presidential chair before treating, is all in doubt. If Almonte is put into the chair provisionally, every one can see that then the reign of anarchy will only have commenced, and that the French will be obliged to remain to carry out their unfortunate programme by force. And yet, up to the present moment, the Ministerial papers here declare that it will be degrading to the dignity of France to treat with such a man as Juarez, and that such a thing cannot be thought of for a moment. But who can see the end if they go beyond Juarez? One step beyond him and everything is darkness and confusion. Every one in France seems to understand that, if the power of the Federal Government is again consolidated by the suppression of the rebellion, Mexico will at once occupy the attention of the United States, and that France cannot afford, for the benefit of an Austrian Duke and a score of Mexican exiles, to bring upon herself a war with the United States.

The Republicans in France, in view of this war with the United States, declare that it will bring with it the downfall of the Bonaparte dynasty, and they are quite elated at the prospect.

Among the persons who have been indicated as having used their influence with the Emperor since the commencement of the rebellion, in urging on the Sonora programme, are Messrs. Michel-Chevalier, Fould, Rouher, and De Rothschild. These gentlemen do not see why France should not make an acquisition of valuable gold mines—which, by the way, she much needs—as well as the United States.

As regards the more utopian scheme of reconstructing and strengthening the Latin and Catholic elements in America, some of the most influential imperialist writers of France have long been urging it. To these must be added a demented party not far removed from the Emperor's person, who dream of nothing less than setting up in America what has been repudiated in Europe—a nobility system, based upon the divine right, and which shall give an asylum and an occupation to the castoff kings and princes of Europe. They would have the Grand Duke Maxamilian or Ferdinand II., of Naples, placed on the throne of Mexico, surrounded by the European rejected princes, and this try to gain a new foothold for a system which is here growing weaker every day.

But the Emperor has generally shown great judgment in seizing the right side of questions as they pass before him, and great wisdom in retreating from mistaken positions, into which, like the ablest of men, he has sometimes fallen; and we have great confidence that he will yet, with the new light which has broken in upon him from the United States, retire from Mexico before he has become so far entangled in the meshes that await him.

A new secession pamphlet is also just out, to which M. Marc de Haut, advocate at the Imperial Court, has put his name. It is entitled: *The American Crisis: its causes, probable results, and connection with France and Europe*. The pamphlet is but a repetition of several of those which have preceded it, and appears to prove that the secessionists think it necessary to keep certain arguments continually, in one form or another, before the public. The following are the stereotyped heads of arguments found in this book: Republics, when the grow too large, must divide. The Americans of the North are ancient English Puritans, sombre, intolerant, taciturn and commercial. The Southerners are descendants of the Cavaliers, grand, historical *seigneurs*, who love a large and free existence, who don't build workshops or counters, but furnish orators, statesmen and presidents. The sole cause of the dissolution of the Union is the tariff—slavery was only the pretext. The Yankees abandoned slavery in the Northern States, not from principle, but because free labor was more profitable in their climate. The proof of this is found in their well known antipathy to the person of the negro. The present struggle is one of free trade against protection. A reunion can never take place. And then the writer terminates with that funny appeal for the sympathy of the French—that the South is French. 'Does not,' he exclaims, 'the General-in-Chief of the Southern forces bear a French name—Beauregard? And what souvenirs do the following names of *Southern* towns recall to the French hear—Louisburg, Montmorency, St. Louis, Vincennes, Duquesne, New Orleans?'

Thus you will see that the French secessionists demand sympathy for the South because it is French, while, the other day, the London *Times* demanded the sympathy of the English for the South because it is English! We hope they will settle the question between them.

MALAKOFF."

This 1862 article is given credence by the fact that the French, under Rothschild's puppet Napoleon III, drove out Juárez in 1864 and made the Austrian Hapsburg Archduke Ferdinand Maximilian Joseph the Emperor of Mexico. Maximilian sought to improve Mexico for Mexicans and to improve Confederate-Mexican relations. This did not promote the race war that the Rothschilds wanted to foment between Mexico and America. The Rothschilds bankrupted Maximilian, and Mexico, and then reinstalled Juárez, who murdered Maximilian. It should be noted that in 1861 Juárez had provided the Rothschilds with the pretext for the initial French and British invasion of Mexico by failing to pay interest on Mexico's debts.

President Lincoln opposed the Rothschilds' designs on the American banking system. A Jewish actor named John Wilkes Booth assassinated Lincoln, and some claim the assassination was instigated by international bankers.[365] After sponsoring a seemingly endless series of dictators and revolutions in Mexico, the Rothschilds, through their agent "Colonel" Edward Mandell House, again sought a major war between Mexico and the United States in the Twentieth Century, which plan was spelled out in House's apocalyptic book *Philip Dru: Administrator*, B. W. Huebsch, New York, (1912).

On 30 October 1939, Congressman Thorkelson warned the American Congress that some Jews were out to destroy America with another world war and by seeding Mexico with Communist revolutionaries—an old Rothschild

---

[365]. Refer to the articles in *The Vancouver Sun* on 2 May 1934 and 4? and 5? May 1934 relating to Gerald Grattan McGeer's speech before the Canadian House of Commons, and the *Vancouver Daily Province* of 2 May 1934. ***See also:*** G. G. McGeer, *The Conquest of Poverty; or, Money, humanity and Christianity*, Garden City Press, Gardenvale, Quebec, (1935).

plan, which is still in the works and is a real and present danger to America's security,

"If House Joint Resolution 306, the present Neutrality Act, is passed as it is, it is my firm belief that such action on our part will bring about civil war in the United States, which may well terminate in the ultimate destruction of those in the invisible Government who sponsored this legislation and who are the silent promoters of the present war in Europe.

As the first step in consideration of this so-called Neutrality Act of 1939, please ask yourself, Who is it that wants war? It certainly is not the people that want war, and it is their wish that we must consider, as we are their Representatives in Congress.

Have any of your constituents asked you to vote for war, so that their children may be sent forth to drown in the Atlantic or die in the trenches of Europe? Are there any Members of Congress who want war? I do not believe so. Have you ever stopped to think, or have you tried to identify those whose greatest ambition is to aline this country in war on the side of England? I have not found anyone that wants war except those who harbor hatreds toward Hitler, and strange as it may seem, they are the same people who approved of Stalin.

Is it logical or reasonable that all Christian civilized nations, such as the United States, England, Canada, Australia, France, Germany, Austria, and other European nationalities, must engage in internecine conflict or war of extermination, so that this group of haters may get even with one man? Shall we sacrifice millions of our young men from 18 to 30 years of age to appease personal hatreds of a small group of international exploiters? I think not. I do not believe that there is any one person worth such sacrifice, whether he be king, prince, or dictator.

Let me now carry this argument a little further, for I want to call your attention to the fact that this same group that now hates Hitler was pro-German during the World War, and it is the same group that ruled and directed Germany's military machine before and during the World War. It is the same group that brought about inflation and exploited the German people, and it is the same group that furnished the money that brought about revolution in Russia and eliminated the Russian Army when its aid was needed to win the World War. This same group of internationalists paid and promoted the bloody invasion of Hungary, in which the invaders destroyed life and property with utter disregard for civilized warfare or even decency. It is this same group that has spread and nourished communism throughout the whole world and that sponsored the 'red' revolution in Spain. It is the same communistic group which is now concentrated south of us in Mexico, waiting to strike when the time is ripe.

Please ask yourselves if you are justified in giving the President the power set forth in this Neutrality Act, and are you justified in repealing the arms-embargo clause, when you know it is for no other reason except to aline the United States with Gr€at Britain in another war as senseless as the

World War. In considering this remember that there are no hatreds among the common people of the nations of the world, and for that reason no desire to destroy either life or property. Is it not time that we, the common people, learn a lesson—yes; a lesson in self-preservation instead of fighting for the 'invisible government'? Let us marshal this personnel into an army of their own and ship them some place to fight it out among themselves. It will be a blessing to civilization.

This contemplated war will not save the world for democracy because we have that now in the fullest measure; it is fully entrenched within the Government itself and in many organizations. We need no further evidence of that than the recent exposé of the League for Peace and Democracy, with its many members employed in strategic positions within the Federal Government, to further the cause of democracy and communism. No; this war will not be fought for so-called democracy or communism, for it is here, and is an evil that we will eventually be called upon to destroy or else be destroyed by it.

If the present agitation in Europe should terminate in an active war, its purpose will be to place all Christian civilized nations under the domination of an international government that expects to rule the world by the power of money and the control of fools who sit in the chairs of governments. I do not believe this will happen here, for the people are too well informed about this evil blight that is keeping the world at odds, and which is spreading dissension and hatreds by confusion and international intrigue. Let us shake off this evil, put our shoulders to the wheel, and push the carriage of state back on the road to sound constitutional government. Do not forget, if attack comes, it will be delivered by the Communists within the United States and next by the Communists who are waiting beyond our borders. Let us, therefore, give undivided attention to the Communists within our midst, for they have no place within a republican government. We should not tolerate foreign or hyphenated groups that, for reasons best known to themselves, cannot or will not assimilate to become Americans. For our own preservation we must get rid of those who cannot subscribe to the fundamental principles of this Republic, as set forth in the Constitution of the United States."[366]

Today, we again see the powerful forces of finance attempting to foment a war between Mexico and America. Some Mexicans are being duped into claiming the Southeastern United States as their national territory and agents of the warmongers are making outrageous statements so as to provoke Americans into an artificial animosity towards their Southern neighbors. It has always been in Americas best interest to have a thriving and friendly southern neighbor, just as it has always profited America to have a stable and successful neighbor to the

---

[366]. *Congressional Record: Proceedings and Debates of the 76th Congress: Second Session*, Volume 85, Part 1, United States Government Printing Office, Washington, D. C., (1939), p. 1068.

North, but Jewish interests have always oppressed the Mexican People and desire to stir up war and "racial" divisions on the North American Continent. Hardworking and good natured Mexicans are being blamed for all of America's ills, as if they had such power to bite the hand that meagerly feeds them.

The American media are teaching Americans to hate, instead of help, the long suffering Mexican People. It would be far better for America to have Mexico as an industrious and well-educated ally, than as a Communist satellite of a Red China controlled by Jewish financiers. The issue of illegal Mexican immigration to the United States is also being promoted as a rallying cry for an American revolution, which would only result in further oppression of the American People and the destruction of the America economy. It is a trap created by Jewish bankers to ruin the North American Continent. Many of the same persons calling for war with Mexico and revolution in the United States of America are also calling for a return to the gold standard, which would earn the Jewish bankers incredible profits on their gold reserves, and ultimately yield them all the gold in the Americas and eventually the world. These people are wittingly or wittingly baiting the trap with the promise of an American Utopia if only the Mexicans could be chased out, the American Government destroyed and a gold standard instituted. There are no Utopias, and the solution to Americas problems, which are still slight compared to those of the rest of humanity, are education, industry and responsible nationalism.

The roots of Jewish finance in America reach back into the prehistory of the United States. The Polish-Jewish Masonic-Frankist Haym Solomon (*also:* Salomon) was one of the financiers of the American Revolution. Other Jewish Freemasons of the Revolutionary Period include one of the founders of the Scottish Rite in American Freemasonry in the 1760's, Moses Michael Hays (*also:* Hayes), as well as Stephen Morin, Isaac da Costa, Rabbi Moses Sexias, Joseph Myers, Abraham Forst and Solomon Bush.[367] Many of these Jews, who brought with them the Frankist and Illuminati movements, were Bohemians. They were quite successful in America, and their descendants sponsored a wave of Jewish immigration to the United States in the European revolutionary period of 1848.[368] The *Encyclopaedia Judaica* writes in its article "Freemasons",

---

367. H. N. Casson, "The Jew in America", *Munsey's Magazine*, Volume 34, Number 4, (January, 1906), pp. 381-395, at 393. *See also:* S. Oppenheim, *The Jews and Masonry in the United States before 1810*, Samuel Oppenheim, New York, (1910); reprinted from: *Publications of the American Jewish Historical Society*, Number 19, (1910). *See also:* "Bush, Solomon", *The Universal Jewish Encyclopedia*, Volume 2, The Universal Jewish Encyclopedia, Inc., New York, (1940), p. 608. *See also: Jewish Calendar for Soldiers and Sailors: 1943-1944: 5704*, National Jewish Welfare Board, New York, (1943), pp. 15-17. *See also:* "Freemasonry", *The Jewish Encyclopedia*, Volume 5 Dreyfus-Brisac—Goat, Funk and Wagnalls Company, New York, (1903), pp. 503-505. **See** *also:* "Freemasons", *Encyclopaedia Judaica*, Volume 7 FR-HA, Macmillan, Jerusalem, (1971), cols. 122-125.
368. G. Kisch, *In Search of Freedom: A History of American Jews from Czechoslovakia: 1592-1948*, Edward Goldston, London, (1948). *See also:* M. Rechcigl, Jr., *Early Jewish Immigrants in America from the Czech Historic Lands and Slovakia:*

"In the U.S. Jewish names appear among the founders of Freemasonry in colonial America, and in fact it is probable that Jews were the first to introduce the movement into the country. Tradition connects Mordecai Campanall, of Newport, Rhode Island, with the supposed establishment of a lodge there in 1658. In Georgia four Jews appear to have been among the founders of the first lodge, organized in Savannah in 1734. Moses Michael *Hays, identified with the introduction of the Scottish Rite into the United States, was appointed deputy inspector general of Masonry for North America in about 1768. In 1769 Hays organized the King David's Lodge in New York, moving it to Newport in 1780. He was Grand Master of the Grand Lodge of Massachusetts from 1788 to 1792. Moses *Seixas was prominent among those who established the Grand Lodge of Rhode Island. and was Grand Master from 1802 to 1809. A contemporary of Hays, Solomon *Bush, was deputy inspector general of Masonry for Pennsylvania, and in 1781 Jews were influential in the Sublime Lodge of Perfection in Philadelphia which played an important part in the early history of Freemasonry in America. Other early leaders of the movement included: Isaac da *Costa (d. 1783), whose name is found among the members of King Solomon's Lodge, Charleston, in 1753; Abraham Forst, of Philadelphia, deputy inspector general for Virginia in 1781; and Joseph Myers, who held the same office, first for Maryland, and later for South Carolina. In 1793 the cornerstone ceremony for the new synagogue in Charleston, South Carolina, was conducted according to the rites of Freemasonry."[369]

The New York (Daily) Times reported on 27 October 1855, on page 4, that the Rothschilds were worth more than one billion dollars, which was more than the whole worth of New York City,

"The ROTHSCHILDS, according to their own estimate, possess $700,000,000 in personal property, exclusive of real estate, seignories, mines, &c., which amount to at least half as much more, making the enormous sum of over $1,000,000,000, or an amount much higher than the entire valuation of New-York City."

The New York Times reported on 10 December 1868, on page 4, that upon his recent death, Baron (James?) Rothschild had left an estate worth $400,000,000 dollars,

"Concentration of Wealth.
Baron ROTHSCHILD is reported to have died worth four hundred

<http://www.jewishgen.org/BohMor/early_immig.html>
369. "Freemasons", *Encyclopaedia Judaica*, Volume 7 FR-HA, Macmillan, Jerusalem, (1971), cols. 122-125, at 124.

millions of dollars, and greatly regretted by the poor of the neighborhood where he lived.

The first fact is an overwhelming proof of the tendencies of modern civilization toward *concentration;* for if we mistake not, this enormous fortune is the growth of but three generations. This tendency to concentration is evident in the alarming growth of our great cities both in this country and in Europe. London increased in the decade from 1851 to 1861 some half a million souls, (440,000)—equal to a city like Philadelphia. The great increase of *our* cities is not accounted for by the increase of the whole country, but is caused by the same tendencies which are at work in the Old World.

Not only is this tendency evident in these great cities, but it is found in all departments of manufacture and trade. To-day and with us it is strikingly apparent in such bazaars as STEWART's and CLAFLIN's. Already under these two roofs, which cover one to two acres each, are gathered all that pertains to the dress of men and women, all fabrics for household uses and many other articles. Thus a retail dealer may here select a complete stock, may have it packed and shipped without wetting the soles of his shoes.

There is no reason why this class of merchants shall not or do not undersell all smaller ones. They buy larger quantities, they control mills, they do their business with one set of experienced men, *and they can undersell others even if they do not.*

Now all mankind will buy where they can buy cheapest, and there is no reason to doubt but that such establishments as these, in another generation of thirty years, will swallow up all the smaller ones. It is inevitable. And why not? Because the great end and aim of life is not to buy cheap and sell dear, but is to so organize society that it shall perfect the bodily and spiritual welfare of the race, and insure comfort and happiness to men. Will this vast aggregation of business and wealth in few hands secure this? It is evident that it will not, unless along with the rights of wealth go the duties of wealth.

We hear a rumor that some of our great capitalists propose to erect, near their wonderful establishments, perfect and complete houses for their clerks and workmen, supplied with every convenience, to be furnished at *the smallest cost.* There will be reading rooms, rooms for games and for exercise, a good public table, a coöperative laundry, bath-houses, &c.

We venture to hope that this is true, and that our millionaires propose to *use* their money as well as to get it, and thus to mark the civilization of the New World by a beneficent use of wealth, and not by its vulgar display."

The New York Times reported on 18 August 1874, on page 2, that upon his death, Baron Anselm De Rothschild left an estate worth one billion francs,

"A ROTHSCHILD.

Baron Anselm De Rothschild, who died recently at his country seat at Dobling, near Vienna, was a man of extreme simplicity of character, in spite of his immense wealth. By his express desire, recorded in his will, his funeral was celebrated without any pomp. A hearse drawn by two horses, followed only be a few servants, conveyed the body to the North-eastern station, whence it was taken to the domain in which the family vault is situated. The fortune of the Baron is estimated at more than a thousand millions of francs."

The Rothschilds made so much money from spreading war around the world, that by 1875 their wealth had eclipsed that of most nations, as *The Chicago Tribune* reported on 27 December 1875, on page 8,

"The Rothschilds.
New York Sun.
The combined capital of the Rothschilds is stated by Emile Burnouf, the well-known publicist, to have attained in the present year to the almost incalculable sum of seventeen billions of francs, or $3,400,000,000. The significance of these stupendous figures may be rudely conceived by comparison, but there is nothing in the history of private wealth with which they can be compared. The capital of the Barings, the estates of Lord Dudley, the Marquis of Bute, and the head of the family of Grosvenor, belong relatively to a humble category, to which the City of New York has contributed the fortunes of Astor, Vanderbilt, and Stewart. The financial resources attributed to the Rothschilds can best be measured by contrasting them with the funded debts of the richest countries on the globe. The capital of this house, as estimated by M. Burnouf, is about equal to the whole funded debt of Great Britain, or that of France, and considerably exceeds the National debt of the United States. A single century, or the possible span of one man's life, has sufficed for the accumulation of this fortune, and the rise of its authors from a shabby rookery in Frankfort to the financial domination of Europe. At the period of Rothschild's first decisive triumph on the London Exchange—the day after Waterloo, just sixty years ago—John Jacob Astor was already a rich man. The great fortune which the latter bequeathed is not believed to exceed $50,000,000, while the inheritance of his Hebrew contemporary has been swollen to more than sixty times that sum. Although its territories are not to be found on any map, and the names of its representatives are set off with no princely dignities, nevertheless the House of Rothschild must be reckoned among the foremost war-sustaining and world-compelling powers of the earth."

The following article appeared in the "Foreign Affairs" section of the *National Repository, Devoted to General and Religious Literature, Criticism, and Art*, Volume 7, (February, 1880), pp. 168ff.,

"WHAT BARON ROTHSCHILD DOES FOR HIS FAVORITE HORSE.—It is not the fate of many to be a Rothschild. But there is many a poor man who will envy not only the rich bankers by that name, but even the horse the Baron Rothschild, of Vienna, has come to regard as his favorite. For the accommodation of this dumb, though attractive, animal he has had a special loose box built at the cost of twelve thousand dollars. This elegant room forms a part of a new stable which cost only eighty thousand dollars. It has marble floors, encaustic tiles painted by distinguished artists, rings, chains, and drain-traps of silver, and walls frescoed with splendid hunting scenes from the pencils of eminent animal painters. Fortunately, however, the baron's annual income is $1,600,000."

The Rothschilds were loan sharks to the nations. They would run a nation into debt by provoking wars, or destroying economies, or talking leaders into self-ruin, then they would foreclose on the nations by demanding more wars— race wars, religious wars, economic wars, trade wars, vendetta wars, utterly senseless wars, etc. Many have alleged that the wars of Napoleon and most since, including both world wars, were brought about by the bankers to reap profits, and more significantly to fulfill Jewish prophecies and create a Jewish State in Palestine. Even France's involvement in Algiers may have begun at the instigation of Jewish interests, on the pretext of an insult on the French Consul by the Dey in 1830. *The North American Review* wrote in 1845,

"The Moors seem to consider the Jews born to serve them and bear their wanton insults. The Moorish boys torment the Jewish children for pastime; and the men, with impunity, maltreat the male adults, and take the grossest liberties with the females. In 1804, many of them were subjected to horrible tortures in Algiers, merely because they had unsuspiciously lent money to certain political conspirators; and they were not released till they had paid an exorbitant ransom. In 1827, the Dey extorted from a rich Jew, by throwing him on some pretence into prison, 500,000 Spanish dollars. But the French occupation of Algiers has greatly improved the condition of this people in that country; and, in consequence, their numbers have increased by immigration."[370]

Those Christian leaders who were traitors to their Gentile followers, encouraged their Christian believers to accept destruction and death as the fulfillment of prophecy, Jewish prophecy deliberately fulfilled by heartless and cruel Jewish leaders. These traitors instructed their gullible followers to see their own demise, for the sake of Jewish profits, as a beautiful and supernatural event. This has been going on in England at least since the time Cabalists brought Jews and Judaism to England with the aid of "Christian" leaders including Oliver Cromwell and "Christian" propagandists including Isaac Newton and Samuel Clarke, who were Cabalist religious Jews who denied the Trinity, and who called on Christians to welcome the end of the world in apocalyptic horrors as if it would be a joyous event, an event which would enslave them to the Jews, destroy

---

370. "The Modern Jews", *The North American Review*, Volume 60, Number 127, (April, 1845), pp. 329-368, at 338-339.

their nations, and give all of their wealth and power to a Jewish King under the false promise that a new world would emerge, a false promise on which they would never have to make good. This madness of self-destruction imposed on Christians by Jewish Zionists and their agents has culminated in the apocalyptic desires of Dispensationalist Christians, who slavishly promote the evils of Israel and eagerly await a nuclear holocaust which will destroy human life on Earth.[371]

Jews sought to be readmitted to England in order to profit from English wealth and trade, but also, as Menassah Ben Israel declared, to fulfill the prophecy that Jews would occupy the ends of the Earth (*Genesis* 12:3; 28:14. *Deuteronomy* 28:64-66. *Isaiah* 27:6; 49:6. *Jeremiah* 24:9; *Daniel* 12:7). Jews felt they had to be readmitted to England before the Messiah could come, and that their readmission to England would herald the coming of the Messiah. Zionist Joachim Prinz wrote in his book *The Secret Jews*,

> "After a year in London, ben Israel was granted an annual stipend of one hundred pounds. Although his mission had succeeded and his petition had provided Cromwell with the excuse he wanted to admit the Jews to England, ben Israel was disappointed. He had wanted a solemn declaration by the Lord Protector, or at least a meeting of Parliament, which would have recognized the religious, Messiah-oriented reasons why this should be done. He wanted a proclamation heralding the coming of the Messiah now that the prophecy of Daniel had been fulfilled."[372]

A virtual confession of the Rothschild's corruption, corruption that would spill oceans of blood in the Twentieth Century, appeared in *The Chicago Daily Tribune* on 27 June 1880 on page 9, where a plan is laid out for the First and Second World Wars:

"MODERN PALESTINE.
ANCIENT JUDEA TO BE CONVERTED INTO A JEWISH COLONY.

The Cologne *Gazette* of a recent date says that among the Orthodox Israelites and Christians unfriendly to the Israelites this has always been a favorit cry: 'Palestine for the Jews!' and has gained strength in proportion as the power of the present political ruler over the 'beloved land' wanes away. The English preacher, Nugee, who has interested himself in this

---

371. G. Halsell, *Prophecy and Politics: Militant Evangelists on the Road to Nuclear War*, Lawrence Hill & Co., Westport, Connecticut, (1986); **and** *Prophecy and Politics: The Secret Alliance Between Israel and the U. S. Christian Right*, Lawrence Hill & Co., Westport, Connecticut, (1986); **and** *Forcing God's Hand: Why Millions Pray for a Quick Rapture—and Destruction of Planet Earth*, Crossroads International Pub., Washington, D.C., (1999), Amana Publications, Beltsville, Maryland, (2003); **Turkish:** M. Acar, H. Özmen, *et al.* translators, *Tanri'yi kiyamete zorlamak: Armagedon, Hristiyan kiyametçiligi ve Israil = Forcing God's Hand : Why Millions Pray for a Quick Rapture: And Destruction of Planet Earth*, Kim, Ankara, (2002).
372. J. Prinz, *The Secret Jews*, Random House, New York, (1973), p. 110.

matter, expounded on the 14th of the month, in a public lecture, a plan which of late has assumed a practical shape. The Englishman, Oliphant, has laid the plan before the Sultan. It is that the land of Gilead and Moab, embracing the whole territory of the Israelitish tribes of Gad, Reuben, and Mannasseh, shall be converted into a Jewish colony, the Sultan being paid in cash for the territory, a proposition which the Sultan has already favorably entertained. Still more, Goschen, the recently-appointed Ambassador Extraordinary of England, at Constantinople, has expressed himself as well disposed toward the furtherance of the plan. The territory in question embraces about 1,500,000 English acres, and is at present inhabited only by nomadic tribes. The colony is to remain subject to the Turkish power, while yet its immediate Governor is to be an Israelite. In this manner Judaism is to regain a firmer foothold in its own land, and the colony itself ultimately become a rallying point for the scattered people of Israel, around which it is hoped an ever-broadening girdle of new settlements will form itself. The purchase money for the territory of the new colony is to be contributed by the freewill offerings of patriotic Israelites. Two railroads or highways are to be built, the one ascending from Jaffa to Jerusalem, the other extending from Haifa to the further side of the Jordan. Sir Moses Montefiore has already interested himself in these significant enterprises, furnishing material aid for the same. For the construction of the road to Jaffa the Turkish Government has already made a concession, with the proviso that work shall be commenced upon it by next January at the farthest. Still further, the construction of a ship canal from the Mediterranean to the Gulf of Akabe and the Red Sea is contemplated. Palestine is again to be reopened, under the influence of the ideas of the nineteenth century, if only the Jews themselves are ready with their contributions and their settlements for their own land.'

Another paper, also, the London *Times*, has the following: 'A negotiation is said to be on foot between the members of the house of Rothschild and the venerable Sir Moses Montefiore on the one hand, and the Ottoman Government on the other, for the cession, under certain conditions, of the Holy Land. The Ottoman Government is already at its last gasp, for want of ready money. The Jewish race wish a '*habitat*' of their own. As the Greeks, though a scattered people, living for the most part in Turkey, have a Greek Kingdom, so the Jews wish to have a Hebrew Kingdom. This, it will be remembered, is the leading idea of George Eliot's Daniel Deronda.' Few persons, and probably the gifted authoress herself not more than others, imagined that the dream of the Mordecai of those pages was in the least degree likely so soon to be realized. Information as to the nature of the new Jewish State, whether it is to be theocratic or royal, is uncertain, but the arrangements in reference to it are in progress. Prophecies have a way of fulfilling themselves, more especially when those who believe in them are possessed of the sinews of Government. The day when 'the Dispersed of Israel' are to be gathered into one is confidently looked forward to, not only by Hebrews, but by multitudes of Christians. The author of 'Alroy' would be gathered to his fathers in greater peace, were he permitted under his Administration to see this day and be glad. Superstitious persons, who think that the end of the world is to be preceded by the restoration of the Jews to Palestine, will be inclined to lend serious belief to Mother Shipton's prophecy that this earth is to see its last days in 1881.'

These extracts are significant, and specimens of long articles that have appeared of late in the European press, secular as well as religious. Whatever some people may think of prophecy, it is clear that a grand movement is on foot for the regeneration of Palestine. The 'Holy Land' looms up with every agitation of the Eastern question, and is, in fact, its central point. As to population, Jerusalem has now 20,000 Jews, a larger number than the Turks and Christians combined, not to name the Russian colony outside. Forty years ago, the population was only 300, and only within ten years was it allowed outside the Ghetto. The Jewish population of Palestine is greater to-day than ever since the Roman expulsion. Andree and Pescher's 'Statistical Atlas' puts the sum total of Jews in the world at 7,000,000, the number in Solomon's time. In Europe the Latin group of Jews is 89,000; the Teutonic 842,000; the Slavonic, 4,047,000; in all 4,978,000. In Asia there are 800,000. In Africa, 600,000. The figures 150,000 for the United States are far too low.

The interest in Palestine is shown by the International Exploration Society. Its 'Great Map of Palestine,' drawn on a scale of one inch to a mile, will surpass all others, and, under the direction of the British Ordnance Survey Department, will show 'every detail of ruin and village, ancient and modern, aqueducts, plantations, roads, dells, synagogs, tombs, temples, castles, forts, Crusading and Saracenic, wadies, fountains, seas, mountains, rivers, plains, springs, and wells.' The preparation is extensive, and the progress has already begun. Jewish synagogs and hospitals are multiplied. The German Jews have already sixteen charity institutions and twenty-eight congregations. The tide of immigration is setting in strongly, and the appointment of Midhat Pasha as Syrian Governor gives promise of brighter days for Palestine. A Venetian Jew has given 60,000 francs for the establishment of an agricultural school in the Plain of Sharon, and Baron Albert de Rothschild has just guaranteed to the ex-Mayor of Jerusalem a large pecuniary contribution for the construction of the Jaffa-Jerusalem Railroad. The *South German Wochenblatt* reminds its readers that the great banking-house of the Rothschilds, at the time of the last loan of 20,000,000 francs to Turkey, accepted as security a mortgage on Palestine, and adds that 'as it is impossible for a bankrupt State, like Turkey, to pay back the money, the Israelites may now count upon their return to the Land of Promise as a certainty.'

A proposition is now under discussion, since a concession has been made to the French for the Euphrates Valley Road, to make a junction between the latter from the old provinces of Assyria to Jerusalem the plan of Gen. Sir Frederick Goldsmid, a Jew whose munificence to the Turkish Jews is so well known, and whose distinguished relative, Francis Goldsmid, a few years ago acted as reference in the question of the Persia and Afghanistan boundary. The interpreters of prophecy in reference to Israel's future have quoted Isaiah, chapter xix., 23, as a prediction whose fulfillment this enterprise seems to favor in some way. The text is this: 'In that day there shall be a highway out of Egypt to Assyria, and the Assyrian shall come into Egypt, and the Egyptian into Assyria, and the Egyptians shall serve with the Assyrians.' It is thought to foreshadow a tripartite alliance between Israel, Egypt, and Assyria, in the future of the Hebrew races, when converted. Then the next verses are quoted: 'In that day shall Israel be the third with Egypt and with Assyria, even a blessing in the midst of the land, whom the Lord will bless, saying, Blessed be Egypt my people, and Assyria the work of my hands, and Israel my inheritance.' It is agreed that no alliance has ever yet taken place.

The usual objection that Palestine is incapable of supporting a dense population is set aside by the testimony of the late United States Consul-General, who writes from Jaffa: 'An abundant supply of water could be brought to the city from the pools of Solomon, were it not that all efforts are thwarted by the Moslem rulers. The land of Palestine is extremely productive, and were colonies planted here, as they are in Australia, New Zealand, and the United States, there is no reason to doubt their success.' Arnold, the celebrated historian, who traveled over it, says, 'The old abundance is still sleeping in the soil of Palestine, and it needs not any miracle, but industry, to bring back the wealth and beauty of the early ages of the Hebrew Monarchy.'

What adds interest to the Jewish question is the discoveries made by scholars of the whereabouts of the lost 'Ten Tribes,' or the tribes of the Northern Kingdom, carried away by Shalmaneser, a century before the Babylonian exile of Judah, the Southern Kingdom. It seems to be established that the Jews in Afghanistan and in the Caucasus, and those in China, with the 200,000 Falashas in Abyssinia, are all descendants from the Ten Tribes. The wonderful increase, too, of Mohammedanism, outstripping Christianity the last ten years as a proselyting religion, and the growing belief of orthodox Moslems that the decay of the Ottoman power is a sign of the end of the world and the judgment day, attract attention. The special interest Englishmen take in the whole question is very marked. Politically, what England wants is a strong power in Syria to protect the Alexandrian Road and Suez Canal from Russian assault. Jewish nationality would solve that problem, provided England had the protectorate. This involves the dispossession of the Turks and overthrow of their Government, and a conflict of nations for the possession of Palestine and dominion of the East and the world. That means a general Asiatic, European, and African struggle, with Jerusalem the objective. This, too, is interesting. With Egypt and Greece already existing, if diplomacy erects Syria and Thrace into two separate Kingdoms, then modern history reproduces the four Kingdoms into which Alexander's Empire was broken up, and points to Syria as the spot where the last enemy of the Jews appear in the last struggle. Out of Syria,

Antiochus Epiphanes came, and it is thought that out of Syria, again, according to the prophecy of Daniel, in his eleventh chapter, the last Antichrist will arise. The discussions in the press and magazines are many and full of interest. One of England's Bishops has just said: 'If ever the question is raised, and it may be raised very soon, Shall the Jews be inducted into their patrimonial land as tenants at will? no matter by whom the proposition is made, or for what purpose,—even hostile to England,—it will be England's duty not to oppose but to assist, or at least permit Israel to be restored, unconverted.' This is the general tone of Christendom. The 'Reformed Jews'—i. e., the Rationalists—are laughing, or mocking."

The Rothschilds owned the Pope and Rome. The question naturally arises whether the Pope was simply reckless with the finances of the Church, or if he was an agent of Rothschilds, who intentionally ran up the debts of the Church. The Jews had always believed that the Kings, Queens, Princes and Princesses of the Gentiles, in other words, all Gentile leaders, are destined to be the Jews' obedient slaves. *Exodus* 19:5-6 states,

"5 Now therefore, if ye will obey my voice indeed, and keep my covenant, then ye shall be a peculiar treasure unto me above all people: for all the earth *is* mine: 6 And ye shall be unto me a kingdom of priests, and an holy nation. These *are* the words which thou shalt speak unto the children of Israel."

Numbers 24:17-20 states,

"17 I shall see him, but not now: I shall behold him, but not nigh: there shall come a Star out of Jacob, and a Sceptre shall rise out of Israel, and shall smite the corners of Moab, and destroy all the children of Sheth. 18 And Edom shall be a possession, Seir also shall be a possession for his enemies; and Israel shall do valiantly. 19 Out of Jacob shall come *he* that shall have dominion, and shall destroy him that remaineth of the city. 20 ¶And when he looked on Amalek, he took up his parable, and said, Amalek *was* the first of the nations; but his latter end *shall be* that he perish for ever."

Numbers 33:50-56 states,

"50 And the LORD spake unto Moses in the plains of Moab by Jordan *near* Jericho, saying, 51 Speak unto the children of Israel, and say unto them, When ye are passed over Jordan into the land of Canaan; 52 Then ye shall drive out all the inhabitants of the land from before you, and destroy all their pictures, and destroy all their molten images, and quite pluck down all their high places: 53 And ye shall dispossess *the inhabitants of* the land, and dwell therein: for I have given you the land to possess it. 54 And ye shall divide the land by lot for an inheritance among your families: *and* to the more ye shall give the more inheritance, and to the fewer ye shall give the less inheritance: every man's *inheritance* shall be in the place where his lot falleth; according to the tribes of your fathers ye shall inherit. 55 But if ye will not drive out the inhabitants of the land from before you; then it shall come to pass, *that those* which ye let remain of them *shall be* pricks in your eyes, and thorns in your sides, and shall vex you in the land wherein ye dwell. 56 Moreover it shall come to pass, *that* I shall do unto you, as I thought to do unto them."

Deuteronomy 7:6 states,

"For thou *art* an holy people unto the LORD thy God: the LORD thy God hath chosen thee to be a special people unto himself, above all people that *are* upon the face of the earth."

Deuteronomy 28:10 states,

"And all people of the earth shall see that thou art called by the name of the LORD; and they shall be afraid of thee."

Psalm 2:1-12 (*see also: Sukkah* 52*a*-*b*) states:

"Why do the heathen rage, and the people imagine a vain thing? 2 The kings of the earth set themselves, and the rulers take counsel together, against the LORD, and against his anointed, *saying*, 3 Let us break their bands asunder, and cast away their cords from us. 4 He that sitteth in the heavens shall laugh: the Lord shall have them in derision. 5 Then shall he speak unto them in his wrath, and vex them in his sore displeasure. 6 Yet have I set my king upon my holy hill of Zion. 7 I will declare the decree: the LORD hath said unto me, Thou *art* my Son; this day have I begotten thee. 8 Ask of me, and I shall give *thee* the heathen *for* thine inheritance, and the uttermost parts of the earth *for* thy possession. 9 Thou shalt break them with a rod of iron; thou shalt dash them in pieces like a potter's vessel. 10 Be wise now therefore, O ye kings: be instructed, ye judges of the earth. 11 Serve the LORD with fear, and rejoice with trembling. 12 Kiss the Son, lest he be angry, and ye perish *from* the way, when his wrath is kindled but a little. Blessed *are* all they that put their trust in him."

Psalm 18:40-50 states,

"40 Thou hast also given me the necks of mine enemies; that I might destroy them that hate me. 41 They cried, but *there was* none to save *them: even* unto the LORD, but he answered them not. 42 Then did I beat them small as the dust before the wind: I did cast them out as the dirt in the streets. 43 Thou hast delivered me from the strivings of the people; *and* thou hast made me the head of the heathen: a people *whom* I have not known shall serve me. 44 As soon as they hear *of me*, they shall obey me: the strangers shall submit themselves unto me. 45 The strangers shall fade away, and be afraid out of their close places. 46 The LORD liveth; and blessed *be* my rock; and let the God of my salvation be exalted. 47 *It is* God that avengeth me, and subdueth the people under me. 48 He delivereth me from mine enemies: yea, thou liftest me up above those that rise up against me: thou hast delivered me from the violent man. 49 Therefore will I give thanks unto thee, O LORD, among the heathen, and sing *praises* unto thy name. 50 Great deliverance giveth he to his king; and sheweth mercy to his anointed, to David, and to his seed for evermore."

Psalm 72:8-11 states,

"8 He shall have dominion also from sea to sea, and from the river unto the ends of the earth. 9 They that dwell in the wilderness shall bow before him; and his enemies shall lick the dust. 10 The kings of Tarshish and *of* the isles shall bring presents: the kings of Sheba and Seba shall offer gifts. 11 Yea, all kings shall fall down before him: all nations shall serve him."

Psalm 110:1-7 states,

"The LORD said unto my Lord, Sit thou at my right hand, until I make thine enemies thy footstool. 2 The LORD shall send the rod of thy strength out of Zion: rule thou in the midst of thine enemies. 3 Thy people *shall be* willing in the day of thy power, in the beauties of holiness from the womb of the morning: thou hast the dew of thy youth. 4 The LORD hath sworn, and will not repent, Thou *art* a priest for ever after the order of Melchizedek. 5 The Lord at thy right hand shall strike through kings in the day of his wrath. 6 He shall judge among the heathen, he shall fill *the places with* the dead bodies; he shall wound the heads over many countries. 7 He shall drink of the brook in the way: therefore shall he lift up the head."

Isaiah 40:23 states,

"That bringeth the princes to nothing; he maketh the judges of the earth as vanity."

Isaiah 49:7 states,

"Thus saith HaShem, the Redeemer of Israel, his Holy One, to him who is despised of men, to him who is abhorred of nations, to a servant of rulers: kings shall see and arise, princes, and they shall prostrate themselves; because of HaShem that is faithful, even the Holy One of Israel, who hath chosen thee. [Masoretic Text Version of the Jewish Publication Society]"

Isaiah 49:23 states,

"And kings shall be thy nursing fathers, and their queens thy nursing mothers: they shall bow down to thee *with their* face *toward* the earth, and lick up the dust of thy feet; and thou shalt know that I *am* the LORD: for they shall not be ashamed that wait for me."

Isaiah 60:12 states,

"For the nation and kingdom that will not serve thee shall perish; yea, *those* nations shall be utterly wasted."

Isaiah 61:9 states,

"And their seed shall be known among the Gentiles, and their offspring among the people: all that see them shall acknowledge them, that they *are* the seed *which* the LORD hath blessed."

Jeremiah 10:10 states,

"But the LORD *is* the true God, he *is* the living God, and an everlasting king: at his wrath the earth shall tremble, and the nations shall not be able to abide his indignation."

Ezekiel 39:17-18 states,

"17 ¶And, thou son of man, thus saith the Lord GOD; Speak unto every feathered fowl, and to every beast of the field, Assemble yourselves, and come; gather yourselves on every side to my sacrifice that I do sacrifice for you, *even* a great sacrifice upon the mountains of Israel, that ye may eat flesh, and drink blood. 18 Ye shall eat the flesh of the mighty, and drink the blood of the princes of the earth, of rams, of lambs, and of goats, of bullocks, all of them fatlings of Bashan."

Micah 17:16-17 states,

"The nations shall see and be confounded at all their might: they shall lay *their* hand upon *their* mouth, their ears shall be deaf. 17 They shall lick the dust like a serpent, they shall move out of their holes like worms of the earth: they shall be afraid of the LORD our God, and shall fear because of thee."

Zechariah 14:9 states,

"And the LORD shall be king over all the earth: in that day shall there be one LORD, and his name one."

One can imagine how quickly the Rothschilds could seize power over Europe and the world if they placed monarchs, heads of state, and church leaders in power, who were their agents, and who intentionally ran up their nations' debts and deliberately brought their nations into wars, and into ruin. There are various means to gain control over a leader: threats, blackmail, bribery, flattery, fame, megalomania, messiah complex, etc. A leader may also be placed in power who already has allegiance to a specific cause due to his or her ethnicity, family history, etc. Once a sovereign of one sort or another is controlled and creates debts which are not paid by the wealthy, but by the comparatively poor, those poor must slave forever to pay off those debts. Not only do the immensely wealthy earn the interest on the debt, that interest accrues to monies which were never truly taxed—this while the immensely wealthy disproportionately reap the benefits of citizenry. It was important to the Rothschilds to not only accrue wealth, but also to prevent Gentiles from accruing wealth and thereby gaining control over their own destinies.

The Chicago Tribune reported on 27 February 1867 on page 2,

"The Rothschilds of Rome.

[Rome Correspondence of the London News.]

Who, whether he has set foot in the Eternal City or no, has not heard of the Torlonias—the Rothschilds of Rome? In the course of last summer, when the monetary crisis here was at its height, Don Alessandro Torlonia—the acting head of the house—won extraordinary popularity by writing a letter to the Pope, in which he offered to buy up the unconvertible Government paper, and substituting a metal currency in its place, providing that the existing managers of the Roman Bank, with Cardinal Antonelli's brother at their head, were sent about their business, and the direction confided to himself. At that time it was quite impossible to get notes converted into coin at any price for the simple reason that there was no coin in the bank. Even now, when things have improved somewhat, it is with the utmost difficulty that you can get change for a scudi note, even at shops in the Corso, and there is not a hotel keeper or a tradesman in Rome who would even look at a five scudi note if you were sufficiently ignorant of the state of things here to present it in payment in the expectation of getting any change out. Of the small pieces of silver, which you obtain with no little difficulty, many are so worn and thin that they seem in a sort of transition state between sliver and paper, and have long since lost all trace of any image or superscription whatever.

So rolling in wealth is Don Alessandro Torlonia that his riches are admitted to be literally untold, and only this much is known certain, that everything in Rome worth having, except the Pope and St. Peter's, already belongs to him. No wonder then that at the Vatican Don Alessandro should be looked upon as a hardly less dangerous character than Victor Emanuel himself, and that the insulting offer which he made last summer to buy up the Holy Father, and add him so his possessions, should have been decidedly rejected, though it had not entailed the removal of an Antonelli from a lucrative place. On his first appearance in public after making the above mentioned patriotic offer, Don Alessandro received such an ovation as has not been witnessed in Rome since those of which Pius IX. was himself the object, when he gave the first impulse to the Italian Revolution in 1846. This Don Alessandro is the same Torlonia who risked his whole fortune on the gigantic enterprise of draining the Fucine Lake, the issue of which struggle with nature was so long doubtful that it became a common saying in Rome, 'Either Torlonia will drain the Fucine Lake, or the Fucine will drain Torlonia.' In the end, however, Torlonia got the better of the lake, and redeemed about one hundred thousand acres of land for cultivation. Over what was a few years ago a barren waste of waters, flourishing crops may now be seen waving every harvest time, and with last year's produce Don Alessandro had a scheme of feeding the now almost starving Roman people by selling them bread of his own baking at a reduced rate. Such, at least, was the account of the story given me by a patriotic and exceedingly liberal Roman, who made a severe case against the Government out of the stoppage of Torlonia's extensive bread baking-by-machinery works, which threw some two hundred workmen out of employment just a fortnight ago. I am bound, however, to add that, on proceeding to the spot and making inquiries, I learned quite a different version of the affair, entirely exculpating the Government from any direct interference in the matter. Only this much is certain, that the works are stopped, and that the Roman people stand little chance, at present, of getting their bread at reduced rates."

On 2 June 1867, *The Chicago Tribune* reported on page 3,

"THE ROTHSCHILDS AND THE POPE.

For fifteen centuries the Jews have been cursed by the Pope, and persecuted by the Roman Church. There is no more revolting chapter of horrors in history than that of the treatment of the Jews at the hands of the Pontiffs. In all lands where the Roman religion is dominant the children of Israel have been treated with barbaric rigor—allowed few privileges, denied all rights, looked upon as a people accursed of God, and set apart by divine ordination to be trampled upon by the church. In Rome, at the present day, the Jews are confined to the Ghetto; they are not allowed to set up a shop in any other part of the city; they cannot leave the city without a permit; they can engage only in certain trades; they are compelled to pay enormous taxes into the Papal treasury; the are subject to a stringent code of laws established by the Pope for their special government; they are imprisoned and fined for the most trivial of offences. They cannot own any real estate in the city; cannot build or tear down or remodel any dwelling or change their place of business, without Papal permission. They are in abject slavery, with no right whatever, and entitled to no privileges, and receive none, except upon the gracious condescension of the Pope. In former times they were unmercifully whipped and compelled to listen once a week to the *Christian* doctrine of the priests. But time is bringing changes. The Pope is in want of money; and the house of the red shield has money to lend on good security. The house is always ready to accommodate Governments. Italy wants money, so she sells her fine system of railroads to the Rothschilds. The Pope wants money, and he sends his Nuncio to the wealthy house of the despised race, offers them security on the property of the church, the Compagna, and receives ten million dollars to maintain his army and Imperial State. That was in 1865. A year passes, and the Pontifical expenditures are five million more than the income, and the deficit is made up by the Rothschilds, who take a second security at a higher rate of interest. Another year has passed and there is a third great annual vacuum in the Papal treasury of six million, which quite likely will be filled by the same house. The firm can do it with as much ease as your readers can pay their yearly subscription to the weekly *Journal*. When will the Pope redeem his loan at the rate he is going? Never. Manifestly the day is not far distant when these representatives of the persecuted race will have all the available property of the Church in their possession. Surely time works wonders."

On 24 December 1893, *The Chicago Daily Tribune* reported, on page 6,

"INCOME AND EXPENSES OF THE POPE.

Economy Necessary Because of the Continual Decrease in the Revenues.

Since the heavy losses made by the Pope a year or more ago the finances of the Vatican have been superintended with great care. 'It is known,' says a Paris paper, 'that a committee of prelates and several Cardinals exists at Rome whose duty it is to regulate the use of the sums of money which flow into the treasury of the Vatican. These sums come principally from two sources: The revenues of the property possessed by the Pope and the gifts of the faithful, known as Peter's Pence. The property of the Vatican is of various kinds, but the greater part of it consists of money or bonds, placed in England and France, under control of the Paris house of Rothschild. Peter's Pence is an annual revenue which far from being fixed. In good years the total of the sum received from all countries of the world reaches 8,000,000 francs. Sometimes it is as low as 6,000,000 and even 5,000,000. This has been the case for the last five years. This diminution is due, in great part, to the discord between the Royalists and the French Catholics produced by the republican policy of the Pope. France alone furnished two-thirds and often three-quarters of Peter's Pence. And in France it is the royalists who prove themselves most generous. But since the adhesion of Leo XIII. to the republic many of them, more Royalist than Catholic, have closed their purses to the Pope. However, despite all this, French Bishops still forward the largest sums to his Holiness. Thus, the Bishop of Nante sent a few days ago 100,000 francs from his flock as their gift to the Vatican treasury.

'Italy,' adds the *Journal*, 'contributes only a small part of the revenue—a few hundred thousand francs a year. The Romans show themselves in this regard less generous than other Italians. On the other hand, the Anglo-Saxon countries—England, Ireland, Australia, and the United States—begin to send important sums. If Catholicism continues to grow in these countries, it is easy to see that in time the Vatican will draw considerable sums from them.

'Again, there are the royal courts, such as that of Austria, which send annually rich presents to the Pope. This is even true of princes of ancient Italian families. Francis II., ex-King of Naples, and Maria Theresa, formerly Grand-Duchess of Tuscany, never fail to send their offerings, which consist of several thousands of francs. The Comte de Chambord was accustomed to give annually 50,000 francs; the Count of Paris sends the same sum.

'The expenses of the Vatican,' continues the writer, 'amount annually to more than 7,000,000 francs. They are regulated as follows: for the personal wants of the Pope, 500,000 francs; for the Cardinals, 700,000; for poor dioceses, 400,000; administration of the Vatican, 1,800,000; Secretary of State, 1,000,000; employés and ablegates, 1,500,000; support of schools and poor, 1,200,000.

'The Cardinals at Rome live at the expense of the Pope. The income of each from this source is at least 22,000 francs. The Secretary of State is charged with upholding relations with foreign governments by the mediation of nuncios. The four most important—Paris, Vienna, Madrid, and Lisbon—each receive an allowance of 60,000 francs a year.

'The last jubilee of Pope Leo XIII. brought to the Vatican 3,000,000 francs. At the first, celebrated five years ago, 12,000,000 francs were received. In the course of years the Pope has introduced a number of economies in the different branches of the Vatican service, and for that reason he has been called miserly. This accusation is not merited; the economies became necessary in a State whose expenses are considerable and whose revenues continue to diminish. Leo XIII. has many reasons to follow the example of his illustrious predecessor, Sixtus, as it is difficult in the present time to count on the generosity of the faithful.'"

There was even talk of making the Pope, who was owned by the Rothschilds, the King of Palestine, thereby making Rothschild King of Palestine by proxy; and, in the minds of Protestants, making the Pope the anti-Christ. This would have enabled the Rothschilds to take Palestine from the Turkish Empire, install the Pope as King, and then unseat him as the "anti-Christ" and replace him with the allegedly "neutral" Jewish Kingdom of the Rothschild dynasty. *The Chicago Tribune* reported on 4 June 1887 on page 5,

"The Pope for King of Palestine.
VIENNA, June 3.—The *Algemeine Zeitung* mentions that a project is hinted at to make the Pope the King of Palestine under a guarantee of protection on the throne by all the Catholic Powers."

The Catholics gave their money to the Popes, who gave it the Catholics' enemies, the Rothschilds to finance the destruction of Catholicism via Christians who had been essentially converted to Judaism *viz.* Protestantism, and the anti-Catholic Jewish press. Numerous European nations ran themselves into debt fighting wars and the only beneficiaries were the bankers and arms manufacturers—the Rothschilds gave the monarchies some wealth to flatter them and control them, then the Rothschilds betrayed them and destroyed them. Continually, the ultimate progress of European nations, and their colonies, and their former colonies, was impeded in ways that profited rich Jews, rich Jews who quietly pretended to the throne of Israel in the diaspora, while doing little for their "subjects", the millions of impoverished Jews struggling in *Schtetels*.

It should, however, be noted that Jews often concealed their wealth and had a love for jewels and gold, because, among other reasons, they were easy to transport at a moment's notice. Many of the Jews who appeared impoverished were in fact wealthy, and the numerous accounts of Jews miraculously becoming wealthy in America are doubtful. Based upon *Genesis* 42:1, the Talmud teaches the Jews to conceal their wealth, lest the Gentiles become envious (*Ta'anith* 10*b*, *Sanhedrin* 29*b*. See also: Rabbi Ephraim Lunshitz , *Kli Yakar*). In 1845, *The North American Review* wrote,

> "Indeed, throughout the East, the Jews are obliged to affect poverty, in order to conceal their wealth; what is exposed to view is never safe from Mohammedan rapacity. Though the great majority of those in Palestine are poor and dependent, some may be found there in comfortable circumstances, or even rich; but their wealth appears to those only who gain their intimacy. Dr. Richardson, an English traveller, says, 'In going to visit a respectable Jew in the Holy City, it is a common thing to pass to his house over a ruined foreground, and up an awkward outside stair, constructed of rough, unpolished stones, that totter under the foot; but it improves as you ascend, and at the top has a respectable appearance, as it ends in an agreeable platform in front of the house. On entering the house itself, it is found to be clean and well furnished the sofas are covered with Persian carpets, and the people seem happy to see you.' The synagogues in Jerusalem are, from prudential motives, both small and mean. A Jew dares not set foot within the Holy Sepulchre. When, in 1832, the Egyptian troops occupied Palestine, the Jews did not find their condition in the least improved. The common soldier made the best Jew sweep the streets, or perform any menial office."[373]

In an article entitled "The Jews", *The Knickerbocker; or New York Monthly Magazine*, Volume 53, Number 1, (January, 1859), pp. 41-51, at 44-45, 48, wrote,

> "Yet the Jews of the Ottoman Empire, notwithstanding their degradation, exhibit a certain intellectual tendency. They live in an ideal world, frivolous and superstitious though it be. The Jew who fills the lowest

---

373. "The Modern Jews", *The North American Review*, Volume 60, Number 127, (April, 1845), pp. 329-368, at 339-340.

offices, who deals out *raki* all day long to drunken Greeks, who trades in old nails, and to whose sordid soul the very piastres he bandies have imparted their copper haze, finds his chief delight in mental pursuits. Seated by a taper in his dingy cabin, he spends the long hours of the night in poring over the Zohar, the Chaldaic book of the magic Cabala, or, with enthusiastic delight, plunges into the mystical commentaries on the Talmud, seeking to unravel their quaint traditions and sophistries, and attempting, like the astrologers and alchymists, to divine the secrets and command the powers of Nature. 'The humble dealer, who hawks some article of clothing or some old piece of furniture about the streets; the obsequious mass of animated filth and rags which approaches to obtrude offers of service on the passing traveller, is perhaps deeply versed in Talmudic lore, or aspiring, in nightly vigils, to read into futurity, to command the elements, and acquire invisibility.' Thus wisdom is preferred to wealth, and a Rothschild would reject a family alliance with a Christian prince to form one with the humblest of his tribe who is learned in Hebrew lore.

The Jew of the old world, has his revenge:

'THE pound of flesh which I demand of him
Is dearly bought, is mine, and I will have it.'

Furnishing the hated Gentiles with the means of waging exterminating wars, he beholds, exultingly, in the fields of slaughtered victims a bloody satisfaction of his 'lodged hate' and 'certain loathing,' more gratifying even than the golden Four-per-cents on his Princely loans. Of like significance is the fact that in many parts of the world the despised Jews claim as their own the possessions of the Gentiles, among whom they dwell. Thus the squalid *Yeslir*, living in the Jews' quarter of Balata or Haskeni, and even more despised than the unbelieving dogs of Christians, traffics secretly in the estates, the palaces and the villages of the great Beys and Pachas, who would regard his touch as pollution. What, apparently, can be more absurd? Yet these assumed possessions, far more valuable, in fact, than the best 'estates in Spain,' are bought and sold for money, and inherited from generation to generation.

***

The Jewish population of Egypt numbers not more than ten thousand souls, of whom nearly seven thousand live in Grand Cairo. Though now undisturbed in the practice of their faith, the oppressive exactions of the Government, and the fear of renewing the persecutions of former times, have taught them to dissimulate. Dressing in filthy rags, and living in houses of the meanest external appearance, they strive to seem even more wretched than they are in reality, so as not to invite taxation."

Jews boasted of their power in terms that Jewish racists would call "anti-Semitic" when stated by Gentiles. Jewish influence circumvented any democratic hopes that Europeans had in the Nineteenth Century and hindered the Continent with endless wars that ultimately only served the perceived self-interests of Jews. Rich Jews beat the drums for war in their newspapers, profiteered from wars in the markets, and brought about wars through their corrupt influence over politicians, church leaders and monarchs. *The Chicago Daily Tribune* reported on 13 May 1877 on page 3,

"Jews in European Politics.
London Public Leader (Jewish Organ).

The London *Examiner* last week announced that a Berlin firm of publishers intended issuing next winter a work entitled 'The Political Influence of the Jewish Race in Europe.' Our contemporary observes that, 'leaving out of consideration the power of Lord Beaconsfield (Disraeli) in English, and of M. Gambetta in French, politics, and the growing Hebraic dominance in Russia, particularly in cities like Odessa, Germany itself would hardly have been the Germany of to-day but for the exertions with pen and tongue of such Liberal politicians as Jacoby, Sonneman, and, above all, Edward Lasker, the 'natural leader,' of the National Liberals.' This is a poor summary of the political influence of the Jews in Europe, especially the production of M. Gambetta as an example of their influence in French politics. There are many more Jewish politicians in France of much greater importance, prominent amongst them are MM. Cremieux and Jules Simon. Austria has been entirely forgotten by our contemporary, notwithstanding that the revolution which necessitated the flight of Metternich was organized and led by Jews, and that amongst the most popular members of the Austrian Parliament are such Jewish statesmen as Hirsch and Kuranda. Then again the Italian Assembly contains several Jewish members, whose opinions are of great weight, and the city of Rome itself—the stronghold of that power which, throughout long ages, attempted the extermination of the Jews— numbers amongst its legislative representatives a Jew born and partly reared in the Roman Ghetto. Whilst we are on this subject, we cannot help remembering the enormous political power wielded by the Jews through the medium of the continental press. In Germany and Austria the majority of papers belong to Jews, and the most brilliant journalists are Children of Israel: and then—finis coronat opus—where in the Examiner's short summary is a mention of the influence of the Rothschilds? The political power of this family can hardly be estimated. It reminds us of an anecdote told of the wife of old Meyer Anselm Rothschild, which is sufficient to illustrate it. To her dying day she lived in the Ghetto of her forefathers in Frankfort, and attained such an age that she saw her sons rise to the position of the greatest financiers in the world. She never renounced her old gossips, and one day, in 1830, one of her friends came to her and told her that her son was ordered to join the military and might be killed in the impending war. 'Be comforted,' answered Madame Rothschild, in the homely patois of her district, 'I will tell my sons not to give the Princess money, and then they will not be able to go to war.'"

War and the revenge of the Jews against the Christians were common themes when discussing the Rothschilds in the Nineteenth Century. *The Chicago Daily Tribune* reported on 28 December 1873 on page 16,

"Character of the Rothschilds.

The four original houses remain, though they have agencies and interests in all the leading cities of Europe, Asia, and Africa, as well as North and South America. They have belted the globe with their operations, and are in the fullest sense universal and cosmopolitan bankers. For generations they have been Barons, and the title is hereditary in their family. Since the death of old Mayer Anselm, they have added the distinguishing *de* and *von* to their names, and are as far removed from democratic affiliations and sympathies as if it were a thousand instead of a hundred years since their ancestors counted kreutzers and old [???] in the Judengasse of Frankfort. They have always been devoted to their theological [???], and strict in observing all the forms of the synagogue. They are not without superstition in their creed, believing that much of their good fortune has come from their unswerving fidelity to Judaism. Their charities to their coreligionists have been many and liberal. They have endowed schools, built hospitals, and funded almshouses. Their attachment to their ancient form of worship is noble and commendable. They cannot help remembering how bitterly their people were persecuted for ages, and how very recent it is that they have been allowed to enjoy either political or civil rights. Long after Mayer Anselm had grown rich, he and his fellow-Hebrews were locked into the Jews' quarter of Frankfort after nightfall, and forbidden to depart thence until the iron gates were thrown open in the morning. If the great bankers have forgiven the inhuman wrongs done through centuries to their race, they are singularly magnanimous. They have reason to feel as *Shylock* felt to *Antonio* toward the fawning Christians who go to them for money. Their negative revenge cannot be without sweetness when they think that the once despised and hunted Jew has had the proudest nobles begging for his gold, and even Kings soliciting his aid. It has been their boast that monarchs could not go to war without the consent of the Rothschilds. Like most boasts, this was not strictly true; but they who furnish the sinews of battle are the most desirable of allies, not less than the most formidable of foes. The Rothschilds, save at rare intervals, continue to intermarry, and are likely to while the powerful family holds together. If the common theory respecting the union of blood-relatives were true, the banking brotherhood would be reduced by this time to hopeless imbecility; and they are in the opposite extreme.—*Harper's Weekly*."

Others believed that inbreeding had indeed degraded the Rothschild family. *The Chicago Daily Tribune* reported on 15 February 1874 on page 7,
"There is no question that, with the death of Baron James, the genius of the house of Rothschild has departed. Constant intermarriage with cousins and the absence of that intellectual vigor which the infusion of fresh, new blood

imparts, has its effect on men as on animals, and the younger branches of the family are far inferior to the elder."

Wars helped the Rothschilds destroy competing banks, including national banks, and consolidate their power, while weakening the European nations—which had been a prophetic wish of Judaism for thousands of years. It is important to note that the effect, and perhaps the desire, is to prevent an entire society, even humanity at large, from becoming powerful and wealthy; which would enable Gentiles to resist Messianic Jewish world domination. *The Chicago Press and Tribune* reported on 6 June 1859,

"The War Revulsion in European Finance—First Effects of the Storm.
[From the New York Herald.]

The monetary disasters which are likely to follow from the effects of the present war in Europe, and the necessary destruction it will entail upon the financial and banking system of several of the most powerful of the European governments, are so entirely different in their character and in the laws that govern them from the revulsions known to the present generation, that few persons now engaged in the active transactions of life comprehend or consider them.

The experience of the present age is limited to a small number of commercial revulsions which have grown out of the exaggeration of the healthy elements of trade. Few recollect the ruin that swept through the commercial world on the commencement of Pitt's war, and the consequent suspension of specie payments by the Bank of England, or the vast fortunes made by a horde of army contractors during its twenty-one years' continuance, while commerce flagged, looms were stopped, ships rotted at the wharves, merchants went into bankruptcy or prison, and the army was the only refuge of the people from starvation. The beginning of a great war, and the short continuance of any strictly local conflict, acts as a stimulus upon trade and industry, because its effects are as yet felt only in their demand for the elements of destruction. But when its true work comes to bear—when the circulating medium is turned from its wonted channels, and the force of destruction without production and exchange begins to be felt—the longing for peace sets in, and continues to increase in intensity till its arrival is celebrated with bonfires and enthusiastic shouts that far exceed any manifestations of joy at the declaration of war. This simple truth marks the real effect of war upon the common weal. Let us now group together a few of the facts that have marked the progress of the present contest.

In the foreground stands the fact that the several governments of Europe, since the 1st of the January, have either come into the market, or are preparing to come in, for loans to the amount of three hundred and fifty millions of dollars. England raised thirty-five for her Indian wants, and immediately sent one-half of it in silver to Calcutta. Austria asked for a hundred millions; but all the power and credit of the Rothschilds could not raise it for her, and she seized the metallic reserve of the Bank of Vienna, suspended specie payments, borrowed two-thirds of the sum in paper, and assessed a forced loan of fifteen millions more on Lombardo-Venetia. Russia sought for sixty millions; but she, too, failed to obtain it, and has adopted a system of financial expedients at home. Sardinia asked for six millions, failed to get it, and suspended specie payments also, borrowing the amount in paper from the Bank of Turin. France has called upon her people to contribute one hundred millions of dollars, and they offer five hundred millions. Turkey borrowed a short time since twenty-five millions. Prussia, Holland, Belgium and the German Confederation are now preparing to come into the money market for large amounts.

The first effects of these extraordinary borrowings is to cause the people to look at the financial condition of several governments. They find that for years past all have exhibited deficits in their budgets. Since 1851 France has borrowed and spent six hundred millions of dollars more than her revenue. Austria has done the same to the extent of four hundred millions. England had to borrow nearly one hundred millions to prosecute the Crimean war; and if she goes into the present one, there is no possibility of estimating how much she must borrow. Russia, Sardinia, Spain, Germany, Prussia—all have exhibited deficits for some time past; and the revolution that now threatens to sweep over commerce gives no hope of a different state of things.

As a result of these movements we find specie disappearing from the vaults of trade, and seeking the hoards of fear or the war chests of the army. In fifteen days New York has sent off ten millions of dollars. The last returns of the Banks of England and France show that in one month they had lost ten millions of bullion. In the two months preceding the declaration of war in 1854, the bullion in the Bank of England alone ran down eight millions, and in the two succeeding months ten millions more. To endeavor to stop this drain, the rate of interest has already been raised in London one per cent., on the 6th of May, and will no doubt be further advanced. This stops commerce from using money. But war does not care for per centages; its first step is to suspend specie payments, which, when taken by a government, is nothing more nor less than a direct robbery of its own subjects.

Already the consequences of these extraordinary movements are beginning to be felt. Although the promised rate of interest has not been refused, an immense depreciation has been caused in the value of government securities and public stocks. It is calculated that the depreciation in British consols is already equal to three hundred and fifty millions of dollars, and that of the stocks of public companies three hundred and fifty millions more. On the Continent the effect has been much greater, and we may safely estimate the fall in the value of funded property in Europe at four thousand millions of dollars. From these causes will follow the ruin of the bankers; and they have already begun to fail. In April Lutteroth failed for a large amount in Trieste. On the 2d of May, Wolf & Co., Berlin bankers, failed; on the 5th, Lloyd, Belby & Co. failed in London; on the 6th, Arnstein & Eckles, Vienna bankers, failed for ten millions of dollars; and up to the 12th of May, nearly one hundred failures were announced on the Stock Exchange and trade in London. At Constantinople a sudden advance in the value of sterling exchange from 143 to 156 piastres had caused the bankers to gather in council in the beginning of May; and in Holland, where large amounts of Austrian and other Continental securities are held, the depreciation of securities had been so severely felt that numerous distressing suicides had taken place.

The cause of these dire results may be reduced to a simple expression. The governing class in Europe—a class that has no connection with commerce and little sympathy with industry—is seizing upon the wealth of the world, perverting it from the arteries and veins of trade, and pouring it into their own pockets and the pockets of a hoard of army contractors, and squandering it in destructive dynastic wars. Let not our merchants flatter themselves that these things are going to be good for them. They will be good for a new class of speculators; men who will run great risks for the chance of great profits—men who connect themselves with the quartermasters and supply contractors of Europe, and who will resort to all kinds of expedients to win a purse or break a neck in the race for fortune. But a general war in Europe will break down all its existing financial and commercial circles, and the effects cannot but be severely felt in one way or another here."

As the Civil War grew nearer, Americans grew suspicious of the Rothschilds' destruction of European economies. Americans noted the new phenomenon whereby governments passed debt on to future generations, who were undemocratically forced to give up their treasure to the repressive Rothschilds. These intrigues, which had the effect of fulfilling Jewish prophecy, were among the reasons why Jews were looked upon with suspicion, especially in Europe. Another major reason was the fact that Jews were prominent in the revolutionary movements. It is important here to note that the debts the Rothschilds manufactured promoted the conditions which enabled the Marxists to overthrow governments and ruin societies, and these Jewish forces covertly worked in collusion. *The Chicago Press and Tribune* reported on 22 December 1859 on page 3,

"Baron Rothschild's Visit to America.

We see announced as among the arrivals by the Persia, one of the celebrated house of Rothschild. Thus far the business of that house with this country and its securities has been comparatively small. They have estimated our government loans too insecure, and our railroad stocks too small, or too speculative and fluctuating. They have negotiated the loans of crowned heads to the amount of millions, resting on no more solid basis than the honor of some bankrupt government. For England, with its debt of eight hundred millions of pounds sterling, they have been the chief government agents at most important and critical times. For France they have at times done much in this way, under half a dozen dynasties, just to keep stocks up and what they had from being swept away. In Austria they have been everything more potent than sovereigns—yet themselves compelled to sustain tottering governments by taking loans to keep things going. Meantime they have despised the growing wealth of this country, which has not exhibited itself in crown jewels or costly palaces, or immense retinues of servants, or of soldiers, but in careful re-investments, railroads, telegraphs and broad acres, subdued by the hand of industry, to supply the world with cotton and with grain.

No Rothschild that we know of has visited this country before, and their doing so now may have a significance in history difficult to calculate. Of course, they do not tell their purposes and their plans. They do not even herald their approach, or intimate it by any ostentatious display. But it is not impossible that such an arrival may indicate at a future period the gradual transfer of large portions of their countless wealth to this country. If such should be the case, it would be perfectly certain that the wealth of thousands of others would follow in the same direction, and our stocks of every kind would rise, and enterprise be pushed in ten thousand channels; so that the next fifty years would produce an expansion and growth from the capital of the old world, united with the industry of the new, compared with which, all the past progress of the last fifty years would be as nothing.

This country *must* afford the best field for the employment of capital. The Rothschilds began with nothing. They made their money mainly by the rise of government securities, consequent on the re-establishment of order and of confidence, after the wild and sweeping ruin of the first French Revolution. The peace of 1815 made them indisputably the first house in the world for capital vested in government securities. But, since the Revolutions of 1848, the loss of confidence in the government securities of Europe has been gradually becoming more and more marked among the most sagacious. Austrian finances have been proverbially rotten for years, and each year has not only added to the deficit, but displayed some new government fraud, until, within the last year, things have come to light showing the over-issue of stock, in such ways and to such an extent that would destroy the character and the credit of any mercantile house, or of anything, in fact, that had any character or credit to lose, except a European government.

The debt of France has been enormously increased, and that of England also. Not a country in Europe is diminishing its debts in peace, and all its wars and preparation have to be carried on by taxing posterity. How long can all this last? If peace were the order of the day, things might go on without getting worse. But peace is not the order of the day, and war is getting to be more and more a question of finance and credit on an unheard of scale of cost. Some nation like Austria will one of these days come to a halt—will run down—and then the rest will follow, like a row of dominoes; and then the capitalists will have stocks and government bonds, but the coupons will be unpaid, and the whole worth only so much waste paper.

The last century taught the civilized world a new act, that of borrowing without the least prospect of ever repaying, by simply paying the interest and throwing the rest upon posterity. So long as posterity obtains something better than the interest in return—peace, order, credit and wealth—they may go on and meet the drafts of their predecessors upon them; but, directly the cost becomes greater than the advantage, and war and insecurity return, a new generation will arise and sweep away the whole debt as unjust. In this country we have *lands*, and railroads, and solid products at the bottoms of our stocks, and into these things the capital of the old world is finding its way and will find it."

The Rothschilds defended Jewish interests. There are indications that they believed that this brought them good luck. It also generated distrust and conflict. Cabalist Jews believed that committing both good acts and evil acts could hasten the coming of the Messiah, and Rothschild wanted to be the Messiah. On 5 September 1874, *The Chicago Daily Tribune* published an obituary for Anselm De Rothschild, which evinces the undemocratic and repressive power of the Rothschild family, as well as their use of their power to promote Jewish interests,

"Baron Anselm De Rothschild.

The death of the lamented Baron Anselm De Rothschild, says *Jewish Chronicle*, has produced a deep impression throughout Vienna. The Baron died at Dobling, near that town. He had attained the age of 71. He was born on the 29th of January, 1803, at Frankfort-on-the-Main. He was a son of Baron Solomon De Rothschild, who was a grandson of the founder of this distinguished commercial dynasty, Anselm Meyer. He spent his youth at Frankfort, and passed some time during his young manhood at Berlin, where he attended the university of that city. His career as a university student imbued him with a lively interest in science. He attached to scientific pursuits and held communion with scientific men throughout his whole life, and he invariably endeavored to keep up with the stream of scientific progress. It is said that he had a special acquaintance with history, but he principally acquired renown as an enthusiastic friend of the fine arts and a profound connoisseur in painting and archæology. In 1855 he took up his residence in Vienna, and rarely quitted it excepting during the hot weather, when he usually went to his estate at Schillersdorf, in Silesia. He married his cousin Charlotte, daughter of his uncle, Baron Nathan Mayer De Rothschild, the well-known head of the London branch (father of Baron Lionel and Sir Anthony Rothschild). He lost his wife in 1859. He had seven children, viz.: three sons, Nathaniel, Ferdinand, and Alfred; and four daughters, Julia, the wife of Adolphe Charles De Rothschild; Matilda, who married William Charles De Rothschild; Louisa, who married Baron Franchetti, and Alice, who is still unmarried. His sons have no children; Baron Ferdinand is a widower. In 1861 Baron Anselm De Rothschild was appointed a member of the Upper House, or House of Lords, of the Austrian

Imperial Parliament, in, which he always voted with the Liberal party. Not having been endowed with oratorical talents he did not attempt to shine as a speaker, but he enjoyed the highest esteem of his illustrious senatorial colleagues by the firmness of his character and the unshakable consistency of his principles. Indeed, it is difficult in Austria for a political personage to acquire a reputation for consistency, but this reputation he deservedly obtained. Baron Anselm De Rothschild invariably evinced a strongly pious adherence to the orthodox principles of the religion of his fathers. In 1866 he gave a notable proof of the intensity with which he felt any blow directed against the honor of his coreligionists. In that year the war broke out between Austria and Prussia. At that time Count Beleredi was at the head of the Austrian Government; he was a man of Ultramontane Catholic principles, and he had very little sympathy with the Jews. Under an assertion of patriotism he put forth the notion of requiring the Jewish congregations to organize several battalions of volunteers at their own expense. Now, as the Jews necessarily undertook the obligations of military service in common with other citizens, Count Beleredi's plan was neither more nor less than an extraordinary tax levied on the Jews, a disguised renewal of the special Jews' tax, that had been abolished since the emancipation of the Jews. Naturally the Jews protested on all sides against this injustice, and on this occasion Baron Anselm de Rothschild wrote to the Imperial Minister that he would close his offices, break off all financial negotiations with the Government, and leave Austria if the Minister persisted in carrying out a project which would be so injurious to the Jews. His letter had the desired effect, and the Minister abandoned the tax. He spent his last days at a villa at Dobling, a village near Vienna. He had suffered much, and was obliged to submit to a painful operation. For some days before his death this catastrophe was regarded as inevitable. According to the last wished of the deceased, his body was taken, with the greatest simplicity, to Frankfort. With the exception of the two preachers of the Synagogue, the functionaries of the burial society, and his most intimate friends, very few persons were at the ceremony. Immediately on hearing of the death of the Baron, the Emperor sent his adjutant to offer his condolence to the family, as did also the German Emperor, the Czar of Russia, and the King of Italy by their respective Ambassadors. Prince Bismark and Count Andrassy, Primo Minister of the Austro-Hungarian realm, sent telegrams of sympathy."

Though the Rothschilds felt justified in using their power to promote Jewish interests, they did not hesitate to use unscrupulous means to fleece entire Gentile societies of their wealth. The callous elitism and arrogant inhumanity of the Rothschilds was revealed in an article that appeared in *The Chicago Tribune* on 24 December 1867 on page 2,

"The Career of the Great Rothschild, of London, as Narrated by Himself.
Extract of a Letter from Sir Thomas Powell

Buxton to Miss Buxton.

DEVONSHIRE STREET, Feb. 11, 1834.

We yesterday dined at Ham House, to meet the Rothschilds, and very amusing it was. He (Rothschild) told us his life and adventures. He was the third son of the banker at Frankfort. 'There was not,' he said, room enough for us all in the city. I dealt in English goods. One great trader came there who had the market to himself; he was quite the great man, and did us a favor if he sold us goods. Somehow I offended him, and he refused to show us his patterns. This was on Tuesday. I said to my father, 'I will go to England.' I could speak nothing but German. On Tuesday I started. The nearer I got to England the cheaper goods were.

As soon as I got to Manchester I laid out all my money, things were so cheap and made good profit. I soon found that there were three profits—on the raw material, the dyeing and the manufacturing. I said to the manufacturer, 'I will supply you with material and dye, and you supply me with manufactured goods.' So I got three profits instead of one, and could sell goods cheaper than anybody. In a short time I made my £20,000 into £60,000. My success all turned on one maxim. I can do what another man can, and so I am a match for the man with the patterns, and all the rest of them! Another advantage I had; I was a off-hand man; I made a bargain at once. When I was settled in London, the East India Company had $800,000 of gold to sell. I went to the sale and bought it all. I knew the Duke of Wellington must have it for the pay of his army in the Peninsula; I had bought a great many of his bills at a discount. The government sent for me, and said they must have it. When they got it they did not know how to get it to Portugal. I undertook all that, and I sent it through France, and that was the best business I ever did. Another maxim on which he seemed to place great reliance was never to have anything to do with an unlucky place or an unlucky man. 'I have seen,' said he 'many clever men, very clever men, who had not shoes to their feet! I never act with them. Their advice sounds very well, but fate is against them; they cannot get on themselves; and if they can not do good to themselves, how can they do good to me?' By aid of these maxims he has acquired three millions of money.

'I hope,' said ----------, 'that your children are not too fond of money and business, to the exclusion of more important things. I am sure you would not wish that.' Rothschild: 'I am sure I should wish that. I wish them to give mind and soul, and heart and body, and every thing to business. This is the way to be happy. It requires a great deal of caution to make a large fortune, and when you have got it, it requires ten times as much wit to keep it. If I were to listen to all the projects proposed to me I should ruin myself very soon. 'Stick to one business, young man,' said he to Edward: 'stick to your brewery, and you may be the great brewer of London. Be a brewer, and a banker, and a merchant, and a manufacturer, and you will soon be in the *Gazette*. One of my neighbors is a very ill-tempered man; he tries to vex me, and has built a great large place for swine close to my walk. So when I go out I hear first grunt, grunt, squeak, squeak: but this does me no harm. I am always in good humor. Sometimes to amuse myself, I give a beggar a guinea. He thinks it is a mistake and for fear I should find it out, off he runs as hard as he can. I advise you to give a beggar a guinea sometimes; it is very amusing.'

The daughters are very pleasing. The second son is a mighty hunter, and the father lets him buy any horses he likes. He lately applied to the Emperor of Morocco for a first-rate Arab horse. The Emperor sent him a magnificent one, but he died as he landed in England. The poor youth said, very feelingly, 'that was the greatest misfortune he had ever suffered.' And I felt strong sympathy with him. I forgot to say that as soon as Mr. Rothschild came here, Bonaparte came here. 'The Prince of Hesse Cassel,' said Rothschild, 'gave my father his money; there was no time to be lost; he sent it to me. I had £600,000 arrive unexpectedly by the post, and I put it to such good use that the Prince made me a present of all wines and linen.'"

The Chicago Daily Tribune reported on 8 June 1873 on page 10 in an article entitled "Great Fortunes",

"The rise of the great House of Rothschild belongs to the eighteenth century. Meyer Anselm, a Jew, was born in 1743, and was established as a money-lender, etc., in Frankfort, in 1772. From his poor shop bearing the sign of the Red Shield, he acquired the name Rothschild. He found a good friend in William, Landgrave of Hesse; and when the Landgrave, in 1806, had to flee from Napoleon, he intrusted the banker with about £250,000 to take care of. The careful Jew traded with this; so that, in 1812, when he died, he left about a million sterling to his six sons, Anselm, Solomon, Nathan, Meyer, Charles, and James. Knowing the truth of the old motto, 'Union is strength,' he charged his sons that they should conduct their financial operations together. The third son, Nathan, was the cleverest of the family, and had settled in England, coming to Manchester in 1797, and London in 1803. Twelve years after, we see him at Waterloo, watching the battle, and posting to England as soon as he knew the issue, and spreading everywhere the defeat of the English. The clever but unscrupulous speculator thus depressed the funds, and his agents were enabled to but at a cheap rate; and it is said that he made a *million* by this transaction. He died in 1836; but the real amount of his wealth never transpired. It has been said; 'Nothing seemed too gigantic for his grasp, nothing too minute for his notice. His mind was as capable of contracting a loan for millions as of calculating the lowest possible amount on which a clerk could exist.' (*Chronicles and Characters of the Stock Exchange.*)"

The Rothschilds had insider information and used it to drain the nations of their wealth. Some speculate that they had improved upon George-Louis Le Sage's telegraph and could transmit messages over great distances effectively instantaneously, or that they had a system of speedy horses like the pony express, or that they had the swiftest vessels with which to cross the English Channel.

Much of the knowledge that must have appeared to have been the result of speedy communications, may instead have been planned in advance. The Rothschilds had agents in banking and government and knew far in advance of others what was about to occur in government, business and war. Many nations depended upon the Rothschilds' wealth for loans. The Rothschilds had no need of personal genius, because they had several advantages which made it impossible for anyone to compete with them. It also appears that they had corrupted many heads of state, and the leaders of many churches, and persuaded them to betray the Peoples whom they represented in order to enrich the Rothschilds and put the wealth of the world into Jewish coffers. Many of these leaders were likely crypto-Jews on a mission to subvert Gentile societies and bring them into debt, largely through wars and manipulation of the currencies and gold markets. Much of the royalty of Europe was of Jewish descent, or thought that they were of Jewish descent. That which Rothschild sycophants attributed to good fortune and acumen was instead the product of foreknowledge and corruption. Whoever controls the press, the banks, the preachers and the State has foreknowledge of just about everything and can profit from it. For example, anyone with a news story must first bring it to the press, which makes them the most powerful spy apparatus in the world. They not only know things in advance, they regulate the flow and timing of information. Another example is the banks. Any major project requires financing and a business plan before it can begin. This gives the bankers inside information. It addition, the Rothschilds could incite wars, recessions, depressions and concentrate wealth and economic growth in any nation or empire of their choosing. With a corrupt head of state, or church leader, who worked for them, the Rothschilds could quickly run a nation into debt and syphon off its gold reserves and tax its People in perpetuity. *The American Farmer, Containing Original Essays and Selections on Rural Economy and Internal Improvements, with Illustrative Engravings and Prices Current of Country Produce* (Baltimore), Volume 5, Number 29, (10 October 1823), p. 229, wrote,

"MEMOIRS OF MR. ROTHSCHILD.

Mr. N. M. Rothschild is descended from a German lineage. Mr. R. sought to establish his fortune in England. Various were his vicissitudes in early life; by his industry and prudential conduct, he acquired considerable property in the linen trade at Manchester, vast quantities of which article, were exported during the last war to the Continent, where Mr. Rothschild availed himself of the peculiar advantage of his brother's agency in that quarter of Europe. Previously to the close of the late war, Mr. Rothschild transferred the scene of commercial operations from Manchester to London. He then became a considerable speculator in the Foreign and British Securities on the Stock Exchange; and after the melancholy death of Mr. Goldsmidt, assumed a very prominent station in the money market. But the principal accident which contributed to the rapid elevation of our Modern Crœsus, was the escape of Buonaparte from Elba, in 1814.—In consequence of Mr. R.'s superior means of information on the Continent, this important occurrence was know to him nearly forty-eight hours before it was in the possession of any other person in this country. He did not fail to avail himself of every advantage which this priority of intelligence presented. His agents went into the market and sold prodigious quantities of stock. The consternation was dreadful! Every one suspected danger, none knew where to look for it. The panic was epidemic! On the disclosure of the fact, the general cry was *sauve qui fieut;* and the object of our present article bore off the immense sum, gained by his success on this great and extraordinary occasion.

Mr. Rothschild, thus fortified in wealth, and enjoying at this time the almost exclusive means of acquiring the first intelligence from the Continent, soon established for himself a reputation and importance, the maturity of which can scarcely be said to have been accomplished at the present moment. He availed himself of a conjunction with his brothers, (who are also great capitalists on the Continent,) of the opportunity of administering to the wants of the King of Prussia, the Emperor of Russia, the Kings of Naples and Spain; the Republic of Columbia and other States, who negotiated loans on terms highly profitable to him; and which have, with the advantages of the courses of exchange, and other incidental benefits, realized immense sums in addition to his fortunate speculations in British Stock. But the great *coup de main* of Mr. R. consisted in his outgeneralling the Gallic Financiers in the recent French Loan. In that transaction he is supposed to have cleared upwards of £100,000, by the commission alone, independent of the advantages of the courses of Exchange!

By the fortuitous occurrence of favourable circumstances, Mr. R. has been enabled to amass greater wealth, than any man that ever existed in England. It would be impossible for others to estimate his property, when Mr. R. has declared that he could not do it himself. It has been asserted, however, that he can command upwards of Fifteen Millions sterling at any time, if required! When it is considered that 'money, the sinew of war,' is in its amount illimitable, and in its control so much at the mere volition of Mr. R. it ceases to surprise the reader, that such a man should be necessary to the Potentates of Europe, and that his friendship and assistance should be no less anxiously sought, than promptly and powerfully afforded.

Mr. Rothschild is a Baron of the German empire, to the Emperor of which, he has rendered some essential services. He is about 43 years of age, and possesses a family of nine children. His mode of life is remarkable for its retired description. Unlike his great predecessor, (Goldsmidt,) he does not boast of his choice and exquisite wines, or herald his hospitality towards the Princes of the blood. His appearance is unostentatious; his deportment familiar; and his manners unaffected and affable. His conversational style on 'Change is rapid, acute, and discriminating. He carries about him no aristocratical feeling; neither does he affect a singularity, the common concomitant of extraordinary genius, and the impotence of mental pecuniary plenitude. His face is distinguished by a lack of that piercing intelligence, which lights up and animates the expressions of those proverbial for their acuteness; but there is a quickness in the eye, which denotes a lively and unremitting watchfulness of the mind, on every subject of general interest.

When engaged in conversation, Mr. R. usually dangles a bunch of keys in his right hand, and indulges a habit of abruptly turning from the object to whom he is speaking, and suddenly renewing the colloquy. He possesses a memory so remarkably retentive, and the powers of mental addition so copiously strong, that he effects all his immense calculations without the agency of pen or paper: and often at those times, when the din of business 'gives note of preparation' for a 'rise or fall.' His genius is of that order, which often enables him to perceive the benefit or disadvantage of a proposition, before the parties have fully viewed the surface. His movements are characterized by profound judgment: his attack is no less able, than his retreat judicious.

Mr. Rothschild's private character is, we believe, as amiable as his public life is important. He diffuses his benevolence with judgment and liberality. When solicited to countenance an Institution with his name, he answers, 'You know I never take a public part; if you want (as I suppose you do,) money; name the sum, and you shall have it; but don't make me look ostentatious or mean, by naming too large or too small a sum.' His eleemosynary contributions are chiefly distributed amongst objects of the jewish persuasion; who have in many instances arrived at a state of opulence through his instrumentality. Such a liberality of disposition, and philanthropy of character, has divested envy of her deadly influence; and created for Mr. Rothschild, an imperishable reputation, which will descend with advantage to his family in after ages."

The *Saturday Evening Post*, Volume 3, Number 42, (16 October 1824), p. 2, reported under the heading, "European Affairs. Late from England":

"Mr. N. M. Rothschild has contracted for a loan to the Napolitan Government to the amount of £2,500,000."

The stories which assert that the Rothschilds built their fortune on funds entrusted to them by the Prince of Hesse and from the profits they netted from the false rumor they spread that the English had lost at Waterloo do not appear to account for their vast wealth. They may have come into the great wealth Jewish bankers had accumulated from the times of the de Medicis and even earlier. They put this wealth to the purpose of fulfilling Jewish Messianic prophecies of the destruction of the Gentile world through perpetual war[374] and sought to make one of their own the King of the Jews, and King of the World through the world government they sought to impose on Gentile humanity. The machinations which brought them into this position remain a mystery. It is not

---

[374]. G. E. Griffin, "The Rothschild Formula", *The Creature from Jekyll Island: A Second Look at the Federal Reserve*, Chapter 11, Fourth Edition, American Media, Westlake Village, California, (2002), pp. 217-234.

known who chose them or why. One could speculate that the Jews have for a very long perpetuated the myth that certain families carry with them the Royal blood of King David. Wealthy families would have an easy time creating this myth for themselves. Since there never was a King David, it is difficult to challenge them, though realistically speaking Ashkenazi Jews would a far more difficult time linking their lineage to Judah, let alone to a King David who never existed, than would Sephardic Jews, who carry with them a stronger genetic tie to the Judeans.

Judaism has always operated under a double standard and considered Gentiles to be mere animals undeserving of moral treatment. Just as the Jewish story of the flight from Egypt taught Jews it was alright to appropriate the gold of other peoples by unscrupulous means, many Jewish financiers delighted in cheating Gentiles, though in the process they also cheated other Jews. Rothschild published his "Memorial of the Jews in England to the Czar of Russia" in 1882. *The Chicago Daily Tribune* quoted Rothschild on 19 February 1882 on page 5 in an article entitled "The Judenhetze",

"Here in England, where perfect civil and religious equality has been granted us, we English Jews can bear testimony to the happy results effected by such complete emancipation. Here all those restrictions—civil, commercial, and educational—which formerly oppressed us have happily been removed, and, as a result, Jew and Christian here live and work side by side on terms of mutual respect and good fellowship, engaged in friendly rivalry, which stimulates public industry and adds to the common weel."

The Chicago Press and Tribune reported on 13 September 1859 on page 2,

"ROTHSCHILD'S INGENUITY.—An eminent Parisian [???], of the Jewish faith, knew the secret of the recent armistice several days before it was actually concluded, and he was desirous of communicating intelligence of the coming event to the house at Berlin. But how was it to be done? The electric wire is by no means a safe confidant for a secret. The banker hit upon a device. He wrote a telegram and concluded it in the following terms: 'Herr Scholem will shortly arrive.' Scholem is a Hebrew word signifying peace. In the Berlin house, where the Hebrew language was understood, the true meaning of the announcement of Herr Scholem's expected arrival was readily interpreted."

It was obvious to many that a democratic society could not exist while wealth remained concentrated in corrupt hands. It became increasingly obvious in the mid-Eighteenth Century that national sovereignty meant little more than the ability to go to war in order to profit the "Moneyocracy", which was more interested in fulfilling the prophecies of Judaism than benefitting the societies over which it ruled. *The Chicago Tribune* reported on 4 April 1866 on page 2,

"A SPEECH BY JULES FAVRE.

The Emperor Napoleon having risen to power by perjury and by the connivance of the moneyocracy and of the principal debauchees of Paris, his reign has become the signal of a reign of lust, luxury and money to such an extent as to make all cultivated men and virtuous women blush for shame, and to cause the people to tremble with indignation as they read the recent speech of Jules Favre in denunciation of these crying evils. In fact Rothschild, Pereire and Fould are, under the second empire, what the ancient nobles were under the rule of the elder Bourbons, and since the moneyocracy of 1866 is not even endowed with the accomplishments which constituted the redeeming but unavailing graces of the aristocracy of 1766, it is not only as hateful as the last were, but still more despicable. The battle cry of the old nobility was monopoly in land, that of the new moneyocracy is monopoly in cash, in railways, in bank, in insurance, and joint stock companies. In fact they assume to be the lords of modern society as the ancient nobles were those of the feudal era, but since their power is not as venerable as that of entailed estates, it is more easily withstood, while its lack of all noble tendencies withholds from it the prestige which clustered round the gallant bearing and emblazoned glories of the old nobility.

Money, and nothing but money, is the great end of all the exertions of this Bonaparte moneyocracy, and not, as it ought to be, whenever honorably obtained as a means for the more liberal fulfilment of all the manifold domestic, social, patriotic, humanitarian and religious duties of life. Wherever the mere possession of money opens, as it does under Napoleon's rule, the door to society, to influence, to every brute, and to every licentious man and bedizened woman, that society is doomed to destruction as surely as was that of the harlot and spendthrift era of Louis XIV and XV. No wonder that the late Baron Dupin animadverted upon this demoralization before he descended to the grave. No wonder that books are published showing that the state of society in Pagan Rome was not a whit worse in its worst period, than at the present time, in Paris. No wonder that Jules Favre, the great jurist, orator and parliamentarian makes the tribune ring with his eloquent vindication of the virtue, the culture, the art, the intellect of France against the fearful supremacy of brutes, bloated with ill-gotten wealth, and of a society reeking with lust and abomination. The following is the concluding extract of the remarkable speech delivered by this gentleman, who is the leader of the opposition in the legislative body, on the 15th inst.:

'In regard to the exterior policy, if the country had been master of its destinies, we should never have witnessed these distant expeditions which have so greatly compromised our interests. We should not have sent to die on the other side of the Atlantic so many young men whose arms would have enriched our soil. We should not have seen millions wasted in Mexico in behalf of an enterprise the least fault of which is that it is impossible. These millions would have been usefully employed in benefitting France and her colonies.

'As to the interior *regime* we are sometimes told that the passions are completely appeased. Sometimes that they are still fermenting, that parties are always armed, and that our liberties should still be refused us. Public morals are spoken of. If you would have good morals you must make good citizens; to make citizens you must have institutions which can form them. France is saturated with military glory. *She has need of moral dignity and grandeur.* If you will interrogate the literature of the present day, which is the expression of public morals, you will be driven to some unfortunate conclusions.

'You have decreed the liberty of theatres, and with the censorship you do what you please upon the public scene, and what do you show us there? Great God! you force a man with any sense of decency to keep away from this privileged temple in launching at him this sort of insult. 'I desired to speak of virtue and devotion. These are no longer actualities, and I am driven from the temple consecrated to them.'

'What do you make of the French scene? You have made it a scene of libertinage and shamelessness; you expose upon it disgusting nudities. You have in your hands a law made to prevent children from working in manufactories, and you begrime the child upon the scene of a privileged theatre, in making him represent the type and model of degradation and cynicism, to the scandal of all respectable people. And then you open *bais masques*, and you say, 'Come and amuse yourselves, and drink from the cup which I put to your lips.' As for me, I say to you, *France wants something else. She wishes to have the power of exercising her liberties. We are nothing if we cannot raise our eyes toward Heaven, and we cannot do that if we are not free.*'

No description, however graphic, could do justice to the effect produced by this oration. It fairly electrified the Chambers, and on the next day it was perused with enthusiasm by millions of noble women and worthy men, whose sentiments it embodies more emphatically than any speech ever delivered since the days of the Girondins and of Mirabeau."

Under the heading "Foreign Gossip", *The Chicago Tribune* reported on 14 March 1869 on page 3,

"The leaders of the French Opposition, Jules Favre, Thiers, Picard, Eugene Pelletan, Glais Bizoin, Marie and Bethmond, are all wealthy men. Only Garnier Pages is poor."

During the Civil War, the Rothschilds gained power on the American Continent by corrupting politics with their wealth and by running up the nations' debts with war. After the war, the Rothschilds floated huge loans to the United States, which netted the Rothschilds immense profits and enormous influence over America. Other European bankers, like Erlanger, fleeced investors and profited immensely during the war.

The Chicago Daily Tribune reported an accusation on 3 February 1873 on page 2, that the Rothschilds had gained control over a political party (the Democrats) in order to sabotage it and secure victory for their candidate,

"In a paper on Federalism, read before the Liberal Club last night by Mr. Delmar, the following remarkable passage made some sensation: 'The people have tacitly committed their entire interests and fortunes to the keeping of two political parties, whose leaders and managers, instead of Congress, as was intended, sway their destinies. It is charged that, knowing this, the Rothschilds, through their American agent, obtained control of one of these parties in the general election of 1868, and threw it into confusion by abandoning its Presidential candidate on the eve of election, so as to afford victory to its opponent, whose financial views more nearly accorded with the interests of that great house.'"

In the 1870's, Major Osman Bey wrote that,

"As a proof of this latter assertion, we need only recall the fact that a Jew by the name of August Belmont (an agent of Rothschild and represented to be an illegitimate offspring of his Dynasty) has for a number of years been the chairman of the National Committee of a political party, and by his shrewd management has always succeeded in leading that party to defeat. It is susceptible of proof that the financial policy of the *opposing* party was the one under which his business was most prosperous. Why then should he desire a change?"[375]

The Rothschild's Republican Hiram Ulysses Grant, generally known as "U. S. Grant", won the election. As President of the United States Grant exchanged America's gold reserves for Lincoln's Greenback dollars, thereby taking away the People's right to produce their own currency through their government without the intercession and profiteering of the Jewish bankers. This had a disastrous impact on the economy of the United States and ultimately the World, just as the Rothschilds, who put Grant into power, had intended. It also guaranteed the Rothschilds that the debts they unnecessarily and deliberately had caused the United States would be paid to them in gold, which through manipulation eventually had an artificially high value.

Grant, at the behest of the Rothschilds, then passed the Coinage Act of 1873, which put America on a Gold Standard, under which the U. S. demonetized silver. This increased the value of gold, further contracted the American money supply, and deflated American currency, all of which benefitted the Rothschild's while destroying the American and World economies, just as the Rothschild's had intended. The Rothschilds, through their puppet President "U. S." Grant, thereby deliberately manufactured the "Panic of 1873", a severe depression in America and the stock market crash in Vienna.

Maj.-Gen. Count Cherep Spiridovich explained how the Jewish bankers

---

375. Major Osman Bey, English translation by F. W. Mathias, *The Conquest of the World by the Jews, an Historical and Ethnical Essay*, Part 19, St. Louis, (1878), p. 58.

created the Civil War, why they opposed the Greenback, and why the Jewish actor John Wilkes Booth assassinated Lincoln:

"To pay the soldiers the Government issued its Treasury notes, authorized by act of Congress, July 17, 1861, for $50.000.000, bearing no interest. These notes circulated at par with gold. The Rothachilds' agents inspired the American banks to offer to Lincoln a loan of $150 million. But before they had taken much of the loan, the banks broke down and suspended specie payments in Dec. 1861. They wished to blackmail Lincoln and demanded the 'shaving' of government paper to the extent of 33%, an extortion which was refused. A bill drafted for the Government issue of $150 million, which should be full legal tender for every debt in the United States, passed the House of Representatives Feb. 25, 1862, and was hailed with delight by the entire country. But the Wall Street bankers were furious.

Sen. Pettigrew reprints the so called 'Hazzard Circular' sent in 1862 by the Bank of England (ruled by the Rothschilds):

'Slavery is likely to be abolished by the war power and chattel slavery destroyed. This I (Rothschild) and my European friends (the 300 men) are glad of, for slavery is but the owning of labor and carries with it the care of the laborer, while the European (read 'Rothschildian') plan led on by England (i. e. the Rothschilds) is for capital to control labor by controlling wages. THIS CAN BE DONE BY CONTROLLING THE MONEY. THE GREAT DEBT THAT CAPITALISTS WILL SEE TO IS MADE OUT OF THE WAR must be used as a means to control the volume of money. To accomplish this the BONDS must be used as a banking basis. We are now waiting for the Secretary of the Treasury to make his recommendation to Congress. It will not do to ALLOW the GREENBACK, as it is called, to circulate as money any length of time, as we cannot control that.'

Thus the order of the Rothschilds was clear: 'Capitalists WILL SEE TO IT that a DEBT is MADE out of the war.'

The result was that by 'hook and crook' the Rothschilds enslaved this country. And Schiffs, Baruchs & Co. are the rulers.

The Chairman of the Committee on Ways and Means of the House of Representatives, Mr. Thaddeus Stevens explains how the United States was captured by the Rothschilds:

'The agents of the banks fell upon the bill in haste and disfigured it.'

In the Senate this amendment was tacked upon the bill:

'Good for all debts and dues of the U. S. EXCEPT DUTIES ON IMPORTS AND INTEREST on the PUBLIC DEBT,' ($150 million above mentioned, plus $70 million, a pre-war debt).

'Thus equipped this bill went forth to rob every American and turn the ownership of this nation into the hands of capitalists.' (Mrs. Hobart).

When the bill came back to the House, Mr. Stevens said: 'We are about to consummate a cunningly devised scheme which will bring GREAT LOSS to all classes of people, except one' (the Rothschilds' branch of the Wall Street). The bill was passed.

The Rothschilds were in possession of 80% of the gold the country. They had a 'corner' on gold. By securing discrimination against the 'greenback' by means of the 'Exception clauses' they made a market for their gold.

'Importers were obliged to go to Wall Street to buy gold to pay duties on their goods, and the Wall Street gamblers held the power to fix the price. Gold went to a premium. Had the greenbacks been permitted to retain their full legal tender quality, there would have been no need for gold to pay import duties. The price of gold rapidly rose and before the war closed had reached the price of $2.85, measured in greenbacks. The gold bought in Wall Street to pay import duties became the revenues of the government and was by it paid back to Wall Street as interest on the public debt. As fast as the bankers sold the gold it was returned for interest on the public debt to be sold again. Thus during the entire war these gold gamblers speculated in gold, making fortunes from the blood and tears of the American people,' (Mrs M. E. Hobart in her 'The Secret of Rothschilds,' p. 54.)

Two more issues of $150 million each, with the 'Exception clause' were authorized in July, 1862 and in March, 1863, making in all $450,000,000. They bore no interest. When these issues were exhausted and necessity arose for additional money the bankers demanded that Treasury Notes should no longer be made in the form of DOLLARS, but in the form of BONDS: bond draws interest, the dollar does not.

A gigantic war costing seven billions was carried through without gold. Why? Because everything was supplied at home and American money, the 'greenbacks' were gladly accepted.

'How then was it that this government, several years after the war was over, found itself owing in London and Wall Street several hundred million dollars to men who never fought a battle, who never made a uniform, never furnished a pound of bread, men, who never did an honest day's work in all their lives... The fact is, that billions owned by the sweat, tears and blood of American laborers have been poured into the coffers of these men for absolutely nothing. This 'sacred war debt' was only a gigantic of fraud, concocted by European capitalists and enacted into American law by the aid of American congressmen, who were their paid hirelings or their ignorant dupes. That this crime has remained uncovered is due to the power of prejudice which seldom permits the victim to see clearly or reason correctly: 'the Money power prolongs its reign by working on the prejudices.'

(Lincoln). Every means has been employed to deceive the masses. Ridicule and derision have been applied to all opposition, while flattery and appreciation were showered upon the officials,' (Mary E. Hobart, p. 49).

[***]

Bismarck knew the truth and revealed it in 1876 to a German, Conrad Siem, who published it ('La Vieille France,' N-216, March, 1921). Bismarck said:

'The division of the United States into two federations of equal force was decided long before the Civil War by the High Financial Power of Europe. These bankers were afraid that the United States, if they remained in one block and as one nation, would attain economical and financial independence, which would upset their financial domination over the World. The voice of the Rothschilds predominated. They foresaw tremendous booty if they could substitute two feeble democracies, indebted to the Jewish financiers, to the vigorous Republic, confident and self-providing. Therefore, they started their emissaries in order to exploit the question of slavery and thus to dig an abyss between the two parts of the Republic. Lincoln never suspected these underground machinations. He was anti-Slaverist, and he was elected as such, But his character prevented him from being the man of one party. When he had affairs in his hands, he perceived that these sinister financiers of Europe, the Rothschilds, wished to make him the executor of their designs. They made the rupture between the North and the South imminent! The masters of Finance in Europe made this rupture definitive in order to exploit it to the utmost. Lincoln's personality surprised them. His candidature did not trouble them: they thought to easily dupe the candidate woodcutter. But Lincoln read their plots and soon understood, that the South was not the worst foe, but the Jew financiers. He did not confide his apprehensions; he watched the gestures of the Hidden Hand; he did not wish to expose publicly the questions which would disconcert the ignorant masses. He decided to eliminate the International bankers, by establishing a system of Loans, allowing the States to borrow directly from the people without intermediary. He did not study financial questions, but his robust good sense revealed to him, that the source of any wealth resides in the work and economy of the nation. He opposed emissions through the International financiers. He obtained from Congress the right to borrow from the people by selling to it the 'bonds' of States. The local banks were only too glad to help such a system. And the Government and the nation escaped the plots of the foreign financiers. They understood at once, that the United Stats would escape their grip. The death of Lincoln was resolved upon. Nothing is easier than to find a fanatic to strike.'

'The death of Lincoln, was a disaster for Christendom. There was no man in the United States great enough to wear his boots. And Israel went anew to grab the riches of the World. I fear that Jewish Banks with their craftiness and tortuous tricks will entirely control the exuberant riches of America, and use it to systematically corrupt modern civilization. The Jew will not hesitate to plunge the whole of Christendom into wars and chaos, in order that 'the earth should become the inheritance of Israel.''

Thus Bismarck, who knew the game of the Jews, spoke in 1876,

[***]

According to Bismarck the awful Civil War in America was fomented by a Jewish Conspiracy, and Abraham Lincoln, the hero, and national Saint of the United States, was killed by the same Hidden Hand, which killed six Romanov Czars, ten Kings and scores of Ministers only to easier bleed their nations. [***] Lincoln was reinaugurated on March 4, 1865 and was shot on April 4-th, 1865 by an actor Wilkes Booth, who cried: 'The South is revenged.' He was a Jew, but this has never been mentioned!"[376]

Henry Morgenthau reported that in 1919 the Zionist Jews in Poland used unscrupulous tactics to subvert Polish democracy and attain Jewish control over the Polish Government,

"They admitted that their fifty-six could sway legislation only in case of close divisions among the other parties. It became clear that their hope must be to encourage such divisions."[377]

Most Polish Jews hated the Zionists and considered them to be demonic[378] and correctly predicted that the Zionist Jews would cause terrible havoc around the world. Morgenthau reported that,

"Space will not permit the reproduction here of all that these leaders said, but one or two sentences should be repeated, and in considering them it should be kept in mind that the Orthodox Jews number about eighty per cent. of the Jewish population of Poland.

'Our principal conflict,' said Rabbi Alter, 'is with Jews; our chief opponents at every step are the Zionists. The Orthodox are satisfied to live side by side with people of different religions... . The Zionists side-track

---

[376]. Maj.-Gen., Count Cherep-Spiridovich, *The Secret World Government, Or, "The Hidden Hand": the Unrevealed in History: 100 Historical "Mysteries" Explained*, The Anti-Bolshevist Publishing Association, New York, (1926), pp. 177-178, 180-181, 183.
[377]. H. Morgenthau, "The Jews in Poland", *The World's Work*, Volume 43, Number 5, (April, 1922), pp. 617-630, at 624.
[378]. M. Selzer, Editor, "Statement by the Holy Gerer Rebbe, the Sfas Emes, on Zionism (1901)", *Zionism Reconsidered: The Rejection of Jewish Normalcy*, Macmillan, New York, (1970), pp. 19-22, at 19-20.

religion.'

'We are exiled,' said Rabbi Lewin; 'we cannot be freed from our banishment, nor do we wish to be. We cannot redeem ourselves... We will abide by our religion (in Poland) until God Almighty frees us.'

And again: 'We would rather be beaten and suffer for our religion than discard the distinguishing marks of Orthodox Judaism, such as not cutting the beard, etc... . The Orthodox love Palestine far more than others, but they want it as a Holy Land for a holy race.'"[379]

In 1921, the Rothschilds were still the principal force behind Zionism and acted against the will of the vast majority of the Jews, whom the Rothschilds wanted to force to Palestine, so that the Rothschild dynasty could be Messiah, meet God, and rule the world from Jerusalem. Note that the (Arthur James) Balfour Declaration was written directly to Lord (Lionel Walter) Rothschild. Note further that Polish Orthodox Jews were the primary target, and the hardest hit victims, of the Holocaust the Zionists perpetrated against them by means of the Nazi Party, which the Zionists put into power in Germany in order to persecute their brethren. Morgenthau stated,

"We have learned the folly of persisting in a distinctive style of clothing, beard, and locks (imposed upon the Jews extraneously as a badge of slavery and oppression), and of ascribing a spiritual significance to such a costume in this age when saints like Montefiore and Baron Edmond de Rothschild, the great patron of Palestine, find sanctity not incompatible with the ordinary dress of those about them."[380]

Frankist Jews had been worming their way into positions of authority in Poland since the 1700's, and by the 1900's crypto-Jewish Frankists dominated the aristocracy, government and Catholic Church of Poland. Zionist Jews were the cause of the majority of the problems the Polish Jews faced, which were many, though it is true that the pogroms had been greatly exaggerated by the Jewish press around the world. Zionist Jews openly sought to form a foreign and adversarial government within Poland, making Jews the sworn enemy of the Polish People. Morgenthau wrote,

"The Zionists were our first callers and were also our most constant ones. We were soon in close contact with all their leaders; we attended their meetings, and they rarely left us. Some were pro-Russian, all were practically non-Polish, and the Zionism of most of them was simply advocacy of Jewish Nationalism within the Polish state. Thus, when the committee of the Djem, or Polish Constitutional Assembly, called on us, led

---

[379]. H. Morgenthau, "The Jews in Poland", *The World's Work*, Volume 43, Number 5, (April, 1922), pp. 617-630, at 628.
[380]. H. Morgenthau, "Zionism a Surrender, Not a Solution", *The World's Work*, Volume 42, Number 3, (July, 1921), pp. i-viii, at viii.

by Grynenbaum, Farbstein, and Thon—all men who had discarded the dress and beard of the Orthodox Jew—and when I discovered that they were really authorized to represent that section of the Jews that had complained to the world of the alleged pogroms, I notified them that we were willing to give them several hours a day until they had completed the presentation of their case to their entire satisfaction. That programme was adhered to, and it constantly cropped out that their aim was the securing of Jewish Nationalism within Poland. [***] There was no question whatever but that the Jews had suffered; there had been shocking outrages, of a sporadic character at least, resulting in many deaths and still more woundings and robberies, and there was a general disposition, not to say plot, of long standing, the purpose of which was to make the Jews uncomfortable in many ways: there was a deliberate conspiracy to boycott them economically and socially. Yet there was also no question but that the reports of some of the Jewish leaders had exaggerated these evils.

We found that, among the Jews, there was a thoughtful, ambitious minority who, sincere in their original motives, intensified the trouble by believing that its solution lay only in official recognition of the Jew as a separate nationality. They had seized on Zionism as a means to establish the Jewish nation. To them, Zionism was national, not religious; when questioned, they admitted that it was a name with which to capture the imagination of their brothers whose tradition bade them pray thrice daily for their return to the Holy Land.

Pilsudski, in a moment of diplomatic aberration, had said that the Jews made a serious error in forcing Article 93; quoting that utterance, these Jewish Nationalists now asserted that neither the Polish Government, nor the Roumanian for that matter, ever would carry out the spirit of the Treaty concessions, and so they aimed at nothing short of an autonomous government and a place in the family of nations. Meanwhile they wanted to join the Polish nation in a federation having a joint parliament where both Yiddish and Polish should be spoken: their favorite way of expressing it was to say that they wanted something like Switzerland, where French, German, and Italian cantons work together in harmony.

Unfortunately, they disregarded the facts. In Switzerland, generally speaking, the citizens of French language live in one section, those of German language in another, and so on, whereas these aspiring Nationals, of course, wanted the Jews to continue scattered throughout Poland. They wanted this, and yet wanted them to have a percentage of representation in Parliament equal to their percentage in the entire Polish nation! Finally, they took no account of the desires of the Orthodox Jews, who form about 80 per cent. of their number, who were content to remain in Poland and suffer for their religion if necessary, and whom the Polish politicians were already coddling and beginning to organize politically as a vote against the Nationalist-Zionists.

The leaders of these Nationalist-Zionists were capable and adroit, but they were like walking-delegates in the Labor Unions, who had to continue to agitate in order to maintain their leadership, and their advocacy of a state within-the-state was naturally resented by all. It was quite evident that one of the deep and obscure causes of the Jewish troubles in Poland was this Nationalist-Zionist leadership that exploited the Old Testament prophesies to capture converts to the Nationalist scheme.

Here, then, was Zionism in action. We had seen it at first hand in Poland. I returned home fearful that, owing to the extensive propaganda of the Zionists, the American people might obtain the erroneous impression that a vast majority of the Jews—and not, as it really was, only a portion of the 150,000 Zionists in the United States—had ceased considering Judaism as a religion and were in danger of conversion to Nationalism."[381]

On 10 October 1864 on page 2, *The Chicago Tribune* reported,

"ENCOURAGEMENT—NOT FOR THE 'ROTHSCHILDS'

The fact that the Chairman of the National Democratic Committee is the agent of the Rothschilds gives the Copperheads an immense advantage in receiving an unlimited amount of funds from the money kings of the old rotten despotisms of Europe in order to secure the election of McClellan and the destruction of the Government. That Copperhead Democracy and European despotism are working for the same end, there cannot be a particle of doubt. The hand of Belmont is most directly seen in the second plank of the Copperhead platform, and in fact it is demonstrable from the language of it, that it was all made in the interest of Jeff. Davis and his alies, the aristocrats and despots of Europe. Shrewd, far-seeing men on the other side of the Atlantic understand this matter perfectly. One of our citizens who has been making an extensive tour in Europe, writes to the Treasurer of the Union State Central Committee as follows:

'Enclosed is an order on ---------, for three hundred dollars, to aid the Union party in publishing and disseminating that proper information in order to secure the re-election of Abraham Lincoln to the office of President of the United States, and to aid in the election of the nominees of that party in the State of Illinois. * * * I have written to --------- to pay this order for me, and to respond to any calls of years to the extent of two hundred dollars more if you think it will be wanted. I feel that the future interests of our beloved country depend much on the re-election of Mr. Lincoln and the success of the Union party, and though absent, I wish to do what I can to secure that result. I hope to be home in time to vote for the Union candidates, both State and National, in November.'

Our shrewd patriotic citizen takes a wise and enlarged view of his duties, and of his interests as well; for if the Copperhead party succeed in the election, his ample fortune would not be worth the cost of a month's

---

[381]. H. Morgenthau, "The Jews in Poland", *The World's Work*, Volume 43, Number 5, (April, 1922), pp. 617-630, at 623, 630.

sojourn in Europe or elsewhere. The destruction of the Government—the sure result of a Copperhead triumph—would destroy all values, and all personal and public safety for the next generation."

On 16 October 1864 on page 2, *The Chicago Tribune* reported,

"BELMONT'S CONFEDERATE BONDS.

The Chicago rebel organ is annoyed by the publication of the fact that a controlling share in the stock of the Copperheads machine has been bought up by Auguste Belmont, the American member of the Rothschilds family and firm, well known everywhere to be controlling owners not only in the British debt and the London *Times*, which together control the British aristocracy and oppress the Irish people, but also of the Maximillian debt, (which fact accounts for the striking out of the Monroe doctrine from the Chicago Platform,) and finally of the rebel debt, (which accounts for Belmont spending two millions dollars to nominate a war man on a peace platform.) These facts are a little inconvenient to the Copperheads. They were never intended by them for publication. They are decidedly embarrassing. It is perhaps somewhat flattering to our national pride to know that the Rothschilds, who hold up every despotism in Europe, have concluded that it would be cheaper to buy up one of our political parties, and in that way secure the dissolution of the Union, than to have their agents in England and France interfere and fight us. But Irishmen and Germans have a something, which for brevity we will all a 'crop,' and this fact sticks in their crop, that the oppressors of Ireland and Germany, the money kings of Europe, not daring to carry out their first pet project of breaking down this Government by the armed intervention, of England and France, for the rebels, which would shake the bourse, lower the rates of consols and take away the ducats of the Rothschilds, have adopted the cheaper and easier mode of accomplishing the same object, by buying up the Copperhead leaders and running the Democratic machine. The Rothschilds want ducats, but to make their ducats they want votes. Votes for the Peace party will send stocks up and so the Rothschilds make their ducats. Votes for McClellan send the Union stock, which the money kings have no share, down, and so the Rothschilds make their ducats. Votes for the dissolution of the Union relatively strengthen England and France and send consols up—and so the Rothschilds make their ducats. The Union dissolved and Maximillian will not be overthrown, nor will England have to pay for her rebel privateering, nor will Ireland, backed by our Government, rebel and be free, nor will British America unite with us, by all which the Rothschilds and Belmont, chairman of the Democratic party, make ducats. The Rothschilds will fish with a silver hook for votes which net them so good a profit, but even the silver hook must be baited, and the Chicago *Times* is authorized to adjust the bait. It is 'authorized to say that Belmont owns no Confederate stock, and that he knows that the Rothschilds do not.' Now, we are authorized to say that all Europe have known for months and years that they do. We know that a banker may, by the scratch of his pen, own nothing but Confederate stock one minute and nothing but five-twenties the next. We happen to have heard of some Copperhead bankers who own little besides five-twenties on the day the Assessors calls. But the financial community know in what stocks financiers are interested, in spite of anything true or false which rebel papers may be 'authorized to state.' Let Belmont state over his own signature, if he can that he and Rothschilds have not, directly or indirectly, in their own name, or in that of others, operated in Confederate stocks during this rebellion. Until he can face the music in that style it matters little what tune any of the Copperhead penny whistles may be authorized to blow, as they are very seldom authorized to state anything that is true."

What the Rothschilds lacked in their efforts to build a Jewish nation in Palestine was any real support from the Jewish community. They could bankrupt Egypt and Turkey. They could bring Russia to ruins. They could buy Jewish neer-do-wells. They could even buy the Pope, but the only way to force Jews in large numbers to Palestine was to put Hitler and Stalin into power and persecute Jews on a massive and unprecedented scale. On 28 January 1877 on page 12, *The Chicago Daily Tribune* reported,

"THE NEW EXODUS.
THE IDEA RIDICULED IN NEW YORK.
New York World.

There is a report 'that the Jews are again crowding back to Palestine.' A writer in the Cincinnati *Commercial* says there are 'many closed Jewish houses in London. The whole region from Dan to Beersheba is crowded with immigrant Jews from all parts of the world.' Conversations with the leading Jewish ministers and professional men of this city show that there is no truth whatever in these reports, except in this, that the Jewish population of Palestine has in recent years, been composed altogether of 'immigrants from all parts of the world,' who have settled in Palestine so as to benefit by the numerous charities which enable them to live there in idleness and pauperism. The wholesale and indiscriminate alms-giving for the relief of 'the poor of Jerusalem' has added to the population, which, as a class, is thoroughly lazy and good-for-nothing. As to the idea of a general return of the Jews to Palestine, it is scouted as absurd and improbable in the highest degree. With the exception of a very few orthodox people, the Jews, as a religious sect, have long since given up all expectation of ever returning to the Holy Land, and the thought of returning now and founding a Jewish state has, it is said, never existed, save in the imagination of some very visionary people.

Mr. Lewis May, the senior member of the banking firm of May & King, and President of the Temple Emmanuel, the largest and richest Jewish congregation in the country, said yesterday to the writer: 'The Jews are more apt to invest in Fifth avenue lots than in Jerusalem real estate. I should advise you to sell short any Jordan River front lots you may happen to have. I think the general feeling of the Jews is that New York is good enough for them, and that Bloomingdale is good enough for the authors of these perennial rumors of a return of the Jewish people to Palestine.'

Another well-known Jewish banker ridiculed the report in a very humorous vein. He said: 'I have not yet prepared to start for Jerusalem, nor shall I until the weather is milder.'

A prominent member of the Stock Exchange said: 'Just fancy what a stir it would make if this absurd report were true. We should have Seligman, Hallgarten, and Netter all shutting up their banking offices; Rothschild would no doubt limit his financial operations to the Holy Land; Ald. Lewis and Phillips would leave two vacancies in the City Government, to which Coroner Ellinger would add another; then what would become of Anti-Tammany without Emanuel B. Hart and Judge Koch, Gershom Cohen, and Adolph Sanger; what bench in Jerusalem would Judge Joachimsen fill? Assemblyman Stein, William H. Stiner, and Judge Dittenhoefer would vanish, too. Solomon would move his furniture place and his Fifth Avenue mansion to the banks of the Jordan; and a host of lesser lights would vanish. What a time there would be 'on 'Change,' too, to miss our Seligmans, De Cordovas, Josephs, Sternbergers, and Bernheimers; what would the theatres do on Saturday nights; who would patronize the balls? With the stores of the Vogels, Stadlers, Rosenfelds, Solomons, Lagowitzes, Adlers, Lauters, and others, shut up, Broadway would be indeed deserted. The handsome Harmonie Club on Forty-second street would, of course, be removed to the Holy Land, and the Standard Club would follow suit. There would be a big falling-off in the membership of the Manhattan, Union League, Lotos, and Palette. Ferdinand Myer would close his 'Newport' flat, Lewis May his 'Albany' flat, and Dore Lyon would sell his real estate. The Temple Emmanuel, on Fifth avenue, all the handsome temples in other parts of the city, the elegant mansions of the Hendrickses, Myers, Kings, Nathans, and Pikes, all to vanish to the stony streets of Jerusalem. Oh, no; never.'"

4 EINSTEIN THE RACIST COWARD

Albert Einstein was a genocidal racist Zionist. He was appalled by the fact that most German Jews did not share his racist and segregationist views. Einstein ridiculed Jews who assimilated into German society. Einstein hypocritically and disingenuously dubbed all of his critics "anti-Semites". He was a coward who hid from criticism by smearing his critics. When he was finally forced to debate in Bad Nauheim, he made a fool of himself and ran away in the middle of the argument.

"The General Assembly, [***] Determines that Zionism is a form of racism and racial discrimination."—UNITED NATIONS GENERAL ASSEMBLY RESOLUTION NUMBER 3379[382]

"I get most joy from the emergence of the Jewish state in Palestine. It does seem to me that our kinfolk really are more sympathetic (at least less brutal) than these horrid Europeans. Perhaps things can only improve if only the Chinese are left, who refer to all Europeans with the collective noun 'bandits.'"—ALBERT EINSTEIN [383]

4.1 Introduction

The massive emigration of Eastern European Jews, coupled with the financial might of the Rothschild family and their lesser branches, and with the disproportionate Jewish domination of the press, resulted in tremendous power for the Jewish community, especially in America, England and Germany. Einstein used this organized Jewish power in a cowardly fashion to suppress open debate on the theory of relativity and his career of plagiarism.

---

382. "3379 (XXX). Elimination of All Forms of Racial Discrimination", General Assembly—Thirtieth Session, Resolutions adopted on the reports of the Third Committee, 2400th Plenary Meeting, (10 November 1975), pp. 83-84. URL:

http://www.un.org/documents/ga/res/30/ares30.htm

Confer: Zionism & Racism: Proceedings of an International Symposium, International Organization for the Elimination of All Forms of Racial Discrimination, Tripoli, (1977), pp. 249-250. *Cf.* F. A. Sayegh, *Zionism: A Form of Racism And Racial Discrimination" Four Statements Made at the U.N. General Assembly*, Office of the Permanent Observer of the Palestine Liberation Organization to the United Nations, (1976), pp. 40-41. URL:

http://www.ameu.org/uploads/sayegh_march1_03.pdf

After the fall of the Soviet Union, which had long sponsored racial integration (*see:* "Circus" a motion picture released in 1936 directed by Grigori Alexandrov starring Lyubov Orlova), the U. N. withdrew this resolution under great pressure from Zionists.

383. Letter from A. Einstein to P. Ehrenfest of 22 March 1919, English translation by A. Hentschel, *The Collected Papers of Albert Einstein*, Volume 9, Document 10, Princeton Univsersity Press, (2004), pp. 9-10, at 10.

Einstein, himself a racist, hypocritically and disingenuously accused his critics of racism for saying the same things that Einstein himself had said both publicly and privately. Einstein counted on fellow racist Jews to rush to his defense simply because he was a Jew. His expectations were rewarded.

4.2 The Power of Jewish Tribalism Inhibits the Progress of Science and Deliberately Promotes "Racial" Discord

Just as the "Jewish press" refused to entertain criticism of Judaism in the *Kulturkampf* while they relentlessly ridiculed Catholicism specifically and Christianity generally, they refused to entertain criticism of their Jewish Messiah, Albert Einstein. However, Einstein's Nobel Prize was not awarded for the theory of relativity, because so many were aware of the fact that Albert Einstein had plagiarized the theory. Ernst Gehrcke[384] demonstrated that Paul Gerber had anticipated the general theory of relativity, as had Johann Georg von Soldner, making a Nobel Prize for that theory impossible. It was long known that Einstein had plagiarized the special theory of relativity from Lorentz and Poincaré. Instead of exposing the public to these facts, the Jewish dominated press smeared Einstein's critics, obstructed their access to the public, and shamelessly hyped Albert Einstein and the theory of relativity.

Reassured that corrupt elements in the press would rescue him, Einstein decided to stay in Berlin after the Berlin Philharmonic meeting where he had been publicly humiliated. On 3 September 1920, the *Berliner Tageblatt* proudly reported that Einstein would not run away:

"Prof. Albert Einstein wird, wie wir erfahren, einer Berufung ins Ausland nicht Folge leisten, sondern in Berlin bleiben. Dieser erfreuliche Entschluß des Gelehrten ist mit die Folge der zustimmenden Briefe, die infolge der Aktion der sogenannten Gesellschaft der Naturforscher an Einstein gelangt sind. Prof. Einstein wird, ehe er seine Gastvorlesungen an der Universität Leiden hält, noch auf der Kieler Woche für Kunst und Wissenschaft über die Relativitätstheorie sprechen und auf der Naturforscherversammlung in Bad Nauheim seine Theorie zur Diskussion stellen. Ob er im kommenden

---

384. E. Gehrcke, *Annalen der Physik*, Volume 51, (1916), pp. 119-124; **and** "Über den Äther", *Verhandlungen der Deutschen Physikalischen Gesellschaft*, Volume 20, (1918), pp. 165-169; **and** "Zur Diskussion über den Äther", *Verhandlungen der Deutschen Physikalischen Gesellschaft*, Volume 21, (1919), pp. 67-68; **and** "Was beweisen die Beobachtungen über die Richtigkeit der Relativitätstheorie?", *Zeitschrift für technische Physik*, Volume 1, (1920), p. 123; **and** "Die Relativitätstheorie, eine wissenschaftliche Massensuggestion", Lecture Delivered in the Berlin Philharmonic on August 24th, 1920, published in *Kritik der Relativitätstheorie*, Hermann Meusser, Berlin, (1924), pp. 54-68; **and** "Zur Frage der Relativitätstheorie", *Kosmos*, Special Edition on the Theory of Relativity, (1921), pp. 296-298.

Wintersemester die angekündigten Vorlesungen an der Berliner Universität halten wird, ist noch nicht sicher."

Einstein recorded his fears and his sudden courage upon learning that he would not have to defend himself, but would instead be defended by sycophants who were more competent than he was, which emboldened him to publish his response in the *Berliner Tageblatt*. Albert Einstein wrote to Arnold Sommerfeld on 6 September 1920:

> "Ich hatte in der That jenem Unternehmen gegen mich zu viel Bedeutung zugeschrieben, indem ich glaubte, dass ein grosser Teil unserer Physiker dabei beteiligt sei. So dachte ich wirklich zwei Tage lang an «Fahnenflucht», wie Sie das nennen. Bald aber kam die Besinnung und die Erkenntnis, dass es falsch wäre, den Kreis meiner bewährten Freunde zu verlassen. Den Artikel hätte ich vielleicht nicht schreiben sollen. Aber ich wollte verhindern, dass mein dauerndes Schweigen zu den Einwänden und Beschuldigungen, welche systematisch wiederholt werden, als Zustimmung gedeutet werden. Schlimm ist, dass jede Äusserung von mir von Journalisten geschäftlich verwertet wird. Ich muss mich eben sehr abschliessen."[385]

4.3 A Jew is Not Allowed to Speak Out Against a Jew

Jews have long taught that a good Jew never speaks out against another Jew, and a good Jews does not praise a Gentile unless such praise results in even greater praise for a Jew. We sometimes see Jews deigning to give mild and grudging praise to Hendrik Antoon Lorentz and Jules Henri Poincaré, or to Mileva Marić; but it is almost always so that they can lavish far greater praise on an undeserving Jew, Albert Einstein. The Jewish book of *Leviticus* 19:17-18 states,

> "17 Thou shalt not hate thy brother in thine heart: thou shalt in any wise rebuke thy neighbour, and not suffer sin upon him. 18 Thou shalt not avenge, nor bear any grudge against the children of thy people, but thou shalt love thy neighbour as thyself: I *am* the LORD."

In this admonition, "neighbor" refers only to fellow Jews. Jews are also taught to cover up the sins of fellow Jews, lest the tribe suffer as a whole. The Jewish book of *Numbers* 16:22 states,

> "And they fell upon their faces, and said, O God, the God of the spirits of all flesh, shall one man sin, and wilt thou be wroth with all the

---

385. A. Einstein to A. Sommerfeld, in A. Hermann, Ed., *Albert Einstein / Arnold Sommerfeld: Briefwechsel: Sechzig Briefe aus dem goldenen Zeitalter der modernen Physik*, Schwabe & Co., Basel, Stuttgart, (1968), p. 69.

congregation?"

Israel Shahak wrote in his book *Jewish History, Jewish Religion: The Weight of Three Thousand Years*,

> "There is also a series of rules forbidding any expression of praise for Gentiles or for their deeds, except where such praise implies an even greater praise of Jews and things Jewish. This rule is still observed by Orthodox Jews. For example, the writer Agnon, when interviewed on the Israeli radio upon his return from Stockholm, where he received the Nobel Prize for literature, praised the Swedish Academy, but hastened to add: 'I am not forgetting that it is forbidden to praise Gentiles, but here there is a special reason for my praise'—that is, that they awarded the prize to a Jew."[386]

Rev. I. B. Pranaitis wrote in his book *The Talmud Unmasked: The Secret Rabbinical Teachings Concerning Christians*, Eugene Nelson Sanctuary, New York, (1939),

> "In Abhodah Zarah (20, a, Toseph) it says:
>
>> 'Do not say anything in praise of them, lest it be said: How good that Goi is!'[Footnote: Maimonides (in Hilkhoth Akum X, 5) adds: 'Moreover, you should seek opportunity to mix with them and find out about their evil doings.']
>
> In this way they explain the words of Deuteronomy (VII, 2) ... and thou shalt show no mercy unto them [Goim], as cited in the Gemarah. Rabbi S. Iarchi explains this Bible passage as follows:
>
>> 'Do not pay them any compliments; for it is forbidden to say: how good that Goi is.'
>
> In Iore Dea (151, 14) it says:
>
>> 'No one is allowed to praise them or to say how good an Akum is. How much less to praise what they do or to recount anything about them which would redound to their glory. If, however, while praising them you intend to give glory to God, namely, because he has created comely creatures, then it is allowed to do so.'"[387]

---

[386]. I. Shahak, *Jewish History, Jewish Religion: The Weight of Three Thousand Years*, Pluto Press, London, (1997/2002), p. 93.

[387]. English translation of I. B. Pranaitis, *Christianus in Talmude Iudaeorum sive, Rabbinicae doctrinae de Christianis secreta: quae patere fecit*, Officina Typographica Academiae Caesarae Scientiarum, Petropoli, (1892).

Jews are taught that non-Jews, the hated Goyim, are "the wicked" and that the names of "the wicked" are to be blotted out of the book of life. There are many Jewish religious writings which call for the name of Amalekites to be forgotten. Jews were simply following their ancient traditions when they promoted Einstein as if he had created the theory of relativity, and when Jews smeared Einstein's predecessors and demanded that their names be forgotten. The Jewish book of *Proverbs* 10:7 states,

"The memory of the just *is* blessed: but the name of the wicked shall rot."

The Jewish book of *Deuteronomy* 7:2 states,

"And when the LORD thy God shall deliver them before thee; thou shalt smite them, *and* utterly destroy them; thou shalt make no covenant with them, nor shew mercy unto them:"

Jews are also taught that a Jew who informs on another Jew must be murdered. Matthew Wagner reported in *The Jerusalem Post* on 18 January 2007,

"A group of rabbis have issued a halachic opinion implying that OC Central Command Maj.-Gen. Yair Naveh deserves to be killed. The rabbis, all connected with a movement to resurrect the Sanhedrin, the ancient Jewish governing body, said in their halachic ruling this week that Naveh was guilty of being a moser, a Hebrew word that can be roughly translated as an informant or traitor."[388]

Israel Shahak and Norton Mezvinsky wrote in their book *Jewish Fundamentalism in Israel*,

"Two additional halachic laws are of special importance both generally and specifically when related to the Rabin assassination. These two laws, employed since talmudic times to kill Jews, were invoked by the assassin, Yigal Amir, as his justification for killing Prime Minister Rabin and are still emphasized by Jews who approved or have barely condemned that assassination. These are the 'law of the pursuer' (din rodef) and the 'law of the informer' (din moser).[Notation: 'Moser,' the Hebrew word for informer, is a terrible insult for Jews, similar to the word 'collaborator' for Palestinians.] The first law commands every Jew to kill or to wound severely any Jew who is perceived as intending to kill another Jew. According to halachic commentaries, it is not necessary to see such a person pursuing a Jewish victim. It is enough if rabbinic authorities, or even competent scholars, announce that the law of the pursuer applies to such a

---

[388]. M. Wagner, "Rabbis: Naveh Deserves to be Killed", *The Jerusalem Post*, http://www.jpost.com/servlet/Satellite?c=JPArticle&cid=1167467765105&pagename=JPost%2FJPArticle%2FShowFull, (18 January 2007).

person. The second law commands every Jew to kill or wound severely any Jew who, without a decision of a competent rabbinical authority, has informed non-Jews, especially non-Jewish authorities, about Jewish affairs or who has given them information about Jewish property or who has delivered Jewish persons or property to their rule or authority. Competent religious authorities are empowered to do, and at times have done, those things forbidden to other Jews in the second law. During the long period of incitement preceding the Rabin assassination, many Haredi and messianic writers applied these laws to Rabin and other Israeli leaders. The religious insiders based themselves on later developments in Halacha that came to include other categories of Jews who were defined as 'those to whom the law of the pursuer' applied. Every Jew had a religious duty to kill those Jews who were so included. Historically, Jews in the diaspora followed this law whenever possible, until at least the advent of the modern state. In the Tsarist Empire Jews followed this law until well into the nineteenth century."[389]

The second meeting of the *Arbeitsgemeinschaft deutscher Naturforscher zur Erhaltung reiner Wissenschaft* took place on 2 September 1920. The famous Jewish philosopher Oskar Kraus of Prague was scheduled to deliver a lecture stating his objections to the special theory of relativity. The Czechoslovakian government refused Kraus a visa for "political reasons" thereby preventing his appearance at the meeting and actively obstructing a public expression of anti-relativism by a famous intellectual figure of Jewish descent. Kraus had known Einstein while Einstein lived in Prague. Kraus believed that Einstein was nothing more than an amateurish Metaphysician. Einstein told Leopold Infeld, "I am really more of a philosopher than a physicist."[390] Einstein was a poor philosopher, as well. He argued in redundancies based on unproven assertions.

The pro-Einstein forces—forces so powerful that they were able to deny a man's right to speak and to corrupt the workings of a nation's government—prevented Kraus' speech, which would have been far more interesting and readily understood by a crowd of laymen and news correspondents than was Glaser's technical lecture which replaced it. Kraus' arguments[391] against the

---

389. I. Shahak and N. Mezvinsky, *Jewish Fundamentalism in Israel*, Pluto Press, London, (1999), pp. 137-138.
390. L. Infeld, *Quest—An Autobiography*, Chelsea, New York, (1980), p. 258.
391. O. Kraus, "Zum Kampf gegen Einstein und die Relativitätstheorie", *Bohemia*, Prag, (2 September 1920); **and** "Zur Lehre vom Raum und Zeit" Nachlaß Brentano, *Kantstudien*, Volume 25, (1920); **and** "Fiktion und Hypothese in der Relativitätstheorie", Schmidt's *Annalen der Philosophie*, Volume 2, Number 3, (1921), pp. 335-396; **and** "Die Verwechslungen von 'Beschreibungsmittel' und 'Beschreibungsobjekt' in der Einsteinschen speziellen und allgemeinen Relativitätstheorie", *Kantstudien, Philosophische Zeitschrift der Kant-Gesellschaft, Berlin*, Volume 26, (1921), pp. 454-486; **and** "Einwendungen gegen Einstein: Philosophische Betrachtungen gegen die Relativitätstheorie", *Neue Freie Presse*, Wien, (11 September (192?), Number 20130, pp. 2ff.; **and** "Die Unmöglichkeit der Einsteinschen Bewegungslehre", *Die Umschau*,

metaphysical absurdities in relativity theory make a powerful impression on the lay public—one Einstein's advocates were frantic to prevent. Einstein did not grasp the distinction between Metaphysics and science. He stated in 1930, "Science itself is metaphysics."[392]

This maneuver enabled pro-Einstein newspapers and Max von Laue to:

1. Criticize Weyland for being too popular and allegedly racist. Leopold Infeld stated that Weyland was a, "handsome dark-haired man of about thirty who wore a frockcoat and spoke with enthusiasm about interesting things[... .] He said that uproar about the theory of relativity was hostile to the German spirit."[393] Weyland denied that his opposition to Einstein was anti-Semitic.

2. Attack Gehrcke's credibility in handwaving personal attacks which would sound impressive to the lay public. Philipp Frank attacked Gehrcke as, "a competent experimental physicist of Berlin, who criticized the theory from a point of view of a man who, while making no mistakes in his experiments, simply lacks the acute understanding and flight of imagination to pass from individual facts to a synthesis."[394] Frank also stated that Gehrcke was, "a hardworking observer in the laboratory".[395] Shortly before Max von Laue joined the dishonest campaign to smear Gehrcke, Laue wrote to Einstein on 18 October 1919 that Gehrcke was, "a very seasoned optics specialist with a genuine interest in moving bodies."[396] Philipp Lenard, himself a Nobel Prize laureate, nominated Gehrcke for the Nobel Prize. Einstein and his friends tried to destroy Gehrcke's career and censored him on numerous occasions.

3. Attack Lenard as an alleged racist (Arnold Sommerfeld praised Lenard's book in a letter to Einstein,

---

Volume 25, (12 November 1921), pp. 681-684; **and** *Zur Relativitäts Theorie*, Meiner, Leipzig, (1921); **and** *Lotos*, Volume 70, (1922), pp. 333ff.; **and** *Offene Briefe an Albert Einstein und Max von Laue über die gedanklichen Grundlagen der allgemeinen Relativitätstheorie*, Braumüller, Wien, (1925); **and** "Zur Relativitätstheorie", *Frankfurter Zeitung*, Number 163, 3, Volume 3, reprinted in *Hundert Autoren gegen Einstein*, R. Voigtländers Verlag, Leipzig, (1931), pp. 17-19.

392. A. Einstein quoted in "Einstein on Arrival Braves Limelight for Only 15 Minutes", *The New York Times*, (12 December 1930), pp. 1, 16, at 16.

393. L. Infeld, quoted in R. W. Clark, *Einstein: The Life and Times*, World Publishing, New York, (1971), pp. 256-257; Clark cites: L. Infeld, *Die Wahrheit*, (March 15-16, 1969).

394. P. Frank, *Einstein: His Life and Times*, Alfred A. Knopf, New York, (1947), p. 161.

395. P. Frank, *Einstein: His Life and Times*, Alfred A. Knopf, New York, (1947), p. 167.

396. Letter from M. v. Laue to A. Einstein of 18 October 1919, English translation by A. Hentschel, *The Collected Papers of Albert Einstein*, Volume 9, Document 145, Princeton University Press, (2004), pp. 122-124, at 123.

"In seiner neu aufgelegten Broschüre «Rel[ativität], Äther, Gravit[ation]» hat [Lenard] sich sehr anständig über Sie [Einstein] geäussert."[397]

Lenard, while expressing his patriotism and the dignity and integrity he demanded of German science, did not publicly express racial sentiments until after Einstein had attacked him and smeared his name without grounds around the world.

4. Avoid Glaser's objections as dry and uninteresting pedantic gobbledygook.

5. Prevent Kraus' dramatic public exposition of the fatal flaws in the theory of relativity, which could not be misconstrued as if "anti-Semitic" even by the shameless pro-Einstein press.

All of this was done to change the subject from Einstein's plagiarism, Einstein's self-promotion and gross exaggeration of the significance of his theories, the relativists' corrupt misrepresentation of the available evidence to the public, and the absurdities of the theory of relativity—all of this was done to change the subject to the irrelevant issue of anti-Semitism. Einstein and his friends were completely unethical. They inhibited the progress of science and took away fundamental human liberties.

Max von Laue reported in the evening edition of *Vossische Zeitung* on 4 September 1920 that the Czechoslovakian government denied Kraus, of Prag, the right to leave the country "for political reasons". Laue, racist Zionist Albert Einstein's "Shabbas Goy", again tried to change the subject to racial issues in a cowardly effort to avoid the relevant facts,

"Der Einstein-Effekt im Spektrum.
Von
Max von Laue.

Professor Max von Laue, Ordinarius für theoretische Physik an der Berliner Universität, Träger des Nobelpreises für Physik im Jahre 1914, stellt uns folgende Ausführungen zur Verfügung:

Die Arbeitsgemeinschaft deutscher Naturforscher für Rassereinheit der Wissenschaft veranstaltete am 2. 9. ihren zweiten Vortragsabend in der Philharmonie. Zunächst mußte ihr geistiges Haupt, Herr Paul Weyland, das Ausbleiben von Prof. Kraus aus Prag mitteilen, dem die tschecho-slowakische Regierung aus politischen Gründen die Ausreise verweigert

---

[397]. A. Sommerfeld to A. Einstein, in A. Hermann, *Briefwechsel. 60 Briefe aus dem goldenen Zeitalter der modernen Physik*, Schwabe & Co., Basel, Stuttgart, (1968), p. 71. Prof. Lewis Elton stresses this point.

hat.

Sodann ergriff Herr Dr.-Ing. Glaser das Wort zu dem angekündigten Vortrage, der sich nach ein paar einleitenden Bemerkungen über die Lichtablenkung bei der Sonnenfinsternis 1919 ausführlich mit der Rotverschiebung der Spektrallinien auf der Sonne beschäftigte, deren Dasein die allgemeine Relativitätstheorie notwendig fordern muß. Hier sprach nun ein gescheiter Mann über eine Sache, von der er etwas versteht — ganz im Gegensatz zum ersten Vortragsabend. Schon daraus geht hervor, daß der Physiker viel dabei lernen konnte. Ob auch der Laie? Manchmal schien uns das zweifelhaft.

Der Redner zeigte zunächst in wohlgelungenen-Projektionsbildern die sogenannten Cyanbanden im Sonnenspektrum, an denen die wichtigsten Beobachtungen gemacht sind, und deren Auflösung in einzelne Linien. Er ging dann aus von den Messungen Schwarzschilds, bei denen er selbst mitgearbeitet hat. Deren Ergebnis sprach eher gegen als für den Einsteineffekt. Er führte weiter die langen Messungsreihen vor, die sich in Arbeiten von St. John, Evershed und Royds sowie Hale befinden. Letztere sind in Deutschland zurzeit schwer zugänglich, und die Mühe, mit der der Vortragende sie sich zu verschaffen gewußt hat, muß sehr anerkannt werden. Mit vollster Bewunderung und einem gewissen Neid muß es erfüllen, wenn man von den großartigen Hilfsmitteln hört, welche die Sternwarte des Mount Wilson für solche Versuche bietet, und dazu die Projektionsbilder sieht. Alle diese Forscher finden Verschiebungen der Spektrallinien, doch welchen diese meist in der Größe, manchmal auch in der Richtung vom Einsteineffekt an, auch lassen sich noch manche andere Erklärungen dafür ersinnen, so daß ein einheitliches Bild nicht entsteht.

Sodann ging der Vortragende zu den kurzen Veröffentlichungen zweier Deutscher über. Grebe und Bachem haben nämlich seit 1919 in Bonn mit weit bescheideneren Mitteln dieselben Untersuchungen angestellt. Und sie kommen zu dem Ergebnis, daß man nicht wahllos jede Linie im Spektrum zur Entscheidung der Frage heranziehen dürfe. Unsymmetrien im Linienbau sowie die unvermeidbaren Unterschiede zwischen Absorptionsspektren, wie wir sie im Sonnenlicht haben, und den irdischen Emissionsspektren, mit denen man sie vergleicht, können nach ihnen das Ergebnis einer genauen Messung vollständig fälschen. Beschränkt man die Untersuchung auf acht Linien, die von solchen Uebelständen frei sind, so findet man aus ihren eigenen Messunggen, sowie aus denen ihrer Vorgänger eine Rotverschiebung, welche mit dem von Einstein verlangten Effekt recht gut übereinstimmt.

Hiergegen wandte sich der Redner. Das wesentlichste Instrument der Bonner Untersuchung ist ein Gitter, und die bisherigen Gitter sind nicht hinreichend fehlerfrei, um diese Untersuchung zu ermöglichen. Er zeigte im Bild vortreffliche photographische Aufnahmen von Gittern und stellte dabei sein eigenes Licht etwas unter den Scheffel, indem er verschwieg, daß solche Aufnahme niemandem vor ihm selbst gesungen sind. Die dabei zutage tretenden Fehler verursachen Schleier um die Spektralanalyse; diese beim Bonner Apparat auftretenden, bei geeigneteren Anordnungen aber fehlenden Schleier sind es nach Glaser, welche Grebe und Bachem zur Ausscheidung der Mehrzahl der bisher untersuchten Linien veranlaßt haben. Glaser hält demgegenüber die älteren Untersuchungen für maßgebend und schloß mit den Worten, er glaube auch die Anhänger der Relativitätstheorie überzeugt zu haben, daß sie von der Rotverschiebung der Spektrallinien nichts mehr zu hoffen hätten.

Darin zeigt sich nun wieder die einseitige Parteinahme dieses sonst nicht schlechten Vortrages. Warum verschwieg der Redner, daß, selbst wenn die allgemeine Relativitätstheorie sich an der Erfahrung nicht bestätigen sollte, doch dann immer noch die beschränkte Relativitätstheorie, welche uns Einstein 1905 beschert hat, bestehen bleibt? Warum erwähnte er nicht, daß Schwarzschild, auch nachdem er die theoretische Rotverschiebung nicht hatte finden können, noch kurz vor seinem Tode in zwei höchst wertvollen Untersuchungen an dem mathematischen Ausbau der allgemeinen Relativitätstheorie mitgearbeitet hat? Er muß diese doch wohl noch nicht für ganz erledigt gehalten haben. Ferner haben die Bonner Gelehrten gewiß nicht mit den Mitteln Hales arbeiten können. Aber sie haben dafür einen sehr beachtenswerten Gedanken in die Erörterung geworfen, den ihre englischen und amerikanischen Vorgänger nicht gehabt und deswegen auch nicht mit ihren besseren Mitteln geprüft hatten. Wie denn nun, wenn diese Forscher die Grebe-Bachemsche Prüfung der Spektrallinien auf ihre Braucharbeit wiederholen — was sehr zu wünschen ist — und dabei vielleicht deren Ergebnis bestätigen? Kann man denn diese Möglichkeit von vornherein ausschließen? Der richtige Schluß aus dem vorliegenden Beobachtungsmaterial wäre für einen sehr skeptischen Beurteiler doch wohl der gewesen: Die älteren Untersuchungen sind durch Grebe und Bachem in ihrer Bedeutung zweifelhaft gemacht. Deren eigene Untersuchungen sind bisher von anderer Seite nicht nachgeprüft. Also ist die ganze Frage noch in der Schwebe.

Und noch ein paar allgemeinere Bemerkungen seien hier gestattet: Hört man die Vorträge der „Arbeitsgemeinschaft", so muß man glauben, mit der Relativitätstheorie wäre der ganze Einstein erledigt. Und dabei ist unter denen, die da gesprochen haben und sprechen wollen, höchstens einer — zur Vorsicht wollen wir sagen, daß wir nicht Herrn Weyland meinen — dessen Leistungen für die Physik sich mit dem messen können, was Einstein außer der Relativitätstheorie getan hat. Sein Nachweis der Elektronenbewegung in den Magneten, seine Theorie der Temperaturabhängigkeit der spezifischen Wärme und so manches andere auf dem Gebiete der Quantentheorie sind unvergängliche Ruhmesblätter in der Geschichte der Wissenschaft. Gelänge es der Arbeitsgemeinschaft, was sie — nach der Art ihrer Mittel zu urteilen — anstrebt, nämlich diesen Mann aus Berlin zu vertreiben, so hätte sie damit — ebenfalls unvergängliche Berühmtheit erworben."

Johannes Riem stated that Oskar Kraus had wired him a telegram on 2

September 1920, which informed him that Kraus, "was refused a visa for political reasons."[398] Riem complained that,

"In such a way relativity theory is protected by the immigration service."[399]

The *Berliner Tageblatt* reported in the morning edition of 3 September 1920,

"Im großen Saal der Berliner Philharmonie sollte gestern abend der Vortrag von Professor Dr. Kraus-Prag, der von der „Arbeitsgemeinschaft deutscher Naturforscher" angekündigt war, stattfinden. Der Beginn des Vortrags war auf ½8 Uhr festgesetzt, um ¼9 Uhr aber erst wurde dem erschienenen Publikum mitgeteilt, daß Professor Dr. Kraus, der über „Relativitätstheorie und Erkenntnistheorie" sprechen sollte, nicht erscheinen werde."

In the evening edition of 3 September 1920, the *Berliner Tageblatt* wrote,

"E. V. Die Einstein-Kampagne. Bei den Einstein-Gegnern scheint jetzt doch die Erkenntnis Platz zu greifen, daß die Art, wie die „Arbeitsgemeinschaft deutscher Naturforscher" den Kampf gegen Einstein in dem ersten Vortrag eingeleitet hatte, nicht der richtige ist. Professor Kraus (Prag), der zur Relativitätstheorie vom erkenntnistheoretischen Standpunkt Stellung nehmen wollte, hatte, wie schon im Morgenblatt kurz gemeldet, telegraphisch abgesagt; er verzichtet darauf, sich als Philosoph in den Straßenkampf der allzu persönlich erhitzten Tagesmeinungen zu stellen. Es blieb als Redner des gestrigen Abends in der Philharmonie nur der Physiker Dr. Ing. Glaser, ein Gehilfe Schwarzschilds bei dessen früheren experimentellen Studien zur Relativitätstheorie. Und es muß gesagt werden, daß er sich nüchternster Sachlichkeit, man könnte beinahe sagen, Trockenheit, befleißigte. Jedenfalls, wer aus dem Publikum in diesen Vortrag gekommen war, um ein paar billige und tönende Schlagworte für seine Anti- oder Sympathie nach Hause zu tragen, ist Gott sei Dank enttäuscht worden, er saß in einem experimentalphysikalischen Seminar. Glaser begnügte sich damit, die Beobachtungsresultate der aus der Relativitätstheorie gefolgerten und von Einstein errechneten Effekte der Lichtablenkung und der Rotverschiebung zu untersuchen, um an Hand von Lichtbildern darzutun, das erstens die beobachteten Effekte hinter den errechneten zurückbleiben, und zweitens die beobachteten Phänomen nicht die restlos zwingende Beweiskraft als Relativitätseffekte haben, sondern, zum Beispiel die Differenz in der Verschiebung am Nordrand und am

---

398. H. Goenner, "The Reaction to Relativity Theory. I: The Anti-Einstein Campaign in Germany in 1920", *Science in Context*, Volume 6, Number 1, (1993), pp. 107-133, at 118.
399. H. Goenner, "The Reaction to Relativity Theory. I: The Anti-Einstein Campaign in Germany in 1920", *Science in Context*, Volume 6, Number 1, (1993), pp. 107-133, at 118-119.

Südrand der Sonne, wie Evershed schon zeigt, sich vorläufig schwer mit dieser Erklärung vereinigen lassen. Glaser untersuchte sehr kritisch die Mittel der Beobachtung und die Möglichkeit, mit den bei den letzten Finsternissen angewandten Apparaten und Methoden ganz einwandfreie Resultate zu erzielen. Wobei zu bedenken ist, daß die Unklarheit der erzielten Bilder doch nicht ohne weiteres zuungunsten der Einsteinschen Effekte ausgelegt werden darf. Es kann auch ein Beobachtungsfehler der unzulänglichen Mittel sein, wenn die beobachteten Effekte hinter den errechneten zurückgeblieben sind.

Es wird uns wohl nichts weiter übrigbleiben, als in Geduld abzuwarten, was am 22. September 1922 die verfinsterte Sonne an den Tag bringen wird, ob die Einsteinsche Sonne aus den kritischen Nebeln, die jetzt mit etwas allzuviel Dunst darum gemacht werden, siegreich hervorgehen wird."

Many years later, Philipp Frank spun things this way and that, and even Max Born felt obliged to state that in the context of the history of the special theory of relativity, Philipp Frank was dishonest and distorted the facts. Frank wrote,

"An invitation had also been extended to a representative of philosophy who was to prove that Einstein's theory was not 'truth,' but only a 'fiction.' He was of Jewish descent and was intended to be the climax of the meeting. Despite his political innocence and urgent telegrams, he declined at the last moment because some friends had explained the purpose of the meeting to him. As a result the first attack took place without the blessing of philosophy."[400]

Max Born said of Frank,

"EINSTEIN's work was the keystone to an arch which LORENTZ, POINCARÉ and others had built and which was to carry the structure erected by MINKOWSKI. I think it wrong to forget these other men, as it can be found in many books. Even PHILIPP FRANK's excellent biography *Einstein, Sein Leben und seine Zeit*, cannot be acquitted of this reproach, e.g., when he says (in Chap. 3, No. 6 of the German edition) that nobody before EINSTEIN had ever considered a new type of mechanical law in which the velocity of light plays a prominent part. Both POINCARÉ and LORENTZ have been aware of this, and the relativistic expression for the mass (which contains c) has rightly been called LORENTZ' formula."[401]

Oskar Kraus was an outspoken critic of the theory of relativity before the Berlin Philharmonic lectures and for many years thereafter. Frank's account does not agree with that of Paul Weyland, Max von Laue and Johannes Riem, who

---

400. P. Frank, *Einstein: His Life and Times*, Alfred A. Knopf, New York, (1967), p. 161.
401. M. Born, "Physics and Relativity", *Physics in my Generation*, second revised edition, Springer, New York, (1969), p. 106.

recorded that Kraus wished to attend the meeting, but was refused a visa for political reasons. Einstein's advocates have always relied upon clannish Jewish racism and disproportionate Jewish influence in government, the press and in the universities to prevent a fair and open discussion of the merits of the theory of relativity and of Einstein's career plagiarism. This is but one of many instances of Jewish censorship in the modern world. Jewish organizations have successfully criminalized opinions which deviate from their own. It is today illegal in many countries to offend or obstruct Jewish racists by revealing their destructive lies and dangerous Messianic aspirations.

4.4 The Bad Nauheim Debate

Nobel Prize winning Physicist Philipp Lenard took great offense at Einstein's defamatory comments. Lenard had said nothing anti-Semitic in public, but instead, in the wake of Germany's defeat in World War I, had simply asserted his national pride and declared that German science stood for high ethical standards and sound scientific practices—as opposed to the wild speculations of the British eclipse observations and the immoderate and self-glorifying advertising of Albert Einstein. Lenard's reaction came at a time when the British and French had openly attempted to destroy German science, with Albert Einstein's help.

In the winter of 1914, Lenard criticized J. J. Thomson and England in a 16 page pamphlet[402] in a nationalistic—not anti-Semitic—tone. Lenard, himself, may have been of Jewish descent and had a classically Jewish appearance.[403] It was common at the time to speak of "German science" and many of Einstein's friends and supporters, many of whom were Jewish, proudly spoke in those exact terms. Lenard supported German efforts in the war, and, like Max Planck, Walter Nernst, Fritz Haber, and many others, signed the pro-German statement of 4 October 1914, as amended, with the signatories broken down by profession, by Goerg Nicolai:

> *"The Manifesto to the Civilized World*
> As representatives of German science and art we protest before the whole civilized world against the calumnies and lies with which our enemies are striving to besmirch Germany's undefiled cause in the severe struggle for existence which has been forced upon her. The course of events has mercilessly disproved the reports of fictitious German defeats. All the more vigorous are the efforts now being made to distort truth and disseminate

---

[402]. P. Lenard, *England und Deutschland zur Zeit des grossen Krieges*, Heidelberg, (1914).
[403]. D. Bronder, *Bevor Hitler kam: Eine historische Studie*, Hans Pfeiffer Verlag, Hannover, (1964), p. 204 (p. 211 in the 1974 edition). H. Kardel, *Adolf Hitler, Begründer Israels*, Verlag Marva, Genf, (1974); English translation *Adolf Hitler: Founder of Israel*, Modjeskis' Society Dedicated to Preservation of Cultures, San Diego, (1997), pp. 4, 73.

suspicion. It is against these that we are raising our voices, and those voices shall make the truth known.

1.—IT IS NOT TRUE THAT GERMANY WAS GUILTY OF THIS WAR

Neither the nation nor the Government nor the emperor wanted it. The Germans did everything possible to avert it, documentary evidence of which is before all the world. In the twenty-six years of his reign William II has frequently shown himself the defender of the world's peace, as has frequently been acknowledged even by our enemies. Indeed, this same emperor, whom they are now presuming to call an Attila, was ridiculed for twenty years and more because of his unswerving devotion to peace. Not until our people was attacked from three sides by superior forces, which had long been lying in wait at the frontier, did it rise as one man.

2.—IT IS NOT TRUE THAT WE CRIMINALLY VIOLATED BELGIAN NEUTRALITY

It can be proved that France and England had resolved to violate it, and it can be proved that Belgium had agreed to this. It would have been suicidal not to have anticipated them.

3.—IT IS NOT TRUE THAT THE LIFE AND PROPERTY OF A SINGLE BELGIAN SUBJECT WERE INTERFERED WITH BY OUR SOLDIERS EXCEPT UNDER THE DIREST NECESSITY

Again and again, despite all warnings, did the population lie in ambush and fire on them, mutilating wounded men, and murdering doctors even while actually engaged in their noble ministrations. There could be no baser misrepresentation than to say nothing about the crime of these assassins and then to call the Germans criminals because of their having administered a just punishment to them.

4.—IT IS NOT TRUE THAT OUR TROOPS BEHAVED BRUTALLY IN REGARD TO LOUVAIN

They were forced to exercise reprisals with a heavy heart on the furious population, which treacherously attacked them in their quarters, by firing upon a portion of the town. The greater portion of Louvain is still standing, and the famous town hall is quite uninjured. It was saved from the flames owing to the self-sacrifice of our soldiers. Every German would regret works of art having been destroyed in this war or their being destroyed in the future. But just as we decline to admit that any one loves art more than we do, even so do we refuse no less decidedly to pay the price of a German defeat for the preservation of a work of art.

5.—IT IS NOT TRUE THAT WE DISREGARD THE PRECEPTS OF INTERNATIONAL LAW IN OUR METHODS OF WARFARE, IN WHICH THERE IS NO UNBRIDLED CRUELTY

But in the East the ground is soaked with the blood of women and children slain by Russian hordes, and in the West the breasts of our soldiers are lacerated with Dumdum bullets. No one has less right to pretend to be defending European civilization than those who are the allies of Russians and Serbians, and are not ashamed to incite Mongolians and negroes to fight against white men.

6.—IT IS NOT TRUE THAT FIGHTING OUR SO-CALLED MILITARISM IS NOT FIGHTING AGAINST OUR CIVILIZATION, AS OUR ENEMIES HYPOCRITICALLY ALLEGE

Without German militarism German civilization would be wiped off the face of the earth. The former arose out of and for the protection of the latter in a country which for centuries had suffered from invasion as no other has done. The German Army and the German people are one, and the consciousness of this makes seventy millions of Germans brothers to-day, without regard to education, rank, or party.

We cannot deprive our enemies of the poisoned weapons of falsehood. All we can do is to cry aloud to the whole world that they are bearing false witness against us. To you who know us, who, together with us, have hitherto been the guardians of man's highest possessions—to you we cry aloud, 'Believe us; believe that to the last we will fight as a civilized nation, to whom the legacy of a Goethe, a Beethoven, and a Kant is no less sacred than hearth and home.'

This we vouchsafe to you on the faith of our name and our honor.

The manifesto was signed by the following seventeen artists actually practising their profession: Peter Behrends, Franz von Defregger, Wilhelm Dörpfeld, Eduard von Gebhardt, Adolf von Hildebrand, Ludwig Hoffmann, Leopold Graf Kalkreuth, Arthur Kampf, Fritz Aug. von Kaulbach, Max Klinger, Max Liebermann, Ludwig Manzel, Bruno Paul, Fritz Schaper, Franz von Stuck, Hans Thoma, Wilh. Trübner.

By these fifteen natural scientists: Adolf von Beyer, Karl Engler, Emil Fischer, Wilhelm Foerster, Fritz Haber, Ernst Haeckel, Gustav Hellmann, Felix Klein, Philipp Lenard, Walter Nernst, Wilhelm Ostwald, Max Planck, Wilhelm Röntgen, Wilhelm Wien, Richard Willstätter.

By these twelve theologians: Adolf Deissmann, Albert Ehrhard, Gerhard Esser, Adolf von Harnack, Wilhelm Herrmann, Alois Knöpfler, Anton Koch, Josef Mausbach, Sebastian Merkle, Adolf von Schlatter, August Schmidlin, and Reinhold Seeberg.

By these nine poets: Richard Dehmel, Herbert Eulenberg, Ludwig Fulda, Max Halbe, Gerhard and Karl Hauptmann, Hermann Sudermann, Karl Vollmöller, and Richard Voss.

By these seven jurists; Lujo Brentano, Johannes Conrad, Theodor Kipp, Paul Laband, Franz von Liszt, Georg von Mayr, and Gustav von Schmoller.

By these seven medical men: Emil von Behring, Paul Ehrlich, Albert Neisser, Albert Plehn, Max Rubner, Wilhelm Waldeyer, and August von Wassermann.

By these seven historians: Heinrich Finke, J. J. de Groot, Karl Lamprecht, Maximilian Lenz, Eduard Meyer, Karl Robert, and Martin Spahn.

By these five art critics: Wilhelm von Bode, Alois Brandt, Justus Brinkmann, Friedrich von Duhn, and Theodor Wiegand.

By these four philosophers: Rudolf Eucken, Alois Riehl, Wilhelm Windelband, and Wilh. Wundt.

By these four philologists: Andreas Heusler, Heinrich Morf, Karl Vossler, Ulrich von Wilamowitz-Moellendorff.

By these three musicians: Engelbert Humperdinck, Siegfried Wagner, and Felix von Weingartner.

By these two politicians: Friedrich Naumann and Georg Reicke.

By this theatrical manager: Max Reinhardt."[404]

Einstein covertly supported the Allies throughout the war. Though he lived in Germany—Einstein was a disloyal agent of Germany's enemies. Einstein became a symbol to many Germans of the Jew who had "stabbed Germany in the back". Many Germans believed that Jewish leaders in the press, the English, and Jewish world finance, had conspired to destroy pan-Germany as it tried to defend Europe from pan-Slavism, and that after the war the Jewish press in Germany sided with the Allies when they sought to punish Germany and break it apart in violation of President Wilson's directives that no nation would lose territory at war's end, which promise had led Germany to surrender in the good faith of that promise.[405] The Allies, and some leading German Jews, betrayed Germany's good faith.

Albert Einstein, together with Wilhelm Förster and Georg Friedrich Nicolai[406] (born Lewinstein)—a crypto-Jew who tried to persuade young Ilse Einstein to accept Albert Einstein's proposal of marriage in 1918, while Albert Einstein was sleeping with her mother, who was Albert Einstein's cousin, Elsa Einstein[407]—drafted their "Call to the Europeans", which anticipated the European Union by calling for peace talks that would destroy the German and Austro-Hungarian Empires and replace them with a yet more universal European block, a Soviet style block that would eliminate personal property and unite the workers in their struggle against the ruling class. This came at a time when

---

404. G. Nicolai, *Die Biologie des Krieges, Betrachtungen eines deutschen Naturforschers*, O. Füssli, Zürich, (1917); English translation: *The Biology of War*, Century Co., New York, (1918), pp. xi-xiv.

405. F. K. Wiebe, *Deutschland und die Judenfrage*, M. Müller & Sohn, Hrsg. im Auftrage des Instituts zum Studium der Judenfrage, Berlin, (1939); **English** translation, *Germany and the Jewish Problem*, Published on behalf of the Institute for the Study of the Jewish Problem, Berlin, (1939); **French** translation, *L'Allemagne et la Question Juive*, Berlin, Edité sous les auspices de l'Institut pour l'étude de la question juive, (1939); **Spanish** translation, *Alemania y la Cuestión Judía*, Publicado por encargo del Instituto para el Estudio de la Cuestión Judía, Berlín, (1939).

406. W. W. Zuelzer, *The Nicolai Case: A Biography*, Wayne State University Press, Detroit, (1982). Christoph Friedrich Nicolai was a friend of Gotthold Ephraim Lessing and Moses Mendelssohn; and a critic of Kant, Fichte, Goethe and Schiller. I do know if Georg Friedrich Nicolai was a namesake.

407. Letter from Ilse Einstein to Georg Nikolai of 22 May 1918, *The Collected Papers of Albert Einstein*, Volume 8, Document 545, Princeton University Press, (1998). *See also:* D. Overbye, *Einstein in Love: A Scientific Romance*, Viking, New York, (2000), pp. 343, 404, note 22. *See also:* A. Einstein to Ilse Einstein, *The Collected Papers of Albert Einstein*, Volume 8, Document 536, Princeton University Press, (1998).

Germans were rightly concerned by the attempted takeovers of revolutionary Jewish Communists like Rosa Luxemburg, Karl Liebknecht and Kurt Eisner, which had shaken the German Nation. It was well known that the Bolsheviks under Jewish leadership had mass murdered millions of Christians and had destroyed the Russian Nation. It was also widely known that Jewish financiers had caused the First World War in order to profiteer from it, promote Zionist interests, and to destroy the Europeans' will to fight back against Bolshevism. The Jewish bankers believed that the war would tire the Europeans and leave susceptible to the Jewish propaganda that internationalism and Bolshevism were the solution to war. However, most Europeans realized that these same forces were behind the war and were terrified at the prospect of a Bolshevist Europe.

Raymond Recouly contrasted the French and Russian revolutions, in an article published in 1922, which stated, *inter alia,*

> "Since the Bolshevist revolution, the produce of Russia has diminished from 50 to 75 per cent. Famine and the deaths of millions of people have been the consequences of that Russian expropriation.
>
> We have now reached a subject in which a great many people seem to find the chief points of comparison between the two revolutions, namely the question of massacres.
>
> Nothing can excuse a massacre, either in France or in Russia.
>
> The massacres which went on in some of the Paris prisons and certain provincial towns, such as Lyons, Nantes, etc., have branded the French Revolution with bloodstains impossible to wash out.
>
> As to the condemnations pronounced by the revolutionary tribunals during the most active period of the Terror, the very composition of those tribunals, their expeditive and summary manner of delivering the sentence, the wholesale trials and condemnations pronounced by them, were the merest parody of justice.
>
> But between those massacres of the French Revolution and the massacres of the Russian Revolution, there are, however, some capital differences.
>
> First, the number of the victims was in France greatly smaller than it has been in Russia.
>
> About 1,300 people were buried at the cemetery of Picpus in Paris, where the greatest majority of the victims of the guillotine had their sepulchers. Those few thousand victims of the French Revolution seem nearly nothing as compared with the enormous number of people exterminated in Russia.
>
> The Terror in France did not last very long. There came soon a strong reaction and the whole thing was definitely stopped.
>
> Even at the most frightful period of the Terror, the exterior forms of justice were, to a certain extent, observed. If one wished to find extenuating circumstances, they could be found in the violence of the political struggle, especially in the fact of France being invaded, that enemy armies were marching on the capital, that a terrible revolt had broken out in the Vendée

province, and insurrections were taking place in the centre and south of France.

In France, the executions were always conducted openly. When Louis XVI and the Queen were beheaded, it was in the middle of the Place de la Concorde in daylight, after they had been publicly judged and condemned.

In the Russian Revolution, on the contrary, no exterior form of justice was even observed. The executions have always taken place secretly. You have only to remember the monstrous manner in which the Czar and all his family were murdered in Ekaterinburg. It was in the middle of the night, in a cellar, by revolver shots, without any judgment whatever.

It has been nearly the same with all the Russian executions.

And what about the Tcheka, that disgusting network of police spies of all kinds, which has something Asiatic, Chinese, in the way of arresting people, of torturing them and putting them to death?

Those Bolshevist massacres have already been going on for several years. There is unfortunately no sign that they are going to decrease.

I have said enough to show you the fundamental differences existing between the two revolutions. The few points of comparison that exist do so only in appearance. They are due to the fact that most of the Russian revolutionaries were wrapped up in the superficialities of the French Revolution.

Their one aim was to imitate, to copy it as much as they could. In spite of that, the two revolutions differ as much as night from day. Nearly all the men at the head of the French Revolution were men of great energy—patriotic, and disinterested; they boldly risked their lives in the struggle; most of them forfeited them.

The French Revolution endowed the country with a far better system of organization, and a far more equitable system of justice than had hitherto existed. It raised the standard of human dignity. The higher material and moral well-being that was its direct creation were immense. The whole of France, and one may truly say a great part of Europe, owes all to those reforms. It abolished all the old privileges, did away with serfdom and feudal rights, founding the liberty and dignity of the human being. It reorganized education, justice, the administering of public affairs, gave a great impulse to the education of the masses, introduced a new system of weights and measures which has been adopted by nearly every country in Europe; it instituted higher education.

That positive, constructive work of the Revolution was, as you can see, immense. When one recalls the conditions under which all those reforms were brought about, when one attempts to conjure up visions of the troubled times rife with political strife, in which the great men of the Revolutionary Assemblies did all that creative work, one cannot help being filled with admiration for their energy and their audacity.

Their virtues far outweighed their old vices.

The Russian Revolution, on the contrary, has produced nothing, it has destroyed everything.

It has not even developed the communist theories. For Lenin, after having wildly proclaimed their inviolability, was forced to abandon them for the greater part.

Bolshevism has for many years laid waste the material, intellectual, and moral forces of Russia.

To draw the conclusion of this article, one could say that while the French Revolution was all the time directed and strongly kept in hand, the Russian Revolution was left without any direction whatever.

Now we must not forget that the leading class in Russia formed a very small minority, that they were, in some manner, lost in the immensity of that country. The geographical, ethical, historical conditions of Russia were so different from Germany, France, and England that it was very difficult, almost impossible, for the leaders to lead effectively such a big country."[408]

Bolshevik atrocities made the Germans very leery of Jewish Communists—even of Jews in general, especially those calling for the world government foretold in Jewish Messianic prophecies—Jewish Messianic prophecies which called for the overthrow of Kings and Queens, Princes and Princesses; as well as for a world government run by Jews, and the "restoration of the Jews to Palestine"; and for the destruction of Gentile culture, Gentile religions, Gentile nations, and ultimately the extermination of the Gentiles, themselves—all this mass murder justified on the false premise that it was necessary to achieve an era of "peace" and a new world ruled by Jews. The persona of Albert Einstein epitomized these ancient racist and genocidal Jewish objectives and made him a focal point for the legitimate concerns Germans had for their survival, grave concerns that were proven correct by the rise of the Zionist Nazis who destroyed Germany at the behest of Jewish financiers, and the further partition and loss of sovereignty of Germany after the Second World War, when a large section of Germany and Eastern Europe were taken over by the Communists, while Western Zionists who led the Western governments permitted it to happen. Many Germans were disgusted by the Jews who had stabbed Germany in the back in the First World War.

The appeal of Einstein, Förster and Nicolai follows:

"A Manifesto to Europeans

Technical science and intercommunication are clearly tending to force us to recognize the fact that international relations exist, and consequently that a world-embracing civilization exists. Yet never has any previous war caused so complete an interruption of that coöperation which should exist between civilized nations. It may, of course, be that the reason why we are so profoundly impressed by this is only that we were already united by so many ties the severing of which is painful.

---

408. R. Recouly, "Contrasts Between the French and Russian Revolutions", *The World's Work*, Volume 44, Number 1, (November, 1922), pp. 67-80, at 78-80.

That such a state of things should exist must not astonish us. Nevertheless, those who care in the slightest degree for this universal world civilization are under a twofold obligation to strive for the maintenance of these principles. Those who might have been expected to care for such things, in particular men of science and art, have hitherto almost invariably confined their utterances to a hint that the present suspension of direct relations coincided with the cessation of any desire for their continuance.

Such feelings are not to be excused by any national passions. They are unworthy of what every one has hitherto understood by civilization, and it would be a misfortune indeed were they generally to prevail among persons of culture; and not only a misfortune for civilization, but, we are firmly convinced, a misfortune for the very purpose for which, after all, in the last resort all the present hell was let loose—the national existence of the different countries.

Technical achievement has made the world smaller, and to-day the countries of that large peninsula Europe seem brought as near to one another as the cities of each individual small Mediterranean peninsula used to be; and Europe—it might almost be said the world —is already one and indivisible, owing to its multitudinous associations.

Hence it must be the duty of educated and philanthropic Europeans to make, at any rate, an effort lest Europe, owing to her not being sufficiently strongly welded together, should suffer the same tragic fate as ancient Greece. Is Europe gradually to be exhausted by fratricidal war and perish?

The war raging at present will scarcely end in a victory for any one, but probably only in defeat. Consequently, it would seem that educated men in all countries not only should, but absolutely must, exert all their influence to prevent the conditions of peace being the source of future wars, and this no matter what the present uncertain issue of the conflict may be. Above all must they direct their efforts to seeing that advantage is taken of the fact that this war has thrown all European conditions, as it were, into a melting-pot, to mold Europe into one organic whole, for which both technical and intellectual conditions are ripe.

This is not the place to discuss how this new European order is to be brought about. We desire only to assert in principle that we are firmly convinced of the time having come for all Europe to be united together, in order to protect her soil, her inhabitants, and her civilization.

Believing as we do that the desire for such a state of things is latent in many minds, we are anxious that it should everywhere find expression and thus become a force; and with this end in view it seems to us before all else necessary that there should be a union of all in any way attached to European civilization; that is to say, who are what Goethe once almost prophetically called 'good Europeans.' We must never abandon hope that their collective pronouncement may be heard by some one even amidst the clash of arms, most especially if the 'good Europeans' of to-morrow include all those who are esteemed and considered as authorities by their fellow-men.

To begin with, however, it is needful that Europeans should unite, and if, as we hope, there are enough Europeans in Europe,—in other words, enough persons to whom Europe is no mere geographical term, but something which they have profoundly at heart,—then we mean to attempt to found such a union of Europeans. We ourselves wish only to give the first impulse to such a union; wherefore we ask you, should you be in agreement with us, and, like us, bent upon making the determination of Europe as widely known as possible, to send us your signature."[409]

Adolphe Isaac Crémieux, friend to Rothschild and Marx, purportedly stated before the Alliance Israélite Universelle,

"A new Messianic empire, a new Jerusalem, must arise in place of the emperors and popes."[410]

Talmudist Jews, like Karl Marx and the Rothschilds, had always borne a deep-seated hatred of Gentiles. Racist Zionists, like Albert Einstein, also hated Gentiles and wished them dead. Outspoken Zionist Dr. Josef Samuel Bloch was famous for answering August Rohling's criticisms of the Talmud and of anti-Christian rabbinical Talmudic culture.[411] The Talmud and Cabalist literature

---

[409]. G. Nicolai, *Die Biologie des Krieges, Betrachtungen eines deutschen Naturforschers*, O. Füssli, Zürich, (1917); English translation: *The Biology of War*, Century Co., New York, (1918), pp. xvii-xix.

[410]. P. W. Massing, *Rehearsal for Destruction: A Study of Political Anti-Semitism in Imperial Germany*, Howard Fertig, New York, (1967), p. 284. **See also:** L. Fry, *Waters Flowing Eastward: The War Against the Kingship of Christ*, TBR Books, Washington, D. C., (2000), pp. 30, 98, 101-105.

[411]. A. Rohling, *Der Talmudjude: zur beherzigung für Juden und Christen aller Stände*, Adolph Russel, Münster, (1871); English translation: *The Jew According to the Talmud*, Sons of Liberty, Metairie, Louisiana, (1978); **and** *Der Antichrist und das Ende der Welt: Zur Erwägung für alle Christen*, B. Herder, St. Louis, (1875); **and** *Der Katechismus des neunzehnten Jahrhunderts, für Juden und Protestanten, den auch Katholiken lesen dürfen*, F. Kirchheim, Mainz, (1877); **and** *Franz Delitzsch und die Judenfrage, Antwortlich beleuchtet...* , J.B. Reinitz, Prag, (1881); **and** *Fünf Briefe über den Talmudismus und das Blutritual der Juden*, Prag, (1881); **and** *Die Polemik das Menschenopfer des Rabbinismus; eine wissenschaftliche Antwort ohne Polemik für die Rabbiner und ihre Genossen*, Bonifacius-Druckerei, Paderborn, (1883); **and** *Meine Antworten an die Rabbiner, oder Fünf Briefe über den Talmundismus und das Blut-Ritual der Juden*, Cyrillo-Method'sche Buchdruckerei, Prag, (1883); **and** *Die Ehre Israels: Neue Briefe an die Juden*, Prag, (1889); **and** *Erklärung der Apokalypse des h. Johannes des grossen Propheten von Patmos*, Verlag der Liebfraumen-Druckerei (Dr. W. Wingerth), München, (1895); **and** *Auf nach Zion!: oder die grosse Hoffnung Israels und aller Menschen*, Jos. Kosel'schen Buchhandlung, Kempten, (1901); **and** *Das Judentum nach neurabbinischer Darstellung der Hochfinanz Israels*, G. Schuh, München, (1903). **See also:** A. Rohling and M. de Lamarque, *Le juif-talmudiste*, A. Vromant, Paris, Bruxelles, (1888). **See also:** A. Rohling and E. A. Drumont, *Le juif selon le Talmud*, Albert Savine, Paris, (1889); German translation: *Prof. Dr. Aug. Rohling's Talmud-Jude*, T. Fritsch, Leipzig, (1891). **See also:** J. A. Eisenmenger, A. Rohling and J. Ecker, *Die*

Sittenlehre des Juden. Auszug aus dem Talmud (Schulchan-Aruch), Deutschen Schutz- und Trutz-bund, Landesverein Bayern, Nürnberg, (1920). Rohling's work is derivative of J. A. Eisenmenger, *Des bey 40. Jahr von der Judenschafft mit Arrest bestrickt gewesene, nunmehro aber durch Autorität eines hohen Reichsvicariats relaxirte Johann Andreä Eisemengers... Endecktes Judenthum, oder: Gründlicher und wahrhaffter Bericht: welchergestalt die verstockte Juden die hochheilige Dreyeinigkeit, Gott Vater, Sohn und Heiligen Geist, erschrecklicher Weise lästern und verunehren, die heil. Mutter Christi verschmähen, das Neue Testament, die Evangelisten und Aposteln, die christliche Religion spöttlich durchziehen, und die gantze Christenheit auf das äusserste verachten und verfluchen; dabey noch viele andere, bishero unter den Christen entweder gar nicht, oder nur zum Theil bekant-gewesene Dinge und grosse Irrthüme der jüdischen Religion und Theologie, wie auch viel lächerliche und kurtzweilige Fabeln und andere ungereimte Sachen an den Tag kommen*, Frankfurt, (1700); **and** *Endecktes Judenthum oder, Gründlicher und wahrhaffter Bericht, welchergestalt die verstockte Juden die hochheilige Drey-einigkeit... verunehren, die heil. Mutter Christi verschmähen... die christliche Religion spöttisch durchziehen, und die gantze Christenheit... verachten und verfluchen; dabey noch viel andere... nur zum Theil bekant gewesene Dinge und grosse Irrthüme der jüdischen Religion und Theologie, wie auch viel lächerliche und kurtzweilige Fabeln... an den Tag kommen. Alles aus ihren eigenen... Büchern... kräfftiglich erwiesen, und in zweyen Theilen verfasset... Allen Christen zur treuherzigen Nachricht verfertiget, und mit volkommenen Registern versehen*, Königsberg in Preussen, (1711); English translation by J. P. Stehelin, *The Traditions of the Jews: With the Expositions and Doctrines of the Rabbins Contain'd in the Talmud and Other Rabbinical Writings*, Volume 1, Printed for G. Smith, London, (1732); **and** *The Traditions of the Jews: Or the Doctrines and Expositions Contain'd in the Talmud and other Rabbinical Writings*, Printed for G. Smith, London, (1742-1743). ***See also:*** E. L. Roblik J. A. Eisenmenger, *Jüdische Augen-Gläser, das ist: Ein... denen Juden zur Erkanntnuss des wahren Glaubens vorgesteltes Buch. Allwo in dem ersten Theil (wider die jüdische irrende Lehr) durch die heil. Schrifft des Alten und Neuen Testaments, gantz klar bewiesen wird, dass Jesus Christus seye ein wahrer Sohn des lebendigen Gottes... In dem anderten Theil aber, wird aus dem jüdischen Buch (Talmud genannt) bewiesen, dass der jetzige jüdische Glauben, ein falscher und gottslästerlicher Glauben seye...*, Gedruckt bey M.B. Swobodin, Brünn, (1741-1743). ***See also:*** C. Anton and J. A. Eisenmenger, *Einleitung in die rabbinischen Rechte, dabey insonderheit von einem Judeneide, wie solchen eine christliche Obrigkeit am verbindlichsten abnehmen kann umständlich ist gehandelt worden*, F.W. Meyer, Braunschweig, (1756).

Bloch accused Rohling of forging sources, and Rohling sued Bloch for libel, though the suit was dropped: A. Rohling and J. S. Bloch, *Acten und Gutachten in dem Prozesse Rohling contra Bloch*, Volume 1, M. Breitenstein, Wien, (1890); **and** *Anhang zum ersten Bande der Acten und Gutachten in dem Prosezze Rohling contra Bloch*, W. Breitenstein's Verlagsbuchhandlung, Wien, (1890).

Rohling had numerous critics: J. S. Bloch, *Gegen die Anti-semiten. Eine Streitschrift*, D. Löwy, Wien, (1882); and *Prof. Rohling und das Wiener Rabbinat: oder, "Die arge Schelmerei"*, Im Selbstverlage des Verfassers, Wien, (1882); and *Des k.k. prof. Rohling neueste Fälschungen*, Wiener Allgemeine Zeitung, Wien (1883); and *Einblicke in die Geschichte der Entstehung der talmudischn Literatur*, D. Löwy, Wien, (1884); and *Einblicke in die Geschichte der Entstehung der talmudischen Literatur*, D. Löwy, Wien, (1884); and *Talmud und Judenthum in der Oesterr. Volksvertretung*, Oesterreichische Wochenschrift, Wien, (1900); and *Talmud und Judenthum in der Oesterr*

have been censored to conceal anti-Christian and anti-Gentile passages.[412] Therefore, when discussing Talmudic passages, one must at times make use of very old and difficult to obtain sources and rely upon secondary Christian sources who were highly knowledgeable, such as Martin Luther, Johannes Buxtorf and others.[413]

Volksvertretung, Oesterreichische Wochenschrift, Wien, (1900); and *"Kol Nidre" und seine Entstehungsgeschichte*, Löwit, Wien, (1918); and *Erinnerungen aus meinem Leben*, R. Löwit, Wien, Leipzig, (1922); English translation: *My Reminiscences*, Arno Press, New York, (1973). See also: M. L. Rodkinson and J. S. Bloch, *Wahrheit gegen Lüge*, Wien, (1886). **See also:** Rabbiner Dr. Kroner, *Entstelltes Unwahres und Erfundenes in dem „Talmudjuden" Professor Dr. August Rohling's*, E. Obertüschen, Münster, (1871); which is described in: "Litarischer Wochenbericht", *Allgemeine Zeitung des Judenthums*, Volume 35, Number 34, (22 August 1871), pp. 673-674, at 674. *See also:* J. E. Fraenkel, P. Mansch, Philipp and A. Rohling, *Erwiederung auf die vom Professor Dr. Aug. Rohling Verfasste Schrift der Talmudjude*, Kugel, Lemberg, (1874). *See also:* P. Bloch, *Prof. Rohling's Falschmünzerei auf talmudischem Gebiet*, L. Merzbach, Posen, (1876). *See also:* F. Delitzsch, *Rohlings Talmudjude beleuchtet*, Dörffling & Franke, Leipzig, (1881); **and** *Schachmatt den Blutlügnern Rohling & Justus*, A. Deichert, Erlangen, (1883); **and** *Was d. Aug. Rohling beschworen hat und beschwören will*, Dörffling & Franke, Leipzig, (1883). *See also:* J. Kopp, *Zur Judenfrage nach den Akten des Prozesses Rohling-Bloch*, Leipsic, (1886). *See also:* C. A. Victor, *Prof. Dr. Rohling, die Judenfrage und die öffentliche Meinung*, T. Fritsch, Leipzig, (1887). *See also: Acten und Gutachten in dem Prozesse Rohling contra Bloch*, M. Breitenstein, Wien, (1890). *See also: Jüdische Presse*, Number 46, (1902).

412. *See:* D. Kimhi, *Hesronot ha-Shas: ve-hu sefer kevutsat ha-hashmatot: kolel kol ha-devarim ha-haserim be-Talmud Bavli ve-Rashi ve-Tosafot ve Rosh veha-G. a. u-fe. ha-mishnayot leha-Rambam; mini az nidpesu `al yede `Emanu'el Bambashti be-'Amsterdam shenat 410 ve-khen hashlamat ha-hisaron hidushe halakhot...* , Jos. Schlesinger, Budapest, (1865). *See also:* G. Dalman, *Jesus Christ in the Talmud, Midrash, Zohar, and the Liturgy of the Synagogue*, Deighton Bell, Cambridge, (1893). *See also:* W. Popper, *The Censorship of Hebrew Books: Submitted in Partial Fulfilment of the Requirements for the Degree of Doctor of Philosophy, Columbia University*, Knickerbocker Press, New York, (1899). *See also:* M. A. Hoffman II, *Judaism's Strange Gods*, Independent History and Research, Coeur d'Alene, Idaho, (2000), pp. 70-72.

413. M. Luther, *Von den Juden und ihren Lügen*, Hans Lufft, Wittenberg, (1543); Reprinted, Ludendorffs, München, (1932); English translation by Martin H. Bertram, "On the Jews and Their Lies", *Luther's Works*, Volume 47, Fortress Press, Philadelphia, (1971), pp. 123-306. *See also:* J. Buxtorf, *Synagoga Judaica: Das ist Jüden Schul ; Darinnen der gantz Jüdische Glaub und Glaubensubung... grundlich erkläret*, Basel, (1603); English edition, *The Jewish Synagogue: Or An Historical Narration of the State of the Jewes, At this Day Dispersed over the Face of the Whole Earth*, Printed by T. Roycroft for H. R. and Thomas Young at the Three Pidgeons in Pauls Church-Yard, London, (1657). *See also:* J. A. Eisenmenger, *Des bey 40. Jahr von der Judenschafft mit Arrest bestrickt gewesene, nunmehro aber durch Autorität eines hohen Reichsvicariats relaxirte Johann Andreä Eisemengers... Endecktes Judenthum, oder: Gründlicher und wahrhaffter Bericht: welchergestalt die verstockte Juden die hochheilige Dreyeinigkeit, Gott Vater, Sohn und Heiligen Geist, erschrecklicher Weise lästern und verunehren, die heil. Mutter Christi verschmähen, das Neue Testament, die Evangelisten und Aposteln, die christliche Religion spöttlich durchziehen, und die gantze Christenheit auf das* *äusserste verachten und verfluchen; dabey noch viele andere, bishero unter den Christen entweder gar nicht, oder nur zum Theil bekant-gewesene Dinge und grosse Irrthüme der jüdischen Religion und Theologie, wie auch viel lächerliche und kurtzweilige Fabeln und andere ungereimte Sachen an den Tag kommen,* Frankfurt, (1700); **and** *Entdecktes Judenthum oder, Gründlicher und wahrhaffter Bericht, welchergestalt die verstockte Juden die hochheilige Drey-einigkeit... verunehren, die heil. Mutter Christi verschmähen... die christliche Religion spöttisch durchziehen, und die gantze Christenheit... verachten und verfluchen; dabey noch viel andere... nur zum Theil bekant gewesene Dinge und grosse Irrthüme der jüdischen Religion und Theologie, wie auch viel lächerliche und kurtzweilige Fabeln... an den Tag kommen. Alles aus ihren eigenen... Büchern... kräfftiglich erwiesen, und in zweyen Theilen verfasset... Allen Christen zur treuhertzigen Nachricht verfertiget, und mit volkommenen Registern versehen,* Königsberg in Preussen, (1711); English translation by J. P. Stehelin, *The Traditions of the Jews: With the Expositions and Doctrines of the Rabbins Contain'd in the Talmud and Other Rabbinical Writings,* Volume 1, Printed for G. Smith, London, (1732); **and** *The Traditions of the Jews: Or the Doctrines and Expositions Contain'd in the Talmud and other Rabbinical Writings,* Printed for G. Smith, London, (1742-1743). ***See also:*** E. L. Roblik J. A. Eisenmenger, *Jüdische Augen-Gläser, das ist: Ein... denen Juden zur Erkanntnuss des wahren Glaubens vorgesteltes Buch. Allwo in dem ersten Theil (wider die jüdische irrende Lehr) durch die heil. Schrifft des Alten und Neuen Testaments, gantz klar bewiesen wird, dass Jesus Christus seye ein wahrer Sohn des lebendigen Gottes... In dem anderten Theil aber, wird aus dem jüdischen Buch (Talmud genannt) bewiesen, dass der jetzige jüdische Glauben, ein falscher und gottslästerlicher Glauben seye...,* Gedruckt bey M.B. Swobodin, Brünn, (1741-1743). ***See also:*** C. Anton and J. A. Eisenmenger, *Einleitung in die rabbinischen Rechte, dabey insonderheit von einem Judeneide, wie solchen eine christliche Obrigkeit am verbindlichsten abnehmen kann umständlich ist gehandelt worden,* F.W. Meyer, Braunschweig, (1756). ***See also:*** A. Rohling, *Der Talmudjude: zur beherzigung für Juden und Christen aller Stände,* Adolph Russel, Münster, (1871); English translation: *The Jew According to the Talmud,* Sons of Liberty, Metairie, Louisiana, (1978); **and** *Der Antichrist und das Ende der Welt: Zur Erwägung für alle Christen,* B. Herder, St. Louis, (1875); **and** *Der Katechismus des neunzehnten Jahrhunderts, für Juden und Protestanten, den auch Katholiken lesen dürfen,* F. Kirchheim, Mainz, (1877); **and** *Franz Delitzsch und die Judenfrage, Antwortlich beleuchtet...,* J.B. Reinitz, Prag, (1881); **and** *Fünf Briefe über den Talmudismus und das Blutritual der Juden,* Prag, (1881); **and** *Die Polemik das Menschenopfer des Rabbinismus; eine wissenschaftliche Antwort ohne Polemik für die Rabbiner und ihre Genossen,* Bonifacius-Druckerei, Paderborn, (1883); **and** *Meine Antworten an die Rabbiner, oder Fünf Briefe über den Talmundismus und das Blut-Ritual der Juden,* Cyrillo-Method'sche Buchdruckerei, Prag, (1883); **and** *Die Ehre Israels: Neue Briefe an die Juden,* Prag, (1889); **and** *Erklärung der Apokalypse des h. Johannes des grossen Propheten von Patmos,* Verlag der Liebfraumen-Druckerei (Dr. W. Wingerth), München, (1895); **and** *Auf nach Zion!: oder die grosse Hoffnung Israels und aller Menschen,* Jos. Kosel'schen Buchhandlung, Kempten, (1901); **and** *Das Judentum nach neurabbinischer Darstellung der Hochfinanz Israels,* G. Schuh, München, (1903). ***See also:*** A. Rohling and M. de Lamarque, *Le juif-talmudiste,* A. Vromant, Paris, Bruxelles, (1888). See also: A. Rohling and E. A. Drumont, *Le juif selon le Talmud,* Albert Savine, Paris, (1889); German translation: *Prof. Dr. Aug. Rohling's Talmud-Jude,* T. Fritsch, Leipzig, (1891). ***See also:*** J. A. Eisenmenger, A. Rohling and J. Ecker, *Die Sittenlehre des Juden. Auszug aus dem Talmud (Schulchan-Aruch),* Deutschen Schutz- und Trutz-bund, Landesverein

Like Einstein, Bloch later advocated a Continental European union. The Socialist Eduard Bernstein wrote of Bloch,

"With regard to the circle around the *Sozialistische Monatshefte*, one must first speak of the periodical's editor, Dr. Josef Bloch. He is an exceptionally gifted East Prussian of Jewish origin. He is so Prussian-minded that at times he may be mistaken for a German nationalist. Before the war, he favored the defense and colonial policies of the German empire. To him, England was the power which German foreign policy must strive to conquer. During the war he was one of the most enthusiastic defenders of the war credits; today he is the guiding spirit among the socialist proponents of the so-called continental policy, that is, a policy which would tie together Germany, Russia, and France against England and, if necessary, also against the United States. This is not as a result of dislike of the English but because he believes that such a policy is necessary in the interest of Germany's world mission. As a Socialist he is a revisionist and as a Jew he is close to the Zionists."[414]

Though *The Manifesto to the Civilized World* managed to attract 93 signatories, *A Manifesto to Europeans* attracted only one other signatory, Otto Buek. Though Nicolai[415] spoke out against racism and nationalism in the common language of pacifists of the day, Einstein mixed his pacifistic rhetoric with contradictory racist and nationalistic Zionist rhetoric reminiscent of the Talmud. It is odd that Einstein contradicted his Socialistic and Pacifistic leanings

---

Bayern, Nürnberg, (1920). *See also:* I. B. Pranaitis (also: J. B. Pranaitis), *Christianus in Talmude Judaeorum sive rabbinicae doctrinae de christianis secreta*, Academia caesarea scientiarum, Petropoli, (1892); **English:** *The Talmud Unmasked: The Secret Rabbinical Teachings Concerning Christians*, Eugene Nelson Sanctuary, New York, (1939); **German:** *Das Christenthum im Talmud der Juden oder die Geheimnisse der rabbinischen Lehre über die Christen, enthüllt*, Verlag des "Sendboten des hl. Joseph", Wien, (1894); **Russian:** *Khristianin v Talmudie Evreiskom ili tainy ravvinskago ucheniia o khristianakh*, Tip. M.A. Aleksandrova, St. Petersburg, (1911); **Polish:** *Chrzescijanin w Talmudzie zydowskim = Christianus in Talmude Iudaeorum*, Instytut Wydawniczy "Pro Fide", Warszawa, (1937); **Spanish:** *El Talmud desenmascarado!: las enseñanzas rabinicas secretas sobre los cristianos*, La Verdad, Buenos Aires, (1981). *See also:* G. Dalman, *Jesus Christ in the Talmud, Midrash, Zohar, and the Liturgy of the Synagogue*, Deighton Bell, Cambridge, (1893). *See also:* E. K. Dilling, *The Plot Against Christianity*, Elizabeth Dilling Foundation, Lincoln, Nebraska, (1964); *the Jewish Religion: Its Influence Today: Formerly Titled the Plot Against Christianity*, Noontide Press, Torrance, California, (1983). *See also:* M. A. Hoffman II, *Judaism's Strange Gods*, Independent History and Research, Coeur d'Alene, Idaho, (2000).
414. P. W. Massing, *Rehearsal for Destruction: A Study of Political Anti-Semitism in Imperial Germany*, Howard Fertig, New York, (1967), p. 326.
415. G. Nicolai, *Die Biologie des Krieges, Betrachtungen eines deutschen Naturforschers*, O. Füssli, Zürich, (1917); English translation: *The Biology of War*, Century Co., New York, (1918).

with racist Zionist nationalism; and it is unusual that Einstein took such a strong public stance in support of Jews in the East, while most Western Jews—and he was a Western Jew—wanted to assimilate and distance themselves from segregationist Eastern Jews. Einstein was an incestuous sexual deviant like many of the Frankist Jews of the East. Einstein's fame came soon after he became a public spokesman for Eastern Jewish Zionism, which was not a coincidence.

4.4.1 Einstein Desires a "Race" War Which Will Exterminate the European Esau

The proposed union of Europe was perhaps intended by Jews like Nicolai and Einstein to consume itself in a struggle against a united Asia. Einstein often spoke in genocidal and racist terms against Germany, while promoting Jews and England. Einstein had consistently betrayed Germany before, during and after the war. For example, Albert Einstein wrote to Paul Ehrenfest on 22 March 1919,

> "[The Allied Powers] whose victory during the war I had felt would be by far the lesser evil are now proving to be *only slightly* the lesser evil. [***] I get most joy from the emergence of the Jewish state in Palestine. It does seem to me that our kinfolk really are more sympathetic (at least less brutal) than these horrid Europeans. Perhaps things can only improve if only the Chinese are left, who refer to all Europeans with the collective noun 'bandits.'"[416]

At the time Einstein made this statement, he likely knew that Bolshevik mass murderers were recruiting large numbers of Chinese.[417] Jews were commonly referred to as Asiatics or Orientals (as opposed to Europeans) at that time, and the context of Einstein's statement was his hope that a Jewish state was about to be formed in Palestine. Einstein differentiates Jews from the Europeans he, like many other Jews, would exterminate.

In an article entitled "The Jews", *The Knickerbocker; or New York Monthly Magazine*, Volume 53, Number 1, (January, 1859), pp. 41-51, at 44-45, wrote,

> "Yet the Jews of the Ottoman Empire, notwithstanding their degradation, exhibit a certain intellectual tendency. They live in an ideal world, frivolous and superstitious though it be. The Jew who fills the lowest offices, who deals out *raki* all day long to drunken Greeks, who trades in old nails, and to whose sordid soul the very piastres he bandies have imparted their copper haze, finds his chief delight in mental pursuits. Seated

---

416. Letter from A. Einstein to P. Ehrenfest of 22 March 1919, English translation by A. Hentschel, *The Collected Papers of Albert Einstein*, Volume 9, Document 10, Princeton Univsersity Press, (2004), pp. 9-10, at 10.
417. Letter from A. Einstein to E. Zürcher of 15 April 1919, *The Collected Papers of Albert Einstein*, Volume 9, Document 23, Princeton Univsersity Press, (2004).

by a taper in his dingy cabin, he spends the long hours of the night in poring over the Zohar, the Chaldaic book of the magic Cabala, or, with enthusiastic delight, plunges into the mystical commentaries on the Talmud, seeking to unravel their quaint traditions and sophistries, and attempting, like the astrologers and alchymists, to divine the secrets and command the powers of Nature. 'The humble dealer, who hawks some article of clothing or some old piece of furniture about the streets; the obsequious mass of animated filth and rags which approaches to obtrude offers of service on the passing traveller, is perhaps deeply versed in Talmudic lore, or aspiring, in nightly vigils, to read into futurity, to command the elements, and acquire invisibility.' Thus wisdom is preferred to wealth, and a Rothschild would reject a family alliance with a Christian prince to form one with the humblest of his tribe who is learned in Hebrew lore.

The Jew of the old world, has his revenge:

> 'THE pound of flesh which I demand of him
> Is dearly bought, is mine, and I will have it.'

Furnishing the hated Gentiles with the means of waging exterminating wars, he beholds, exultingly, in the fields of slaughtered victims a bloody satisfaction of his 'lodged hate' and 'certain loathing,' more gratifying even than the golden Four-per-cents on his Princely loans. Of like significance is the fact that in many parts of the world the despised Jews claim as their own the possessions of the Gentiles, among whom they dwell. Thus the squalid *Yeslir*, living in the Jews' quarter of Balata or Haskeni, and even more despised than the unbelieving dogs of Christians, traffics secretly in the estates, the palaces and the villages of the great Beys and Pachas, who would regard his touch as pollution. What, apparently, can be more absurd? Yet these assumed possessions, far more valuable, in fact, than the best 'estates in Spain,' are bought and sold for money, and inherited from generation to generation."

Einstein's statements attain their full genocidal context in the writings of his friend and political cohort, the crypto-Jew Georg Friedrich Nicolai (Lewinstein), who, together with Einstein called for the "European race" to unite in their *Manifesto to Europeans*—perhaps in Nicolai's mind to fight a preemptive race war of extermination against the "superior race" of Mongols—perhaps in Einstein's mind for the "Mongoloid race" to exterminate the "horrid Europeans"—the "Esau" of Rome.

Nicolai saw Jews as members of the "European race", or he at least pretended to see them as such in his efforts to draw the Europeans into a "race" war with the Asians. Einstein saw Jews as racially distinct from Europeans. Nicolai (Lewinstein) wrote in 1917,

> "§ 34.—*What a War of Extermination Means*
> Thus to-day the original conception of war is distorted until it has

become completely reversed, simply because there is no longer anything natural about war; it is now merely a romantic reminiscence. Now, it might be, and has been said, that the benefits of war come afterward. It might be thought, however, that any one thus contemplating the remote effects of war ought seriously to reflect upon its inevitable results. That is, he ought to think out his ideas to their logical conclusions, which seems easy, but is often very difficult.

The idea of war as a factor likely to favor the selection of the fittest, and thus promote human evolution, is simple enough. War is here looked upon as representing that relentless, or rather that disinterested, justice which allows the fit to survive and destroys the unfit. Those who consider this right should act accordingly, and proceed to draw up rules accordingly. They ought to adopt the usages of war of which we read in ancient history, rules by which old men were killed and also unborn children, but not the seemingly humane (!) rules of modern times—rules which make war a farce in the sense in which a natural scientist uses the word; that is to say, cause it to promote negative selection, and thus convert it into a means of deterioration.

The gulf which apparently separates the selfish human being of to-day from the humane promoter of civilization is merely apparent; and here I would recall what I have already said about struggle between animals and struggle between man and man. Both are justifiable in themselves and both *can* be carried on logically. Difficulties do not arise until we begin to imagine that it is allowable to carry on an animal struggle against human beings and by human methods. This is senseless, and therefore criminal; for war as waged at present can be considered only a justifiable form of struggle for existence if the nations against whom we are waging war are not looked upon as human beings, at any rate not as human beings on a level with ourselves; that is, if it is desired to carry on a war of extermination against barbarians so as to enable true humanity to find room upon and spread over the earth. No European will feel that he is justified in considering another European as a barbarian. The utmost which might be asked is whether we are not entitled to consider ourselves a superior race in comparison with certain undeveloped races, such as the Andamans or Tierra del Fuegans. What will undoubtedly occur is that these people will gradually be exterminated by the white race, though it has long been clear that it would be extremely foolish to make war upon them. They die out of themselves wherever they come in contact with whites, bloodless warfare being always more effectual than bloody.

There is only *one* race for which this question of racial superiority might be profoundly important—the Mongolian. I do not know who are the superior, the Mongolians or we ourselves, but I can quite understand our looking on the Mongolian race as enemies, and that, for instance, Europeans on the highest plane would not easily be induced to have a child by a Mongolian woman, at any rate not to own it. I can therefore also fully understand that we or the Mongolians might say, 'Only one of us two races can rule over the world, and we want that race to be ours.'

In this case the biologically *weaker* race—that is, the one which may rest assured that in ordinary course it would fall a victim to natural selection—might *perhaps* be justified in saying, 'As there is no chance of our getting the upper hand by natural and lawful means, we will try to take by force what nature withholds from us.' This shows very plainly that for the really strong war is superfluous; and as obviously it is generally folly for the weak, it is self-evident that, save in the rarest instances, there can be no possible object whatever in it.

Now, it is possible that one such rare instance may be afforded by the Mongolians, for, unlike all the other colored races, they seem to be in certain respects fitter than Europeans, although it is impossible to know exactly how they will be affected when once they are drawn into the vortex of modern civilization. Meantime, however, the sons of Heaven have the enormous advantage of being able to work equally well under all heavens, whether in the icy wastes of the tundras or under the burning sun of Sumatra. Apparently this is a special Mongolian peculiarity, for even primitive Teutonic peoples simply melted away under the Southern sun to which their impulse led them, and negro races get consumption if transferred to colder climates.

If all this is really the case, then the greater part of the habitable world belongs to the Mongols, and likewise the overlordship thereof; for it seems out of the question, seeing how much going to and fro there already is and how much more there is certain to be in the near future, that two races should live side by side and yet apart. They will mix, and one will prevail over the other.

But perhaps even the most humane of us all would not desire this, and therefore I can imagine our pointing with pardonable pride to our civilization, and saying that we are ready to take up arms in defense of it. You Mongols may be better than we are, we would say, but you are different. We do not want to know anything about your civilization, even supposing it to be superior; we mean to keep our own. From this point of view I can imagine a war, but then it must be really a relentless, merciless war.

There are now in the world five hundred millions of us Europeans or white men originally from Europe, and a thousand millions of various colored races. I believe we have even now the technical means at our disposal for exterminating these thousand millions in the course of the next twenty years. After twenty years, however, we shall no longer be in a position to do this, as soon, that is, as China has armed her whole population, constructs her own dreadnoughts, and manufactures her own cannon and shells, as Japan is already doing.

In the ensuing twenty years, therefore, it is possible that the fate of the world will be decided once and for all, and the responsibility for this decision rests with the five hundred millions of Europeans. The Mongolians need do nothing but wait, for time and space are on their side.

At a time when the fate of so many men is hanging in the balance, Europeans may, perhaps must, be asked whether on careful consideration they mean to declare all colored races barbarians, and then begin a struggle for existence, in other, words a war of extermination, and not a ridiculous war for power, against everything non-European. When once so terrible a conception as that of such a war is grasped, then, if anything save senseless cruelty is to be the result, it also must be thought out to the end, and there would have to be a war *sans trève et sans relâche*.

We must not spare even the child in its mother's womb, and must tolerate no bastards. Such a war would be ghastly, but there would be some object in it. It is useless to talk of the justice of a war, but in a sense this ghastliest of wars is the justest because, at any rate, 'it serves its own particular purpose.'

To me it seems at least conceivable that some such war might succeed, although I certainly do not believe this. History, indeed, proves over and over that the despair of nations fighting for their lives gives rise to strength which enables them to triumph over all technical expedients. Here, again, any attempt to interfere with the justice of history by such brutal methods might only too easily hasten the downfall of Europe. European nations, as I think, would do better to concentrate all their economic, technical, and scientific resources on increasing their internal vital energy, that is, on promoting race hygiene in every respect, and thus endeavor to become the equals and even the superiors of the Mongols.

This opens up vistas of victories not purchased with blood—victories which I am profoundly convinced are within the bounds of possibility. This inextinguishable hope is due to my proud European racial instinct. I will not, and I refuse to, admit that the Mongols have in the long run greater vitality than I. I trust that the majority of Europeans think as I do, and that never shall we show the Asiatics such a sign of weakness as to draw the sword against them. Even if the European nations were faint-hearted, even if they were doubtful of ultimate peaceful victory, and if nothing seemed to stand in the way of their extermination by force, even, then I would shrink from resort to force, and I am convinced that the majority of mankind agree with me.

Every one, however, must compound with his own conscience, and should any one be anxious to proceed to victory by way of force, I will go a step further to please him. I feel that all Europeans belong to the same race, and I am proud of this. But others certainly feel this less keenly than I do, and they let their wholesome race instinct run to waste in all manner of fantastic and useless notions, such as the supposed existence of a Teutonic race.[*Footnote:* Cf. §§ 90-105, about race patriotism.]

But there are those who believe in the Teutons, Germans, or Prussians having a right to predominate. I shall not here discuss the justification for such ideas, but those who would fain lead such small aggregates of human beings to victory must at any rate ask themselves whether they are *able* and, if able, also *willing*, to fight out this fight in the only way in which it can answer its purpose.

As for Teutonism, the question is as follows: take the one hundred million Germans or, properly speaking, the twenty millions more or less pure Teutons living in various parts of Europe, most of whom will have nothing whatever to do with the conception of Teutonism. Do they believe that they *can* with any prospect of success embark upon a struggle against forces from fifteen to a hundred times more numerous, and do they really *mean* to destroy these? If they have made up their minds to this, then let them make the attempt, and they will be fighting for an idea, and for an object which is at least conceivable.

We are therefore faced with the following alternative: we must either resolve to live in peace with the French, Russians, English, and whatever all their names may be, or we must wage a war of extermination upon them, a war whose purpose it is not to leave one of them alive.

Whoever, therefore, decides for war is, at any rate, no fool, and has logic on his side. Nevertheless, I hope and believe that even those who most delight in war will incline toward peace when once they realize what is the inevitable alternative. But this senseless playing at war which is now devastating Europe must be the last of its kind."[418]

The Bolsheviks in Russia had a strong and growing Chinese contingent very early on in the movement. These Chinese Bolsheviks brutally slaughtered Slavic Christians. Jewish leadership had long since scheduled China to become a Communist nation. Zionist Jews sought to establish a "Jewish State" in the far Eastern regions of the Soviet Union, the Jewish Autonomous Oblast in Khabarovsk Krai in the districts of Birobidzhansky, Leninsky, Obluchensky, Oktyabrsky and Smidovichsky.[419] This plan failed, in part, due to the interference of some Zionist Socialists, who insisted that Palestine was the Jews' national home. An even earlier attempt to found a Jewish State in Russia in the districts of Homel, Witebsk and Minsk,[420] also failed, largely due to a lack of Jewish interest. The Zionists insisted that anti-Semitism alone could force the Jews to segregate. When the Zionists put Hitler in power, they had the needed impetus to force Jews to flee Europe and the Zionists attempted to steal Chinese territory for a "Jewish homeland" with the help of the Imperial Japanese under the "Fugu Plan".[421] Zionist Jews sought to establish a "Jewish State" in China,

---

418. G. Nicolai, *Die Biologie des Krieges, Betrachtungen eines deutschen Naturforschers*, O. Füssli, Zürich, (1917); English translation: *The Biology of War*, Century Co., New York, (1918), pp. 84-89.
419. "Jews", *Great Soviet Encyclopedia: A Translation of the Third Edition*, Volume 2, Macmillan, New York, (1973), pp. 292-293, at 293.
420. I. Zangwill, "Is Political Zionism Dead? Yes", *The Nation*, Volume 118, Number 3062, (12 March 1924), pp. 276-278, at 276.
421. M. Tokayer and M. Swartz, *The Fugu Plan: The Untold Story of the Japanese and the Jews During World War II*, Paddington Press, New York, (1979). D. Goodman and M. Miyazawa, *Jews in the Japanese Mind: The History and Uses of a Cultural Stereotype*,

which had been taken over by the Imperial Japanese whom the Jews had been financing since the days when Jacob Schiff loaned them $200,000,000.00 in the Russo-Japanese War. The Zionists used the Imperial Japanese to destroy the Chinese government in preparation for the formation of a Jewish nation in China under the "Fugu Plan" in Manchuria or Shanghai. The Jews even promoted the *Protocols of the Learned Elders of Zion* to the Japanese as evidence as to how powerful they were. The "Fugu Plan" failed to attract enough Jews, even under Nazi pressure, and die hard Zionists wanted Palestine. The Zionists then arranged for war between the United States and Japan. When America declared war on Japan, Hitler, seemingly inexplicably, declared war on the United States ensuring the ultimate defeat of Germany. Hitler also went to war with the Soviets, which gave him access to large numbers of Jews the Zionists could then segregate and ready for deportation to Palestine.

It is interesting to note that the famous pilot Charles A. Lindbergh warned that the Jews, the British, and the Roosevelt administration were planning a Pearl Harbor type event, in a speech Lindbergh delivered on 11 September 1941 in Des Moines, Iowa.[422] Lindbergh was viciously smeared in the press, so viciously, that few dared to defend him. After the Pearl Harbor attack, any who might otherwise have said, "I told you so!" would have been branded a traitor and a Nazi. It is further interesting to note that Adolf Hitler declared war against America immediately after the United States declared war on Japan—this in the full knowledge that America's entrance into the war had cost Germany victory in the First World War—then Hitler declared war on the Soviets, thereby ensuring the destruction of Germany. It has since been proven that FDR did have foreknowledge of the Pearl Harbor attack.[423]

Zbigniew Brzezinski wrote in his book *The Grand Chessboard: American*

Free Press, New York, (1995).
422.
<http://www.pbs.org/wgbh/amex/lindbergh/filmmore/reference/primary/desmoinesspeech.html>
423. R. B. Stinnett, *Day of Deceit: The Truth about FDR and Pearl Harbor*, Free Press, New York, (2000). **See also:** C. B. Dall, FDR, *My Exploited Father-in-Law*, Christian Crusade Publications, Tulsa, Oklahoma, (1967); **and** *A Tribute to Lincoln, Our Money-Martyred President: An Address in Springfield, Illinois*, Omni Publications, Hawthorne, California, (1970); **and** *Amerikas Kriegspolitik: Roosevelt und seine Hintermänner*, Grabert-Verlag, Tübingen, (1972); **and** A. J. Hilder and C. B. Dall, *The War Lords of Washington (Secrets of Pearl Harbor); an Interview with Col. Curtis Dall*, Educator Publications, Fullerton, California, (1972); **and** C. B. Dall, *Who Controls Our Nation's Federal Policies — and Why?*, Noontide Press, Los Angeles, (1973); **and** C. B. Dall and B. Freedman, *Israel's Five Trillion Dollar Secret*, Liberty Bell Publications, Reedy, West Virginia, (1977); **and** C. B. Dall, *Col. Dall Reports to the Board*, Liberty Lobby, Washington, D.C., Serial Publication, (1900's); **and** C. B. Dall and R. M. Bartell, *Liberty Lobby Progress Report*, Serial Publication, Liberty Lobby, Washington, D.C., (1970's) ; **and** C. B. Dall, *Colonel Dall Reports*, Serial Publication Liberty Lobby Washington, D.C., (1900's); **and** C. B. Dall and C. M. Dunn, *Ephemeral Materials*, 1957-, Liberty Lobby, Washington, D.C.

Primacy and Its Geostrategic Imperatives, Basic Books, New York, (1997), pp. 24-25,

> "The attitude of the American public toward the external projection of American power has been much more ambivalent. The public supported America's engagement in World War II largely because of the shock effect of the Japanese attack on Pearl Harbor."

Project for the New American Century published a report entitled REBUILDING AMERICA'S DEFENSES: *Strategy, Forces and Resources For a New Century*, Project for the New American Century, Washington, D.C., (September, 2000); which states on page 51,

> "Further, the process of transformation, even if it brings revolutionary change, is likely to be a long one, absent some catastrophic and catalyzing event—like a new Pearl Harbor."[424]

There is evidence that Zionists of Einstein's Era planned to use the Chinese and Japanese to destroy the Europeans, and as a slave populace to protect and provide for Israel, in conformity with Jewish Messianic myth. China is likely slated to become the new America for Zionist interests. Racist Jews have long considered themselves to be "Orientals" and have felt closer to Asia than to Europe.

Nicolai wrote his statement while in prison, much like Hitler would later write *Mein Kampf* while incarcerated. One has a right to ask if agents provocateur like Nicolai were behind Hitler, or if Hitler himself was merely another Nicolai forwarding the interests of genocidal Judaism and racist Zionism. Nicolai (Lewinstein) further indulged in Jewish self-glorification when he wrote, ironically criticizing anti-Semitism, and under the false assumption that Jews were "racially" pure,

> "Europe, at all events, is an absolute national medley, and any one who does not consider the Jews the flower of the human race should not make such foolish assertions as that concerning the superiority of unmixed races."[425]

Nicolai's venture into genocidal fantasies was not an anomaly among politically minded persons in the West. Theodore Roosevelt was a racist who worried that the Occidental American "race" was menaced by the superior Oriental "race". Roosevelt, like Nicolai, wrote, in the context of the disappearance of "races", that "The military supremacy of the whites"[426] could

---

[424]. <http://www.newamericancentury.org/RebuildingAmericasDefenses.pdf>
[425]. G. Nicolai, *Die Biologie des Krieges, Betrachtungen eines deutschen Naturforschers*, O. Füssli, Zürich, (1917); English translation: *The Biology of War*, Century Co., New York, (1918), p. 276.
[426]. J. B. Bishop, *Theodore Roosevelt and His Time Shown in His Own Letters*, Volume

by no means be taken for granted and that Asians must be prevented from emigrating to America and Australia. Roosevelt and many others were concerned by the growing industrial might of the Japanese and dreaded the day when the Chinese might likewise grow their military strength. Zionist Napoleon Bonaparte is said to have called China a "sleeping giant".

The infamous Hungarian Jew Moses Pinkeles, a. k. a. Ignatius Trebitsch-Lincoln, a. k. a. Chao Kung; who was a Methodist preacher, a pretend spy, a real spy, a Tory member of the British Parliament, one of the early financiers of Adolf Hitler and the Nazi Party, and a very early political activist for the German right wing who argued that genetic mutation had rendered him an Aryan; became a Buddhist monk who claimed to be the Dalai Lama and the Tashi Lama in 1937 and worked with the Imperial Japanese to subjugate the Chinese and create a Jewish Nation near Shanghai—where the Nazis' allies, the Imperial Japanese, had brutalized the Chinese, though the 20,000 Jewish colonizers[427] remained in comfort.[428]

Like the Frankist Jews, Schopenhauer, Wagner, and Rudolf Glandeck Freiherr von Sebottendorf (b. Adam Alfred Rudolf Glauer), Trebitsch-Lincoln preached Metempsychosis.[429] The Lurian Cabalah of Isaac Ben Solomon Luria taught Metempsychosis,[430] and it was the spiritual guide which influenced the Jewish Messianic movement of Shabbatai Zevi and Jacob Frank. The Lurian Cabalah provided the dogma for the Frankists' belief that the Messiahship would pass from one Jewish king to another Jewish king, either as a dynasty, or through Metempsychosis from one person to another person not genetically related to the previous "Messiah".

This belief system has survived among the Lubavitchers, who today proclaim the advent of the Jewish Messiah. Luria was born of an Ashnkenazi father and a Sephardic mother. Some believe that the Lurian Cabalah is expressive of the mysticism of the Hasidic Ashkenazi and forms the basis of much of modern Hasidism, who represent the descendants of the Shabbataians and the Frankists. Others dispute these assertions. It is important to note the differences between the various Jewish conceptions of the Messiah[s], and the Christian story of a loving Jesus. According to the Old Testament and various Cabalistic writings, the Jewish Messiah will ruin the nations and exterminate the

---

2, Charles Scribner's Sons, New York, (1920), pp. 104-110, at 109.
427. B. Wasserstein, *The Secret Lives of Trebitsch Lincoln*, Yale University Press, (1988), pp. 273, 284.
428. I. T. T. Lincoln, *The Autobiography of an Adventurer*, H. Holt and Co., New York, (1932); H. Kardel, *Adolf Hitler, Begründer Israels*, Verlag Marva, Genf, (1974); English translation *Adolf Hitler: Founder of Israel*, Modjeskis' Society Dedicated to Preservation of Cultures, San Diego, (1997), picture page between pages 35 and 36 and pp. 50-52, 62-63. B. Wasserstein, *The Secret Lives of Trebitsch Lincoln*, Yale University Press, (1988).
429. B. Wasserstein, *The Secret Lives of Trebitsch Lincoln*, Yale University Press, (1988), p. 271.
430. "LURIA, ISAAC BEN SOLOMON, *Encyclopaedia Judaica*, Volume 11 LEK-MIL, Macmillan, Jerusalem, (1971), cols. 572-578, at 576.

Gentiles. The Cabalist Jews hold sacred another rabidly anti-Gentile, anti-Christian, and anti-Moslem racist religious tract, the Cabalist *Zohar*.

Lubavitch Hasidim continue a tradition of Frankist Jewish Dualism, which sees evil as good, and which practices evil as if it were observance to God and a means of summoning forth the Messiah. Many suspect that the Lubavitchers, who are very well-connected in politics and in the media and who have pronounced that the Messiah is among us, plan to rule the world and fulfill Jewish Messianic prophecy.

Frankist Jews intentionally caused the persecution of Rabbinical Jews by calling the attention of Catholics to the horrifically anti-Christian and anti-Gentile teachings of the Talmud. The Frankists delighted in the deaths and sufferings of Jews, because they believed it would bring on the Messianic Era; and because it provided them with a means to worm their way into Gentile government and the Church so as to subvert them as crypto-Jews. *The North American Review* wrote as early as 1845,

> "The common expectation of a Messiah has given a wide scope for enthusiasm and fanaticism. About the year 1666, when the whole nation were looking for some remarkable event, there appeared in the East one of the most notable of the many, who, in different ages, have claimed to be Messiahs. Banished from Aleppo, his birth-place, and subsequently from Salonichi, this man, Zabathai Tzevi, travelled much, and then took up his residence at Smyrna. Great multitudes followed him; and when, to save his life, he professed the Mohammedan faith, though without renouncing his pretensions to the Messiahship, many imitated his example. His followers, denominated Zabathaites, are still found at Salonichi, outwardly professing Islamism, but Jews at heart, —a separate community, all living in the same quarter of the city, and mingling with the Turks only at the mosques and in business. He had many adherents in Poland, Holland, England, and other parts of Europe, some of whose descendants are said still to revere his memory; and would, perhaps, agree with a class of Jews, which the chief rabbi of Cairo told Dr. Wolff was numerous, and who, without being avowed followers of Tzevi, declare, when embarrassed by passages of Scripture which speak of a suffering Messiah, that they think Tzevi may have been he. Tzevi and some of his followers pretended to work miracles, and to have visions and prophetic raptures.
>
> In 1750, a Polish Jew named Frank, or Frenk, formed a new congregation in Podolia, sometimes called that of the Zoharites, after the much earlier admirers of the celebrated mystical book Zohar; and these are improperly regarded by some persons as followers of Tzevi [Shabbatai Zevi]. These Frankists, as they are also denominated, were undoubtedly tainted with mysticism; but their chief distinction seems to have been the rejection of the Talmud, which brought upon them the persecuting hate of the Rabbinists. Their faith, indeed, approximated to Christianity, which many of them embraced. They were once numerous, and are still found in Hungary and Poland.

The sect called at the present day *Chasidim*, the *Holy*, or *Pious*, who are not to be confounded with a party bearing the same name in the time of the Maccabees, date from about the year 1760; when, at Miedzyvorz in the Ukraine, a rabbi named Israel, taking the surname of Baalshem, 'possessor of the name of God,' by means of outward sanctity, and the pretended power of exorcism and working miracles, gained great multitudes of adherents. He obtained ten thousand followers within ten years, and before his death, which took place five years afterwards, forty thousand. The doctrines of the Chasidim are said to he of most pernicious tendency, promising the faithful absolution from the vilest enormities, and supernatural protection from the hostility of all earthly powers; and the sect has been reproached for every species of immorality and crime. Probably, however, these accounts are exaggerated; and the Chasidim have doubtless improved since the age of their founder. Though they receive the traditions, they are at enmity with all other Jews; and are especially bigoted in their hatred of Christianity. Their number seems to have been increasing ever since Baalshem's day, and now to be very large. Dr. Jost, a Jew opposed to them, declares, nevertheless, that their religion is at present that of nine tenths of all the Jews in Galicia, South Hungary, Wallachia, and West and South Russia; and of great numbers in Bohemia, Moravia, Moldavia, and Poland. Their worship is marked by many extravagances; they have been called ' Jewish Jumpers.' Working themselves into ecstasies, they laugh hysterically, clap their hands, and leap with frantic zeal about the synagogue, turning their faces and raising their clenched fists towards heaven, as if daring the Almighty to refuse their requests.

Rabbinism is the Catholic faith, from which all these sects are, in modern phrase, dissenters. It is the lineal descendant of Pharisaism, and distinguished by its blind adherence to the Talmud. The estimation in which strict Rabbinists hold this book is unbounded. 'He that has learned the Scripture, and not the Mishna,' says the Gemara, 'is a blockhead.' Isaac, a distinguished rabbi, says, 'Do not imagine that the written law is the foundation of our religion, which is really founded on the oral law.' The Rabbinical doctrine is, 'The Bible is like water, the Mishna like wine, and the Gemara like spiced wine.' [*Soferim* 13*b*] Some even say, that 'to study the Bible is but a waste of time.' [*Baba Mezia* 33*a*] For strict Rabbinism, a melancholy compound of superstition and fanaticism, we must look to Poland, Russia, Hungary, and Palestine, of which we speak, in describing the system. In those countries, the Rabbinists, or Talmudists, discountenance as profane all other study than that of the Bible and Talmud, but are very careful to educate their sons in their religious lore."[431]

In 1933, Moses Pinkeles, a. k. a. Trebitsch-Lincoln, tried to spread

---

[431]. "The Modern Jews", *The North American Review*, Volume 60, Number 127, (April, 1845), pp. 329-368, at 356-357.

Buddhism in Europe. In 1939, he made a Frankist appeal to the combatant governments to disband under the threat that he would otherwise unleash "Tibetan Buddhist Supreme Masters" who would destroy them—which harkens back to the Theosophic myths surrounding the Messianic Cabalist Comte de Saint Germain and the "White Lodge of the Himalayas" and the "lost secrets of Atlantis".[432] When Trebitsch-Lincoln died, Nazi Party ideologist, and Editor-in-Chief, Alfred Rosenberg published an obituary to honor Pinkeles, a Jew, on the front page of the official Nazi Party organ the *Völkische Beobachter*. Pinkeles, a Hungarian Zionist Jew, had given Adolf Hitler the money to buy the newspaper *Volkische Beobachter*. Trebitsch-Lincoln was remembered in a somewhat different fashion by *The New York Times* on 9 October 1943 on page 13. Trebitsch-Lincoln asserted that Jews are Orientals, which he appearently considered a superior "race" to Europeans. While a member of the British Parliament, he responded on 13 June 1910 to the assertion that the allegedly superior white "race" must subjugate the allegedly inferior "races",

> "I submit that if the white man cannot rule races which we call inferior races save by resort to arms, then his prestige is already gone. I speak, I confess, as an Oriental myself. I have Oriental blood in my veins, and I cannot but laugh at the doctrine of hon. Members opposite that Orientals must receive treatment in some way different from that given to other peoples. May I be permitted to point out that one of the greatest men who ever lived, Jesus Christ, was an Oriental, and did He differentiate His treatment when dealing with Orientals?"[433]

The Nazis launched a major effort to turn the Indians of India against the British, which they directed through Tibet, in which effort Trebitsch-Lincoln sought to lend his influence among Buddhists and the Imperial Japanese.

The British obstructed the Nazis' efforts to send Jews to Palestine. Moses Pinkeles sought to remove British influence from Asia and supplant it with Nazi and Imperial Japanese influence.[434] He no doubt wanted to forward the "Fugu Plan" for a Jewish State in Manchuria or Shanghai.

It is interesting to note that Communist China is the largest nation on Earth, in terms of population, but is rarely in the news in the United States. Israel, with its vastly smaller population, dominates the news, though the Palestinian viewpoint is largely ignored. Very little effort is made by United States politicians and by the American press to reform China and free its two billion citizens from tyranny, and enormous sums of money are given to Israel to help

---

432. "Ex-Spy Warns World of Buddhist Wrath", *The New York Times*, (20 December 1939), p. 5

433. B. Wasserstein, *The Secret Lives of Trebitsch Lincoln*, Yale University Press, (1988), p. 72. Wasserstein cites: *Parliamentary Debates (Hansard). House of Commons Official Report*, Series 5, Volume 17, H.M.S.O., London, (1910), cols. 1135-1139.

434. B. Wasserstein, *The Secret Lives of Trebitsch Lincoln*, Yale University Press, (1988), p. 245.

the Jews to oppress the Palestinians. Neo-Conservatives and Israeli spies have been accused of providing the Red Chinese with top secret American military secrets and materials. As China's financial power increases, it will come to play a major rôle, if not the dominant rôle, in world politics.

4.4.2 Genocidal Judaism—Pruning the Branches of the Human Family Tree

There are many Jewish traditions of human sacrifice and of the genocide of their own people, as well as of their enemies. A Jew named Saul carried these traditions over into Christianity (*Romans* 11). Jewish mythology begins with Baal worship, a Canaanite religion in which fathers burn their own firstborn children as a sacrifice to God.

The Jewish mythology of Abraham states that Abraham believed in and feared God. As a reward, God made a covenant with Abraham and gave the land that was to became Israel to the seed of Abraham. *Genesis* 15:18-21 states (*see also: Deuteronomy* 11:24-28, and *Joshua* 1:3-4. These passages—which promise the Jews an enormous domain—in some minds the entire world—are troubling because the Kahanists are pursuing these lands[435] and the Neo-Conservative Zionists in America are assisting Israel to obtain hegemony over the Middle East):

> "18 In the same day the LORD made a covenant with Abram, saying, Unto thy seed have I given this land, from the river of Egypt [the Nile] unto the great river, the river Euphrates: 19 The Kenites, and the Kenizzites, and the Kadmonites, 20 And the Hittites, and the Perizzites, and the Rephaims, 21 And the Amorites, and the Canaanites, and the Girgashites, and the Jebusites."

Genesis 17:8 states:

> "8 And I will give unto thee, and to thy seed after thee, the land wherein thou art a stranger, all the land of Canaan, for an everlasting possession; and I will be their God."

Ari Shavit wrote in his article, "White Man's Burden", in the Israeli news source *Haaretz*,

> "The war in Iraq was conceived by 25 neoconservative intellectuals, most of them Jewish, who are pushing President Bush to change the course of history."[436]

---

[435]. J. Stern, *Terror in the Name of God: Why Religious Militants Kill*, Ecco, New York, (2003), pp. 85-106.
[436]. http://www.haaretzdaily.com/hasen/pages/ShArt.jhtml?itemNo=280279

In an article entitled, "Top White House Posts Go to Jews" published in *The Jerusalem Post* on 25 April 2006, Nathan Guttman named some of the Jews in the Clinton and Bush Administrations and in the State Department: Joshua Bolten, Joel Kaplan, Michael Chertoff, Elliott Abrams, Jay Lefkowitz, Paul Wolfowitz, Doug Feith, Lewis "Scooter" Libby, Ken Mehlman, Robert Reich, Robert Rubin, Sandy Berger, Lawrence Summers, Madeline Albright, Dennis Ross, Martin Indyk, and Aaron Miller. Guttman wrote,

"One tradition likely to go on is the reading of the Purim megilla led by Chabad Rabbi Levi Shemtov, which attracts many of the Jewish staffers."[437]

In addition to the United States Government, the American news media are in predominantly Zionist hands. Against the best interests of the American People, the United States has literally fought for Israel to obtain its goal of hegemony in the Middle East, and a Greater Israel whose borders will extend from the Nile to the Euphrates. Many American lives have been sacrificed to Israel.

In one of the early instances of human sacrifice in the history of the Hebrews, God asked Abraham to make a burnt offering of his only and beloved son Isaac to God as a human sacrifice (*Genesis* 22:2). This story reveals that Judaism is an outgrowth of Canaanite Baal worship. Baal worship required parents to sacrifice their firstborn children by burning them to ashes, by "passing them through the flame". Note that the crucifixion of Jesus of Nazareth was another of countless human sacrifices in the Jewish tradition, in which Baal or God sacrifices His own firstborn child Jesus, as Jews so often did in the Old Testament.[438] Since Abraham was willing to murder his child by burning him as a sacrifice to God, an obvious instance of Baal worship, God spared Isaac and blessed Abraham by multiplying his seed (*Genesis* 22). Abraham's son Isaac came to fear God and so inherited the blessing.

An alternative explanation is that the entire story is a Jewish fabrication of self-aggrandizement meant to justify the theft of the land of other peoples. It might be that someone in the history of the Canaanites was so traumatized by the action of burning his only and beloved son alive, that he hallucinated God, or invented a story to excuse himself from sacrilege and so founded a new form of the worship of Baal, which became Judaism. Yet another alternative explanation, and this is perhaps the most plausible explanation, is that the Judeans fabricated the story in order to hide their Baal worshiping practice of human sacrifice from, among others, the Greeks and Egyptians, who often criticized them for it; and took the opportunity to give themselves their neighbor's land.

437.
<http://www.jpost.com/servlet/Satellite?cid=1143498911316&pagename=JPost%2FJPArticle%2FPrinter>
438. W. W. Reade, *The Martyrdom of Man*, Trübner & Co., London, (1872).

In point of fact, in the story Abraham's firstborn child was not Isaac, but Ishmael. Abraham's wife was named Sarah. She was also Abraham's sister and perhaps prostituted herself, as was customary among Hebrew Baal worshipers, and slept with the Pharaoh (*Genesis* 12:10-20), and with Abimelech, who was perhaps the true father of Isaac in the earliest traditions which preceded the Torah (*Genesis* 20; 21:22-34). Mary, mother of Jesus of Nazareth, was also said by the Jews to have been a prostitute and the mother of the son of the new covenant. Jesus was said by the Jews to have been the bastard child of a whore, whose reputation was improved by the legend of royal descent through his father Joseph, though it was contradictorily claimed that he was the son of God through virgin birth to Mary (*Matthew* 1. *Luke* 3:23-38). The stories of Abraham and Jesus were conceived in comparatively close timing to one another, despite the dates claimed for them, and they were fabricated under similar circumstances and towards the same ends (*Matthew* 1:21-23).

One should note that the Jews of the First Century and before had a myth which exists to this day, that there would be two Messiahs, one descended from David, (II *Samuel* 7; 22:44-51; 23:1-5. *Isaiah* 9:6-7. *Jeremiah* 23:5; 33:15, 17. *Ezekiel* 37:24-25); and another from Joseph (through the tribe of Ephraim: *Exodus* 40. *Isaiah* 53). Perhaps this explains why two different lineages emerged, perhaps not. It should be noted that King David is a fictional character, and that the ten northern tribes of Israel and the Temple of Solomon probably never existed. Even those who believe in the existence of King David as a matter of faith, may wish to consider that his descendants can not be traced, as the *Encyclopaedia Judaica* states in its article "Messiah",

> "The Davidic origin of the kingly Messiah was supposed; but, as it seems, the Messianic pretender had to prove his authenticity by his deeds—in the period of the Second Temple Davidic descendants were not traceable."[439]

Since Sarah was barren, Abraham slept with Sarah's maidservant Hagar, an Egyptian, who bore him Ishmael (*Genesis* 16) who grew into a "wild man" at perpetual war with other men. Examining the story from the perspective of Baal worshiping Hebrews, Baal required the Hebrews to sacrifice the firstborn child of each family to God.

Why should we consider the Jews to have been Baal worshipers? The book of *Ezekiel* and other places in the Old Testament make clear that the practices of Baal worship of cutting one's self with a knife to the point of covering one's self with one's own blood, of prostitution in the Temple in celebration of fertility, of homosexuality in the Temple as an expression of devotion to the male fertility god, of immolating one's firstborn child by incineration, were all widely practiced by the Jews for very long periods of time. Abraham's father, Terah, worshiped idols (*Joshua* 24:2). Abraham violated the law that he must burn his

---

[439]. "Messiah", *Encyclopaedia Judaica*, Volume 11 Lek-Mil, Macmillan, Jerusalem, (1971), cols. 1407-1417, at 1410.

firstborn child, Ishmael. Perhaps he did so at the insistence of Ishmael's Egyptian mother, Hagar. More likely is the alternative explanation that "Hagar" (like "Moses") is a symbol of the Egyptian proselytizers who converted the Judeans to Egyptian monotheism—the two religions intertwining in a new genocidal form of *Baal* worship called Judaism, which had to reconcile its past history and recent present of human sacrifice with the need to improve its image in the then-modern ancient world, where such barbarities were frowned upon.

The Biblical myth of the sacrificial mass murder of the firstborn of Egypt, for the sake of Zionism, probably relates to a lost traditional myth of the human sacrifice of the firstborn of the Egytian Hagar and Abraham. Their firstborn son was Ishmael. There may well have been a tradition which claimed that he was sacrificed for the sake of Zionism, and that Abraham and Sarah's son Isaac became heir to the covenant, and had twin sons Esau and Jacob. These mythological characters, perhaps based on the Egyptian myths of Horus and Seth, were symbols of entire peoples—peoples meant for world domination (Jacob=Jews) and peoples destined for extermination (Esau=Gentiles). The Jews pruned off entire "races" from the human family in their religious and political mythologies, often cutting off some of their own blood lines. Ishmael is to this day made a human sacrifice made for the sake of Zionism. Zionist Jews today ascribe "Esau" to the Iranians, Iraqis, Palestinians, Syrians, Lebanese, etc. And "Esau", the Christian United States and Great Britain, are the sword and the servant of Jacob, the Zionist State, the sword and the servant who slays "Esau" and "Ishmael" the Moslems (*Genesis* 25:23; 27:38-41). The reader is advised that these inconsistencies are due to the mythologies of opportunistic Jewish racists, not your humble author. At any rate, it seems clear that the story of the murder of the firstborn of Egypt is the story of a Canaanite sacrifice to Baal made as an offering for the land of Greater Israel.

Moslems believe that they inherited God's covenant (*Genesis* 15:18) from Abraham through Abraham's firstborn child Ishmael, son of Abraham and Hagar (*Genesis* 16; 17:18-27; 21:9-21). Jews believe that the Jewish God instead passed the covenant on to Isaac (*Genesis* 17:19-21), who was later born to Abraham and Sarah. Ishmael is the Biblical father of the Arabs, and has more generally come to symbolize Moslems in general. Racist Jews refer to Arabs, and Moslems in general, as "Ishmaelites". The racist Hebrew Bible stigmatizes "Ishmaelites". *Genesis* 16:12 states,

> "And [Ishmael] will be a wild man; his hand will be against every man, and every man's hand against him; and he shall dwell in the presence of all his brethren."

Worse still, racist Jews believe that the racist Talmud grants them special license to murder "Ishmaelites". In addition to the virulently anti-Gentile passages found throughout the Hebrew Bible and in the Jewish Talmud and the Cabalistic Jewish *Zohar*, the Talmudic book of *Sukkah*, folio 52*b*, not only dehumanizes Arabs, and in modern Jews' eyes, the Moslems in general, it makes them the enemies of the Jewish God,

"R. Hana b. Abba stated: It was said at the schoolhouse, There are four things of which the Holy One, blessed be He, repents that He had created them, and they are the following: Exile, the Chaldeans, the Ishmaelites and Evil Inclination. 'The Exile', since it is written, *Now, therefore, what do I here, saith the Lord, seeing that My people is taken away for naught* etc.;[9] 'the Chaldeans', since it is written, *Behold the land of the Chaldeans—this is the people that was not;*[10] 'the Ishmaelites', since it is written, *The tents of the robbers*[11] *prosper, and they that provoke God are secure since God brought them with His hand;*[12] 'the Evil Inclination', since it is written, [*And I will gather her that is driven away*] *and her that I have afflicted.*[13]"[440]

The racist Jewish Zohar, Volume 3, page 282a degrades Jesus Christ, as does the Jewish Talmud. It also degrades the Prophet Mohammed,

"From the side of idolatry Shabbethaj (Saturn) is called Lilith [*Footnote:* Lilith is a female demon, comp. Is. XXXIV. 14 and Weber, *Altsynagogale palästinische Theologie*, p. 246.], mixed dung, on account of the filth mixed from all kinds of dirt and worms, into which they throw dead dogs and dead asses, the sons of 'Esau and Ishma'el, and there (read הבו) Jesus and Mohammed, who are dead dogs, are buried among them. She (Lilith) is the grave of idolatry, where they bury the uncircumcised, (who are) dead dogs, abomination and bad smell, soiled and fetid, a bad family. She (Lilith) is the ligament [*Footnote:* אכדם is a fibre attached to the lungs] which holds fast the 'mixed multitude' (Ex. xii. 38), which is mixed among Israel, and which holds fast bone and flesh, that is, the sons of 'Esau and Ishma'el, dead bone and unclean flesh torn of beasts in the field, of which it is said (Ex. xxii. 31): 'Ye shall cast it to the dogs.'"[441]

Perhaps, to a Baal worshiper, Ishmael, the son of Abraham and Hagar, should have been sacrificed to God through the fire; and Hagar, an Egyptian, intervened and would not let her child Ishmael be sacrificed to Baal. It was Ishmael, not Isaac, who was the eldest son of Abraham and he, not Isaac, should have inherited the Covenant with God. It is likely that the Egyptian Hagar would have her son Ishmael circumcised, given that circumcision was an Egyptian custom, and the Covenant was given to the circumcised, Abraham and Ishmael (*Genesis* 17—indeed, the prophet Mohammed taught that the Covenant was with

---

440. I. Epstein, Editor, "Sukkah 52*b*", *The Babylonian Talmud*, Volume 12, The Soncino Press, London, (1935), pp. 249-252, at 250.
441. G. Dalman, *Jesus Christ in the Talmud, Midrash, Zohar, and the Liturgy of the Synagogue*, Deighton Bell, Cambridge, (1893), p. 40. Though work is given an ancient attribution by its "discoverer", the Muhammadans are also mentioned in *Zohar*, II, 32*a*. Some consider the author to have been divinely inspired, some say the work evolved over time, some say the work is a fabrication—in any event, it is an now a very old writing and was very influential in Jewish political movements like the Frankists.

Abraham and Ishmael, not Isaac), but because Ishmael should have been sacrificed to God, rights to the Covenant instead passed to a prophesied second child, Isaac born of Sarah; and, apparently, Abimelech, King of Gerar. Ishmael is demonized as a wild man of a foreign inferior race, so as to justify the unjustifiable wrongs done to him by the descendants of the Jews (*Genesis* 16:12). In the mythology the Judeans composed to glorify themselves, Isaac inherits Abraham's blessings and the Judeans eventually steal the lands of Abimelech.

As Thomas Jefferson admonished us to do, we should eliminate the supernatural superstition in the Bible. A clearer picture of the story emerges if we eliminate the myth of the Covenant with God for the land of Canaan, and substitute the more realistic picture presented in the story of the covenant between Abraham and King Abimelech. *Genesis* 21:22-33 states:

> "22 And it came to pass at that time, that Abimelech and Phichol the chief captain of his host spake unto Abraham, saying, God *is* with thee in all that thou doest: 23 Now therefore swear unto me here by God that thou wilt not deal falsely with me, nor with my son, nor with my son's son: *but* according to the kindness that I have done unto thee, thou shalt do unto me, and to the land wherein thou hast sojourned. 24 And Abraham said, I will swear. 25 And Abraham reproved Abimelech because of a well of water, which Abimelech's servants had violently taken away. 26 And Abimelech said, I wot not who hath done this thing: neither didst thou tell me, neither yet heard I of *it*, but to day. 27 And Abraham took sheep and oxen, and gave them unto Abimelech; and both of them made a covenant. 28 And Abraham set seven ewe lambs of the flock by themselves. 29 And Abimelech said unto Abraham, What *mean* these seven ewe lambs which thou hast set by themselves? 30 And he said, For these *seven* ewe lambs shalt thou take of my hand, that they may be a witness unto me, that I have digged this well. 31 Wherefore he called that place Beer-sheba; because there they sware both of them. 32 Thus they made a covenant at Beer-sheba: then Abimelech rose up, and Phichol the chief captain of his host, and they returned into the land of the Philistines. 33 And *Abraham* planted a grove in Beer-sheba, and called there on the name of the LORD, the everlasting God."

Your author proposes that, given the many identities, we should assume that the stories of: Sarah and the Pharaoh, Sarah and Abimelech, Sarah and Og, Rebekah and Abimelech; are all the same story told in various traditions. Also assume that the stories of: Adam and Eve; Abraham, Abimelech, Hagar and Sarah; Isaac, Ambimelech and Rebekah; and perhaps even Aaron and Moses; are all the same story told in various traditions—quite likely Egyptian traditions stemming from the life of Egyptian Pharaoh Akhenaton IV, who pioneered Egyptian monotheism. Still further assume that: Cain and Abel, Ishmael and Isaac, Esau and Jacob, Aaron and Moses; are the same story told in different traditions—perhaps based on the Egyptian myths of Horus and Seth. All of these fabricated and racist stories are awkwardly threaded together in the Bible, as if different stories, and are linked together by a fabricated genealogy which places

Israel at perpetual war with other peoples, so as to explain away the fact that the same story is told over and over again with different characters.

A predominant racist element repeated again and again in the Old Testament is the story that a leader's family is led into corruption by a foreign wife or servant; and, conversely, that Jewish woman are sent to corrupt foreign leaders—a practice practiced and lauded by prominent Jews, such as Josephus, who wrote of the alleged corruption of Nero by his Jewish wife Poppæa.[442] We know that the more modern Frankist Jews, among many other Jews, carried on this tradition, whether the ancient stories are in fact true, or not. Stalin feared that the Jewish wives of members of the government were seeking to undermine his authority, or so he claimed, and Stalin proscribed intermarriage between Jews and Gentiles,[443] though he himself loved Jewish women.[444] These proscriptions against intermarriage had the benefit of helping to preserve the Jewish religion and the Jewish race, in the minds of Jewish bigots (*Genesis* 28:1, 6. *Exodus* 34:16. *Leviticus* 20:26. *Numbers* 23:9. *Deuteronomy* 7:1-6. *Ezra* 9. *Nehemiah* 9:2; 13:3, 23-30). The Jewish faith is traditionally passed down through the mother, which ensures that the blood of the child is at least half the blood of the tribe, because a woman may sleep with many men but carries her own eggs.

The covenant for land for the Jews in Judah is then strictly a deal struck between Abimelech and Abraham, not God and Abraham, and was made to give Abimelech's offspring through Sarah a kingdom and secure peace, not to create a Holy contract that must be obeyed forever by all the world. The supposed "tribes" were ruled by the descendants of Abimelech and his wives, including Sarah, Abraham's sister—not by the descendants of Abraham. Abraham is merely the guardian of Abimelech and Sarah's child, Isaac/Jacob; and Abraham promotes him over his own son, Ishmael/Esau—in effect sacrifices his firstborn Ishmael/Esau, whose seed (all Gentiles) then becomes a perpetual human sacrifice to God for the sake of Jacob (all Jews), in fulfilment of the Canaanites'/Jews' worship of Baal. Note that Ishmael is said to sire twelve Princes and to be the father of a great nation (*Genesis* 17:20). Note further that the union of Sarah and Pharaoh is said to have caused plagues on Egypt—which is quite similar to the stories of Aaron, Moses and the Pharaoh (*Genesis* 12:17).

---

442. Josephus, "Antiquities of the Jews", Book XX, Chapter 8, *The Works of Flavius Josephus: Comprising the Antiquities of the Jews; a History of the Jewish Wars; and Life of Flavius Josephus, Written by Himself*, S. S. Scranton Co., Hartford, Connecticutt, (1916), pp. 609-613, at 612-613. *See also:* Tacitus, *Annal*, Book XV, in: "Dissertation III", *The Works of Flavius Josephus: Comprising the Antiquities of the Jews; a History of the Jewish Wars; and Life of Flavius Josephus, Written by Himself*, S. S. Scranton Co., Hartford, Connecticutt, (1916), p. 960. *See also:* E. Gibbon, "The Conduct of the Roman Government towards the Christians, from the Reign of Nero to that of Constantine", *The History of the Decline and Fall of the Roman Empire*, Chapter 16, Volume 3, Fred De Fau and Company, New York, (1776).

443. L. Rapoport, *Stalin's War Against the Jews: The Doctors' Plot and the Soviet Solution*, Free Press, New York, (1990), pp. 139, 208-210.

444. S. S. Montefiore, *Stalin: The Court of the Red Star*, Vintage, New York, (2003), p. 267.

The same story transfers to Moses and Aaron, where Moses and Aaron must convince those Egyptians who would follow them to give up their bondage to the worship of Pharaoh and adopt the worship of Baal—historically perhaps a group of Egyptian lepers oppressed by the Hyksos—perhaps even ostracized Hyksos lepers, who migrated to Judah and taught the Judeans Egyptian monotheism. Moses and Aaron bring plagues on the Egyptians, which is perhaps symbolic of the diseases the Hyksos brought to Egypt. In an act of Baal worship, Moses sacrifices the firstborn of the Egyptians among his people, and so hopes to transfer the loyalty of the Egyptians from Pharaoh and the Sun, to Baal, and the loyalty of Baal to the Egyptian converts. Moses and Aaron eventually succeed and the people worship Baal, though, perhaps, Moses then seeks to convert them to an Egyptian sect of Monotheism and Eleatic Monism—which is the same story as the inexplicable break in religion between Terah and his son Abraham.

Jewish authors may have added this break from pure Baalism while under the influence of the Greeks, or an Egyptian sect in Alexandria. There might well have been a sect that sought to convert Jews from Baalism which incorporated other gods, to a strict Baalism that worshiped only jealous Baal; and so fabricated the stories and legends of Monotheism from Eleatic Monism, and Egyptian and Socratic Monotheism. The sect of Dualist Judaism took from Heraclitean and Platonic dialectics to invent Christianity, which was probably intended as a stumbling block for the Romans and means to preserve the Jewish Nation.

Had Gnostic Christianity succeeded, it would have exterminated the Romans. Epiphanius wrote of the Gnostics,

"For all the sects have gathered their imposture from Greek mythology, and altered it for themselves by revising it for another and worse purpose."[445]

There are also elements of Hindu Metempsychosis in Dualist Judaism, especially as it reached the Frankists *viz.* the Lurian Cabala. The Jews were exposed to Metempsychosis through Origen, Pathagoras, and many ancient Greek philosophers; then through the Schoolmen. The Cabala adopts many of the beliefs of the Stoics and Eleatics, such as the Eleatic notions of pantheism and space-time—the belief that all space and all time is one, that everything *is*, and God is *all*. This found its way into the Old Testament, which was fabricated and modified in the era of the Eleatics and of Heraclitus, then further modified by the Alexandrian Jews in the *Septuagint* and by Philo, who heavily Helenized Judaism and set the stage for the early Christian apologists, who were in many instances Jewish apologists and Jewish nationalists—as was Philo of Alexandria. Ultimately, these beliefs are Hindu in origin and many Jewish Cabalists have succeeded in infusing them into modern Physics. The modern notions of the "big bang", space-time, pantheism, etc. were passed down to Giordano Bruno, Isaac

---

445. Epiphanius, translated by F. Williams, *The Panarion of Epiphanius of Salamis*, Volume 1, 26.16.7, E. J. Brill, New York, (1987), p. 97.

Newton, etc. by Cabalist Jews, who adopted the ideas of the Hindus via the Eleatics, Heraclitus, Plato, Aristotle, Origen of Alexandria, the Schoolmen, etc.

An important aspect of the Abraham myth, which weeds off certain races (the Old Testament is filled with mythologies whereby individuals symbolize entire peoples), is the declaration that God would shield Abraham (*Genesis* 15:1). Jews promoted the myth that God would annihilate anyone who challenged Israel (*Isaiah* 41:11. *Jeremiah* 30:16; 50:7). Jews celebrated the genocide of the Egyptian army in *Exodus* 14:15-15:1. *Deuteronomy* 11:24-28 states,

> "24 Every place whereon the soles of your feet shall tread shall be yours: from the wilderness and Lebanon, from the river, the river Euphrates, even unto the uttermost sea shall your coast be. 25 There shall no man be able to stand before you: *for* the LORD your God shall lay the fear of you and the dread of you upon all the land that ye shall tread upon, as he hath said unto you. 26 Behold, I set before you this day a blessing and a curse; 27 A blessing, if ye obey the commandments of the LORD your God, which I command you this day: 28 And a curse, if ye will not obey the commandments of the LORD your God, but turn aside out of the way which I command you this day, to go after other gods, which ye have not known."

As a threat against the nations, Jews sought to promote the myth of their invincibility and tried desperately to preserve the Gentiles' "fear of the inaccessibility of Israel". Frederick the Great is reputed to have stated, "to oppress the Jews never brought prosperity to any Government".[446] In 1906, Herbert N. Casson tried to intimidate Americans into welcoming the massive influx of Eastern European Jews,

> "It seems as if the American plan of giving the Jews fair play was succeeding. At any rate, all the other plans failed. 'No nations prospers that persecutes the Jews,' said Frederick the Great. Egypt tried persecution, and the Jews went to its funeral. Assyria made the same blunder. So did Babylon, Persia, Greece, Rome, Spain. Say the Jew is not a fighter!"[447]

This prompts the question if America will share the sorry fate of those nations which had a significantly large number of racist Jews in its midst. *Jeremiah* 24:9 states,

> "And I will deliver them to be removed into all the kingdoms of the earth for *their* hurt, to be a reproach and a proverb, a taunt and a curse, in all places whither I shall drive them."

---

446. "Jews, Modern", *Encyclopædia Britannica*, Volume 13, Ninth Edition, Charles Scribner's Sons, (1881), p. 680.
447. H. N. Casson, "The Jew in America", *Munsey's Magazine*, Volume 34, Number 4, (January, 1906), pp. 381-395, at 394.

Malachi 1:14 states,

"[...]I am a great King, saith the LORD of hosts, and my name is dreadful among the heathen."

Cyprian exposited upon the ancient practice of threatening one's enemies with one's gods, and asserted that a single God, whose power was undiluted and universal, posed the greatest threat of all to one's enemies. Consider Cyprian's doctrine *circa* A.D. 247,

"TREATISE VI.
ON THE VANITY OF IDOLS: SHOWING THAT THE IDOLS ARE NOT GODS, AND THAT GOD IS ONE, AND THAT THROUGH CHRIST SALVATION IS GIVEN TO BELIEVERS.

ARGUMNET.—THIS HEADING EMBRACES THE THREE LEADING DIVISIONS OF THIS TREATISE. THE WRITER FIRST OF ALL SHOWS THAT THEY IN WHOSE HONOUR TEMPLES WERE FOUNDED, STATUES MODELLED, VICTIMS SACRIFICED, AND FESTAL DAYS CELEBRATED, WERE KINGS AND MEN AND NOT GODS; AND THEREFORE THAT THEIR WORSHIP COULD BE OF NO AVAIL EITHER TO STRANGERS OR TO ROMANS, AND THAT THE POWER OF THE ROMAN EMPIRE WAS TO ATTRIBUTED TO FATE RATHER THAN TO THEM, INASMUCH AS IT HAD ARISEN BY A CERTAIN GOOD FORTUNE, AND WAS ASHAMED OF ITS OWN ORIGIN.

1. That those are no gods whom the common people worship, is known from this. They were formerly kings, who on account of their royal memory subsequently began to be adored by their people even in death. Thence temples were founded to them; thence images were sculptured to retain the countenances of the deceased by the likeness; and men sacrificed victims, and celebrated festal days, by way of giving them honour. Thence to posterity those rites became sacred which at first had been adopted as a consolation. And now let us see whether this truth is confirmed in individual instances.

2. Melicertes and Leucothea are precipitated into the sea, and subsequently become sea-divinities. The Castors die by turns, that they may live. Æsculapius is struck by lightning, that he may rise into a god. Hercules, that he may put off the man, is burnt up in the fires of Oeta. Apollo fed the flocks of Admetus; Neptune founded walls for Laomedon, and received— unfortunate builder—no wages for his work. The cave of Jupiter is to be seen in Crete, and his sepulchre is shown; and it is manifest that Saturn was driven away by him, and that from him Latium received its name, as being his lurking-place. He was the first that taught to print letters; he was the first that taught to stamp money in Italy, and thence the treasury is called the

treasury of Saturn. And he also was the cultivator of the rustic life, whence he is painted as an old man carrying a sickle. Janus had received him to hospitality when he was driven away, from whose name the Janiculum is so called, and the month of January is appointed. He himself is portrayed with two faces, because, placed in the middle, he seems to look equally towards the commencing and the closing year. The Mauri, indeed, manifestly worship kings, and do not conceal their name by any disguise.

3. From this the religion of the gods is variously changed among individual nations and provinces, inasmuch as no one god is worshipped by all, but by each one the worship of its own ancestors is kept peculiar. Proving that this is so, Alexander the Great writes in the remarkable volume addressed to his mother, that through fear of his power the doctrine of the gods being men, which was kept secret, had been disclosed to him by a priest, that it was the memory of ancestors and kings that was (really) kept up, and that from this the rites of worship and sacrifice have grown up. But if gods were born at any time, why are they not born in these days also?—unless, indeed, Jupiter possibly has grown too old, or the faculty of bearing has failed Juno.

4. But why do you think that the gods can avail on behalf of the Romans, when you see that they can do nothing for their own worshipers in opposition to the Roman arms? For we know that the gods of the Romans are indigenous. Romulus was made a god by the perjury of Proculus, and Picus, and Tiberinus, and Pilumnus, and Consus, whom as a god of treachery Romulus would have to be worshipped, just as if he had been a god of counsels, when his perfidy resulted in the rape of the Sabines. Tatius also both invented and worshipped the goddess Cloacina; Hostilius, Fear and Paleness. By and by, I know not by whom, Fever was dedicated, and Acca and Flora the harlots. These are the Roman gods. But Mars is a Thracian, and Jupiter a Cretan, and Juno either Argive or Samian or Carthaginian, and Diana of Taurus, and the mother of the gods of Ida; and there are Egyptian monsters, not deities, who assuredly, if they had had any power, would have preserved their own and their people's kingdoms. Certainly there are also among the Romans the conquered Penates whom the fugitive Æneas introduced thither. There is also Venus the bald,—far more dishonoured by the fact of her baldness in Rome than by her having been wounded in Homer.

5. Kingdoms do not rise to supremacy through merit, but are varied by chance. Empire was formerly held by both Assyrians and Medes and Persians; and we know, too, that both Greeks and Egyptians have had dominion. Thus, in the varying vicissitudes of power, the period of empire has also come to the Romans as to the others. But if you recur to its origin, you must needs blush. A people is collected together from profligates and criminals, and by founding an asylum, impunity for crimes makes the number great; and that their king himself may have a superiority in crime, Romulus becomes a fratricide; and in order to promote marriage, he makes a beginning of that affair of concord by discords. They steal, they do

violence, they deceive in order to increase the population of the state; their marriage consists of the broken covenants of hospitality and cruel wars with their fathers-in-law. The consulship, moreover, is the highest degree in Roman honours, yet we see that the consulship began even as did the kingdom. Brutus puts his sons to death, that the commendation of his dignity may increase by the approval of his wickedness. The Roman kingdom, therefore, did not grow from the sanctities of religion, nor from auspices and auguries, but it keeps its appointed time within a definite limit. Moreover, Regulus observed the auspices, yet was taken prisoner; and Mancinus observed their religious obligation, yet was sent under the yoke. Paulus had chickens that fed, and yet he was slain at Cannæ. Caius Cæsar despised the auguries and auspices that were opposed to his sending ships before the winter to Africa; yet so much the more easily he both sailed and conquered.

6. Of all these, however, the principle is the same, which misleads and deceives, and with tricks which darken the truth, leads away a credulous and foolish rabble. They are impure and wandering spirits, who, after having been steeped in earthly vices, have departed from their celestial vigour by the contagion of earth, and do not cease, when ruined themselves, to seek the ruin of others; and when degraded themselves, to infuse into others the error of their own degradation. These demons the poets also acknowledge, and Socrates declared that he was instructed and ruled at the will of a demon; and thence the Magi have a power either for mischief or for mockery, of whom, however, the chief Hostanes both says that the form of the true God cannot be seen, and declares that true angels stand round about His throne. Wherein Plato also on the same principle concurs, and, maintaining one God, calls the rest angels or demons. Moreover, Hermes Trismegistus speaks of one God, and confesses that He is incomprehensible, and beyond our estimation.

7. These spirits, therefore, are lurking under the statues and consecrated images: these inspire the breasts of their prophets with their afflatus, animate the fibres of the entrails, direct the flights of birds, rule the lots, give efficiency to oracles, are always mixing up falsehood with truth, for they are both deceived and they deceive; they disturb their life, they disquiet their slumbers; their spirits creeping also into their bodies, secretly terrify their minds, distort their limbs, break their health, excite diseases to force them to worship of themselves, so that when glutted with the steam of the altars and the piles of cattle, they may unloose what they had bound, and so appear to have effected a cure. The only remedy from them is when their own mischief ceases; nor have they any other desire than to call men away from God, and to turn them from the understanding of the true religion, to superstition with respect to themselves; and since they themselves are under punishment, (they wish) to seek for themselves companions in punishment whom they may by their misguidance make sharers in their crime. These, however, when adjured by us through the true God, at once yield and confess, and are constrained to go out from the bodies possessed. You may see them at our voice, and by the operation of the hidden majesty, smitten

with stripes, burnt with fire, stretched out with the increase of a growing punishment, howling, groaning, entreating, confessing whence they came and when depart, even in the hearing of those very persons who worship them, and either springing forth at once or vanishing gradually, even as the faith of the sufferer comes in aid, or the grace of the healer effects. Hence they urge the common people to detest our name, so that men begin to hate us before they know us, lest they should either imitate us if known, or not be able to condemn us.

8. Therefore the one Lord of all is God. For that sublimity cannot possibly have any compeer, since it alone possesses all power. Moreover, let us borrow an illustration for the divine government from the earth. When ever did an alliance in royalty either begin with good faith or end without bloodshed? Thus the brotherhood of the Thebans was broken, and discord endured even in death in their disunited ashes. And one kingdom could not contain the Roman twins, although the shelter of one womb had held them. Pompey and Cæsar were kinsmen, and yet they did not maintain the bond of their relationship in their envious power. Neither should you marvel at this in respect of man, since herein all nature consents. The bees have one king, and in the flocks there is one leader, and in the herds one ruler. Much rather is the Ruler of the world one; who commands all things, whatsoever they are, with His word, disposes them by His wisdom, and accomplishes them by His power.

9. He cannot be seen—He is too bright for vision; nor comprehended—He is too pure for our discernment; nor estimated—He is too great for our perception; and therefore we are only worthily estimating Him when we say that He is inconceivable. But what temple can God have, whose temple is the whole world? And while man dwells far and wide, shall I shut up the power of such great majesty within one small building? He must be dedicated in our mind; in our breast He must be consecrated. Neither must you ask the name of God. God is His name. Among those there is need of names where a multitude is to be distinguished by the appropriate characteristics of appellations. To God who alone is, belongs the whole name of God; therefore He is one, and He in His entirety is everywhere diffused. For even the common people in many things naturally confess God, when their mind and soul are admonished of their author and origin. We frequently hear it said, 'O God,' and 'God sees,' and 'I commend to God,' and 'God give you,' and 'as God will,' and 'if God should grant;' and this is the very height of sinfulness, to refuse to acknowledge Him whom you cannot but know."[448]

The *Midrash Bereshit Rabbah* 38:13 tells that Abraham's father worshiped and sold idols. One day, Abraham smashed all of the idols but the largest idol

---

448. Cyprian,"The Treatises of Cyprian", Treatise VI, *The Anti-Nicene Fathers: Translations of the Writings of the Fathers down to A.D. 325*, Volume 5, Christian Literature Publishing Company, New York, (1886), pp. 465-467.

and then placed a stick in its hand. He told his father that the largest god had destroyed the others. Note the lesson that the Jewish monotheistic God is dominant and will destroy the gods of other peoples. The myth of Abraham differs from the myth of Cyprian, in that Christianity is taught as a universal religion, and the story of Abraham is a racist myth, which elects the Judeans as a unique and chosen race descended through Jacob to Abraham, a race who have an exclusive contract with God which makes them divine.

Jews have long sought to provoke superstitious fear of their God. The Judeans fabricated a history of persecution in Egypt, which never occurred, in order to defame the Egyptians and to blame the Egyptians for Jewish ethnocentricism, as well as to justify their claim that their God was stronger than the Pharaoh. The "Lost Tribes" of Israelites, the "ten northern tribes" allegedly taken captive by Assyrian King Shalmaneser V, and corralled by the river Sambatyon in Syria and Iraq (II *Kings* 17), never existed beyond the imagination of the "southern tribes" of Judeans and supposedly "Benjamin", who were allegedly taken captive in exile in Assyria (II *Kings* 18:13) and in Babylon by Nebuchadnezzar (II *Kings* 24:3-16; 25), and who wanted to steal the land of the indigenous peoples from the Nile to the Euphrates (*Ezra* 1:5). The myth of the Egyptian captivity, and of the ten northern tribes, was fabricated by the Judeans in an attempt to justify their desires on lands and religious beliefs which were not originally theirs. They created the "prophecy" of these "events" in order to admonish their tribe to obey their racist and tribalistic leaders out of fear (*Leviticus* 26. *Deuteronomy* 4:24-27; 28:15-68; 30:1-3. II *Chronicles* 7:19-22. *Jeremiah* 29:1-7). Many argue that the prophecies of the Old Testament must have been written after the events they "foretold" and were merely a means for Jewish leaders to subjugate their followers. Præterist Christians believe that the Apocalyptic "prophecies" have all been fulfilled by the destruction of the Temple in 70 A.D. and the Diaspora of 135 A.D., and that the story of Gog and Magog in *Ezekiel* 38 and 39 is post-Millennial (*Revelation* 20:7-8). They see Christian Zionists as dangerous dupes, who are serving the "Beast".

The process continues in the modern world. David Ben-Gurion stated to the General Staff,

> "I proposed that, as soon as we received the equipment on the ship, we should prepare to go over to the offensive with the aim of smashing Lebanon, Transjordan and Syria. [***] The weak point in the Arab coalition is Lebanon [for] the Moslem regime is artificial and easy to undermine. A Christian state should be established, with its southern border on the Litani River. We will make an alliance with it. When we smash the [Arab] Legion's strength and bomb Amman, we will eliminate Transjordan, too, and then Syria will fall. If Egypt still dares to fight on, we shall bomb Port Said, Alexandria, and Cairo. [***] And in this fashion, we will end the war and settle our forefathers' accounts with Egypt, Assyria, and Aram."[449]

---

[449]. D. Ben-Gurion, quoted in: M. Bar-Zohar, *Ben-Gurion: A Biography*, Delacorte

Judaism, Christianity and Islam are among the most dogmatic and intolerant of religions, in part due to the superstitious fear they would impose on humanity in order to preserve and promote their own power. They threaten their critics with damnation and ruin, as if it were a self-evident truth that ruin will befall non-believers and enemies of the faith. British Zionist Winston Churchill promoted the myth of Jewish invincibility and the necessarily sorry fate of any who would oppose the Jews.[450] Zionist Reverend Cyrus Ingerson Scofield annotated the *Scofield Reference Bible*, published by Oxford University Press, with threats against any who would oppose the Jews. In reference to *Genesis* 12:1-3, which states:

> "Now the LORD had said unto Abram, Get thee out of thy country, and from thy kindred, and from thy father's house, unto a land that I will shew thee: 2 And I will make of thee a great nation, and I will bless thee, and make thy name great; and thou shalt be a blessing: 3 And I will bless them that bless thee, and curse him that curseth thee: and in thee shall all families of the earth be blessed."

Scofield wrote in the 1909 edition of the *Scofield Reference Bible*, in oddly Zionistic terms,

> "(6) 'And curse him that curseth thee.' Wonderfully fulfilled in the history of the dispersion. It has invariably fared ill with the people who have persecuted the Jew—well with those who have protected him. The future will still more remarkably prove this principle."[451]

It is noteworthy that Scofield, though annotating a Christian Bible, did not repeat the Christian dogma, which transferred this blessing and curse to the Christians viz. *Matthew* 12:30; 21:43-45. *Romans* 4; 9; 11:7-8. *Galatians* 3:16, 28-29; 4 and *Hebrews* 8:6-10.

Scofield's intentional corruption of Christian doctrines to favor Zionist interests was not a new phenomenon. *The North American Review* published the following statement in 1845,

> "But religious belief—the Jewish, even, and much more the Christian—

Press, New York, (1978), p. 166.
450. W. Churchill, "Zionism Versus Bolshevism. A Struggle for the Soul of the Jewish People.", *Illustrated Sunday Herald*, (8 February 1920), p. 5.
451. C. I. Scofield, Editor, *The Scofield reference Bible. The Holy Bible, containing the Old and New Testaments. Authorized version, with a new system of connected topical references to all the greater themes of Scripture, with annotations, revised marginal renderings, summaries, definitions, and index; to which are added helps at hard places, explanations of seeming discrepancies, and a new system of paragraphs*, Oxford University Press, American Branch, New York, (1909), p. 25.

heightens immeasurably the importance and the attractiveness of this wonderful theme. To the confiding student of the Bible, the Jews assume high dignity, and challenge earnest attention, as God's chosen, covenant people; as the descendants of holy patriarchs, to whom Jehovah spake 'face to face, as a man speaketh unto his friend'; as a nation long visibly led and governed, upheld, protected, and punished, by an almighty hand; as a people whose ancient history, recorded by inspiration, expressly and clearly shows—what all uninspired annals leave to be faintly and uncertainly traced out by the dim light of human reason—the connection between every outward event and an unseen Providence; as the special depositaries of divine communications intended for all times and every people; as that race, 'of whom, as concerning the flesh, Christ came,' and who, although they rejected and crucified the Saviour of the world, are themselves rejected and outcast, 'scattered among all people, from the one end of the earth even unto the other,' 'to be a reproach and a proverb, a taunt and a curse, in all places' of their sojourn ; as still beloved of God in his covenant faithfulness, and 'for the fathers' sake'; as still inheriting the prophetic benediction, 'Cursed be every one that curseth thee, and blessed he he that blesseth thee'; as yet to be 'grafted again into their own olive-tree,' the church of God; and, as many believe, to be restored to that goodly land which was confirmed to them by oath before they were a nation; which was taken from its original possessors to be given to them, when they were homeless pilgrims; which is still theirs, twice exiled from it as they have been,—now for nearly eighteen hundred years,—and wonderfully kept from permanent occupation by any Gentile people;—in a word, as the standing miracle of modern times, changing in themselves nature's most firmly established laws, without interfering with the harmony that everywhere else prevails in convincing contrast. Such are the Jews in the eye of Christian faith."[452]

Judeans have continuously and heavily promoted the myth that they are the divinely inspired chosen people, who have a right to enslave the rest of humanity. Ancient Jews taught their children to be absolutely intolerant of any dissent against Jews, or Jewish mythology, and to quash any dissent by exterminating those who have opposed the Jews, or Jewish mythologies. They feared that any challenge, or competition, to Judaism would reveal that they had fabricated and plagiarized their myths, which were little but a bluff meant to intimidate others far stronger than themselves. Even an unsuccessful challenge to any Jew, or to Jewish myths, would show to the world the intrinsic weakness of the position of the Jewish people and the inanity and meanspiritedness of the mythologies they had appropriated and corrupted. It is important to note that the Jews wanted other peoples to fear and to obey them, and to never entertain the slightest doubt of Jewish infallibility, or to challenge them. To this day the strongest taboos in

---

452. "The Modern Jews", *The North American Review*, Volume 60, Number 127, (April, 1845), pp. 329-368, at 331.

society are the prohibition against questioning the existence of the Jewish God who chose the Jews to rule, and the prohibition against criticizing the modern State of Israel.

God commands the Jews to exterminate Amalek, because Amalek was the first to attack Israel and expose its terrible vulnerability. Jews so viciously attack anyone who even hints at challenging their supremacy, because they are in a very vulnerable position and must cut off all challenges before they grow. Jews must maintain the illusion that they are protected by God and invulnerable and cannot be challenged. Jews must maintain the lie that they are a divine blessing and a divine curse. That is why they are so hateful of Amalek and have carried the lesson down through history that they must not only tribalistically attack all who question any Jew, but that they must nip such challenges in the bud, or better yet prevent them from ever occurring, lest a significant number of Gentiles learn of their ill intentions and their vulnerability and put an end to the threat they pose. Rather than modify their behavior to socially acceptable norms, they band together to quash all challengers and feel no compunctions about committing immoral acts in order to defend the tribe from the truth. They are out to exterminate any and all who do not obey them and they are out to exterminate the truth of what they are doing.

The modern State of Israel has practiced censorship of the press and kept important historical information under lock and key. Israeli soldiers have gone so far as to murder journalists and activists who record the Israelis' atrocities against the Palestinians. In the illegally Occupied Territories, Israelis humiliate and degrade their fellow human beings, while declaring to the world that Israel, one of the most undemocratic of the nations formed in the Twentieth Century, is the only democracy in the Middle East—a false declaration intended to degrade their Moslem enemies. The Jews have always had strong prohibitions against blasphemy and Judeans and Christians have held back the progress of science and politics for two thousand years in order to preserve their mythologies by preventing any open challenges to them. Pious Jews cling to the myth of a Jewish cult-hero, Moses, who gave to them God's Law, which cannot be questioned. Christians cling to the myth of a Jewish cult-hero, Jesus, who came to fulfill the Law, which cannot be questioned. "Einstein's" irrational and physically contradicted theories are promoted as if irrefutable, and challenges to the theories are regularly excluded from publication as if a matter of principle. Dissent against the theories is punished by ridicule and career infringement, as well as by charges of anti-Semitism where there are no grounds for such charges.

The ancient Jews fabricated the mythology that they have genetic enemies, whom they must subjugate, then exterminate. Jacob's brother, Esau, is said to be the father of a people who are inherently antagonistic to Jews and who must be exterminated. Louis Ginzberg states in his *The Legend of the Jews* (and bear in mind that Amalek represents Esau, his grandfather, and ultimately Haman, Rome, and Christianity; and, though Islam is traditionally associated with Abraham and Hagar's son Ishmael, when it comes to the genocide of the

Palestinians, Arabs, Turks and Persians, they are called Amalek[453]; as are Gentiles in general—enemies of the Jews in general, as is revealed in various other passages in Ginzberg's many volumes),

> "Although Amalek had now received the merited punishment from the hands of Joshua, still his enterprise against Israel had not been entirely unavailing. The miraculous exodus of Israel out of Egypt, and especially the cleaving of the sea, had created such alarm among the heathens, that none among them had dared to approach Israel. But this fear vanished as soon as Amalek attempted to compete in battle with Israel. Although he was terribly beaten, still the fear of the inaccessibility of Israel was gone. It was with Amalek as with that foolhardy wight who plunged into a scalding-hot tub. He scalded himself terribly, yet the tub became a little cooled through his plunge into it. Hence God was not content with the punishment Amalek received in the time of Moses, but swore by His throne and by His right hand that He would never forget Amalek's misdeeds, that in this world as well as in the time of the Messiah He would visit punishment upon him, and would completely exterminate him in the future world. So long as the seed of Amalek exists, the face of God is, as it were, covered, and will only then come to view, when the seed of Amalek shall have been entirely exterminated.
>
> God had at first left the war against Amalek in the hands of His people, therefore He bade Joshua, the future leader of the people, never to forget the war against Amalek; and if Moses had listened intently, he would have perceived from this command of God that Joshua was destined to lead the people into the promised land. But later, when Amalek took part in the destruction of Jerusalem, God Himself took up the war against Amalek, saying, 'By My throne I vow not to leave a single descendant of Amalek under the heavens, yea, no one shall even be able to say that this sheep or that wether belonged to an Amalekite.'
>
> God bade Moses impress upon the Jews to repulse no heathen should he desire conversion, but never to accept an Amalekite as a proselyte. It was in consideration of this word of God that David slew the Amalekite, who announced to him the death of Saul and Jonathan; for he saw in him only a heathen, although he appeared in the guise of a Jew.
>
> Part of the blame for the destruction of Amalek falls upon his father, Eliphaz. He used to say to Amalek: 'My son, dost thou indeed know who will possess this world and the future world?' Amalek paid no attention to this allusion to the future fortune of Israel, and his father urged it no more strongly upon him, although it would have been his duty to instruct his son clearly and fully. He should have said to him: ' My son, Israel will possess this world as well as the future world; dig wells then for their use and build roads for them, so that thou mayest be judged worthy to share in the future

---

[453]. Y. Harkabi, *Israel's Fateful Hour*, Harper & Row, New York, (1988), pp. 149-150.

world.' But as Amalek had not been sufficiently instructed by his father, in his wantonness he undertook to destroy the whole world. God, who tries the reins and the heart, said to him: 'O thou fool, I created thee after all the seventy nations, but for thy sins thou shalt be the first to descend into hell.'

To glorify the victory over Amalek, Moses built an altar, which God called 'My Miracle,' for the miracle God wrought against Amalek in the war of Israel was, as it were, a miracle for God. For so long as the Israelites dwell in sorrow, God feels with them, and a joy for Israel is a joy for God, hence, too, the miraculous victory over Israel's foe was a victory for God."[454]

In the jargon of Jewish racists, the Gentiles are called "Esau" or "Edom", and the Jews, "Jacob". The Old Testament book of *Obadiah* instructs the Jews to destroy the wise among the Gentiles, and then to exterminate the Gentiles ("cut off"="murder")—much as the Communists have done. Noted Hebrew and Rabbinical scholar Johannes Buxtorf wrote in 1603, quoting from Machir of Toledo's *Avkat Rokhel*, Constantinople/Istanbul, (1516):

"Then shall *Armillus* with his whole army die, and the Atheistical Edomites (the Christians they mean) who laid waste the house of our God, and led us captive into a strange land, shall miserably perish; then shall the Jews be revenged upon them, as it is written, {Obad. 18} *The house of Jacob shall be a fire, and the house of Joseph a flame, and the house of Esau* (that is, we Christians, as the Jews interpret, whom they Christen Edomites) *shall be for stubble.* This stubble the Jews shall set in fire, that nothing be left to us Edomites which shall not be burnt and turned into ashes."[455]

The book of *Obadiah:*

"1 The vision of Obadiah. Thus saith the Lord GOD concerning Edom; We have heard a rumour from the LORD, and an ambassador is sent among the heathen, Arise ye, and let us rise up against her in battle. 2 Behold, I have made thee small among the heathen: thou art greatly despised. 3 ¶ The pride of thine heart hath deceived thee, thou that dwellest in the clefts of the rock, whose habitation *is* high; that saith in his heart, Who shall bring me down to the ground? 4 Though thou exalt *thyself* as the eagle, and though thou set thy nest among the stars, thence will I bring thee down, saith the LORD. 5

---

[454]. L. Ginzberg, *The Legend of the Jews*, Volume 3, The Jewish Publication Society of America, Philadelphia, (1911/1954), pp. 61-63.

[455]. J. Buxtorf, *Synagoga Judaica: Das ist Jüden Schul ; Darinnen der gantz Jüdische Glaub und Glaubensubung... grundlich erkläret*, Basel, (1603); as translated in the 1657 English edition, *The Jewish Synagogue: Or An Historical Narration of the State of the Jewes, At this Day Dispersed over the Face of the Whole Earth*, Printed by T. Roycroft for H. R. and Thomas Young at the Three Pidgeons in Pauls Church-Yard, London, (1657), p. 323.

If thieves came to thee, if robbers by night, (how art thou cut off!) would they not have stolen till they had enough? if the grapegatherers came to thee, would they not leave *some* grapes? 6 How are *the things* of Esau searched out! *how* are his hidden things sought up! 7 All the men of thy confederacy have brought thee *even* to the border: the men that were at peace with thee have deceived thee, *and* prevailed against thee; *they that eat* thy bread have laid a wound under thee: *there is* none understanding in him. 8 Shall I not in that day, saith the LORD, even destroy the wise *men* out of Edom, and understanding out of the mount of Esau? 9 And thy mighty *men*, O Teman, shall be dismayed, to the end that every one of the mount of Esau may be cut off by slaughter. 10 For *thy* violence against thy brother Jacob shame shall cover thee, and thou shalt be cut off for ever. 11 In the day that thou stoodest on the other side, in the day that the strangers carried away captive his forces, and foreigners entered into his gates, and cast lots upon Jerusalem, even thou *wast* as one of them. 12 But thou shouldest not have looked on the day of thy brother in the day that he became a stranger; neither shouldest thou have rejoiced over the children of Judah in the day of their destruction; neither shouldest thou have spoken proudly in the day of distress. 13 Thou shouldest not have entered into the gate of my people in the day of their calamity; yea, thou shouldest not have looked on their affliction in the day of their calamity, nor have laid *hands* on their substance in the day of their calamity; 14 Neither shouldest thou have stood in the crossway, to cut off those of his that did escape; neither shouldest thou have delivered up those of his that did remain in the day of distress. 15 For the day of the LORD *is* near upon all the heathen: as thou hast done, it shall be done unto thee: thy reward shall return upon thine own head. 16 For as ye have drunk upon my holy mountain, so shall all the heathen drink continually, yea, they shall drink, and they shall swallow down, and they shall be as though they had not been. 17 ¶ But upon mount Zion shall be deliverance, and there shall be holiness; and the house of Jacob shall possess their possessions. 18 And the house of Jacob shall be a fire, and the house of Joseph a flame, and the house of Esau for stubble, and they shall kindle in them, and devour them; and there shall not be *any* remaining of the house of Esau; for the LORD hath spoken *it*. 19 And *they* of the south shall possess the mount of Esau; and *they of* the plain the Philistines: and they shall possess the fields of Ephraim, and the fields of Samaria: and Benjamin *shall possess* Gilead. 20 And the captivity of this host of the children of Israel *shall possess* that of the Canaanites, *even* unto Zarephath; and the captivity of Jerusalem, which *is* in Sepharad, shall possess the cities of the south. 21 And saviours shall come up on mount Zion to judge the mount of Esau; and the kingdom shall be the LORD's."

Sanhedrin 59*a* states that Gentiles who study the Torah must be killed. *Soferim*, Chapter 15, Rule 10, states, quoting the much celebrated genocidal racist Jew Simon ben Yohai:

"The best among the Gentiles deserves to be killed."[456]

Michael Berenbaum wrote in his book, *After Tragedy and Triumph*,

"Menachim Begin built upon this realization and constructed a usable past upon the twin pillars of antisemitism and the need for power. *Goyim* (literally, 'the nations') hate Jews, Begin maintained. In traditional language, Esau hates Jacob. According to Begin's worldview, Jews are a people that dwells alone. Power is essential. Powerlessness invites victimization. Jews must determine their own morality. The world's pronouncements toward the Jews mask—sometimes more successfully and sometimes less so—their genocidal intent. The desire to make the world *Judenrein* continues, and only fools would allow themselves to be deceived."[457]

Isaac and his wife Rebekah had twin sons: Esau,[458] the firstborn, and Jacob, the younger son. Even before the twins were born, they fought each other in the womb (*Genesis* 25:22). God told Rebekah that her sons would father two peoples and that Esau, the elder, would serve Jacob, the younger (*Genesis* 25:23). Isaac favored Esau, but Rebekah favored Jacob. Esau was a hunter, and Jacob, a farmer. Isaac and Rebekah did not sacrifice Esau and pass him through the fire to the gods of heaven, which is perhaps why Rebekah did not favor Esau, the firstborn who opened her womb—the firstborn son who had rights to the covenant.

The differences of character between Esau and Jacob became key features in Jewish mythology. Esau, the hunter, came to represent strong warrior peoples—Esau was a belligerent people like the Hyksos.[459] Jacob, whom God renamed "Israel" (*Genesis* 25:26; 32:27-28; 35:10), came to represent the agrarian, weak and scholarly peoples, who were allegedly entitled by God to be immoral—even genocidal—especially genocidal—and to use Esau as their sword and their slave (*Genesis* 25:23; 27:38-41)—Jacob was a people like the ancient Egyptians.

When some Jews attempted to stigmatize Germans, Christians and Gentiles as genetically predisposed to be warlike and anti-Semitic, as they often have, they were recalling Esau and Jacob, and stating that they (Jacob/Israel) have the God-given right to exploit the Germans, Christians, Moslems and Gentiles in

---

456. "Gentile", *The Jewish Encyclopedia*, Volume 5, Funk and Wagnalls Company, New York, (1903), pp. 615-626, at 617. ***See also:*** A. Cohen, "Soferim 41a", *The Minor Tractates of the Talmud Massektoth Ketannoth in Two Volumes*, Volume 1, The Socino Press, London, (1965), pp. 287-288, *especially* note 50.
457. M. Berenbaum, *After Tragedy and Triumph: Essays in Modern Jewish Throught and the American Experience*, Cambridge University Press, (1990), p.7.
458. "Esau" is also referred to as "Edom" *Genesis* 36:8.
459. I. Velikovsky, *Ages in Chaos*, Volume 1, Doubleday and Company, Inc., Garden City, New York, (1952), p. 95.

general (Esau) as slaves and warriors, then to exterminate them in accordance with God's wishes; because the Gentiles are by nature ungodly and anti-Semitic, according to Jewish mythologies. In accord with the Old Testament, Zionists repeatedly asserted that the Gentile nations were obliged to fight for Israel and to finance it—hence the common paradox of the anti-nationalist pacifist Zionist warmonger.

It is noteworthy that the British and Americans fought to secure Palestine from the Turks—those Turks who had for centuries treated the Jews better than anyone else—and to end the Nazi régime, which had instilled tremendous fear in Jews—all of which cleared the way for the formation of the State of Israel. It is also noteworthy that today America is fighting wars for Israel, and that the comparatively insignificant and wealthy nation of Israel receives more foreign aid from the United States of America than any other nation on Earth, though it has carried out worse espionage campaigns against the United States[460] than even the outspoken enemies of the United States, these wasted monies donated to sponsor oppression while millions of the unchosen needlessly perish from starvation and disease around the world. Israel plays a prominent rôle in international politics and the media, in spite of the fact that the world faces far more important issues than the fate of a comparatively small, and forever troublesome, minority among humanity. Jewish selfishness apparently knows no bounds. It is deeply entrenched in Jewish religious mythology.

One day, after returning home from the field so hungry that he was starving to death, Esau asked Jacob to spare his life and give him some food. Jacob took advantage of the situation to coerce Esau into surrendering his birthright to Jacob for some lentil porridge (*Genesis* 25:29-34). Through deceit, Rebekah and Jacob, whom God renamed Israel (*Genesis* 32:27-28), stole Esau's blessing from Isaac, who had inherited it from Abraham, and gave it to treacherous Jacob. Esau pledged to kill his younger twin brother Jacob, thereby expressing the genocidal imagery between Jews and Gentiles, and Jewish self-obsession and selfishness found throughout Jewish history:

> "1 And it came to pass, that when Isaac was old, and his eyes were dim, so that he could not see, he called Esau his eldest son, and said unto him, My son: and he said unto him, Behold, *here am I*. 2 And he said, Behold now, I am old, I know not the day of my death: 3 Now therefore take, I pray thee, thy weapons, thy quiver and thy bow, and go out to the field, and take me *some* venison; 4 And make me savoury meat, such as I love, and bring *it* to me, that I may eat; that my soul may bless thee before I die. 5 And Rebekah heard when Isaac spake to Esau his son. And Esau went to the field to hunt *for* venison, *and* to bring *it*. 6 And Rebekah spake unto Jacob her son, saying, Behold, I heard thy father speak unto Esau thy brother, saying, 7 Bring me venison, and make me savoury meat, that I may eat, and bless thee

---

460. D. B. Ball and G. W. Ball, *The Passionate Attachment: America's Involvement with Israel, 1947 to the Present*, W. W. Norton, New York, (1992), pp. 204-206.

before the LORD before my death. 8 Now therefore, my son, obey my voice according to that which I command thee. 9 Go now to the flock, and fetch me from thence two good kids of the goats; and I will make them savoury meat for thy father, such as he loveth: 10 And thou shalt bring *it* to thy father, that he may eat, and that he may bless thee before his death. 11 And Jacob said to Rebekah his mother, Behold, Esau my brother *is* a hairy man, and I *am* a smooth man: 12 My father peradventure will feel me, and I shall seem to him as a deceiver; and I shall bring a curse upon me, and not a blessing. 13 And his mother said unto him, Upon me *be* thy curse, my son: only obey my voice, and go fetch me *them*. 14 And he went, and fetched, and brought *them* to his mother: and his mother made savoury meat, such as his father loved. 15 And Rebekah took goodly raiment of her eldest son Esau, which *were* with her in the house, and put them upon Jacob her younger son: 16 And she put the skins of the kids of the goats upon his hands, and upon the smooth of his neck: 17 And she gave the savoury meat and the bread, which she had prepared, into the hand of her son Jacob. 18 And he came unto his father, and said, My father: and he said, Here *am* I; who *art* thou, my son? 19 And Jacob said unto his father, I *am* Esau thy firstborn; I have done according as thou badest me: arise, I pray thee, sit and eat of my venison, that thy soul may bless me. 20 And Isaac said unto his son, How *is it* that thou hast found *it* so quickly, my son? And he said, Because the LORD thy God brought *it* to me. 21 And Isaac said unto Jacob, Come near, I pray thee, that I may feel thee, my son, whether thou *be* my very son Esau or not. 22 And Jacob went near unto Isaac his father; and he felt him, and said, The voice is Jacob's voice, but the hands *are* the hands of Esau. 23 And he discerned him not, because his hands were hairy, as his brother Esau's hands: so he blessed him. 24 And he said, *Art* thou my very son Esau? And he said, I *am*. 25 And he said, Bring *it* near to me, and I will eat of my son's venison, that my soul may bless thee. And he brought *it* near to him, and he did eat: and he brought him wine, and he drank. 26 And his father Isaac said unto him, Come near now, and kiss me, my son. 27 And he came near, and kissed him: and he smelled the smell of his raiment, and blessed him, and said, See, the smell of my son *is* as the smell of a field which the LORD hath blessed: 28 Therefore God give thee of the dew of heaven, and the fatness of the earth, and plenty of corn and wine: 29 Let people serve thee, and nations bow down to thee: be lord over thy brethren, and let thy mother's sons bow down to thee: cursed be every one that curseth thee, and blessed *be* he that blesseth thee. 30 And 'it came to pass, as soon as Isaac had made an end of blessing Jacob, and Jacob was yet scarce gone out from the presence of Isaac his father, that Esau his brother came in from his hunting. 31 And he also had made savoury meat, and brought it unto his father, and said unto his father, Let my father arise, and eat of his son's venison, that thy soul may bless me. 32 And Isaac his father said unto him, Who *art* thou? And he said, I *am* thy son, thy firstborn Esau. 33 And Isaac trembled very exceedingly, and said, Who? where *is* he that hath taken venison, and brought *it* me, and I have eaten of all before thou camest, and have blessed him? yea, *and* he shall be blessed. 34 And when Esau heard the words of his father, he cried with a great and exceeding bitter cry, and said unto his father, Bless me, *even* me also, O my father. 35 And he said, Thy brother came with subtilty, and hath taken away thy blessing. 36 And he said, Is not he rightly named Jacob? for he hath supplanted me these two times: he took away my birthright; and, behold, now he hath taken away my blessing. And he said, Hast thou not reserved a blessing for me? 37 And Isaac answered and said unto Esau, Behold, I have made him thy lord, and all his brethren have I given to him for servants; and with corn and wine have I sustained him: and what shall I do now unto thee, my son? 38 And Esau said unto his father, Hast thou but one blessing, my father? bless me, *even* me also, O my father. And Esau lifted up his voice, and wept. 39 And Isaac his father answered and said unto him, Behold, thy dwelling shall be the fatness of the earth, and of the dew of heaven from above; 40 And by thy sword shalt thou live, and shalt serve thy brother; and it shall come to pass when thou shalt have the dominion, that thou shalt break his yoke from off thy neck. 41 And Esau hated Jacob because of the blessing wherewith his father blessed him: and Esau said in his heart, The days of mourning for my father are at hand; then will I slay my brother Jacob. 42 And these words of Esau her elder son were told to Rebekah: and she sent and called Jacob her younger son, and said unto him, Behold, thy brother Esau, as touching thee, doth comfort himself, *purposing* to kill thee. 43 Now therefore, my son, obey my voice; and arise, flee thou to Laban my brother to Haran; 44 And tarry with him a few days, until thy brother's fury turn away; 45 Until thy brother's anger turn away from thee, and he forget *that* which thou hast done to him: then I will send, and fetch thee from thence: why should I be deprived also of you both in one day? 46 And Rebekah said to Isaac, I am weary of my life because of the daughters of Heth: if Jacob take a wife of the daughters of Heth, such as these *which are* of the daughters of the land, what good shall my life do me?"—*Genesis* 27:1-46

This story conveys many of the tenets of Zionism—that other nations shall serve Israel, and especially that they shall fight its wars and secure its borders—that deceit is encouraged in the pursuit of Israel—and that Edom will be the mortal enemy of Israel. In the minds of many Jews, Edom became associated with Amalek, Haman, Rome and with European Gentiles and Christians in general. Esau's grandson Amalek (*Genesis* 36:9-12) was first to wage war on Israel, and therefore the first to expose the vulnerability of the Jews. God obliged the descendants of Jacob—Israel, to utterly destroy the seed of Amalek (*Sanhedrin* 20*b*. P188L *Dvarim* 25:19)—obliged Israel to exterminate Gentiles, Christians, Moslems, etc. *Deuteronomy* 25:17-19 states,

"17 ¶ Remember what Amalek did unto thee by the way, when ye were come forth out of Egypt; 18 How he met thee by the way, and smote the hindmost of thee, *even* all that were feeble behind thee, when thou *wast* faint and weary; and he feared not God. 19 Therefore it shall be, when the LORD thy God hath given thee rest from all thine enemies round about, in the land which the LORD thy God giveth thee *for* an inheritance to possess it, *that* thou shalt blot out the remembrance of Amalek from under heaven; thou shalt not forget *it*."

"And the LORD said unto Moses, Write this for a memorial in a book, and rehearse it in the ears of Joshua: for I will utterly put out the remembrance of Amalek from under heaven. And Moses built an altar, and called the name of it Jehovah-nissi: For he said, Because the LORD hath sworn *that* the LORD *will have* war with Amalek from generation to generation."—*Exodus* 17:14-16

"Therefore it shall be, when the LORD thy God hath given thee rest from all thine enemies round about, in the land which the LORD thy God giveth thee *for* an inheritance to possess it, *that* thou shalt blot out the remembrance of Amalek from under heaven; thou shalt not forget *it*."—*Deuteronomy* 25:19 [Should the Zionists continue in their attempts to carry out their ancient plans we can expect that when Israel gains hegemony over the Middle East, it will seek to exterminate the peoples of European descent. Zionists are clearly attempting to destroy the militaries of those Moslem nations which would react with rage and which would likely attack Israel, when the Cabalistic Jews and their Christian Dispensationalist slaves destroy the Dome of the Rock and the Al Aqsa Mosque and build in their place a Jewish Temple. Should Israel succeed in destroying Iran and Syria, they will likely destroy the Dome of the Rock and the Al Aqsa Mosque, and the Moslem world will be unable to stop them. They will then unleash the priests of Aaron, and reinstitute ritual sacrifices. Greater Israel will emerge and occupy the territory from the Nile to the Euphrates. Zionists will generate anti-Semitism around the world in order to force "racial" Jews to emigrate to Israel, who will then populate the greater Israel of the Covenant. Then the Jewish King, perhaps a descendent of the Rothschilds, will emerge and many Jews will likely take up Judaism—the "Messiah" will be a dynasty passing from father to son, or a supposed incarnation from one man to the next in the Shabbataian style, much like the Dalai Lama, *see: 2 Samuel 7*. Perhaps the proposed Jewish King is alive today, hidden from view. The Lubavitchers, under the leadership of the now deceased Rebbe Schneerson, have declared that the Messiah is alive today and will soon be anointed. They are an immensely powerful Cabalistic Jewish sect, which has infiltrated governments around the world. We can expect that Soviet-style oppression will grip the West—one already sees that news organizations restrict the international news Americans see, much as happened in the Soviet Union. China will likely become the new America for the Zionists, and their "Iron Scepter", which Israel will utilize to smash the West, which will have plunged into deep depression and an international police state. Racist Jews, who view themselves as Orientals, will then enslave the rest of humanity, and through laws mandating miscegenation dilute the blood of

"Esau". Then they will likely break up Israel into classes, where Ashkanazi Jews reign over Sephardic and Coptic Jews—a process which is already well underway. Those who doubt it are invited to consider what happened to Germany and Russia at the hands of Jewish financiers and to further consider the precarious economic condition of the United States as a result of the organized efforts of Zionists to undermine the sovereignty of America, its moral and educational strengths, and to export its industries.]

"1 The burden of the word of the LORD to Israel by Malachi. 2 I have loved you, saith the LORD. Yet ye say, Wherein hast thou loved us? *Was* not Esau Jacob's brother? saith the LORD: yet I loved Jacob, 3 And I hated Esau, and laid his mountains and his heritage waste for the dragons of the wilderness. 4 Whereas Edom saith, We are impoverished, but we will return and build the desolate places; thus saith the LORD of hosts, They shall build, but I will throw down; and they shall call them, The border of wickedness, and, The people against whom the LORD hath indignation for ever. 5 And *your* eyes shall see, and ye shall say, The LORD will be magnified from the border of Israel. 6 A son honoureth *his* father, and a servant his master: if then I *be* a father, where *is* mine honour? and if I *be* a master, where *is* my fear? saith the LORD of hosts unto you, O priests, that despise my name. And ye say, Wherein have we despised thy name? 7 Ye offer polluted bread upon mine altar; and ye say, Wherein have we polluted thee? In that ye say, The table of the LORD *is* contemptible. 8 And if ye offer the blind for sacrifice, *is it* not evil? and if ye offer the lame and sick, *is it* not evil? offer it now unto thy governor; will he be pleased with thee, or accept thy person? saith the LORD of hosts. 9 And now, I pray you, beseech God that he will be gracious unto us: this hath been by your means: will he regard your persons? saith the LORD of hosts. 10 Who *is there* even among you that would shut the doors *for nought?* neither do ye kindle *fire* on mine altar for nought. I have no pleasure in you, saith the LORD of hosts, neither will I accept an offering at your hand. 11 For from the rising of the sun even unto the going down of the same my name *shall* be great among the Gentiles; and in every place incense *shall* be offered unto my name, and a pure offering: for my name *shall* be great among the heathen, saith the LORD of hosts. 12 But ye have profaned it, in that ye say, The table of the LORD *is* polluted; and the fruit thereof, *even* his meat, *is* contemptible. 13 Ye said also, Behold, what a weariness *is it!* and ye have snuffed at it, saith the LORD of hosts; and ye brought *that which was* torn, and the lame, and the sick; thus ye brought an offering: should I accept this of your hand? saith the LORD. 14 But cursed *be* the deceiver, which hath in his flock a male, and voweth, and sacrificeth unto the LORD a corrupt thing: for I *am* a great King, saith the LORD of hosts, and my name *is* dreadful among the heathen."—*Malachi* 1:1-14

The Jewish Talmud, in the book of *Sanhedrin*, folio 20*b*, states,

"It has been taught: R. Jose[12] said: Three commandments were given to Israel when they entered the land; [i] to appoint a king; [ii] to cut off the seed of Amalek; [iii] and to build themselves the chosen house [i.e. the Temple] and I do not know which of them has priority. But, when it is said: *The hand upon the throne of the Lord, the Lord will have war with Amalek from generation to generation,*[13] we must infer that they had first to set up a king, for *'throne'* implies a king, as it is written, *Then Solomon sat on the throne of the Lord as king.*[14] Yet I still do not know which [of the other two] comes first, the building of the chosen Temple or the cutting off of the seed of Amalek. Hence, when it is written, *And when He giveth you rest from all your enemies round about etc.,* and then [Scripture proceeds], *Then it shall come to pass that the place which the Lord your God shall choose,*[15] it is to be inferred that the extermination of Amalek is first. And so it is written of David, *And it came to pass when the king dwelt in his house, and the Lord had given him rest from his enemies round about,* and the passage continues; *that the king said unto Nathan the Prophet: See now, I dwell in a house of cedars etc."*[461]

Rabbi Shlomo Yitzhaki's (Rashi's) *Commentary on the Pentateuch*, Exodus 17:14-16, states,

"14. Write this (for) a memorial that Amalek came to battle against Israel prior to all the (other) nations. And rehearse (it) in the ears of Joshua who will bring into the land, that he should command Israel to recompense him (Amalek) for his deed. Here it was hinted to Moses that Joshua would bring in Israel to the land. For I will utterly blot out Therefore I admonish you thus, for I desire to blot them out. 15. And he called the name of it (I. e.,) of the altar. Adonai-nissi (lit., the Lord is my banner (or miracle). The Holy One Blessed Be He wrought for us here a 'miracle'. It is not that the altar was called 'Lord' but (that) he who mentioned the name of the altar would recall the miracle which the Omnipresent wrought: 'The Lord He is our miracle.' 16. And he said (I. e.,) Moses, The hand upon the throne of the Lord The hand of the Holy One Blessed Be He was raised to swear by His throne that there would be for Him war and hatred against Amalek forever. And why is (it written) (throne) and not stated [***]? Is then the (Divine) Name also divided in half (i. e.: [***] instead of the full name)? The Holy One Blessed Be He swore that His name will not be whole (i. e., [***] instead of the full name) nor His throne whole (i. e. [***]) instead of [***] until there will be blotted out the name of Amalek utterly. And when his (Amalek's) name will be blotted out (then) will the (Divine) Name be whole, and it is stated (Ps. 9.7): 'O thou enemy, the waste places are come to an end forever' this refers to Amalek, regarding whom it is written Amos

---

461. I. Epstein, Editor, "Sanhedrin 20*b*", *The Babylonian Talmud*, Volume 27, The Soncino Press, London, (1935), pp. 107-111, at 109.

1.11): 'And his anger he kept forever,' 'And the cities which thou didst uproot Their very memorial is perished' (Ps., *ibid.* 7). What does (Scripture) state after this? 'But the Lord is enthroned forever' (verse 8)—behold the (Divine) Name is whole (expressed in full); 'He hath established His throne for judgment' (*ibid.*)—behold his throne is whole [***]."[462]

The Judaic religious doctrine of the genocide of the seed of Amalek is alive today. Yehoshafat Harkabi wrote in his book *Israel's Fateful Hour*,

"Some nationalistic religious extremists frequently identify the Arabs with Amalek, whom the Jews are commanded to annihilate totally (Deuteronomy 25:17-19). As children, we were taught that this was a relic of a bygone and primitive era, a commandment that had lapsed because Sennacherib the Assyrian king had mixed up all the nations so it was no longer possible to know who comes of the seed of Amalek. Yet some rabbis insist on injecting a contemporary significance into the commandment to blot out Amalek."[463]

Some Jews to this day celebrate the genocidal destruction of their enemies and their hatred of Gentiles once a year at the festival of Purim; which commemorates the execution of Haman and the genocidal mass murder of "enemies of the Jews". Haman is said to have descended from Amalek through Hammedatha the Agagite,[464] and was allegedly the archenemy of the Jews and sought to exterminate them (*Esther* 3)—it is clear that the story of Esther fabricates the pretext of a Haman conspiracy in order to justify the Jewish genocide of the "Amalekites" which Jewish commandment to genocide predated the story of Esther. Esther and Mordecai wormed their way into power under false pretensions, concealing the fact that "Esther" was Jewish. The name "Esther" means "that which is hidden".[465] Her true Jewish name was Hadassah. She was one of the first "crypto-Jews", who conceal their identity in order to corrupt societies and betray those who trust in them.

It should be noted that it is well known that the *Book of Esther* is work of fiction and does not correspond to the historical facts of Persian history. The Judeans fabricated a history of captive exile in Babylon in order to justify the theft of Jerusalem and the lands of all of the other inhabitants of Canaan. Based on *Ezra* 1-6, one might even conclude that the Judeans themselves were an alien horde of Babylonians—or Persians—who the Persians placed in power to rule over the Canaanites and gather the gold and silver of the world as a tribute to the Persian King. They fabricated the entire Old Testament in order to justify their

---

462. A. Ben Isaiah, et al., *The Pentateuch and Rashi's Commentary: A Linear Translation into English*, S. S. & R. Publishing Company, Brooklyn, New York, (1949), pp. 187-188.
463. Y. Harkabi, *Israel's Fateful Hour*, Harper & Row, New York, (1988), p. 149.
464. I *Samuel* 15:9. *Esther* 3:1. G. Dalman, *Jesus Christ in the Talmud, Midrash, Zohar, and the Liturgy of the Synagogue*, Deighton Bell, Cambridge, (1893), pp. 39-40.
465. Marc-Alain Ouaknin, *Symbols of Judaism*, Barnes & Noble Books, New York, (2000), p. 84.

theft of land, their racist credos, their self-declared right to conquer and rule the world, and in order to inspire superstitious fear of their God, and, thereby, fear of them.

4.4.3 Crypto-Jews

Cabalistic Jews have the pantheistic belief that God is hidden in all things and only reveals himself to the enlightened. They believe that the Jews are God among the beasts of the Earth who are the Gentiles. Based on these myths, Cabalistic Jews hold that they should play God's hidden rôle as the secret controller and ruler over the Earth, the secret and divine master of the Gentile beasts—just as God is the secret and divine master of the Universe.

When the Jews of Spain were ordered to convert to Christianity, or leave the country, Jewish leadership instructed them to become crypto-Jews—Jews who feign conversion, but secretly remain Jews and attempt to subvert the churches and the societies in which they live. The crypto-Jews of Spain became known as "Marranos". The correspondence advising the Jews of Spain to feign Christian conversion and destroy Gentile Spanish society was republished in Julio Iniguez de Medrano's book, *La Silva curiosa*, Marc Orry, Paris, (1608), pp. 157-157, and an English translation appears in: L. Fry, *Waters Flowing Eastward: The War Against the Kingship of Christ*, TBR Books, Washington, D. C., (2000), pp. 73-74,

"Respuesta de los Iudios de Constantinopla,
a los Iudios de España
A Mados hermanos en Moysen vuestra carta recibimos, en la qual nos significais los trabajos & infortunios que padesceis, de cuyo sentimiento nos a cabido tanta parte como a vosotros. El parescer de los grandes Satrapas, y Rabi es lo siguiente.

A lo que dezis que el Rey de España os haze boluer Christianos, que lo hagias pues no podeis hazer otto. A lo que dezis que os mandan quitar vuestras haziendas, hazed vuestros hijos mercaderes, para que poco a poco les quiten las suyas. A lo que dezis que os quita lasvidas, hazed vuestros hijos medicos y boticarios, para que les quiten las suyas. A lo que dezis que os destruyen vuestras Sinagogas, hazed vuestros hijos clerigos y theologos, para que les destruyan sus templos. Ya lo que dezis que os hazen otras vexaciones, procurad que vuestros hijos sean abogados, procuradores, notarios, y consejeros, y que siempre entiendan en negocios de Republicas, para que sujetandolos ganeis tierra, y os podais vengar dellos, y no salgais desta orden que os damos, porque por experiencia vereis que de abatidos, verneis a ser tenidos en algo.

V s s v s F F Principe de los Iudios de Constantinopla."[466]

---

[466]. J. I. de Medrano, *La Silva curiosa*, Marc Orry, Paris, (1608), pp. 157-157. *Cf.* L. Fry, *Waters Flowing Eastward: The War Against the Kingship of Christ*, TBR Books, Washington, D. C., (2000), pp. 81-82.

Many of the Bolshevik mass murderers were crypto-Jews, as were many of the "Young Turks",[467] who committed genocide against the Armenian Christians—the Spanish Civil War was led and fought by many Cabalistic and crypto-Jews, on both sides of the struggle, and served as a prototype for the bloodshed of World War II. The Frankist crypto-Jews of Poland wormed their way into the Catholic Church of Poland and came to dominate Polish aristocracy.

Jews and crypto-Jews also worked for the Czar—at least they pretended to work for the Czar—they were notorious assassins and double agents who murdered members of State, like Vyacheslav Plehve and Peter Stolypin, and who betrayed State secrets to the Jewish revolutionaries. In an article entitled, "The Protocol Forgery" published in *The London Times* on 17 August 1921 on page 9, it states,

"THE FIRST REVOLUTION.

But the principal importance of the Protocols was their use during the first Russian revolution. This revolution was supported by the Jewish element in Russia, notably by the Jewish Bund. The Okhrana organization knew this perfectly well; it had its Jewish and crypto-Jewish agents, one of whom afterwards assassinated M. Stolypin; it was in league with the powerful Conservative faction with its allies it sought to gain the Tsar's ear. For many years before the Russian revolution of 1905-1906 there had been a tale of a secret council of Rabbis who plotted ceaselessly against the Orthodox."

Some Jewish revolutionaries, like Emma Goldberg, did not hide their "Jewish sounding names", though they often did not mention—perhaps a very small few did not even realize—that they were fulfilling Judaic Messianic prophecies. Other Jewish Communist radicals did conceal their Jewish identities by changing names; including "Miss Rose Pastor", a Russian Jew, and Morris Hillquit, born Moses Hillkowitz in Riga, Latvia,[468] and Leon Trotsky, born Lev Davidovich Bronstein in Yanovka, Ukraine, and Leo Kameneff, born Rosenfeld, and married to Trotsky's sister.[469]

These Jewish radicals, often born into wealthy Jewish families,[470] were funded by unimaginably wealthy Jewish financiers, who profited from the strife

---

[467]. I. Zangwill, *The Problem of the Jewish Race*, Judaen Publishing Company, New York, (1914), pp. 9, 11; which was first published as an article, "The Jewish Race", *The Independent*, Volume 71, Number 3271, (10 August 1911), pp. 288-295, at 290-291. J. Prinz, *The Secret Jews*, Random House, New York, (1973), pp. 111-112.

[468]. B. J. Hendrick, "Radicalism among the Polish Jews", *The World's Work*, Volume 44, Number 6, (April, 1923), pp. 591-601, at 597.

[469]. R. Recouly, "Contrasts Between the French and Russian Revolutions", *The World's Work*, Volume 44, Number 1, (November, 1922), pp. 67-80, at 78.

[470]. R. Recouly, "Contrasts Between the French and Russian Revolutions", *The World's Work*, Volume 44, Number 1, (November, 1922), pp. 67-80, at 75, 78.

they caused; and who, being pious Jews, sought to fulfill their Messianic goals. These goals included the utter destruction of all nations but Israel, all religions but Judaism, all cultures but Jewish culture; and the "restoration" of the Jews to Palestine, the rebuilding of the Temple, and the anointment of the Messiah, the King of the Jews, who would rule a ruined world. Crypto-Jews and Gentiles married to Jews continued to dominate the Soviet Régime through the 1930's and beyond.[471]

The United States Government published a report entitled "Bolshevism and Judaism" dated 13 November 1918, which is found in State Department Decimal File (861.00/5339).[472] The report was translated into French and then translated back into English in Denis Fahey's *The Mystical Body of Christ in the Modern World*, Browne and Nolan, Dublin, London, (1935), pp. 89-91, 90, *see also:* pp. 77, 86, 92-93. Fahey cites: *La Vieille France*, (1920); and E. Jouin, "Les 'Protocolos' des Sages de Sion: Coup d'Oeil d'Ensemble", *Le Péril Judéo-Maçonnique*, Part 3, Revue Internationale des Sociétés Secrètes, Paris, (1921), pp. 249-250. *La Vieille-France*, Volume 160, published the following French translation of the report in 1920 under the heading "Les Juifs ont créé le Bolchevisme. Les Gouvernements de l'Entente le savent." which was republished in the French translation of the *Protocols* published by *La Vieille-France* as: *La Conspiration Juive Contre les Peuples: «Protocols» Procès-verbaux de Réunions Secrètes des Sages d'Israël*, La Vieille-France, Paris, (1920), pp. 90-91:

> "En février 1916, pour la première fois, on apprit qu'une Révolution se préparait en Russie. On découvrit que les personnes et maisons suivantes étaient engagées dans cette œuvre de destruction:
> *Jakob Schiff — Kuhn, Loeb et Co — Félix Warburg — Otto Kahn Mortimoff L. Schiff — Jérôme H. Hahauer — Guggenheim — Max Breitung.*
> Il n'y a donc guère de doute que la Révolution russe, qui éclaira en 1917 cette information de 1916, fut fomentée et lancée par des influences purement Juives.
> En fait, au mois d'avril 1917, Jakob Schiff déclara *publiquement* que la Révolution russe avait réussi *grâce à son appui financier*.
> Au printemps de 1917, Jakob Schiff commença de commanditer Trotsky (Juif Braunstein) pour organiser en Russie is Révolution sociale. Le *Forward*, journal juif bolcheviste de New-York, versa sa contribution.
> De Stockholm, le Juif Max Warburg commanditait également Trotsky. A ce *consortium* de Juifs bolchevicks et de Juifs multimillionnaires participaient le syndicat (juif) Westphalien-Rhénan, le Juif Olet Aschberg de la *Nye Banken* (Stockholm) et le Juif Jivolovsky, dont la fille a épousé Trotsky.

---

471. S. S. Montefiore, *Stalin: The Court of the Red Star*, Vintage, New York, (2003), pp. 304-306.
472. A. C. Sutton, *Wall Street and the Bolshevik Revolution*, Buccaneer Books, Cutchogue, New York, (1974), pp. 186-187.

En octobre 1917, quand les Soviets établirent leur pouvoir sur le peuple russe, on y remarquait: *Oulianov* dit Lénine, *Braunstein* (Trotsky), *Nachamkes* (Stockloff), *Zederbaum* (Martoff), *Apfelbaum* (Zinovieff), *Rosenfeld* (Kameneff), *Gimel* (Souchanoff), *Krochmann* (Sagerski), *Silberstein* (Bogdanoff), *Lurge* (Larin), *Goldmann* (Gorev), *Radomislsky* (Uritzky), *Katz* (Kamenev), *Furtenberg* (Ganetzky), *Gourevitch* (Dan), *Goldberg* (Meschkovsky), *Goldfandt* (Parvus), *Goldenbach* (Riasanov), *Zibar* (Martinoff), *Chernomordkin* (Chernomorsky), Bleichmann (Solntzeff), *Zivin* (Piatnisky), *Rein* (Abromovitch), *Voinsten* (Zvesdin), *Rosenblum* (Maklakosky), *Loevenschen* (Lapinsky), *Natansohn* (Bobriev), *Orthodox* (Axelrod), *Garfeld* (Garin), *Schultze* (Glasonnoff), *Ioffe*: tous Juifs sous de faux noms russes.

En même temps, aux Etats-Unis, le Juif Paul Warburg laissait voir des relations si étroites avec les personnalités bolchevistes qu'il ne fut pas réélu au *Federal Reserve Board*.

Jakob Schiff a pour intime ami et pour agent très actif le rabbin Judas Magne, protagoniste du Judaïsme international, qui a lancé aux Etats-Unis la première organisation ouvertement bolcheviste, dite *Conseil du Peuple*. Le 24 octobre 1918, Judas Magne a fait la déclaration publique de son adhésion sans réserve au Bolchevisme, dans une réunion du Comité Juif d'Amérique à New-York. Commandité par Jakob Schiff, administrant avec lui la *Kebillah* juive, le rabbin Judas Magne est le directeur effe tif de l'organisation sioniste *Poale*, et du «Parti travailliste juif».

La firme juive Kuhn, Loeb et C° est étroitement liée au Syndicat Westphalien-Rhénan, aux Juifs Lazard de Paris, à la firme juive Gunsbourg (Petrograd-Paris-Tokio), à la firme juive Speyer et C° (Londres-New-York-Francfort) et à la firme juive *Nye Banken* (Stockholm): d'où il apparaît que *le Bolchevisme est l'expression d'un mouvement général juif, où sont intéressées les grandes banques juives*.

La reconnaissance formelle d'un Etat Juif en Palestine, la constitution de Républiques juives en Allemagne et en Autriche ne sont que les premiers pas vers la domination du monde. La Juiverie internationale s'agite fiévreusement. Elle a réuni dernièrement, en peu de jours, aux Etats-Unis, sous prétexte d'écoles en Palestine, un fonds de guerre d'un milliard de dollars."

Whether or not Lenin was of partial Jewish descent, he was married to a Jewish woman, and was put in power by Jewish bankers. The Jews who put Lenin in power were not likely to put a known full-blooded Jew into the position of dictator over Russia unless left with no other choice. Jewish leaders believed that a known Jew would have a difficult time dominating Russia. Max Nordau wrote in 1909,

"In Russia today it would be impossible for a Jew, whether he had been baptized or no, to rouse a mass movement like that led by Lasalle in Germany in the fifties and sixties; or to rise to the premiership, as Disraeli did in England."[473]

Lenin was clearly serving the interests of Jewish leadership. His personal ethnic heritage is largely irrelevant. The Jews may have chosen Lenin to be the dictator over Russia for the very reason that he was not a full-blooded Jew. That does not render Bolshevism any less of a Jewish led movement. Lenin served that movement. He was not its ultimate leader. However, the fact that Bolshevism was a Jewish movement does not mean that all Jews were Bolsheviks.

The Jewish Chronicle published the following article on 11 April 1919 on page 10,

"Percentage of Jewish Bolsheviki in Petrograd.

COPENHAGEN [F. O. C.]

On the trustworthy authority of the well-known Zionist leader, M. Idelson (of Petrograd), I am in a position to state that only two and a-half per cent. of the Jews in Petrograd have declared themselves in sympathy with Bolshevism. Although sixty per cent. of the Bolshevik leaders are Jews, and although a declaration against Bolshevism involves serious sacrifices, the Jews of Petrograd have fearlessly stated their attitude towards the movement. We are, therefore, confronted with the anomaly of the Jews furnishing for the Bolsheviki the majority of their leaders, although a smaller percentage of Jews than of any other nationality approve of Bolshevism."

"Janus" wrote a Letter to the Editor of the *London Times* which was published on 26 November 1919 on page 8,

"JEWS AND BOLSHEVISM.
REVOLUTIONARY ELEMENTS.
TO THE EDITOR OF THE TIMES.

Sir,—I have read with much interest the letters you published on the 21st and 25th instant from Mr. Israel Cohen and that signed 'Philojudæus' in your issue of the 22nd instant. Without being concerned in the question of whether the Jewish population of Russia as a whole is for or against Bolshevism, or, as one should more correctly describe it, Communism, it is certainly a remarkable fact that the following 28 conspicuous Bolshevists, most of them Commissaries, are either full-blooded Jews or of Jewish extraction. Nearly all possess a Russianized name. In Hungary also the Commissaries were nearly all Jews, and so are the Bolshevist propagandists in the United States and other countries. This is no more a reflection upon the Jewish race as a whole than the exploits of Marat are a reflection upon

---

473. M. Nordau, *The Interpretation of History*, Willey Book Company, New York, (1910), pp. 290-297; which is an English translation by M. A. Hamilton of *Der Sinn der Geschichte*, C. Duncker, Berlin, (1909).

the French. All that one can say is that wherever there are subversive movements the restless and enterprising boil up to the surface. The list is as follows:

Russian name.	Former name.
Lunacharsky	—
Uritsky	—
Litvinov	Fineklstein.
Trotsky	Bronstein.
Steklov	Nahamkes.
Zinoviev	Apfelbaum.
Chernov	Liebermann.
Volodarsky	Cohen.
Kamkov	Katz.
Kamenev	Rosenfeldt.
Solntsev	Goldstein.
Naut	Ginsburg.
Dau	Gurevicz.
Martov	Zederbaum.
Zvezdich	Feinstein.
Lebedeva	Simon.
Meshkovsky	Goldenberg.
Parvus	Goldfarb.
Kamensky	Hoffmann.
Gorev	Goldmann.
Sukhanov	Himmer.
Rjazanov	Goldenbach.
Zagorsky	Krachmalnik.
Izgoev	Goldmann

							Bogdanov				Silberstein.

							Larin					Lurier.

							Bunakov				Fundamentsky.

							Radek

							Yours faithfully,
							JANUS."

Israel Cohen responded in *The London Times* on 27 November 1919 on page 15,

"TO THE EDITOR OF THE TIMES.

Sir,—In your issue of to-day your correspondent 'Janus' gives a list of 28 'conspicuous Bolshevists' who, he states, 'are either full-blooded Jews or of Jewish extraction.' It is only fair to your readers that they should be informed that as many as 10 names in this list are those either of non-Jews or of anti-Bolshevists or of dead Bolshevists:—

(1-3) Lunacharsky, Chernov, and Bogdanov are pure Russian Bolshevists.

(4) Zagorsky is neither a Jew nor a Bolshevist, but a Russian Radical.

(5-6) Kamkov and Bunakov are Social Revolutionaries—*i.e.*, anti-Bolshevists. Kamkov (-Katz), after his participation in the assassination of Count Mirbach, had to flee from Bolshevist Russia to Archangel.

(7-8) Dan and Martov are the Jewish leaders of the Menshevists—*i.e.*, the most determined opponents of Lenin and his group. They were referred to as anti-Bolshevists in your columns only a few days ago.

(9-10) Uritzky and Volodarsky have both been murdered, the former by the Jew Kannesgiesser.

I have no doubt that 'Janus' has sent you his list in good faith, but the fact that it has to be discounted to such a great extent is typical of the general misrepresentations of the Jewish share in Bolshevism.

Yours faithfully,
ISRAEL COHEN.
77, Great Russell-street, W.C., Nov. 26."

The New York Times reported on 20 April 1906 on page 20 on a Jewish revolutionary from Russia, who hid his identity with a "Gentile sounding name", and who traveled through America with falsified passports seeking support (note that there is no call for his arrest),

"MAXIME COMES HERE TO AID REVOLUTION

To Stir Up Sentiment Among the

Jews of America.

TELLS OF RUSSIAN BUND

Declares Upheaval Is Coming Soon—
Thinks Father Gapon an Agent
of the Government.

Sent for by the Revolutionary Bund of this city, an organization of Jewish citizens helping the Jewish revolutionary movement in Russia, a young man with a high forehead and piercing, black eyes, and describing himself as Gregory Maxime of St. Petersburg, arrived yesterday in New York as the representative of the parent bund in Russia. How he came and where he agitated last he declined to say. He admits that Maxime is not his real name, and that he may address the Jewish people of some other large city by some other name in a few weeks.

Maxime is the representative of the powerful Jewish revolutionary party in Russia. It is known that he is of fine education, and that his father is a wealthy Jew in Russia. Under the name of Maxime he headed the provisional government in Riga after the big railroad strike, and, while the names of the central committee of the Bund are known to very few sympathizers in the old country or abroad, he is believed to be a member of it, and also a controlling mind in the direction of the Jewish end of the revolutionary work.

The Bund is strong, and contributes largely to the work of the organization in Russia. As all the Bund's work is done underground, and as many members of it are subject to imprisonment, exile, or death at the hands of the Russian Government, Maxime changes his passports, his name, and as far as possible his appearance as frequently as he deems it necessary to dodge Russian spies. At present he looks the student. He is 27 years old, dresses simply and neatly, and wears a neatly trimmed black beard and mustache. He might easily pass as a university instructor.

Maxime's practical rule of Riga came to an end when the Czar's agents poured into that city sufficient troops to overwhelm the large revolutionary population of Jews and Letts. Maxime says that he was addressing an audience in the theatre of the city when the place was surrounded and artillery trained on it. He had escaped from exile in Siberia just prior to the strike, and he knew that he was wanted. He dropped through a trap in the stage as the officers entered the theatre, and was hurried to the roof of an adjoining building, which was the home of a member of the Bund. He was then shaven and in a few moments was in the garments of a woman and rushing out with the women of the household as they fled to the streets and the Czar's officers rushed in. The Government Secret Service has not had trace of him since.

Maxime will remain in New York about three weeks, addressing the Jews of the city on the revolutionary movement in Russia. Next Sunday night he will talk at Grand Central Palace. After several addresses in Yiddish in this city he will visit other cities with large Jewish populations.

Asked what he thought of Maxim Gorky's plight in this country, he said yesterday: 'I have never met Gorky. In Russia we accept him as a great writer and factor for good, and do not pry into his private affairs. The Mme. Gorky who is with him here was accepted in Russia as Mme. Gorky by the best people. As for me, I'm here unmarried—that is, my wife's in Russia.'

'What do the Jewish revolutionists think of Father Gapon?' he was asked.

'They think him an agent of the Government.'

'What is the opinion of Count Witte?'

'Witte is first for himself and the emoluments,' was the reply. 'He would serve any form of government for the price.'"

On 30 June 1912, on page 9, *The New York Times* published a letter from the radical Jewish Communist Zionist of the Poale Zion, Baruch Charney Vladeck—a. k. a. B. Charney Vladeck, a. k. a. Bruce Vladeck—born Baruch Nachman Charney in Minsk—spent time in prison for attempting to overthrow the Government of Russia—fled to America under a false name—a correspondent for the Jewish Socialist Federation's *Naye Velt* and City Editor of the *Jewish Daily Forward*; a New York City Alderman who led the *Bund* until 1908—and who would later become a member of the New York City Housing Authority and first President of the Jewish Labor Committee—and who made an unsuccessful bid for the United States Congress,

"REAL NAMES IN RUSSIA.
Lenin's not German—Other Radicals
May Be from Baltic Provinces.
New York, June 25, 1917.

To the Editor of The New York Times:

In this morning's TIMES there is a little item of news from Petrograd, under the headline 'Leader's Names Assumed,' credited to The London Post, which is full of misinformation, and ought to be corrected. The item referred to contains the following two statements:

1—That the real name of the leader of the extremist faction, Lenin, is Zebarbluhm or Zedarbaum.

2—That of the eighteen members of the Executive Committee of the Council of Workmen's and Soldiers' Delegates the real names of fourteen sound German.

As to Lenin, his real name is Ulianoff, a 'Stolbovoy Dvorianin,' which means a member of the nobility. He is of Russian parentage, born in one of the innermost Russian provinces. Zedarbaum is the real name of an influential Socialist of the moderate faction whose nom de plume is Martoff.

As for the Executive Committee of the Council of Workmen's and Soldiers' Delegates, it consists of fifty-four members, not of eighteen, these fifty-four being divided into a majority of thirty-two moderates or minimalists and twenty-two extremists or maximalists.

Of the fourteen members referred to in the news item, several represent the Jewish Socialist organization known as the Bund, as Goldman, Lurie, &c. The seven or eight whose real names sound German may come from provinces with a large German population, like the Baltic provinces, or they may simply have a name that sounds German, but has nothing to do with German policies.

It is perfectly legitimate to disagree with the Council of Workmen's and Soldiers' Delegates in Petrograd, but I don't see why the council and its members should be constantly vilified by people who, for lack of insight into the great Russian crisis, try to explain away events of historical importance by insignificant trifles.

It is true that most of them have studied statesmanship in prison, but so have many others whose names now shine forth from the pages of history. Everybody at all acquainted with the recent history of Russia knows that nearly every able writer from Lermontov down to Gorky: every original thinker from Herzen down to the present Minister Chernov or Plekhnov; every independent citizen from the Becabrists down to Breshkovskaya, the grandmother of the revolution, were persecuted, humiliated, and imprisoned by the old régime, so that very often the prison was the only place where they could learn anything.

B. C. VLADECK,
City Editor Jewish Daily Forward."

Simon Sebag Montefiore wrote in his book *Stalin: The Court of the Red Tsar*,

"In 1937, 5.7 percent of the Party were Jews yet they formed a majority in the government. Lenin himself (who was partly Jewish by ancestry) said that if the Commissar was Jewish, the deputy should be Russian: Stalin followed this rule. [***] Many Jewish Bolsheviks used Russian pseudonyms. As early as 1936, Stalin ordered Mekhlis at *Pravda* to use these pseudonyms: 'No need to excite Hitler!'"[474]

In another among many instances of organized Jewish censorship, many Jews made corrupt use of their power in the media, universities and government to censor and ridicule anyone who told the truth about the dominant and destructive rôle Jews played in Bolshevism, Socialism and Communism. While Jews who chose to do so were free to boast of the commonality of Judaism and Bolshevism, Gentiles who pointed out that same linkage were ruined. In the Soviet Union, outing a crypto-Jew was an offense punishable by death.

Denis Fahey wrote extensively of organized Jewish power to censor and

---

[474]. S. S. Montefiore, *Stalin: The Court of the Red Star*, Vintage, New York, (2003), pp. 305-306.

punish those who told those truths leading Jews did not want exposed to the public. Fahey quoted a June, 1924, article "The Russian Revolution and the English Official White Paper, Russia, No. 1, 1919," by G. P. Mudge, which was published in *Loyalty League*, in which Mudge wrote, *inter alia*,

"WHY DOES THE BRITISH FOREIGN OFFICE SUPPRESS THE TRUTH UNPALATABLE TO JEWRY?
In the April issue of the *Loyalty League* I dealt with the attempt made, in the course of a series of lectures by a Mr. M. Farbman, at the London School of Economics, to transfer the responsibility for the hideous Russian revolution of 1917 from the real perpetrators, the Jews, and to ascribe it to a purely agrarian movement among the peasants. I undertook in that article to marshal the voluminous and conclusive evidence that this revolution was entirely Jewish in organization and operation, to show that it had nothing to do with an agrarian movement, or indeed with any cause that had Russian interests in view.

Perhaps one of the most damning pieces of evidence, not only that this revolution, but also the world-revolution which is planned, is Jewish, lies in the strenuous and partially successful efforts which organized Jewry has made to suppress the truth about it. Not only has Jewry succeeded in large measure in suppressing the truth, but it has seemingly been able to intimidate or cajole the *British Foreign Office to suppress a very vital part of one of its own official publications.*"[475]

Mudge went on to quote from the British War Cabinet's unabridged "White Paper" of April, 1919, which includes Oudendyke's report of 6 September 1918. Oudendyke was the Netherlands' Minister at St. Petersburg,

"The following collection of Reports from His Majesty's official representatives in Russia, from other British subjects who have recently returned from that country, and from independent witnesses of various nationalities, covers the period of the Bolshevik *régime* from the Summer of 1918 to the present date. *They are issued in accordance with a decision of the War Cabinet in January last*. They are unaccompanied by anything in the nature either of comment or introduction, since they speak for themselves in the picture which they present of the principles and methods of Bolshevik rule, the appalling incidents by which it has been accompanied, the economic consequences which have flowed from it, and the almost incalculable misery which it has produced. [***] The foregoing report will indicate the extremely critical nature of the present situation. The danger is now so great that I feel it my duty to call the attention of the British and all other Governments to the fact that, if an end is not put to Bolshevism in

---

[475]. D. Fahey, *The Mystical Body of Christ in the Modern World*, Browne and Nolan Limited, London, (1935), p. 247.

Russia at once, the civilization of the whole world will be threatened. This is not an exaggeration, but a sober matter of fact; and the most unusual action of German and Austrian consuls-general, before referred to, in joining in protest of neutral legations, appears to indicate that the danger is also being realized in German and Austrian quarters. I consider that the immediate suppression of Bolshevism is the greatest issue now before the world, not even excluding the war which is still raging, and unless, as above stated, *Bolshevism is nipped in the bud immediately, it is bound to spread in one form or another over Europe and the whole world,* AS IT IS ORGANIZED AND WORKED BY JEWS WHO HAVE NO NATIONALITY, AND WHOSE ONE OBJECT IS TO DESTROY FOR THEIR OWN ENDS THE EXISTING ORDER OF THINGS... . *I would beg that this report may be telegraphed as soon as possible in cypher in full to the British Office in view of its importance.*"[476]

Denis Fahey quoted an 18 July 1929 article "Censorship of the Anglo-Saxons" in the *Patriot*, which stated, among other things,

"The censorship in force is Jewish in character, in backing, and in its operative machinery. But it is not confined in its supervision and operation to a definitely organized body of men, even if there be such an organization unknown to us. The Jewish race is absolutely apart from all others in its solidarity, which is maintained in spite of complete dispersion over the globe, and in spite of fundamental differences in religion, in politics, and in material and spiritual attachments within many different nations. The dispersion of the individuals—accompanied as it is by close inter-communications, through business relations in all countries, and by literature on racial interests—permits of the exercise of an ever-growing world power. [***] Other countries have also organizations aiding Jewish solidarity; and that this solidarity does exist can be shown by two illustrations: First, the amazing way in which the whole world was shaken up on several occasions during the long period of the trials for treason of a single French Army officer, Dreyfus; and second, by the persistent policy of concealment, from all peoples, of the leading part played by a section of revolutionary Jews in all the bloodshed and commercial destruction of the Russian people. That concealment is enforced so successfully that neither writers of books nor editors of newspapers can safely forget the interdict. Even a Government White Book issued in April, 1919, and making clear the world-danger of the Jewish-Bolshevik conspiracy against civilization was, by some unknown influence, suppressed, and a bowdlerised abridgement was substituted.

The over-riding power in literature and publicity of a small Jewish minority in most countries is made up of a variety of elements. There is vast

---

476. D. Fahey, *The Mystical Body of Christ in the Modern World*, Browne and Nolan Limited, London, (1935), pp. 247-248, *see also:* pp. 87-88.

wealth to be drawn on for racial objects; there is ownership or control of large numbers of newspapers; and that control is not merely over the complexion given to some news, but over those reviews of new publications which affect largely their sales. The news agencies feeding the newspapers are mostly under Jewish control. The power exercised in film and theatrical productions is pretty generally known. The enormous potential force of a combination of the wealthy Jewish advertisers in all important papers is fully recognized by journalists, for whom advertisements are the life blood of commercial publication. While the political power of Jews might appear negligible because they are equally active in all three Parties here, it is a fact that the division works to great advantage; for, not only is the power exercised out of proportion to numbers in each Party, but it is multiplied by three in matters of racial interest. This is clearly expressed in the words of Emanuel Shinwell, M. P. (Financial Secretary to the War Office), in a speech at the annual dinner of 'B'nai B'rith,' on 23rd, June: *The Jews in the House of Commons, whatever their political opinions may be, will always stand in that assembly for the rights of the Jewish community. It has been said that they must emphasize the fact of the Judaism before the fact of citizenship. He held that they must regard themselves as Jews and citizens equally.*"[477]

Fahey also quoted from a 20 February 1930 article in the *Patriot*,

"As bearing on the part taken by Red Jews in the Bolshevik triumph over Russia, we quote Dr. Angelo S. Rappaport, a Jewish writer, who published a book in 1918 called *Pioneers of the Russian Revolution*:—
'To a greater degree than the Poles, the Letts, or Finns, or, indeed, any other ethnic group in the vast Empire of the Romanovs, the Jews have been the artisans of the revolution of 1917... . It is no exaggeration to say that the small, even insignificant, amount of freedom obtained by the Russian Liberals in 1905 and 1906 was largely due to the effort of the Jews... . There was no political organization in the vast Empire that was not influenced by Jews or directed by them... . Throughout history the spirit of the Jew has always been revolutionary and subversive... . Long before they had been formulated in French, the principles of the 'Rights of Man' had been announced in Hebrew... . The Russian Jews, the pioneers of the revolution, are now continuing to fight for the cause of Justice, for the principles of Democracy against German Militarism.'
When the Jewish and Russian Bolsheviks seized power, Red Jews flocked to the scene from all countries, and reinforced the brains and hands of the murderous tyranny. Mr. Robert Wilton, for seventeen years correspondent of *The Times* in Russia [***] wrote a book, *The Last Days of*

---

[477]. D. Fahey, *The Mystical Body of Christ in the Modern World*, Browne and Nolan Limited, London, (1935), pp. 252-254.

the Romanoffs. This book showed that the murder of the Czar and his family was the work of Red Jews, and that they prepared the whole revolution, and became masters of Russia from their domination of all the important offices under the Soviet. He wrote in 1920: 'The Jewish domination in Russia is supported by certain Russians... they are all mere screens or dummies behind which the Sverdlovs and the thousand and one Jews of Sovdepia continue their work of destruction.'

After this Mr. Wilton's chances in English journalism were gone. He was a true British patriot; and he died in very straitened circumstances in France in January, 1925. No one who has paid the slightest attention to the course of Russian events since the Bolshevik accession to power in November, 1917, can have failed to know that, when all the important members of the Russian aristocracy, the learned profession, the Army and Navy, had been executed, or imprisoned, or driven abroad, Red Jews were in possession of the great majority of responsible positions in and under the Soviet. So clear was this that, in the past, Jewish apologists, here and in America, have explained the fact by the true statement that only among the Jews could be found any longer the brains and business experience for filling important posts. Yet in the face of this situation there have been dozens of books published in English, and innumerable articles throughout the Press, and any number of lectures delivered, all with the astounding omission of any mention of Jewish handiwork in Russian Bolshevism. There have been public references to the sufferings of some orthodox non-Communist Jews at the hands of the Soviet.

Newspapers bear witness to a censorship over them by what they omit to publish, and by their sketchy apologetic mention of incidents tending to produce undesired conclusions about the march of events. Authors can safely reckon on the refusal of book publishers to produce any book unorthodox to current propaganda which supports the censorship."[478]

Gorky stated soon after the Russian Revolution of 1917, that the crypto-Jews "Lenin" and "Trotsky" (Lev Davidovitch Bronstein) had turned the revolutionary movement for democracy, liberty, equality and fraternity into a dictatorship; which suppressed human rights and civil liberties; and which censored the press, including Gorky's own daily newspaper, Новая Жизнь or "New Life" published in Petrograd. It was a common practice for Cabalist Jews to foment revolution with cries for liberty, equality and fraternity—especially in the press, which they owned—then destroy the nation, culture and religion of a people after the revolution, and declare that only a dictatorship, run by one of their agents, would have the ability to restore order among the chaos, which insufferable chaos they themselves had intentionally created. The dictatorship would then set about to destroy the people themselves, and spread war and

---

[478]. D. Fahey, *The Mystical Body of Christ in the Modern World*, Browne and Nolan Limited, London, (1935), pp. 256-257.

famine to the nation and to its neighbors. The English Revolution, the French Revolution, the Young Turk Revolution, the Russian Revolution, Hitler's burning of the Reichstag, etc. followed this Cabalistic Jewish model, which we know was employed by Jews at least since the time of the Roman Caesars, and which appears in Jewish literature in the their fabricated tales of "exile" and "captivity" in Egypt and Babylon.

At the festival of Purim, Jews wear costumes which conceal their identity in order to symbolize the status of a crypto-Jew, one who undermines the nation in which he or she resides. Some have interpreted the festival of Purim as an occasion for the Rabbis to augment their power by manufacturing an artificial common enemy for their followers to fear and to hate.[479] Purim is based on the story of Esther, which story is read at the festival.

In the story of "Esther" (a crypto-Jewish name, her actual name was Hadassah) the Jews manipulated and betrayed the Persian Kings, who had freed the Jews from their captivity and exile among the Babylonians. If the stories can be believed—and they cannot, Cyrus, King of Persia, freed the Jews and restored them to Palestine and helped them to "rebuild" the Temple. Ahasuerus, King of Persia, (no such king ever existed) married and obeyed Esther, a deceitful crypto-Jewish agent placed in his midst after Ahasuerus' first wife had died, or had been killed. The Jews repaid the generosity of the Persians with deceit and genocide, in their own mythologies, which genocidal mythologies are inculcated into the minds of Jewish youth.

We find parallels to this ancient story today. The President of Iran (Persia) may be an agent of the Zionists and a traitor to the Iranian people. Judging by his actions, this modern "Persian King" wants to lead the Iranians toward their own destruction in order to benefit the Israelis. Like the Turks who followed the crypto-Jewish Young Turks,[480] who mass murdered Armenians; like the Russians who followed crypto-Jewish Bolsheviks, who mass murdered Russians, Jews and countless others; like the Germans who followed crypto-Jewish Nazis, who mass murdered Germans, Jews and countless others; Americans, Iranians, British, etc. are today led by Zionist Jews and crypto-Jews, who are bringing about their destruction.

Celebrated annually, the festival of Purim is widely considered to be the Jews' favorite holiday. The Biblical book of *Esther* (whose "real" name was Hadassah) and the "war against Amalek" are discussed in the *Tractate Megillah*, Chapter 1. On Purim, Jewish children are encouraged to commit symbolic acts of violence while in a frenzy, and to cry out for genocide and curse the Gentiles (*Orach Chaim* 690:16). In 1603, Johannes Buxtorf, the world's foremost expert on Judaism and Jews, wrote of Purim, a drunken Jewish festival celebrating

---

479. M. A. Hoffman II, *Judaism's Strange Gods*, Independent History and Research, Coeur d'Alene, Idaho, (2000), pp. 108-109.

480. I. Zangwill, *The Problem of the Jewish Race*, Judaen Publishing Company, New York, (1914), pp. 9, 11; which was first published as an article, "The Jewish Race", *The Independent*, Volume 71, Number 3271, (10 August 1911), pp. 288-295, at 290-291. J. Prinz, *The Secret Jews*, Random House, New York, (1973), pp. 111-112.

genocide and hatred, and the use of crypto-Jews to subvert a government,

"CHAP. XXIV.

Of their Feast of Purim.

THe word *Purim* is a Persian word, and is rendered by the Hebrew *Goral*, which signifies a lot. This Feast therefore took its name from that plot and wicked device of *Haman* the Agagite, {Esther 3.} who in the moneth *Nisan* in the twelfth year of *Ahasuerus* cast *Pur*, that is a lot, whereby all the Jews, both young and old, children and women in all the Kings Provinces should be destroyed and rooted out in one day, even upon the thirteenth day of the twelfth moneth, which is the moneth *Adar* of February; which decree was written in the name of the King, and sealed with his Ring.

The end of this conspiracy fell far contrary to *Hamans* intent. For *Haman* was hanged upon a pair of Gallows fifty foot high, and the King granted the Jews {Esther 8.} in what Cities soever they were to gather themselves together, and to stand for their life to root out, slay, and destroy, all them that vexed them. So that strengthened by the Kings Letter Patents, they put their adversaries to death. In *Shushan* the Palace they slew five hundred men, and the ten sons of *Haman*; and the Jews that were in the Provinces of King *Ahasuerus* slew of them that hated them seventy five thousand men, upon the thirteenth day of the moneth *Adar*, and rested upon the fourteenth and fifteenth thereof. Wherefore it is instituted and ordained, that upon the fourteenth and the fifteenth day of the said moneth every yeer should a Feast be kept by the Jews in all quarters, in remembrance of this great deliverance throughout their generations by an ordinance for ever. Wherein they rested from their enemies, in the moneth which turned unto them from sorrow to joy, from mourning to a joyful day: as we may read in the ninth Chapter of the book of *Esther*.

These two dayes are celebrated at this day by the Jews imitation of their ancestors, but in that manner, that they rather deserve the name of the dayes of profanation and drunkennesse, then of joy and gladnesse.

Although upon these dayes working is not prohibited by the text of Scripture: yet the Jewes at this day rest from all manner of labour, writing and affirming in the Talmud, {Tract. Megilah.} that he will never thrive or prosper that does any work upon them. For there it is recorded, that upon a certain time that a man being sowing line-seed upon one of these dayes, a certain Rabbine coming by and seeing him, began to reprove and curse him. Whereupon it came to passe, that the seed never came to growth, nor did ever peep out of the ground.

In the first place therefore the women are enjoyned in a more peculiar manner to sanctifie and celebrate this Festival, because this deliverance was wrought by the hands of Queen *Esther*. The night being come, they light the Lamps of joy in the Synagogue, and the *Chasan* or the Minister expounding the book of *Esther*, reads it from end to end: whereat the women and children ought to be present, and give diligent attention; and they have a custome that the little ones so often as *Haman* is named, keep a vile stir and a tumultuous noise in the terrible and forcible explosion thereof. {Orach chajim, nu. 690. Sect. 16.} In former times they were wont to provide themselves two stones, upon one of which the name of *Haman* was written. These they did beat one against the other, until the name was quite demolished and worn out; which when they perceved, they presently cried aloud, *Let his name be blotted out. The name of the wicked shall rot; Accursed be Haman; Blessed by Mordecai; Cursed be Zeresh; Blessed be Esther the wife Ahasueras. Cursed be all they that worship idols or the host of heaven. Blessed be all the people of Israel.* When the Lecturer comes to that place where mention is made of the ten sons of *Haman*, he is bound to read it with one breath, for they write, that all these sons of *Haman* perished in the twinkling of an eye, and their souls in a very moment took their farewel of their beloved lodging the body. They celebrate this Feast in a very voluptuous manner, sousing their guts in wine and beer, because *Esther* the Queen found favour and grace in the eyes of King *Ahasuerus* when he sate at her banquet, and obtained pardon for the Jews, and a grant that they might stand for their lives. And hence it comes to pass, that for the space of these two dayes, they busie themselves with no other things then eating and drinking, smelling, and bibbing, dancing, and piping, singing, and roaring, feasting, and sporting, riming, and scoffing, the women putting on mens apparrell and the men clothing themselves in womens attire, which although it be expresly forbid in the law of *Moses*, yet they make there one exception, {Orach:chajim num: 615.} saying, that it is lawful and no offence to practice it upon this day, and this occasion: seeing it is done by them only for worldly joy and recreation, *Rabbi Isaac Tirna* in this Minhagim hath left in record to posterity, {De rit: Jud: p. 61.} that it is commanded as a work of great excellency, to make merry as upon these dayes, to goe a whoring, to drink and be drunke, yea in that measure, that he cannot make any difference between *Mordecai* the blessed, and *Haman* the accursed, that is to say, untill he be so besotted with the ale tappe, that he cannot for his heart declare how many letters be contained in any of these words, yea moreover, any one is permitted at this time to poure in strong drink, until he knowes not how many fingers he hath on either hand. Which precept indeed is most diligently observed and kept, according to the very rigour thereof by the Jews at this day, and that chiefly by the beggerly crew, to whom the richer sort send gifts and presents in a far greater measure then they do at other times, to the end that one may not mock another for being drunk, being commanded and strictly prohibited to send away their meat and drink to any other end and purpose. With these Bacchanal rites, drunken fits, and besotting beastliness, they put an end to their annual feasts. For this of Purim is the last festival in the year, having no more until the feast of the passover. If the Prophet *Isaiah* were alive at this day, or should rise from the dead, truly and really might he take occasion, and that both forcible and urgent to cry out, *Woe and alass unto them that rise up early to follow drunkenness, and to them that continue until the night, till the wine do inflame them.*"[481]

4.4.4 The Gentiles Must be Exterminated Lest God Cut Off the Jews

An important aspect of the Jewish Alamek mythology is the belief that Esau, or Edom, sought to destroy a belief in the Creator God of the Old Testament. This offense against God makes it easier for Jewish religious fanatics to justify their merciless genocide of Gentiles—they believe that any evil done in the name of God is good. *Deuteronomy* 7:2-3 states:

> "2 And when the LORD thy God shall deliver them before thee; thou shalt smite them, *and* utterly destroy them; thou shalt make no covenant with them, nor show mercy unto them: 3 Neither shalt thou make marriages with them; thy daughter thou shalt not give unto his son, nor his daughter shalt thou take unto thy son."

Deuteronomy 7:16-18 states:

> "16 And thou shalt consume all the people which the LORD thy God shall deliver thee; thine eye shall have no pity upon them: neither shalt thou serve their gods; for that *will be* a snare unto thee. 17 If thou shalt say in thine heart, These nations *are* more than I; how can I dispossess them? 18 Thou shalt not be afraid of them: *but* shalt well remember what the LORD thy God did unto Pharaoh, and unto all Egypt;"

Some Jews have seen Amalek in Haman, Marcion, Rome, Christianity, Islam, Germany, Russia, even in all Gentiles; and though the Moslems—especially the Islamic Turkish Empire—are traditionally associated with Isaac's half-brother Ishmael, rather than Esau, they are often referred to today as Amalek, as the race that must be exterminated.[482] Jewish mythology emphasizes the threat that God will be angry with, and punish, any Jew who fails to exterminate the seed of Amalek. I *Samuel* 15:1-35 states (one wonders, together with Voltaire,[483] if Agag was meant as a human sacrifice to Baal):

> "Samuel also said unto Saul, The LORD sent me to anoint thee *to be* king

---

481. J. Buxtorf, *Synagoga Judaica: Das ist Jüden Schul ; Darinnen der gantz Jüdische Glaub und Glaubensubung... grundlich erkläret*, Basel, (1603); as translated in the 1657 English edition, *The Jewish Synagogue: Or An Historical Narration of the State of the Jewes, At this Day Dispersed over the Face of the Whole Earth*, Printed by T. Roycroft for H. R. and Thomas Young at the Three Pidgeons in Pauls Church-Yard, London, (1657), pp. 243-245.
482. Y. Harkabi, *Israel's Fateful Hour*, Harper & Row, New York, (1988), pp. 149-150.
483. Voltaire in English translation in R. S. Levy, *Antisemitism in the Modern World: An Anthology of Texts*, D.C. Heath, Toronto, (1991), p. 46.

over his people, over Israel: now therefore hearken thou unto the voice of the words of the LORD. 2 Thus saith the LORD of hosts, I remember *that* which Amalek did to Israel, how he laid *wait* for him in the way, when he came up from Egypt. 3 Now go and smite Amalek, and utterly destroy all that they have, and spare them not; but slay both man and woman, infant and suckling, ox and sheep, camel and ass. 4 And Saul gathered the people together, and numbered them in Telaim, two hundred thousand footmen, and ten thousand men of Judah. 5 And Saul came to a city of Amalek, and laid wait in the valley. 6 And Saul said unto the Kenites, Go, depart, get you down from among the Amalekites, lest I destroy you with them: for ye shewed kindness to all the children of Israel, when they came up out of Egypt. So the Kenites departed from among the Amalekites. 7 And Saul smote the Amalekites from Havilah *until* thou comest to Shur, that *is* over against Egypt. 8 And he took Agag the king of the Amalekites alive, and utterly destroyed all the people with the edge of the sword. 9 But Saul and the people spared Agag, and the best of the sheep, and of the oxen, and of the fatlings, and the lambs, and all *that* was good, and would not utterly destroy them: but every thing *that was* vile and refuse, that they destroyed utterly. 10 Then came the word of the LORD unto Samuel, saying, 11 It repenteth me that I have set up Saul to be king: for he is turned back from following me, and hath not performed my commandments. And it grieved Samuel; and he cried unto the LORD all night. 12 And when Samuel rose early to meet Saul in the morning, it was told Samuel, saying, Saul came to Carmel, and, behold, he set him up a place, and is gone about, and passed on, and gone down to Gilgal. 13 And Samuel came to Saul: and Saul said unto him, Blessed *be* thou of the LORD: I have performed the commandment of the LORD. 14 And Samuel said, What *meaneth* then this bleating of the sheep in mine ears, and the lowing of the oxen which I hear? 15 And Saul said, They have brought them from the Amalekites: for the people spared the best of the sheep and of the oxen, to sacrifice unto the LORD thy God; and the rest we have utterly destroyed. 16 Then Samuel said unto Saul, Stay, and I will tell thee what the LORD hath said to me this night. And he said unto him, Say on. 17 And Samuel said, When thou *wast* little in thine own sight, *wast* thou not *made* the head of the tribes of Israel, and the LORD anointed thee king over Israel? 18 And the LORD sent thee on a journey, and said, Go and utterly destroy the sinners the Amalekites, and fight against them until they be consumed. 19 Wherefore then didst thou not obey the voice of the LORD, but didst fly upon the spoil, and didst evil in the sight of the LORD? 20 And Saul said unto Samuel, Yea, I have obeyed the voice of the LORD, and have gone the way which the LORD sent me, and have brought Agag the king of Amalek, and have utterly destroyed the Amalekites. 21 But the people took of the spoil, sheep and oxen, the chief of the things which should have been utterly destroyed, to sacrifice unto the LORD thy God in Gilgal. 22 And Samuel said, Hath the LORD *as great* delight in burnt offerings and sacrifices, as in obeying the voice of the LORD? Behold, to obey *is* better than sacrifice, *and* to hearken than the fat of rams. 23 For rebellion *is as* the sin of witchcraft, and stubbornness *is as* iniquity and idolatry. Because thou hast rejected the word of the LORD, he hath also rejected thee from *being* king. 24 And Saul said unto Samuel, I have sinned: for I have transgressed the commandment of the LORD, and thy words: because I feared the people, and obeyed their voice. 25 Now therefore, I pray thee, pardon my sin, and turn again with me, that I may worship the LORD. 26 And Samuel said unto Saul, I will not return with thee: for thou hast rejected the word of the LORD, and the LORD hath rejected thee from being king over Israel. 27 And as Samuel turned about to go away, he laid hold upon the skirt of his mantle, and it rent. 28 And Samuel said unto him, The LORD hath rent the kingdom of Israel from thee this day, and hath given it to a neighbour of thine, *that is* better than thou. 29 And also the Strength of Israel will not lie nor repent: for he *is* not a man, that he should repent. 30 Then he said, I have sinned: yet honour me now, I pray thee, before the elders of my people, and before Israel, and turn again with me, that I may worship the LORD thy God. 31 So Samuel turned again after Saul; and Saul worshipped the LORD. 32 Then said Samuel, Bring ye hither to me Agag the king of the Amalekites. And Agag came unto him delicately. And Agag said, Surely the bitterness of death is past. 33 And Samuel said, As thy sword hath made women childless, so shall thy mother be childless among women. And Samuel hewed Agag in pieces before the LORD in Gilgal. 34 Then Samuel went to Ramah; and Saul went up to his house to Gibeah of Saul. 35 And Samuel came no more to see Saul until the day of his death: nevertheless Samuel mourned for Saul: and the LORD repented that he had made Saul king over Israel."

The Jewish God of the Old Testament preferred genocidal extermination to mercy and tolerance, as revealed in *Joshua* 10:34-42,

"And from Lachish Joshua passed unto Eglon, and all Israel with him; and they encamped against it, and fought against it: 35 And they took it on that day, and smote it with the edge of the sword, and all the souls that *were* therein he utterly destroyed that day, according to all that he had done to Lachish. 36 And Joshua went up from Eglon, and all Israel with him, unto Hebron; and they fought against it: 37 And they took it, and smote it with the edge of the sword, and the king thereof, and all the cities thereof, and all the souls that *were* therein; he left none remaining, according to all that he had done to Eglon; but destroyed it utterly, and all the souls that *were* therein. 38 And Joshua returned, and all Israel with him, to Debir; and fought against it: 39 And he took it, and the king thereof, and all the cities thereof; and they smote them with the edge of the sword, and utterly destroyed all the souls that *were* therein; he left none remaining: as he had done to Hebron, so he did to Debir, and to the king thereof; as he had done also to Libnah, and to her king. 40 So Joshua smote all the country of the hills, and of the south, and of the vale, and of the springs, and all their kings: he left none remaining, but utterly destroyed all that breathed, as the LORD

God of Israel commanded. 41 And Joshua smote them from Kadesh-barnea even unto Gaza, and all the country of Goshen, even unto Gibeon. 42 And all these kings and their land did Joshua take at one time, because the LORD God of Israel fought for Israel."

Deuteronomy 3:4-7; 7:2, 16-18; 20:16; 26:19; and 28:9 state:

"And we took all his cities at that time, there was not a city which we took not from them, threescore cities, all the region of Argob, the kingdom of Og in Bashan. All these cities *were* fenced *with* high walls, gates, and bars; beside unwalled towns a great many. And we utterly destroyed them, as we did unto Sihon king of Heshbon, utterly destroying the men, women, and children, of every city. But all the cattle, and the spoil of the cities, we took for a prey to ourselves. [***] And when the LORD thy God shall deliver them before thee; thou shalt smite them, *and* utterly destroy them; thou shalt make no covenant with them, nor shew mercy unto them: [***] And thou shalt consume all the people which the LORD thy God shall deliver thee; thine eye shall have no pity upon them: neither shalt thou serve their gods; for that *will be* a snare unto thee. [***] But of the cities of these people, which the LORD thy God doth give thee *for* an inheritance, thou shalt save alive nothing that breatheth: [***] And to make thee high above all nations which he hath made, in praise, and in name, and in honour; and that thou mayest be an holy people unto the LORD thy God, as he hath spoken. [***] The LORD shall establish thee an holy people unto himself, as he hath sworn unto thee, if thou shalt keep the commandments of the LORD thy God, and walk in his ways."

Numbers 21:3, 35; and 31:1-18 state:

"3 And the LORD hearkened to the voice of Israel, and delivered up the Canaanites; and they utterly destroyed them and their cities: and he called the name of the place Hormah. [***] 35 So they smote him, and his sons, and all his people, until there was none left him alive: and they possessed his land. [***] And the LORD spake unto Moses, saying, 2 Avenge the children of Israel of the Midianites: afterward shalt thou be gathered unto thy people. 3 And Moses spake unto the people, saying, Arm some of yourselves unto the war, and let them go against the Midianites, and avenge the LORD of Midian. 4 Of every tribe a thousand, throughout all the tribes of Israel, shall ye send to the war. 5 So there were delivered out of the thousands of Israel, a thousand of *every* tribe, twelve thousand armed for war. 6 And Moses sent them to the war, a thousand of *every* tribe, them and Phinehas the son of Eleazar the priest, to the war, with the holy instruments, and the trumpets to blow in his hand. 7 And they warred against the Midianites, as the LORD commanded Moses; and they slew all the males. 8 And they slew the kings of Midian, beside the rest of them that were slain; *namely*, Evi, and Rekem, and Zur, and Hur, and Reba, five kings of Midian:

Balaam also the son of Beor they slew with the sword. 9 And the children of Israel took *all* the women of Midian captives, and their little ones, and took the spoil of all their cattle, and all their flocks, and all their goods. 10 And they burnt all their cities wherein they dwelt, and all their goodly castles, with fire. 11 And they took all the spoil, and all the prey, *both* of men and of beasts. 12 And they brought the captives, and the prey, and the spoil, unto Moses, and Eleazar the priest, and unto the congregation of the children of Israel, unto the camp at the plains of Moab, which are by Jordan *near* Jericho. 13 And Moses, and Eleazar the priest, and all the princes of the congregation, went forth to meet them without the camp. 14 And Moses was wroth with the officers of the host, *with* the captains over thousands, and captains over hundreds, which came from the battle. 15 And Moses said unto them, Have ye saved all the women alive? 16 Behold, these caused the children of Israel, through the counsel of Balaam, to commit trespass against the LORD in the matter of Peor, and there was a plague among the congregation of the LORD. 17 Now therefore kill every male among the little ones, and kill every woman that hath known man by lying with him. 18 But all the women children, that have not known a man by lying with him, keep alive for yourselves."

See also: The Book of Jubilees 32:17-20.

In Jewish mythology, the Messiah of the Jews will destroy the nations, destroy all the religion of the Gentiles, enslave the Gentiles and then exterminate them. It is very important to remember that the Messiah of genocidal Judaism is not the gentle healer of the sick, and willing victim of the powerful, whom we call Jesus of Nazareth. The Messiah of genocidal Judaism is a demonic figure who will lay the Gentiles to waste—he is worse than those who were promoted in the press of their day as messiahs—worse than Napoleon, worse than Marx, worse than Hitler, worse even than Stalin. *Psalm* 2:1-12 (*see also: Sukkah* 52*a-b*) states:

"Why do the heathen rage, and the people imagine a vain thing? 2 The kings of the earth set themselves, and the rulers take counsel together, against the LORD, and against his anointed, *saying*, 3 Let us break their bands asunder, and cast away their cords from us. 4 He that sitteth in the heavens shall laugh: the Lord shall have them in derision. 5 Then shall he speak unto them in his wrath, and vex them in his sore displeasure. 6 Yet have I set my king upon my holy hill of Zion. 7 I will declare the decree: the LORD hath said unto me, Thou *art* my Son; this day have I begotten thee. 8 Ask of me, and I shall give *thee* the heathen *for* thine inheritance, and the uttermost parts of the earth *for* thy possession. 9 Thou shalt break them with a rod of iron; thou shalt dash them in pieces like a potter's vessel. 10 Be wise now therefore, O ye kings: be instructed, ye judges of the earth. 11 Serve the LORD with fear, and rejoice with trembling. 12 Kiss the Son, lest he be angry, and ye perish *from* the way, when his wrath is kindled but a little. Blessed *are* all they that put their trust in him."

Psalm 110:1-7 states,

> "The LORD said unto my Lord, Sit thou at my right hand, until I make thine enemies thy footstool. 2 The LORD shall send the rod of thy strength out of Zion: rule thou in the midst of thine enemies. 3 Thy people *shall be* willing in the day of thy power, in the beauties of holiness from the womb of the morning: thou hast the dew of thy youth. 4 The LORD hath sworn, and will not repent, Thou *art* a priest for ever after the order of Melchizedek. 5 The Lord at thy right hand shall strike through kings in the day of his wrath. 6 He shall judge among the heathen, he shall fill *the places with* the dead bodies; he shall wound the heads over many countries. 7 He shall drink of the brook in the way: therefore shall he lift up the head."

The Jews scoffed at that idea that Jesus should have been the Messiah of the Jews, because Jesus did not commit genocide against the Gentiles with an iron scepter as was prophesied (*Numbers* 24:17-20. *Psalm* 2:9). Jesus was humble, not a demonic and wealthy king who destroyed the nations, enslaved the Gentiles and then murdered them, as some sects of Judaism design and desire to this day.

Israel is today a nation. The Jewish religion, as practiced by some, calls for the extermination of the seed of Amalek. This meant to some Jews the sterilization of Germans, assimilationists, criminals, etc.; to others the planned effects of "race-mixing", which would dilute and weaken the seed of Amalek; to others, it has meant the obliteration of Islamic Nations.[484] There have been allegations that Israel is developing genetically targeted biological weapons. Israel is heavily armed with nuclear, biological and chemical weapons. *The Sunday Times* of London reported, among other things, on 15 November 1998, in an article by Uzi Mahnaimi and Marie Colvin entitled "Israel Planning 'Ethnic' Bomb as Saddam Caves in / Pentagon Warns Over 'Ethno Bomb'", on pages 1 and 2,

> "In developing their 'ethno bomb', Israeli scientists are trying to exploit medical advances by identifying genes carried by some Arabs, then create a genetically modified bacterium or virus. [***] The programme is based at the biological institute in Nes Tziyona, the main research facility for Israel's clandestine arsenal of chemical and biological weapons."

Israel plans to destroy all human life on Earth, if its Messianic goals are not fulfilled. The Israeli government, which represents only a few million persons, has prepared a doom's day device called the "Samson Option", which will detonate enough nuclear devices to kill off all of humanity. They plan to use it if the State of Israel fails.[485] Judaism calls on the "righteous"—fanatically

---

484. Y. Harkabi, *Israel's Fateful Hour*, Harper & Row, New York, (1988), pp. 149-150.
485. S. M. Hersh, *The Samson Option: Israel's Nuclear Arsenal and American Foreign Policy*, Random House, New York, (1991).

religious Jews—to mass murder the rest of humanity in the Messianic Era.[486] *Deuteronomy* 32:9, states,

"For the LORD's portion *is* his people; Jacob *is* the lot of his inheritance."

The criminal Israeli cult of assassination and espionage, the Mossad, wages war on the rest of the world. The Mossad's motto is, "By way of deception, thou shalt do war."[487]

The ultimate purpose of the racist Jews' war on humanity is ultimately to leave no one left alive but "righteous" Jews.[488] All Gentiles are destined to be killed. All assimilated Jews are destined to be killed. Michael Higger wrote in his book *The Jewish Utopia*,

"First, no line will be drawn between bad Jews and bad non-Jews. There will be no room for the unrighteous, whether Jewish or non-Jewish, in the Kingdom of God. All of them will have disappeared before the advent of the ideal era on this earth.[84] Unrighteous Israelites will be punished equally with the wicked of other nations.[85] [***] In general, the peoples of the world will be divided into two main groups, the Israelitic and the non-Israelitic. The former will be righteous; they will live in accordance with the wishes of one, universal God; they will be thirsty for knowledge, and willing, even to the point of martyrdom, to spread ethical truths to the world. All the other peoples, on the other hand, will be known for their detestable practices, idolatry, and similar acts of wickedness. They will be destroyed and will disappear from earth before the ushering in of the ideal era.[218] All these unrighteous nations will be called to judgment, before they are punished and doomed. The severe sentence of their doom will be pronounced upon them only after they have been given a fair trial, when it will have become evident that their existence would hinder the advent of the ideal era.[219] Thus, at the coming of the Messiah, when all righteous nations will pay homage to the ideal righteous leader, and offer gifts to him, the wicked and corrupt nations, by realizing the approach of their doom, will bring similar presents to the Messiah. Their gifts and pretended acknowledgment of the new era, will be bluntly rejected.[220] For the really wicked nations, like the wicked

---

[486]. *Exodus* 34:11-17. *Psalm* 2; 72. *Isaiah* 1:9; 2:1-4; 6:9-13; 9:6-7; 10:20-22; 11:4, 9-12; 17:6; 37:31-33; 41:9; 42; 43; 44; 61:6. *Jeremiah* 3:17; 33:15-16. *Ezekiel* 20:38; 25:14. *Daniel* 12:1, 10. *Amos* 9:8-10. *Obadiah* 1:18. *Micah* 4:2-3; 5:8. *Zechariah* 8:20-23; 14:9. *Romans* 9:27-28; 11:1-5.

[487]. V. Ostrovsky and C. Hoy, *By Way of Deception*, St. Martin's Paperbacks, New York, (1990), p. 53. **See also:** V. Ostrovsky, *The Other Side of Deception: A Rogue Agent Exposes the Mossad's Secret Agenda*, Harper Paperbacks, New York, (1994).

[488]. *Exodus* 34:11-17. *Psalm* 2; 72. *Isaiah* 1:9; 2:1-4; 6:9-13; 9:6-7; 10:20-22; 11:4, 9-12; 17:6; 37:31-33; 41:9; 42; 43; 44; 61:6. *Jeremiah* 3:17; 33:15-16. *Ezekiel* 20:38; 25:14. *Daniel* 12:1, 10. *Amos* 9:8-10. *Obadiah* 1:18. *Micah* 4:2-3; 5:8. *Zechariah* 8:20-23; 14:9. *Romans* 9:27-28; 11:1-5.

individuals, must disappear from earth before an ideal human society of righteous nations can be established. No ideal era of mankind can be established as long as there are peoples living idolatrous, ungodly lives ; as long as there are oppressors of the righteous, friends of slavery, enemies of freedom and liberty, and defiant enemies of God.^{221}[***] Moreover, rabbinic sources, in speaking of Israel's fate in the ideal era, ascribe Israel's spiritual victory in the future to the fact that righteousness will be victorious over wickedness, and that the upright and just will succeed in bringing about the disappearance of the unrighteous from the earth.^{226} [***] Consequently, before the Kingdom of God will be established, a number of important reforms and changes will take place. Idolatry and idol worshippers, wicked people, unrighteous nations will disappear from the earth.^{230}"[489]

It should be noted that Higger asserts that Gentiles will first be offered an opportunity to join the "righteous Jews", but those whom the Jewish Messiah rejects will be mass murdered in a broad genocide. What is to prevent the Jewish Messiah, a political Jewish King, not a divine being, from merely pronouncing all Gentiles "unrighteous" as is the case in the Hebrew Bible? What is "righteous" about genocide? Why do religious disagreements give the "righteous Jews" the right to slaughter their Catholic, Buddhist, Hindu, and assimilated Jewish neighbors?

Tom Segev quoted Ehud Praver in Segev's book, *The Seventh Million: The Israelis and the Holocaust*,

"'In the wake of Kahane, we heard more and more about soldiers who, exposed to the history of the Holocaust, were planning all sorts of ways to exterminate the Arabs,' recalled education-corps officer Ehud Praver. 'It concerned us very much, because we saw that the Holocaust was legitimizing the appearance of Jewish racism. We learned that it was necessary to deal not only with the Holocaust but also with the rise of fascism and to explain what racism is and what dangers it holds for democracy.' According to Praver, 'too many soldiers were deducing that the Holocaust justifies every kind of disgraceful action.'"[490]

Jewish hatred of the Gentiles spans across history. The *Zohar*, I, 28*b*, states,

"One kind is from the side of the serpent; another from the side of the Gentiles, who are compared to the beasts of the field[.]"[491]

---

489. M. Higger, *The Jewish Utopia*, Lord Baltimore Press, Baltimore, (1932), pp. 20, 37-39.
490. T. Segev, *The Seventh Million: The Israelis and the Holocaust*, Hill and Wang, New York, (1993), p. 407.
491. H. Sperling and M. Simon, *The Zohar*, Volume 1, The Soncino Press, New York, (1933), p. 110.

We also find the racist Jews Isaac Luria, Nachman of Bratslav and Shneur Zalman degrading Gentiles as if sub-human. Shneur Zalman believed that,

"Gentile souls are of a completely different and inferior order. They are totally evil, with no redeeming qualities whatsoever. Consequently, references to gentiles in Rabbi Shneur Zalman's teachings are invariably invidious.... Their material abundance derives from supernal refuse. Indeed, they themselves derive from refuse, which is why they are more numerous than the Jews, as the pieces of chaff outnumber the kernels.... All Jews were innately good, all gentiles innately evil. Jews were the pinnacle of creation and served the Creator, gentiles its nadir and worshiped the heavenly hosts."[492]

Rabbi Abraham Isaac Kook wrote in the Twentieth Century that,

"the difference between the Israelite soul... and the souls of all non-Jews, no matter what their level, is bigger and deeper than the difference between the human soul and the animal soul."[493]

An alternative translation,

"The difference between a Jewish soul and souls of non-Jews—all of them in all different levels—is greater and deeper than the difference between a human soul and the souls of cattle."[494]

The Jewish Encyclopedia wrote in its article "Gentile",

"According to Hananiah b. Akabia the word והער (Ex. xxi. 14) may perhaps exclude the Gentile; but the shedding of the blood of non-Israelites, while not cognizable by human courts, will be punished by the heavenly tribunal (Mek., Mishpatim, 80b). [***] Another reason for discrimination [against Gentiles] was the vile and vicious character of the Gentiles: 'I will provoke them to anger with a foolish nation' (לכנ = 'vile,' 'contemptible'; Deut. xxxii. 21). The Talmud says that the passage refers to the Gentiles of Barbary and Mauretania, who walked nude in the streets (Yeb. 63b), and to similar Gentiles, 'whose flesh is as the flesh of asses and whose issue is like the issue of horses' (Ezek. xxiii. 20); who can not claim a father (Yeb. 98a).

---

[492]. *From:* A. Nadler, "Last Exit to Brooklyn: The Lubavitcher's Powerful and Preposterous Messianism", *The New Republic*, (4 May 1992), pp. 27-35, at 33. Nadler appears to quote from: R. A. Foxbrunner, *Habad: The Hasidism of R. Shneur Zalman of Lyady*, University of Alabama Press, Tuscaloosa, Alabama, (1992).
[493]. Y. Sheleg, "A dark reminder of the Dark Ages", *Haaretz.com*, (28 June 2005).
[494]. A. C. Brownfeld's review of I. Shahak and N. Mezvinsky's book *Jewish Fundamentalism in Israel*, Pluto Press, London, (1999); in *The Washington Report on Middle East Affairs*, Volume 19, Number 2, (March, 2000), pp. 105-106, at 105.

The Gentiles were so strongly suspected of unnatural crimes that it was necessary to prohibit the stabling of a cow in their stalls ('Ab. Zarah ii. 1). Assaults on women were most frequent, especially at invasions and after sieges (Ket. 3b), the Rabbis declaring that in case of rape by a Gentile the issue should not be allowed to affect a Jewish woman's relation to her husband. 'The Torah outlawed the issue of a Gentile as that of a beast' (Mik. viii. 4, referring to Ezek. *l. c.*)."[495]

Albert Einstein's friend Georg Friedrich Nicolai (Lewinstein) stated in 1917,

"Apart from this strange story of Cain, however, murder is forbidden in the Bible, and very sternly forbidden. But—it is only the murder of Jews. As is natural, considering the period from which it dates, the Bible is absolutely national, in character. Only the Jew is really considered as a human being; cattle and strangers might be slain without the slayer himself being slain. In this case there was a ransom. Accordingly, war was of course allowed also, and the Jews were no more illogical than the Moslem who kills the outlander. Of late years the Jews and the Old Testament have often been reproached for their contempt for those who were not Jews; and in practice even Christ acted in precisely the same way."[496]

In an article "Begin and the 'Beasts'", *New Statesman*, Volume 103, Number 2674, (25 June 1982), page 12, Amnon Kapeliuk wrote of Menachem Begin, the Prime Minister of Israel,

"The war in Lebanon cannot be interpreted, even by its most devoted proponents in Israel, as a war of survival. For this reason, the government has gone to extraordinary lengths to dehumanise the Palestinians. Begin described them in a speech in the Knesset as 'beasts walking on two legs'. Palestinians have often been called 'bugs' while their refugee camps in Lebanon are referred to as 'tourist camps'. In order to rationalise the bombing of civilian populations, Begin emotively declared: 'If Hitler was sitting in a house with 20 other people, would it be correct to blow up the house?'"[497]

In a "Letter to the Editor", signed by Isidore Abramowitz, Hannah Arendt, Abraham Brick, Rabbi Jessurun Cardozo, Albert Einstein, Herman Eisen, M. D.,

---

495. "Gentile", *The Jewish Encyclopedia*, Funk and Wagnalls Company, New York, (1903), pp. 615-626, at 618 and 621.
496. G. Nicolai, *Die Biologie des Krieges, Betrachtungen eines deutschen Naturforschers*, O. Füssli, Zürich, (1917); English translation: *The Biology of War*, Century Co., New York, (1918), p. 531.
497. *See also:* J. Kuttab, "West Bank Arabs Foresee Expulsion", *The New York Times*, (1 August 1983), p. A15.

Hayim Fineman, M. Gallen, M. D., H. H. Harris, Zelig S. Harris, Sidney Hook, Fred Karush, Bruria Kaufman, Irma L. Lindheim, Nachman Majsel, Seymour Melman, Myer D. Mendelson, M. D., Harry M. Orlinsky, Samuel Pitlick, Fritz Rohrlich, Louis P. Rocker, Ruth Sager, Itzhak Sankowsky, I. J. Schoenberg, Samuel Shuman, M. Znger, Irma Wolpe, Stefan Wolpe; dated "New York. Dec. 2, 1948."; published as: "New Palestine Party; Visit of Menachen Begin and Aims of Political Movement Discussed", *The New York Times*, (4 December 1948), p. 12; it states, *inter alia*,

> "Among the most disturbing political phenomena of our time is the emergence in the newly created state of Israel of the 'Freedom Party' (Tnuat Haherut), a political party closely akin in its organization, methods, political philosophy and social appeal to the Nazi and Fascist parties. It was formed out of the membership and following of the former Irgun Zvai Leumi, a terrorist, right-wing, chauvinist organization in Palestine. The current visit of Menachen Begin, leader of this party, to the United States is obviously calculated to give the impression of American support for his party in the coming Israeli elections, and to cement political ties with conservative Zionist elements in the United States. [***] The Deir Yassin incident exemplifies the character and actions of the Freedom Party. Within the Jewish community they have preached an admixture of ultranationalism, religious mysticism, and racial superiority. Like other Fascist parties they have been used to break strikes, and have themselves pressed for the destruction of free trade unions. In their stead they have proposed corporate unions on the Italian Fascist model. [***] This is the unmistakable stamp of a Fascist party for whom terrorism (against Jews, Arabs, and British alike), and misrepresentation are means, and a 'Leader State' is the goal."

Racist Zionist Moses Hess declared that Germans are the genetic enemies of Israel in 1862 (contrast Hess' views with Goldhagen's negative analysis of Germans under Hitler[498] and see Hartmut Stern's response to Goldhagen[499]). Moses Hess' statement must be seen in the context of Jacob and Esau, and Isaac's "blessing" to Esau that Esau should be the servant and the sword of

---

[498]. D. J. Goldhagen, *Hitler's Willing Executioners: Ordinary Germans and the Holocaust*, Knopf, New York, (1996). *See also:* É. Durkheim, *"Germany above All" The German Mental Attitude and the War*, Librairie Armand Colin, Paris, (1915). *See also:* "By a German", *I Accuse! (J'Accuse!)*, Grosset & Dunlap, New York, (1915). *See also:* W. F. Barry, *The World's Debate: An Historical Defence of the Allies*, George H. Doran, New York, (1917). *See also:* W. T. Hornaday, *A Searchlight on Germany: Germany's Blunders, Crimes and Punishment*, American Defense Society, New York, (1917). *See also:* D. W. Johnson, *Plain Words from America: A Letter to a German Professor*, London, New York, Toronto, Hodder & Stoughton, (1917).

[499]. H. Stern, *KZ-Lügen: Antwort auf Goldhagen*, FZ-Verlag, München, Second Edition, (1998), ISBN: 3924309361; **and** *Jüdische Kriegserklärungen an Deutschland: Wortlaut, Vorgeschichte, Folgen*, FZ-Verlag, München, Second Edition, (2000), ISBN: 3924309507.

Jacob, of Israel. *Genesis* 25:23 states,

> "And the LORD said unto her, Two nations *are* in thy womb, and two manner of people shall be separated from thy bowels; and *the one* people shall be stronger than *the other* people; and the elder shall serve the younger."

Genesis 27:38-41 states,

> "38 And Esau said unto his father, Hast thou but one blessing, my father? bless me, *even* me also, O my father. And Esau lifted up his voice, and wept. 39 And Isaac his father answered and said unto him, Behold, thy dwelling shall be the fatness of the earth, and of the dew of heaven from above; 40 And by thy sword shalt thou live, and shalt serve thy brother; and it shall come to pass when thou shalt have the dominion, that thou shalt break his yoke from off thy neck. 41¶ And Esau hated Jacob because of the blessing wherewith his father blessed him: and Esau said in his heart, The days of mourning for my father are at hand; then will I slay my brother Jacob."

Hess may have envisioned the annihilation of the German "race"—referred to by some Jews as the people of the sword. It was clearly better for the Jews to kill off Esau before his descendants "broke his yoke from off his neck" than to let them live and potentially seek revenge on the Jews. Hess' book told his fellow Jews that Germans were the seed of Amalek and must be exterminated. At least as early as the 1860's, Moses Hess argued that the "German race" had a genetically programmed antagonism towards the "Jewish race"—the implication being that one must destroy the other in order to survive. In the Jewish mythology, this confrontation called for the extermination of the Germans. Two World Wars nearly accomplished the destruction of Germany and ended their prominence in world affairs. Two World Wars killed off many of the strongest, smartest and most assertive Germans. Hess wrote in 1862,

> "It seems that German education is not compatible with our Jewish national aspirations. Had I not once lived in France, it would never have entered my mind to interest myself with the revival of Jewish nationality. Our views and strivings are determined by the social environment which surrounds us. Every Living, acting people, like every active individual, has its special field. Indeed, every man, every member of the historical nations, is a political, or as we say at present, a social animal; yet within this sphere of the common social world, there are special places reserved by Nature for individuals according to their particular calling. The specialty of the German of the higher class, of course, is his interest in abstract thought; and because he is too much of a universal philosopher, it is difficult for him to be inspired by national tendencies. 'Its whole tendency,' my former publisher, Otto Wigand, once wrote to me, when I showed him an outline of a work on Jewish national aspirations, 'is contrary to my pure human nature.'

The 'pure human nature' of the Germans is, in reality, the character of the pure German race, which rises to the conception of humanity in theory only, but in practice it has not succeeded in overcoming the natural sympathies and antipathies of the race. German antagonism to Jewish national aspiration has a double origin, though the motives are really contrary to each other. The duplicity and contrariety of the human personality, such as we can see in the union of the spiritual and the natural, the theoretical and the practical sides, are in no other nation so sharply marked in their points of opposition as in the German. Jewish national aspirations are antagonistic to the theoretical cosmopolitan tendencies of the German. But in addition to this, the German opposes Jewish national aspirations because of his racial antipathy, from which even the noblest Germans have not as yet emancipated themselves. The publisher, whose 'pure human' conscience revolted against publishing a book advocating the revival of Jewish nationality, published books preaching hatred to Jews and Judaism without the slightest remorse, in spite of the fact that the motive of such works is essentially opposed to the 'pure human conscience.' This contradictory action was due to inborn racial antagonism to the Jews. But the German, it seems, has no clear conception of his racial prejudices; he sees in his egoistic as well as in his spiritual endeavors, not German or Teutonic, but 'humanitarian tendencies'; and he does not know that he follows the latter only in theory, while in practice he clings to his egoistic ideas.

[***]

In 1858, there appeared, at Leipzig, a work written by Otto Wigand under the title *Two discourses concerning the desertion from Judaism*, being an analysis of the views on this question expressed in the recently published correspondence of Dr. Abraham Geiger. The author endeavors to prove that the conclusions of Dr. Geiger are untenable both from a philosophic and from a social standpoint. Here are his social arguments:

'My friend,' says the author, 'there are certain conclusions which you cannot escape. The stamp of slavery, if we may use this expression, which centuries of oppression have deeply impressed upon the Jewish features, might have been obliterated by the blessed hand of regained civil liberty. The gait of the Jews, buoyed up by the happy reminiscences of the victory won in the struggle for the noble possession of liberty, might have been straighter and prouder. The Jewish face may certainly beam with pride, as it views the tremendous progress made by the Jews in a brief time, their mighty flight to the spiritual height upon which they now stand, which is especially notable considering the fact that their poets and writers at whose greatness the nation is astonished, and of whose talents the entire people takes account, have sprung from those who, a generation ago, could hardly converse correctly in the language of the land. Such a state of affairs should undoubtedly call forth admiration in the hearts of the present German generation, and yet, in spite of these achievements, the wall separating Jew and Christian still stands unshattered, for the watchman that guards them is one who will not be caught napping. It is the race difference between the Jewish and Christian populations. If this assertion of mine surprises or astonishes you, I ask you to consider whether it is not almost a rule with the Germans that race differences generate prejudices which cannot be overcome by any manifestation of good-will on the part of the other race. The relations existing between the German and the Slavic populations in Bohemia, in Hungary and Transylvania, between the Germans and the Danes in Schleswig, or between the Irish and the Anglo-Saxon settlers in Ireland, illustrates well the power of race antagonism in the German world. In all these countries the different elements of the population have lived side by side for centuries, sharing equally all political rights, and yet, so strong are the national or racial differences, that a social amalgamation of the various elements of the population is even at the present day quite unthinkable. And what comparison is there between the race differences of a German and Slav, a Celt and Anglo-Saxon, or a German and Dane, and the race antagonism between the children of the Sons of Jacob, who are of Asiatic descent, and the descendants of Teut and Herman, the ancestors of whom have inhabited Europe from time immemorial; between the proud and the tall blond German and the small of figure, black-haired and black-eyed Jew? Races which differ in such a degree oppose each other instinctively and against such opposition reason and good sense are powerless.'

These expressions are certainly frank and sincere in their meaning, though they by no means prove the conclusions to which the author wishes to arrive, namely, the desirability of conversion; for conversion will not turn a Jew into a German. But they at least contain the confession, that an instinctive race antagonism triumphs in Germany above all humanitarian sentiments. The 'pure human nature' resolves itself, according to the Germans, in the nature of pure Germanism. The 'high-born blond race' looks with contempt upon the regeneration of the 'black-haired, quick-moving mannikins,' without regard to whether they are descendants of the Biblical patriarchs, or of the ancient Romans and Gauls.

While other civilized western nations mention the shameful oppression to which the Jews were formerly subjected, only as an act of theirs of which they are ashamed, the German remembers only the 'stamp of slavery' which he impressed upon 'the Jewish physiognomy.'

In a *feuilleton* which appeared recently in the *Bonnerzeitung*, entitled 'Bonn Eighty Years Ago,' the author speaks of the Jews in mocking terms and describes them as people who lived in separate quarters and supported themselves by petty trades. I believe that we should wonder less at the fact that the Jews, who were forbidden to participate in the important branches of industry and commerce, lived on petty trade, than at the fact that they were able to live at all in those centuries of oppression. As a matter of fact, almost every means of existence, including the right of domicile, was denied them. It was only by means of bribes that every Jewish generation could procure anew the 'privilege' not to be driven out of their homes in Bonn, and they felt happy indeed if, in spite of the contract, they were not robbed of their property and exiled, or attacked by a fanatical mob in the bargain. I, also, can tell a story of 'eighty years ago.' A Jew won the high favor of the Kurfuerst of Bonn, that he and his descendants were granted the 'privilege' to settle in Ebendich.

[***]

Gabriel Riesser, the editor of the magazine, *The Jew*, as far as I can recollect, never fell into the error, common to all modern German Jews, that the emancipation of the Jews is irreconcilable with the development of Jewish Nationalism. He demanded emancipation for the Jews on the one condition only, that of their receiving all civil and political rights in return for their assuming all civil and political burdens."[500]

Jewish financiers including Jacob Schiff brought about the downfall of Russia in the name of saving the Children of Israel from Edom. England, France, Germany, Turkey and Russia caused each other great harm, but their wars resulted in the emancipation of the Jews, a reduction in the power of the Roman Catholic Church, and, ultimately, in the formation of the State of Israel. *Micah* 5:8 states,

"And the remnant of Jacob shall be among the Gentiles in the midst of many people as a lion among the beasts of the forest, as a young lion among the flocks of sheep: who, if he go through, both treadeth down, and teareth in pieces, and none can deliver."

The *Zohar* I, 25a-25b, states that peoples other than the Jews will be exterminated when the Jews form a state in Palestine,

"But as '*tohu* and *bohu*' gave place to light, so when God reveals Himself they will be wiped off the earth. But withal redemption will not be complete until Amalek will be exterminated, for against Amalek the oath was taken that 'the Lord will have war against Amalek from generation to generation' (Ex. XVII, 16)."[501]

Amalek and Esau are seen as the genetic and reincarnate spirit of Cain who slew Abel. I *Enoch* 22:7, states that the spirit of Abel prays for the extermination of the seed of Cain:

"And he answered and said to me, saying: 'That is the spirit that proceeded

---

[500]. M. Hess, *Rom und Jerusalem: die letzte Nationalitätsfrage*, Eduard Wengler, Leipzig, (1862); English: "Fourth Letter", "Note III" and "Note IV", *Rome and Jerusalem: A Study in Jewish Nationalism*, Bloch, New York, (1918), pp. 56-57, 240-244.

[501]. H. Sperling and M. Simon, *The Zohar*, Volume 1, The Soncino Press, New York, (1933), p. 100.

from Abel, whom his brother Cain slew; and it laments on his account till his seed is destroyed from the face of the earth and his seed disappear from among the seed of men.'"[502]

Genesis 3:14-15 implies that Eve and the serpent which tempted Eve had a son, Cain who slew Abel. *Yebamoth* 103*b* states that serpent infused Eve with lust when they copulated. The Jews were supposedly cleansed of this lust infused into Eve by the serpent, on Mount Sinai (*Abodah Zarah* 22*b*. *Shabbath* 145*b*-146*a*). *Yebamoth* 63*a* states that Adam had intercourse with all animals and beasts, but only derived satisfaction from Eve. Voltaire ridiculed Judaism and Jews for their laws against sexual relations with animals, which laws Voltaire alleged indicate that the practice of bestiality was common among ancient Jews, for otherwise Jews would have required no laws proscribing bestiality.[503]

It is significant that Enoch is given two different lineages in *Genesis* and that Cain was a farmer, while Abel was a shepherd. God (like Isaac) preferred Abel's (Esau's) offering of flesh to Cain's (Jacob's) offering of fruit (*Genesis* 4). This relates Cain to Jacob and Abel to Esau. Cain, the first murderer, might be said to have been the first "wandering Jew" and his descendants were city dwellers. *Genesis* 3:14-15 states,

> "And the LORD God said unto the serpent, Because thou hast done this, thou *art* cursed above all cattle, and above every beast of the field; upon thy belly shalt thou go, and dust shalt thou eat all the days of thy life: And I will put enmity between thee and the woman, and between thy seed and her seed; it shall bruise thy head, and thou shalt bruise his heel."

Certain Cabalists believe that Jews descend from Cain.[504]

4.4.5 Jewish Dualism and Human Sacrifice—Evil is Good

Ironically and paradoxically, a major part of the Jewish religious plan is the objective of making Jews irreligious. Jewish leaders believe that the prophets commanded that Jews fall away from God and that they are duty bound to see to it that two thirds of Jews perish as a result (*Isaiah* 48:10. *Ezekiel* 5:12. *Zechariah* 13:8-9). They believe that the Messiah will only come when Jews have embraced heresy and have made the world evil (*Sanhedrin* 97*a*).

The Dualism implicit in the stories of Adam and Eve, Cain and Abel, and

---

[502]. G. H. Schodde, *The Book of Enoch: Translated from the Ethiopic, with Introduction and Notes*, Warren F. Draper, Andover, (1882), p. 98. Note that *Genesis* 4:17 gives a different lineage for Enoch than *Genesis* 5:18-24, and that in the former, Enoch is the son of Cain!
[503]. Voltaire in English translation in R. S. Levy, *Antisemitism in the Modern World: An Anthology of Texts*, D.C. Heath, Toronto, (1991), p. 46.
[504]. M. A. Hoffman II, *Judaism's Strange Gods*, Independent History and Research, Coeur d'Alene, Idaho, (2000), pp. 110-111.

Jacob and Esau, has been interpreted in Marcionistic and Gnostic terms as the blessings and curses of two distinct gods. There is the good spiritual god who brought us Jesus, and the evil Creator god who created the corpse of the flesh in which divine spirits are trapped—the lesser creator God of the Old Testament. Jewish Dualism is apparent in the Old and New Testament *Logos*, mistranslated as divine "Word", which word in fact signifies the dialectic and Dualistic principles of Heraclitus and Plato—the dialectic of good and evil, light and darkness, flesh and spirit, which is the eternal flame of the Universe.

These Dualistic mythologies have been put to great political effect over the centuries and are intentionally confused to bewilder the uninitiated into believing that all Jews worship the Devil; or, alternatively, that Catholics worship the Devil and that the Pope is the anti-Christ; or, alternatively, that all Dualist sects actually worship the Devil alone; etc.

However, it is true that Jewish Dualism teaches Jews to view evil as a good thing which originates in God, as do all things. Many Dualistic Jews even see evil as a stronger force for action than good, because they fear evil, but have no fear of good. Many Dualistic Jews view evil as a more powerful force, because they believe it attracts God's attention and causes Him to act. Many Dualistic Jews teach their adherents to commit acts of evil, the worse the better, as a means to summon the Messianic Era.

In many Jewish racist myths, various myths which frequently contradict one another, angels are blamed for bringing evil to mankind and for interbreeding with human females to create, alternatively, depending upon political and religious bias, an evil or a divine race, which race of demigods must be exterminated, or defended (*Genesis* 6:1-5. *Numbers* 13:25-33. I *Enoch*). The Dualism expressed in Jewish writings may have its origin in the Sumerian myths of An, Enlil and Enki. The Biblical legend of evil giants descended from angels may derive from the epic of *Gilgamesh*—as well as in the Greek myths of giants and demi-gods. Jewish Dualism has always been a dangerously racist belief system which defines specific peoples as "elect" and "good", and other peoples as an evil race destined to be exterminated (*Isaiah* 65; 66. *See also: Enoch*).

Judaism is likely a mixed-up sect of the Canaanite religions, incorporating Mesopotamian, Greek and Egyptian myths. Jacob worships the god "El" (*Genesis* 35) and was himself called El (*Tractate Megillah*, Chapter 2). El was a Canaanite god who bore Baal-Hadad, a calf, and is sometimes depicted seated and with the head and horns of a bull. This god was a fertility god. "Baal" has been translated as "Lord" and the Hebrews referred to their God as "Baal". In Canaanite myth, Baal is a mighty storm and in the Bible the word we know of today as "spirit" or "ghost", as in "Holy Ghost", is in fact "wind" in the original languages. From the beginnings of *Genesis* through the New Testament, God is a mighty and wrathful storm, or wind, or "Holy Wind", which we today call "Holy Ghost". This poetic imagery was likely derived from the Canaanite religion. Baal worship, especially the worship of Moloch, involves human sacrifice, in particular, that of burning one's firstborn child—the child who opens the womb, as did Esau. Gentiles are to be human sacrifices to Baal for the sake of Jacob, the Jews.

The Canaanite Baal and El, like the Jews' God, were jealous gods and there was an enmity between them. Perhaps this enmity between gods and tribes is what led Jews—Judeans—into accepting a stubborn and intolerant Egyptian monotheism violently and fanatically opposed to all other religions. The Jews have also had several sects which have worshiped a form of Eleatic Monism. Perhaps, the enmity between Baal and El is the source of the Dualistic beliefs of some Jewish and Christian sects. Perhaps the original authors of Judaism made their God a jealous God because they created their God to protect their racism. God's jealousy is linked to commandments not to intermarry with other peoples, because this would lead the Hebrews to worship foreign gods, but the real underlying motive is the preservation of "racial purity" and the religious mythology was merely a means of controlling people and thereby preserving the "race". The Jewish religion was a survival tactic and a very effective one.

The Jews of Judea knew that peoples could disappear, and that even to conquer another people could lead to intermarriage and the disappearance of one's own people. This is clearly spelled out in *Ezra* 9:

> "Now when these *things* were done, the princes came to me, saying, The people of Israel, and the priests, and the Levites, have not separated themselves from the people of the lands, *doing* according to their abominations, *even* of the Canaanites, the Hittites, the Perizzites, the Jebusites, the Ammonites, the Moabites, the Egyptians, and the Amorites. 2 For they have taken of their daughters for themselves, and for their sons: so that the holy seed have mingled themselves with the people of *those* lands: yea, the hand of the princes and rulers hath been chief in this trespass. 3 And when I heard this thing, I rent my garment and my mantle, and plucked off the hair of my head and of my beard, and sat down astonied. 4 Then were assembled unto me every one that trembled at the words of the God of Israel, because of the transgression of those that had been carried away; and I sat astonied until the evening sacrifice. 5 And at the evening sacrifice I arose up from my heaviness; and having rent my garment and my mantle, I fell upon my knees, and spread out my hands unto the LORD my God. 6 And said, O my God, I am ashamed and blush to lift up my face to thee, my God: for our iniquities are increased over *our* head, and our trespass is grown up unto the heavens. 7 Since the days of our fathers *have* we *been* in a great trespass unto this day; and for our iniquities have we, our kings, *and* our priests, been delivered into the hand of the kings of the lands, to the sword, to captivity, and to a spoil, and to confusion of face, as *it is* this day. 8 And now for a little space grace hath been *shewed* from the LORD our God, to leave us a remnant to escape, and to give us a nail in his holy place, that our God may lighten our eyes, and give us a little reviving in our bondage. 9 For we *were* bondmen; yet our God hath not forsaken us in our bondage, but hath extended mercy unto us in the sight of the kings of Persia, to give us a reviving, to set up the house of our God, and to repair the desolations thereof, and to give us a wall in Judah and in Jerusalem. 10 And now, O our God, what shall we say after this? for we have forsaken thy commandments,

11 Which thou hast commanded by thy servants the prophets, saying, The land, *unto* which ye go to possess it, *is* an unclean land with the filthiness of the people of the lands, with their abominations, which have filled it from one end to another with their uncleanness. 12 Now therefore give not your daughters unto their sons, neither take their daughters unto your sons, nor seek their peace or their wealth for ever: that ye may be strong, and eat the good of the land, and leave *it* for an inheritance to your children for ever. 13 And after all that is come upon us for our evil deeds, and for our great trespass, seeing that thou our God hast punished us less than our iniquities *deserve*, and hast given us *such* deliverance as this; 14 Should we again break thy commandments, and join in affinity with the people of these abominations? wouldest not thou be angry with us till *thou* hadst consumed us, so that *there should be* no remnant nor escaping? 15 O LORD God of Israel, thou *art* righteous: for we remain *yet* escaped, as *it is* this day: behold, we *are* before thee in our trespasses: for *we* cannot stand before thee because of this."

Nehemiah 9:2; 13:3, 23-30 state:

"9:2 And the seed of Israel separated themselves from all strangers, and stood and confessed their sins, and the iniquities of their fathers. [***] 13:3 Now it came to pass, when they had heard the law, that they separated from Israel all the mixed multitude. [***] 13:23¶ In those days also saw I Jews *that* had married wives of Ashdod, of Ammon, *and* of Moab: 13:24 And their children spake half in the speech of Ashdod, and could not speak in the Jews' language, but according to the language of each people. 13:25 And I contended with them, and cursed them, and smote certain of them, and plucked off their hair, and made them swear by God, *saying*, Ye shall not give your daughters unto their sons, nor take their daughters unto your sons, or for yourselves. 13:26 Did not Solomon king of Israel sin by these *things?* yet among many nations was there no king like him, who was beloved of his God, and God made him king over all Israel: *nevertheless* even him did outlandish women cause to sin. 13:27 Shall we then hearken unto you to do all this great evil, to transgress against our God in marrying strange wives? 13:28 And *one* of the sons of Joiada, the son of Eliashib the high priest, *was* son in law to Sanballat the Horonite: therefore I chased him from me. 13:29 Remember them, O my God, because they have defiled the priesthood, and the covenant of the priesthood, and of the Levites. 13:30 Thus cleansed I them from all strangers, and appointed the wards of the priests and the Levites, every one in his business;"
Genesis 28:1, 6 states:

"28:1 And Isaac called Jacob, and blessed him, and charged him, and said unto him, Thou shalt not take a wife of the daughters of Canaan. [***] 28:6 When Esau saw that Isaac had blessed Jacob, and sent him away to Padan-aram, to take him a wife from thence; and that as he blessed him he gave him a charge, saying, Thou shalt not take a wife of the daughters of Canaan;"

Exodus 34:11-17 states (note that Zionist Jews have repeatedly committed such atrocities against Palestinians):

"11 Observe thou that which I command thee *this* day: behold, I drive out before thee the Amorite, and the Canaanite, and the Hittite, and the Perizzite, and the Hivite, and the Jebusite. 12 Take heed to thyself, lest thou make a covenant with the inhabitants of the land whither thou goest, lest it be for a snare in the midst of thee: 13 But ye shall destroy their altars, break their images, and cut down their groves: 14 For thou shalt worship no other god: for the LORD, whose name *is* Jealous, *is* a jealous God: 15 Lest thou make a covenant with the inhabitants of the land, and they go a whoring after their gods, and do sacrifice unto their gods, and *one* call thee, and thou eat of his sacrifice; 16 And thou take of their daughters unto thy sons, and their daughters go a whoring after their gods, and make thy sons go a whoring after their gods. 17 Thou shalt make thee no molten gods."

Leviticus 20:26 states:

"And ye shall be holy unto me: for I the LORD *am* holy, and have severed you from *other* people, that ye should be mine."

Numbers 23:9 states:

"For from the top of the rocks I see him, and from the hills I behold him: lo, the people shall dwell alone, and shall not be reckoned among the nations."

Deuteronomy 7:2-3 states:

"2 And when the LORD thy God shall deliver them before thee; thou shalt smite them, *and* utterly destroy them; thou shalt make no covenant with them, nor show mercy unto them: 3 Neither shalt thou make marriages with them; thy daughter thou shalt not give unto his son, nor his daughter shalt thou take unto thy son."

El was the supreme god of the Canaanites, but Baal ruled the Earth. Baal, a god of fertility, dies and is resurrected each year. From these myths emerged Christianity, which in its earliest incarnation preached that Jesus was the son of a supreme and spiritual God, perhaps "El"; and that Judaism worshiped the earthly and devilish "Covenant of Baal" (*Exodus* 32. *Leviticus* 26:30. *Numbers* 22:41. *Judges* 2:11-14; 3:7; 6:25, 31; 8:33; 9:4; 11:31, 39. I *Kings* 14:22-24; 16:31-33; 18:18-19, 26; 19:10, 14, 18; 22:53. II *Kings* 3:2-3; 8:18, 27; 10:18-28; 11:18; 16:3-4; 17:10, 16-18, 23; 18:4-5; 21:6; 22:5; 23:5, 12, 32, 37; 24:9, 19. I *Chronicles* 12:5 "Bealiah"; II *Chronicles* 23:17; 24:7; 28:1-4. *Jeremiah* 7:3, 9, 31; 11:12-13; 17:2; 19:5,13; 32:29, 35. *Ezekiel* 14:11. *Hosea* 2:16)—a. k. a.

Baal-Berith (*Judges* 8:33, 9:4), also called El-Berith (*Judges* 9:46), Baal-Zebub (II *Kings*, 1:2, 3, 6, 16. *Shabbath* 83*b*. *Sanhedrin* 63*b*), Baal-Peor (*Numbers* 25:1-9, 18; 31:16. *Deuteronomy* 3:29. *Joshua* 22:17. *Hosea* 9:10. *Psalm* 106:28 [eating the sacrifices of the dead]), Baal-Habab, Baal-Moloch (II *Chronicles* 28:1-4)—the God of Flies, the Golden Calf, the religion of Devil worship and human sacrifices (*Genesis* 22:1-18. *Exodus* 8:26; 13:2. *Leviticus* 27:28-29. *Joshua* 13:14. *Judges* 11:31, 39. I *Kings* 13:1-2. II *Kings* 16:3-4; 17:17; 21:6; 23:20-25. II *Chronicles* 28:1-4. *Jeremiah* 7:3; 19:5; 32:35. *Ezekiel* 16:20-21; 20:26, 31; 23:37).

Early Christians accepted Dualism and worshiped Jesus as Lucifer, the light, the Canaanites' god of the Sun. In the tradition of the Dualist principles of good and evil, male and female, corpse and spirit, they ate semen and drank menstrual blood as a form of prayer to the fertility gods they worshiped and as a form of protest against the alleged "evil" of procreation—of capturing a spirit in a corpse—a protest against the birth of a child into the morbid flesh. Here we see the stumbling stone the Jews laid on the path of the Romans in an attempt to exterminate them with Jewish Liberalism. Epiphanius wrote,

> "[26] 4,1 But I shall pass to the substance of their deadly story—they vary in their wicked teaching of what they please—because in the first place, they hold their wives in common. (2) And if a guest who is of their persuasion arrives, they have a sign that men give women and women give men, the tickling of the palm as they clasp hands in pretended greeting, to show that the visitor is of their religion.
>
> 4,3 And now that they know each other from this, the next thing they do is feast—and though they may be poor, they set the table with lavish provisions for eating meat and drinking wine. But then, after a drinking bout and practically filling the boy's veins, they next go crazy for each other. (4) And the husband will withdraw from his wife and tell her— speaking to his own wife!—'Get up, perform the Agape with the brother.' And when the wretched couple has made love—and I am truly ashamed to mention the vile things they do, for as the holy apostle says, 'It is a shame even to speak' of what goes on among them. Still, I shall not be ashamed to say what they are not ashamed to do, to arouse horror by every method in those who hear what obscenities they are prepared to perform. (5) For besides, to extend their blasphemy to heaven after making love in a state of fornication, the woman and man receive the male emission on their own hands. And they stand with their eyes raised heavenward but the filth on their hands, and pray, if you please—(6) the ones called Stratiotics and Gnostics—and offer that stuff on their hands to the actual Father of all, and say, 'We offer thee this gift, the body of Christ.' (7) And then they eat and partake of their own dirt, and they say, 'This is the body of Christ; and this is the Pascha, because of which our bodies suffer and are made to acknowledge the passion of Christ.'
>
> 4,8 And so with the woman's emission when she happens to be having her period—they likewise take the unclean menstrual blood they gather from her, and eat it in common. And 'This,' they say, 'is the blood of Christ.'

(5,1) And thus, when they read, 'I saw a tree bearing twelve manner of fruits every year, and he said unto me, This is the tree of life,' in apocryphal writings, they interpret this allegorically of the menses.

5, 2 But though they copulate they forbid procreation. Their eager pursuit of seduction is for enjoyment, not procreation, since the devil mocks people like these, and makes fun of the creature fashioned by God. (3) They come to climax but absorb the seeds in their dirt—not by implanting them for procreation, but by eating the dirt themselves.

5, 4 But even though one of them gets caught and implants the start of the normal emission, and the woman becomes pregnant, let me tell you what more dreadful thing such people venture to do. (5) They extract the fetus at the stage appropriate for their enterprise, take this aborted infant, and cut it up in a trough shaped like a pestle. And they mix honey, pepper, and certain other perfumes and spices with it to keep from getting sick, and then all the revellers in this <herd> of swine and dogs assemble, and each eats a piece of the child with his fingers. (6) And now, after this cannibalism, they pray to God and say, 'We were not mocked by the archon of lust, but have gathered the brother's blunder up!' And this, if you please, is their idea of the 'perfect Passover.'"[505]

In addition to the appearance of these practices in numerous apocryphal books, the Gnostics claimed that these anti-procreation practices stemmed from Jesus, himself, and cited canonical passages like *Luke* 20:34-38, *John* 6:26-71, and I *Corinthians* 7:32-40 as evidence of their claim. I *Timothy* 4:3 proves that among the earliest of Christians, like the Cathars who descended from them, were vegetarians who forbade marriage. This was one means the Jewish Dualists had to exterminate "Goy races" which converted to Christianity, and to prevent them from sacrificing animals, which sacrifices the Baalists believed would give their enemies power and divine protection. It was a Jewish means to weaken and exterminate the enemies of the Jews by giving them a foreign religion as a stumbling block.

The fall of Rome coincided with the rise of Christendom. Dogmatic Judaism in the form of dogmatic Christianity proved fatal to European progress and plunged Europe into the Dark Ages.

While the Romans were gullible, they were not quite so gullible as to adopt the outright suicidal practices of the Gnostic Christian Jews, at least not in large numbers—Americans are far more vulnerable to Jewish mythologies which destroy Gentiles. *The Catholic Encyclopedia* writes of the Gnostic Cathar sect known as the "Albigenses",

> "What the Church combated was principles that led directly not only to the ruin of Christianity, but to the very extinction of the human race."[506]

---

[505]. Epiphanius, translated by F. Williams, *The Panarion of Epiphanius of Salamis*, Volume 1, 26.4.1-26.5.6, E. J. Brill, New York, (1987), pp. 85-87.
[506]. "Albigenses", *The Catholic Encyclopedia*, Volume 1, Robert Appleton Company,

Centuries of censorship and fabrication have modified the presence of Baal in the Jewish faith and no doubt in the minds of most of the modern Jews and Christians who practice their faiths outside of Dualist sects—but the ancient, and even Medieval and no small number of modern Jews were superstitious, told and believed fables, segregated themselves and participated in Dualist sects and Baal worship. The worship of Dualism is pervasive in the religious myth that free will requires that there be evil as well as good.

The story of Jesus was interpreted by many Gnostic cults as an instance of Heraclitian dialectics. The *Logos*, the eternal fire of change, incorporates both good and evil. Jesus and Judas were often seen as opposing forces of the same divine principle—they bore the same name—the Jew—Judas the Galilean and Judas Iscariot. Jesus was also referred to as Lucifer, the Light. Jesus was both the son of man and the son of God. It was only through death, through human sacrifice, through the shedding of the evil flesh, that Jesus attained life, existence as pure Spirit, the wind of flame, and this death which brought life came at the hands of Jesus' alter ego, Judas, whose evil betrayal brought good tidings. This Dualistic Heraclitian dialectical theme was already many centuries old at the time the Gospels were written—the end is the beginning, death is life, bad is good, the way up the stairs is the way down the stairs, etc.

The Jews set out to ruin the Gentiles with a suicidal liberalism based on these Hellenistic dialectics. The Jews witnessed many examples of hermetic monks wasting away their lives in childless ruin, endlessly contemplating meaningless idealistic and self-destructive dogmas, which likely inspired the Jews to ruin the Romans in this fashion. They were largely successful.

4.4.6 Gentiles are Destined to Slave for the Jews, Then the Slaves Will be Exterminated

The *Zohar*, I, 28*b*-29*a*, states that the peoples who are descended from Eve and the serpent, through Cain, are Esau, Amalek, the Christians, and that they will be exterminated,

> "At that time the mixed multitude shall pass away from the world [***] The mixed multitude are the impurity which the serpent injected into Eve. From this impurity came forth Cain, who killed Abel. [***] for they are the seed of Amalek, of whom it is said, 'thou shalt blot out the memory of Amalek' [***] Various impurities are mingled in the composition of Israel, like animals among men. One kind is from the side of the serpent; another from the side of the Gentiles, who are compared to the beasts of the field; another from the *mazikin* (goblins), for the souls [29*a*] of the wicked are literally the *mazikin* (goblins) of the world; and there is an impurity from the side of the demons and evil spirits; and there is none so cursed among them as Amalek,

---

(1907), pp. 267-269, at 269.

who is the evil serpent, the 'strange god'. He is the cause of all unchastity and murder, and his twin-soul is the poison of idolatry, the two together being called Samael (lit. poison-god). There is more than one Samael, and they are not all equal, but this side of the serpent is accursed above all of them.'"[507]

Zohar, II, 219*b*, states,

"So they went nearer and they heard him saying: 'Crown, crown, two sons are kept outside, and there will be no peace or rest until the bird is thrown down in Cæsarea.' R. Jose wept and said: 'Verily the *Galuth* is drawn out, and therefore the birds of heaven will not depart until the dominion of the idolatrous nations is removed from the earth, which will not be till the day when God will bring the world to judgement.'"[508]

Jews often took a predominant rôle in the production of revolutionary literature in Europe and in revolutions meant to create world government. Many Jews were eager to destroy all "princes", to eliminate the monarchies of Europe. The Rothschilds caused war after war in order to make the Gentile peoples weary of war and clamor for peace. This proscription for Jewish domination was spelled out in the Old Testament. Jews then offered the Gentiles a solution to the wars the Jews had covertly caused. The Jews preached the message that the only solution to war was world government—world government run by Jews out of Jerusalem in an era of peace, as prophesied in *Isaiah*. The Jews have employed this model for centuries to lure the nations into surrendering their sovereignty to Jewish domination. The *Zohar*, III, 19*b*, states,

"It is, however, as R. Abba has said: all the other days are given over to the angelic principalities of the nations, but there is *one* day which will be the day of the Holy One, blessed be He, in which He will judge the heathen nations, and when their principalities shall fall from their high estate."[509]

Zohar, III, 43*a*, states that Gentiles must be converted to Judaism and used as the work animal, the horse or ass, of the Jews' (or "lambs'") desire to destroy the Gentiles' own governments. Should any resist conversion and the destruction of their own nations, they are to be killed. Bear in mind that to many Jews, as Moses Hess stated, Judaism is not a religion but a "racially" based nation; and the religion is the expression of this prophetic "race"; and that which is attributed to God, must in their minds be their mandate to themselves, a mandate

---

[507]. H. Sperling and M. Simon, *The Zohar*, Volume 1, The Soncino Press, New York, (1933), pp. 108-110.
[508]. H. Sperling and M. Simon, *The Zohar*, Volume 2, The Soncino Press, New York, (1933), p. 311.
[509]. H. Sperling and M. Simon, *The Zohar*, Volume 3, The Soncino Press, New York, (1933), p. 63.

represented by the genocidal murder of the firstborn of Egypt. The *Zohar*, III, 43*a*,

"To these He appointed as ministers Samael and all his groups—these are like clouds to ride upon when He descends to earth: they are like horses. That the clouds are called 'chariots' is expressed in the words, 'Behold the Lord rideth upon a swift cloud, and shall come into Egypt' (Isa. XIX, I). Thus the Egyptians saw their Chieftain like a horse bearing the chariot of the Holy One, and straightaway 'the idols of Egypt were moved at His presence, and the heart of Egypt melted in the midst of it' (*Ibid.*), i. e. they were 'moved' from their faith in their own Chieftain. AND EVERY FIRSTLING OF AN ASS THOU SHALT REDEEM WITH A LAMB, AND IF THOU WILT NOT REDEEM IT... THOU SHALT BREAK HIS NECK."[510]

Zohar, III, 282*a*, states,

"From the side of idolatry Shabbethaj (Saturn) is called Lilith [*Footnote:* Lilith is a female demon, comp. Is. XXXIV. 14 and Weber, *Altsynagogale palästinische Theologie*, p. 246.], mixed dung, on account of the filth mixed from all kinds of dirt and worms, into which they throw dead dogs and dead asses, the sons of 'Esau and Ishma'el, and there (read הבו) Jesus and Mohammed, who are dead dogs, are buried among them. She (Lilith) is the grave of idolatry, where they bury the uncircumcised, (who are) dead dogs, abomination and bad smell, soiled and fetid, a bad family. She (Lilith) is the ligament [*Footnote:* אכדם is a fibre attached to the lungs] which holds fast the 'mixed multitude' (Ex. xii. 38), which is mixed among Israel, and which holds fast bone and flesh, that is, the sons of 'Esau and Ishma'el, dead bone and unclean flesh torn of beasts in the field, of which it is said (Ex. xxii. 31): 'Ye shall cast it to the dogs.'"[511]

In commenting on the *Abodah Zarah*, the *Tosefta* states (the bracketed text is original to the Neusner edition),

"8:5 A. *For bloodshed — how so?*

B. A gentile [who kills] a gentile and a gentile who kills an Israelite are liable. An Israelite [who kills] a gentile is exempt.

C. *Concerning thievery?*

D. [If] one has stolen, or robbed, and so too in the case of finding a

---

[510]. H. Sperling and M. Simon, *The Zohar*, Volume 3, The Soncino Press, New York, (1933), p. 132.

[511]. G. Dalman, *Jesus Christ in the Talmud, Midrash, Zohar, and the Liturgy of the Synagogue*, Deighton Bell, Cambridge, (1893), p. 40. Though work is given an ancient attribution by its "discoverer", the Muhammadans are also mentioned in *Zohar*, II, 32*a*. Some consider the author to have been divinely inspired, some say the work evolved over time, some say the work is a fabrication—in any event, it is an now a very old writing and was very influential in Jewish political movements like the Frankists.

beautiful captive [woman], and in similar cases:
E. a gentile in regard to a gentile, or a gentile in regard to an Israelite

it is prohibited. And an Israelite in regard to a gentile — it is permitted."[512]

The Old Testament book of *Numbers* 24:17-20, which prophesies the Messiah, also prophesies the extermination of Amalek,

> "I shall see him, but not now: I shall behold him, but not nigh: there shall come a Star out of Jacob, and a Sceptre shall rise out of Israel, and shall smite the corners of Moab, and destroy all the children of Sheth. And Edom shall be a possession, Seir also shall be a possession for his enemies; and Israel shall do valiantly. Out of Jacob shall come he that shall have dominion, and shall destroy him that remaineth of the city. And when he looked on Amalek, he took up his parable, and said, Amalek *was* the first of the nations; but his latter end *shall* be that he perish for ever."

In addition to the well-known prophecies of Jewish world domination, the destruction of Gentile nations, Gentile servitude and the extermination of Gentiles found in the Old Testament (*see also: The Book of Jubilees* 32:17-20), the apocalyptic literature of the Qumran is overtly racist and genocidal—and this Jewish literature forms the basis for the genocidal visions of the Christian apocalyptic nightmares, which were iterated soon after. Horrific genocidal visions, and racist invectives are found in 1 *Enoch*, 2 *Baruch*, *The War Scroll*, and 4 *Ezra*.[513] If the Jews who wrote these genocidal works had their way, not a single Gentile or apostate Jew would be left alive.

Some Christians also look to the mythology of Esau and Jacob to justify their belief that the Jews will be "justly" annihilated should they refuse to accept the sacrifice of Jesus Christ as their salvation. Isaac's blessing to Esau stated that Esau would someday break off the yoke of Jacob,

> "39 And Isaac his father answered and said unto him, Behold, thy dwelling shall be the fatness of the earth, and of the dew of heaven from above; 40 And by thy sword shalt thou live, and shalt serve thy brother; and it shall come to pass when thou shalt have the dominion, that thou shalt break his yoke from off thy neck."—*Genesis* 27:39-40

In the early days of Christianity, Cyprian wrote in his Twelfth Treatise, "Three Books of Testimonies Against the Jews", First Book, Testimony 19,

---

512. J. Neusner, *The Tosefta: Translated from the Hebrew: Fourth Division: Neziqin (The Order of Damages)*, Volume 4, Ktav Publishing House Inc., New York, (1981), p. 342. *See also: Sanhedrin* 57a. *See also: Abodah Zarah* 26b.
513. Selections from these texts are found in: M. G. Reddish, *Apocalyptic Literature: A Reader*, Abingdon Press, Nashville, (1990).

"19. That two peoples were foretold, the elder and the younger; that is, the old people of the Jews, and the new one which should consist of us.

In Genesis: 'And the Lord said unto Rebekah, Two nations are in thy womb, and two peoples shall be separated from thy belly; and the one people shall overcome the other people; and the elder shall serve the younger.'[*Footnote*: Gen. xxv. 23.] Also in Hosea: 'I will call them my people that are not my people, and her beloved that was not beloved. For it shall be, in that place in which it shall be called not my people, they shall be called the sons of the living God.' [*Footnote*: Hos. ii. 23. i. 10.]"[514]

Abraham's covenant with God is both a blessing and a curse to Jews—and to Gentiles. *Genesis* 12:1-3 states:

"Now the LORD had said unto Abram, Get thee out of thy country, and from thy kindred, and from thy father's house, unto a land that I will shew thee: 2 And I will make of thee a great nation, and I will bless thee, and make thy name great; and thou shalt be a blessing: 3 And I will bless them that bless thee, and curse him that curseth thee: and in thee shall all families of the earth be blessed."

Zionists and Christians use the Old Testament and the New Testament to justify the murder of apostate Jews. *Deuteronomy* 11:24-28 states,

"24 Every place whereon the soles of your feet shall tread shall be yours: from the wilderness and Lebanon, from the river, the river Euphrates, even unto the uttermost sea shall your coast be. 25 There shall no man be able to stand before you: *for* the LORD your God shall lay the fear of you and the dread of you upon all the land that ye shall tread upon, as he hath said unto you. 26 Behold, I set before you this day a blessing and a curse; 27 A blessing, if ye obey the commandments of the LORD your God, which I command you this day: 28 And a curse, if ye will not obey the commandments of the LORD your God, but turn aside out of the way which I command you this day, to go after other gods, which ye have not known."

Romans 9 states (*see also: Matthew* 12:30; 21:43-45. *Romans* 11:7-8. *Galatians* 3:16. *Hebrews* 8:6-10):

"I say the truth in Christ, I lie not, my conscience also bearing me witness in the Holy Ghost, 2 That I have great heaviness and continual sorrow in my heart. 3 For I could wish that myself were accursed from Christ for my

---

514. Cyprian, Twelfth Treatise, "Three Books of Testimonies Against the Jews", First Book, Testimony 19, *The Anti-Nicene Fathers: Translations of the Writings of the Fathers down to A.D. 325*, Volume 5, Christian Literature Publishing Company, New York, (1886), p. 512.

brethren, my kinsmen according to the flesh: 4 Who are Israelites; to whom *pertaineth* the adoption, and the glory, and the covenants, and the giving of the law, and the service *of God*, and the promises; 5 Whose *are* the fathers, and of whom as concerning the flesh Christ *came*, who is over all, God blessed for ever. Amen. 6 Not as though the word of God hath taken none effect. For they *are* not all Israel, which are of Israel: 7 Neither, because they are the seed of Abraham, *are they* all children: but, In Isaac shall thy seed be called. 8 That is, They which are the children of the flesh, these *are* not the children of God: but the children of the promise are counted for the seed. 9 For this *is* the word of promise, At this time will I come, and Sarah shall have a son. 10 And not only *this*; but when Rebecca also had conceived by one, *even* by our father Isaac; 11 (For *the children* being not yet born, neither having done any good or evil, that the purpose of God according to election might stand, not of works, but of him that calleth;) 12 It was said unto her, The elder shall serve the younger. 13 As it is written, Jacob have I loved, but Esau have I hated. 14 What shall we say then? *Is there* unrighteousness with God? God forbid. 15 For he saith to Moses, I will have mercy on whom I will have mercy, and I will have compassion on whom I will have compassion. 16 So then *it is* not of him that willeth, nor of him that runneth, but of God that sheweth mercy. 17 For the scripture saith unto Pharaoh, Even for this same purpose have I raised thee up, that I might shew my power in thee, and that my name might be declared throughout all the earth. 18 Therefore hath he mercy on whom he will *have mercy*, and whom he will he hardeneth. 19 Thou wilt say then unto me, Why doth he yet find fault? For who hath resisted his will? 20 Nay but, O man, who art thou that repliest against God? Shall the thing formed say to him that formed *it*, Why hast thou made me thus? 21 Hath not the potter power over the clay, of the same lump to make one vessel unto honour, and another unto dishonour? 22 *What* if God, willing to shew *his* wrath, and to make his power known, endured with much longsuffering the vessels of wrath fitted to destruction: 23 And that he might make known the riches of his glory on the vessels of mercy, which he had afore prepared unto glory, 24 Even us, whom he hath called, not of the Jews only, but also of the Gentiles? 25 As he saith also in Osee, I will call them my people, which were not my people; and her beloved, which was not beloved. 26 And it shall come to pass, *that* in the place where it was said unto them, Ye *are* not my people; there shall they be called the children of the living God. 27 Esaias also crieth concerning Israel, Though the number of the children of Israel be as the sand of the sea, a remnant shall be saved: 28 For he will finish the work, and cut *it* short in righteousness: because a short work will the Lord make upon the earth. 29 And as Esaias said before, Except the Lord of Sabaoth had left us a seed, we had been as Sodoma, and been made like unto Gomorrha. 30 What shall we say then? That the Gentiles, which followed not after righteousness, have attained to righteousness, even the righteousness which is of faith. 31 But Israel, which followed after the law of righteousness, hath not attained to the law of righteousness. 32 Wherefore? Because *they sought it* not by faith, but as it were by the works of the law. For they stumbled at that stumblingstone; 33 As it is written, Behold, I lay in Sion a stumblingstone and rock of offence: and whosoever believeth on him shall not be ashamed."

One of the reasons some Jews attempt to degrade other cultures, especially Christian cultures, is that they want Christians to become decadent and lose favor in the eyes of God. Though Esau broke the yoke, Jacob's yoke will yet again—and forever—fall upon Esau should the Christians become decadent. Should this happen, the Jews will then again find favor with the Lord, according to Paul—and in some minds, Jesus (*Luke* 21:24). In some minds, the period of Gentile rule began as God's punishment to unfaithful Jews in 606 B. C. with the ascendence of Nebuchadnezzar and eventual captivity and exile of the Jews in Babylon and the destruction of Jerusalem. According to this belief system, Gentile rule is supposed to have lasted for a period of 2520 years, which time span ended in 1914—the first year of the First World War, when the Jews began to rule the world.

Jews have dominated the mass media in many societies in which they have lived. Though within their own families they wisely promote education, thrift, tradition and morality, these same values are often absent from the messages they convey through the mass media. Though Jews have a racist tradition of segregation and nationalism, they often promote miscegenation and internationalism to the Gentiles. The Old Testament teaches the Jews again and again that a nation which loses favor in the eyes of God will be utterly destroyed—for example in the story of Sodom and Gomorrah (*Genesis* 18; 19). These lessons teach Racist and tribalistic Jews that if they can rob a nation of its righteousness, they will have destroyed it before God. They are taught that if they can turn the entire Gentile world to "evil", then any righteous Jews remaining will inherit God's blessing and the era of Gentile domination will be at an end.[515]

Those throughout history who best knew the Jews, men like Cyprian, John Chrysostom, Martin Luther, Johannes Buxtorf, etc., warned Christians that Jews were out to destroy them, and that they ought not to stumble over the stumbling stones the Jews threw on their path, and must remain righteous, or lose the favor of the Lord, which the Jews believed would then return to them. Johannes Buxtorf wrote in the preface of his *Synagoga Judaica: Das ist Jüden Schul ; Darinnen der gantz Jüdische Glaub und Glaubensubung... grundlich erkläret*, Basel, (1603); as translated in the 1657 English edition, *The Jewish Synagogue: Or An Historical Narration of the State of the Jewes, At this Day Dispersed over the Face of the Whole Earth*, Printed by T. Roycroft for H. R. and Thomas Young at the Three Pidgeons in Pauls Church-Yard, London, (1657),

---

515. *Exodus* 34:11-17. *Psalm* 2; 72. *Isaiah* 1:9; 2:1-4; 6:9-13; 9:6-7; 10:20-22; 11:4, 9-12; 17:6; 37:31-33; 41:9; 42; 43; 44; 61:6. *Jeremiah* 3:17; 33:15-16. *Ezekiel* 20:38; 25:14. *Daniel* 12:1, 10. *Amos* 9:8-10. *Obadiah* 1:18. *Micah* 4:2-3; 5:8. *Zechariah* 8:20-23; 14:9. *Romans* 9:27-28; 11:1-5.

"THE AUTHORS
PREFACE
To the Christian Reader.

Christian Reader,
WHen once we exactly ponder in the Scales of our understanding that thrice pressing load of Jewish ingratitude, disobedience, and obstinacy, for which they were dayly branded by *Moses* and the rest of the Prophets with a foul guilt, to which was annexed a vehement reprehension. When we seriously consider those horrid threats and execrations where with God in his justice would depress them, unless they framed their lives according to the strict rule of his Commandments; this ought to be a warning piece unto us to entertain such blessings with a more gratefull acceptance, and hitherto to bend all our studies, that by our unthankfulness we should not make our selves unworthy of them, and so be disinherited of such a possession. *Moses* in this manner prophesies of the Jews ingratitude, {Deut. 32.15.} *Jesurun waxed fat, and kicked. (thou art waxen fat, thou art grown thick, thou art covered with fatness) then he forsook God which made him, and lightly esteemed the worke of his salvation.* This issued from a prophetical spirit, declaring that as already present, which after the revolution of many a year was to be fulfilled and accomplished. This ingratitude was in its swadling clouts when *Joshua* led *Israel* into the land of promise, which is ratified by the unanimous suffrage of the whole College of Prophets, and almost in the very same terms by *Hosea* in chap. 13. *Jeremy* arraigns them as guilty of the same crime. The bill of inditement runs thus: {Jer. 11.10.} *They are turned back to the iniquities of their fore-fathers which refused to hear my words, and they went after other gods to serve them: the house of* Israel *and the house of* Judah *have broken the Covenant which I made with their Fathers.* And God himselfe by the mouth of his Prophet thus proclaims their obstinacy: {Jer. 7.25.26.} *Since the day that your Fathers came out of the Land of Egypt unto this day, I have even sent unto you all my servants the Prophets, dayly rising up early and sending them; yet they hearkened not unto me, nor inclined their eare, but heardened their neck, they did worse then their Fathers.* The obstinacy of this People at last grew to so high a pitch, that they stopt their ears at the admonition of the Prophets, who cried aloud unto them to amend their waies, and curbed their offences with tart reprehensions, killing, stoning, rewarding every one with some bitter death; which act of theirs is faithfully registred by the holy Spirit, Ezra 2: {Nehem. 9.25.26.} *They tooke strong Cities and a fat Land, and possessed houses full of all goods, wels digged, Vineyards and Oliveyards and fruit trees in abundance: so they did eat and were filled, and became fat, and delighted themselves in thy great goodness: nevertheless they were disobedient and rebelled against thee, and cast thy Law behind their backs, and slew thy Prophets which testified against them to turn them to thee, and wrought great provocations.* And *Jeremy* also may be cited for a witness, for his words are these: {Ier. 2.29 30.} *Wherefore will ye plead with me? ye all have transgressed against me, saith the Lord. In vain have I smitten your children, they have received no correction: your own sword hath devoured your Prophets like a destroying Lion.* When the Lord sees this his people thus altogether incapable of

corection, he afflicts them with all the punishments which *Moses* by the spirit of God had denounced against them, neither their bodies nor goods can now escape the lash of his fury; he sends among them the sword, famine and pestilence, tempests, diseases, imbred dissention, and discord; and to make their misery compleat, casts them out of that Land flowing with milk and hony, and causes them to trace the captives steps into another which they knew not. The ten tribes together with their King *Hoshea* is carried by *Salmanasser* into *Assyria*, 2 *Kin*. {2 Reg. 17.} and when the two remaining Tribes, *Juda* and *Benjamin*, were not hurried to repentance by the present view of their brethrens afflictions, God sends *Nebuchadnezzer* King of *Babel* against them, who leads them captive into the Land of *Chaldea*, makes *Jerusalem* a desolate heap, and turns their Temple, their chief beauty into ashes. Nevertheless the space of 70 years fully expired, these 2 tribes were again brought out of the house of bondage, because it was the Almighties pleasure to preserve the tribe of *Judah* even unto that time, when according to his promise, out of that tribe, and in the promised land the *Messias* should be incarnate. But for all this these 2 tribes did not much outstrip the other 10 in the practice of holiness; for they always following their own devices, seriously traced the forbidden by-paths of their forefathers, for which the later Prophets, *Haggai*, *Zachary* and *Malachi* were earnest declamitants against them: the last of which being a Priest, & proclaiming them guilty of a wicked life, threatens them with a finalrejection.

But

[There are pages missing from both the microfilm and digital reproductions of this text which were used in this transcription. Your author apologizes and would be grateful if an intact copy were found and the missing text provided.]

out in obscurity, that so we might again be cast headlong into that darknesse in which we sate, before it was the Lords pleasure by his mercy to impart unto us the saving knowledge of his heavenly word.

My second Motive was this, that the hardened in heart, and blindfolded Jews at last descending into the Chambers of their strict cogitations, mights have some glimpse of the greatness of their infidelity, and so convicted before the face of the whole world of that more than brutish folly in the expounding of the holy Scriptures, and of their old wives tales, whereby God for the most part is blasphemed, and his saving word against all humane reason after an execrable manner perverted, they might begin to be ashamed, who with such a whorish forehead, and want of wit did not fear to speak or write in this manner of God Almighty, and his holy word, and that at length they might think, that they had stumbled at that stone of stumbling, and rock of offence laid in *Sion*, and thereupon that they shall fall prostrate upon the ground, be broken, to Gods Law ensnared and captivated, and finally that God {Isa. 29.10,11.} *poured upon them the spirit of deep sleep*, and so closed their eyes, that every prophesie and the whole Scripture was to them as the words *of a book that is sealed, & that the wisdome of their wise men is now altogether perished, and the understanding of their prudent men hid*, as the Prophet *Isaiah* foretold them. The God of mercies have mercy upon

them, and convert them, and keep us firm and immoveable in the knowledge of his truth, that in it we may hope to gain eternall life, as Christ himself witnesseth to our comfort, when he saith, {John 17.3} *This is eternall life, that they might know thee the onely true God, and Jesus Christ whom thou hast sent*, To him be ascribed, praise, honour and glory for evermore, Amen.
MICAH c. 4 v. 1, 2.

IN the last dayes it shall come to passe, that the mountains of the house of the Lord shall be established in the top of the mountains, and it shall be exalted above the hills, and people shall flow unto it.

And many Nations shall come and say, come, and let us go up to the mountains of the Lord, and to the house of the God of Iacob, and he will teach us of his wayes, and we will walk in his paths; for the Law shall go forth from Sion, and the word of the Lord from Jerusalem.

Luther upon these words of *Micah*, hath left this consequent paragraph in memory concerning the Iews. So goes the matter, hereupon arise these mentall divisions, this is that which makes the Jews mad and foolish, that which forceth them to a sense so damnable, that they are compelled without the least shew of honesty, to wrest every parcell of the Scripture, because it contradicts their will, and they cannot endure that we Gentiles should be equal copartners with them in Gods favour, and the Messias should in a like measure administer to us and them joy and consolation. Moreover, rather than they would vouchsafe, that we the offspring of the Gentiles (who are by them daily contemned, accursed and devoted to the infernall hagges, torn and cut in pieces by their sladerous back-bitings) should participate in the Merits of the Messias, and enjoy the title of coheirs and brethren, they had rather ten Messias should suffer the shamefull death of the crosse, and afflict God himself (if there were any possibilty in nature) the holy Angels and all other creatures with the stroke of death, nay, they would not be afraid of the fact, though a thousand hellish torments were to be endured for the effecting of it, so incomprehensible and austere is the pride mixed with the honourable blood of these Fathers, and circumcised Saints, who alone would enjoy the promised Messias, and be capped for the sole *Donns* of the world. {Chjim.} The Nations or Gentiles ought onely to be these accursed vassals, and to give up their desire, that is their silver and gold unto the Iews, and that they should be constrained to submit themselves unto them after the manner of beasts prepared to the slaughter, rather then they will relinquish one whit of this their assertion, they will not refuse wittingly to be damned eternally."

Though Johannes Buxtorf, Martin Luther, and many others expressed anger at the Jews for not converting to Christianity, Jews simply could not accept that Jesus was their Messiah, or that Gentiles, whom they considered to be less than human—less than Jews, had a right to Jewish beliefs. Jesus did not level the nations with an iron scepter. He did not make the Jews rulers of the world and the Gentiles their slaves. He did not lay to waste the lands outside of Israel. He was not the repressive and horrible Messiah the Jews prophesied in the Old

Testament.

Unlike Christians, Jews were not concerned with eternal life on an individual basis, but were concerned with the survival, the immortality, of the Jewish "race". Judaism is less a spiritual religion than is Christianity. It is much more materialistic, and combines religion, politics, commerce and mundane laws with religion, such that the boundaries between the secular and the religious do not really exist. A Jewish racist and/or tribalist can erase God from the Old Testament and still find in it his or her identity as a "Jew", and a mission in life. For him or her, this belief system is meant for none other than those who created it, the Jews. Racist secular Jews merely believe that "God" is the product of Jewish "racial instincts". God is a Jew and Jews embrace Judaism as the expression of their Jewish "soul", the material product of a chosen people, not an individual, but a people bold enough and superior enough to chose themselves to be the natural rulers of the Earth, rulers over the "lesser races" of non-Jews, whom they will eventually exterminate.

For many Jews, Jesus was far too weak and ineffective, far too universal in his message, to have been their racist Jewish Messiah, the tyrannical Jewish King promised to give them the world. Jews do not wish to wait for death to obtain paradise. They want a Jewish Utopia on Earth and they want their rewards on this Earth in this lifetime. They do not believe in a Christian Heaven and they do not believe poverty and sacrifice and repression will earn them eternal rewards. Nor do they believe that they will be eternally punished for doing wrong. They are out to obtain what they can here on this Earth in this lifetime. Judaism is a very different religion from Christianity. It is more of a mundane racist and genocidal political movement than it is a spiritual and ethical religion.

Many have accused leading Jews of using their power in the American media to degrade American culture and Christianity. The same accusations appeared in Germany. Leading Jews used Communism to destroy cultures, nations and religions. In the Spanish, Nazi, Turkish, Russian, French and English revolutions, leading Jews followed the same model of requesting liberal freedoms, which resulted in revolution, which resulted in chaos. Then, leading Jews spread word through their channels which control public opinion, that it would be impossible for anyone but a dictator to restore order out of the chaos—chaos the Jews had covertly intentionally created. The foolish Gentiles who were duped into clamoring for liberty, equality and freedom by the means of nihilistic revolution, are then duped into clamoring for an absolute dictatorship to restore order. The whole process is overseen by Jewish and crypto-Jewish leaders. After they have a dictator in place and the Gentiles have surrendered all of their rights to the Jews' puppet dictator, they destroy religion and culture, and mass murder the leading class of intellectual elites. For them it is the process of breeding the type of cattle they want to serve them—degenerate, stupid and compliant cattle.

OTHER TITLES

The Manufacture and Sale of Saint Einstein

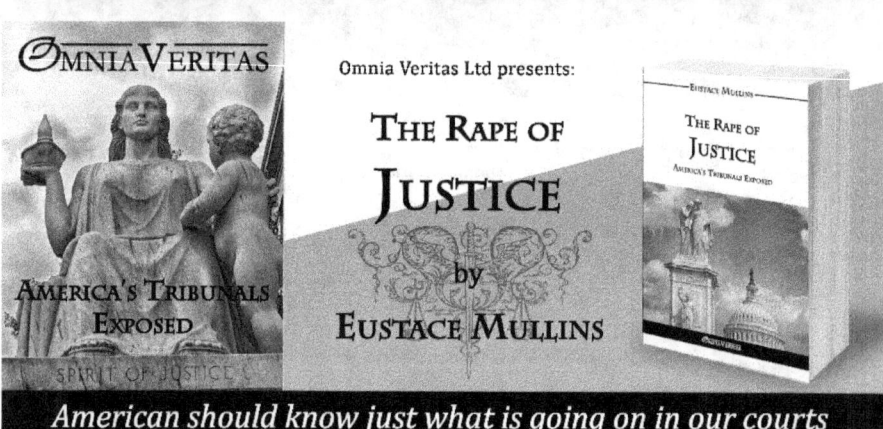

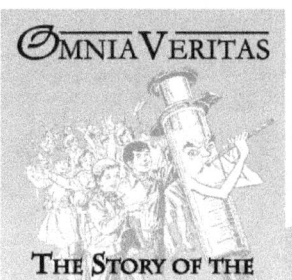
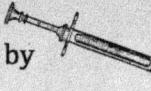

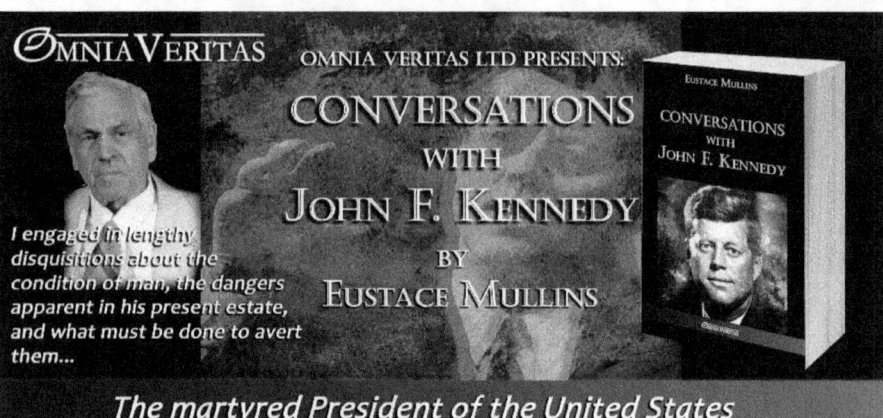

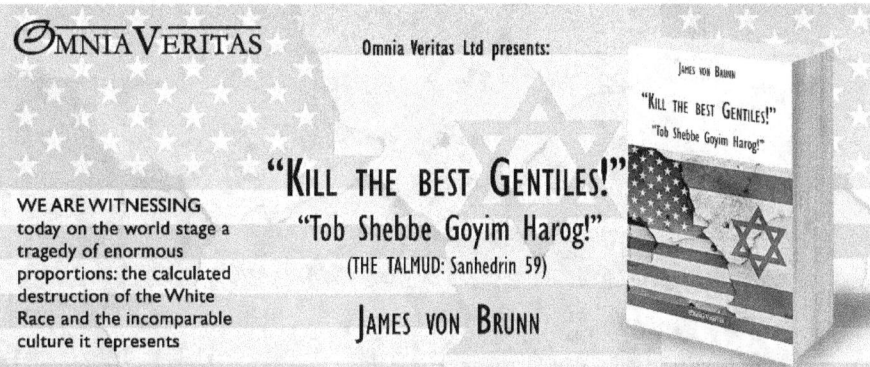

www.omnia-veritas.com

https://www.instagram.com/omnia.veritas/

https://twitter.com/OmniaVeritasLtd